# ENDOCYTOBIOLOGY III

ANNALS OF THE NEW YORK ACADEMY OF SCIENCES
Volume 503

# ENDOCYTOBIOLOGY III

*Edited by John J. Lee and Jerome F. Fredrick*

28666

The New York Academy of Sciences
New York, New York
1987

Cover designed by Norman Tempia. On a strand of DNA, a prokaryote is captured by another cell and eventually becomes established as an organelle of the capturing or "host" cell and the endosymbiosis is now integrated into the overall cellular and organellar DNA.

### Library of Congress Cataloging-in-Publication Data

Endocytobiology III.

    (Annals of the New York Academy of Sciences, 0077-8923 ; v. 503)
    Papers presented at the Third International Colloquium on Endocytobiology held June 10-12, 1986 in New York, N.Y. and sponsored by the New York Academy of Sciences.
    Bibliography: p.
    1. Cell organelles—Congresses. 2. Eukaryotic cells—Congresses. 3. Symbiosis—Congresses. 4. Evolution—Congresses. 5. Micro-organisms—Evolution—Congresses.
    I. Lee, John J. II. Fredrick, Jerome F.
III. International Colloquium on Endocytobiology (3rd : 1986 : New York, N.Y.) IV. New York Academy of Sciences.
V. Series.
Q11.N5 vol. 503     500 s     87-15304
[QH573]
ISBN 0-89766-402-7
ISBN 0-89766-401-9 (pbk.)

PCP
*Printed in the United States of America*
**ISBN 0-89766-402-7 (Cloth)**
**ISBN 0-89766-401-9 (Paper)**
**ISSN 0077-8923**

ANNALS OF THE NEW YORK ACADEMY OF SCIENCES

Volume 503
July 9, 1987

# ENDOCYTOBIOLOGY III[a]

Editors and Conference Organizers
JOHN J. LEE and JEROME F. FREDRICK

## CONTENTS

[a]This volume is the result of a conference entitled Endocytobiology III held from June 10 to
June 12, 1986 in New York, N.Y., and sponsored by The New York Academy of Sciences.

**Financial assistance was received from:**

- OFFICE OF NAVAL RESEARCH GRANT NO. N00014-86-G-0077

# Introductory Remarks

JOHN J. LEE

*Department of Biology*
*City College*
*City University of New York*
*Convent Avenue at 138th Street*
*New York, New York 10031*

On behalf of the International Organizing Committee consisting of Drs. Jerome Fredrick, D. C. Smith, T. Cavalier-Smith, H. E. A. Schenk, L. N. Edmunds, W. Schwemmler, F. J. R. Taylor and myself, it is a pleasure to introduce this volume resulting from the Third International Colloquium on Endocytobiology.

This conference, in a way, recognizes the maturing of ideas and concepts that were still quite controversial 10 years ago when my coorganizer, Dr. Jerome F. Fredrick, first thought about organizing the predecessor (or ancestor in terms of advancement) conference—"Origins and Evolution of Intracellular Eukaryotic Organelles" (January 22-24, 1980) sponsored by the New York Academy of Sciences. Drs. Schwemmler and Schenk organized the Endocytobiology (I and II) Colloquia in Tubingen to keep open the lines of communication between the multidisciplines that were contributing to endocytobiological advancement. It is a tribute to them and their colleagues on the International Committee that an international society has been formed and that two new journals dedicated to research in this field, *Symbiosis* and *Endocytobiosis and Cell Research,* have become established.

As we start this volume, we must pause for a moment to gain perspective. What is the rationale for the special scientific focus on endocytobiology? The answer is simple and all of us as participants know it. Scientific specializations tend to isolate us into traditional fields and technical approaches. Often new developments outside our traditional focus escape our immediate attention. The aim of this symposium is to bring together researchers in quite a number of diverse fields who are working on various aspects related to the origin and integration of cellular organelles (e.g., mitochondria, chloroplasts) and more general aspects of invertebrate and plant endosymbiotic systems (e.g., corals and zooxanthellae, insects and endosymbiotic bacteria). The organizers have taken great care not to allow this colloquium to become sets of minispecialist meetings at the expense of cohesive examination of the mechanisms and processes of integration and evolution of genetically distinct entities at the cellular and organismic levels. If we look back to 1980 we were asking questions aimed at broadly rarifying the "endocytobiological theory." In this volume our focus will be quite different; we will discuss evidence for coevolution and integration of endosymbiotic partners. I know we can look forward to a vigorous and constructive interchange of ideas and expansions of our intellectual horizons.

It goes without saying that many persons have worked behind the scenes to make this conference possible and to guide it along its present lines. After the initial conference proposal was drawn up and approved by the international committee cited above, many of our peers made helpful suggestions. We are particularly grateful to Drs. Lynn Margulis and J. W. Hastings for their help during this early process. The

Conference Committee of the New York Academy of Sciences and anonymous reviewers made additional suggestions. A subcommittee of the NYAS Conference Committee consisting of Drs. Walter N. Scott, Chair, Jacqueline Messite, and Constance Sancetta helped us finalize the program. As usual, Mrs. Ellen A. Marks, the Conference Director, and her staff worked busily to make things go smoothly. Ms. Barbara Parker was our key contact person.

# An Overview of the Status of Evolutionary Cell Symbiosis Theories

F. J. R. "MAX" TAYLOR

*Departments of Botany and Oceanography*
*University of British Columbia*
*Vancouver, British Columbia, Canada V6T 1W5*

The history of phylogenetic studies is strewn with the wrecks of broken theories.[72]

P. H. A. Sneath, 1974

The history of the natural sciences is strewn with the corpses of intricately organized theories.[73]

P. Kitcher, 1982

## INTRODUCTION

The purpose of this presentation is to provide a primarily historical background for an area of enquiry that is, indeed, littered with discarded theories. This broad outline will be supplemented by the more detailed papers that follow. My assigned task is to focus on the fundamental role of symbiosis in cell evolution, and so there will be little more than brief mention of the nonsymbiotic processes that were also involved. The paper by Cavalier-Smith which follows in this volume provides more information on the latter aspects.

Symbiosis has been thought to be involved in cell evolution in several different ways: the primary origins of eukaryotic organelles, such as plastids and mitochondria; the possibility of multiple primary origins for the same organelles; and the secondary acquisition of organelles by one eukaryote from another. Each of these, and their biological implications, will be examined in turn.

## SYMBIOSIS AND THE PRIMARY ORIGINS OF EUKARYOTIC ORGANELLES

In the past decade or so, a true Kuhnian minirevolution has occurred in cell biological theory. A group of hypotheses for the origin of eukaryotes, collectively known as the serial endosymbiosis theory (SET),[1] or similar title, has been transformed from an amusing ingenuity into a respectable alternative and is now a preferred explanation for the origins of plastids and mitochondria.[2-4] In essence it can be stated

1

that the origin of eukaryotic cells involved at least two symbioses with a fermentative host, probably an Archaebacterium, which first acquired a purple nonsulfur bacterial cytobiont which became the mitochondrion and, subsequently, a cyanobacterium which became the plastid, other cell components having arisen by gradual differentiation. Extentions of this that have not become established include the symbiotic origins of the flagellum and nucleus (see below) and the involvement of multiple symbiotic events for each organelle. As a consequence the paradigm of eukaryotic cell evolution has changed fundamentally, from one of gradual differentiation to one of abrupt organellar acquisition by symbiosis; from gradualism to punctuated equilibrium of a very distinct variety. Historical, evolutionary theories cannot be proven in the strict sense, because of the impossibility of repeated observation; however, it is often possible to specify observations that would contradict them, thus remaining fundamentally scientific. The change referred to here has come about because accumulating data, particularly molecular, have resulted in a quirky set of hypotheses postulating the symbiotic, foreign (xenogenous) origin of two common eukaryotic organelles being much more consistent with them than a conventional, more conservative view. It should be self-evident that the guiding principle throughout all these enquiries into cell evolution is that the simplest explanation is always that a cell has produced all its components by evolutionary differentiation (autogenously) unless there are compelling reasons to doubt this.

Elsewhere, I have detailed the history of this change, and the key observations (see also TABLE 1) that brought it about.[5] It is well known that Schimper first proposed a symbiotic cyanobacterial origin of plastids, based on simple resemblances and the first establishment of their continuity, as early as 1883.[6] This notion, explicitly developed by Mereschkowsky between 1905 and 1920,[7-9] remained dormant for more than 40 years. It is important to realize that, until the major distinction of prokaryotes from eukaryotes, based on positive features rather than absences in prokaryotes, was recognized in the 1960s,[10] there were few events in eukaryogenesis to explain. Eukaryotes were simply thought to have grown larger from an aerobic, nonphotosynthetic prokaryotic stock, the organelles arising from membrane envelopments. Until the 1970s,[11,12] there were no attempts to detail how this might have arisen; it was simply an autogenous *assumption,* rather than a theory. The fundamental choices, and some of their implications, are illustrated in FIGURE 1.

Similarly, although the regular existence of mitochondria as eukaryotic organelles was known by the 1920s, their true function was not appreciated until the 1930s (for a review see Reference 13), and their continuity in all cases only in the 1960s. It is hardly appropriate, therefore, to give Altmann and Portier credit for early symbiotic proposals for mitochondrial origins, as some recent authors have done, when these early cell biologists had completely erroneous ideas about the organelles.[14,15] Altmann thought that they were "bioblasts," the primordial vital entities, surrounded by inert cell substance. Portier's claim that he could culture them was evidently the fruit of bacterial contamination. A further oddity was Mereschkowsky's proposal that nuclei arose from the symbiosis of bacteria with the anucleate moneran ancestral to the eukaryotes[8] (he did *not* propose that this was how mitochondria arose, as has been claimed).[16] Wallin's proposals in the 1920s, precipitated by the furor surrounding Portier's claims, were the first explicitly directed to the idea that mitochondria, although true eukaryotic organelles, originated as symbiotic bacteria.[17,18] Furthermore, he proposed that this type of symbiosis was a hitherto unrecognized form of major evolutionary novelty. This form of speciation was dubbed "symbionticism" (more recently discussed in Reference 19).

Another organelle for which a symbiotic origin was proposed is the eukaryotic flagellum. The first to make this suggestion appears to have been Kozo-Polyansky,[20]

**TABLE 1.** A Simplified Chronology of Major Events in the Recognition of the Involvement of Symbioses in Cell Evolution (See Text for Details)

| Decade | Plastids | Mitochondria | Observations |
|--------|----------|--------------|--------------|
| 1880 | Schimper: first symbiotic proposal | | Morphology, permanence |
| 1980 | | (Altmann: "bioblasts") | (Primordial entities) |
| 1900 | | | Photosynthetic cytobioses recognized |
| | Mereschkowsky: three papers explicitly developing an origin for plastids in symbiotic cyanobacteria. | | First plastic genetics (Correns) |
| 1910 | | | |
| 1920 | | (Portier: bacteria) | (cultivation claimed) |
| | | Wallin: organelles, but of symbiotic bacterial origin; Symbionticism | Functions of organelles gradually established |
| 1930 | | | |
| 1940 | | | Antibiotics developed |
| 1950 | Provasoli: streptomycin cures *Euglena* of plastids | Ephrussi: effects of acriflavin | First mitochrondrial mutants established |
| 1960 | Ris: TEM demonstration of cDNA | Nass and Nass: mtDNA seen with TEM | Organellar DNA Symbiotic flagellum proposal (Sagan) |
| 1970 | Margulis: synthesis Piggot and Carr: hybridization cDNA/ rRNAs Schwartz and Dayhoff | Margulis: synthesis Schwartz and Dayhoff | First nucleic acid comparisons Amino acid sequencing RNA sequencing |
| 1980 | Many | Many | Introns Promiscuous DNA |

although the idea is now inextricably linked to Margulis, who introduced it independently in a paper on the origins of mitosis in 1967 (as Sagan).[21] This extension of the SET does not seem to be supported by sufficient data to warrant acceptance over an autogenous origin (see the papers by To and Bermudes and Margulis in this volume). Although microtubule-like structures have been found in some spirochetes and cyanobacteria, basal bodies/centrioles can form without the presence of other basal bodies or centrioles in the cytoplasm. DNA and possibly RNA are absent. New centriolar formation may involve RNA (with a reverse-transcriptase postulated as

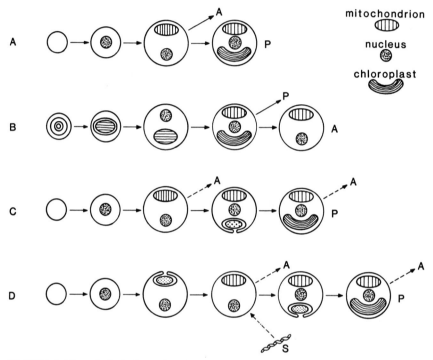

**FIGURE 1.** The principal alternatives in the mechanisms for eukaryogenesis (redrawn from Taylor).[1] (A) The fully autogenous development from a nonphotosynthetic ancestor, the Urzoan. (B) Fully autogenous development from a photosynthetic, cyanobacterium-like Uralga. (C) Autogenous mitochondrial development in a purple nonsulfur bacterium-like ancestor, with symbiotic acquisition of the plastid from a cyanobacterium. (D) The serial endosymbiosis theory (see text for details). A, animals (alternative derivations); P, plants; S, Margulis's extension, in which a spirochete gives rise to the flagellum.

carrying information to the nucleus). These features[22] do not support the idea that microtubular organizing centers arose from the permanent incorporation of ectosymbiotic spirochetes, although it can be argued that all the organellar DNA has now been incorporated into the nucleus. If symbiosis had to be involved, a viral origin might fit the data better,[19] although it still seems simpler to consider microtubular systems as being of autogenous origin in the ancestral eukaryote,[19,22] with eukaryotic flagella a complex derivative, rather than source, structure.

During the 40-year "eclipse period" for the hypotheses, there were several indications that things might not be all that they seemed with eukaryotic cell evolution. Evidence for "extrachromosomal inheritance" factors accumulated which pointed towards semiindependant genetic properties of plastids, indicated as early as 1909 by the pattern of flower color inheritance on the variegated "four o'clock *Mirabilis*" plant by C. Correns,[23] and beginning in the late 1940s for mitochondria (the classic studies of respiration-deficient "petite" mutants of yeast by Boris Ephrussi and his colleagues).[24] The action of antibiotics developed after the Second World War indicated a functional similarity between these organelles and bacteria,[25] as mutagens or lethal agents, including the classic "bleaching" of *Euglena* by streptomycin, permanently "curing" it of its plastids, also in the late 1940s.[26]

Many examples of obligate intracellular symbiosis were being discovered that revealed that the postulated symbiotic events were at least feasible,[30] and the difficulty of deciding if some cyanellae (the photosynthetic blue-green bodies within the cell), such as those of *Cyanophora paradoxa,* are plastids or symbiotic cyanobacteria,[27] is still particularly provocative. Indeed, one can find contemporary analogues for nearly all the participants and events postulated by the SET. Considering that the events in question are thought to have occurred some 1500 million years ago, this seems to be extraordinarily fortuitous. While illustrating that such mechanisms and prototypes, as postulated by the SET, are reasonable, these analogues do not prove that the theory is correct.

I have previously attributed the serious revival of interest largely to the conclusive demonstration of DNA in plastids[28] and mitochondria[29] in the 1960s. Another critical factor was Margulis' active and provocative espousal of the hypotheses in a synthetic form, within a broad geological context, in a germinal book in 1970.[30] It is one of the ironies of scientific history that her own original contribution to the question, i.e., the origin of "9+2" microtubular structures by a further symbiosis (see below and in this volume), has not been widely accepted. However, her book represents the first synthesis of the separate hypotheses for each organelle into a general theory of eukaryotic cell evolution (unnamed and not presented as such at the time), marshalling the available supportive evidence and indicating predictions that could be, and were, tested by cell biologists.

It must be extremely irritating to the properly cautious scientist when hypotheses that seem facile and too "neat" to be true, turn out to be more or less correct. This is much the case with the SET, recognized and named as a general theory with multiple constituent hypotheses in 1974.[1] The crude resemblances between plastids, mitochondria, and prokaryotes became quantitatively measurable relative to the nucleocytoplasm from which they were supposed to have arisen, as macromolecular sequencing techniques grew in sophistication from amino acid sequencing of cytochromes and ferredoxins,[31] to nucleic acid hybridization,[32] to oligonucleotide cataloguing,[33] to partial and complete ribosomal nucleic acid sequencing,[34] to secondary and tertiary structure comparisons[35] (for further details, see Reference 45 and the articles by Delihas, Erdmann *et al.,* and Sogin in this volume).

Essentially, the pattern observed in crude DNA-RNA hybridization comparisons by Pigott and Carr, published in 1972,[32] has been reinforced as more sophisticated studies have revealed greater and more complete detail, including the recognition of more or less conserved regions of the molecules. They found that plastid DNA hybridized considerably more with the rRNA of free-living cyanobacteria (to more than 40%, the same magnitude between different genera of cyanobacteria) than with the comparable, surrounding cytoplasmic rRNA (less than 1%), the reverse of what might be expected if plastids had become differentiated from cytoplasm, long after the divergence from cyanobacteria, but almost better than supporters of the SET might hope for.

The strength of evidence required to convince skeptics varies considerably. The strong general morphological, physiological, and metabolic biochemical resemblances of plastids to cyanobacteria were evident in the mid-1960s. By the mid-1970s the sequencing of proteins and ribosomal RNAs had confirmed Pigott and Carr's picture, sufficient for Woese to assert that "the case for the [symbiotic origin of the] chloroplast is a clearcut one, and it has now been proven."[36]

The case for mitochondria was complicated by the relatively small amounts of DNA present (interpreted by SET adherents as being the product of an earlier symbiotic incorporation and consequently longer integration and greater interdependence), combined with much greater diversity than plastids at the molecular level.[37] Nevertheless, a pattern essentially similar to plastids emerged, if one filtered out the "noise" produced by the oddity of some mitochondrial features, including departure from the so-called universal code (see review by Gellissen and Michaelis, this volume). These peculiarities, not present in contemporary bacteria, must have evolved subsequent to the primordial symbiotic events. In the 1970s strong links between purple nonsulfur (PNS) bacteria particularly to the so-called alpha group,[38] and mitochondria were discovered, initially as a result of cytochrome c and ferredoxin sequencing.[31] Free-living PNS members possess membrane invaginations remarkably like the tubular[39] or flat cristae of mitochondria (FIGURE 2). It was particularly striking when the nonphotosynthetic bacterium *Paracoccus denitrificans*, which showed enough metabolic similarities to mitochondria to be dubbed a "free-living mitochondrion"[40] (see also John in this volume) was subsequently shown to group with the alpha nonsulfur bacteria with rRNA sequencing.

To some,[35] the SET was established for both plastids and mitochondria when Schwartz and Dayhoff published their "tree" based on a combination of cytochrome, ferredoxin, and 5SrRNA sequences in 1978.[31] Despite criticism of their methodological assumptions,[41,42] the general clusters exhibited by their tree have held up as further organisms and molecules have been studied in greater detail. For example, tRNA comparisons initially seemed to yield equivocal results,[43] but secondary structure information has once again supported the prokaryote-like nature of mitochondrial tRNAs.[44]

As it became evident that the similarities between plastids, mitochondria, and free-living prokaryotes (the review by Gray and Doolittle covers much of the relevant data)[45] were more than superficial, opposition to the SET focused on the fact that many of the organellar proteins, particularly of mitochondria, are coded for in the nucleus. The strongest, most fully documented opposition to the symbiotic origin of mitochondria came from Raff and Mahler,[35] who also made much of the unique features and of mitochondrial diversity. If the SET is correct, then there must have been a wholesale transfer of genes from the organelles to the nucleus,[35] since it is straining credulity to assume their fortuitous prior presence in the latter. By the late 1970s many examples of apparent gene migration, dubbed "promiscuous DNA", had been found[46,47] (see also Gellissen and Michaelis, this volume) and this form of opposition faded away.

A renewed attack on the symbiotic origin of the mitochondrion came after the discovery of introns in the genome of some mitochondria,[48] because this was the first major feature that mitochondria shared with the eukaryotic cytoplasm that was not known to be present in prokaryotes. When first discovered, it was assumed that, as introns are a regular feature of eukaryotes, and eukaryotes are more advanced than prokaryotes, introns must have some advanced function. However, if introns are retained primitive features, as proposed by Darnell[49] and Doolittle[50], then they could have been lost from mitochondria in parallel with their loss from most prokaryotes. Animal mitochondria do not have introns, having minimal amounts of DNA and showing indications of much more rapid evolutionary change than plant or fungal mitochondria.[37] Intron-like segments have been found also in a few plastids.[51] No introns are known yet from purple nonsulfur bacteria or cyanobacteria, and so, if the SET is correct, the loss must have occurred after the early symbiotic events. Alternatively, the organellar introns may be products of the nuclear genome, having moved in the opposite direction from most of the gene migration.

Early suggestions for the selective advantage conferred by the acquisition of the protomitochondrion were based on the assumption that it arose as an aerobic respirer in an anaerobic host.[30] This raised the problem of how the twain would meet.[1] However, arguments could be presented for an advantage even if both were aerobic and the PNS connection raised a new and intriguing possibility. The initial protomitochondria could have been nonoxygenic, photosynthetic anaerobes (like most PNS bacteria) in an anaerobic host, much as suggested by Dickerson et al. and Woese.[52,36] Elsewhere I have drawn attention to a property of PNS bacteria which is very interesting in this regard: they are facultative aerobes, surviving in the presence of oxygen but losing

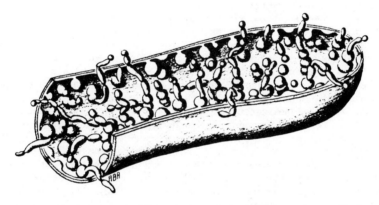

**FIGURE 2.** Artist's reconstruction of the membrane projections in a purple nonsulfur bacterium, *Rhodospirillum rubrum*. (From Holt and Marr, with permission.)[71]

their ability to photosynthesize.[53] Thus, if they were trapped inside the host and exposed to rising environmental oxygen levels, they would convert to respiratory organelles. The later cyanobacterial symbiosis re-created the original symbiotic pairing, but in an aerobic situation.

It must be recalled that there are major problems with the autogenous or partially autogenous alternatives (FIGURE 1) which transcend molecular comparisons. An autogenous origin from a nonphotosynthetic host (the Urzoan) is extremely improbable because it requires the evolution of photosystems I and II, with all their pigments and enzymes twice:[1] once in cyanobacteria and again in eukaryotes when plastids arose. Although one can get around this by deriving eukaryotes autogenously from a cyanobacteria-like ancestor (the Uralga), one must then explain why mitochondria are more like purple nonsulfur (PNS) bacteria.[45] The reverse is the problem if one begins with a PNS ancestor. If one supports an autogenous origin of mitochondria, but a xenogenous symbiotic origin for plastids, then one must explain the many similarities between the two organelles (both surrounded by two membranes confluent with endoplasmic reticulum, both containing circular, naked DNA and ribosomes, both always arising from preexisting organelles, having similar proton translocation geometry, etc.) as being due to parallelism. The surrounding cytoplasm should show a relationship to PNS bacteria, which is not the case so far.

At present, many molecular biologists might consider any further attempts to "prove" the role of symbiosis in the origin of plastids and mitochondria as "flogging a dead horse," and I am inclined to agree with them, bearing in mind that the molecular trees cannot yet be interpreted as absolutely precise indications of organellar relationships, as witnessed by trees in which amphibia arise before fish[54] and so forth. They can be influenced by several factors including differing rates of evolution in the same molecules after divergence, number and variety of determinations, and methodological differences, including alignments and tree making.[75] However there appears to be little reason to doubt that they are roughly correct for the organisms included. Certainly, there are those who will continue to doubt the SET, or even the involvement of symbiosis at all, which is as it should be. However, if current evidence is insufficient, it is difficult to imagine what new observations would convince them.

This does not mean that important questions do not remain. One that appears to be nearing resolution is the stock from which the eukaryotic nucleocytoplasmic "host" (Urkaryote) arose. To Woese and Fox,[55] its divergence was not from prokaryotes, but from a more ancient, extinct, ill-defined *progenote* stock, and they pictured a trifurcation of lineages: the Eubacteria, the Archaebacteria, and the Eukaryotes (FIGURE 3), initially defined on the basis of $16S$ rRNA sequencing. This implied that eukaryotes were no more similar to either the Eubacteria or the Archaebacteria. However, if features other than $16S$ rRNA are taken into account, some members of the Archaebacteria show features that might be expected in the Urkaryote: *Thermoplasma acidophila* lacks a cell wall, has a histone-like protein which is associated with DNA nucleosome formation, some similarities (and differences) in its initiator tRNA, it has a b-type cytochrome, its cell membrane can change configuration (but not truly phagocytose), and it has an actomyosin system inhibitable by cytochalasin B (see Searcey, this volume). Other Archaebacteria also show some resemblances to eukaryotes, including *Halobacterium,* which has some initiator tRNA similarities, a ferredoxin like Metaphytes and a red pigment extraordinarily like rhodopsin in the Metazoa. The lipids of Archaebacteria are markedly different from both the pro- and eukaryotes in having ether linkages, rather than esterifications; and so, if they are ancestral to the eukaryotes, this would have to have evolved after their divergence. Despite their peculiarities, the Archaebacteria are the only prokaryotes to exhibit any Urkaryote-like features, leading me to believe that Van Valen and Maiorana[56] and others[74] are fundamentally correct in their derivation of the eukaryotes from an Archaebacteria-like host. Searcey, and also Doolittle (this volume), provide further assessments of the possible role of Archaebacteria in eukaryotic evolution.

A further question that awaits resolution is whether there are any eukaryotic stocks that have never had plastids in them. At first sight it seems obvious that many of the "protozoan" groups diverged before the origin of plastids. However, many nonphotosynthetic groups appear to be more closely related to photosynthetic groups than to each other,[57] such as ciliates to dinoflagellates or bodonids and trypanosomes to euglenoids, and so the matter is not so simple. If there was only a single symbiosis which led to plastids (see below), then they would appear to have lost them, whereas if there were multiple symbioses, different lines could become independently and partially "infected." Perhaps some amebae, slime molds, or fungi have never been photosynthetic and others have lost it. I have difficulty conceiving of the elaborately multiflagellated trichomonads and hypermastigids (see Cavalier Smith, this volume) inhabiting the guts of insects and other low-oxygen metazoan environments as primordially amitochondrial, having lurked free living in some anaerobic oasis for millions of years until their highly specialized metazoan hosts evolved. We need a good biochemical "handle" on this question.

## MULTIPLE PRIMARY ORIGINS FOR ORGANELLES?

With the symbiotic origin of plastids and mitochondria more or less established, attention has shifted to further symbiotic influences. For example, have the organelles in question arisen by similar processes more than once? Mereschkowsky first proposed this notion,[7,8] although it is Raven's paper on multiple plastid origins that is best

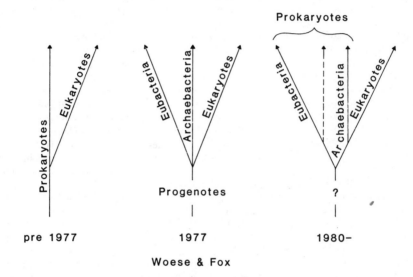

**FIGURE 3.** Alternative phylogenies of the early divergence of eukaryotes: changes in recent years. Archaebacteria may be on *both* sides of the divergence.[74]

known.[58] Subsequently, multiple origins for mitochondria have also been proposed.[59,60]

Proponents of these views use them to explain the diversity of biochemical or structural features found in these organelles. Thus, plastids have been characterized as being of three or more basic types, recognizable primarily by their pigments and, in particular, their chlorophylls: *a* only ("reds"), *a* + *b* ("greens"), and *a* + *c* ("goldens/browns"). Each of these is postulated to have arisen from a different cyanobacteria-like stock. As the cyanobacteria already had chlorophyll *a* and phycobilins similar to those found in the red algae, the ancestral stock for red algal plastids is considered to be essentially similar to contemporary cyanobacteria. However, what of the postulated green and golden/brown stocks?

The discovery of *Prochloron,*[16,61] an oxygenic prokaryote with chlorophylls *a* and *b* and lacking phycobilins, caused a revival of the multiple-origin-of-plastids notion, because it seemed to represent the missing "green" Chlorophyte-Metaphyte plastid stock. Indeed, to Margulis, this discovery "has made it virtually certain that P. H. Raven . . . is correct concerning the multiple origin of plastids."[62] For reasons indicated below, this is a distinct overstatement. Margulis and Obar subsequently presented *Heliobacterium* as a present-day counterpart to the golden/brown "chrysoplast" stock,[63] although the absence of either chlorophyll *a* or *c*, or any photosynthesis, for that matter, makes its candidacy very unconvincing. The combined presence of phycobilins with chlorophylls *a* and *c* in the cryptomonads, plus the fact that chlorophyll *c* is more like a chlorophyllide than a fully developed chlorophyll, lacking the phytol chain (FIGURE 4), argue for the origin of *c* in a phycobilin-containing-stock as a side branch in chlorophyll synthesis.

Further data that could be taken to support multiple plastid origins arose from the comparisons of amino acid sequences of cytochrome c and 2Fe-2S ferrodoxins,[31] and the nucleotide sequences of 5$S$ and 16$S$ rRNAs.[33,64] In the "trees" that resulted, plastids occurred on several branches and the 16$S$ data indicated a somewhat greater similarity between the plastid of *Porphyridium,* a red alga, and some free-living cyanobacteria than between red and *Euglena* plastids. More recently, comparisons of additional 5$S$ and 4.5$S$ rRNA sequences resulted in trees in which higher plant plastids cluster with cyanobacteria, but not with euglenoid or chlamydomonad plastids, and yet all the plastids are of the *a* + *b* type.[64] Euglenoid accessory pigments are markedly different from *Chlamydomonas* or Metaphytes, but the latter two are very similar and are the only photosynthetic organisms to produce starch within their plastids.

If plastids have not had multiple origins, then chlorophyll *b* must have evolved independently at least twice, in view of its presence in *Prochloron*. As noted elsewhere,[53] this is not as drastic an event as to make it unlikely to have evolved more than once, for it simply requires the replacement of a methyl group with an aldehyde on ring II (FIGURE 4), perhaps with more than one step involved, however. With increasing information, *Prochloron* has become less and less like a free-living green plastid and more and more evidently simply an anomalous cyanobacterium.[65]

Proponents of multiple plastid origins have differed as to whether additional biochemical or structural differences among existing plastids arose from postsymbiotic divergence or the involvement of still further prokaryotic stocks. Such differences include the presence of only one form of chlorophyll *c* ($c_2$) versus two ($c_1, c_2$) which also differ in only one position on ring II (FIGURE 4), the radically different xanthophylls of the euglenoid *a* + *b* plastid from that of the Chlorophytes, the intraplastid location of starch in the Chlorophytes and Metaphytes versus its extraplastidic location in other starch-producing lower eukaryotes, and the presence of phycobilins within, rather than external to, the thylakoids of the *a* + $c_2$ cryptomonads. In addition, secondary symbioses have also been invoked (see below), based primarily on the number of membranes surrounding the plastids. For example, the scheme presented by Whatley involves three primary and four secondary symbioses to account for present plastid diversity.[66] The most extreme form can be found in Mirabdullaev's scenario, involving six primary and four secondary symbioses, with minimal postsymbiotic divergence.[67]

I have argued elsewhere that, with the exception of cyanelles (most of which appear to be obligately symbiotic cyanobacteria), existing plastids have had a single primary origin, their diversity being due to postsymbiotic divergence.[53,56] The principal basis for this is the consistency in apparent phylogenetic patterns between plastids, mitochondrial cristal morphology, and other features such as flagellar hairs or storage products, which are strongly indicative of coevolution. On the other hand, the cyanelles of today appear to be transforming into plastids. Why should this only be so now and 1500 million years ago? What of the interim?[1]

Polyphylety has also been suggested for mitochondria, based primarily on the form of the cristae or the presence or absence of introns. In protists the cristal form is usually highly group conservative, with flattened or tubular/vesicular forms as the basic dualism. Basic mitochondrial cristal morphology was first used to determine gross affinities in protistan phylogeny in the mid-1970s. Their conservativeness in some groups had been noted as early as 1959 by Manton,[68] but mitochondrial cristae were not used to assist in determining affinities until the study of the cryptomonad symbiont of *Mesodinium rubrum* by Taylor *et al.* 10 years later.[69]

Stewart and Mattox have taken the fundamental cristal dualism to be indicative of a biphyletic origin.[59,60] The presence of intron-like segments in the mitochondrial DNA of yeasts and other nonflagellated fungi but absence in animal mitochondria

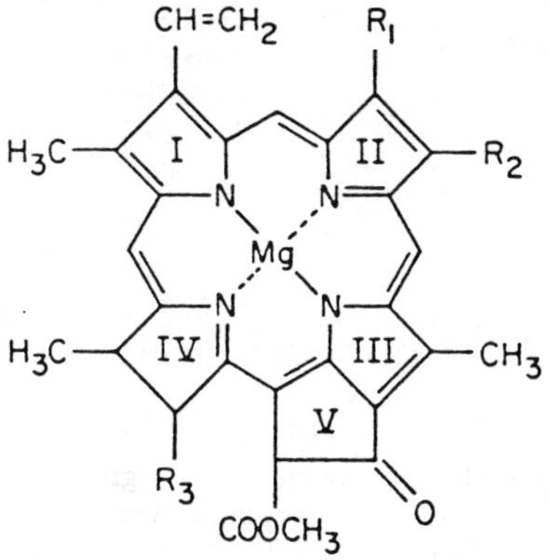

|        | $R_1$ | $R_2$ | $R_3$ |
|--------|-------|-------|-------|
| Chl a  | $CH_3$ | $CH_2CH_3$ | $(H_2C)_2$ — PHYTOL |
| Chl b  | CHO | $CH_2CH_3$ | $(H_2C)_2$ — PHYTOL |
| Chl $c_1$ | $CH_3$ | $CH_2CH_3$ | $HC=CH_2$ — COOH |
| Chl $c_2$ | $CH_3$ | $CH=CH_2$ | $HC=CH_2$ — COOH |

**FIGURE 4.** The structure of the eukaryotic chlorophylls (courtesy of B. Prézelin).

(see above), and comparisons of 16$S$ and 23$S$ rRNAs, have also been taken to indicate biphyletys, but they are different from those suggested by cristal morphology. This, taken together with the fact that it can vary considerably in a few protists, notably within the same species of amebae, some trypanosomes or yeasts, depending on environmental oxygen levels, and within the tissues of a single metazoan, complicates the multiple origin case. Mitochondrial crista differences between different amebae and heliozoans do appear to be the result of polyphylety,[56] but as a result of divergence after the original mitochondrial event.

It is possible to construct a conservative phyletic "tree" of the protists[56] in which the primary crystal form is flat, with tubular (or vesicular—short, swollen tubular) being derivative. This pattern is consistent with plastid and flagellar characteristics (with cryptomonads being an interesting anomaly—see below), suggesting coevolution of all three systems. It is difficult to obtain molecular information on the tubular cristal type,[60] the mitochondria of the ciliate *Tetrahymena* still being the only one studied at the appropriate level. Unfortunately, the genetic system of this ciliate is so unusual (including the presence of linear DNA)[37] that its use as an exemplar for tubular cristal mitochondria is highly questionable. The intron question has been discussed above.

An important point is that, if mitochondria had a polyphyletic origin, why are they of near-universal occurrence in eukaryotes? It is true that some protists, such as the giant ameba *Pelomyxa,* and a variety of symbiotic gut flagellates lack mitochondria (see above). These organisms permanently inhabit very low oxygen environments in which mitochondria would be of little use. It seems reasonable to suggest that they lost them, although the possibility of *Pelomyxa* being of premitochondrial stock has been proposed.[30] Further, unlike the situation with plastids, if PNS bacteria gave rise to all mitochondria, this could occur only under anaerobic conditions.

Taking all this into consideration the cases for multiple symbioses involving the same organelles do not seem strong enough to maintain at this time.


## SECONDARY SYMBIOTIC ORIGINS OF EUKARYOTIC ORGANELLES


There are many examples of obligate intracellular symbioses, particularly among protists, in which the endosymbiont (cytobiont;[19] endocytobiont[4]) has become an integral part of its host. As long as its nucleus continues to control the functions and destiny of its organelles, we still think of the cytobiont as a foreign entity. The study of contemporary symbioses indicates that there is a strong tendency for dependence to increase, with "degeneration" of the cytobiont and loss of genetic material from both.[53] In the early 1970s the possibility was raised that these integrative processes might proceed to the point where the nucleus of the cytobiont becomes nonfunctional and is lost, the host cell now "farming" its acquired organelles.

No instances of this actually happening are known, although there are some well-known examples of extreme integration involving photosynthetic cytobionts, such as the ciliate host/cryptomonad cytobiont of *Mesodinium rubrum*[19,53,69] and the dinoflagellates *Peridinium balticum* and *Kryptoperidinium* (*"Glenodinium"*) *foliaceum* with chrysomonad or diatom-like cytobionts.[19] In these cases the cytobionts ramify to such an extent within the hosts, separated by only a single membrane, that their presence was not clearly recognized without electron microscopy. In most populations of *M. rubrum* the cytobiont has fragmented into many "islands" of cytoplasm containing mitochondria and plastids, with only one island containing a cryptomonad nucleus.

The discovery of the "nucleomorph" in cryptomonads and the enigmatic *Chlorachnion reptans,* plus the subsequent demonstration of DNA inside it (see Ludwig and Gibbs, this volume) is perhaps the most compelling evidence for a secondary symbiotic origin of plastids. It would provide a solution to the paradoxical apparent relationships of the group,[56] with most of the cell resembling other flagellates with flattened mitochondrial cristae, but the plastid containing a unique combination of pigments (chlorophyll *c* and phycobilins). There are no free-living organisms that have plastids like those of the inner compartment of cryptomonads, and so the group must have become extinct subsequent to the symbiosis having occurred. The topology of the putative cytobiont is peculiar, being surrounded by two membranes (chloroplast endoplasmic reticulum, confluent with the nuclear envelope). Various interpretations have been placed on this. Cavalier-Smith's interpretation of the process by which this arose seems most reasonable to me.[70]

Some authors have invoked secondary symbioses several times during protist phylogeny (see above) but, apart from the the specific cases discussed above, the evidence is, once again, not strong enough to consider them more compelling than simple divergence from one or two photosynthetic eukaryotic ancestors. Furthermore, its invocation does not provide any new insights other than cryptomonad affinities.

Although a number of instances of protists temporarily maintaining foreign plastids from their digested food are known[1] (see also Stoecker, this volume), there are no known cases in which this "chloroplast maintenance" has persisted longer than approximately six weeks.

# CONCLUSIONS

A consequence of the change in paradigm for the origins of plastids and mitochondria is that their primordial state was one of independence. Consequently, interest focuses on the processes of integration and regulation that appear to have taken place. Studies of contemporary symbioses become valuable indicators of the types of processes that can be expected to have taken place. Given the rapidity with which some symbiotic relationships can progress (see Jeon, this volume), one cannot help being struck by the fact that the symbioses involved in eukaryotic evolution occurred approximately 1500 million years ago:[62] they are the oldest symbioses on earth, and it seems extraordinary to me that the cytobionts should retain sufficient individuality to still be recognizable as such. Equally surprising, stasis notwithstanding, is the fact that so many of the putative cast of free-living characters of the SET are still present. A priori one might expect that, since the events occurred so long ago, they would have become extinct. The SET would still be defensible even if no cyanelles were known, if cyanobacteria were much more different from plastids than they are, and if purple nonsulfur bacteria lacked crista-like membrane protrusions. It would be better still if the the surviving organellar ancestral stocks possessed introns. No living Archaebacteria possess all the features required for the "host" Urkaryote, but is that surprising? There seem to be enough eukaryote-like features scattered among members of the nonmethanogenic Archaebacteria to justify the derivation of the Urkaryote from an extinct member of the group, particularly in the Thermoproteales.[74]

As I have noted on several previous occasions,[19,53] while the SET has acquired wide acceptance, some of its other consequences are not fully realized. Cell biologists accept that the SET is the most probable explanation, but nevertheless still teach the classical "cell theory" (modified to allow for viruses). Eukaryotic and prokaryotic "cells" are *not* equivalent: the former are polygenomic multiples of prokaryotic "cells:" they are at different levels of organization.[1] Similarly, the mechanism of organellar acquisition, "inheritable symbiosis," i.e., phagotrophy followed by retention, genetic interdependence, and endless (so far) transmission to daughter cells due to cytoplasmic sharing, seems to be a fine example of the "evolution of acquired characteristics," even if not of the type Lamarck had in mind and without future goals.

When I first became intrigued by this evolutionary topic nearly 20 years ago I had no idea how increasingly definable and data rich it was to become. There are many evolutionary questions for which opposite views can be equally justifiably or unjustifiably advocated and where personal influence and rhetoric are of major importance in the championing of one view over the other. However, this has certainly not been a principal feature of the SET. Its predictions have been upheld by increasingly powerful molecular biological tools, and rejection can only be by denial of the value of the heart of modern biology, molecular sequencing. To the historian of science, the SET controversy offers an excellent model of the causal relationships of observations and ideas, inertia and momentum, pontification and timidity. The story is worth following: it has been fascinating so far.

## REFERENCES

1. TAYLOR, F. J. R. 1974. Implications and extensions of the Serial Endosymbiosis Theory of the origin of eukaryotes. Taxon 23: 229-258.
2. GRAY, M. W. 1983. The bacterial ancestry of plastids and mitochondria. Bioscience 33: 693-699.
3. ALBERTS, B., D. BRAY, J. LEWIS, M. RAFF, K. ROBERTS & J. D. WATSON. 1983. Molecular Biology of The Cell. Garland Publishing Inc. New York, N.Y.
4. SCHWEMMLER, W. 1984. Reconstruction of Cell Evolution: A Periodic System. CRC Press, Inc. Boca Raton, Fl.
5. TAYLOR, F. J. R. 1980. The stimulation of cell research by endosymbiotic hypotheses for the origin of eukaryotes. In Endocytobiology. W. Schwemmler & H. E. A. Schenk, Eds.: 917-942. Walter de Gruyter & Co. Berlin, FRG.
6. SCHIMPER, A. F. W. 1883. Uber die Entwicklung der Chlorophyllkorner und Farbkorner (1 Teil). Bot. Zeit. 41: 105-114.
7. MERESCHKOWSKY, C. 1905. Über Natur und Ursprung der Chromatophoren im Pflanzenreiche. Biol. Centr. 25: 593-604.
8. MERESCHKOWSKY, C. 1910. Theorie der zwei Plasmaarten als Grundlage der Symbiogenesis, einer neuen Lehre von der Entstehung der Organismen. Biol. Centralbl. 30: 278-303, 321-347, 353-367.
9. MERESCHKOWSKY, C. 1920. La plante considérée comme une complex symbiotique. Bull. Soc. Nat. Sci. Ouest France 6: 17-21.
10. STANIER, R. Y. & C. B. VAN NIEL. 1962. The concept of a bacterium. Arch. Mikrobiol. 42: 17-35.
11. CAVALIER-SMITH, T. 1975. The origin of nuclei and of eukaryotic cells. Nature 256: 463-468.
12. TAYLOR, F. J. R. 1976. Autogenous theories for the origin of eukaryotes. Taxon. 25: 377-390.
13. ERNSTER, L. & G. SCHATZ. 1981. Mitochondria: a historical review. J. Cell Biol. 91: 227s-255s.
14. ALTMANN, R. 1890. Die Elementarorganismen und ihre Beziehung zu den Zellen. Veit & Co. Leipzig, Germany.
15. PORTIER, P. 1918. Les Symbiotes. Masson et Cie. Paris, France.
16. LEWIN, R. 1981. Prochloron and the theory of symbiogenesis. Ann. N.Y. Acad. Sci. 361: 325-329.
17. WALLIN, I. E. 1922. On the nature of mitochondria. Am. J. Anat. 30: 203-229, 451-471.
18. WALLIN, I. E. 1927. Symbionticism and the Origin of Species. Bailliere, Tindall & Cox. London, England.
19. TAYLOR, F. J. R. 1979. Symbionticism revisited: a discussion of the evolutionary impact of intracellular symbioses. Proc. R. Soc. London Ser. B 204: 267-286.
20. KOZO-POLYANSKY, B. M. 1924. New Principles of Biology. (In Russian.) (As cited in OPARIN, A. I. 1938. The Origin of Life. MacMillan. London, England.)

21. SAGAN, L. 1967. The origin of mitosing cells. J. Theor. Biol. **14:** 225-274.
22. WHEATLEY, D. N. 1982. The Centriole: a Central Enigma of Cell Biology. Elsevier Biomedical Press. Amsterdam, Holland.
23. CORRENS, C. 1909. Vererbungsversuche mit blas (gelb) grunen und bunt blattrigen sippen bei *Mirabilis, Urtica* und *Lunaria.* Z. Vererbungs. **1:** 291-329.
24. EPHRUSSI, B., H. HOTTINGUER & A.-M. CHIMENES. 1949. Action de l'acriflavine sur les levures. I. La mutation "petite colonie". Ann. Inst. Pasteur Paris **76:** 351-367.
25. EBRINGER, L. 1978. Effects of drugs on chloroplasts. *In* Progress in Molecular and Subcellular Biology, F. E. Hahn, Ed.: 271-350. Springer Verlag. Berlin, FRG.
26. PROVASOLI, L., S. H. HUTNER & A. SCHATZ. 1948. Streptomycin-induced chlorophyllless races of *Euglena.* Proc. Soc. Exp. Biol. Med. **69:** 279-282.
27. KIES, L. 1980. Morphology and systematic position of some endocyanomes. *In* Endocytobiology. W. Schwemmler & H. E. A. Schenk, Eds.: 7-19. Walter de Gruyter. Berlin, FRG.
28. RIS, H. 1961. Ultrastructure and molecular organization of genetic systems. Can. J. Genet. Cytol. **3:** 95-120.
29. NASS, M. M. K. & S. NASS. 1963. Intramitochondrial fibers with DNA characteristics. I. Fixation and electron staining reactions. J. Cell Biol. **19:** 593-611.
30. MARGULIS, L. 1970. Origin of Eukaryotic Cells. Yale University Press. New Haven, Conn.
31. SCHWARTZ, R. M. & M. O. DAYHOFF. 1978. Origins of prokaryotes, eukaryotes, mitochondria, and chloroplasts. Science **199:** 395-403.
32. PIGOTT, G. H. & N. CARR. 1972. Homology between nucleic acids of blue-green algae and chloroplasts of *Euglena gracilis.* Science **175:** 1259-1261.
33. BONEN, L. & W. F. DOOLITTLE. 1975. On the prokaryotic nature of red algal chloroplasts. Proc. Nat. Acad. Sci. USA **72:** 2310-2314.
34. PHILLIPS, D. O. & N. G. CARR. 1980. Molecular approaches to the endosymbiotic hypothesis. Ann. N.Y. Acad. Sci. **361:** 298-311.
35. WOLTERS, J. & V. A. ERDMANN. 1984. Comparative analyses of small ribosomal RNAs with respect to the evolution of plastids and mitochondria. Endocytol. Cell Res. **1:** 1-23.
36. WOËSE, C. R. 1977. Endosymbionts and mitochondrial origins. J. Mol. Evol. **10:** 93-96.
37. GRAY, M. W. 1982. Mitochondrial genome diversity and the evolution of mitochondrial DNA. Can. J. Biochem. **60:** 157-171.
38. VILLANUEVA, E., K. R. LUEHRSEN, J. GIBSON, N. DELIHAS & G. E. FOX. 1985. Phylogenetic origins of the plant mitochondrion based on a comparative analysis of 5S ribosomal RNA sequences. J. Mol. Evol. **22:** 46-52.
39. PFENNIG, N. & H. G. TRUPER. 1983. Taxonomy of phototrophic green and purple bacteria: a review. Ann. Microbiol. **134**(B): 9-20.
40. JOHN, P. & F. R. WHATLEY. 1975. *Paracoccus denitrificans* and the evolutionary origin of the mitochondrion. Nature **254:** 495-498.
41. DEMOULIN, V. 1979. Protein and nucleic acid sequence data and phylogeny. Science **205:** 1036-1038.
42. UZZELL, T. & C. SPOLSKY. 1980. Two data sets: alternative explanations and interpretations. Ann. N.Y. Acad. Sci. **361:** 481-499.
43. PARTHIER, B. 1982. Transfer RNA and the phylogenetic origin of cell organelles. Biol. Zbl. **101:** 577-596.
44. WREDE, P. 1986. Evolution of the mitochondrial tRNA genes. Endocytol. Cell. Res. **3:** 1-27.
45. GRAY, M. W. & W. F. DOOLITTLE. 1982. Has the endosymbiont hypothesis been proven? Microbiol. Rev. **46:** 1-42.
46. FARRELLY, F. & R. A. BUTOW. 1983. Rearranged mitochondrial genes in the yeast nuclear genome. Nature **301:** 296-301.
47. STERN, D. B. & D. M. LONSDALE. 1982. Mitochondrial and chloroplast genomes of maize have a 12-kilobase DNA sequence in common. Nature **299:** 698-702.
48. MAHLER, H. R. 1983. The exon:intron structure of some mitochondrial genes and its relation to mitochondrial evolution. Int. Rev. Cytol. **82:** 1-97.
49. DARNELL, J. E. 1978. Implications of RNA:RNA splicing in evolution of eukaryotic cells. Science **202:** 1257-1260.

50. DOOLITTLE, W. F. 1978. Genes in pieces: were they ever together? Nature **272:** 581-582.
51. ALLET, B. & J.-D. ROCHAIX. 1979. Structure analysis at the ends of the intervening DNA sequences in the chloroplast 23S ribosomal genes of *C. reinhardii.* Cell **18:** 55-60.
52. DICKERSON, R. E., R. TIMKOVICH & R. ALMASSY. 1976. The cytochrome fold and the evolution of bacterial energy metabolism. J. Mol. Biol. **100:** 473-491.
53. TAYLOR, F. J. R. 1979. Symbionticism revisited: a discussion of the evolutionary impact of intracellular symbioses. Proc. R. Soc. London Ser. B **204:** 267-286.
54. HORI, H. & S. OSAWA. 1979. Evolutionary change in 5S RNA secondary structure and a phylogenic tree of 54 5S RNA species. Proc. Nat. Acad. Sci. USA **76:** 381-385.
55. WOESE, C. R. & G. E. FOX. 1977. Phylogenetic structure of the prokaryotic domain: the primary kingdom. Proc. Nat. Acad. Sci. USA **74:** 5088-5090.
56. VAN VALEN, L. M. & V. C. MAIORANA. 1980. The archaebacteria and eukaryotic origins. Nature **287:** 248-250.
57. TAYLOR, F. J. R. 1978. Problems in the development of an explicit hypothetical phylogeny of the lower eukaryotes. BioSystems **10:** 67-89.
58. RAVEN, R. H. 1970. A multiple origin for plastids and mitochondria. Science **169:** 641-645.
59. STEWART, K. D. & K. R. MATTOX. 1980. Phylogeny of phytoflagellates. *In* Phytoflagellates. E. R. Cox, Ed.: 433-462. Elsevier/North Holland Inc. New York, N.Y.
60. STEWART, K. D. & K. R. MATTOX. 1984. The case for a polyphyletic origin of mitochondria: morphological and molecular comparisons. J. Mol. Evol. **21:** 54-57.
61. LEWIN, R. A. & N. W. WITHERS. 1975. Extraordinary pigment composition of a prokaryotic alga. Nature **256:** 735-737.
62. MARGULIS, L. 1981. Symbiosis in Cell Evolution. W. H. Freeman & Co. San Francisco, Calif.
63. MARGULIS, L. & R. OBAR. 1985. *Heliobacterium* and the origins of chrysoplasts. BioSystems **17:** 317-325.
64. GEORGE, D. G., L. T. HUNT & M. O. DAYHOFF. 1983. Sequence evidence for the symbiotic origins of chloroplasts and mitochondria. *In* Endocytobiology II. H. E. A. Schenk & W. Schwemmler, Eds.: 845-861. Walter De Gruyter. Berlin, FRG.
65. STACKEBRANDT, E. 1983. A phylogenetic analysis of *Prochloron. In* Endocytobiology II. H. E. A. Schenk & W. Schwemmler, Eds.: 921-932. Walter De Gruyter. Berlin, FRG.
66. WHATLEY, J. M. & F. R. WHATLEY. 1981. Chloroplast evolution. New Phytol. **87:** 233-247.
67. MIRABDULLAEV, I. M. 1985. The evolution of plastids and origin of cyanobacteria. Zhurn. Obshch. Biol. **46:** 483-490.
68. MANTON, I. 1959. Electron microscopical observations on a very small flagellate: the problem of *Chromulina pusilla* Butcher. J. Mar. Biol. Ass. U.K. **38:** 319-333.
69. TAYLOR, F. J. R., D. J. BLACKBOURN & J. BLACKBOURN. 1969. Ultrastructure of the chloroplasts and associated structures within the marine ciliate *Mesodinium rubrum* (Lohmann). Nature **224:** 819-821.
70. CAVALIER-SMITH, T. 1985. The kingdom Chromista: origin and systematics *In* Progress in Phycological Research **3.** Biopress Ltd. Bristol, England.
71. HOLT, S. C. & A. G. MARR. 1965. Location of chlorophyll in *Rhodospirillum rubrum.* J. Bacteriol. **89:** 1402-1412.
72. SNEATH, P. H. A. 1974. Phylogeny of microorganisms. Symp. Soc. Gen. Microbiol. **24:** 1-39.
73. KITCHER, P. 1982. Abusing Science: the Case against Creationism. MIT Press. Cambridge, Mass.
74. ZILLIG, W., R. SCHNABEL & K. G. STETTER. 1985. Archaebacteria and the origin of eukaryotic cytoplasm. Curr. Top. Microbiol. Immunol. **114:** 1-18.
75. ROTHSCHILD, L. J., M. A. RAGAN, A. W. COLEMAN, P. HEYWOOD & S. A. GERBI. 1986. Are rRNA sequence comparisons the Rosetta Stone of phylogenetics? Cell **47:** 640.

# The Origin of Eukaryote and Archaebacterial Cells

T. CAVALIER-SMITH

*Department of Biophysics, Cell and Molecular Biology*
*King's College London*
*26-29 Drury Lane*
*London WC2B 5RL, United Kingdom*

## INTRODUCTION AND SUMMARY

Ultrastructural and molecular data have recently so rejuvenated the study of organismic diversity that we may soon have a clear understanding of the overall phylogeny of the living world, and even of the major steps in its diversification. Of these, the transition from the prokaryote to the eukaryote cell is certainly the most profound,[1] so much so that Prokaryota and Eukaryota are now generally ranked as superkingdoms (TABLE 1). To explain the origin of the eukaryote cell one has to determine the properties of the most primitive eukaryote, identify its likely prokaryotic ancestor, and explain in detail how the latter evolved into the former. That is the object of this paper.

Identifying the most primitive eukaryote depends upon a proper understanding of the diversity and phylogenetic relationships within the most primitive eukaryote kingdom, the Protozoa. TABLE 2 indicates the protozoan classification and nomenclature that will be used in this paper: those protozoa recently separated as the subkingdom Archezoa are of special significance for early eukaryote evolution. All four archezoan phyla, though fully eukaryotic, completely lack any trace of mitochondria; the first three of them, especially, have a variety of other apparently primitive characters suggesting that they are living representatives of the earliest phases of eukaryote evolution. I argue that the most primitive eukaryote was a phagotrophic archezoan, with no chloroplasts, no mitochondria, no microbodies, and no stacked smooth cisternae forming a Golgi dictyosome, but possessing a single cilium with a sheaf of rootlet microtubules surrounding the single nucleus that divided by a closed mitosis in which the ciliary kinetosome was attached to the centrosome. The present day Mastigamoebea closely fit this description and may be "living fossils." I suggest that this first eukaryote had a single chromosome and could form a resting cyst or spore in which (as in modern Anaxostylea) the ciliary axoneme and rootlets were not disassembled, and that with the origin of sexual syngamy allopolyploidy led to the formation of a cell with two chromosomes and two dissimilar (i.e., "anisokont") cilia. This ancestral two-chromosomed anisokont perfected a primitive one-step meiosis and was the ancestor not only of the amitochondrial Metamonada and Parabasalia but also of all eukaryotes with mitochondria and chloroplasts, which it acquired by endosymbiosis of purple nonsulfur bacteria[6] and cyanobacteria[7] respectively. The

17

amitochondrial Pelobiontea and Microspora probably evolved from Mastigamoebea by the loss of cilia.

The host that acquired mitochondria was, I have argued,[4,6,7] an early member of the Anaxostylea that had a gullet and two anisokont cilia with paraxial rods and nontubular mastigonemes. During the conversion of the symbionts into organelles,[8] they diversified considerably to produce the great diversity of chloroplasts[7] and mitochondria[3,6] as outlined in FIGURE 1 with special reference to the variety of

TABLE 1. The Seven-Kingdom Classification[2-5]

| Kingdom | Subkingdom | Branch | Number of Phyla |
|---|---|---|---|
| **Superkingdom 1. Prokaryota** | | | |
| 1. Eubacteria | 1. Negibacteria[a] | 1. Murnebacteria[a] | 8 |
| | | 2. Planctobacteria[a] | 1 |
| | 2. Posibacteria[a] | | 1 |
| 2. Archaebacteria (= Metabacteria) | | | 3 |
| **Superkingdom 2. Eukaryota** | | | |
| 1. Protozoa | 1. Archezoa | | 4 |
| | 2. Mitozoa | 1. Miozoa[a] | 2 |
| | | 2. Allozoa[a] | 4 |
| | | 3. Radiozoa[a] | 1 |
| | | 4. Sarcozoa[a] | 3 |
| | | 5. Euglenozoa | 1 |
| | | 6. Choanozoa | 1 |
| 2. Chromista | 1. Cryptomonada | | 1 |
| | 2. Chromophyta | | 2 |
| 3. Plantae | 1. Biliphyta | | 1 |
| | 2. Viridiplantae | | 5 |
| 4. Fungi | | | 4 |
| 5. Animalia | 1. Phytozoa[a] (i.e., Porifera, Placozoa, Cnidaria) | | 3 |
| | 2. Bilateria | 1. Lophophorata | 3 |
| | | 2. Deuterostomia | 3 |
| | | 3. Spiralia | 14 |
| | | 4. Ctenophora | 1 |
| | | 5. Pseudocoelomata | 3 |

[a] New taxa and names proposed here. Posibacteria comprise the gram-positive bacteria (Firmicutes) and the mycoplasmas (Mollicutes); Negibacteria comprise all those bacteria with a second lipoprotein outer membrane in addition to the plasma membrane. Murnebacteria are *mure*in-containing *ne*gibacteria; Planctobacteria are negibacteria with no murein sacculus. For Miozoa, Allozoa, Radiozoa, and Sarcozoa see TABLE 2.

mitochondrial cristae. At this time, the ciliary transition region also underwent its basic diversification[4] and the Golgi membranes became stacked to form dictyosomes, associating with cytoskeletal elements in ways distinctive for particular protist groups. On this view the symbiotic origin of mitochondria and chloroplasts was associated with a radical diversification of eukaryote protists,[13] but not with the origin of the first ciliated eukaryote which occurred purely autogenously. The basic reason for the above scenario is the phylogenetic sense it makes of protist diversity as outlined in FIGURE 1. This paper concentrates on the origin of the first eukaryote and of protozoan diversity; diversification within the kingdoms Plantae,[7] Chromista,[4] and Fungi[5] has been discussed previously.

It has been suggested that a direct transition from prokaryote to eukaryote is impossible[15] and that eukaryotes are as old as prokaryotes[15-18] and diverged independently from a primitive precellular organism of ill-specified properties, named the progenote.[15-18] However, as the fossil record strongly indicates that eukaryotes did not evolve before about 960-1450 million years (My) ago[19] but that prokaryotes go back at least 3500 My,[19,20] I am not attracted by that hypothesis, and consider that the ancestor of eukaryotes must have been a well-developed prokaryote.[1] Of the various bacterial groups, only two show sufficient affinity with eukaryotes to be considered seriously as possible ancestors: (1) the aerobic archaebacteria such as *Halobacterium* and *Thermoplasma* (TABLE 3); and (2) the aerobic eubacteria of the actinomycete subdivision of the posibacteria (TABLE 4).

The major objection to an archaebacterial ancestry for eukaryotes is the peculiar isoprenoid ether-linked membrane lipids,[49] which replace the typical fatty acid based ester-linked phospholipids of eubacteria and eukaryotes. The simplest way of reconciling this fundamental difference with the marked resemblances to eukaryotes, which seem too great for coincidence, is to suppose that eukaryotes evolved from an early ancestor of modern archaebacteria that had already evolved the features shown in TABLE 3 but still possessed the ordinary ester-linked phospholipids of their eubacterial ancestry. This, of course, raises the question of the origin of archaebacteria themselves, conventionally shrouded in the mists around the mysterious progenote.[15,17]

I shall argue (1) that the archaebacteria evolved from the actinomycete branch of the posibacteria as a result of a mutation that caused the loss of their cell wall by preventing the synthesis of muramic acid;[1] (2) that the unusual archaebacterial membrane and wall chemistry, on the one hand, and the eukaryote cytoskeleton and membrane sterols, on the other,[50] evolved as divergent ways of coping with the resulting dangerously fluid cell surface of this ancestral mutant; (3) that by developing a cytoskeleton rather than a new cell wall, eukaryotes, unlike archaebacteria, were able to evolve phagocytosis[50,51] and sex;[50] (4) that phagocytosis necessarily led to the evolution of the nucleus,[50] mitosis,[50] cilia, and endomembrane system,[50] and some time afterwards led to the symbiotic origin of mitochondria,[6,13] chloroplasts,[7,13] and microbodies.[13]

## THE RELATIONSHIP BETWEEN EUBACTERIA, ARCHAEBACTERIA, AND EUKARYOTES

My thesis (FIGURE 2b) that eukaryotes evolved from a transient intermediate between posibacteria and the ancestral archaebacterium also goes counter to the conventional wisdom[15] that there is no direct connection between the two bacterial kingdoms and that they also evolved independently from an early "progenote" (FIGURE 2a). The idea that the three groups are not directly related (FIGURE 2a) arose because of the very low similarity coefficients ($S_{AB}$ values) calculated from their small subunit ribosomal RNA (srRNA) oligonucleotide catalogues. But it has been shown that low $S_{AB}$ values grossly underestimate the degree of sequence homology, and thus are "almost meaningless":[27] no important conclusion can be based on them.

TABLE 2. Classification of the Kingdom Protozoa

Subkingdom 1. Archezoa[3] (Mitochondria Universally Absent)

| Phylum | Class | Examples |
|---|---|---|
| 1. Archamoebae | 1. Mastigamoebea[a] (ciliated archamoebae) | *Mastigina, Mastigella,* those Mastigamoebae with no mitochondria |
| | 2. Pelobiontea[a] (nonciliated archamoebae) | *Pelomyxa palustris, Entamoeba* |
| 2. Microspora | 1. Metchnikovellea | *Metchnikovella* |
| | 2. Microsporidea | *Nosema* |
| 3. Metamonada | 1. Anaxostylea | Diplomonadida, Retortamonadida |
| | 2. Axostylaria | Oxymonadida |
| 4. Parabasalia[e] | 1. Trichomonadea[a] | Trichomonads |
| | 2. Hypermastigea[a] | Hypermastigotes |

Subkingdom 2. Mitozoa[4] (Mitochondria Present)[b]

| Branch | Phylum | Examples |
|---|---|---|
| 1. Miozoa[a] (1-step meiosis; ampulliform mitochondrial cristae) | 1. Dinozoa | classes Peridinea[a] (free-living "dinoflagellates" with exonuclear spindles and few histones), Syndinea, Oxyrrhea[a] (i.e., *Oxyrrhis:* intranuclear spindle; histone-rich) |
| | 2. Sporozoa[c] | classes Gregarinea, Haematozoa (e.g., malaria), Coccidea |
| 2. Allozoa[a] (2-step meiosis; cristae rounded tubules, often curved) | 1. Proterozoa | Opalinida, Proteromonadida, Pseudodendromonadida, Plasmodiophorida, Cercomonadida, *Reckertia* |
| | 2. Infusoria[b] (=Ciliophora= Heterokaryota) | "ciliates" and suctorians |
| | 3. Mesozoa | |
| | 4. Mycetozoa | classes Protostelea, Dictyostelea, Myxogastrea |
| 3. Radiozoa[a] (cristae rounded or flattened tubules) | 1. Radiozoa[a] | subphyla Acantharia,[a] and Radiolaria[a] (classes Spumellaria, Nassellarea, Phaeodarea) |
| 4. Sarcozoa[a] | 1. Sarcodina | 3 subphyla: Amoebozoa[a] (classes Lobosea, Acarpomyxea, Xenophyophorea), Rhizopoda[a] (classes Filosea, Granuloreticulosea, Chlorarachnea), and Heliozoa |
| | 2. Myxozoa | Myxosporidia (incl. Actinosporidia) |

TABLE 2 (*continued*)

| Subkingdom 2. Mitozoa[4] (Mitochondria Present)[b] | | |
| --- | --- | --- |
| Branch | Phylum | Examples |
| | 3. Ascetospora | Haplospondia, Paramyxa |
| 5. Euglenozoa[a] (meiosis unknown;[d] discoidal mitochondrial cristae) | 1. Euglenozoa | classes Euglenoidea,[a] Kinetoplastidea[a] (i.e., trypanosomes and bodonids), Pseudocileata, Acrasea sensu strictu |
| 6. Choanozoa[a] (meiosis unknown; nondiscoidal, flattened, platelike cristae) | 1. Choanozoa | Class Choanomonadea[a] (sole order Choanoflagellida) |

[a] New branches, phyla, subphyla, or classes: except for the Miozoa, Allozoa, Peridinea, Oxyrrhea, Amoebozoa, Rhizopoda (present sense), which are novel, these taxa have previously been recognized at different levels.

[b] A few infusoria symbiotic in animal guts apparently lack mitochondria.

[c] Recent evidence suggests that at least 1 sporozoan has 2-step meiosis.

[d] Recent evidence suggests that trypanosomes must have meiosis at some stage of the life cycle.

[e] Possibly degenerate mitozoans,[13] not true archezoa.

Total srRNA sequences now available show a high degree of homology in sequence (over 50%) and deduced secondary structure between all three groups,[26] which therefore could not have diverged from each other as earlier proposed in the earliest phases of ribosome evolution. The eubacterial and archaebacterial sequences and predicted secondary structures resemble each other much more than they do those of eukaryotes,[26] and the eukaryotic rRNA is more similar to the archaebacterial than to the eubacterial rRNA, which favors the phylogenies in FIGURE 2b or 2c rather than those of FIGURE 2a or 2d. Phylogenies shown in FIGURES 2c and 2d are unlikely for the reasons summarized in the legends, so the overall most satisfactory one is that of FIGURE 2b which treats the eubacteria as the most primitive cells.

Other characters also show that archaebacteria and eubacteria could not have diverged in the earliest phases of evolution (TABLE 5), and clearly show that archaebacteria are fundamentally prokaryotes and not a "third type" of organism. Certainly the most fundamental division within prokaryotes is that between archaebacteria and eubacteria (TABLE 1), but it need not have happened early in evolution. The fossil record in the form of stromatolites shows that eubacteria go back at least 3500 My,[52] whereas there is no such evidence for archaebacteria which may be much more recent: indeed authors skeptical of their archaism call them Metabacteria.[27]

Additional evidence against a divergence between prokaryotes and eukaryotes before DNA replication, repair, transcription, and translation all became efficient lies in (1) the extensive sequence homology of the largest RNA polymerase subunit in bacteria, yeast, and *Drosophila;*[55,56] (2) the mutual recognition[62] of the signal sequences involved in protein secretion; (3) the ability of certain posibacterial and archaebacterial plasmids to replicate in eukaryotes (significantly for my present thesis no negibacterial plasmids are known to do this),[63] and (4) the sensitivity of translation to anisomycin in both archaebacteria and eukaryotes.[24]

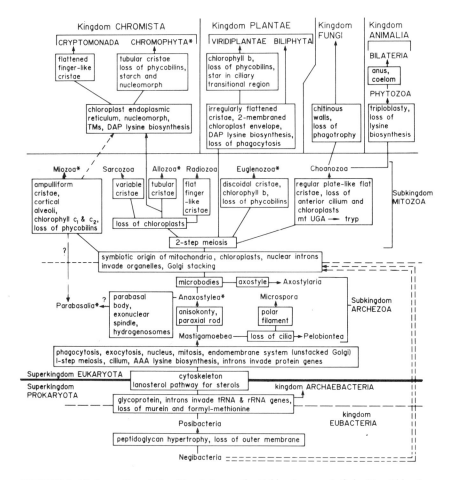

**FIGURE 1.** Phylogenetic relationships between the 7 kingdoms and their 10 subkingdoms (capitals). For the subkingdoms Eubacteria and Mitozoa their separate branches are also shown, and in the Archezoa distinct phyla or classes, because of their special phylogenetic importance. Major innovations are shown in boxes. Taxa containing some members with ciliary paraxial rods, which probably originated with the Anaxostylea,[4] are shown by an asterisk. The ancestral eukaryote lysine biosynthetic pathway via an α-amino-adipic acid (AAA) intermediate has been retained by the Euglenoidea and Fungi, whereas the prokaryotic lysine biosynthetic pathway with diamino-pimelic acid (DAP) apparently replaced it in the kingdoms Chromista and Plantae, having entered the eukaryote cell with the symbiotic cyanobacterial genome;[1] unfortunately the pathway used by most protozoa is not known. Likewise I suggest that the lanosterol pathway for sterol biosynthesis (fungi and animals) was the ancestral eukaryote condition, and that the cycloartenol pathway of chloroplasts came in with the symbiotic cyanobacteria. Mitochondrial cristae are considered to have originated independently[6] in about 9 lines of evolution following the conversion of a symbiotic purple nonsulfur bacterium into an organelle: in Miozoa, Allozoa, Radiozoa, Sarcozoa, Cryptomonada, Chromophyta, Plantae, Euglenozoa, and Opisthokonta (i.e., the clade comprising the Choanozoa, Fungi, and Animalia). Ampulliform cristae are straight, elongated, and unbranched invaginations of the inner mitochondrial membrane, circular in cross-section, with a lumen 5-10 times the thickness of the membrane; the tubular cristae of the Allozoa are typically slenderer (lumen usually 2-4 times the thickness of the membrane) and

[**Note added in proof.** Woese and Wolfe[106] have recently advocated an independent origin of eukaryotes and eubacteria from archaebacteria as in FIGURE 2c. Though vastly preferable to Woese's earlier suggestion of independent origins of all three lineages from a progenote (FIGURE 1a),[15,15a] as it is much more compatible with the extensive similarities summarized in TABLES 3-5, the new proposal has the disadvantages mentioned in the legend to FIGURE 2. But I accept their thesis that the ancestral archaebacterium was thermophilic; specifically, I suggest, thermacidophilic since this provides a selective force for the replacement of membrane fatty acid glycerol esters by isoprenoidal glycerol di- and especially tetraethers, which are more resistant to acid hydrolysis as well as being more rigid in heat. The fact that these have not been replaced in derived mesophilic archaebacteria by fatty acid esters makes their assumption of such replacement independently in the ancestral eukaryote and eubacterium highly implausible. Also one should stress that all thermophilic archaebacteria are aerobic or anaerobic respirers, unlikely ancestral phenotypes for the first bacteria in early nonoxidizing atmospheres since the necessary oxidized terminal electron acceptors would not have become widely available until after the much later evolution of oxygenic photosynthesis. Woese's earlier suggestion that the first cells were anaerobic

---

are often curved and sometimes branched; the tubular cristae of the Chromophyta usually contain a dense central filament in the lumen[9] and probably evolved independently of the allozoan tubular cristae during the early diversification of the Chromista.[4] Others have proposed Axostylaria or Parabasalia, rather than the Anaxostylea, as the hosts for the presumptive mitochondrial symbionts;[10] however the axostyle and preaxostyle seem complex structures without parallel in the Mitozoa; and Axostylaria, like the also uniquely complex Parabasalia, are exclusively gut symbionts of higher animals. By contrast the Anaxostylea and Archamoebae both have free-living as well as symbiotic members and, for this reason, as well as their less specialized structure, are more likely to represent primitive transitional eukaryotes. Moreover a gullet, widespread in mitochondrial protists, and arguably a uniquely evolved structure,[4,7] is found in the Archezoa only in the Anaxostylea (in which one of the ciliary roots also seems to prefigure the multilayered structure of certain Viridiplantae and Euglenoidea). The very early divergence from other eukaryotes of intracellular parasitic Microspora is strongly supported by their unique 70S ribosomes, with 16S and 23S rRNA like prokaryotes, in which, unlike all other eukaryotes so far studied, there is no distinct 5.8S rRNA, the homologous sequences being part of the 23S rRNA molecule[11] (as in all bacteria). Similar studies are needed of all major Archezoan groups. The three major endosymbioses important in eukaryote evolution are shown by dashed lines: two involving prokaryotic symbionts taken up by an early biciliated anaxostylean to form mitochondria[6] and chloroplasts;[7] and one in which such an early eukaryotic photosynthesizer (which like Dinozoa had already acquired chlorophyll $C_1$ and $C_2$, but which had not lost phycobilins) was taken up by an early ancestor of the proteromonads (that had not yet evolved cristae of stable morphology) to form the cryptomonad nucleomorph and chromist chloroplast endoplasmic reticulum.[4] At the same time the tubular somatonemes of this eochromist were transferred to the cilia to become tubular mastigonemes (TMs).[4] Here I adopt the long-established name Cryptomonada for the chromist subkingdom and phylum containing the Cryptophyceae, instead of the misleading name Cryptophyta used previously,[2-4,7] because the cryptomonads are virtually all unicellar monads, some being nonphotosynthetic (either with leucoplasts or even no plastids at all), and are nothing like plants. Because of their single-membraned envelope, if microbodies originated symbiotically,[12] they probably did so from a posibacterial or archaebacterial rather than a negibacterial symbiont.[13] The double-envelope parabasalian hydrogenosomes[14] might in principle have originated symbiotically from any of the three types of bacteria, or alternatively by reduction from mitochondria with the loss of mitochondrial DNA.[13] If the latter were proven, parabasalia should be transferred to the Mitozoa (branch Miozoa) from the Archezoa (in which they are anomalous, as the only group with a stacked Golgi, which however because of its unique association with the cross-striated parabasal filament might have originated independently of the Golgi stacking in mitozoan dictyosomes).

TABLE 3. Resemblances between Archaebacteria and Eukaryotes Not Shared with Eubacteria

1. Introns in rRNA and tRNA genes.[a,21]
2. Protein synthesis invariably initiated by methionine, not formyl-methionine[15,17] (mitochondria and chloroplasts, like their negibacterial ancestors, do use formyl-methionine).
3. Murein (peptidoglycan containing muramic acid) absent.[22]
4. True $N$-asparagine-linked glycoproteins present.[b,23]
5. Antigenic similarities in RNA polymerase.[24,24a]
6. Closer sequence similarities for the ribosomal A proteins.[25]
7. Closer sequence similarities for small subunit rRNA.[26]
8. Closer sequence similarities for 5S rRNA.[27]
9. Protein synthesis elongation factor susceptible to adenosine diphosphate ribosylation by diphtheria toxin.[28]
10. A more substantial projecting "bill" on the small ribosomal subunits (both ribosomal subunits have the simplest form in eubacteria: various extra lobes and bulges are found in eukaryotes and in some but not all archaebacteria).[29]
11. Chloramphenicol and many other antibiotic inhibitors of eubacterial protein synthesis do not inhibit at least some archaebacteria.[30]
12. Anisomycin inhibits protein synthesis.[24]
13. Retinene-containing photoreceptor proteins.[31]
14. Aphidicolin inhibits replicating DNA polymerase.[24a]
15. CCA 3' terminus of tRNA not coded by gene; added posttranslationally.[24a]
16. Some similar tRNA modification.[24a]

[a] The intron in the thymidylate synthase gene of phage T4 is not in a bacterial gene; from its sequence similarity with mammalian thymidylate synthase and dissimilarity from that of *Escherichia coli,*[32] the gene was clearly evolutionarily derived from the mammalian host of *E. coli* and not from *E. coli* itself, and thus is not an exception to the rule that introns are totally absent in eubacterial genes.

[b] A few posibacteria have glycosylated envelope proteins which might on closer investigation prove to be true glycoproteins:[33] if so this would strengthen the idea of a posibacterial ancestry for archaebacteria. The presence of histone-like proteins and "actin-like" proteins in *Thermoplasma*[34] has also been used to suggest a link with eukaryotes,[34,35] but similar proteins have also been found in eubacteria.[36,37]

photosynthesizers—the probable ancestral phenotype of eubacteria[68]—is much more plausible and compatible with a simple origin for the negibacterial envelope from an inside-out cell (FIGURE 3a), so as to produce the first true (eubacterial) cell.]

## THE LOSS OF MUREIN AND ITS CONSEQUENCES

The peptidoglycan murein is an alternating copolymer of $N$-acetylglucosamine (NAG) and $N$-acetyl muramic acid covalently cross-linked by peptides to form a rigid bag round the eubacterial cell. It plays a fundamental role not only in maintaining the shape, integrity, and viability of the cell despite osmotic fluctuations but also in cell growth, cell division, and DNA segregation. This is dramatically shown by the effect of disrupting its synthesis by penicillin or degrading it by lysozyme: the wall-less cells thus produced are osmotically fragile, have very low viability in natural conditions, and even when protected by agar grow and divide in a chaotic manner, producing cells far larger and far smaller than usual containing several or no copies of the chromosome.[64,65] Both evaginations and invaginations of surface membranes are found. Because the peptide cross-links in murein are between muramic acid residues, a mutation blocking muramic acid synthesis or assembly would prevent the cross-linking and have the same phenotypic effect as penicillin or lysozyme; it would be so traumatic for the cell that it could survive only through secondary mutations giving the cell envelope stability in other ways. Neither archaebacteria nor eukaryotes make or contain muramic acid, but both make and use NAG (which in eubacteria is the direct biosynthetic precursor of muramic acid). In fact NAG is a key constituent of asparagine-linked glycoproteins which are universal constituents of the eukaryote cell surface and form the cell wall of the archaebacterium *Halobacterium*[23] but are so far unknown in eubacteria.

Moreover there is a definite similarity in the methods of biosynthesis and secretion of peptidoglycans and such glycoproteins.[66] In both cases a key role is played by an isoprenoid carrier in the membrane [for glycoproteins the isoprenoid alcohol dolichol, in the endoplasmic reticulum (ER); and for peptidoglycan the phosphorylated isoprenoid alcohol undecaprenol phosphate, in the plasma membrane.] Dolichol holds nascent oligosaccharide precursors covalently attached to it via NAG on the noncytoplasmic face of the membrane to facilitate their covalent attachment to proteins, also via the NAG residue. In peptidoglycan synthesis the muramic acid is attached to the undecaprenol phosphate on the cytoplasmic face of the membrane, the amino

TABLE 4. Resemblances between Posibacteria and Eukaryotes

1. Pyruvate dehydrogenase sequence and 3-D structure closer than to negibacteria.[38]
2. Citrate synthase structure and regulatory mode closer than to negibacteria.[39,40]
3. Succinate thiokinase size closer than to negibacteria.[40]
4. Fatty acid synthetase is a complex macromolecular assembly in a few actinomycete posibacteria, as in *Euglena,* animals, and fungi.[41] [In all negibacteria and the clostridial branch of posibacteria it consists of separate soluble molecules, as it does also in green plants (Viridiplantae) which probably obtained theirs from the symbiotic (negibacterial) cyanobacteria that became chloroplasts.]
5. Aerobic desaturation of fatty acids (universal in eukaryotes) is found in prokaryotes only in certain posibacteria.[42]
6. Some posibacteria have rifamycin-insensitive RNA polymerases[43] like all archaebacteria and eukaryotes.[24]
7. Both actinomycetes and fungi produce *N*-penicillin and cephalosporins.[44]
8. Secretion of soluble extracellular digestive enzymes is common in both groups.[45]
9. Some posibacteria (unlike virtually all negibacteria except cyanobacteria) apparently make sterols (universal in eukaryotes).[42]
10. The only prokaryotes where cell fusion to form stable diploids (as in eukaryote syngamy) has been observed are *Bacillus subtilis*[46,47] and (artificially with protoplasts) certain actinomycetes.[101]
11. The method of exosporulation (producing desiccation-resistant spores) in some actinomycetes[103] could simply be converted to sexual syngamy and zygospore formation following loss of the murein vegetative cell wall.
12. Conjugational pheromones of *Bacillus* resemble sexual pheromones[48] of yeast.
13. Unlike negibacteria both lack a second outer membrane outside the plasma membrane.
14. The actinomycete posibacterium *Streptomyces* is the only prokaryote known to produce chitin;[48a] that it uses the same UDP-*N*-acetylglucosamine precursor as for murein supports my thesis that eukaryote chitin synthesis evolved directly from murein synthesis.[1]
15. A few posibacterial (but no negibacterial) plasmids can replicate in eukaryotes.[63]

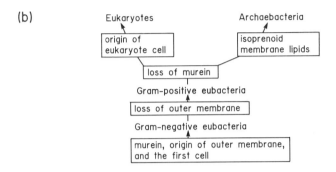

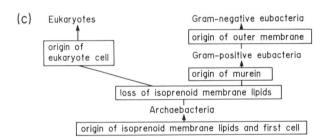

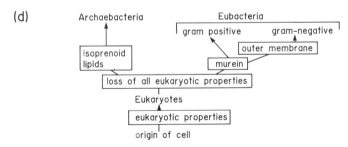

**FIGURE 2.** Possible phylogenetic relationships between eubacteria, archaebacteria, and eukaryotes. (a) Independent descent[15] from a primitive "progenote" that had not yet evolved DNA, large proteins, or a defined cell structure[15a] (b-d). Direct conversion of one cell type into another but with a different group chosen as ancestral: (b) The eubacteria as ancestors of both archaebacteria and eukaryotes; in the form advocated here, this assumes that the first cell was a gram-negative eubacterium that originated as shown in FIGURE 3a and that posibacteria originated from negibacteria as shown in FIGURE 3b. (c) The archaebacteria as ancestors: this postulates loss of isoprenoid membrane lipids in the ancestor of eukaryotes and eubacteria, for which there is no evidence or obvious reason, and the late origin of the outer membrane of the gram-negative cell for which there is neither evidence nor a plausible mechanism. The much greater ecological and bioenergetic diversity of the eubacteria than the archaebacteria, strongly suggesting that

acids are added and then the NAG, and the complex shuttles across the membrane for polymerization of the peptidoglycan on its outer face.

I suggest that in the posibacterial common ancestor of eukaryotes and archaebacteria, muramic acid was lost by a mutation preventing either its biosynthesis or its incorporation into peptidoglycan.[1] Either the same or a separate mutation then caused the enzymatic machinery to transfer NAG to cell surface proteins, including those of the S-layer, instead of to the muramopeptide (FIGURE 3c). I argue that such a mutational shift from the peptidoglycan murein to glycoprotein was the key step in the origin of both the archaebacteria and the eukaryotes, and that virtually all the features that they share and in which they differ from eubacteria can be understood as the direct or indirect consequences of this drastic change in cell envelope chemistry and organization. Because of the features that they share, and their new envelope chemistry, I refer to the archaebacteria and eukaryotes collectively as the Neomura, which I regard as a true clade. The inhibition of the synthesis of both glycoproteins and posibacterial (but not negibacterial) peptidoglycan by tunicamycin[66] supports my thesis that the neomuran ancestor was a posibacterium.

The major differences between eukaryotes and archaebacteria can also be explained as an indirect consequence of the loss of murein. The first major difference is in their plasma membrane lipids. Eukaryotes, unlike archaebacteria and most eubacteria, have sterols as well as ordinary phospholipids that are esters of fatty acids. Archaebacteria unlike both eukaryotes and eubacteria have ether-linked isoprenoidal phospholipids. Isoprenoidal lipids and sterols have the same basic function: making the fluid bilayer of the plasma membrane more rigid, which would have been of key importance in rescuing the ancestral wall-less neomuran from an early demise. Both are synthesized from the isoprenoid squalene, known to be present in the posibacteria suggested here as the ancestors of the neomura. I suggest that the biosynthesis of sterols by the cycloartenol cyclization and of ether-linked isoprenoid lipids evolved independently in two different but closely related offshoots of their wall-less common ancestor, and both were selected because they reduced undesirable uncontrolled invaginations, evaginations, and fragmentations of the cell membrane by increasing its rigidity. Thereafter eukaryotes and archaebacteria diverged, but by then their common ancestor had already acquired the common neomuran characters shown in TABLE 3.

The second major difference between the two evolutionary branches, and that with the most profound long-term evolutionary consequences, was the evolution in the sterol-containing presumptive-eukaryote line of a cytoskeleton and consequently[50] phagocytosis and all the other features unique to eukaryotes (TABLE 6, FIGURE 3d). The evolution of an internal cytoskeleton provides an entirely new way of preserving the cell against osmotic shock, of allowing controlled growth and division, and a rigid apparatus for DNA segregation. By making all these possible for the first time in the absence of a rigid exoskeletonlike external cell wall, the evolution of an endoskeleton enabled phagocytosis to evolve. Phagocytosis inevitably led to the origin of the nucleus, endoplasmic reticulum, and the invasion of protein-coding genes by introns. By re-

---

they evolved first and had filled nearly every potential prokaryote niche before the archaebacteria had evolved, also makes this hypothesis implausible. Stromatolites and isotopic data suggest that photosynthetic eubacteria were present as early as $3.5 \times 10^9$ years ago.[52] (d) Eukaryotes as ancestors: this makes the unlikely assumptions that the first cell was eukaryotic and that all its complex characters (TABLE 6) not only evolved directly from the prebiotic molecular soup but also were all later lost to yield the first prokaryote. The fossil record gives no evidence whatever for eukaryotes before $1.5 \times 10^9$ years ago[19,20] and nothing convincing before $0.96 \times 10^9$ years ago,[19,20] so we may reject this possibility.

evolving a prokaryotic type of rigid exoskeleton instead of a eukaryotic type of endoskeleton, archaebacteria could not evolve phagocytosis or sex and therefore remained fundamentally prokaryotic both in way of life and in cell architecture, though were able to invade certain extreme environments (hot acid, alkali, or extreme salt) in which their envelope stability and chemistry gave them a marked advantage over preexisting eubacteria.

Since the high internal salt concentration of those methanobacteria most closely related to the halobacteria[67] is more likely to be vestigial than a predaptation, I suggest that methanobacteria evolved from a halobacterium by the evolution of the special

TABLE 5. Cellular and Molecular Characters Shared by Archaebacteria and Eubacteria

1. Genome of DNA associated with histone-like proteins.
2. Plasma membrane.
3. Proton-driven DCCD-sensitive plasma membrane ATP synthetase.
4. Cytochromes b and c.
5. 70S ribosomes with Shine-Dalgarno initiation mechanisms, and rRNA molecules with 60%-70% sequence homology and basically the same secondary structure and gene order.
6. Fatty acids (rare but present in archaebacteria).
7. Isoprenoids.
8. Bacterial flagella.
9. All advanced eukaryotic features shown in TABLE 6 absent (including introns in protein-coding genes).
10. Coenzyme F (Archaebacteria and the posibacteria *Streptomyces* only).
11. Gas vacuoles.
12. Many coenzymes (e.g., acetyl-CoA, flavins, NAD/P, pantothenate, thiamine, pyridoxine, paraaminobenzoate).
13. tRNA with up to 60% homology and modified in same way at 11 identical sites.
14. Restriction nucleases.
15. Homologous ferredoxins.
16. Superoxide dismutase.
17. Quinones.
18. Hydrogenase.
19. Circular plasmids.
20. Tailed bacteriophages.
21. Glycogen.
22. Dozens of enzymes of intermediary metabolism including glycolysis, most of Krebs cycle, amino acid and nucleotide synthesis.
23. Typical sugars and *N*-acetyl glucosamine.
24. Pili and fimbriae (but homology not demonstrated).

enzymes and coenzymes that make methanogenesis possible; this enabled them to colonize anaerobic niches rich in $CO_2$ and $H_2$. Some of them retained cytochromes, and most retained protein or glycoprotein walls: but one branch evolved a new type of peptidoglycan called pseudomurein because it lacks muramic acid (which could not be reevolved), which would have helped protect them in environments rich in proteases (e.g., rumens).

The natural loss of murein to generate a major new group of organisms seems to have occurred on about five occasions during the history of life (FIGURES 3c-3f), and is a very rare and unusual event that could only be successful if associated with other major cellular changes. Only two groups of eubacteria lack murein: the mycoplasmas, which clearly evolved from murein-walled posibacteria,[58,59] and the *Planctomyces* group of negibacteria. The mycoplasmas (FIGURE 3c) survived without a cell wall by undergoing three major changes: (1) extreme reduction in cell (and genome) size, which would give their surface greater stability; (2) by using sterols made by their eukaryote hosts to make their membranes more rigid; (3) by evolving membrane-skeleton proteins that presumably confer on them their shape and capacity to divide and segregate DNA accurately. But as mycoplasmas all depend on their eukaryote hosts for osmotic protection and (with one exception) for sterols, they could not have evolved till after the origin of eukaryotes, and therefore could not be ancestral to them as sometimes suggested. However, they are important in showing that murein loss by a posibacterium can lead to the origin of a radically different cell type, as here proposed for the origin of the Neomura. How *Planctomyces* manage without peptidoglycan is not clear; they may have lost it, as I suspect, or may be primitively without it. *Planctomyces* however like all negibacteria have an outer membrane as well as a plasma membrane. Loss of peptidoglycan without losing the negibacterial outer membrane is perfectly possible, as shown by negibacterial spheroplasts and by wall-less negibacterial L-forms which typically retain both membranes. I have argued that this also occurred during the origin of chloroplasts[7] (FIGURE 3e) and mitochondria[6,13] (FIGURE 3f), the most clearly demonstrated cases of natural murein loss by negibacteria. Though negibacterial L-forms without an outer membrane can be created in the laboratory, there is no case where this is known to have occurred in nature. The conversion of a negibacterium into a posibacterium probably involved not the loss of murein but its hypertrophy (FIGURE 3b), probably in a spore-forming negibacterium such as *Sporomusa* which has clear 16S rRNA affinities with posibacteria.[69]

## EARLY EVOLUTION OF THE NEOMURA

Prior to their separation through evolving different envelope chemistries, archae-bacteria and eukaryotes must have undergone a common phase of evolution in which they acquired the various characters shown in TABLE 3. Ten out of 16 of these involve ribosomes (or tRNA) which have to interact via the signal mechanism with the membrane when synthesizing integral membrane proteins or secreting extracellular proteins. It is therefore likely that the major changes in the cell surface occurring in the early neomuran would entail coadaptive changes in the ribosomes and the signal mechanism. Most of the changes however could well be neutral or quasi-neutral rather than coadaptive with membrane changes, and may have occurred at an unusual rate because of the major destablization caused by the loss of the cell wall. Because the replication and segregation machinery are associated with the membrane, and because of the instability of the membrane, all sorts of major and minor errors in replication, segregation, recombination, and repair would be likely. This sort of trauma would be very likely to derepress transposases and lead to a bout of transpositions and to an unprecedented multiplication of selfish transposable elements. New mutations partially rescuing the traumatized cell could carry many neutral or slightly harmful ones to fixation by the hitchhiking effect. The drastic cell surface changes would eliminate many of the viruses previously infecting the cell by removing or altering their receptors. All in all a drastic period of genetic upheaval would occur. Such accelerated quasi-neutral evolution would change molecules like 16S rRNA much more rapidly than before or since, invalidating the molecular clock assumptions (they cannot be correct

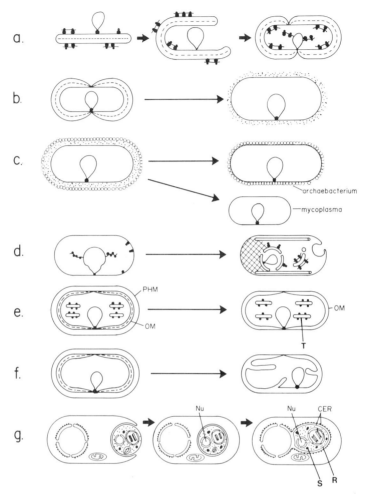

**FIGURE 3.** Eight major transitions in cell evolution. (a) The origin of the first cell—a gram-negative bacterium (possibly an anaerobic photosynthetic green bacterium)[53] with lipoprotein outer membrane as well as a plasma membrane: an "inside-out cell"[53,54] with attached ribosomes and chromosome and good access to small molecule precursors in the prebiotic soup, after becoming able to transport and store them reversibly inside the cisternae, can fold up and fuse with itself to enclose its chromosome and ribosomes inside the cell without starving.[53] Even the peptidoglycan murein layer (dashed line between the inner and outer membranes), DNA replication, and accurate segregation and division may have evolved in the inside-out cell phase.[53] It is much harder to envisage a viable cell being formed merely by the accidental enclosure of chromosomes and ribosomes by a single membrane. (b) The evolution of the posibacterial cell: a mutation causing hypertrophy of the murein layer made the wall so thick that outer membrane lipids and proteins could no longer be transferred directly from the plasma membrane, causing the permanent loss of the outer membrane as the cell grew.[57] (c) Evolution of the archaebacteria and mycoplasmas[57] from a gram-positive bacterium by the mutational loss of murein. The archaebacteria though losing the murein retained the outer surface layer (S-layer)[33] of crystalline protein as their new wall (or in some cases evolved pseudomurein as well); after the evolution of rigidifying covalently bonded isoprenoid lipids, one genus (*Thermoplasma*) could lose the

for srRNA divergences between prokaryotes and eukaryotes for they place the origin of life earlier than the origin of the universe)[70] and making it appear that eukaryotes, archaebacteria, and eubacteria diverged much earlier than they really did.

One particular consequence of this upheaval was I suggest the origin of introns from transposable elements to create split genes.[71] Perhaps the simplest mechanism for this is via RNA-based transposable elements that multiplied by using reverse transcriptase to generate DNA copies[72] and which were also capable of self-splicing.[73] The association of reverse transcriptase with eukaryotic transposable elements is now well established;[74] I suggest it began during the early neomuran genetic upheaval that preceded the origin of eukaryotes. Reverse transcriptase itself may have evolved as a mutant DNA polymerase. I suggest that such a mutant polymerase was initially selected because it rescued the cell from a lethal mutation. At a time of exceptionally high mutation rates, damaged copies of any gene could be replaced from undamaged messengers by reverse transcription and recombination, so reverse transcriptase would be highly advantageous, and only became a nuisance and restricted to "selfish" genetic symbionts after things had stabilized with the formation of the archaebacterial envelope or of the eukaryotic cytoskeleton, nucleus, and segregation mechanism. Because of the tight coupling between transcription and translation in prokaryotes, one might think that introns could invade protein-coding genes only after the origin of the nucleus and the functional and topographic separation of the transcriptional RNA processing

---

protein wall. Mycoplasmas evolved similarly much later in eukaryote hosts, from which they acquired sterols to help maintain surface rigidity despite loss of the murein and S-layer, and became greatly reduced in size. (d) Origin of the eukaryote cell from an early archaebacterium lacking isoprenoidal membrane lipids, by the origin of the cytoskeleton (microtubules, actin microfilaments, and intermediate filaments),[50] and of phagocytosis which internalized the DNA and ribosomes to form the nucleus and rough endoplasmic reticulum.[50] (e) Origin of the chloroplast from a cyanobacterium phagocytosed by an early eukaryote:[7] in the ancestor of plants and most algae the food vacuole (phagosomal) membrane (PHM) surrounding the symbiotic cyanobacterium probably burst and was lost prior to its conversion into a chloroplast,[7,8] so the outer membrane of the chloroplast is derived from the outer membrane (OM) of the cyanobacterium;[7,8] only in the euglenoids and most peridineans was the food vacuole membrane retained and incorporated into the envelope as the outermost of its three concentric membranes.[7,8] The peptidoglycan layer of the cyanobacterial wall (dashed line) was lost in all chloroplasts except those of *Cyanophora* and one or two other Glaucophyceae. All chloroplasts could have diverged from a single cyanobacterial ancestor; the Biliphyta retained phycobilisomes [dark blobs on the thylakoids (T)] whereas other algae lost them and evolved thylakoid stacking and either chlorophyll b or c.[7] Acquisition of the symbiont membranes was as important as the acquisition of their DNA. The permanent loss of the phagosomal membrane was presumably a membrane mutation[4] not caused by a DNA mutation. (f) Origin of mitochondria from a symbiotic purple nonsulfur bacterium[6,60,61] phagocytosed by an early eukaryote. As in most chloroplasts the peptidoglycan and the phagosome membrane were lost.[6] Cristae evolved to increase the amount of respiratory machinery, but with divergent morphologies in different lines of protozoan evolution.[3,6] (g) Origin of the chromistan chloroplast endoplasmic reticulum (CER) and the cryptophyte nucleomorph (Nu) by the phagocytosis and conversion into organelles of an early photosynthetic eukaryote by another early eukaryote that had already acquired a mitochondrion but not yet evolved cristae of stable morphology.[4,7,8] The Chromophyta (e.g., brown algae and diatoms) probably evolved from the Cryptomonada by the loss of the nucleomorph,[4,7] and of periplastidal ribosomes (R) and starch (S). The accidental fusion of the phagosomal membrane with the nuclear envelope and consequent acquisition of ribosome receptors could have been another heritable "membrane mutation" neither initiated nor maintained by any change in the DNA;[4] the permanent association of the CER and nuclear envelope in some but not other chromist groups, however, must probably have been caused by changes in cytoskeletal proteins.

TABLE 6. Unique Features of Eukaryotes Never Found in Prokaryotes

(a) Universal nuclear characters
    1. Nuclear envelope; total separation of transcription and translation.
    2. Nuclear pore complexes with active transport of mRNA and ribosomal subunits; sequence-specific targeting of nuclear proteins.
    3. Nuclear lamina (and ? matrix).
    4. Mitotic spindle of tubulin-containing microtubules (sometimes purely cytoplasmic).
    5. Nucleolus.
    6. Split protein-coding genes.
    7. Separate RNA polymerases for rRNA (pol I), mRNA (pol II) and tRNA and other small RNAs (pol III).
    8. 5′ guanosine-capped and 3′-polyadenylated obligately monocistronic mRNA.
    9. Plural replicons (i.e., many replication units per chromosome).
    10. C-Value paradox: genome size directly proportional to nuclear and cell volume.
    11. Plural linear chromosomes with specialized telomeres.
    12. Centromeres.
(b) Universal cytoplasmic characters
    13. Actin microfilaments.
    14. Myosin/ATPases (? also kinesin ATPase) that mediate intracellular movements and cytoplasmic contractility.
    15. Intermediate filaments.
    16. Exocytosis.
    17. Endocytosis; clathrin coated vesicles.
    18. Rough endoplasmic reticulum (RER); "smooth" vesicle budding from RER.
    19. $Ca^{++}$ as secondary messenger with calmodulin and inositol phosphate controls.
    20. Lysosomes.
    21. Golgi cisternae (not identified in Metamonada and Archamoebae, but assumed to be present),[a] and smooth vesicle fusion with and budding from Golgi membranes.
    22. Signal recognition particle with 7SL RNA; ribophorins (both identified only in animals, but assumed to be universal).[b]
    23. Capacity to harbor cellular endosymbionts.
(c) Important nonuniversal characters (character 24 suggested to have evolved in neomuran ancestor, 25-30 during origin of eukaryotes; others later)
    24. Capacity to harbor retroviruses and transposons with reverse transcriptase (may be universal).
    25. Nucleosomes + 4-5 histones (absent, or at least grossly reduced, in Peridinea where thymine is largely replaced by 5′ OH methyluracil).
    26. Chromatin condensation during mitosis (absent in a disparate variety of protists).
    27. Centrosomes (apparently absent in higher plants, though indistinct centrosomal material may in fact be present).
    28. Cilia and kinetosomes.
    29. Meiotic pairing; synaptonemal complex.
    30. Phagocytosis.
    31. 80S ribosomes with 5.85s, 18s and 28s rRNA, (like bacteria Microspora have 70s ribosomes with 16s and 23s rRNA without separate 5.8s rRNA).
    32. Two-step meiosis; chiasmata (at least 3 protozoan phyla—Dinozoa, Sporozoa, Metamonada—only have nonchiasmate one-step meiosis; the Parabasalia have both one-step meiosis and a unique nonchiasmate two-step meiosis).
    33. Mitochondria with mitochondrial DNA and ribosomes (absent in subkingdom Archezoa, a few Spizomycete fungi and a few Infusoria).
    34. Microbodies (peroxisomes and glyoxysomes) (absent in Archamoebae, Microspora, Anaxostylea, Kinetoplastidea).
    35. Chloroplasts, with chloroplast DNA and 70s ribosomes. (Absent from animals, fungi, most Protozoa, and many chromists.)
    36. Centrioles (absent in most nonciliated eukaryotes, and others such as Infusoria).
    37. Hydrogenosomes (in Parabasalia and anaerobic Infusoria and Spizomycetes).

**TABLE 6** (*continued*)

38. Glycosomes (in Kinetoplastidea only).
39. Chloroplast endoplasmic reticulum (kingdom Chromista only).
40. Tubular ciliary mastigonemes (kingdom Chromista only).
41. Dictyosomes (stacked Golgi cisternae—absent from all Archezoa except Parabasalia and from Eufungi and Allomycetes).
42. Breakdown of nuclear envelope during mitosis (i.e., open mitosis—most Protozoa, most fungi and some other protists have a closed mitosis in which the nuclear envelope persists and divides).
43. Breakdown of mitotic spindle during interphase (absent in a few protists).
44. Ameboid movement.
45. Extrusomes.
46. Heterochromatin.
47. AAA lysine biosynthetic pathway.
48. Lanosterol steroid biosynthetic pathway (replaced in green plants by cycloartenol pathway shared by cyanobacteria).

[a] I now suspect that a true Golgi, topologically distinct from the ER, may be absent in these phyla, and even in Microspora.
[b] Either or both could be absent in Archezoa, and need to be specifically sought.

machinery on the one hand and the translational machinery on the other; this is a reason for doubting the idea that they evolved early in evolution to facilitate the evolution of new proteins. It may also partially explain why introns are found only in tRNA and rRNA in Archaebacteria; however introns with a definite secondary structure at the 5′ end would be able to fold up immediately so as to prevent faulty translation. Self-splicing introns of this type are compatible with prokaryotic genetics, and may one day be found in archaebacteria; probably the $T_4$ intron is self-splicing.[32]

## THE ORIGIN OF EUKARYOTES: CYTOSKELETON AND NUCLEUS

The most fundamental and radical molecular innovation in the origin of eukaryotes was the origin of the cytoskeleton.[1,50] A cross-linked actin gel pervading the whole cell and attached to the plasma membrane helps to prevent osmotic swelling and lysis. But since the cell must grow it must also have proteins able to sever and/or depolymerize actin filaments,[75] just as a eubacterial cell has murein hydrolases to sever links in the peptidoglycan to allow growth. This makes it more vulnerable to osmotic shocks which cause excessive swelling or partial puncturing of the cell, which could be counteracted by active contraction of the gel. Even bacteria pump calcium out of their cells, so a calcium-sensitive protein like calmodulin (like actin one of the best conserved universal eukaryote proteins)[76] would immediately respond to leakiness caused by excessive swelling or local puncturing; coupling its response to activation of contraction would ensure the adaptively effective response to localized or generalized stretching or damage to the plasma membrane. Bacteria may already have had a suitable $Ca^{++}$-binding protein to do this, and many proteins would have the potential to mutate so as to polymerize into filaments: probably many more than the "actinlike" proteins already identified in bacteria. Mechanochemical ATPases such as myosin to mediate contraction could have arisen from a prokaryotic protein such as a DNA helicase which is also a mechanochemical ATPase (TABLE 7).

Just as actin filaments are suitable for withstanding tension, microtubules are the most economical way of resisting compressive or shearing forces. I suggest that almost at the outset they became laterally attached to the plasma membrane and served to prevent the contractile actin gel or environmental shearing forces from chopping off bits of nonnucleated cell. By the joint action of actin filaments and microtubules, the cell could be stabilized from damage by external forces, be able to grow and to divide by longitudinal fission (FIGURE 4). Many proteins, even hemoglobin, probably are potentially able to mutate into a suitable microtubular form; intracellular microtubules, presumably unrelated to eukaryotic tubulin-based microtubules, have even been observed in bacterial L-forms as well as in normal bacteria. The replacement of the bacterial segregation mechanism by a premitotic system based on microtubules was the essential prerequisite that led to the evolution of plural replicons (FIGURE 4c), which in turn led to great increases in genome size.[83]

The third cytoskeletal component, intermediate filaments, probably evolved slightly later, but the intermediate filament family of proteins must date back at least to the

TABLE 7. Analogies between DNA Helicase I of *E. coli* and Myosin

1. Both are ATPases activated by a linear structurally polarized macromolecular substrate (DNA and F-actin respectively).
2. Both are exceptionally large proteins that form ordered aggregates (helicase I monomer molecular weight 180,000).
3. These aggregates use energy from ATP to move unidirectionally along the linear substrate.
4. As they move along the substrate they can do useful mechanical work (DNA strand separation and intracellular movements respectively).
5. Both F-actin and DNA contain adenine nucleotides and both have a strong binding affinity for pancreatic DNAase, raising the possibility of some similarity in their binding mechanisms for myosin and helicases respectively.

origin of the nucleus. This is because the nuclear lamins[85] which mediate the attachment of the nuclear envelope to the chromosomes, are members of the intermediate filament family,[86] and the origin of their capacity to assemble as a lamina on a chromosomal template must have been the key molecular step in the origin of the nucleus[87] following the internalization of the chromosome attachment site by phagocytosis (FIGURE 5). Intermediate filaments also serve as anchoring sites for ribosomes;[96] the novel properties of eukaryote ribosomes (extra proteins and extra loops inserted into rRNA—presumably for the extra proteins to attach to) may be connected partly with this novel attachment to the cytoskeleton, partly with novel features of their assembly in the nucleolus in association with uniquely eukaryotic nucleolar matrix proteins, partly with their active transport through the nuclear pores (obviously not occurring in prokaryotes), and partly with the changes in the cotranslational protein secretion mechanism to be discussed later. For all these coadaptive reasons, one expects the eukaryote rRNA to have evolved exceptionally rapidly during the origin of the eukaryotes and to have become more radically different from the ancestral eubacterial rRNA than did the archaebacterial rRNA. One must not construe the undoubted radical differences as evidence for an early divergence despite the fossil evidence that prokaryotes are somewhere between two-and-a-half and four times as old as eukaryotes.

Evolution of the nucleus must have gone on hand in hand with the origin of mitosis (FIGURE 5); the latter would have led to the origin of linear chromosomes as outlined in FIGURE 6. The later diversification of mitosis (FIGURE 7) also depended mainly on changes in the association with the nuclear envelope of cytoskeletal elements; the origin of histones and nucleosomes can also be attributed to the evolution of mitosis and intracellular cytoskeleton-based motility.[1] Most of the unique features of the eukaryotic genetic system can be interpreted as secondary or tertiary consequences of prior changes in the cytoskeletal system, which is therefore the fundamental thing that has made eukaryotes so uniquely different from bacteria.

## THE ORIGIN OF CAPPED MESSENGERS, SPLIT PROTEIN-CODING GENES, AND THE NUCLEOLUS

These fundamental properties of eukaryotes are not arbitrary differences from prokaryotes, as they have often seemed, but necessary prerequisites, or consequences, of the effective exclusion of ribosomes from the nucleus (FIGURE 5b). In prokaryotes messengers are typically short-lived and degraded by 5' exonucleases often even before their synthesis is complete. Exclusion of ribosomes from the nucleus would have been possible only if this could be prevented, which I argue is why guanosine 5' capping evolved to prevent messengers from degradation before they could be translated. 3'OH polyadenylation and the attachment of mRNA-binding proteins would also have helped to prevent less specific degradation by 3' exonucleases and endonucleases and thus increase messenger lifetimes, especially for messengers needed throughout the cell cycle (unnecessary for short-lived ones like histone messengers).

Once greater messenger stability allowed ribosomes to be excluded from the nucleus, there would be no risk of their translating unspliced premessengers. Therefore transposable introns would inevitably invade protein-coding genes without doing any significant harm irrespective of whether or not they could fold up in such a way as to prevent faulty translation; increased messenger stability would allow ample time for splicing. But in prokaryotes such invasion, especially by non-self-splicing introns, might lead to synthesis of harmful unspliced proteins and so be more strongly selected against. It seems likely that the first introns were self-splicing and that the non-self-splicing mechanism of nuclear mRNA[97] evolved from them only after the origin of the nucleus.

A secondary or incidental consequence of capping was the provision of a novel structure that could be recognized by ribosomes as binding sites for the initiation of protein synthesis. But the evolution of stable ribosome-binding to caps would have two major consequences: (1) Improvements to the cap-binding mechanisms of ribosomes, I suggest, reduced their capacity to initiate on internal ribosome-binding sites, thus mutations replacing these by 5' cappable termini would be favored; eventually cistrons would all thereby become monocistronic. Further improvements in cap binding would no longer be constrained by necessity to recognize also internal sites, and soon eukaryote ribosomes would have become unable to translate polycistronic messengers. (2) The prokaryotic Shine-Dalgarno sequences would no longer be necessary for the binding of ribosomes to messengers, and might even be harmful by delaying the movement of the small ribosomal subunit from the cap site to the initiation site, so were soon lost by deletion.

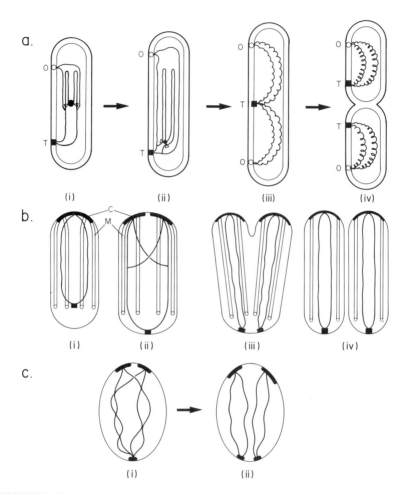

**FIGURE 4.** The origin of mitotic spindles and the eukaryote cell cycle. (a) The bacterial DNA segregation mechanism is thought[77] to depend on attachment of the chromosome terminus (T) and one origin (O) to the cell wall at opposite poles (i), movement of a sister origin to the opposite pole by DNA helicase (triangles) and its attachment there (ii), detachment of the terminus and its pulling by DNA gyrase-induced DNA supercoiling to the cell equator (iii), duplication of the terminus and its attachment site (prokinetochore) and septation between the two prokinetochores (iv). Concomitantly, growth of the rigid cell wall moves the two polar DNA attachment sites farther apart. (b) Following the loss of the rigid murein wall the basic features of the eukaryote cell cycle probably evolved before the nucleus: (i) The replicon origin-attachment proteins (or other proteins bound to them) evolved into microtubule nucleating centers and became primitive centrosomes (c) with the microtubules (m) attached to them at their minus ends and thus all having the same molecular polarity as in modern centrosomes;[78] the microtubules would stabilize the plasma membrane by lateral attachments to it as in euglenozoan and sporozoan protozoa. (ii) From the start centrosome duplication would be temporally and probably causally coupled to the initiation of DNA replication, as it is in eukaryotes,[79] and as replicon origin attachment sites probably are in bacteria, so this fundamental feature of the eukaryote cell cycle would be directly inherited from bacteria;[80] centrosome duplication would produce two parallel half-spindles as in protozoan pleuromitosis;[81] (iii) After completion of

Increased messenger stability during the evolution of the nucleus would potentially allow economies in mRNA transcription frequency by lowering promoter/RNA polymerase binding affinities. But selection for larger cell size associated with the switch to phagotrophic feeding would require greatly increased transcription rates for rRNA and tRNA genes. This "conflict" was resolved partly by selection of tandem duplications of ribosomal genes and the evolution of a localized nucleolar matrix to maximize the rate of transcription and processing of rRNA, and its assembly into ribosomal subunits, and partly by duplicating the RNA polymerase gene which allowed RNA polymerase I to be tuned for maximal transcriptional rates of long rRNA transcripts (partly by means of long RNA polymerase binding sequences), RNA polymerase II for minimal transcriptional rates for mRNA genes, and RNA polymerase III for the most efficient rapid transcription of short tRNA genes with internal promoters in addition to or instead of upstream promoters.

## ORIGIN OF PHAGOCYTOSIS AND THE ENDOMEMBRANE SYSTEM

Loss of the cell wall by a gram-positive bacterium that secreted digestive enzymes into the medium and actively absorbed the products would allow the evolution of phagocytosis, as outlined in FIGURE 8. A central assumption is that the prokaryotic cotranslational protein insertion and secretion mechanism at the plasma membrane became transformed into that of the eukaryotic rough endoplasmic reticulum (RER). Removal of the ribosomes and chromosome from the plasma membrane to form the RER and nuclear envelope required (1) their internalization by phagocytosis,[50] and (2) evolution of smooth vesicle budding from the RER and their fusion with the plasma membrane by exocytosis; such budding acted as a selective "valve" preventing the chromosome and ribosome receptors from ever returning to the cell surface after they were phagocytosed.

---

replication and decatenation of the two daughter DNAs by a type II DNA topoisomerase,[82] the contractile actin gel cleaves the cell longitudinally between the two half-spindles as in pleuro-mitosis to produce two daughter cells each with a half-spindle of cortical microtubules attached to an apical centrosome, and each with a single circular chromosome. Histones and nucleosomes probably evolved at this stage so as to fold the chromosomes more compactly and prevent DNA fragmentation[1] by the rapidly contracting cleavage furrow; this would allow energy wasting active negative supercoiling by DNA gyrase to be dispensed with. (I suggested that histones were secondarily lost in Peridinea after the evolution of nuclei and exonuclear spindles which would protect the DNA from the contractile machinery.)[1] (c) Once the basic premitotic division mechanism shown in (b) evolved it would no longer be necessary for sister origins to be actively separated by the DNA helicase as in (a), and therefore for the chromosome to have but a single replicon origin and terminus: mutations creating new origins throughout the chromosome would no longer be selected against and plural replicons would inevitably evolve (i), S-phase could become much shorter than the cell cycle even in rapidly growing cells. If not present in the bacterial ancestor, a mechanism to prevent more than one replication per cell cycle would have had to evolve at this stage. Plural replicons would allow the increase of genome size[83] and noncoding DNA and of intragenomic populations of selfish, neutral, and useful genetic symbionts (integrated viruses and transposable elements),[84] but introns would find it hard to spread into mRNA genes till after the origin of the nucleus. The former bacterial chromosome origins and termini later became centromeres and telomeres (FIGURES 5 and 6).

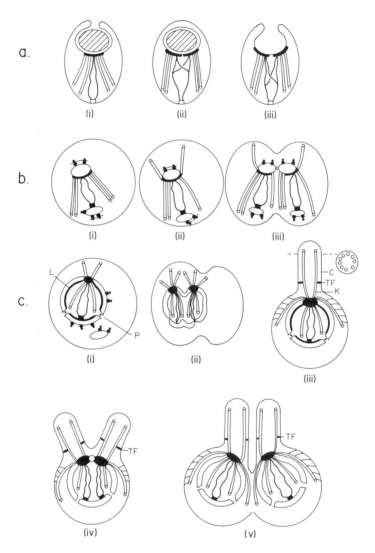

**FIGURE 5.** The origin of the nucleus, mitosis, and cilia. (a) Phagocytosis involving membrane containing the attachment site for the chromosome origins (i, ii) would separate the centrosome from the cell surface, but in early stages be reversible after digestion of the prey by refusion with the plasma membrane (iii). (b) Following the evolution of more selective exocytotic refusion as shown in FIGURE 8biv, phagocytosis involving both the origin- and the terminus-attachment site (i) would inevitably separate the chromosome from the cell surface, but it would remain attached to internal RER cisternae, which could then evolve into the nucleus as shown in (c). The internalized centrosome would be free to nucleate microtubules on both faces (ii) and its attached membrane divide as before (iii). (c) The internalized RER/centrosome/chromosome/ microtubule complex could be converted into a nucleus with a vegetatively persistent half-spindle and aster (as in just a few protozoa) by the evolution of a nuclear lamina (L) by mutation(s) causing the polymerization of proteins able to bind to DNA (or chromatin) and then, and only then, to RER membrane, and by the evolution of pore complexes (P) to prevent the nuclear

The budding of smooth vesicles from the rough endoplasmic reticulum is a universal eukaryotic process of such fundamental importance that it needs a specific name: I suggest ercytosis. The maintenance of the fundamental topological, chemical, and functional distinctiveness of the plasma membrane and the endoplasmic reticulum (ER, of which the nuclear envelope is an integral part) depends upon the existence of both ercytosis and exocytosis. Growth of the eukaryotic plasma membrane depends on (1) biosynthesis of its lipids and proteins in the ER, (2) ercytosis, which selectively buds off smooth vesicles, which lack not only attachment sites for DNA and ribosomes, but also the ER-located lipid-making enzymes, and (3) exocytosis which inserts these vesicles into the plasma membrane. Both ercytosis and exocytosis are absent from bacteria, since their membrane lipids are made by enzymes embedded in the plasma membrane, and proteins and lipids are both inserted directly into the plasma membrane (in cyanobacteria they must presumably also be directly inserted into the thylakoids, which unlike all other bacterial cytoplasmic membranes are topologically distinct from the plasma membrane). The evolution of the three cytotic mechanisms, ercytosis, exocytosis and endocytosis (including phagocytosis), together with the origin of the cytoskeleton, comprised the fundamental innovations that created the eukaryotic cell.

The Golgi apparatus, however, in which additional cytotic processes (i.e., vesicle fusions and buddings) occur, may not have evolved in the first eukaryote. Its apparent absence from the three most primitive archezoan phyla suggests that it evolved later,

---

envelope and lamina thus formed from totally cutting off the nucleoplasm from the cytoplasm, which would have been lethal. Initially pore complexes could be purely passive channels allowing ribosomes, polymerases, mRNA, and so on to pass freely in and out. Later the concentration benefits of compartmentalization would favor mutants making them more selective: active transport mechanisms for mRNA and nascent ribosomal subunits, and of sequence-specific nuclear-targeting mechanisms,[88] and the exclusion of mature ribosomal subunits from the nucleus would have evolved at this stage and required distinctive modifications to ribosomal RNA and proteins. Division of the nucleus could at first occur by the very same premitotic mechanism of FIGURE 4b—not a single change would be necessary. The only new problem would be that cytokinesis (by actomyosin furrowing as before) would be uncoordinated with nuclear division so that binucleate and enucleate cells would sometimes be produced (ii). This could be avoided simply by attaching the cytoplasmic centrosomal microtubules to the plasma membrane (iii). After centrosome duplication, the daughter centrosomal microtubules would become similarly attached (iv): the nuclear and cytoplasmic cleavage furrows would thereby be mechanically constrained to pass always between the daughter centrosomes and parallel to the cortical microtubules (v), as occurs in protozoan monads such as euglenoids, and thus invariably produce daughters each with a single nucleus. I assume that initially exactly 9 cytoplasmic microtubules become orthogonally attached to the plasma membrane by nine proteinaceous attachments (TF) which evolved into the modern ciliary transitional fibers; the region distal to them evolved into a protocilium (C), and that proximal to them into the kinetosome (basal body, K). Dynein, I suggest, had evolved earlier from a duplicated myosin gene and had been selected to mediate sliding of cortical microtubules so as to propel the cell by a euglenoid or axostyle-like motility. B-Tubules and kinetosomal C-tubules would have been added later[89] with the evolution of nontubulin protofilament protein. The remaining cytoplasmic centrosomal microtubules became microtubular ciliary roots[89] and/or subpellicular microtubules. Thus mutation in a single protein to create the 9 transitional fibers would simultaneously create a protocilium with microtubular roots and an accurate nuclear segregation mechanism, i.e., a primitive form of intranuclear pleuromitosis which could readily evolve into the more advanced pleuromitosis that is widespread in protozoa and probably the most primitive type of mitosis.[81]

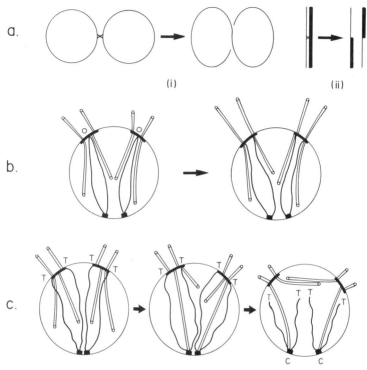

**FIGURE 6.** The origin of linear chromosomes, telomeres, and centromeres. (a) For a circular chromosome an odd number of sister-chromatid exchanges produces a dimer (i) as occurs commonly in bacterial plasmids, but for a linear chromosome it results in crossing over (ii). I argue that bacteria can resolve such dimers at the termination of replication but that the eukaryote segregation and replication mechanisms prevent this, and that prevention of dimerization following their evolution (FIGURE 4) was the selective force for the origin of linear chromosomes and telomeres. In bacteria the gradual separation of sister origins by the DNA helicase (FIGURE 4) and their folding into spatially distinct nucleoids by DNA gyrase will "terminalize" chiasmata by pulling them to the single chromosome terminus where the topoisomerase II can presumably resolve a dimer into 2 monomers as well as decatenate the dimers. But with the evolution of pleural replicons (FIGURE 4c) and simultaneous folding of daughter chromosomal domains at many points on the chromosome, chiasma terminalization would not in general occur and there would be no localized way of determining whether the overall number of exchanges was odd or even, and about half the offspring would produce dimers and experience lethal DNA breakage at cell division. This could be avoided by evolving telomeres and an end replication mechanism:[50,90] initially it could be quite inaccurate, for it would be more advantageous than retaining circular chromosomes so long as it produced more than 50% viable products. (b) Origin of telomeres by cutting the origin (o) into two, and evolution of prokaryotic origin-attachment proteins into eukaryote telomere-attachment proteins. The telomere replication mechanism[91] works by the addition of extra terminal repeats of a sequence with the general formula $C_{2-4}$ $(T)_{0-1}$ $A_{1-4}$ $(CA)_{0-3}$. Possibly such repeats evolved from the CA rich part of the sequence TTATACAAA which is repeated four times at the *Escherichia coli* replicon origin[93] and which is thought to be the binding site for the replication initiation protein coded by gene *dnaA*. $C_2A_4$ and $C_4A_3$ are also the RNA primer initiation sites for DNA replication in human mitochondria;[93a] as these are probably inherited from their bacterial ancestor it suggests that $C_xA_y$ sites were widely used to initiate bacterial replication. Cutting the origin in the middle would give two repeats at each telomere and allow both to replicate, but for continued viability means for adding further repeats

at the time of the simultaneous symbiotic origin of mitochondria, chloroplasts, and peroxisomes,[13] which would have stimulated the evolution of a more sophisticated sorting system for intracellular smooth vesicles so as to prevent the incorporation of plasma membrane, lysosomal, and secretory proteins into the three new organelles. Some of the differences between the bacterial and eukaryotic signal mechanisms for the insertion of proteins into membranes (e.g., its greater dependence on ribosome attachment to the membrane in the ER, and the origin of the signal recognition particle and docking protein), as well as the origin of 80S ribosomes, may also have occurred at this time so as to prevent mistakes in intracellular protein targeting in the considerably more complex system of cytoplasmic organelles than was present in the premitochondrial archezoan eukaryotes.

If, as argued here, phagocytosis originated in a preeukaryote with cortical microtubules, it must have occurred between them or have been accompanied by their disassembly where the prey was adsorbed. Its origin would therefore have stimulated two divergent pathways of cytoskeletal evolution: (1) formation of a localized gullet or cytostome specialized for phagocytosis, as in Anaxostylea, from which all protist gullets may be directly descended: (2) internalization of cortical microtubules to form axostyles and/or ciliary rootlets, thus allowing phagotrophy over the whole surface as in trichomonads and Mastigamoebea. Only after the nucleus evolved could this have led to the (polyphyletic) evolution of ameboid motion without disastrously shearing the DNA.

---

to repair the 5′ gaps left by the RNA primer must have evolved. An attractive alternative possibility is that the telomere replication mechanism evolved first in a linear virus and was only subsequently taken over by the host cell;[90a] this proposal receives some support from the discovery in infusorian nuclei of transposable elements[92,92a] (well established to be related to retroviruses)[92b] with termini homologous to telomeric repeats. Chromatin diminution in hypotrich infusoria itself may be mediated by reverse transcription[92c] and may therefore be another cellular property acquired from a viral symbiont.[90a] The viral and chromosome origin hypotheses for the origin of telomeres are not necessarily mutually exclusive, since ends of linear viruses are themselves replicon origins, which quite probably themselves originally evolved from host chromosomal or plasmid replicon origins. (c) Evolution of centromeres from the prokaryotic replicon termini by the attachment to them of the + end of one of the centrosomal microtubules (as in modern mitosis). This was necessary to allow telomeres (T) to be freed from the centrosome and nuclear envelope during mitosis, for otherwise chromosome breakage would still be caused by an odd number of chromatid exchanges. Temporary freeing of the telomeres would also avoid the problem of the centromere being segregated into the opposite daughter from the telomeres of the same chromosome. The centromere splitting mechanism I suggest evolved from the prokinetochore splitting mechanism of bacteria (FIGURE 4a). From the beginning the microtubule attachment sites on the two sister termini must have been oriented at an angle and not parallel to each other, so as to avoid sister chromatids being attached to microtubules from the same centromere. These changes produce a closed intranuclear pleuromitosis indistinguishable from that of many protozoa, which is thought to be the most primitive extant form.[81] More advanced types of mitosis can readily be produced from it (FIGURE 7). Originally I attributed the origin of linear chromosomes to the production of odd numbers of chiasmata by the origin of sexual recombination,[50] but sister chromatid exchange would have caused the same problems even earlier; however sex itself (FIGURE 9) may have started to evolve even before all the changes shown in FIGURES 4-6 were complete, and therefore may have provided an additional, albeit minor, selective force in the same direction.

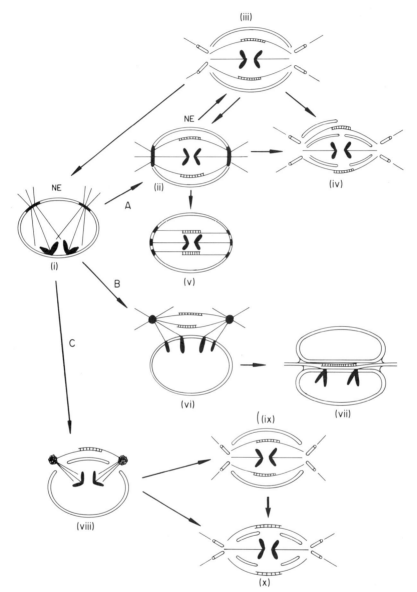

**FIGURE 7.** Mitotic diversification. The ancestral form was probably closed intranuclear pleuromitosis with two half-spindles arranged side by side (i; NE = nuclear envelope), found both in Archezoa (standard for Microspora and Oxymonadida) and Mitozoa (Radiozoa, trypanosomes, and Granuloreticulosea) and in at least one primitive green alga (Plantae); it evolved in three directions: (A) By forming a central spindle by cross connections between endonuclear nonkinetochore microtubules to make pole-to-pole spindle fibers, it yielded a closed intranuclear orthomitosis (ii), the other very widespread form in Protozoa,[81] also common in the derived kingdoms Fungi, Chromista, and Plantae. (B) By freeing the centrosomes from direct contact with the nuclear envelope and forming an exonuclear central spindle by cross connections between

# THE ORIGIN OF SEX AND ANISOKONTY

As sex is epigenetically linked to sporulation in most protists, and as sporulation in both prokaryotes[103] and protists[102] is so genetically complex, I suggest that the sporulation genetic program was conserved during the origin of the eukaryotes[103a] and sexuality grafted onto it when the loss of the prokaryotic cell wall and the evolution of the cytoskeleton allowed cells to fuse more readily. FIGURE 9 outlines how exospore formation in an actinomycete posibacterium could be converted after vegetative wall loss into eukaryote syngamy and zygospore formation. The mutation initating syngamy may have been initially selected because it allowed the formation by allopolyploidy of the first anisokont eukaryote and a marked improvement in predatory efficiency (FIGURES 10 and 11).

---

centrosome microtubules to form an exonuclear pleuromitosis as in Parabasalia and Dinozoa of the class Syndinea (vi). (C) By developing polar fenestrae in the envelope and allowing (as in B) a closer association of spindle poles with the kinetosomes (now centrioles) to form semiopen pleuromitosis as in most sporozoa. The origin of orthomitosis (ii, ix, x) involves not only the cross-linking of microtubules of opposite polarity to form pole-to-pole spindle fibers but also a change in the orientation of the two binding sites for microtubules on the sister centromeres: i.e., increasing the angle between them to 180°, which further reduces the chance of sisters becoming attached to the same centromere (with unmodified pleuromitosis this error would have become more frequent as the haploid chromosome number increased). Semiopen ortho-mitosis could have arisen both from closed endonuclear orthomitosis (iii) and from semiopen pleuromitosis (ix); with the loss of cilia and the kinetosome/centriole it could readily revert to closed endonuclear mitosis (i) or (ii); this has clearly happened during fungal evolution.[5] Open orthomitosis with the nuclear envelope totally rearranged[94] so as to lie parallel to spindle fibers (iv, x), as in animals, higher green plants, and some protists, could originate in three ways: from semiopen pleuromitosis (viii), semiopen orthomitosis (iv, ix), and from closed orthomitosis (ii). Though the ancestral form for animals, it has clearly evolved more than once in the kingdoms Protozoa, Plantae, Fungi, and Chromista, possibly from all three sources. Its selective advantage, I proposed,[87,94] was to shorten the diffusion paths for $Ca^{++}$, used for control of mitosis, in larger nuclei. The unique pluricentrosomal endonuclear spindle of the dinozoan *Oxyrrhis*[95] (v) could have evolved from closed orthomitosis simply by fragmentation of the centrosomes after they ceased to be associated with kinetosome/centrioles. The equally unique noncentric exonuclear spindle of the dinozoan class Peridinea (vii) could readily have evolved from the Syndinean type by the loss of centromeres and centrioles (vi); because this mitotic type is unknown in Archezoa, and because it and the ciliary apparatus, cell cortex, and hydroxymethylcytosine-containing DNA of the Peridinea seem so specialized,[1] I consider it to be the most derived rather than the most primitive form of mitosis. Though there are differences between exonuclear mitosis in the Parabasalia and Syndinea (e.g., the presence of transnuclear cytoplasmic channels in the latter, as in the Peridinea), it is possible that they have a common origin: if this is so then it would imply that the Parabasalia are not primitively mitochondrionless and strengthens the conjecture[13] that their hydrogenosomes evolved from degenerate mitochondria, and necessitates their transfer from the Archezoa to the Mitozoa; the fact that they are the only Archezoa with a permanent dictyosome (present in all Mitozoa) would also support such transfer. rRNA sequences from Dinozoa and Archezoa might resolve this important issue. Many of the mitotic forms shown occur both with and without centrioles; the acentriolar variant results from the polyphyletic loss of cilia and kinetosomes (e.g., flowering plants, Zygnematalean green algae, and nearly all Amoebozoa, and Myxozoa) or their great intracellular distance from spindle poles and or multiplication in numbers (e.g., Peridinea, Infusoria).

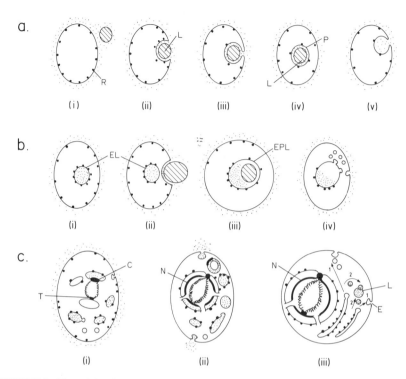

**FIGURE 8.** The origin of phagocytosis and the cellular endomembrane system (RER, nuclear envelope, Golgi, and lysosomes). (a) In the ancestral L-form, secreting enzymes at random into the environment (i), many enzymes would be wasted and many of the digestion products not absorbed. A mutant surface glycoprotein causing adhesion to the walls of bacterial prey (ii) would cause the L-form partially to surround the prey (L) thus increasing efficiency of absorption of digestion products. Increased size of the prokaryote would allow its surface completely to surround the prey (iii); accidental fusion of the lips to produce a closed phagosome (i.e., food vacuole, P) (iv) would allow complete absorption of all the digestion products, after which the phagosome membrane could refuse with the plasma membrane (v); mutations modifying proteins or lipids of the outer leaflet of the membrane bilayer so as to promote such fusion would improve efficiency of phagocytosis, so long as their effect was causally linked (possibly allosterically) to adhesion to the prey. But much secreted enzyme would still be wasted. R = membrane-bound ribosomes making secretory proteins. (b) This waste would be reduced by two possibly concurrent changes: (i) delay in refusion of the phagosome allowing secreted enzyme to accumulate in its lumen to form a primitive lysosome which I call the erlysosome (EL) because, having ribosomes in its outer surface, it was a precursor also of the rough endoplasmic reticulum (RER). (ii) Fusogenic mutations affecting the proteins in the cytoplasmic leaflet of the erlysosome would be especially favored if they caused them to bind preferentially to the cytoplasmic face of integral membrane glycoproteins allosterically altered by adhesion of their outer surface to the prey. This would directly liberate the erlysosomal digestive enzymes into the forming phagosome to form an erphagolysosome (EPL) (iii). (iv) Evolution of the budding from the erlysosome of smooth vesicles, which would fuse constitutively with the plasma membrane, would prevent the excessive membrane internalization caused by steps (i-iii) and at the same time allow the erlysosomal ribosomes to continue secreting enzyme into the EL lumen rather than the external medium, and erlysosome fusion to become totally specific to the phagosome. (c) Such specificity coupled with the specificity of the vesicle budding were the key steps in eukaryogenesis: henceforth if centrosomes (C), chromosome termini (T) or ribosomal attachment sites were internalized by

## α-AMINO-ADIPIC ACID (AAA) PATHWAY FOR LYSINE BIOSYNTHESIS

This pathway probably evolved from the degradative pathway in bacteria which involves AAA,[41] and was substituted for the prokaryotic diamino-pimelic acid (DAP) pathway during the early evolution of eukaryotes.[1,2,57] I earlier suggested that the loss of murein (and with it the need for DAP in the tetrapeptides that typically cross-link the muramic acid residues in murein) was what allowed the total loss of DAP by the ancestral eukaryote during the changeover. Yet many posibacteria use lysine rather than DAP in their murein tetrapeptides, but still use the DAP pathway; thus loss of DAP from the wall, though probably necessary, is insufficient in itself to account for the changeover. I now suggest that the evolution of phagocytosis was the decisive secondary cause of the changeover. Firstly, the prey would provide such an abundant new source of lysine that mutations inactivating the DAP pathway could have spread by random drift or hitchhiking. Secondly, it might positively have favored the changeover if the ancestral eukaryote was a synthesizer, like some actinomycete posibacteria and fungi, of N-penicillin, cephalosporins, or other β-lactam antibiotics containing AAA as a major constituent.[44]

In actinomycetes, which use the DAP pathway to make lysine, excess lysine stimulates the formation of antibiotic, presumably by increasing the AAA produced by its degradation.[44] Though advantageous to a posibacterial saprotroph for which other bacteria are competitors for nutrients, this would be disadvantageous for a phagotroph for which they are valuable food. In fungi however, which use AAA for making lysine, excess lysine reduces the output of β-lactam antibiotics, presumably by feedback inhibition of AAA formation;[44] this pattern of regulation would be more advantageous for a phagotroph—especially one that was also a facultative saprotroph as is likely for the first eukaryote—since it would not inhibit the growth of the prey. However this argument depends on the assumption that fungi inherited the enzymes

---

phagocytosis (i and FIGURE 5) they could not be returned to the surface by exocytosis. Fusion of vesicles around the chromosome mediated by lamin proteins and the evolution of nuclear pores and transenvelope RNA and protein transport would produce the nucleus (ii, N). At some stage mutations making enzyme secretion obligately cotranslational (as now appears so for ribophorins) ensured that as soon as all ribosome receptors were internalized by phagocytosis (ii) ribosomes could never again colonize the plasma membrane, and would reduce the wastage of enzyme. But some would still be lost via smooth vesicle exocytosis until a sorting process evolved to allow the budding from the erlysosome of two sorts of vesicle: (1) those lacking lysosomal enzymes and lysosome receptors and able to fuse constitutively with the plasma membrane (E), and (2) those with enzymes and phagosome receptors and able to fuse only with each other to form true lysosomes (L) without ribosomes on their surface. The latter vesicles are distinguished in animal cells by the exposure of mannose 6-phosphate residues by Golgi deglycosylases; but as shown in iii this process might have evolved even before the smooth membranes and the glycosylation process became topologically segregated from the RER in a distinct Golgi apparatus; whether this segregation never happened in the mitochondrionless Metamonada and Archamoebae, in which no Golgi apparatus has been reported, or whether like higher fungi they possess unnoticed unstacked Golgi cisternae is not known. It is equally unknown at what stage in eukaryote evolution the Golgi apparatus became subdivided into topologically distinct subcompartments interconnected by smooth vesicle budding and fusion.[98]

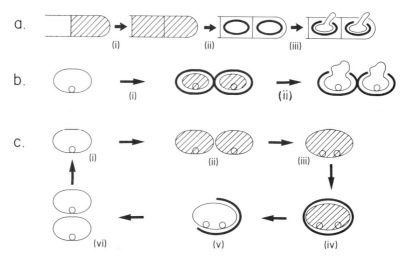

**FIGURE 9.** Sporulation and the origin of sex. (a) The ancestor of eukaryotes and archaebacteria is assumed to have been a posibacterium of the actinomycete subdivision, which (in response to nutrient depletion), like some modern actinomycetes, [103] formed two exospores per sporangium, each with a resistant wall, which germinated when conditions improved. (b) Loss of murein would make the vegetative cells and the two presumptive spore cells naked and potentially able to fuse with each other but would not disrupt sporulation (i) or spore germination (ii) or the epigenetic program controlling them, so a single vegetative cell would still differentiate specifically into two spores. (c) A mutation rendering a preexisting membrane glycoprotein fusogenic would be selected against if it were in a gene with a promoter expressed also in vegetative cells, as it would cause indiscriminate fusion of sister cells. But such a mutation in a gene already under the control of a sporulation promoter would not be harmful. It would simply cause the two prespore cells to fuse to form a "diploid" cell (iii) which would surround itself by a wall to produce a zygospore (iv). When the zygospore germinated the gene would not be expressed and no further fusions would occur. Provided syngamy originated after the origin of centrosomal microtubules, then on return to the vegetative dividing phase a division furrow would probably be formed between each centrosome at the next division thus returning the cell immediately to the haploid state. In the archaebacterial line because of the absence of centrosomes and cytokinesis by actomyosin cleavage every cell fusion would double their ploidy, so as in posibacteria sexual fusion could not evolve. Meiosis would be unnecessary until the cell acquired two dissimilar chromosomes (FIGURE 10); though the sexual life cycle shown here is autogamous, it is not obligately so, and fusion with nonsister cells would become much more frequent with the evolution of ciliary motility, and one would have a homothallic sexual haploid life cycle with gametogenesis, syngamy, zygospore formation, and reduction during zygospore germination, as in many protists.

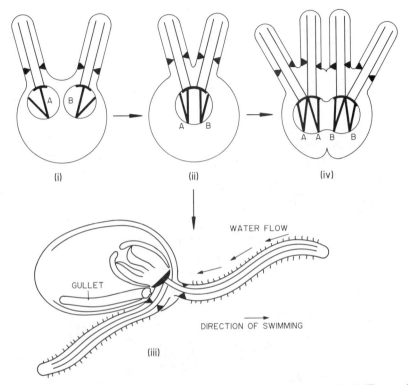

**FIGURE 10.** Allopolyploidy and the origin of anisokonty. After cilia had evolved (FIGURE 5), syngamy (FIGURE 9) would produce a biciliated cell (i). Failure in reduction during spore germination would create a "polyploid" cell with two chromosomes and two cilia: such a biciliated polyploid would be stable if two parental centrosomes became fused or reorganized into a single structure (ii) which duplicated as a unit. If the cilia of the two strains (A and B) that fused had already diverged structurally and functionally then the resulting allopolyploid would be preadapted for the evolution of anisokonty: the anterior cilium would become specialized for pulling and the posterior cilium for pushing the cell forward, and the water current generated by their coordinated beat would direct bacteria into a gullet underlying the posterior cilium (iii). But the anisokont state could only be maintained if mitotic segregation regularly produced ABs rather than AAs or BBs. If the chromosome were still circular, sister origins would be preferentially attached to the same centrosome and AAs and BBs preferentially produced (iv); but if, as I suggest, linear chromosomes with a microtubule-binding centromere had already evolved (FIGURE 6), ABs would be automatically produced and the anisokont would be genetically stable. However sexual fusion between such two- chromosomed anisokonts would provide a powerful selective force for the origin of meiotic pairing (FIGURE 11). Such an asymmetric biciliated monad, as the first anaxostylean metamononad, would have been ancestral to all ciliated eukaryotes other than Archamoebea; their characteristic mode of feeding would have been to swim along bacteria-rich surfaces as do modern euglenozoa. Ciliary paraxial rods and nontubular mastigonemes probably evolved in this early anaxostylean and like sex were independently lost in many higher protists.

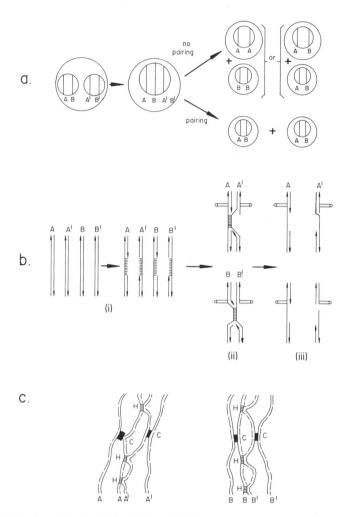

FIGURE 11. The origin of meiosis, (a) Sexual fusion between two-chromosomed anisokonts would produce zygotes with four chromosomes; in the absence of pairing these would segregate randomly (unless the nuclei failed to fuse). Initially random segregation would be relatively harmless because all chromosomes would have the same genes; it would merely have an equal chance of producing two anisokont (AB) or two isokont (AA and BB) offspring: the latter, though less efficient, would not inevitably die. But as chromosomes A and B inevitably diverged by mutation and chromosomal rearrangement, a proportion of the offspring would be lethally genetically deficient in the absence of chromosome pairing. Pairing probably evolved to reduce the proportion of such offspring (and of less efficient isokonts) rather than to facilitate recombination. Until pairing became efficient the haploid chromosome number could probably not increase beyond two (a number still found in the Parabasalia and Dinozoa, both with apparently primitive achiasmate meiosis),[81] because the proportion of defective offspring produced by reduction without pairing increases sharply with the number of different chromosomes. Typical two-step chiasmate meiosis probably evolved only after the symbiotic origin of mitochondria by the evolution of premeiotic S-phase and delayed replication of Zyg DNA.[104] (b) The simplest origin for pairing is for one complementary polynucleotide chain from each homologue to be

for β-lactam biosynthesis from actinomycetes via their protozoan ancestors. If instead they acquired them directly through genetic transformation by actinomycete plasmids, it would seem more likely that the first phagotroph would have simply lost any capacity for antibiotic and lysine synthesis possessed by its actinomycete ancestors. In this case the AAA pathway might have originated distinctly after the origin of phagotrophy. But the presence of the AAA pathway in *Euglena* suggests that it must have evolved before the simultaneous origin of chloroplasts and mitochondria,[13] for if it had not it would have been easier for the host to acquire the DAP pathway from the symbiont (as apparently happened in the Chromista and Plantae)[1,2,7] than to evolve a new AAA pathway. The postulated choanomonad common ancestor of animals and fungi[5] must have had the AAA pathway; unfortunately what lysine pathway, if any, is present in the Choanozoa, or in the Archezoa, Dinozoa, and Archaebacteria, is unknown despite their key importance for understanding the phylogeny of lysine biosynthesis.

# ENVOI

The present theory of eukaryote and archaebacterial origins is more detailed and explicit than any previous ones. Yet it gives a much more gradual and less hazardous evolution of mitosis, cilia, and sex than do previous theories, and it rationalizes many more of the unique genetic and cytoplasmic features of eukaryotes. It is a purely autogenous theory of the origin of eukaryotes; symbiosis is necessary only to explain the origin of mitochondria, chloroplasts, chloroplast endoplasmic reticulum, and perhaps microbodies and peroxisomes. It should be tested by comparative molecular studies of the Archezoa, Posibacteria, and Archaebacteria.

---

digested in the centromere region by site-specific nucleases (i) so as to allow DNA hybrids to form (ii), which will automatically occur preferentially between homologous chromosomes; if hybridizing sites were appropriately positioned near the centromeric microtubule-binding sites they could ensure that these were oriented so as to attach to microtubules from opposite centrosomes. The fact that the duration of meiotic prophase is directly proportional to genome size[105] strongly supports the thesis that meiotic pairing is still mediated by DNA hybridization,[104] and therefore the simple molecular mechanism for the origin of Mendelian segregation proposed here. It yields a one-step meiosis with no chiasmata as found in some Archezoa and in Dinozoa and Sporozoa.[81] The gap left in the DNA after segregation (iii) can be repaired after division. (c) Typical two-step meiosis probably evolved after the symbiotic origin of mitochondria by the evolution of a premeiotic S-phase and either delayed replication or one-strand nuclease digestion of Zyg DNA which would allow hybrid DNA (H) and therefore chiasmata to form at many points on the chrososome and thus increase the diversity of meiotic products; at the same time centromere replication must be delayed or one sister centromere (C) masked (? by synaptonemal complex proteins) till the second division. For clarity hybrid DNA is only shown between two of the four chromatids, but in practice the hybrids could occur randomly between any of the four homologous chromatids.

## ACKNOWLEDGMENTS

I thank the Australian National University for a Visiting Fellowship, the Royal Society for a Traveling Fellowship, Des Clark-Walker for stimulating discussions, and Val Rawlings for typing.

### REFERENCES

1. CAVALIER-SMITH, T. 1981. The origin and early evolution of the eukaryote cell. Symp. Soc. Gen. Microbiol. **32:** 33-84.
2. CAVALIER-SMITH, T. 1981. Eukaryote kingdoms: seven or nine? BioSystems **14:** 461-481.
3. CAVALIER-SMITH, T. 1983. A 6-kingdom classification and a unified phylogeny. In Endocytobiology II. W. Schwemmler & H. E. A. Schenk, Eds.: 1027-1034. de Gruyter. Berlin, FRG.
4. CAVALIER-SMITH, T. 1986. The kingdom Chromista: origin and systematics. In Progress in Phycological Research. F. E. Round and D. J. Chapman, Eds. **4:** 309-347. Biopress Ltd. Bristol, England.
5. CAVALIER-SMITH, T. 1987. The origin of Fungi and Pseudofungi. In Evolutionary Biology of Fungi. A. D. M. Rayner, C. M. Brasier & D. Moore, Eds. Symp. Br. Mycol. Soc. **13.** Cambridge University Press. Cambridge, England. (In press.)
6. CAVALIER-SMITH, T. 1983. Endosymbiotic origin of the mitochondrial envelope. In Endocytobiology II. W. Schwemmler & H. E. A. Schenk, Eds.: 265-279. de Gruyter. Berlin, FRG.
7. CAVALIER-SMITH, T. 1982. The origins of plastids. Biol. J. Linn. Soc. **17:** 289-306.
8. CAVALIER-SMITH, T. & J. LEE. 1985. Protozoa as hosts for endosymbioses and the conversion of symbionts into organelles. J. Protozool. **32:** 376-379.
9. MARKEY, D. R. & R. T. WILCE. 1976. The ultrastructure of reproduction in the brown alga *Pylaiella littoralis.* II. Zoosporogenesis in the unilocular sporangia. Protoplasma **88:** 147-173.
10. ROBERTS, K. R., K. D. STEWART & K. R. MATTOX. 1982. The flagellar apparatus of *Chilomonas paramecium* (Cryptophyceae) and its comparison with certain zooflagellates. J. Phycol. **17:** 159-167.
11. VOSSBRINCK, C. R. & C. R. WOESE. 1986. Eukaryotic ribosomes that lack a 5.8s rRNA. Nature **320:** 287-288.
12. DE DUVE, C. 1982. Peroxisomes and related particles in historical perspective. Ann. N.Y. Acad. Sci. **386:** 1-4.
13. CAVALIER-SMITH, T. 1987. The simultaneous symbiotic origin of mitochondria, chloroplasts, and microbodies. Ann. N.Y. Acad. Sci. (This volume.)
14. BENCHIMOL, M. & W. DE SOUZA. 1983. Fine structure and cytochemistry of the hydrogenosome of *Tritrichomonas foetus.* J. Protozool. **30:** 422-455.
15. WOESE, C. R. 1982. Archaebacteria and cellular origins: an overview. Zbl. Bakt. Hyg. 1 Abt. Orig. C **3:** 1-17.
15a. WOESE, C. R. 1983. The primary lines of descent and the universal ancestor. In Evolution from Molecules to Men. D. S. Bendall, Ed.: 209-233. Cambridge University Press. Cambridge, England.
16. DOOLITTLE, W. F. 1980. Revolutionary concepts in evolutionary cell biology. Trends Biochem. Sci. **5:** 146-149.
17. DOOLITTLE, W. F. & C. J. DANIELS. 1985. Prokaryotic genome evolution: what we might learn from the Archaebacteria. In Evolution of Prokaryotes. K. H. Schleifer & E. Stackebrandt, Eds.: 31-44. Academic Press. London, England.
18. DARNELL, J. E. 1978. Implications of RNA:RNA splicing in evolution of eukaryotic cells. Science **202:** 1257-1260.

19. KNOLL, A. H. 1983. Biological interactions and precambrian eukaryotes. *In* Biotic Interactions in Recent and Fossil Benthic Communities. M. J. S. Tevesz & P. L. McCall, Eds.: 251-283. Plenum Press. New York, NY.

20. KNOLL, A. H. 1985. Patterns of evolution in the Archaean and Proterozoic eons. Paleobiology **11:** 53-64.

21. KJEMS, J. & R. A. GARRETT. 1985. An intron in the 23s ribosomal RNA gene of the archaebacterium *Desulfurococcus mobilis.* Nature **318:** 675-677.

22. KANDLER, O. 1982. Cell wall structures and their phylogenetic implications. Zbl. Bakt. Hyg. I. Abt. Orig. **C3:** 149-160.

23. MESCHER, M. F. & J. L. STROMINGER. 1975. Purification and characterisation of a prokaryotic glycoprotein from the cell envelope of *Halobacterium salinarium.* J. Biol. Chem. **251:** 2005-2014.

24. ZILLIG, W., R. SCHNABEL & K. O. STETTER. 1985. Archaebacteria and the origin of the eukaryotic cytoplasm. Curr. Topics Microbiol. Immunol. **114:** 1-18.

24a. ZILLIG, W., R. SCHNABEL, F. GROPP, W. D. REITER, K. STETTER & M. THOMM. 1985. The evolution of the transcription apparatus. FEMS Symp. **29:** 45-72.

25. OTAKA, E., T. OOI, T. KUMAZAKI & T. ITOH. 1985. Examination of protein sequence homologies. II. Six *Escherichia coli* L7/L12-type ribosomal 'A' protein sequences from eukaryotes and metabacteria, contrasted with those from prokaryotes. J. Mol. Evol. **22:** 342-350.

26. OLSEN, G. J., N. R. PACE, M. NUELL, B. P. KAINE, R. GUPTA & C. R. WOESE. 1985. Sequence of the 16S rRNA gene from the thermacidophilic archaebacterium *Sulfolobus solfataricus* and its evolutionary implications. J. Mol. Evol. **22:** 301-307.

27. HORI, H., T. ITOH & S. OSAWA. 1982. The phylogenetic structure of the metabacteria. Zentralbl. Bakteriol. Hyg. [C]**3:** 18-30.

28. KESSEL, M. & F. KLINK. 1982. Identification and comparison of eighteen archaebacteria by means of the diphtheria toxin reaction. Zbl. Bakt. Hyg., I Abt. Orig. **C3:** 140-148.

29. LAKE, J., M. W. CLARK, E. HENDERSON, S. P. FAY, M. OAKES, A. SCHEINMAN, J. P. THORNBER & R. A. MAH. 1985. Eubacteria, halobacteria and the origin of photosynthesis: the photocytes. Proc. Nat. Acad. Sci. USA **82:** 3716-3720.

30. CAMMARANO, P., A. TEICHNER, P. LONDEI, M. ACCA, B. NICOLAUS, J. L. SANZ & R. AMILS. 1985. Insensitivity of archaebacterial ribosome to protein synthesis inhibitors. Evolutionary implications. EMBO J. **4:** 811-816.

31. STOECKENIUS, W. 1985. The rhodopsin-like pigments of halobacteria: light-energy and signal transducers in an archaebacterium. Trends Biochem. Sci. **10:** 483-486.

32. BELFORT, M., J. PEDERSEN-LANE, D. WEST, K. EHRENMAN, G. MALEY, F. CHU & F. MALEY. 1985. Processing of the intron-containing thymidylate synthase (td) gene of phage T4 is at the RNA level. Cell **41:** 375-382.

33. SLEYTR, U. B. & P. MESSNER. 1983. Crystalline surface layers on bacteria. Annu. Rev. Microbiol. **37:** 311-339.

34. SEARCY, D. G., D. B. STEIN & G. R. GREEN. 1981. A mycoplasma-like archaebacterium possibly related to the nucleus and cytoplasm of eukaryotic cells. Ann. N.Y. Acad. Sci. **361:** 312-323.

35. VAN VALEN, L. M. & V. C. MAIORANA. 1980. The archaebacteria and eukaryotic origins. Nature **287:** 248-250.

36. THOMM, M., K. O. STETTER & W. ZILLIG. 1982. Histone-like proteins in eu- and archaebacteria. Zbl. Bakt. Hyg. I. Abt. Orig. **C3:** 128-139.

37. MANILOFF, J. 1981. Cytoskeletal elements in mycoplasmas and other prokaryotes. BioSystems **14:** 305-312.

38. HENDERSON, C. E., R. N. PERHAM & J. T. FINCH. 1979. Structure and symmetry of B. *stearothermophilus* pyruvate dehydrogenase multienzyme complex and implications for eukaryote evolution. Cell **17:** 85-93.

39. PULLEN, A. M., N. BUGDEN, M. J. DANSEN & D. W. HOUGH. 1985. Citrate synthase: an immunological investigation of interspecies diversity. FEBS Lett. **182:** 163-166.

40. WEITZMAN, P. D. J. 1985. Evolution in the citric acid cycle. FEMS Symp. **29:** 253-275.

41. RAGAN, M. A. & D. J. CHAPMAN. 1978. A Biochemical Phylogeny of Protists. Academic Press. London, England.

42. NES, W. R. & W. D. NES. 1980. Lipids in Evolution. Plenum. New York, N.Y.

43. CUNDLIFFE, E., R. H. SKINNER & J. THOMPSON. 1984. Resistance to antibiotics in antibiotic-producing organisms. *In* Biological, Biochemical and Biomedical Aspects of Actinomycetes. L. Ortiz-Ortiz, L. F. Bojalil & V. Yakoleff, Eds.: 303-314. Academic Press. Orlando, Fla.

44. MENDELOVITZ, S., F. T. KIRNBERG & N. M. MAGUL. 1984. Feedback regulatory steps involved in diaminopimelic acid and lysine biosynthesis in *Streptomyces clavigerus. In* Biochemical and Biomedical Aspects of Actinomycetes. L. Ortiz-Ortiz, L. F. Bojalil & V. Yakoleff, Eds.: 357-366. Academic Press. Orlando, Fla.

45. MEZES, P. S. F. & J. O. LAMPEN. 1985. Secretion of proteins by Bacilli. *In* Molecular Biology of the Bacilli. D. A. Dubnau, Ed. 2: 151-183. Academic Press. Orlando, Fla.

46. LANDMAN, O. E. & R. A. PEPIN. 1982. *Bacillus subtilis* mating system: dikaryotic recombinants and supression of the prototrophic phenotype. *In* Molecular Cloning and Gene Regulation in Bacilli. A. T. Ganesan, S. Chang & J. A. Hoch, Eds.: 25-39. Academic Press. New York, N.Y.

47. CHEN, Z., S. F. WOJCIK & N. E. WELKER. 1986. Genetic analysis of *Bacillus stearothermophilus* by protoplast fusion. J. Bacteriol. **165:** 994-1001.

48. CLEWELL, D. B., B. A. WHITE, Y. IKE & F. Y. AN. 1984. Sex pheromones and plasmid transfer in *Streptococcus faecalis. In* Microbial Development. R. Losick & L. Shapiro, Eds.: 133-149. Cold Spring Harbor Laboratory. Cold Spring Harbor, N.Y.

48a. SMUCKER, R. A. 1984. Biochemistry of the *Streptomyces* spore sheath. *In* Biological, Biochemical and Biomedical Aspects of Actinomycetes. L. Ortiz-Ortiz, L. F. Bojalil & V. Yakoleff, Eds.: 171-177. Academic Press. Orlando, Fla.

49. DE ROSA, M., A. GAMBACORTA & A. GLIOZZI. 1986. Structure, biosynthesis, and physicochemical properties of Archaebacterial lipids. Microbiol. Rev. **50:** 70-80.

50. CAVALIER-SMITH, T. 1975. The origin of nuclei and of eukaryotic cells. Nature **256:** 463-468.

51. STANIER, R. Y. 1970. Some aspects of the biology of cells and their possible evolutionary significance. Symp. Soc. Gen. Microbiol. **20:** 1-38.

52. SCHOPF, J. W., Ed. 1983. Earth's Earliest Biosphere: Its Origin and Evolution. Princeton University Press. Princeton, N.J.

53. CAVALIER-SMITH, T. 1985. Introduction: the evolutionary significance of genome size. *In* The Evolution of Genome Size. T. Cavalier-Smith, Ed.: 1-36. Wiley. Chichester, England.

54. BLOBEL, G. 1980. Intracellular membrane topogenesis. Proc. Nat. Acad. Sci. USA **77:** 1496-1500.

55. ALLISON, L. A., M. MOYLE, M. SHALES & C. J. INGLES. 1985. Extensive homology among the largest subunits of eukaryotic and prokaryotic RNA polymerases. Cell **42:** 599-610.

56. BIGGS, J., L. L. SEARLES & A. L. GREENLEAF. 1985. Structure of the eukaryotic transcription apparatus: features of the gene for the largest subunit of Drosophila RNA polymerase II. Cell **42:** 611-621.

57. CAVALIER-SMITH, T. 1980. Cell compartmentation and the origin of membranous eukaryote organelles. *In* Endocytobiology: Endosymbiosis and Cell Biology. W. Schwemmler & H. E. A. Schenk, Eds.: 831-916. de Gruyter. Berlin, FRG.

58. WOESE, C. R., J. MANILOFF & L. B. ZABLEN. 1980. Phylogenetic analysis of the mycoplasmas. Proc. Nat. Acad. Sci. USA **77:** 494-498.

59. WOESE, C. R., E. STACKEBRANDT & W. LUDWIG. 1985. What are mycoplasmas: the relationship of tempo and mode in bacterial evolution. J. Mol. Evol. **21:** 305-316.

60. GRAY, M. W. & W. F. DOOLITTLE. 1982. Has the endosymbiosis hypothesis been proven? Microbiol. Rev. **46:** 1-42.

61. YANG, D., Y. OYAIZU, H. OYAIZU, G. J. OLSEN & C. R. WOESE. 1985. Mitochondrial origins. Proc. Nat. Acad. Sci. USA **82:** 4443-4447.

62. RAPOPORT, T. O. & M. WIEDMANN. 1985. Application of the signal hypothesis to the incorporation of integral membrane proteins. Curr. Topics Membr. Transp. **24:** 1-63.

63. MEILE, L., T. LEISINGER & J. N. REEVE. 1985. Cloning of DNA sequences from *Methanococcus vaniellii* capable of autonomous replication in yeast. Arch. Microbiol. **143:** 253-255.

64. MADOFF, S. 1982. The L-forms of bacteria. *In* The Prokaryotes. M. P. Starr, H. Stolp, H. G. Truper, A. Balows & H. G. Schegel; Eds. **2:** 2225-2237. Springer. Berlin, FRG.

65. PATTERSON, S. N. & R. W. GILPIN. 1982. Basic biology of cell wall-deficient bacteria. *In* Cell Wall-Deficient Bacteria: Basic Principles and Clinical Significance. G. J. Domingue, Ed.: 1-58. Addison-Wesley. Reading, Mass.

66. STODDART, R. W. 1984. The Biosynthesis of Polysaccharides. Croom Helm. London, England.

67. MATHESON, A. T. & M. YAGUCHI. 1982. The evolution of the archaebacterial ribosome. Zbl. Bakt. Hyg. I Abt. Orig. **C3:** 192-199.

68. WOESE, C. R., E. STACKEBRANDT, T. J. MACKE & G. E. FOX. 1985. A phylogenetic definition of the major eubacterial taxa. Syst. Appl. Microbiol. **6:** 143-151.

69. STACKEBRANDT, E., H. POHLA, R. KROPPENSTEDT, H. HIPPE & C. R. WOESE. 1985. 16S rRNA analysis of *Sporomusa, Selenomonas,* and *Megasphaera:* on the phylogenetic origin of gram-positive Eubacteria. Arch. Microbiol. **143:** 270-276.

70. HASEGAWA, M., Y. IIDA, T. YANO, F. TAKAIWA & M. IWABUCHI. 1985. Phylogenetic relationships among eukaryote kingdoms inferred from ribosomal RNA sequences. J. Mol. Evol. **22:** 32-38.

71. CAVALIER-SMITH, T. 1985. Selfish DNA and the origin of introns. Nature **315:** 283-284.

72. HICKEY, D. A. & B. F. BENKEL. 1985. Splicing and the evolution of introns. Nature **316:** 582.

73. CECH, T. R. 1985. Self-splicing RNA: implications for evolution. Int. Rev. Cytol. **93:** 3-22.

74. ERREDE, B., M. COMPANY & R. SWANSTROM. 1986. An anomalous TY1 structure attributed to an error in reverse transcription. Mol. Cell Biol. **6:** 1334-1338.

75. WEEDS, A. 1982. Actin-binding proteins—regulators of cell architecture and motility. Nature **296:** 811-816.

76. VAN ELDIK, L. J., J. G. ZENDEGUI, D. R. MARSHAK & D. M. WATTERSON. 1982. Calcium binding proteins and the molecular basis of calcium action. Int. Rev. Cytol. **72:** 1-61.

76a. ABDEL-MONEM, M., H. LAUPPE, J. KARTENBECK, H. DÜRWALD & H. HOFFMANN-BERLING. 1977. Enzymatic unwinding of DNA. III. Mode of action of *Escherichia coli* DNA unwinding enzyme. J. Mol. Biol. **111:** 667-685.

77. CAVALIER-SMITH, T. 1987. Bacterial DNA segregation: its motors and positional control. J. Theor. Biol. (In press.)

78. EUTENEUER, U. & J. R. MCINTOSH. 1980. The polarity of midbody and phragmoplast microtubules. J. Cell Biol. **87:** 509-515.

79. BYERS, B. & L. GOETSCH. 1974. Duplication of spindle plaques and integration of the yeast cell cycle. Cold Spring Harbor Symp. Quant. Biol. **38:** 123-131.

80. KORNBERG, A. 1980. DNA Replication. Freeman. San Francisco, Calif.

81. RAIKOV, I. 1982. The Protozoan Nucleus: Morphology and Evolution. Springer. Vienna, Austria.

82. WANG, J. C. 1985. DNA topoisomerases. Annu. Rev. Biochem. **54:** 665-697.

83. CAVALIER-SMITH, T. 1985. DNA replication and the evolution of genome size. *In* The Evolution of Genome Size. T. Cavalier-Smith, Ed.: 211-251. Wiley. Chichester, England.

84. CAVALIER-SMITH, T. 1985. Selfish DNA, intragenomic selection and genome size. *In* Evolution of Genome Size. T. Cavalier-Smith, Ed.: 253-265. Wiley. Chichester, England.

85. KROHNE, G. & R. BENAVENTE. 1986. The nuclear lamins: a multigene family of proteins in evolution and differentiation. Exp. Cell Res. **162:** 1-10.

86. MCKEON, F. D., M. W. KIRSCHNER & D. KAPUT. 1986. Homologies in both primary and secondary structure between nuclear envelope and intermediate filament proteins. Nature **319:** 463-468.

87. CAVALIER-SMITH, T. 1982. Evolution of the nuclear matrix and envelope. *In* The Nuclear Envelope and the Nuclear Matrix. (Wistar Symposium.) G. G. Maul, Ed. **2:** 307-318. Liss. New York, N.Y.

88. DINGWALL, C. 1985. The accumulation of proteins in the nucleus. Trends Biochem. Sci. **10:** 64-66.

89. CAVALIER-SMITH, T. 1982. The evolutionary origin and phylogeny of eukaryote flagella. Soc. Exp. Biol. Symp. **35:** 465-493.

90. CAVALIER-SMITH, T. 1974. Palindromic base sequences and replication of eukaryote chromosome ends. Nature 250: 467-470.
90a. CAVALIER-SMITH, T. 1983. Genetic symbionts and the origin of split genes and linear chromosomes. In Endocytobiology II. H. E. A. Schenk & W. Schwemmler, Eds.: 29-45. de Gruyter. Berlin, FRG.
91. GREIDER, C. W. & E. H. BLACKBURN. 1985. Identification of a specific telomere terminal transferase activity in Tetrahymena extracts. Cell 43: 405-413.
92. CHERRY, J. M. & E. M. BLACKBURN. 1985. The internally located telomeric sequences in the germ line chromosomes of Tetrahymena are at the end of transposon-like elements. Cell 43: 747-758.
92a. HERRICK, G., S. CARTINHOUR, D. DAWSON, D. ANG, R. SHEETS, A. LEE & K. WILLIAMS. 1985. Mobile elements bounded by $C_4A_4$ telomeric repeats in Oxytricha fallax. Cell 43: 759-768.
92b. VARMUS, H. E. 1985. Reverse transcriptase rides again. Nature 314: 583-584.
92c. LIPSS, H. J. 1985. A reverse transcriptase like enzyme in the developing macronucleus of the hypotrichous ciliate Stylonychia. Curr. Genet. 10: 239-243.
93. ZYSKIND, J. W., D. W. SMITH, Y. HIROTA & M. TAKANAMI. 1981. The consensus sequence of the bacterial origin. ICN-UCLA Symp. Mol. Cell. Biol. 21: 26.
93a. CHANG, D. D. & D. A. CLAYTON. 1985. Priming of human mitochondrial DNA replication occurs at the light-strand promoter. Proc. Nat. Acad. Sci. USA 82: 351-355.
94. HEPLER, P. K. & S. M. WOLNIAK. 1984. Membranes in the mitotic apparatus: their structure and function. Int. Rev. Cytol. 90: 169-238.
95. TRIEMER, R. E. 1982. A unique mitotic variation in the marine dinoflagellate Oxyrrhis marina (Pyrrhophyta). J. Phycol. 18: 399-411.
96. NELSON, W. J. & P. TRAUB. 1981. Fractionation of the detergent-resistant filamentous network of Ehrlich ascites tumour cells. Eur. J. Cell Biol. 23: 250-251.
97. ROGERS, J. H. 1985. The origin and evolution of retroposons. Int. Rev. Cytol. 93: 187-279.
97a. RODRIGUEZ-BOULAN, E., D. E. MISEK, D. V. DE SALAJ, P. J. I. SALAS & E. BARD. 1985. Protein sorting in the secretory pathway. Curr. Topics. Membranes Transp. 24: 251-294.
98. KOZAK, M. 1983. Comparison of initiation of protein synthesis in procaryotes, eucaryotes, and organelles. Microbiol. Rev. 47: 1-45.
98a. DUNPHY, W. G. & J. E. ROTHMAN. 1985. Compartmental organization of the Golgi stack. Cell 42: 13-21.
99. BENSON, S. A., M. N. HALL & T. J. SILHAVY. 1985. Genetic analysis of protein export in Escherichia coli K12. Annu. Rev. Biochem. 54: 101-134.
100. VON HEIJNE, G. 1985. Ribosome-SRP-signal sequence interactions. The relay helix hypothesis. FEBS Lett. 190: 1-5.
101. CHATER, K. F. 1984. Morphological and physiological differentiation in Streptomyces. In Microbial Development. R. Losick & L. Shapiro, Eds.: 89-115. Cold Spring Harbor Laboratory. Cold Spring Harbor, N.Y.
102. ESPOSITO, R. E. & S. KLAPHOLZ. 1981. Meiosis and ascospore development. In The Molecular Biology of the Yeast Saccharomyces: Life Cycle and Inheritance. J. N. Strathern, E. W. Jones & J. R. Broach, Eds.: 211-287. Cold Spring Harbor Laboratory. Cold Spring Harbor, N.Y.
103. BLAND, C. E. & J. N. COUCH. 1981. The family Actinoplanaceae. In The Prokaryotes. M. P. Starr, M. Stolp, H. G. Truper, A. Balows & H. G. Schlegel, Eds. 2: 2004-2010. Springer. Berlin, FRG.
103a. DAWES, I. W. 1981. Sporulation in evolution. Symp. Soc. Gen. Microbiol. 32: 85-130.
104. HOTTA, Y., S. TABATA, L. STUBBS & H. STERN. 1985. Meiosis-specific transcripts of a DNA component replicated during chromosome pairing: homology across the phylogenetic spectrum. Cell 40: 785-793.
105. BENNETT, M. D. 1974. The duration of meiosis. In Cell Cycle in Development and Differentiation. M. Balls & F. S. Billet, Eds.: 111-131. Cambridge University Press. Cambridge, England.
106. WOESE, C. R. & R. S. WOLFE. 1985. Archaebacteria: the Urkingdom. In The Bacteria. Archaebacteria. C. R. Woese & R. S. Wolfe, Eds. 8: 561-564. Academic Press. New York, N.Y.

# The Simultaneous Symbiotic Origin of Mitochondria, Chloroplasts, and Microbodies

## T. CAVALIER-SMITH

*Department of Biophysics, Cell and Molecular Biology*
*King's College London*
*26-29 Drury Lane*
*London WC2B 5RL, United Kingdom*

## INTRODUCTION

The purpose of this paper is not to discuss the continually growing evidence for the symbiotic origin of both chloroplasts and mitochondria.[1] It is to argue that many uncritically accepted features of the conventional serial endosymbiosis theory[2,3] are so seriously in error that they need to be replaced by a more realistic symbiotic theory. I shall first state and then justify the seven major changes needed:

1. The host was not an imaginary ameboid prokaryote, but a real and fully eukaryotic nonameboid, biciliated anaxostylean metamonad protozoan.
2. Mitochondria and chloroplasts each originated only once, and not several times, from prokaryote symbionts, i.e., from a purple nonsulfur photosynthetic bacterium and a cyanobacterium respectively.
3. The chloroplast endoplasmic reticulum of the eukaryote kingdom Chromista[4] originated only once by the endosymbiosis of an early photosynthetic eukaryote.
4. Mitochondria and chloroplasts did not have a purely symbiotic origin, but are chimeric structures incorporating proteins from the host as well as the symbiont.
5. The outer membrane of mitochondria[5] and plant and chromist chloroplasts[6] (which all have a two-membraned envelope) originated not from the host phagosomal membrane but from the outer membrane of the negibacterial symbiont; only the outer membrane of the chloroplasts of the protozoan phyla Euglenozoa and Dinozoa (which have a three-membraned envelope) represents the host phagosomal membrane.[6]
6. The selective advantage of the gene transfer from symbiont to nucleus, was not to "increase control" or to make use of sexual recombination, but to increase the metabolic and energetic efficiency of the organelle by reducing the space taken up by ribosomes and to a lesser extent DNA; neutral and "selfish" gene transfers of DNA into the organelles from the rest of the cell have also occurred, e.g., of introns.[7]
7. The conversion of symbiotic negibacteria into mitochondria and chloroplasts did not occur serially at distantly separated times but simultaneously in the

same host cell about 960 million years ago some time after the origin of the first efficient phagotrophic eukaryote.

I shall refer to my present unified view of the endosymbiotic origin of these two organelles as the simultaneous endosymbiosis theory, but stress two things: (1) The seven theses are logically independent; one or more could prove incorrect without invalidating the others. (2) This is not a symbiotic theory of the origin of eukaryotes; I remain strongly opposed to Mereschkowsky's defunct symbiotic theory of the origin of the nucleus,[8] to recent attempts to resuscitate it,[9,10] and to Margulis' postulates[2,3] (i) that the symbiotic origin of mitochondria was a prerequisite for the origin of phagocytosis and the eukaryote endomembrane system, and (ii) that the ecto- (not endo-) symbiotic origin of cilia from spirochaetes or spiroplasmas was a prerequisite for the origin of nuclear division by mitosis. Any of these very different theories is correctly called a symbiotic theory of eukaryote origin, but for none of them is there a scrap of even semisolid evidence; nor have their authors even begun to show how symbiosis helps to explain the origin of the detailed structure of cilia, the nucleus, or any of the universal eukaryote characters;[11] they have merely ignored the problems, which have been tackled in some detail by the autogenous theory of eukaryote origin.[11] It is high time that those who, with good reason, accept the symbiotic origin of mitochondria and chloroplasts[1] cure themselves of the grossly confusing habit of referring to this as the "symbiotic origin of eukaryotes," which it emphatically is not.

Though all biologists know that most eukaryotes lack chloroplasts, few are aware that many have no mitochondria either. The falsehood that all eukaryotes have mitochondria has been repeated so often that the importance of the more than a thousand species of protozoa with no mitochondria at all has only recently been recognized by their treatment as a separate subkingdom, the Archezoa,[12] containing four separate phyla: Archamoebae, Metamonada, Microspora, and Parabasalia. In my view it is only the existence of such fully eukaryotic phagotrophs that makes a symbiotic origin of mitochondria mechanistically plausible; no bacteria, not even the predatory ones,[13] can take up or harbor other living cells in their cytoplasm, and to suppose that any ever did is to stray into the realms of science fiction. It was the purely autogenous origin of phagotrophic eukaryotes that first made cellular endosymbiosis possible, or rather so easy as immediately to assume immense evolutionary importance. A major new argument for the symbiotic origin of mitochondria and chloroplasts comes from recent studies of the mechanism of posttranslational import of cytoplasmically synthesized proteins into the two organelles.[14] I shall argue that this mechanism could have evolved more simply by the reversal of the transenvelope protein export mechanism of a prokaryotic symbiont than by autogenous development. Similar considerations suggest that microbodies too may have had a chimeric endosymbiotic origin, which I shall also briefly discuss.

## MONERAN MYTH VERSUS METAMONAD REALITY

Haeckel originally invented the Monera as amebae without a nucleus,[15] which he thought were the most primitive forms of life and which he also thought he had actually discovered. However it is now clear that such organisms do not exist, and probably never did. Nonetheless Mereschkowsky adopted the nineteenth century idea of a symbiotic origin for the nucleus in an ameboid moneran host and extended the idea to the symbiotic origin of the chloroplast.[8] Onto such ideas Margulis[2,3] grafted Wallin's symbiotic theory[16] of the origin of mitochondria and cilia ("cilia" also embraces and supersedes "eukaryotic flagella")[17] to form the serial endosymbiosis theory, and also adopted Copeland's highly inappropriate use of Monera for the kingdom Bacteria.[18]

But no bacteria are capable of ameboid locomotion. Though Haeckel's energetic popularizations firmly placed amebae in the popular mind as primitive organisms—how often we hear the phrase "from ameba to man"—less speculatively inclined protistologists have more often considered ciliated monads to be the most primitive eukaryotes.[19] This is because of the clear phylogenetic evidence, amply confirmed by electron microscopy, that amebae have originated polyphyletically from monads on numerous separate occasions; at heart the nature of the first eukaryote and the nature of the hosts for the symbiotic origin of organelles—not necessarily the same thing—are phylogenetic problems and not matters for a priori speculation.

The main difficulty with the idea of a monad ancestry for all eukaryotes was the idea that the nonciliated red algae evolved autogenously from cyanobacteria.[20] Now that Mereschkowsky's theory that only their plastids did so, by endosymbiosis, is strongly substantiated, a monad ancestry becomes highly plausible. In the accompanying paper I argue that a simultaneous autogenous origin for the cilium and nuclear division by mitosis is the simplest explanation for both, and that the first eukaryote was a monad with a single cilium and chromosome and cortical microtubules serving also as ciliary rootlets.[11] But the nature of the host(s) for organelle endosymbioses is a separate question best approached by phylogenetic analysis. Using ultrastructural characters and a few molecular ones I initially tried to reconstruct a common ancestor for all eukaryotes with chloroplasts,[6] and concluded that it was a biciliated monad with two anterior but unequal (i.e., anisokont) cilia and a number of distinctive ultrastructural characters. Subsequently I have done the same for all eukaryotes with mitochondria, and somewhat to my surprise, found that the same biciliated monad can serve as a common ancestor for them also.[4,11] The characters of the common ancestor are shown in TABLE 1.

Of the four archezoan phyla, any or all of which may be primitively amitochondrial, the Microspora have no cilia and are obligate intracellular parasites and so could not have been the host, while the Archamoebae have only one or no cilia and so probably diverged before the origin of the biciliated anisokont state,[11] which must have been a key step in early eukaryote evolution. The Metamonada and Parabasalia however are clearly both derived from an anisokont common ancestor with one trailing cilium and one or more anterior cilia per kinetid. The Metamonada are more probable candidates for the host for seven reasons: (1) They lack the specialized Golgi dictyosome and associated striated fiber that together form the parabasal body, which is unique to Parabasalia. (2) They lack double-membraned hydrogenosomes which may possibly represent degenerate mitochondria. (3) They clearly had biciliated anisokont ancestors as shown by the biciliated character of each of the two kinetids of the Axostylaria, whereas even the most primitive Parabasalia have more than one anterior cilium (and so may have evolved from a similarly endowed metamonad). (4) They have an endonuclear rather than a specialized exonuclear spindle. (5) They are not all gut symbionts; some are free living. (6) Some have cytostomes, gullets, and cortical microtubules. (7) Some even have a microtubular root with struts like a multilayered structure. Since the latter three characters are found only in the metamonad class Anaxostylea (comprised of the orders Diplomonadida and Retortomonadida), they are the most likely hosts; the other metamonad class Axostylaria (sole order Oxymonadida) also has a paracrystalline preaxostyle and (usually contractile) axostyle found in no other organisms—as they are all symbionts in arthropod guts they probably evolved from a biciliated anaxostylean long after the origin of mitochondria and chloroplasts. Thus I conclude that the eukaryote host that converted purple bacteria and cyanobacteria into mitochondria and chloroplasts was a biciliated anaxostylean metamonad protozoan.

The 70s ribosomes of the Microspora,[21] and their primitive rRNA sequences[31] suggesting a branching of the eukaryote tree before the origin of mitochondria, strongly support my theory that Archezoa are genuinely primitive.


## SIMULTANEOUS VERSUS SERIAL ENDOSYMBIOSIS


Margulis suggested that mitochondria originated well before chloroplasts,[2,3] presumably because they are found in most eukaryotes and chloroplasts only in a minority. But this tells us nothing about the relative timing of their origin. She also asserted that their uptake was a prerequisite for the origin of endocytosis, exocytosis, and the eukaryote endomembrane system (including the nuclear envelope) on the grounds

TABLE 1. Reconstructed Characters of the Eomitozoan

1. Nucleus with endonuclear mitotic spindle and achiasmate one-step meiosis; zygospore.
2. Rough endoplasmic reticulum, lysosomes, exocytosis, and phagocytosis.
3. Two dissimilar cilia with paraxial rods and nontubular mastigonemes; laterally attached towards the anterior end of the cell, one pointing forwards and pulling the cell, and one pointing backwards and pushing; kinetosomes oriented at an angle of 90% or more, and attached closely to the nucleus.
4. Gullet with cytostome opening near ciliary bases to phagocytose bacteria wafted there by the ciliary currents.
5. Microtubular rootlet with lateral struts similar to the multilayered structure (MLS) of green plants running alongside the gullet.
6. Cortical microtubules converging on kinetosomes; ill-developed membranous cortical alveoli.
7. Golgi dictyosome evolving but not yet stable in position.
8. Primitive mitochondria in process of evolving cristae.
9. Primitive chloroplasts still with phycobilisomes, and envelopes in a state of flux.

that (1) sterols were needed for membrane budding and fusion because they make membranes more flexible, and (2) sterols entered eukaryotes only with mitochondria because that is where they are now made. But on the contrary, sterols in fact make membranes more stable;[22] they are not needed for membrane fusion and budding, for Oomycetes[23] achieve this without them (as do archezoa without mitochondria), and they are made in the endoplasmic reticulum,[24] not mitochondria!

Stanier, the other early protagonist of serial endosymbiosis, suggested that chloroplasts came first,[25] because oxygenic photosynthesis probably evolved before aerobic respiration. But, as he himself first stressed, only eukaryotes can harbor cellular endosymbionts; his argument is therefore invalid unless eukaryote phagotrophy evolved before aerobic respiration. Aerobic respiration probably evolved soon after oxygenic photosynthesis (at least 2000 million years ago),[26] long before eukaryotes (1450-960 million years ago). Clearly it was the availability of the host, not the symbionts, that was the important factor. As soon as an efficient phagotroph evolved, all kinds of symbiont could readily be taken up; it is most unlikely that photosynthetic ones would be taken up appreciably earlier or later than respiratory ones. Thus both versions of the serial endosymbiosis theory are ill-founded.

In fact cyanobacteria and purple nonsulfur bacteria are both able to respire aerobically and to photosynthesize. I have argued that the reason why cyanobacteria became chloroplasts and purple bacteria became mitochondria, and not the reverse, is that they were taken up by the very same metamonad host and evolved into organelles simultaneously and synergistically.[5] Purple bacteria cannot photosynthesize in the presence of oxygen, and so lost photosynthesis. Cyanobacteria cannot control respiration separately from photosynthesis, as both share the same electron transport chain; as soon as the resources of the host and two symbionts were pooled by inserting energy translocators into their envelopes, the symbiotic cyanobacterium could lose respiration and rely in the dark on the more efficient purple bacterial respiration (purple bacteria but not cyanobacteria have a complete Krebs cycle). With the evolution of envelope translocators for intermediary metabolites the three-cell chimera would be highly flexible and energetically and nutritionally efficient. In the light it could photosynthesize in both aerobic and anaerobic conditions, in the light or dark it could respire aerobically using metabolites acquired either phagotrophically or photosynthetically; it could also exist anaerobically by phagotrophy in the dark like its metamonad ancestors. In the light it could recycle its own oxygen and carbon dioxide in the same cell. It would be so successful that it would rapidly multiply and inevitably diversify to produce all the major nonarchezoan phyla before any competitor could evolve mitochondria and chloroplasts in the same manner. Some lines of evolution would lose chloroplasts, and become the purely phagotrophic phyla of the protozoan subkingdom Mitozoa, including our own ancestors—the Choanozoa. But for phototrophs the value of having mitochondria was so great that they were never lost. Only two phyla contain members that have clearly lost mitochondria: Infusoria (Protozoa) and Archemycota[27] (Fungi).

The smaller number of genes in mitochondrial DNA than in chloroplast DNA is sometimes taken as evidence that mitochondria evolved earlier and have therefore had a longer time for gene transfer to the nucleus. But a comparison between different phyla of genome size in chloroplasts [typically 100-200 kilobases (kb)] and of gene numbers in mitochondria (8-20 or so proteins) shows that no major transfers have occurred since the divergence of the phyla over 500 million years ago. Clearly most DNA was transferred in a burst of quantum evolution during the origin of the organelles, since when there has been near stasis with respect to gene distribution. Other factors than elapsed time must have determined how many genes were retained by the former symbionts. As their relative genome sizes tell us nothing about relative times of origin, we shall have to look to gene sequences evolving in a more clocklike fashion to test the simultaneous theory further.

## DOUBLE DOUBLE, TOIL AND TROUBLE

Schnepf's idea that the double-membraned envelope of the mitochondrion and chloroplast was a chimeric structure[28] has long been an accepted tenet of the symbiotic theory; the inner membrane was the symbiont's plasma membrane and the outer membrane the phagosome (food vacuole) membrane. But his idea predated the demonstration that the negibacterial envelope itself consists of two membranes and that of the choroplast in Euglenoidea and Dinozoa of three. As purple and cyanobacteria are both negibacteria with two separate membranes, I proposed that the double envelope of the symbionts evolved directly into those of the organelles and that the phagosome membrane was entirely lost,[5,6] except in the Euglenoidea and Dinozoa where it did indeed become the third outermost membrane. I thus assume the retention of the negibacterial outer membrane and periplasmic space throughout the evolutionary history of the organelles. This newer theory's chief merits are (1) numerous examples are known of loss of the phagosome membrane, which liberates symbionts into the cytoplasm; (2) not a single case is known in the whole of evolutionary history where a free-living or symbiotic negibacterium has clearly lost its outer membrane as well as its peptidoglycan layer, as Schnepf's view requires; (3) the outer membrane of the organelle double envelopes closely resembles that of negibacteria in its permeability mediated by proteins called porins,[29] and is quite unlike the phagosome membrane; (4) the growth and division of the two membranes are tightly coupled, probably with the help of focal adhesions between them present in both organelles and negibacteria; (5) to abandon the last two properties by losing the outer membrane and then to reevolve exactly the same properties by modifying the nondividing phagosome membrane is very much more than double the toil .and trouble of simply losing the phagosome membrane.

## DNA TRANSFERS: FUNCTIONAL, NEUTRAL, AND SELFISH

Cells contain such a variety of DNA-splicing and repair enzymes that the incorporation of symbiont DNA into the nucleus no longer presents any mechanistic problems; with several copies of each symbiont per cell, one could burst, liberating its DNA into the cytoplasm from where it could readily enter the nucleus. Mitochondrial genes[30] and even whole mitochondrial genomes have been found in the nucleus, and chloroplast genes in the mitochondrion.[32] There has been a tendency to look for a function for such relatively recent transfers, but perhaps there are none, and they are merely inevitable accidents—examples of neutral rather than selfish or functional evolution. The real question is why during the origin of the organelles there was such a massive transfer of genes into the nucleus. There are, very roughly, about 1000 nuclear genes for mitochondrial and about 1000 for chloroplast proteins. Though some or even many of these may have been host genes from the beginning, hundreds of them must be transferred symbiont genes.

Some have suggested that transfer gives better control; but control is merely a longer and more fashionable word than god to invoke when you do not understand the problem. Sex uses the same amount of ink as god but is more popular. For organelle genes to enjoy its seductive but dubious advantages they must be put in the nucleus.[33] But the truth I fear is less romantic: economy and efficiency. There are hundreds or thousands of copies of each organelle genome in most cells, but only one or two in the nucleus; if the transferred genes were put back into the organelles this would increase the amount of cytoplasmic DNA 10- to 50-fold. But it would not decrease nuclear genome size one iota, for this evolves in direct proportion to cell volume;[34] one would merely replace mitochondrial genes by noncoding DNA—in my view such DNA has a useful, functional skeletal role in the nucleus but not in the organelles.[35] Thus the first selective advantage of gene transfer is to economize on the overall amount of cellular DNA, by at least a factor of two. The second advantage may be even more important: to increase the metabolic efficiency of the organelles by reducing the volume occupied by DNA and ribosomes. The mitochondrial matrix especially is a virtually solid mass of enzymic protein, over 50% by mass;[36] such high concentrations have clearly evolved to maximize the rates and efficiency of biosynthesis. To insert a thousand extra genes, and still worse the numerous bulky ribosomes to make the encoded proteins, into the matrix would decrease the enzyme concentration severalfold. Conversely their original transfer to the nucleus allowing the use of cytoplasmic ribosomes, allowed a marked increase in energetic and biosynthetic efficiency; the principles of compartmentation—specialization simultaneously allowing higher concentrations and throughput and greater economy—apply equally to symbiotic and autogenous modes of origin for membranous organelles.

Such selection for economy and efficiency amply explains the massive gene transfer to the nucleus. But note that for both selective forces there is a law of diminishing returns; as more and more DNA is transferred, the relative cost of not transferring the increasingly little that remains becomes smaller and smaller. If some genes positively need to remain or simply cannot be transferred, the process will not go to completion; when as seems to have been so for the last 700 million years or so only a small residue remains, the selective forces against the insertion of extra bits of DNA into organelle genomes will be relatively slight. This will be so as long as the extra DNA is not translated at a high level and therefore causes no significant increase in the numbers of bulky ribosomes; this is certainly the case for the extra DNA in some fungal mitochondria[37] and probably is for the even higher amounts in some angiosperms. Such DNA can be expected to fluctuate in amounts in ways depending less on selective forces than on varying mutation pressures, which in turn will depend on the particular sequences present in relation to the DNA-manipulating enzymes that happen to be present;[38] mutations in either DNA substrates or the enzymes' own genes will change the pattern of neutral or selfish evolution observed.

## The Ins and Outs of Introns

Obvious examples [7] of the insertion of selfish DNA into organelle genomes are the introns found in both chloroplasts[39] and mitochondria.[37] As they are of the RNA self-splicing type, it would be particularly easy for them to have migrated from the nucleus into the organelles; no accompanying processing machinery would be necessary. The rival dogma that they were present in the symbionts prior to uptake[1,40] holds no phylogenetic water. Introns are unknown in purple or cyanobacteria or in any other eubacterium, and there is not a shred of evidence that they were ever present. Their distribution in the organelles does not support a descent from a common ancestor. It is commonly supposed that the presence of introns in fungal mitochondria but their absence in animal mitochondria means that the latter lost them. But animals did not evolve from fungi! Both evolved from a protozoan, probably a choanociliate.[27] The closest protozoa to their common ancestor for which mitochondrial DNA sequences are known are the trypanosomes (Phylum Euglenozoa), which like animals and fungi, but unlike plants, use UGA to specify tryptophan in their genetic code;[41] they have no introns despite having much larger amounts of noncoding mtDNA than some fungi that do have them. This suggests that fungal introns may have invaded their mitochondria well after the origin of fungi and that animal ancestors may never have had them. Many introns in fungal mtDNA are known to be optional: the genes function equally well without them.[37] Some are mobile: one optional intron in yeast can invade homologous chromosomes lacking it by directed gene conversion[42]—a fine example of a selfish genetic symbiont; one in *Podospora* can excise, multiply as a plasmid, and enter the nucleus.[43] Such direct gene conversion readily explains the homologous insertion points for the homologous introns in protozoan nuclear and fungal mitochondrial rRNA following the transfer of rDNA genes from one location to the other; therefore, identical insertion points are not "strong evidence"[40] for an autogenous origin for mtDNA.

The gene for cytochrome oxidase subunit I always has introns in fungi, but lacks them in plants; subunit II, the only plant gene in mtDNA so far with an intron, never has one in fungi;[37] thus introns may have invaded plant mitochondria independently. A similar pattern is found in chloroplasts. Their introns are probably optional, for the same gene in different species may either possess or lack them.[7] Green plant chloroplasts have only a handful of introns, whereas *Euglena* has about 100;[39] when the same gene is compared in *Euglena* and green plants,[7] the introns are in different positions, again suggesting independent insertions in the two groups (which I classify in different kingdoms).[11]

### Changing the Code at Mitochondrial Palace

The code changes must have followed and cannot possibly have preceded, the transfer of the bulk of the mtDNA to the nucleus, for in the latter case disasters would follow translation by the universal code used by cytoplasmic ribosomes. This disposes of the notion that the new code was really an ancient one found in the symbiont before its uptake.[44] The use of UGA for tryptophan in mycoplasmas must be a convergent response to high AT/GC ratios and associated changes in codon usage.[45] Even in mtDNA this may have happened more than once, though there is at present no reason to think it did. The change to a two-by-two reading pattern by reducing the number of tRNAs would help further to reduce the space-wasting clutter in the matrix. The differences in the code between different groups[46] show that they have evolved within eukaryotes. Many of the minor changes may be purely neutral.

## PROTEIN IMPORT: ORGANELLES AS CHIMERAS

Merely to transfer symbiont genes to the nucleus is useless. They must be transcribed, transcripts processed and translated into proteins, and the proteins imported back into the mitochondrion or chloroplast, before the gene copy remaining in the organelle can be deleted.[5,6] If the transcriptional and translational machinery of the eukaryotic host had diverged from that of the prokaryotic symbionts this would present problems, the more serious the greater the divergence. In higher eukaryotes the ribosomes are so different from those of bacteria that it may be impossible for their numerous bacterial symbionts to be converted into organelles by the transfer of functional genes to the nucleus. But if metamonads retain 70s ribosomes, as they may prove to, they could much more easily convert bacteria into organelles. The major problem then would have been the import of the cytoplasmically made proteins back into the organelle.

This is achieved in modern cells by receptor proteins on the organelle envelope that bind to a peptide at the N-terminal end of the imported protein known as the transit peptide.[14] In many, but not all, cases the transit peptide is cleaved from the imported protein by one or both of two distinct transport peptidases located in the periplasmic space or organellar matrix respectively. Unlike the analogous mechanism for the rough endoplasmic reticulum, involving cleaved signal peptides, the mechanism is posttranslational and requires energy: ATP for chloroplasts and a membrane potential maintained by proton pumping for mitochondria. In the latter respect the mitochondrial system resembles the more primitive signal mechanism of respiring bacteria,[47] which can be either co- or posttranslational. Transit peptides are more basic and generally longer (over 50 amino acids) than signal peptides, and would have had to be added to the N-terminal end of every one of the many hundreds of symbiont genes transferred to the nucleus. This is one of several major reasons for doubting that either organelle evolved more than once. That the mechanism was addition to rather than mutation of the N-terminal end of the protein is shown by such extension of ATPase subunit 9 precursor when nuclearly coded, but not when mitochondrially coded.[48] Moreover cleavage of a peptide mutated from the original end of a protein would commonly destroy or impair its function.

Transit peptides could be added to a gene either by inserting a copy of the DNA encoding a preexisting transit peptide immediately upstream of the start codon or by a mutation creating a new ribosome initiation site closer to the promoter and modifying the base sequence appropriately in the region between it and the old initiation site. The latter mechanism is not as unlikely as it first seems, for initially the transit peptide need not work very efficiently because the gene copy in the organelle could make as much product as needed. How then could improved transit peptides be selected? Assuming feedback control on transcription within the organelle, the more gene product imported the less would need to be made internally, and with fewer RNA polymerases, mRNAs, and ribosomes the more space, energy, and materials for matrix enzymes. Even if each mutation made only a slight improvement, a cell with such improvements in hundreds of simultaneously evolving transit peptides could be at a marked advantage over one with none.

The foregoing presupposes the presence of an import mechanism that discriminates between proteins destined for the mitochondrion and those destined for the chloroplast, and in each case between those to go into the outer membrane, the periplasmic space, the inner membrane, the matrix, and for the chloroplast also the thylakoid membrane and intrathylakoid space. All imported molecules must be labeled (not necessarily exclusively via their transit sequences) for their specific destinations. Such a sophisticated mechanism could hardly evolve de novo or guided by only weak selective forces. I therefore suggest that it evolved by relatively slight modifications of the mechanisms used by the two symbionts when free living to direct their proteins to the right site.[47] Consider first a periplasmic protein like cytochrome c (FIGURE 1a). In a bacterium it must interact with a receptor on the matrix face of the inner membrane, and be translocated across that membrane into the periplasm; for it to reach the same destination when it is made in the cytoplasm following the transfer of its gene to the nucleus it has to bind to a receptor located now on the cytoplasmic face of the outer membrane and be translocated across that membrane instead (FIGURE 1b). In both cases it has to traverse but a single membrane. The key difference is simply the location of the receptor.

A receptor could be transferred from the inner membrane to the outer membrane simply by transferring its gene into the nucleus; if it were a protein capable of self-insertion into a membrane, it would automatically insert itself into the outer membrane and enable periplasmic proteins coded by genes newly transferred into the nucleus to enter the periplasm. Or rather, it would allow those proteins not requiring a membrane potential to propel them across a membrane to do so. This proviso is made because the outer membrane lacks proton pumps and is too permeable to allow a membrane potential to develop. A bacterial protein requiring such a potential (as most appear to) would have to be modified so that it did not. If its specific structure made this impossible it could only enter the periplasm by making use of the membrane potential

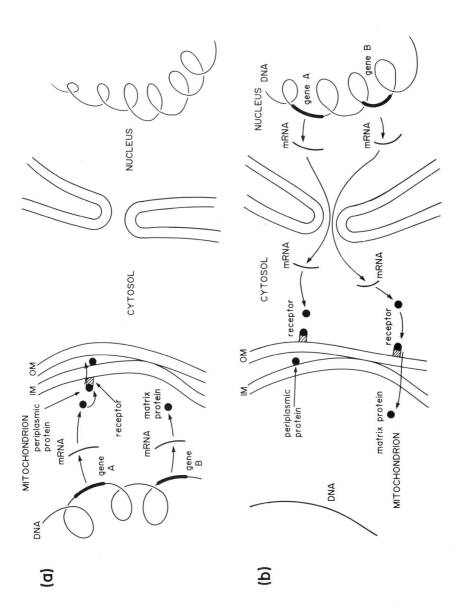

**FIGURE 1.** Origin of the mitochondrial protein import system for genes transferred to the nucleus. For explanation see text. (a) Before gene transfer to the nucleus. (b) After gene transfer to the nucleus.

to pull it first across both membranes from the cytoplasm to the matrix, and then back across the inner membrane into the periplasm. Cytochrome $b_2$ and cytochrome c peroxidase do indeed enter the periplasm in this backhanded fashion.[49] Moreover they are proteolytically processed in two stages: first by a matrix transit peptidase and then by a different one in the periplasm. The latter I suggest was directly derived from the bacterial periplasmic signal peptidase, so the second stage of transport from the matrix to the periplasm exactly recapitulates the complete process of the ancestral bacterial symbiont. What was new was the evolution of the matrix peptidase to allow the import of proteins across both membranes into the matrix (FIGURE 1b); this peptidase cleaves the transit peptide not only from periplasmic proteins but also from matrix proteins, which would not have had to cross any membranes in the bacterial ancestor. For these proteins also a receptor had to be inserted into the outer membrane. Inner membrane proteins typically share the same mechanism and are cleaved by the matrix peptidase. The receptors for proteins that cross the inner membrane, temporarily or permanently, differ from those for cytochrome c, which is not cleaved.

The existence of two distinct mechanisms for periplasmic proteins, and especially of the botched up two-stage mechanism, which so clearly betrays its bacterial ancestry, strongly supports the symbiotic theory. On the autogenous theory one might expect all periplasmic proteins to be transported directly across the outer membrane alone. At least this is so for the plasmid sequestration theory,[50] according to which such proteins would never have been coded by organelle DNA. But in the cluster clone theory,[51] all genes of the parent cell would initially have been present, so the embryonic organelles would mimic whole bacteria; but if one has to start organelle evolution by in effect creating two whole bacteria autogenously in the cytoplasm of another, it would seem far simpler to take them up ready made by phagocytosis. Especially so as cyanobacteria and purple bacteria are predifferentiated in the exact way necessary to yield separate chloroplasts and mitochondria, and already have an envelope transport mechanism that could be readily modified as outlined here. In the plasmid theory a protein transport mechanism would have to evolve virtually instantaneously with the closure of the organelle from the cytoplasm:[52] before then it would have no advantages; afterwards would be too late to prevent death. On the cluster clone theory there would appear to be no initial selective advantage for compartmentation, and therefore no selective force for the evolution of a complex protein transport mechanism.

On the symbiotic theory the first step in the conversion of symbiont to organelle was the tapping of the symbiont's energy supply by the evolution of a nuclear-coded adenine nucleotide carrier for the mitochondrial envelope[5] and analogous carriers for the chloroplast, clearly of immense selective advantage to the host. As this must have occurred prior to the origin of the matrix peptidase, it is highly significant that the adenine nucleotide carrier, unlike other inner membrane proteins, remains uncleaved during its insertion into the inner membrane,[53] but nonetheless requires the inner membrane potential for its transport. It is probable that the outer membrane receptors for the transport of proteins into or across the inner membrane evolved to help direct such carriers to their correct location.[5] As they may have evolved before any gene transfer they are more likely to have evolved from host than from symbiont proteins.[5] That the host was probably more concerned to get organic photosynthesate from the cyanobacterium than ATP is suggested by the presence in chloroplast envelopes of organic phosphate rather than adenine nucleotide carriers. Differentiation between the receptors and transit sequences of the two organelles may therefore have occurred to stabilize this difference even before any gene transfer. Thus any symbiont receptors whose genes were transferred to the nucleus as suggested above could make use of these preexisting organelle specific receptors. So also could transferred outer membrane proteins like porin, which as they would no longer need to be pulled through the inner membrane to reach their destination would no longer require a membrane potential or removal of the transit peptide during their direct insertion into the outer membrane, as is the case.[49]

Thus the symbiotic theory can account in detail for the origin of the protein transport mechanism. But as soon as it evolved it would allow not only proteins from transferred symbiont genes but also those from the eukaryote host to be inserted into the organelles merely by the addition of a transit peptide.Thus the organelles are probably chimeras of host and symbiont genes. For chloroplasts the contribution from the host was probably relatively slight, for the host was nonphotosynthetic and would have lacked most of the relevant genes. But as the host was probably a facultative aerobe, it would probably already have most of the proteins now present in the mitochondrion, so they could be transferred into it and allow deletion of the homologous symbiont genes without any gene transfer being necessary. Thus the smaller number of genes remaining in mtDNA than in chlDNA may not be because more genes were transferred to the nucleus but because fewer such transfers were necessary to allow their deletion. The major respiratory chain proteins and the rRNA probably did not originate in this way, for they show such striking affinities with the purple bacteria.[1] But for the soluble matrix proteins no such affinities have been demonstrated. In fact pyruvate dehydrogenase and two key Krebs cycle enzymes, citrate synthase and succinate thiokinase, have equally clear affinities with posibacterial enzymes;[1,11] as there are other reasons also for thinking that the eukaryote host evolved from an intermediate between posibacteria and archaebacteria,[11] this definitely supports the chimera idea. Each mitochondrial protein will have to be examined in detail for its affinities with purple bacteria on the one hand and posibacteria/archaebacteria on the other to determine whether it came from host or symbiont. Some proteins of course may originally, as such, not have been present in either, for example subunits III-VIII of cytochrome oxidase, which are absent in bacteria.

The identification of the posibacteria/archaebacterial intermediate as the ancestor of eukaryotes,[11] and of the Anaxostylea as the ancestors of all eukaryotes with mitochondria,[4] when added to the earlier evidence that chloroplasts evolved from whole cyanobacteria[1] and that the respiratory and genetic components of mitochondria evolved from purple nonsulfur bacteria,[1] greatly strengthens the phylogenetic evidence for the symbiotic origin of both organelles, as does recent evidence that the respiratory machinery of cyanobacteria is sufficiently different from that of mitochondria as to rule out an autogenous mitochondrial origin from a cyanobacterial ancestor.[54] Likewise my reinterpretation of the origin of the outer membrane,[5,6] and present proposals for the origin of the protein transport mechanism, provide a much more plausible and detailed mechanism than hitherto for the conversion of bacterial symbionts into organelles.

## PEROXISOMES: AND NOW WE ARE FOUR?

It used to be thought that peroxisomes, and the developmentally related glyoxysomes, evolved[55] and develop by budding from the endomembrane system, like lysosomes. But it is now clear that they multiply instead, like mitochondria and chloroplasts, by growth and division,[56] and that their matrix proteins are inserted posttranslationally and not by the cotranslational rough endoplasmic reticulum (RER) signal mechanism used by lysosomes.[57] This immediately raised the possibility that peroxisomes also had a symbiotic origin.[58] If they did there were two major differences from chloroplasts and mitochondria: (1) As peroxisomes are bounded by only one membrane, the symbiont probably also would have been, and therefore must have been a posibacterium or archaebacterium rather than a negibacterium,[11] so the symbiont would have been relatively more closely related to the host. Therefore the latter might even provide all the matrix proteins during the conversion of symbiont into organelle. (2) As there is only one bounding membrane the biophysics of the insertion of symbiont proteins into that membrane would be relatively closely similar before and after the transfer of its genes into the nucleus. There would thus be little or no barrier to the transfer of all its membrane protein genes to the nucleus, and it could lose all its DNA like $\rho_0$ yeast mitochondria, which can still grow, divide, and take up proteins posttranslationally. It has been proposed that the reason why about half a dozen inner membrane proteins remain universally coded by mtDNA is that their structure cannot be modified so as to enable them to be pulled through two membranes.[59] Though historical accident must also be involved in the retention of genes in the organelles (to account for the differential retention of other proteins in different species), this idea is rather attractive. The need to insert certain proteins from the matrix side to ensure their correct polarity is another factor. But so long as this constraint was absent, or could be satisfied by modifying the membrane proteins, the symbiont would be expected to lose all its DNA.

It would so effectively have covered its tracks that it may prove impossible to gain phylogenetic evidence for the above scenario by showing that certain peroxisomal proteins are markedly more similar to ones in some as yet unspecified posi- or archaebacterium than in that which future research may show to be closest to the first eukaryote. The key role of the symbiont would have been to provide a compositionally distinctive outer membrane into which organelle-specific receptors could thereby specifically be inserted and one which is already topologically distinct from other cellular compartments and already capable of perpetuating this distinctiveness by growth and division. Such a distinct compartment can however arise purely autogenously—witness the Golgi apparatus, which grows and divides but which almost certainly evolved by budding and differentiation from the ER.[11,55] However, the difference is that all Golgi proteins are channeled to it via the ER signal mechanism. It is the posttranslational mechanism that suggests, but does not establish, a symbiotic origin.

The other thing pointing, albeit nondecisively, in the same direction is the absence of peroxisomes in the Archezoa, suggesting that eukaryotes may first have acquired them as symbionts at the same time as mitochondria and chloroplasts. Peroxisomes are present in all eukaryotes with mitochondria. I suggest that the close and complementary cooperation observed in modern plants between mitochondria, chloroplasts, cytosol, and peroxisomes/glyoxysomes first evolved in the ancestral mitozoan by the simultaneous conversion of all three symbionts into organelles in a common cytoplasm. Thus the peroxisome was not the respiratory organelle of the first eukaryotes as de Duve once suggested. In the Archezoa the Microspora, Archamoebae, and Anaxostylea have no microbodies of any kind; the Parabasalia have hydrogenosomes, and the Axostylaria have biochemically unstudied microbodies which I guess may also be hydrogenosomes.

## HYDROGENOSOMES: PLURAL ORIGINS?

Parabasalian hydrogenosomes have two bounding membranes,[60] and no DNA.[61] Whatley et al. suggested they evolved from symbiotic Clostridia (posibacteria), and that the outer membrane represents the phagosome membrane.[62] If Parabasalia never had mitochondria, this would seem the most plausible origin, for the reasons given for peroxisomes. But if the Parabasalia once had mitochondria but lost the respiratory chain and DNA on entering the anaerobic niche of animal guts, then the residual mitochondrion (very like that of $\rho_0$ yeasts) could very simply be converted into a hydrogenosome; all that would be needed would be to add mitochondrial transit sequences to existing cytoplasmic enzymes many of which exist in the cytosol in other Archezoa.[63] Two further points support such conversion of a mitochondrion into a hydrogenosome. Unlike all other Archezoa the Parabasalia have a permanent Golgi dictyosome, as do all Mitozoa, and an exonuclear spindle; both making it easier for them to have evolved from a primitive mitozoan than from a metamonad. As their mitosis closely resembles that of the parasitic Syndinea (phylum Dinozoa), I suggest that they evolved from an anisokont (not a dinokont) syndinean by the conversion of mitochondria into hydrogenosomes, by the multiplication of anterior cilia, and by the internalization of cortical microtubules to form the axostyle and pelta and thereby allow phagocytosis over the whole cell surface.

The spizomycete rumen fungi[64] have clearly lost mitochondria, and have hydrogenosomes, but with a single bounding membrane.[65] As it seems unlikely that they would have acquired posibacteria by phagocytosis, I suggest they converted their peroxisomes into hydrogenosomes by the addition of peroxisomal transit sequences to cytosolic hydrogenosomal enzymes. Glycosomes, unique to trypanosome protozoa,[66] could have evolved similarly by the attachment of peroxisomal transit sequences to glycolytic enzymes.

The number of bounding membranes in the hydrogenosomes of anaerobic Infusoria ( = Ciliophora) is unclear,[67] as therefore is their most probable origin—from mitochondria, peroxisomes, both, or neither. But from their patchy distribution in eukaryotes and structural diversity it seems probable that hydrogenosomes had a polyphyletic origin, which I assert was not the case of chloroplasts, mitochondria, and peroxisomes.

## MITOZOAN DIVERSIFICATION AND ORGANELLE COEVOLUTION

My overall thesis is that mitochondria, chloroplasts, and microbodies were converted from bacterial symbionts into organelles approximately simultaneously and coadaptively in a single anisokont, biciliated, and fully eukaryotic anaxostylean protozoan cell. I earlier proposed that this cell had only shortly before evolved as the first biciliated monad[4] and that diversification of the ciliary transition region[4] (so important in eukaryote phylogeny) as well as in the structure of mitochondrial cristae and mtDNA,[5] and of the pigmentation, thylakoid arrangement, and envelope structure of chloroplasts,[6] occurred in its immediate descendants. Since permanent Golgi dictyosomes are absent in all Archezoa except the Parabasalia (whose archezoan status as we have just seen is suspect), but present in all Mitozoa, I suggest that stacking of Golgi cisternae to form dictyosomes also originated in this very same "eomitozoan" cell. Such stacking would involve novel interactions between the Golgi membranes and the cytoskeleton. In different protist classes or phyla the dictyosomes are often attached to different cytoskeletal elements (e.g., a striated fiber in Parabasalia and the chromist class Chrysophyceae, but in no other groups) and located at different cellular sites. As these distinctive features are rather well conserved within groups I suggest that they also arose during the rapid diversification of the immediate descendants of the eomitozoan: when dictyosome attachments to the cytoskeleton first evolved they were fairly fluid, but as they were improved and stabilized (often differently in different diverging lines) they rather soon ceased to evolve significantly and have remained much the same ever since. The mitotic and meiotic systems were also undergoing their most fundamental diversification at the very same time (Reference 11, Figures 1 and 6).

This major burst of quantum evolution approximately 950 million years ago must have occurred after the origin of the Anaxostylea, which I suggest could have occurred so soon after the origin of the first eukaryote as to be geologically apparently instantaneous. Because of the new insights presented here into the origin of protein import into symbiotic organelles I no longer think there need have been a geologically significant delay[5] between the origin of eukaryotes and of mitochondria, chloroplasts, and microbodies. But this does not mean that the whole process was so easy that a plural origin for any of these organelles is at all probable. As there is no phylogenetic evidence requiring more than one origin,[4-6] rapid and unequal divergence from a single ancestor, the eomitozoan, remains the simplest hypothesis. A good test for it will be to examine the organellar protein import mechanisms for each group for which an independent symbiotic origin is claimed. I predict that they will prove so similar as to confirm a single origin for each organelle.

Mereschkowsky's colorful theory of the polyphyletic symbiotic origins of chloroplasts[8] has however become almost a dogma for modern symbiotologists;[1,3,62] supposedly red, green, brown, yellow, and so on, chloroplasts arose independently from correspondingly colored bacteria. The theory was boosted by the discovery of *Prochloron,* but 16s rRNA sequences[68] support my earlier thesis that its resemblances to green plant and euglenoid chloroplasts are convergent.[6] The recent discovery of a free-living prochlorobacterium[69] will allow sequence studies of thylakoid protein genes to clinch the issue one way or another. Despite arguing for a single origin for chloroplasts, I am also convinced that Greenwood was correct in suggesting that a very different type of symbiosis, involving a photosynthetic eukaryotic symbiont, was involved in the origin of the cryptomonad nucleomorph and chloroplast endoplasmic reticulum.[70] But as the symbiont already had a chloroplast, albeit I argue in an early stage of evolution,[4] this was not an independent symbiotic origin of chloroplasts, as sometimes wrongly stated, but the symbiotic uptake of a preexisting chloroplast and surrounding plasma membrane by a different eukaryote cell. I have elsewhere argued in detail that the subkingdom Chromophyta evolved from an early cryptomonad before either group had fully stabilized their chloroplast or mitochondrial structure,[4] i.e. during the earliest phase of mitozoan diversification. Thus the relatively recently recognized kingdom Chromista (comprised of the Cryptomonada and Chromophyta) originated and split into subkingdoms also about 950 million years ago, as for similar reasons I argue did the kingdom Plantae.[6] Our own kingdom, Animalia, arose later (about 680 million years ago),[71] as probably did the related kingdom Fungi,[27] both I argued from choanociliate protozoa.[27]

# ACKNOWLEDGMENTS

I thank the Australian National University for a Visiting Fellowship, the Royal Society for a Traveling Fellowship, and Des Clark-Walker for stimulating discussions.

## REFERENCES

1. GRAY, M. W. & W. F. DOOLITTLE. 1982. Microbiol. Rev. **46:** 1-42.
2. SAGAN, L. 1967. J. Theor. Biol. **14:** 225-275.
3. MARGULIS, L. 1981. Symbiosis in Cell Evolution. Freeman. San Francisco, Calif.
4. CAVALIER-SMITH, T. 1986. *In* Progress in Phycological Research. F. E. Round & D. J. Chapman, Eds. **4:** 309-347. Biopress Ltd. Bristol, England.
5. CAVALIER-SMITH, T. 1983. *In* Endocytobiology II. W. Schwemmler & H. E. A. Schenk, Eds.: 265-279. de Gruyter. Berlin, FRG.
6. CAVALIER-SMITH, T. 1982. Biol. J. Linn. Soc. **17:** 289-306.
7. CAVALIER-SMITH, T. 1985. Nature **315:** 283-284.
8. MERESCHKOWSKY, C. 1910. Biol. Centralbl. **30:** 278-303, 321-347, 353-367.
9. WOESE, C. R. 1982. *In* Evolution from Molecules to Man. D. S. Bendall, Ed.: 209-233. Cambridge University Press. Cambridge, England.
10. LAKE, J. A. 1983. Prog. Nucleic Acid Res. Mol. Biol. **30:** 163-194.
11. CAVALIER-SMITH, T. 1987. Ann. N.Y. Acad. Sci. (This volume.)
12. CAVALIER-SMITH, T. 1983. *In* Endocytobiology II. W. Schwemmler & H. E. A. Schenk, Eds.: 1027-1034. De Gruyter. Berlin, FRG.
13. GUERRERO, R., C. PEDRÓS-ALIO, I. ESTEVE, J. MAS, D. CHASE & L. MARGULIS. 1986. Proc. Nat. Acad. Sci. USA **83:** 2133-2137.
14. REID, G. A. 1985. Curr. Topics Membr. Transp. **74:** 295-336.
15. HAECKEL, E. 1866. Generelle Morphologie der Organismen. G. Reimer. Berlin, FRG.
16. WALLIN, I. E. 1927. Symbionticism and the Origin of Species. Ballière, Tindall & Cox. London, England.
17. CAVALIER-SMITH, T. 1986. BioScience **36:** 293-294.
18. COPELAND, H. F. 1956. The Classification of Lower Organisms. Pacific Books. Palo Alto, Calif.
19. PASCHER, A. 1914. Ber. Dtsch. Bot. Ges. **32:** 136-160.
20. ALLSOPP, A. 1969. New Phytol. **68:** 591-612.
21. VOSSBRINCK, C. R. & C. R. WOESE. 1986. Nature **320:** 287-288.
22. YEAGLE, P. L. 1985. Biochem. Biophys. Acta **822:** 267-287.
23. ELLIOTT, C. G. 1977. Adv. Microbiol. Physiol. **15:** 121-173.
24. RILLING, H. C. & L. T. CHAYET. 1985. *In* Sterols and Bile Acids. H. Danielsson & J. Sjövall, Eds.: 1-39. Elsevier. Amsterdam, Holland.
25. STANIER, R. Y. 1970. Symp. Soc. Gen. Microbiol. **20:** 1-38.
26. SCHOPF, J. W., Ed. 1983. Earth's Earliest Biosphere: Its Origin and Evolution. Princeton University Press. Princeton, N.J.
27. CAVALIER-SMITH, T. 1987. *In* Evolutionary Biology of Fungi. A. D. Rayner, C. M. Brasier & D. Moore, Eds. Symp. Br. Mycol. Soc. **13**. Cambridge University Press. Cambridge, England. (In press.)
28. SCHNEPF, E. 1964. Arch. Mikrobiol. **49:** 112-131.
29. BENZ, R. 1985. Crit. Rev. Biochem. **19:** 145-190.
30. FARRELLY, F. & R. A. BUTOW. 1983. Nature **301:** 296-301.
31. VOSSBRINCK, C. R., T. J. MADDOX, S. FRIEDMAN, B. A. DEBRUNNER-VOSSBRINK & C. R. WOESE. 1987. Nature **326:** 411-414.
32. STERN, D. B. & D. M. LONSDALE. 1982. Nature **299:** 698-702.
33. THORNLEY, A. L. & A. HARINGTON. 1981. J. Theor. Biol. **91:** 515-523.
34. CAVALIER-SMITH, T. 1985. *In* The Evolution of Genome Size. T. Cavalier-Smith, Ed.: 105-184. Wiley. Chichester, England.
35. CAVALIER-SMITH, T. 1985. *In* The Evolution of Genome Size. T. Cavalier-Smith, Ed.: 1-36. Wiley. Chichester, England.
36. SERE, P. A. 1981. Trends Biochem. Sci. **6:** 4-7.
37. SEDEROFF, R. R. 1984. Adv. Genet. **22:** 1-108.
38. CLARK-WALKER, G. D. 1985. *In* The Evolution of Genome Size. T. Cavalier-Smith, Ed.: 277-297. Wiley. Chichester, England.
39. KOLLER, B. & H. DELIUS. 1984. Cell **36:** 613-622.

40. MAHLER, H. R. 1983. Int. Rev. Cytol. **82:** 1-98.
41. SIMPSON, L., A. M. SIMPSON, V. DE LA CRUZ, N. NECKELMANN & M. MUHICH. 1985. *In* Achievements and Perspectives of Mitochondrial Research. E. Quagliariello, E. C. Slater, F. Palmieri, C. Saccone & A. M. Kroon, Eds. **2:** 99-110. Elsevier. Amsterdam, Holland.
42. JACQUIER, A. & B. DUJON. 1985. Cell **41:** 383-394.
43. OSIEWACZ, H. D. & K. ESSER. 1984. Curr. Genet. **8:** 299-305.
44. YAMAO, F., A. MUTO, Y. KAWAUCHI, M. IWAMI, S. IWAGAMI, Y. AZUMI & S. OSAWA. 1985. Proc. Natl. Acad. Sci. USA **82:** 2306-2309.
45. JUKES, T. H. 1985. J. Mol. Evol. **22:** 361-362.
46. BREITENBERGER, C. A. & U. L. RAJBHANDARY. 1985. Trends Biochem. Sci. **10:** 478-483.
47. BENSON, S. A., M. N. HALL & T. J. SILHAVY. 1985. Annu. Rev. Biochem. **54:** 101-134.
48. MICHEL, R., E. WACHTER & W. SEBALD. 1979. FEBS Lett. **101:** 373-376.
49. HAY, R., P. BOHNI & S. GASSER. 1984. Biochem. Biophys. Acta **779:** 65-87.
50. MAHLER, R. & R. A. RAFF. 1975. Int. Rev. Cytol. **43:** 1-124.
51. BOGORAD, L. 1975. Science **188:** 891-898.
52. CAVALIER-SMITH, T. 1980. *In* Endocytobiology: Endosymbiosis and Cell Biology. W. Schwemmler & H. E. A. Schenk, Eds.: 831-916. De Gruyter. Berlin, FRG.
53. ZIMMERMAN, R., U. PALUCH, M. SPRINZ & W. NEUPERT. 1979. Eur. J. Biochem. **19:** 247-252.
54. SANDMANN, G., S. SCHERER & P. BÖGER. 1984. *In* Compartments in Algal Cells and Their Interaction. W. Wiessner, D. G. Robinson & R. C. Starr, Eds.: 207-217. Springer. Berlin. FRG.
55. CAVALIER-SMITH, T. 1975. Nature **256:** 463-468.
56. VEENHUIS, M., J. P. VAN DIJKEN & W. HARDER. 1983. Adv. Microbiol. Physiol. **24:** 1-82.
57. LAZAROW, P. G. & Y. FUJIKI. 1985. Annu. Rev. Cell Biol. **1:** 489-530.
58. DE DUVE, C. 1982. Ann. N.Y. Acad. Sci. **386:** 1-4.
59. HEIJNE, G. 1986. FEBS Lett. **198:** 1-4.
60. BENCHIMOL, M. W. & W. DE SOUZA. 1983. J. Protozool. **30:** 422-455.
61. MULLER, M. 1985. J. Protozool. **32:** 559-563.
62. WHATLEY, J. M., P. JOHN & F. R. WHATLEY. 1979. Proc. R. Soc. London Ser. B. **204:** 165-187.
63. MULLER, M. 1980. Symp. Soc. Gen. Microbiol. **30:** 127-142.
64. HEATH, B., T. BAUCHOP & C. G. ORPIN. 1983. Can. J. Bot. **61:** 295-307.
65. YARLETT, N., C. G. ORPIN, E. A. MUNN, N. C. YARLETT & C. A. GREENWOOD. 1986. J. Biochem. **236.** (In press.)
66. OPPERDOES, F. R. 1982. Ann. N.Y. Acad. Sci. **386:** 543-545.
67. YARLETT, N., A.C. HANN, D. LLOYD & A. WILLIAMS. 1985. Comp. Biochem. Physiol. **74B:** 357-364.
68. SEEWALDT, E. & E. STACKEBRANDT. 1982. Nature **295:** 618-620.
69. BURGER-WIERSMA, T., M. VEENHUIS, H. J. KORTHALS, C. C. M. VAN DER WIEL & L. R. MUIR. 1986. Nature **320:** 262-264.
70. GREENWOOD, A. D., H. B. GRIFFITHS & U. J. SANTORE. 1977. Br. Phycol. J. **12:** 119.
71. GLAESSNER, M. F. 1984. The Dawn of Animal Life. Cambridge University Press. Cambridge, England.

# The Evolutionary Significance of the Archaebacteria

W. FORD DOOLITTLE

*Department of Biochemistry*
*Dalhousie University*
*Halifax, Nova Scotia B3H 4H7, Canada*

## INTRODUCTION

Several independent lines of reasoning led, eight years ago, to a drastic reformulation of views about the evolutionary relationships among major groups of living things. In particular, it became reasonable to believe (1) that earlier claims for the endosymbiotic origin of plastids and mitochondria were at last amply confirmed by molecular data of a variety of kinds, (2) that the hosts for these endosymbiotic events were not, as previously argued,[1,2] descendant from some otherwise typically prokaryotic lineage which had the unusual ability to engulf other prokaryotes; but instead that the nuclear-cytoplasmic components of modern eukaryotic cells comprise an entirely distinct ancient lineage, which diverged from all known prokaryotic lineages at the very dawn of cellular life, and (3) that within the prokaryotes themselves there is an equally profound evolutionary discontinuity, which separates eubacteria from archaebacteria. These radical notions have not, I believe, been as yet fully accepted by the community of molecular biologists and cell biologists, and many of their implications remain to be explored.

## THE PROBLEM OF THE "HOST" IN THE ENDOSYMBIONT HYPOTHESIS FOR THE ORIGIN OF EUKARYOTIC CELLS

In 1962, Stanier and van Niel gave formal articulation to the growing understanding that there are, in a structural sense, two profoundly different types of cellular organization exhibited by living organisms: prokaryotic and eukaryotic.[3] As Woese and Fox have pointed out,[4] this distinction is in fact taxonomic, not phylogenetic—but it has been often taken in the evolutionary sense. In particular, the notion that eukaryotes evolved *from* prokaryotes emerged automatically from (1) the recognition that eukaryotic cellular structure is in many ways more complex than prokaryotic cellular structure and (2) the long-standing assumption that primitive entities must be simpler in structure than advanced ones.[5] One could even place a rough date on the "prokaryote-eukaryote transition" from the first appearance of large cells in the fossil record, some 1400 million years ago.[6]

In 1967 and 1971, Margulis assembled arguments and evidence supporting the hypothesis, advanced some three quarters of a century earlier,[7-9] that the eukaryotic cell is essentially chimeric—plastids and mitochondria having been initially acquired as bacterial endosymbionts by some protoeukaryotic "host" cell(s), perhaps at the time of the prokaryote-eukaryote transition seen in the fossil record. Efforts to test this hypothesis by molecular sequence comparisons using ribosomal and transfer RNAs began in the middle 1970s. We can now conclude from these studies that plastids unquestionably arose from cyanobacterial (or cyanobacterial-like) endosymbionts, while mitochondria came from gram-negative eubacteria (for review, see Reference 10). Most recently, Yang *et al.* have shown strong sequence similarity between the ribosomal RNA genes of wheat mitochondria and purple bacteria related to *Agrobacterium* (see Reference 11). Purple bacteria had earlier been inferred to be closely related to mitochondria from cytochrome *c* sequence data.

Less clearly specified in discussions of the endosymbiont hypothesis during the 1970s was the nature of the host in these ancient symbioses. Margulis has recently summarized such discussions, which concerned mainly the likely metabolic and structural capabilities of the ancestral protoeukaryote.[12] Ignored were two important areas of inquiry. First, there were few molecular sequence data for homologous proteins or nucleic acids coded for by eukaryotic nuclear, organellar, and prokaryotic genomes. Most proteins for which both eukaryotic and prokaryotic versions were known were in fact mitochondrial-genome encoded. What we needed were clear quantifiable data tying the eukaryotic nuclear-cytoplasmic component to some *particular* prokaryotic lineage—eukaryotes cannot have arisen somehow from all the prokaryotes as a class, and what comparative protein data there were showed that eukaryotic nuclear-gene-encoded molecules were distinct from any of their prokaryotic homologues.[10] Second, the manifold differences between eukaryotic nuclear and prokaryotic molecular biologies required better explanation. All eukaryotes differ from all prokaryotes (or at least all eubacteria) in a number of very basic features of genome structure, genome replication, and gene expression. On the other hand, all eubacteria are very similar in such features as genome organization, gene structure, sequence signals for the initiation and termination of transcription and translation, and the like, so that most probably these traits were already in place in the last common ancestor of all eubacteria, which one can infer from the prokaryotic fossil record to have been thriving at least 3.5 billion years ago.[6] If eukaryotic nuclear genomes arose from the genomes of one of these eubacterial lineages, they must then have undergone drastic alterations in form and function—for reasons that are truly not obvious.

## NEW VIEWS OF THE RELATIONSHIP BETWEEN PROKARYOTES AND EUKARYOTES

In 1977, two developments made it possible to entertain an alternative view for the evolutionary relationship between prokaryotes and eukaryotes, and opened up a whole new set of ways for understanding the evolution of gene and genome structure. The first was the discovery of intervening sequences (introns) in eukaryotic and eukaryotic viral genes (see Reference 13). It was very difficult to rationalize the presence of these structures, averaging some 5 to 10 times the length of the coding sequences they interrupt, as an improvement on the more "primitive" continuous and completely functional gene structure known for prokaryotes. Most introns serve no function in gene expression and its control, and although evolution of new genetic function may be facilitated by their presence,[13] this does not constitute selection pressure strong enough to have motivated the insertion of introns at or just after the presumed "prokaryote-eukaryote transition." Their presence in eukaryotes and absence from prokaryotes (or at least from eubacteria) can thus not be explained within the classical prokaryote → eukaryote evolutionary scenario. Darnell and I suggested that, instead, introns were present in the genomes of the very first cells, that prokaryotes (or at least eubacteria) and eukaryotes diverged very soon after the appearance of those first cells, and that introns have been subsequently lost in prokaryotes (eubacteria?) but have been partially retained in the genomes of eukaryotic lineages.[14,15] Strong confirmation for this supposition comes from recent comparative studies of the structures of homologous genes in prokaryotes and eukaryotes. Pyruvate kinase, glyceraldehyde phosphate dehydrogenase, triose-phosphate isomerase, and alcohol dehydrogenase are all very ancient enzymes and the kinase and dehydrogenases contain the mononucleotide-binding domain identified by Rossman and collaborators as highly conserved in structure and function among a variety of prokaryotic and eukaryotic proteins which are, otherwise, largely nonhomologous in structure. Prokaryotic and eukaryotic versions of these enzymes show sufficient structural similarity to support the conclusion that they evolved from a common ancestral enzyme in the common ancestor to eukaryotes and prokaryotes. Genes for the eukaryotic homologues have now been sequenced.[14–16] They contain introns that delimit coding regions corresponding to structural domains of the corresponding proteins. This can be taken to mean that these genes were assembled from coding regions already bounded by introns in the last common ancestor of prokaryotes and eukaryotes, and thus that prokaryotes (or at least eubacteria) have indeed since lost those introns. So introns are in fact a primitive genomic feature, retained in eukaryotic nuclear genomes.

The second development also dates to 1977. It is the first appearance of data from Woese's laboratory that allowed quantitative measurement of evolutionary relationships between prokaryotic genomes and (unlike cytochromes $c$) both organellar and nuclear genomic compartments of eukaryotic cells.[17,18] These measurements were initially based on partial ribosomal RNA sequences obtained through the use of existing RNA technology, but now have been extended to complete gene sequences. They show that all eubacteria are much more closely related to all other eubacteria than any is related to any eukaryotic nucleus, so that there are no prokaryotes to which eukaryotic nuclei are closest, as we would expect if eukaryotic nuclear genomes evolved only 1.4 billion years ago from the genomes of some prokaryotic lineage.

These developments, together with recent remarkable results obtained with the self-splicing ribosomal RNA intron of Tetrahymena,[19] support a consistent grand evolutionary scheme: (1) Self-replicating RNA molecules arose by chance from monomeric and oligomeric precursors synthesized abiotically. RNAs that by chance have an appropriate sequence (comparable to that of the Tetrahymena intron) can catalyze their own replication, without proteinaceous enzymes. Such molecules will evolve towards greater efficiency and stability by Darwinian natural selection. (2) When, by whatever mechanism, RNA came to code for protein, the phenotype-genotype coupling was established and self-replicating RNAs coding for small stable subunits of protein structure and function evolved by natural selection. (3) Such small coding RNAs became linked together by some variant of self-splicing. The noncoding spaces between them became introns, and splicing would have become necessary for the production of larger proteins with multiple structural and functional domains (like modern proteins). (4) At some stage, intron-containing RNA genomes were reverse transcribed into DNA genomes, containing introns. This stage would be comparable to that described by Woese as the "progenote."[20] Progenotic genomes would have shown many other primitive traits, in addition to introns. Eukaryotes and prokaryotes (at least eubacteria) diverged at this stage, and many of the differences between them in terms of gene and genome structure, genome replication, and gene expression and regulation reflect different solutions to problems as yet unsolved in the last common ancestor, and so solved differently in eukaryotic nuclear and prokaryotic (eubacterial) evolutionary lineages.

## THE SIGNIFICANCE OF THE ARCHAEBACTERIA

Woese's ribosomal RNA sequence studies produced an additional major surprise. Certain previously recognized but (from the molecular point of view) little-studied bacteria—methanogens, halobacteria, *Thermoplasma,* and certain other thermophiles—were different from the rest of the bacteria, as different on this quantitative basis as the latter were from eukaryotic nuclei. These unusual bacteria were named the *archaebacteria* by Woese and Fox,[17,18] and the vast array of other, well-studied species became, in their terms, the *eubacteria.* By now, there are a great many other molecular and biochemical traits that define the archaebacteria as a separate coherent group, not obviously closer to eubacteria than to eukaryotes. For instance, archaebacterial tRNA and rRNA genes sometimes have introns; archaebacterial RNA polymerases are complex in subunit composition and immunologically related to eukaryotic polymerases (as W. Zillig will discuss elsewhere in this volume); archaebacterial promoters are neither eukaryotic nor eubacterial in structure; archaebacterial messages may sometimes lack 5' leaders; archaebacterial ribosomes, although bearing 16S, 23S, and 5S rRNAs, are unique in structure and spectrum of antibiotic sensitivity; translation in archaebacteria is initiated with methionyl-, not formylmethionyl-tRNA and is sensitive to diphtheria toxin; and archaebacterial cell envelopes lack peptidoglycan and exhibit ether-linked, rather than ester-linked, lipids (for review, see Reference 21).

What all this means in the larger context of this volume is that we now have still a third perspective on cellular evolution. If the evolutionary lineage that led to the nucleus and cytoplasm of eukaryotic cells indeed diverged from that leading to modern eubacteria at the progenote stage, when molecular machineries for replication, transcription, and translation were still poorly evolved, then decisions as to which eukaryotic molecular features are primitive and which are advanced require special kinds of information which are so far available only for introns. The archaebacteria will exhibit still a third set of evolutionary solutions to problems separately solved in the other two "primary kingdoms." Knowing those solutions, we should be in a much better position to guess at the primitive state.

## WHAT WE KNOW SO FAR

Dennis has most recently reviewed the available molecular data[22] (see also Reference 23 for references). No large picture yet emerges from them, but we can draw several simple conclusions. For instance:

1. In most archaebacteria, genes for 16S, 23S, and 5S ribosomal RNAs are linked, in that order, and probably cotranscribed from promoters 5' to the 16S gene. This similarity to eubacterial gene organization extends to the finding of alanine tRNA genes between the 16S and 23S coding regions in some species, almost a certain indicator of homology predating the eubacterial-archaebacterial divergence. Tentatively, one might suggest that the linked 18S and 28S plus unlinked 5S organization of most eukaryotes is a derived condition.

2. Several tRNA genes in *Sulfolobus solfataricus* and at least one tRNA gene in *Halobacterium volcanii* bear introns adjacent to their anticodon regions, in the position where most intron-containing eukaryotic tRNA genes are interrupted. These observations suggest that introns in these positions are primitive, and perhaps predate the derivation of most tRNA species from a limited number of ancestors during the evolutionary expansion of the genetic code.

3. Some protein-coding genes in methanogenic species appear to be transcribed as parts of polycistronic messages, suggesting that archaebacterial ribosomes can initiate internally, and that this condition might be primitive. However, transcript analysis of the halobacterial bacterioopsin gene and a linked protein-coding gene suggest that, at least for these species, there is no effective Shine-Dalgarno recognition mechanism for the initiation of protein synthesis.

4. A variety of potential promoter sequences have been identified by comparisons of sequences 5' to coding regions of rRNA, tRNA, and protein-coding genes. These are different one from another and from eubacterial and eukaryotic promoter sequences.

5. Comparisons of the subunit structures of archaebacterial, eubacterial, and eukaryotic RNA polymerases yield much that is surprising—this has been covered elsewhere in this volume by W. Zillig.

6. Certain species of halobacteria harbor large numbers of transposable elements. In structure, these resemble eubacterial insertion sequences, but in number and distribution they behave like eukaryotic middle-repetitive DNA, and impart an unusual degree of plasticity to genome structure in these species.

None of these findings by itself gives a very clear picture of early events in the evolution of genome structure in general. All of them make it clear, however, that we are moving in the right direction to achieve that goal. The comparative approach has been as yet little used in evolutionary molecular biology, except in deducing structure-function relationships for macromolecules. With it, we should be able to provide an evolutionary-historical picture of the development of basic features of replication, transcription, and translation.

## REFERENCES

1. MARGULIS, L. 1970. Origin of Eukaryotic Cells. Yale University Press. New Haven, Conn.
2. CAVALIER-SMITH, T. 1975. Nature 256: 463-467.
3. STANIER, R. Y. & C. B. VAN NIEL. 1962. Arch. Mikrobiol. 42: 17-35.
4. WOESE, C. R. & G. E. FOX. 1977. J. Mol. Evol. 10: 1-6.
5. MOROWITZ, H. J. & D. C. WALLACE. 1973. Ann. N.Y. Acad. Sci. 225: 62-73.
6. KNOLL, A. H. 1985. Paleobiology 11: 53-64.
7. SAGAN, L. 1967. J. Theor. Biol. 14: 225-275.
8. SCHIMPER, A. F. W. 1883. Bot. Ztg. 41: 105-114.

9. MERESCHOWSKY, K. C. 1905. Bull. Soc. Nat. Sci. Ouest **6:** 17-98.
10. GRAY, M. W. & W. F. DOOLITTLE. 1982. Microbiol. Rev. **46:** 1-42.
11. PACE, N. R., G. J. OLSEN & C. R. WOESE. 1986. Cell **45:** 325-326.
12. MARGULIS, L. 1981. Symbiosis in Cell Evolution. Freeman. San Francisco, Calif.
13. GILBERT, W. 1978. Nature **271:** 501.
14. LONBERG, N. & W. GILBERT. 1985. Cell **40:** 81-90.
15. STONE, E. M., K. N. ROTHBLUM & R. J. SCHWARTZ. 1985. Nature **313:** 498-500.
16. DUESTER, G., H. JORNVALL & G. W. HATFIELD. 1986. Nucleic Acids Res. **14:** 1931-1941.
17. WOESE, C. R. & G. E. FOX. 1977. Proc. Nat. Acad. Sci. USA **74:** 5088-5090.
18. FOX, G. E., L. J. MAGRUM, W. E. BALCH, R. S. WOLFE & C. R. WOESE. 1977. Proc. Nat. Acad. Sci. USA **74:** 4537-4541.
19. CECH, T. R. 1986. Proc. Nat. Acad. Sci. USA **83:** 4360-4363.
20. WOESE, C. R. 1981. *In* Archaebacteria. O. Kandler, Ed.: 1-17. Fischer. Stuttgart, FRG.
21. WOESE, C. R. & R. S. WOLFE. 1985. The Bacteria, **8:** Archaebacteria. Academic Press. New York, N.Y.
22. DENNIS, P. 1986. J. Bacteriol. **168:** 471-478.
23. DOOLITTLE, W. F. 1985. Trends Genet. **1:** 268-269.

# Eukaryotic Traits in Archaebacteria

## Could the Eukaryotic Cytoplasm Have Arisen from Archaebacterial Origin?

WOLFRAM ZILLIG

*Max-Planck Institute for Biochemistry*
*D-8033 Martinsried, Federal Republic of Germany*

The urkingdom of the archaebacteria[1,2] consists of three major subdivisions or branches: (1) the methanogens plus extreme halophiles, (2) the sulfur-dependent *Thermoproteales* + *Sulfolobales,*[3-5] and (3) the recently recognized sulfur-dependent *Thermococcales,*[6,7] which appear to fall between the other two branches (FIGURE 1). The phylogenetic depth of this kingdom, i.e., the phylogenetic distances between its origin and its present representatives, as measured by comparisons of 16S rRNA sequences, appears larger than those of the other two kingdoms, eubacteria and eucytes (= "eukaryotic cytoplasm" = "truly eukaryotic" fraction of the composite eukaryotes). On the other hand, the distance between its origin and the branching point between the three kingdoms appears particularly small indicating its primeval character.[8,9]

All archaebacteria share certain feature designs which are either (1) unique, such as the ether lipids[10] or the small genome size;[11] or (2) "eubacterial," such as the occurrence of operons,[12-14] Shine-Dalgarno sequences and ribosome binding sites,[12,14,15] class II restriction enzymes[16] and the prokaryotic organization, i.e., absence of a nucleus; or (3) "eukaryotic," such as the occurrence of introns in tRNA genes[17,18] (G. Wich and A. Böck, unpublished data), the absence of chlorophyll-dependent phototrophy and of a mureine sacculus, the insensitivity of DNA-dependent RNA polymerases to rifampicin or streptolydigin,[19] and of the ribosome to streptomycin or chloramphenicol,[20] the sensitivity of DNA polymerase to aphidicolin,[21] the ADP ribosylation of EFII by diphtheria toxin.[22] However, the designs of certain features are (4) divergent in different branches of the archaebacterial kingdom. Such divergent feature designs in the different branches of the archaebacteria include, for example, different RNA polymerase types,[23] different promoter organization (G. Wich and A. Böck, unpublished data), different 5S rRNA secondary structures,[24,25] differences in the sequences of initiator tRNAs,[26] in rDNA organization,[27] in the number of proteins in ribosomal subunits,[20,28] in tRNA intron frequency, different types of viruses,[29] and different modes of cell division.[30]

The designs of most of the divergent features appear more "eukaryotic" in the sulfur-dependent *Thermoproteales* + *Sulfolobales* than in the methanogens + halophiles. The other way round, these features appear more eubacterial in methanogens + halophiles than in sulfur-dependent archaebacteria. This observation has prompted the assumption that the two "modern kingdoms" might have arisen within the archaebacteria, possibly in different branches of them.[30] That is, the archaebacteria appeared paraphyletic, not comprising two important lineages descending from them, the eubacteria and the eukaryotes.

The comparison of major portions of the total sequences of 10 16S rRNAs from archaebacteria with each other and with 16S rRNAs from eubacteria and eukaryotes has, however, yielded a most probable unrooted tree in which all three kingdoms appear monophyletic.[8] Yet the distance from the deepest bifurcation in the archaebacteria to the trifurcation between the three kingdoms is smaller than some distances between branching points within the archaebacterial kingdom itself.

The consideration of the distribution of qualitative characteristics (traits) thus

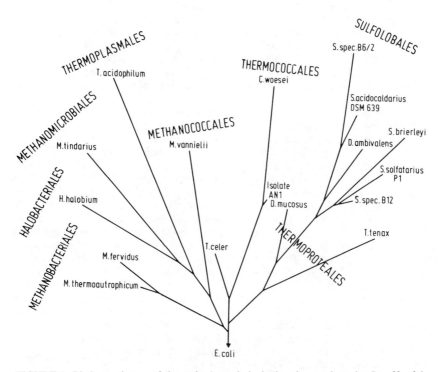

**FIGURE 1.** Phylogenetic tree of the archaebacteria including the novel species *Desulfurolobus ambivalens*[33] and *Pyrococcus woesei,*[7] derived from rRNA-DNA cross-hybridization data. (For details see Reference 7.)

suggests the paraphyletic nature of the archaebacteria, i.e., the descent of eubacteria and eukaryotes from different archaebacterial ancestors whereas the 16S rRNA sequence comparison appears to prove the monophyletic nature of all three kingdoms. How can this paradox be resolved?

First, the descents of traits are not always as clear as they appeared at first glance. For example, the ether lipids are unique to the archaebacteria. If one does not want to assume two independent events of origination of ester lipids in the lineages leading to eubacteria and eukaryotes, one must suppose that both modern kingdoms originated by a bifurcation of one lineage which had previously separated from the archaebacteria.

A similar hint comes from the distribution of sensitivities to certain antibiotics within the kingdoms.[31] Some of these are shared by eubacteria and eukaryotes, but not archaebacteria. *Thermoproteales + Sulfolobales* which are insensitive to most antibiotics differ strongly from methanogens + halophiles which share some responses with eubacteria and a few others with eukaryotes. However, single nucleotide exchanges suffice to eliminate sensitivity to some antibiotics, which weakens the significance of such evidence.

Though there is no doubt that the sulfur-dependent archaebacteria (and *Thermoplasma*) have DNA-dependent RNA polymerases of the same type as the eukaryotes, the possibility exists that the eukaryotic RNA polymerases A(I), B(II), and C(III) are highly conserved, whereas the methanogens + halophiles evolved a derived type independent of the eubacteria. The conservation of different fractions of the antigenic determinants of the original sulfur-dependent archaebacterial B component in the B components of eukaryotic and the $\beta$ subunits of eubacterial RNA polymerases and the different B→B'+B'' splits seen in the RNA polymerases of different orders of methanogens and halophiles have certainly resulted by evolution, but must not necessarily be interpreted as proving different archaebacterial origins of eukaryotes and eubacteria.

Similarly, the suggestive strength of the existence of different 5S rRNA secondary structures in different branches of the archaebacteria remains subject to doubt until the validity of the proposed structures has been settled.[24,25]

Yet, the coincidence in the distributions of alternative designs of a number of features leaves one dissatisfied, if one wants to take the distinct monophyletic nature of each of the three kingdoms as derived from rRNA sequence data for granted. Is the most probable unrooted tree relating the three kingdoms the real one? The resolution of distances between different phyla of different kingdoms appears insufficient. Usually, these distances scatter closely around an average interkingdom distance. Yet, the distance of *Thermoproteus tenax* (and, to a lesser extent, *Sulfolobus solfataricus* and *Thermococcus celer*) from different eukaryotes appears significantly smaller than that of the *Methanobacteriales, Methanomicrobiales,* and especially the *Halobacteriales* from the eukaryotes.[8] This is particularly striking in view of the large average interkingdom distance, which is probably not much smaller than the distance of two random sequences.[32] On the other hand, the archaebacteria clearly appear closer to the eubacteria than to the eukaryotes. The largest average distance is seen between eubacteria and eukaryotes. The strength of the proposal of distinct monophyletic origins of the three kingdoms depends on the significance of the distance measurement between the lowest archaebacterial bifurcation and the interkingdom trifurcation point. With the present information this distance is hard to assess.

Thus, the question of which of the two controversial models is correct remains unanswered until sufficient evidence for one or the other possibility has been obtained. More and independent sequence data, e.g., of different sufficiently conservative and large enough "fundamental" molecules, like the large components of DNA-dependent RNA polymerases or of enzymes of basic metabolism, are required.

The proposed archaebacterial origin of the eucyte thus is unproven at the moment. Yet, the observation that many eukaryotic feature designs appear "archaebacterial" or even similar to the corresponding ones in sulfur-dependent archaebacteria sheds light on a fundamental difference between the modes of evolution of eukaryotes and eubacteria. In view of the distribution of antigenic determinants in the B components of DNA-dependent RNA polymerases of representatives of the three kingdoms and of the major branches of the archaebacterial kingdom, we can safely assume that the extremely thermophilic anaerobic sulfur-dependent archaebacteria, possibly the *Thermococcales* more than the *Thermoproteales,* are closest to the common origin. The archaebacterial feature designs in eukaryotes thus appear primeval or primitive suggesting that eukaryotes have evolved by diversification rather than by streamlining from origins possibly not in the same lineage as but resembling sulfur-dependent archaebacteria.

[**Note added in proof:** The name of the organism termed *Calduplex woesei* in this paper has been changed into *Pyrococcus woesei.*]

## REFERENCES

1. WOESE, C. R. & G.E. FOX. 1977. Phylogenetic structure of the prokaryotic domain: the primary kingdoms. Proc. Nat. Acad. Sci. USA **74:** 5088-5090.
2. FOX, G. E., E. STACKEBRANDT, R. B. HESPELL, J. GIBSON, J. MANILOFF, T. A. DYER, R. S. WOLFE, W. E. BALCH, R. S. TANNER, L. J. MAGRUM, L. B. ZABLEN, R. BLAKE-MORE, R. GUPTA, L. BONEN, B. J. LEWIS, D. A. STAHL, K. R. LUEHRSEN, K. N. CHEN & C. R. WOESE. 1980. The phylogeny of prokaryotes. Science **209:** 457-463.
3. ZILLIG, W., K. O. STETTER, W. SCHÄFER, D. JANEKOVIC, S. WUNDERL, I. HOLZ & P. PALM. 1981. *Thermoproteales:* a novel type of extremely thermoacidophilic anaerobic archaebacteria isolated from Icelandic solfataras. Zbl. Bakt. Hyg. I. Abt. Orig. C2: 205-227.
4. TU, J., D. PRANGISHVILLI, H. HUBER, G. WILDGRUBER, W. ZILLIG & K. O. STETTER. 1982. Taxonomic relations between archaebacteria including 6 novel genera examined by cross hybridization of DNAs and 16S rRNAs. J. Mol. Evol. **18:** 109-114.
5. WOESE, C. R., R. GUPTA, C. M. HAHN, W. ZILLIG & J. TU. 1984. The phylogenetic relationships of three sulfur dependent archaebacteria. System. Appl. Microbiol. **5:** 97-105.
6. ZILLIG, W., I. HOLZ, D. JANEKOVIC, W. SCHÄFER & W. D. REITER. 1983. The archaebacterium *Thermococcus celer* represents a novel genus within the thermophilic branch of the archaebacteria. System. Appl. Microbiol. **4:** 88-94.
7. ZILLIG, W., I. HOLZ, H.-P. KLENK, J. TRENT, S. WUNDERL, D. JANEKOVIC, E. IMSEL & B. HAAS. 1987. *Pyrococcus woesei,* sp. nov., an ultra-thermophilic marine archaebacterium, representing a novel order, *Thermococcales.* System. Appl. Microbiol. **9:** 62-70.
8. WOESE, C. R. & G. J. OLSEN. 1986. Archaebacterial phylogeny: perspectives on the urkingdoms. System. Appl. Microbiol. **7:** 161-177.
9. HUMMEL, H., M. JARSCH & A. BÖCK. 1986. Unique antibiotic sensitivity of protein synthesis in archaebacteria and the possible structural basis. Microbiology: 370-374.
10. LANGWORTHY, T. A., T. G. TORNABENE & G. HOLZER. 1982. Lipids of archaebacteria. Zbl. Bakt. Hyg. I. Abt. Orig. C3: 228-244.
11. SEARCY, D. G. & F. R. WHATLEY. 1982. Zbl. Bakt. Hyg. I. Abt. Orig. C3: 245-257.
12. ALLMANSBERGER, R., C. BOLLSCHWEILER, U. KONHEISER, B. MÜLLER, E. MUTH, G. PASTI & A. KLEIN. 1986. Arrangement and expression of methyl CoM reductase genes in *Methanococcus voltae.* System. Appl. Microbiol. **7:** 13-17.
13. HAMILTON, P. T. & R. N. REEVE. 1985. Structure of genes and an insertion element in the methane producing archaebacterium *Methanobrevibacter smithii.* Mol. Gen. Genet. **200:** 47-59.
14. REITER, W.-D., P. PALM, A. HENSCHEN, F. LOTTSPEICH, W. ZILLIG & B. GRAMPP. 1987. Identification and characterization of the genes encoding three structural proteins of the *Sulfolobus* virus-like particle SSVI. Mol. Gen. Genet. **206:** 144-153.
15. REEVE, J. N., P. T. HAMILTON, G. S. BECKLER, C. J. MORRIS & C. H. CLARKE. 1986. Structure of methanogen genes. System. Appl. Microbiol. **7:** 5-12.
16. KESSLER, C., T. S. NEUMAIER & W. WOLF. 1985. Gene **33:** 1-102.
17. KAINE, B. P., R. GUPTA & C. R. WOESE. 1983. Putative introns in tRNA genes of prokaryotes. Proc. Nat. Acad. Sci. USA **80:** 3309-3312.
18. DANIELS, C. J., S. E. DOUGLAS & W. F. DOOLITTLE. 1986. Genes for transfer RNAs in *Halobacterium volcanii.* System. Appl. Microbiol. **7:** 26-29.

19. GROPP, F., W. D. REITER, A. SENTENAC, W. ZILLIG, R. SCHNABEL, M. THOMM & K. O. STETTER. 1986. System. Appl. Microbiol. **7**: 95-101.
20. SCHMID, G., T. PECHER & A. BÖCK. 1982. Properties of the translational apparatus of archaebacteria. Zbl. Bakt. Hyg. I. Abt. Orig. **C3**: 209-217.
21. FORTERRE, P., C. ELIE & M. KOHIYAMA. 1984. Aphidicolin inhibits growth and DNA synthesis in halophilic archaebacteria. J. Bacteriol. **159**: 800-802.
22. KESSEL, M. & F. KLINK. 1982. Identification and comparison of eighteen archaebacteria by means of the diphtheria toxin reaction. Zbl. Bakt. Hyg. I. Abt. Orig. **C3**: 140-148.
23. SCHNABEL, R., M. THOMM, R. GERARDY-SCHAHN, W. ZILLIG, K. O. STETTER & J. HUET. 1983. Structural homology between different archaebacterial DNA-dependent RNA polymerases analyzed by immunological comparison of their components. EMBO J. **2**: 751-755.
24. FOX, G. E., K. R. LUEHRSEN & C. R. WOESE. 1982. Archaeabacterial 5S ribosomal RNA. Zbl. Bakt. Hyg. I. Abt. Orig. **C3**: 330-354.
25. HUYSMANS, E. & R. DE WACHTER. 1986. The distribution of archaebacterial 5S rRNA sequences in phenetic hyperspace: implications for early biotic evolution. *In* Archaebacteria '85. O. Kandler & W. Zillig, Eds. Gustav Fischer Verlag. Stuttgart & New York.
26. KUCHINO, Y., M. HIARA, Y. YABUSAKI & S. NISHIMURA. 1982. Initiator tRNAs from archaebacteria show common unique sequence characteristics. Nature **298**: 684-685.
27. NEUMANN, H., A. GIERL, J. TU, J. LEIBROCK, D. STAIGER & W. ZILLIG. 1983. Organization of the genes for ribosomal RNA in archaebacteria. Mol. Gen. Genet. **192**: 66-72.
28. CAMMARANO, P., A. TEICHNER & P. LONDEI. 1986. Intralineage heterogeneity of archaebacterial ribosomes, evidence for two physicochemically distinct ribosome classes within the third urkingdom. System. Appl. Microbiol. **7**: 137-146.
29. ZILLIG, W., F. GROPP, A. HENSCHEN, H. NEUMANN, P. PALM, W.-D. REITER, M. RETTENBERGER, H. SCHNABEL & S. YEATS. 1986. Archaebacterial virus host systems. System. Appl. Microbiol. **7**: 58-66.
30. ZILLIG, W., R. SCHNABEL & K. O. STETTER. 1985. Archaebacteria and the origin of the eukaryotic cytoplasm. Curr. Top. Microbiol. Immunol. **114**: 1-18.
31. BÖCK, A. & O. KANDLER. 1985. Antibiotic sensitivity of archaebacteria. *In* The Bacteria. C. R. Woese & R. S. Wolfe, Eds. **8**: 525-544. Academic Press, Inc. Orlando, Fla.
32. ECKENRODE, V. K., J. ARNOLD & R. B. MEAGHER. 1985. Comparison of the nucleotide sequence of soybean 18S rRNA with the sequences of other small-subunit rRNAs. J. Mol. Evol. **21**: 259-269.
33. ZILLIG, W., S. YEATS, I. HOLZ, A. BÖCK, M. RETTENBERGER, F. GROPP & G. SIMON. 1986. *Desulfurolobus ambivalens,* gen. nov. sp. nov., an autotrophic archaebacterium facultatively oxidizing or reducing sulfur. System. Appl. Microbiol. **8**: 197-203.

# Are Centrioles Semiautonomous?

LELENG P. TO

*Goucher College*
*Department of Biological Sciences*
*Dulaney Valley Road*
*Towson, Maryland 21204*

Centrioles are unique to eukaryotes.[1-7] They are small, barrel-shaped structures (about 0.2 $\mu$m in diameter and 0.3-0.5 $\mu$m long) that occur in the cytoplasm of every eukaryotic cell capable of forming cilia or flagella. Higher plants, yeasts, and some protists do not form flagella or cilia, and these eukaryotes have never been found to contain centrioles.

Centrioles and kinetosomes are interchangeable terms. In all cells that have them, centrioles are associated with mitotic or meiotic spindles. As kinetosomes (also called basal bodies), centrioles are located at the bases of cilia and flagella. During mitosis in unicellular green algae, *Chlorella*, and *Chlamydomonas*, the kinetosomes associated with flagella become detached and take up polar positions like centrioles.[7-10] In ciliated vertebrate cells, the centriole pair migrates from its normal location near the nucleus to the apical region of the cell and appears to form several electron-dense satellites. The satellites become kinetosomes and migrate to the cell membrane to initiate ciliary generation.[11] Interchangeability of kinetosomes and centrioles was also demonstrated by inducing aster formation in vertebrate and invertebrate eggs with microinjection of nonvertebrate kinetosomes.[12,13] Since centrioles and kinetosomes are interchangeable, centriole will be used here as the term for both.

The electron microscope has revealed that centrioles contain a 9+0 array of microtubules. This microtubular arrangement is similar to that of the ciliary and flagellar 9+2 complex except centrioles lack the pair of central microtubules and the nine around the centriolar circumference consist of triplets linked with electron-dense elements. Except for a cartwheel-like density pattern at their proximal ends, centrioles appear structureless in the center. Spindle pole centrioles are usually present as pairs positioned perpendicularly to each other. An osmiophilic pericentriolar matrix (PCM) surrounds each centriole pair. Each centriole pair together with the associated PCM is called a centrosome.[14,15]

The ninefold symmetry of centrioles is found in diverse groups of eukaryotes, suggesting that centrioles remained highly conserved through 2 billion years of eukaryotic evolution.[5] Only a few variations in centriole structure have been found. Atypical centrioles vary only in the details of triplet arrangement. Some centrioles with construction different from the ninefold symmetry, however, have been reported.[16-19]

During DNA duplication at interphase, the two parent centrioles move apart slightly, and each produces a small budlike procentriole, separated from the parent centriole by a small space. The growing centrioles extend outward at right angles from the parent centrioles. During mitosis, each pair of duplicated centrioles becomes surrounded by an aster of microtubules. The separated centrioles are moved to opposite poles of the nucleus by lengthening microtubules of the developing spindle. Spindle microtubules form with centrioles as the focal point, but never touch the centrioles. Instead, they terminate in the PCM.[5,20]

Based on light microscope and functional studies of centriole "reproduction," Mazia proposed that centrioles replicate by a generative mechanism,[21] as opposed to a fission mechanism. This generative mechanism is thought to include three stages: (1) a replicative phase leading to the formation of a small, amorphic, electron-dense "germ" or "seed," also called the procentriole; (2) the growth or elongation phase during which a mature centriole is formed under the influence of the procentriole; (3) the separation phase during which the mature parent and daughter centrioles move away from each other. The resulting daughter centriole contains no material from the parent centriole.

Went interpreted cleavage patterns of developing sand dollar eggs treated with actinomycin D, 5-bromodeoxyuridine (BudR), chloramphenicol, and mercaptoethanol as consistent with Mazia's generative mechanism for centriole duplication.[22] All four chemicals disrupted normal cleavage and centriole duplication, but the time of interference by chloramphenicol or mercaptoethanol was distinctly and consistently different from that observed with actinomycin or BudR. Mercaptoethanol and chloramphenicol were thought to disrupt the growth phase of centriole formation. Since actinomycin inhibits DNA-dependent RNA synthesis, the effect of actinomycin on sand dollar egg cleavage pattern suggested the involvement of DNA in the replicative phase.

The parent centriole is normally surrounded by PCM which serves as the microtubule-organizing center (MTOC) for spindle microtubules. It was previously thought that the daughter centriole does not become associated with PCM until it becomes a parent in the next cell cycle.[23] To investigate whether a centriole must experience parenthood before association with PCM, Keryer et al. induced multipolar mitosis in Chinese hamster ovary cells by colcemid treatment.[24] After recovery from colcemid block, 40% of the cells divided into three. Antitubulin immunofluorescence and high-voltage electron microscopy (HVEM) showed that centrioles were distributed among the three spindle poles in a 2:1:1 or 2:2:0 pattern. The 2:1:1 pattern suggests that parenting is not a prerequisite for association with pole function. The 2:2:0 pattern indicates that centrioles, per se, are not necessary for pole function. The investigators proposed that centrioles may help in properly organizing and distributing PCM which is needed for formation of a functional spindle pole.

Using HVEM, Sluder and Rieder examined the ultrastructure of serial sections of mercaptoethanol-induced tetrapolar sea urchin eggs,[20] which eventually cleave into four daughter cells with a monopolar spindle containing only one centriole. In 12 cells examined, no cases of acentriolar spindle poles were found. Spindle poles that would have given rise to monopolar spindles, at the next mitosis, had only one centriole, while spindle poles that would have given rise to bipolar spindles had two centrioles. Therefore, the number of centrioles at a pole is directly related to the reproductive capacity of that pole. These findings and the observations that isolated centrioles induce asters in a somewhat dose-dependent manner suggested to the authors that centrioles may play a causal role in spindle pole formation and reproduction.

In cells not known to have centrioles, spindle microtubules also terminate in densely staining material. Centrioles, therefore, are not necessary for the generation of the spindle apparatus. Dietz prevented the migration of centrioles and asters so that after cell division two lines of cells were produced: one containing two pairs of centrioles and one lacking centrioles.[25] In the next cell division, the centrioleless cells formed spindles and divided normally, suggesting that centrioles are not necessary for spindle generation even in cells that normally contain centrioles. A *Drosophila* cell line has no visible centrioles, again indicating that centrioles or the 9+0 component of centrioles is not necessary for cell division.[26-27] However, mitotic spindles that contain centrioles at each pole are highly focused and have elaborate astral microtubules. In contrast, mitotic spindles that contain no centrioles are less focused at the poles and have no astral array.[1,28] In cells lacking centrioles, microtubules also terminate in osmiophilic material similar to the pericentriolar matrix. Such osmiophilic material, in higher plants, binds to antibodies against PCM.[88]

The absence of centrioles in some mitotic cells, the apparent restriction of centrioles to cells capable of forming cilia or flagella, and Dietz's observations have led some cell biologists to suggest that centriolar association with spindle and astral arrays reflects a mechanism to ensure that a pair of centrioles is distributed to each daughter cell to later organize the elaboration of cilia or flagella.[3,4] The apparent de novo formation of centrioles in eukaryotes prior to development of an axonemal motile stage is consistent with this concept. "De novo" generation of centrioles has been reported in ameboflagellates, algae, and lower plants.[29-38] In all these cases, centriole appearance was coordinated with transformation to a motile phase or with the development of motile gametes. In one case, after flagellar assembly, the centriole disintegrates but the flagellum remains functional. This suggests that centrioles may be necessary for flagellar formation but not for flagellar function.[39]

In the kinetosomal mode, centrioles act as MTOCs in a fashion different from centrosomes. Axonemal doublet microtubules are directly continuous with two of each centriolar triplet. As part of the centrosome, centrioles may also be involved in nonaxonemal cell movement. Using an ultraviolet laser, Berns and his colleagues destroyed newt eosinophil centrioles without damaging the PCM.[40-42] With intact centrioles, eosinophils exhibited directional motility; however, after laser disruption of centrioles, cell movement decreased in rate and motility was uncoordinated and directionless.

Tubulin is the major centriolar protein.[43,44] There are also reports of $\alpha$-actinin,[45] calmodulin,[46] ATPase,[47] and RNA polymerase II[48] in centrioles. Two-dimensional gel analysis of centriolar proteins has shown, in addition to tubulin, at lease 15 different resolvable protein species in centrioles.[49] Nickel and silicon have been reported to be associated with centrioles.[50,51] The lumen of centrioles contain RNA and protein.[52-55] The pericentriolar cloud also contains RNA.[56] The presence of DNA in centrioles and the pericentriolar material has not been established. In insect spermatids, a dense granular structure is closely associated to the posterior region of the nucleus and surrounds the spermatid centriole and the initial portion of the flagellum. This structure is called the centriolar adjunct and also contains RNA and proteins.[57-60]

The nature and function of centriolar RNA are unknown. Centriolar RNA appears to be necessary for the induction of mitotic aster formation. Injection of RNAase-treated centrioles into echinoderm eggs failed to generate the astral array associated with the injection of centrioles into such cells.[54] In contrast, Hirano did not observe loss of aster induction activity in sperm centrioles treated with higher concentrations of RNAases[61] than those used by Heidemann and his colleagues.[54] After injection with centrioles, Hirano activated the eggs by normal insemination. This makes Hirano's observations difficult to interpret. The observed cleavage of inseminated eggs may have been initiated from sperm asters rather than from the injected centrioles. What is clear from Hirano's report is that RNAase treatment of inseminated eggs did not cause aster formation.

RNA seems to be necessary also for the microtubule nucleating capacity of the pericentriolar matrix. Digestion with RNAase specifically destroyed the ability of the pericentriolar cloud to nucleate microtubule assembly, while DNAase had no effect on either structure or nucleating ability of centrosomes.[56] It appears that centriole-associated RNA(s) may play a role in organizing spindle poles.

The presence of RNA and protein in the centriole lumen suggests that centriole RNA may exist as a ribonucleoprotein complex. Ribosomes represent the most abundant ribonucleoprotein assemblies in cells. Kallenbach interpreted the appearance of intracentriolar granules in the electron microscope as putative evidence for 70S ribosomes.[62] The sizes of intracentriolar granules were closer to those of mitochondrial than to cytoplasmic ribosomes. If this is the case, then special ribosomes may be needed for the synthesis of various centriolar and/or axonemal proteins. The granules of insect spermatid centriolar adjuncts were also similar to ribosomes in size.

The unusual mode of duplication of centrioles, the apparent continuity of generations of centrioles, the precise coordination of centriolar migration with mitosis, and the presence of RNA in centriolar lumens inevitably raise the question of centriole semiautonomy. If centrioles are semiautonomous, RNA in the lumen may be replicated by a reverse transcriptase or an RNA replicase. Although nonviral RNA replicases have been reported in uninfected healthy plant cells,[63–67] there is no report of RNA replicases in cells with centrioles. The binding of fluorescent α-amanitin (FAMA) to centrioles during mitosis is noteworthy.[48] α-Amanitin possesses a high affinity to eukaryotic RNA polymerase II, but the affinity of FAMA is 18 times lower than that of α-amanitin. The accumulation of fluorescent amatoxin in centrioles may reflect the presence of another protein with high affinity to amatoxins. In addition, RNA polymerase II is known to use RNA templates *in vitro.*[68]

Citing the disruption of cleavage patterns by actinomycin D and BudR, TTP synthesis in ammonia-treated enucleated fragments of sea urchin eggs, and the presence of intracentriolar RNA as evidence, Went proposed that a reverse transcriptase may be involved in the replicative phase of centriole generation.[22,69] Went's hypothesis is not consistent with studies that demonstrated that DNA synthesis is not necessary for centriole duplication in *Stentor coeruleus* and Chinese hamster ovary cells.[70–73] The possible functions of centriole RNA and evidence consistent with proposed functions are summarized in TABLE 1.

Margulis has proposed that, like chloroplasts and mitochondria, centrioles may be self-replicating semiautonomous organelles which evolved from prokaryotes, probably spirochetes, symbiotic to eukaryotic cells.[74,75] There is no direct evidence to support this view, but the discovery of tubulin-like proteins in spirochetes is consistent with Margulis' hypothesis. Using western blot techniques, Bermudes *et al.* reported

TABLE 1. Possible Role of Centriole RNA

|  | Evidence |
|---|---|
| Ribosomes | EM appearance and sizes |
| Direct assembly of centriole, axonemes, and/or pericentriolar material | Cortical heredity of ciliate kinetid pattern RNAase sensitivity of aster organization |
| Template reverse transcription | Echinoderm egg cleavage pattern disrupted by actinomycin and BUDR TTP synthesis in enucleated eggs RNA in centrioles |
| RNA transcription | RNA polymerase II in centriole lumen *In vitro,* RNA polymerase II able to use RNA as template DNA synthesis is not necessary for centriole duplication Presence of RNA in centriole lumen |

TABLE 2. Origin of Centriole

| | Evidence |
|---|---|
| **Exogenous** | |
| Margulis —centriole-axoneme complex evolved from ancestral spirochetes symbiotic to eukaryotes. | Tubulinlike proteins in spirochetes such as cold/warm cycle purification, taxol-enhanced purification yields, and cross-reactivity to antitubulin antibodies. *Pillotina* and *Hollandina* spirochetes have 25-nm microtubules. |
| **Endogenous** | |
| Hartman —centrioles evolved from part of host to mitochondria and chloroplasts. The host has an RNA genome. RNA is thought to precede DNA as genetic material. | |
| Dustin   —centriole-axoneme complex evolved by direct filiation from ancestral cyanobacteria (hypothetical uralga) or from ancestral anisokont protozoa. | 23-nm striated microtubules, with 13 subunits, in the cytoplasm of cyanobacteria. |
| Jensen | |
| Pickett-Heaps | |
| Wheatley | |
| Cavalier-Smith | |

that guinea pig antitubulin antibody recognized S1 protein from spirochetes of the genus *Spirochaeta*.[76] This immunoreactive S1 protein resembles tubulin in some other aspects: it is purified by warm-cold cycling and purification yields are enhanced greatly by taxol, a microtubule-stabilizing drug. Taylor suggested that if centrioles have an exogenous origin, a virus may be the appropriate ancestor.[77] In contrast to Margulis' view, Hartman suggested that the presence of RNA in centrioles and other MTOCs may indicate that the "host" for the endosymbiotic formation of eukaryotic cells was a primitive RNA-based organism.[78] Also in contrast to Margulis' hypothesis on the exogenous origin, Cavalier-Smith,[79] Dustin,[80] Jensen,[81] Pickett-Heaps,[3-4] and Wheatley[1] favor an endogenous origin of centrioles/kinetosome-axonemal complex by direct filiation from an ancestral anisokont or a cyanobacterial ancestor of eukaryotes. This ancestor has been called the uralga. Different hypotheses on the origin of centrioles are summarized in TABLE 2.

Two important observations that seem to contradict the notion of centriolar semiautonomy are the absence of demonstrable genetic control of centrioles and the apparent de novo generation of centrioles in various cells. The absence of genetic control may be explained by the great selection against mutations affecting centrioles, and by the paucity of our knowledge about centriole RNA. Isolation of centriole RNA and centriole mutants may show an extranuclear linkage group unique to centrioles. In test crosses to 33 markers, four uniflagellar positional mutants of *Chlamydomonas reinhardii* isolated by Huang et al.[82] did not show linkage to any of 16 established nuclear linkage groups, yet these mutants show linkage to 4 other mutations which are also unmapped with respect to linkage group. The 4 other mutations also affect flagellar assembly. Apparent lack of nuclear linkage may be due to incomplete mapping of nuclear genes or to the extranuclear origin of the linkage group giving rise to positional uniflagellar mutants.

Unless demonstrated by test tube assembly of components transcribed and translated by the nucleocytoplasmic system, de novo synthesis and assembly of centrioles cannot be assumed and the possibility of semiautonomy of centrioles cannot be dismissed. De novo appearance of centrioles has been reported, but the nature of the initial nucleating "seed" for the presumed self-assembly of centrioles is unknown. Unfertilized eggs of many animal species that lack centrioles appear to utilize the sperm centriole for the first mitotic division. Under extreme ionic imbalance or electrical stimulation, unfertilized animal eggs can generate varying numbers of centrioles.[73] These observations indicate that some precursor to centrioles may exist in the cytoplasm of unfertilized eggs. Studies of de novo centriole appearance in transforming ameboflagellates or gametogenesis of motile sperm suggest that in most cases, centriole formation involves the production of ill-defined fibrogranular material.[1] The composition of this ill-defined precursor and whether the precursor contains nucleic acids are unknown. The absence of visible centrioles at the poles may mean that the $9+0$ microtubular array is missing, but other centriole components may be present and functional.[20] The subcellular origin of the component(s) of this precursor has not been investigated.

The molecular events leading to the formation of new centrioles are poorly understood. Biochemical analysis of centrioles is hampered by the lack of a rich source of centrioles. The abundance of ribosomes in the cytoplasm complicates the purification of centriolar RNA. My laboratory worked on the large heterotrichous ciliate, *Stentor coeruleus.* Treatment with shedding agents, such as sucrose or urea, can induce *Stentor* cells to release their oral membranellar bands.[83–85] Each membranellar band consists of cilia and 15,000-20,000 centrioles.[71] After removal or dilution of the shedding agent, *Stentor* cells remain viable and can regenerate their oral membranellar bands.

To determine whether RNA is polymerized in centrioles, *in vitro* incorporation of tritiated UTP by isolated oral membranellar bands was examined.[86] The use of oral membranellar bands gave uninterpretable results, possibly because of mitochondrial and bacterial contamination. Cytochemical staining for succinic acid dehydrogenase, a mitochondrial marker enzyme, indicated that isolated oral membranellar bands were very rich in mitochondria. Although oral membranellar bands were obtained from "axenic" *Stentor* cultures that did not show bacterial growth on nutrient agar or the presence of bacteria in gram stains, the cultures may have contained bacteria that do not grow on nutrient agar. Attempts to repeat the experiment with isolated centrioles or to isolate centriolar RNA were unsuccessful because we could not obtain enough centrioles from "axenically" grown *Stentor* cells. For these studies, 2000 stentor cells were picked aseptically from "axenic" cultures and isolated by sucrose-heavy water gradient ultracentrifugation. In spite of precautions against RNAases, no RNA band was detected in isolated centrioles. A much greater number of cells is apparently needed for isolation of detectable levels of RNA. *Stentor* cells do not multiply rapidly, and the axenic cultures grew more slowly than nonaxenic cultures. The low number of cells and the time required to collect such cells led us to conclude that *Stentor coeruleus* is not the appropriate system for purifying centrioles.

Other ciliates, such as *Oxytricha* and *Euplotes,* may be more appropriate sources of kinetosomes. *Oxytricha* can be cultivated in 50 liter fermentation vats to obtain a density of 8000 cells/ml.[89,90] Subsequently, these hypotrichous ciliates can be separated from coexisting prokargotes by filtration and concentrated in a continuous flow centrifuge. Large-scale cultivation of *Euplotes* has not been reported, but more genetic information is available on this ciliate.[91–94]

Recently, a kidney cell line (A6) from *Xenopus laevis* was observed to form multiple motile cilia.[87] Prior to this, cell lines had been observed to make only a single nonmotile $9+0$ primary cilium. Cell lines, including A6 cells, do not form multiple motile cilia when propagated in plastic dishes. After subculture onto collagen or millipore filter bottom cups, however, A6 cells form multiple motile cilia at cellular apices. A pure cell line eliminates the problems of bacterial contamination and slow growth associated with *Stentor* cultures. Currently, we are characterizing the growth conditions for ciliogenesis of this cell line. If a large proportion of A6 cells become ciliated and if ciliogenesis in A6 cells is preceded by generation of multiple new centrioles as observed in mammalian tracheal epithelium, then A6 cells, like *Oxytricha* or *Euplotes,* may provide a feasible system for the biochemical characterization of centrioles. Until such a system can be found, centriole biogenesis will remain, in the words of Wheatley,[1] a central enigma in cell biology.

## ACKNOWLEDGMENTS

I am grateful to (1) Dr. Lynn Margulis for advice, encouragement, and moral support, (2) Dr. Helen Habermann for comments on this paper, (3) Susan Platt and Ann Karczeski for help with manuscript preparation, and (4) Goucher College and Research Corporation for financial support.

## REFERENCES

1. WHEATLEY, D. N. 1982. The Centriole: a Central Enigma of Cell Biology. Elsevier. New York, N.Y.
2. PETERSON, S. P. & M. W. BERNS. 1980. Int. Rev. Cytol. **64:** 81-106.
3. PICKETT-HEAPS, J. D. 1971. Cytobios **3:** 205-221.
4. PICKETT-HEAPS, J. D. 1974. Biosystems **6:** 37-48.
5. FULTON, C. 1982. Cell **30:** 341-343.
6. WOLFE, J. 1970. J. Cell Sci. **6:** 151-192.
7. WILSON, E. B. 1928. The Cell in Development and Heredity. The McMillan Co. New York, N.Y.
8. ATKINSON, A. W., JR., B. E. S. GUNNING, P. C. L. JOHN & W. McCULLOUGH. 1971. Nature New Biol. **234:** 24-25.
9. COSS, R. A. 1974 J. Cell Biol. **63:** 325-329.
10. JOHNSON, U. G. & K. R. PORTER. 1968. J. Cell Biol. **38:** 403-425.
11. SOROKIN, S. P. 1968. J. Cell Sci. **3:** 207-230.
12. HEIDEMANN, S. R. & M. W. KIRSCHNER. 1975. J. Cell Biol. **67:** 105-117.
13. MALLER, J., D. POCCIA, D. NISHIOTA, P. KIDD, J. GERHART & H. HARTMAN. 1976. Exp. Cell Biol. **99:** 185-194.
14. MCINTOSH, J. R. 1983. Mod. Cell Biol. **2:** 115-142.
15. PAWELETZ, N., D. MAZIA & E. FINZE. 1984 Exp. Cell Res. **152:** 47-56.
16. CLEVELAND, L. R. 1963. Function of flagellate and other centrioles in cell reproduction. *In* The Cell in Mitosis. Levine, Ed.: 3-53. Academic Press. New York, N. Y.
17. DALLAI, R. & M. MAZZINE. 1983. J. Ultrastruct. Res. **82:** 19-26.
18. GATENBY, J. B. 1961. J. R. Microscop. Soc. **79:** 299-317.
19. PHILLIPS, D. M. 1967. J. Cell Biol. **33:** 73-92.
20. SLUDER, G. & C. L. RIEDER. 1985. J. Cell Biol. **44:** 117-132.
21. MAZIA, D. 1961. *In* The Cell. J. Brachet & A. E. Mirsky, Eds. **3:** 77. Academic Press. New York, N. Y.
22. WENT, H. A. 1977. Exp. Cell Res. **108**(1): 63-74.
23. RIEDER, C. L. & G. G. BORISY. 1982. Biol. Cell. **44:** 117-132.

24. KERYER, C., H. RIS & G. G. BORISY. 1983. J. Cell Biol. **98:** 2222-2229.
25. DIETZ, R. 1966. Heredity **19:** 161-166.
26. DEBEC, A. 1984. Exp. Cell Res. **151:** 236-246.
27. DEBEC, A., A. SZOLLOSI & D. SZOLLOSI. 1982. Biol. Cell. **44:** 133-138.
28. SCHATTEN, G., C. SIMERLY & H. SCHATTEN. 1985. Proc. Nat. Acad. Sci. USA **82:** 4152-4156.
29. FULTON, C. & A. D. DINGLE. 1971. J. Cell Biol. **51:** 826-836.
30. FULTON, C. 1970. Science **167:** 1269-1270.
31. FULTON, C. 1971. In Origin and Continuity of Cell Organelles. J. Reinert & H. Urspring, Eds.: 170-221. Springer-Verlag. Berlin, FRG.
32. PICKETT-HEAPS, J. D. & L. C. FOWKE. 1969. Aust. J. Biol. Sci. **22:** 857-894.
33. PICKETT-HEAPS, J. D. & L. C. FOWKE. 1970. J. Phycol. **6:** 189-215.
34. DINGLE, A. D. 1970. J. Cell Sci. **7:** 463-487.
35. MIZUKAMI, I. & J. GALL. 1966. J. Cell Biol. **29:** 97-111.
36. MOSER, J. W. & G. L. KREITNER. 1970. J. Cell Biol. **44:** 454-458.
37. TURNER, F. R. 1968. J. Cell Biol. **37:** 370-393.
38. HOFFMAN, E. J. & I. MANTON. 1962. J. Exp. Bot. **13:** 443-449.
39. FAWCETT, D. W. 1975. Dev. Biol. **44:** 394-436.
40. BERNS, G. S. & M. W. BERNS. 1982. Exp. Cell Res. **142:** 103-109.
41. BERNS, M. W., J. AIST, J. EDWARDS, K. STRAHS, et al. 1981 Science **213:** 505-513.
42. KOONCE, M. P., R. A. CLONEY & M. W. BERNS. 1984. J. Cell Biol. **98:** 1999-2010.
43. CHU, L. K. & J. E. SISKEN. 1977. Exp. Cell. Res. **107:** 71-77.
44. PEPPER, D. A. & B. R. BRINKLEY. 1979. J. Cell Biol. **82:** 585-591.
45. GORDON, R. E., B. P. LANE & F. MILLER. 1980. J. Histochem. Cytochem. **28:** 1189-1197.
46. GORDON, R. E., K. B. WILLIAMS & S. PUSZKIN. 1982. J. Cell Biol. **95:** 57-63.
47. ANDERSON, R. G. W. 1977. J. Cell Biol. **74:** 547-560.
48. WULF, E., F. A. BAUTZ, H. FAULSTICH & T. WIELAND. 1980. Exp. Cell Res. **130:** 415-420.
49. ANDERSON, C. W., J. F., ATKINS & J. J. DUNN. 1976. Proc. Nat. Acad. Sci. USA **73:** 2752-2756.
50. KOVACS, P. & Z. DARVAS. 1982. Acta Histochem. **71:** 169-173.
51. SCHAFER, P. W. & J. A. CHANDLER. 1970. Science **170:** 1204-1205.
52. DIPELL, R. V. 1976. J. Cell Biol. **69:** 622-637.
53. HARTMAN, H., J. P. PUMA & T. GURNEY. 1974. J. Cell Sci. **16:** 241-260.
54. HEIDMANN, S. R., G. SANDER & M. W. KIRSCHNER. 1977. Cell **10**(3): 337-350.
55. RIEDER, C. L. 1979. J. Cell Biol. **80**(1): 1-9.
56. PEPPER, D. A. & B. R. BRINKLEY. 1980. Cell Motility **1:** 1-15.
57. TAFFAREL, M. & P. ESPONDA. 1980. Mikroskopie **36:** 35-42.
58. BACETTI, B., R. DALLAI & F. ROSATI. 1969. J. Mikroskopie **8:** 249-262.
59. GASSNER, G. 1970. J. Cell Biol. **47:** 69a.
60. LINDSEY, J. N. & J. J. BIESELE. 1974. Cytobios **10:** 199-220.
61. HIRANO, K. 1982. Dev. Growth Diff. **24:** 273-281.
62. KALLENBACH, R. J. 1983. Biosci. Rep. **3:** 1155-1162.
63. CHIFFLOT, S., P. SOMMER, D. HARTMANN, C. STUSSI-GARAUD & L. HIRTH. 1980. Virology **100:** 91-100.
64. DUDA, C. T. 1979. Virology **92:** 180-189.
65. IKEGAMI, M. & H. FRAENKEL-CONRAT. 1978. Proc. Nat. Acad. Sci. USA **75:** 2122-2124.
66. MASSON, A. 1979. Ann. Phytopathol. **11:** 492-506.
67. ROMAINE, C. P. & M. ZAITLIN. 1978. Virology **86:** 241-253.
68. DIENER, T. O. 1983. Adv. Virus Res. **28:** 241-283.
69. WENT, H. A. 1977. J. Theor. Biol. **68**(1): 95-100.
70. YOUNGER, K. R., S. BANERJEE, J. K. KELLEHER, M. WINSTON & L. MARGULIS. 1972. J. Cell Sci. **11:** 621-637.
71. KURIYAMA, R. & G. G. BORISY. 1981. J. Cell Biol. **91:** 814-821.
72. KURIYAMA, R., S. DASGUPTA & G. G. BORISY. 1983 J. Cell Biol. **97:** 190a.
73. RATTNER & PHILLIPS. 1973. J. Cell Biol. **57:** 359.
74. MARGULIS. L. 1981. Symbiosis in Cell Evolution. W. H. Freeman & Co. San Francisco, Calif.

75. MARGULIS, L. & D. BERMUDES. 1985. Symbiosis **1:** 101-124.
76. BERMUDES, D., R. OBAR & G. TZERTZINIS. 1986. Abstract submitted to the American Society of Microbiology Meetings, Washington, D. C., March 23-28.
77. TAYLOR, F. J. R. 1979. Trans. R. Soc. London Ser. B **204:** 267-286.
78. HARTMAN, H. 1984. Speculations Sci. Technol. **7:** 77-82.
79. CAVALIER-SMITH, T. 1982. Symp. Soc. Exp. Biol. **35:** 465-493.
80. DUSTIN, P. 1983. Treballs Soc. Catalana Biol. **35:** 15-63.
81. JENSEN, T. E. 1984. Cytobios **39:** 35-62.
82. HUANG, B., Z. RAMANIS, S. K. DUTCHER & D. J. L. LUCK. 1982. Cell **29:** 745-753.
83. TARTAR, V. 1960. The Biology of Stentor. Pergamon Press. New York, N. Y.
84. SHIGENAKA, Y., K. ITO, M. KANEDA & T. YAMAOKA. 1979. J. Electron Microsc. **28(2):** 73-82.
85. MALONEY, M. S. & B. R. BURCHILL. 1977. J. Cell Physiol. **93:** 363-374.
86. KHODADOUST, M. 1985. Isolation and Regeneration of Stentor coeruleus centrioles. Senior Independent Research Thesis. Goucher College. Towson, Md.
87. HANDLER, J. S., A. S. PRESTON & R. E. STEELE. 1984. Fed. Proc. **43:** 2221-2224.
88. CLAYTON, L., C. M. BLACK & C. W. LLOYD. 1985. J. Cell Biol. **101:** 319-324.
89. LAUGHLIN, T. J., J. M. HENRY, E. F. PHARES, M. V. LONG & D. E. OLINS. 1983. J. Protozool. **30:** 63-64.
90. LAUGHLIN, T. J., J. M. HENRY, E. F. PHARES, M. V. LONG & D. E. OLINS. 1982. Development of methods for the large scale culture of *Oxytricha nova. In* From Gene to Protein: Translation into Biotechnology. Ahmad *et al,* Eds.: 539. Academic Press. New York, N.Y.
91. KLOBUTCHER, L. A., M. T. SWANTON, P. DONINI & D. M. PRESCOTT. 1981. Proc. Nat. Acad. Sci. USA **78:** 3015-3019.
92. LALOE, F. 1979. Arch. Zool. Exp. Gen. **120:** 109-130.
93. NANNEY, D. L. 1977. Cell-cell interactions in ciliates: evolutionary and genetic constraints. *In* Microbial Interactions. Reissig, Ed.: 351-397. Halsted Press/Wiley. New York, N.Y.
94. KUNG, C. 1976. Membrane control of ciliary motions and its genetic modification. *In* Cell Motility. Goldman, *et al.,* Eds.: 941-948. Cold Spring Harbor Laboratory. Cold Spring Harbor, N.Y.

# Origins of the Plant Chloroplasts and Mitochondria Based on Comparisons of 5S Ribosomal RNAs[a]

NICHOLAS DELIHAS [b] AND GEORGE E. FOX [c]

[b]Department of Microbiology
School of Medicine
State University of New York at Stony Brook
Stony Brook, New York 11794

[c]Department of Biochemical and Biophysical Sciences
University of Houston
University Park
Houston, Texas 77004

Dedicated to the Memory of Dr. Robert van Tubergen who contracted Multiple Sclerosis and whose Promising Scientific Career was Sadly Curtailed by this Disease

The problem of the origin of the organelles has been addressed in the past. The evolution of the organelles has been a major topic in a previous symposium of the New York Academy of Sciences.[1] In addition, before many of the macromolecular data were known, Dr. Lynn Margulis very simply and eloquently described symbiotic relationships that suggested possible organelle origins.[2]

In this paper, we provide macromolecular comparisons utilizing the 5S ribosomal RNA structure to suggest extant bacteria that are the likely descendants of chloroplast and mitochondria endosymbionts. The genetic stability and near universality of the 5S ribosomal gene allows for a useful means to study ancient evolutionary changes by macromolecular comparisons. The value in current and future ribosomal RNA comparisons is in fine tuning the assignment of ancestors to the organelles and in establishing extant species likely to be descendants of bacteria involved in presumed multiple endosymbiotic events.

In assessing the phylogenetic origins of the organelles, the relatedness of 5S RNAs from organelles and eubacteria has been analyzed from several points of view:

1. Residue-to-residue comparisons.
2. Comparisons of segments of the RNA sequence.

[a]Work supported by National Science Foundation Grant DMB 85-02213 to N.D. and National Aeronautics and Space Administration Grant NSG-7440 to G.E.F.

92

3. Signature analysis, i.e., individual base positions, groups of contiguous positions, or secondary structural features that are characteristic of a particular phylogenetic group.
4. Overall conformation of the RNA.

In addition, the relatedness of sequences has been analyzed by use of a chi-square analysis of similarities and differences in sequences by the procedure known as correspondence analysis.[3] The analysis recognizes patterns of change in nucleotide sequence and produces a clustering of related 5S RNA species in a two-dimensional projection.

The availability of over 320 5S RNA sequences allows for a comprehensive phylogenetic analysis. In cases where the above-mentioned comparative analyses were applied to 5S RNAs of organelles and bacteria, consistent correlations have appeared.

## 5S RNA STRUCTURE

A consensus universal secondary structural model of the 5S RNA has been formulated based on comparative sequence analyses[4–10] and enzymatic and chemical probes of the higher-order structure.[11] In addition, the universally conserved positions have been mapped.[11] The universal positions are those positions that are highly conserved (found in approximately 90% or more of known sequences of 5S RNAs) in all classes of organisms and organelles, i.e., eubacteria, archaebacteria, eukaryotes, chloroplasts, and mitochondria. The 5S RNA comparisons also reveal structural subclasses of RNAs and distinct patterns in sequence and higher-order structure that are exhibited by 5S RNAs of related organisms. There are conserved sequences specific to the 5S RNAs of either the eubacterial or eukaryotic 5S RNAs. In addition, major differences between the 5S RNAs of these two phylogenetic divisions are nucleotide insertions and deletions that are found between given universal positions. For example, there are three nucleotide positions between the universal position U40 and G44 (see Reference 11 for universal numbering system) in eubacterial 5S RNAs and two positions in eukaryotic 5S RNA. Other well-characterized chain-length differences between universal positions are found in the stem of helices IV and V.

Deletion and insertion signatures have been used to assess phylogenetic relationships. For example, the presence of a eubacterial chain-length signature in the 5S cytoplasmic RNAs of the enkaryotic red algae (e.g., *Porphyra*) leads to the proposal of the retention of this eubacterial signature during a possible evolutionary transition from eubacteria to eukaryotes.[11]

## CYANOBACTERIAL AND GREEN PLANT PLASTID 5S RNA CORRELATIONS BASED ON PRIMARY AND HIGHER-ORDER STRUCTURES

The overall nucleotide sequence homologies between the chloroplast 5S RNAs and those of the eubacteria reveal that the closest identities of the chloroplast RNAs are to the 5S RNAs of the cyanobacteria.[12] Data on primary and higher-order structure reveal that the cyanobacterial and chloroplast 5S RNAs form a structural subclass of the eubacterial 5S RNAs. All cyanobacterial 5S RNAs have the deletion, insertion, and base-substitution signatures specific to those of the eubacterial 5S RNAs but in addition have characteristics that differ from those of 5S RNAs of most other organisms. Some signatures are specific only to the cyanobacterial and green plant chloroplast 5S RNAs and are therefore unique to the RNAs from these sources; for example, the insertion between positions 30 and 31 (see Reference 11 for universal numbering system). Other signatures, which are not unique, such as the A residue at position 102 nevertheless are found primarily in the cyanobacterial/plant chloroplast 5S RNAs. In addition to the above-mentioned properties, strains II and III of the cyanobacterium *Synechococcus lividus* also share a unique deletion signature (between positions 35-39) with the green plant and photosynthetic protist chloroplast 5S RNAs. This deletion is not found in the 5S RNAs of *Anacystis nidulans,* Prochloron or any of the other 320 known 5S RNA sequences. We suggest that the sharing of unique or rare signatures is indicative of a close phylogenetic relationship between the bacterial organism and the organelle. The signatures shared by the cyanobacterial and chloroplast 5S RNAs are summarized in TABLE 1 and also depicted in FIGURE 1 where the moss chloroplast 5S RNA[13] is used as a model.

Studies on the higher-order structure of the 5S RNAs reveal that the overall conformations of the RNAs of *S. lividus* and the spinach chloroplasts are very similar. This is based on the similarity in electrophoretic mobility on polyacrylamide gels of native, renatured, and denatured 5S RNAs from these sources and the marked differences in mobilities of renatured and denatured 5S RNA from *E. coli* compared to the mobility of the cyanobacterial/chloroplast 5S RNAs.[12] In addition, probes of the higher-order structure by partial nuclease digestion and chemical modification appear to indicate that the spinach chloroplast and *E. coli* 5S RNAs differ in conformation.[14] Thus, primary and higher-order structural comparisons indicate that the 5S RNAs of cyanobacteria and plant chloroplasts form a common structural subclass.

The correspondence analysis as applied to the 5S RNA[3] clusters the chloroplast RNA close to the RNAs of the cyanobacteria and distal to the other bacterial 5S RNAs in a two-dimensional plot. This is in agreement with the similarities in primary and higher-order structures between the cyanobacterial and chloroplast 5S RNAs and

TABLE 1. Signatures Common to the Cyanobacterial and Chloroplast 5S RNAs

| Signature | Position[a] | Characteristics |
|---|---|---|
| Nucleotide insertion | Between 30-31 | Unique to known cyanobacteral, plant chloroplast, and cyanelle 5S RNAs |
| Nucleotide deletion | Between 35-39 | Unique to known *S. lividus,* plant and protist chloroplast, and cyanelle 5S RNAs |
| Base substitution | A100, A102 | Uncommon in eubacterial 5S RNAs |
| Base pairing helix IV | C·G71 | Uncommon in eubacterial 5S RNAs |
| Base pairing helix V | U U81 | Uncommon in eubacterial 5S RNAs. Characteristic of cyanelle and several cyanobacterial and chloroplast 5S RNAs |

[a] Numbering system used is that shown for the universal 5S RNA structure (see Reference 11).

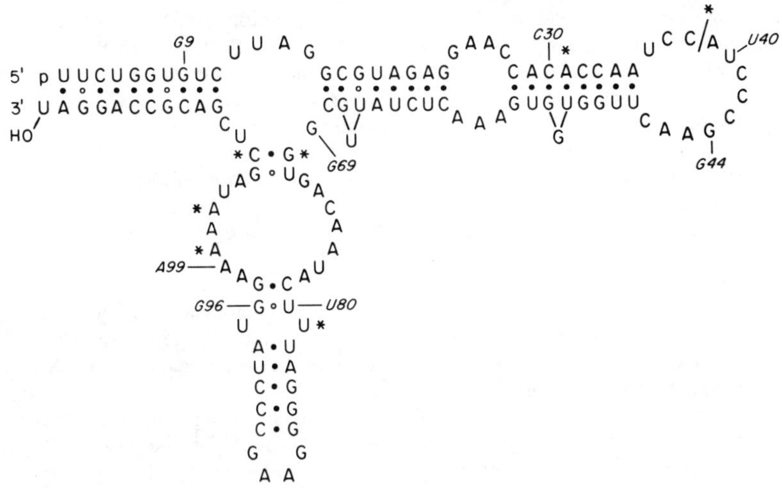

```
                          G9   U U A G                G A A C   C30        U C C/A  U40
                                                                  *                *
     5'  p U U C U G G U G U C     G C G U A G A G     C A C A C C A A         C
         • • • • • • • • •   o     • • o • • • • •     • • • • • • • •         C
     3'  U A G G A C C G C A G     C G U A U C U C  A A A G U G U G G U U   C A A G  C
     HO'                      C   G  \/ U            A        \/          C        G44
                               U   U  U  G69              G
                              *C • G*
                            U A G o U G A
                          * A            C
                            A            A
                          * A            A
                     A99 —— A  A G • C A  U
                      G96 —— G o U —— U80
                              U      U*
                            A • U
                            U • A
                            C • G
                            C • G
                            C • G
                            G     G
                              A A
```

FIGURE 1. Secondary structural model and sequence of the chloroplast 5S RNA from the moss *M. polymorpha*.[13] Asterisks indicate the unique or uncommon positions shared by *S. lividus* and the moss and other chloroplast 5S RNAs. Several universal positions along with the universal numbering system are also shown.

the evidence for a separate subclass of 5S RNA that were mentioned above. An exception that has been revealed by correspondence analysis however is that of the *Clostridium pasteurianum* 5S RNA, which segregates close to the cyanobacterial and plant chloroplast 5S RNAs in the two-dimensional maps (C. Mannella, personal communication). A comparative analysis of the plant chloroplast and *C. pasteurianum* 5S RNAs at each base position reveals that there is an unusual number of matches in 25 positions that are known to be highly variable in eubacterial 5S RNA. *C. pasteurianum* shares none of the deletion or insertion signatures that are specific to the chloroplast and cyanobacterial 5S RNAs. It is assumed that homologies at highly variable positions are a result of sequence convergence and not necessarily a reflection of evolutionary relatedness.


## COMPARISON OF 5S RNAS OF *CYANOPHORA PARADOXA* CYANELLES AND THE CHLOROPLASTS OF *EUGLENA GRACILIS*: A POSSIBLE EXAMPLE OF SEPARATE ENDOSYMBIOTIC EVENTS INVOLVING DIFFERENT PROGENITOR ORGANISMS


Examples of 5S RNA structures that are either very similar or very divergent compared to the cyanobacterium *S. lividus* and other cyanobacterial 5S RNAs are found in the organellar 5S RNAs of the plastids of the protozoa *Cyanophora paradoxa* and *Euglena gracilis*. Earlier work showed the prokaryotic nature of the *C. paradoxa* cyanellar ribosomal RNAs.[15] The sequence of the 5S RNA from *C. paradoxa* cyanelles has recently been determined.[16] Maxwell, Liu, and Shively have shown that this sequence has all the signatures characteristic of the *S. lividus* 5S RNAs, which includes the unique deletion between positions 35-39, the insertion between position 30 and 31, adenine substitutions at residues 100 and 102, and the U U non-Watson-Crick pairing in helix V (FIGURE 2). In addition, its overall sequence homology, compared to the plant chloroplast and eubacterial 5S RNAs, is closest to the 5S RNAs of strains II and III of *S. lividus* (78% and 77% correlations, respectively). The high identity in overall nucleotide sequence and the sharing of unique 5S RNA signatures suggest *S. lividus* as a close relative of the *C. paradoxa* cyanelle endosymbiont. The work of Maxwell *et al.* supports the hypothesis presented by Margulis[2] that *C. paradoxa* cyanelles are a product of a relatively recent endosymbiotic event and an example of a present-day intermediate in chloroplast evolution.

In contrast, the 5S RNA of *Euglena gracilis* chloroplasts has several unusual properties (FIGURE 2). With respect to overall nucleotide sequence homology, it differs greatly from that of the known cyanobacterial, green plant chloroplast, and cyanellar 5S RNAs (TABLE 2). It does have some signatures that are specific to the 5S RNAs of *S. lividus* and the plant chloroplasts, which implies a degree of relatedness to the RNAs from these sources, but it lacks other group-specific signatures. For example, although it has the deletion signature between positions 35-39, the base-substitution signatures (A100 and A102), and the unusual U U pairing in helix V, it does not have the insertion between positions 30 and 31 specific to known cyanobacterial, plant chloroplast, and cyanellar 5S RNAs. The *Euglena gracilis* 5S RNA also has several unique or highly uncommon base substitutions. It is the only 5S RNA that is known to have a base substitution ( U) for the universal position G69, and it does not have the conserved pyrimidines at positions 35 and 36 or several conserved G/Y/ Y/G base pairs in helices IV and V (FIGURE 2). Its unusual 5S RNA sequence may be due to genetic drift, or it arose from an endosymbiosis involving an organism that is distantly related to the extant cyanobacteria whose nucleotide sequences are known. The signature analysis implies that the putative cyanobacterial endosymbiont of *Euglena* represents a separate branch of the cyanobacterial phylogenetic tree. The determination of more 5S RNA sequences from an array of cyanobacteria may reveal if there are extant organisms that have the unusual structural properties found in the 5S RNA of *Euglena gracilis* chloroplasts.

It is important to note from the work of Maxwell *et al.*[16] that the cytoplasmic 5S RNAs of *C. paradoxa* and *E. gracilis* are relatively close in overall sequence homology (79%) and that the *C. paradoxa* 5S RNA sequence is closer to that of *E. gracilis* compared to other known 5S RNAs from eukaryotic sources. Thus these protists appear phylogenetically closely linked. The plastids of these two related organisms may have originated by two different endosymbiotic events which occurred at very different times during the evolution of these organisms.

TABLE 2. Overall Sequence Homology between *Euglena gracilis* Chloroplast 5S RNA and Other RNAs

| Source of 5S RNA | Percent Sequence Homology with 5S RNA of *Euglena* Chloroplast[a] |
|---|---|
| *Anacystis nidulans* | 52 |
| *Synechococcus lividus* III | 59 |
| Spinach chloroplasts | 53 |
| *Cyanophora paradoxa* cyanelles | 62 |
| *Ochromonas* chloroplasts | 58 |

[a] Based on the sequence of *Euglena* 5S DNA.[17]

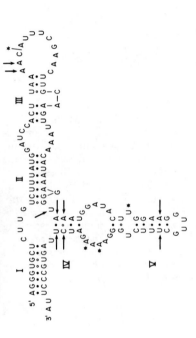

**FIGURE 2.** Nucleotide sequence of the 5S RNA of (A) *Euglena gracilis*,[17] (B) *Cyanophora paradoxa*,[16] and (C) *Synechococcus lividus* III.[18] The arrows point to unusual base substitutions peculiar to the *Euglena gracilis* 5S RNA sequence. Asterisks indicate unique or uncommon positions shared by the chloroplast, cyanellar, and *S. lividus* 5S RNAs.

Additional 5S RNA sequences from another photosynthetic protist, that of *Ochromonas*, a chrysophyte, have been determined (R. Devereux, A. R. Loeblich III, and G. E. Fox, unpublished data). The *Ochromanas* plastid 5S RNA shows a closer homology to that of the green plant chloroplast (72% with spinach chloroplast; 73% with moss chloroplast) than to the *Cyanophora* cyanelle (68%) or the *Euglena* plastid (58%) 5S RNAs. Again, this suggests polyphyletic origins for the organelles of these photosynthetic protists. In contrast to the divergent plastid 5S RNA sequences found among the protists, the known green plant plastid 5S RNA sequences reveal few differences (less than 10 residues per molecule) and some sequences are identical (e.g., from tobacco and spinach).

## MITOCHONDRIAL ORIGINS

At approximately the same time, Dr. Carl Woese and co-workers[19] and our laboratories[20] independently proposed that members of the alpha subdivision of the purple photosynthetic bacteria[21] are phylogenetically related to the green plant mitochondria. Our data, which are based on 5S RNA comparisons, show that both the overall sequence homologies (TABLE 3) and homologies in selected regions of the RNA (namely, regions that have a high concentration of universally conserved positions, TABLE 4) show a closer homology of the plant mitochondrial 5S RNA to the alpha-subdivisional bacteria than to the beta or gamma subdivisions. The alpha subdivision includes the species *Rhodobacter sphaeroides, Paracoccus denitrificans, Rhodospirillum rubrum,* and *Agrobacterium tumefaciens.* This high homology is especially evident in the selected regions that have a high concentration of conserved residues. In these regions there is approximately a 90% identity of the wheat mitochondrial 5S RNA with most alpha-subdivisional species. This contrasts with a 70% homology of the wheat mitochondrial RNA with the 5S RNA of *P. cepacia* (part of the beta

TABLE 3. Nucleotide Differences of 5S RNAs from the Purple Nonsulfur Bacteria and Wheat Mitochondria

| | P.d. | Rb.s. | R.r. | A.t. | Th.t. | Ps.C. | Rc.g. | E.c. |
|---|---|---|---|---|---|---|---|---|
| *P. denitrificans* | — | — | — | — | — | — | — | — |
| *Rb. sphaeroides* | 8 | — | — | — | — | — | — | — |
| *R. rubrum* | 22 | 25 | — | — | — | — | — | — |
| *A. tumefaciens* | 34 | 35 | 22 | — | — | — | — | — |
| *Th. thermophilus* | 45 | 43 | 29 | 33 | — | — | — | — |
| *Ps. cepacia* | 44 | 45 | 39 | 44 | 47 | — | — | — |
| *Rc. gelatinosa* | 42 | 40 | 40 | 48 | 41 | 13 | — | — |
| *E. coli* | 41 | 41 | 43 | 43 | 35 | 40 | 33 | — |
| *T. aestivum* mitochondria | 60 | 60 | 61 | 64 | 64 | 67 | 68 | 68 |

TABLE 4. Nucleotide Differences in Regions of High Homology between 5S RNAs of Purple Bacteria and the Wheat Mitochondrion[a]

| Species | Differences from *T. aestivum* Mitochondrion |
|---|---|
| *Rb. sphaeroides* | 5 |
| *P. denitrificans* | 3 |
| *R. rubrum* | 5 |
| *A. tumefaciens* | 11 |
| *Th. thermophilus* | 8 |
| *Rc. gelatinosa* | 16 |
| *Ps. cepacia* | 18 |
| *E. coli* | 15 |

[a] Sequences represent a total of 51 nucleotide positions.[20]

subdivision of the purple photosynthetic bacteria) in these same selected regions. These regions represent approximately one-half of the molecule. In addition, helices II and III of the molecule reveal structural properties that are similar between the plant mitochondria and the alpha-subdivisional 5S RNAs. The comparisons are based on the wheat mitochondrial 5S RNA sequence. The mitochondrial 5S RNAs from four species of plants have been determined, and these sequences are nearly identical. The conclusions relate to all known plant mitochondrial 5S RNA sequences.

On the other hand, an analysis of another segment of the wheat mitochondrial 5S RNA reveals a striking difference in homology compared to the alpha-subdivisional 5S RNAs. The 5' end (first 24 nucleotides) has only a 30-35% sequence homology with the equivalent 5' region of the alpha-subdivisional 5S RNAs. In sharp contrast to this, the next 28 nucleotides (positions 25-52 of the wheat mitochondrial 5S RNA) show an approximate 90% identity. The remaining portion of the molecule (approximately the 3' half) does not have long stretches that show an extreme divergence in homology. To explain the unusual disparity at the 5' end, we propose that a gene rearrangement took place within the ancestral plant mitochondrion or the progenitor bacterial species whereby the first 24 nucleotide positions of the 5' end of the 5S ribosomal gene was replaced by a genetic segment from another part of the genome. This putative rearrangement produced an aberrant 5S RNA structure (e.g., only 6 base pairs in helix I instead of the usual 9-10) and yet a molecule that can still function.

There is extensive evidence for sequence rearrangements in the plant mitochondrial genome in both the 5S structural gene and the spacer regions.[22] Rearrangements involving the 3' end of the mitochondrial 5S gene have been observed in soybean cells grown in culture.[23] In addition the ribosomal gene order in the plant mitochondrial DNA differs from the ribosomal gene organization in the eubacteria.[24] Considering the extensive evidence for ribosomal gene rearrangements in plant mitochondria, it is not likely that the anomalous 24-base sequence at the 5' end of the mitochondrial 5S gene arose by a sequence rearrangement. A phylogenetic tree of the purple photosynthetic bacteria based on overall sequence homology and the proposed origin of the plant mitochondrion is schematically shown in FIGURE 3. Our data are consistent with the work of others,[26,27] which suggests that *Paracoccus denitrificans* and related species are descendants of the plant mitochondrial endosymbiont.

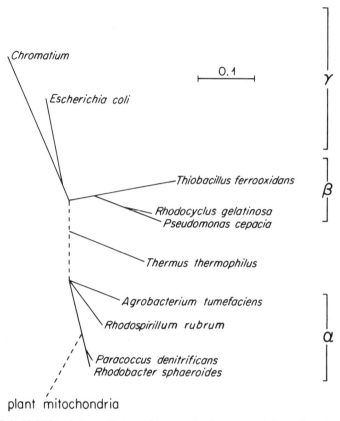

plant mitochondria

**FIGURE 3.** 5S RNA phylogenetic tree of the purple photosynthetic bacteria and plant mito-chondria. Adapted from Stahl *et al.*[25] and Villanueva *et al.*[20] Since the close homologies between the 5S RNAs of the plant mitochondria and the alpha subdivision are based on only given segments of the RNA, the phylogenetic relationship of the mitochondria is shown schematically as a dashed line. The scale bar represents 0.1 nucleotide changes per sequence position.

## CONCLUSIONS

The 5S RNAs from the cyanobacteria, chloroplasts, and cyanelles form a common structural subclass. With respect to primary structure, certain nucleotide insertions and deletions are unique to the RNAs of these sources; in particular, the RNAs from the strains of the cyanobacteria *Synechoccocus lividus,* the green plant and photosyn-thetic protist chloroplasts, and the *Cyanophora paradoxa* cyanelles (Maxwell *et al.*)[16] all share a specific deletion which is unique to the RNAs of these sources. Analysis of denatured and renatured 5S RNA conformations by gel electrophoresis suggests that the cyanobacterial and chloroplast RNAs form a common higher-order structural subclass. The extreme diversity found in the sequences of some of the chloroplast 5S

RNAs of the photosynthetic protists can be explained by polyphyletic origins of the organelles of these organisms.

A comparison of mitochondrial and eubacterial 5S RNA sequences reveals sequence homologies for the plant mitochondrial RNAs that are closest to the alpha 3 subgroup of the purple photosynthetic bacteria (e.g., *Rhodobacter sphaeroides, Paracoccus denitrificans*). Comparisons of particular segments of the RNAs show an unusually low sequence homology at the 5' end of the molecule. This may be attributable to gene rearrangements within the mitochondrial or progenitor genome.

## ACKNOWLEDGMENT

We thank Janet Andersen for constructive criticism of the manuscript.

## REFERENCES

1. FREDRICK, J. F., Ed. 1981. Origins and Evolution of Eukaryotic Intacellular Organelles. Ann. N.Y. Acad. Sci. **361.**
2. MARGULIS, L. 1970. Origins of Eukaryotic Cells. Yale University Press. New Haven, Conn.
3. MANNELLA, C. A., J. FRANK & N. DELIHAS. 1985. Interrelatedness of 5S RNA sequences studied by correspondence analysis. Biophys. J. **47:** 224a.
4. NISHIKAWA, K. & S. TAKEMURA. 1974. Structure and function of 5S ribosomal ribonucleic acid from *Torulopsis utilis.* J. Biochem. Tokyo **76:** 935-947.
5. FOX, G. E. & C. R. WOESE. 1975. 5S RNA secondary structure. Nature **256:** 505-507.
6. BOHM, S., H. FABIAN & H. WELFLE. 1981. Universal secondary structures of prokaryotic and eukaryotic ribosomal 5S RNA derived from comparative analysis of their sequences. Acta Biol. Med. Germany **40:** k19-k24.
7. STAHL, D. A., K. R. LUEHERSEN, C. R. WOESE & R. R. PACE. 1981. An unusual 5S rRNA from *Sulfolobus acidocaldarius,* and its implications for a general 5S rRNA structure. Nucleic Acids Res. **9:** 6129-6137.
8. DELIHAS, N. & J. ANDERSEN. 1982. Generalized structures of the 5S ribosomal RNAs. Nucleic Acids Res. **10:** 7323-7344.
9. DE WACHTER, R., M. CHEN & A. VANDENBERGHE. 1982. Conservation of secondary structure in 5S ribosomal RNA: a uniform model for eukaryotic, eubacterial and organelle sequences is energetically favorable. Biochemie **64:** 311-329.
10. FOX, G. E. 1985. The structure and evolution of archaebacterial ribosomal RNAs. *In* The Bacteria. I. C. Gunsalus, Ed. **8:** 257-310. Academic Press. New York, N.Y.
11. DELIHAS, N., J. ANDERSEN & R. P. SINGHAL. 1984 Structure and evolution of the 5S ribosomal RNAs. Prog. Nucleic Acid Res. Mol. Biol. **31:** 161-190.
12. DELIHAS, N., J. ANDERSEN & D. BERNS. 1985. Phylogeny of the 5S ribosomal RNA from *Synechococcus lividus.* II. The cyanobacterial/chloroplast 5S RNAs form a common structural class. J. Mol. Evol. **21:** 334-337.
13. YAMANO, Y., K. OHYAMA & T. KOMANO. 1984. Nucleotide sequences of chloroplast 5S RNA from cell suspension cultures of the liverworts *Marchantia polymorpha* and *Jungermannia subulata.* Nucleic Acids Res. **12:** 4621-4624.
14. PIELER, T., M. DIGWEED, M. BARTSCH & V. A. ERDMANN. 1983. Comparative structural analysis of cytoplasmic and chlororplastic 5S rRNA from spinach Nucleic Acids Res. **11:** 591-604.
15. SIEBENS, H. C. & R. K. TRENCH. 1978. Aspects of the relation between *Cyanophora*

*paradoxa* (Korschikoff) and its endosymbiotic cyanelles *Cyanocyta korschikoffiana* (Hall and Claus). Proc. R. Soc. Lond. Ser. B. **202:** 463-472.

16. MAXWELL, E. S., J. LIU & J. M. SHIVELY. 1986. Nucleotide sequence of *Cyanophora paradoxa* cellular and cyanelle-associated 5S ribosomal RNAs: the cyanelle as a potential intermediate in plastid evolution. J. Mol. Evol. (In press.)

17. KARABIN, G. D., J. O. NARITA, J. R. DODD & R. B. HALLICK. 1983. *Euglena gracilis* chloroplast ribosomal RNA transcription units. J. Biol. Chem. **258:** 14790-14796.

18. DELIHAS, N., W. ANDRESINI, J. ANDERSEN & D. BERNS. 1982. Structural features unique to the 5S ribosomal RNAs of the thermophilic cyanobacterium *Synecococcus lividus* III and the green plant chloroplasts. J. Mol. Biol. **162:** 721-727.

19. YANG, D., Y. OYAIZU, H. OYAIZU, G. J. OLSEN & C. R. WOESE. 1985. Mitochondrial origins. Proc. Nat. Acad. Sci. USA **82:** 4443-4447.

20. VILLANUEVA, E., K. R. LUEHRSEN, J. GIBSON, N. DELIHAS & G. E. FOX. 1985. Phylogenetic origins of the plant mitochondrion based on a comparative analysis of 5S ribosomal RNA sequences. J. Mol. Evol. **22:** 46-52.

21. WOESE, C. R., E. STACKEBRANDT, T. J. MACKE & G. E. FOX. 1985. A phylogenetic definition of the major eubacterial taxa. System. Appl. Microbiol. **6:** 143-151.

22. BRENNICKE, A., S. MOLLER & P. A. BLANZ. 1985. The 18S and 5S ribosomal RNA genes in *Oenothera* mitochondria: sequence rearrangements in the 18S and 5S rRNA genes. Mol. Gen. Genet. **198:** 404-410.

23. MORGENS, P. H., E. L. GRABAU & R. F. GESTLAND. 1984. A novel soybean mitochondrial transcript resulting from a DNA rearrangement involving the 5S rRNA gene. Nucleic Acids Res. **12:** 5665-5694.

24. BONEN, L. & M. W. GRAY. 1980. Organization and expression of mitochondrial genome of plants. I. The genes for wheat mitocondrial ribosomal and transfer RNA: evidence for an unusual arrangement. Nucleic Acids Res. **8:** 319-335.

25. STAHL, D. A., D. J. LANE, G. J. OLSEN & N. R. PACE. 1984. Analysis of hydrothermal vent-associated symbionts by ribosomal RNA sequences. Science **224:** 407-411.

26. JOHN, P. & F. R. WHATLEY. 1975. *Paracoccus denitrificans* and the evolutionary origins of the mitochondria. Nature **254:** 495-498.

27. DICKERSON, R. E. 1980. Cytochrome c and the evolution of energy. Sci. Am. **242:** 136-153.

# Evolution of Organisms and Organelles as Studied by Comparative Computer and Biochemical Analyses of Ribosomal 5S RNA Structure[a]

VOLKER A. ERDMANN, JÖRN WOLTERS, TOMAS
PIELER, MARTIN DIGWEED, THOMAS SPECHT,
AND NORBERT ULBRICH

*Department of Chemistry*
*Institute for Biochemistry*
*Free University of Berlin*
*Thielallee 63*
*1000 Berlin 33 (Dahlem), Federal Republic of Germany*

## INTRODUCTION

Since the discovery of ribosomal 5S RNA in 1963,[1] this molecule has been the object of numerous investigations, in which attempts were made to elucidate its structure and function[2–6] and, more recently, its evolution.[7–16]

The ribosomal 5S RNA is an essential constituent of the large ribosomal subunits of prokaryotes and eukaryotes. Currently the sequences of 216 eukaryotic, 109 eubacterial, 11 archaebacterial, 11 plastidic, and 4 mitochondrial 5S rRNAs are known.[17,18] The molecules are 120 nucleotides long and have a molecular weight of approximately 40,000 Daltons. Modified nucleotides are usually not observed, with the exception of some in eukaryotic species.

We have been investigating the structure, function, and evolution of 5S rRNA for a number of years using the following methods: chemical and enzymatic modifications, physical characterizations, 50S ribosomal subunit reconstitution experiments, functional analysis, and computer evaluation of 5S rRNA sequences. In this communication we are summarizing our chemical and biochemical structural studies and will show that the results obtained are in line with those derived from computer analysis of 5S rRNA sequences and that a combination of these methods is ideally suited for the investigation of the evolution of organisms and organelles.

[a]The financial support by the Deutsche Forschungsgemeinschaft (Sfb 9/B5) and the Fonds der Chemischen Industrie e.V. is gratefully acknowledged.

# COMPUTER ANALYSIS OF 5S rRNA: ALIGNMENT OF 5S rRNA SEQUENCES AND PROPOSALS FOR SECONDARY STRUCTURAL MODELS

For the alignment of the RNA sequences, the 5S rRNAs were divided into the following four groups: eubacteria including plastids, mitochondria, archaebacteria, and eukaryotes. Computer programs (ALIGNSTAT or ALIGNSEC[19] were used for the alignment of the 5S rRNA sequences, and they were purchased as a software package, named SAGE, from Technoma GmbH, Heidelberg, Germany.[17]

As recently shown,[17] the results of the alignment permitted us to construct minimal structural models for the following 5S rRNA types: eubacteria, eukaryotes, and halophilic-methanogenic archaebacteria. The corresponding secondary structural models are shown in FIGURES 1-3.

The minimal structural models shown in FIGURES 1-3 are in agreement with earlier reported structural data and molecular models as recently discussed.[17] The newly introduced nomenclature for the 5S rRNA structures is based on an earlier proposal and now widely accepted nomenclature for tRNAs.[17,20] The numbering of the nucleotides is from the 5' to 3' end of the 5S rRNA, and the numbers 1 through 120 for the *Escherichia coli* 5S rRNA sequence are used as a basis for all 5S rRNAs.[11] Thus, the conserved G in position 44 of *E. coli* 5S rRNA will also be G44 in eukaryotes and archaebacteria. Positions occurring in other species and not in *E. coli* are indicated by a decimal fraction. The five helices are designated by capital letters A through E, and the loops and single-stranded sequences by the small letters a through e. In referring, for example, to helix B and loop b, one may also refer to segment B, segment B', segment b, and segment b', indicating the corresponding 5' and 3' portions of the segments in the molecule (see FIGURES 1-3).

# CHEMICAL MODIFICATION AND ENZYME DIGESTION OF 5S rRNA

Enzymatic digestion has been used for nearly 15 years in the study of 5S rRNA structure. Early efforts were restricted to RNases, which show a *preference* for single-stranded RNA, and were coupled to laborious and often intrinsically ambiguous sequencing techniques. Over the past few years this important approach has been improved through the availability of RNases with a *specificity* for single- or double-stranded RNA[21,22] and the development of rapid and precise sequencing methods using end-labeled RNA.[23,24] FIGURE 4 shows an example of a state-of-the-art structure examination of 5S rRNA isolated from spinach chloroplasts. Cleavage sites of the single-strand-specific nuclease S1 and the double-strand-specific RNase CSV[25] alternate through the sequence and allow localization of paired and unpaired regions within the molecule. Interpretation of such cleavage patterns is dependent upon previous evaluation of the enzymes by establishing their behavior with model molecules such as synthetic homopolymers and the structurally well-characterized tRNA.[25,26] Summarizing the results of such studies with the enzymes S1 and CSV one can say that S1 nuclease (a) is specific for single-stranded nucleic acid, (b) shows no base specificity, (c) has no minimum required length of single-stranded nucleic acid, (d) does not normally cut regions involved in tertiary interactions, even if these are not of a standard

Watson-Crick base-pair nature, unless disrupted by denaturing agents (e.g., urea, heat), (e) does not recognize unpaired but intercalated bases; and CSV RNase (a) is specific for double stranded RNA, (b) has no base specificity, (c) requires a minimum length of two base pairs in regular helices, (d) may recognize some tertiary interactions

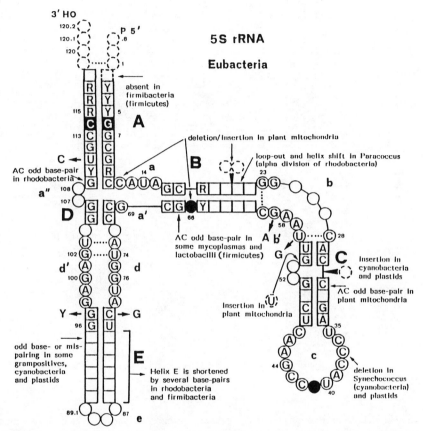

**FIGURE 1.** Minimal model of eubacterial 5S rRNA secondary structure. Squares indicate conserved base pairing; circles unpaired nucleotides. Dotted lines indicate possible helix extensions. Filled squares and circles indicate positions that are unique to one group. Bases indicated in the minimal models are supposedly ancestral for the respective group. Different bases in the majority of species are marked with an arrow. Hypervariable positions remain blank. The occurrence of AC and UU odd base pairs is also marked. (FIGURES 1-3 are reproduced from Reference 17 with permission from the publisher.)

even if they do not involve Watson-Crick base pairing, (e) can cleave helices even if they are "disturbed" by unpaired loopouts or odd base pairs.

The mechanisms of nuclease recognition and cleavage are unclear. Particularly double-stranded nucleic acid presents a complex substrate with two definitive recognition sites: the major and minor grooves. Examination of tRNA cleavage patterns

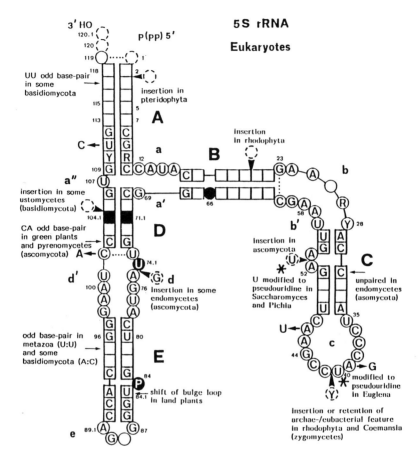

**FIGURE 2.** Minimal model of eukaryote 5S rRNA secondary structure. (See FIGURE 1 for details.)

suggest that the minor groove is the primary recognition site since helices in which the major groove is filled with bases from other parts of the molecule are cut. Furthermore, the uniform and symmetric disposition of hydrogen bond acceptors makes it a good environment for a non-sequence-specific nuclease such as CVS. Single-stranded RNA might be considered a simpler substrate. However, the resistance of polypurine single-stranded regions of 5S rRNA suggests that absence of base pairing alone is not sufficient for S1 cleavage, conformation within the molecule as a whole may allow some single-stranded regions to mimic double-stranded structures.

Examination of the availability of bases in RNA for chemical reaction allows their status to be determined. Depending upon which atom or group of the base is attacked one can monitor either secondary (base-pairing) or tertiary (hydrogen-bonding, general steric disposition) structure.

FIGURE 5 illustrates one simple base-modification system in which the N-1 of adenines is oxidized by monoperphthalic acid.[27,28] Since N-1 of adenine is involved in a hydrogen bond wich N-3 of uracil in the Watson-Crick A:U base pair, resistance to modification indicates involvement in secondary structure. This modification system is complemented by the modification of adenine with diethyl pyrocarbonate which attacks N-7 in the imidazole ring.[29] Since this atom is not involved in hydrogen bonding but does contribute to stacking stability, its resistance to modification may indicate either base pairing or an ordered, stacked arrangement of unpaired bases.

Modification of non-hydrogen-bonded atoms of a base may be coupled to determination of pairing status by compound modification systems such as that illustrated in FIGURE 6.[30] Here the C-6 of cytidines, which is not involved in base pairing and which probably contributes insignificantly to stacking, is sulfonated in the presence of bisulfite. The consequent electronic rearrangement favors hydrolytic deamination

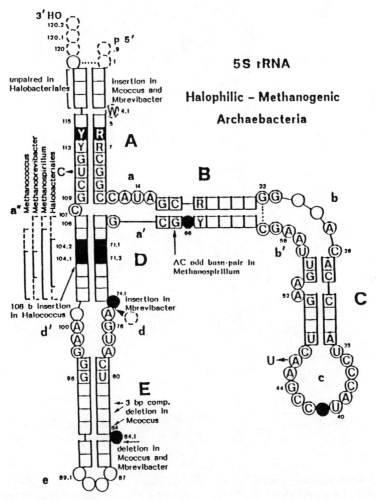

**FIGURE 3.** Minimal model of halophilic-methanogenic archaebacterial 5S rRNA secondary structure. (See FIGURE 1 for details.)

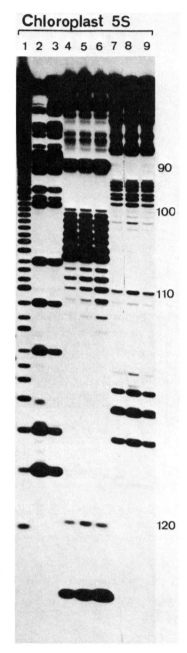

**FIGURE 4.** Probing of RNA secondary structure by use of structure-specific enzymes. 5S rRNA isolated from spinach chloroplast ribosomes has been labeled at its 3'- end with 5'-$^{32}$P-pCp and T$_4$ RNA ligase. Sequence correlation is achieved by limited alkaline hydrolysis (lane 1) and the G-specific enzymatic RNA sequencing reaction (lanes 2 and 3). Digestion with the single-strand-

at C-4 unless the amino group is otherwise stabilized, for example by hydrogen bonding to O=C-6 of guanine in the G:C base pair. The sulfonation of an RNA molecule proceeds rapidly while the deamination requires longer periods, 24 hours for completion in the case of 5S rRNA. Elimination of the sulfonate groups from modified RNA returns cytidines to their original structures but leaves uridines where deamination occurred. These are detected on sequencing gels as "new" uridine bands.

All these chemical modification systems have been applied to model molecules such as tRNA[29] and to 5S rRNA[27,28] particularly under increasingly denaturing conditions.[26] The results of such experiments allow the following summary of modification behavior of RNA:

a. The reactions are base specific.
b. Monitors of base pairing allow estimation of the strength of a base pair.
c. Weak base pairs such as those at helix termini or neighboring destabilizing structures (e.g., bulge loops) are detected.
d. Steric factors may influence modification.
e. Ions bound to the RNA may shield bases from modification.

In addition to bases, other components of nucleic acids may be probed by chemical modification. One such system is the alkylation of phosphates in the phosphodiester backbone.[31] As with the direct monitors of base pairing, resistance of the phosphate to modification indicates involvement of the group in other interactions, although steric effects probably are more pronounced than in base-modification reactions. This is suggested by clusters of less accessible phosphates in contrast to the few individual, inaccessible phosphates found in the same molecule—the latter presumably involved in hydrogen bonding to other groups.

## BIOCHEMICAL ANALYSIS OF 5S RIBOSOMAL RNA SECONDARY/TERTIARY STRUCTURE

Application of the methods discussed in the preceding paragraph on various phylogenetically representative 5S rRNA molecules results in the data displayed for eubacterial and plastid 5S rRNA in FIGURE 7, for eukaryotic 5S rRNA in FIGURE 8, and for archaebacterial 5S rRNA in FIGURE 9.

Positive experimental evidence for the universal helices A, B, C, and E comes from the cleavage studies with the double-strand-specific CSV ribonuclease. The regular digestion patterns with an intense, continuous series of signals (i.e., FIGURE

specific nuclease $S_1$ (lane 4, 100 U $S_1$; lane 5, 200 U $S_1$; and lane 6, 500 U $S_1$) was carried out in 35 $\mu$l buffer $S_1$ (100 mM NaCl, 5 mM $ZnCl_2$, 30 mM NaAc pH 4.9, 5% glycerol) for 30 minutes at 25°C and stopped by ethanol precipitation. Hydrolysis with the double-strand-specific ribonuclease CSV (lane 7, 0.1 U CSV; lane 8, 0.2 U CSV; and lane 9, 0.5 U CSV) was performed in 100 $\mu$l buffer CSV (50 mM Tris-HCl pH 6.8, 50 mM NaCl, 5 mM $MgCl_2$) for 30 minutes at 25°C and stopped by ethanol precipitation. Samples were directly loaded onto 25% sequencing gels.

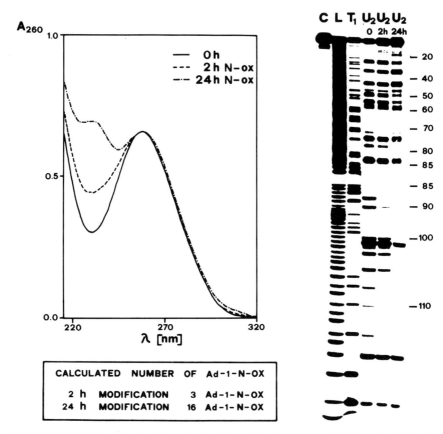

**FIGURE 5.** Modification of adenines in the N-1 position by monoperphthalic acid (MPPA). *Xenopus laevis* oocyte 5S rRNA, 2.5 nmol (100 μg), was incubated in 75 μl 0.4 M NaPO$_4$ pH 7 at 20°C for 2 and 24 hours; reactions were stopped by Sephadex G 50 chromatography in H$_2$O. The number of modified nucleotides can be calculated from the respective UV spectra (A) due to the absorption maximum of AMP-1N oxide at 232 nm. Quantitative analysis of these data performed as described[28] estimates the number of modified adenines per molecule 5S rRNA as 3 after 2 hours and 16 after 24 hours reaction with MPPA. Identification of adenine 1-N oxides relative to the 5S rRNA primary sequence is achieved by use of the adenine-specific RNase U2 sequencing reaction (B), since modified nucleotides are no longer recognized by the enzyme. Results for *Xenopus laevis* oocyte 5S rRNA are indicated in FIGURE 8.

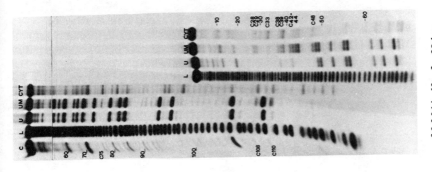

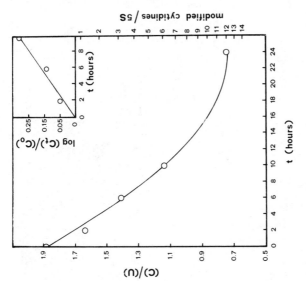

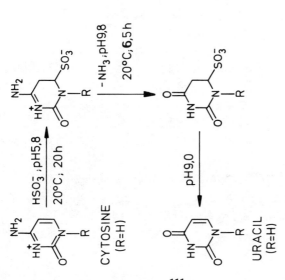

**FIGURE 6.** The bisulfite-mediated deamination of cytosine. The reaction is shown on the left: RNA is incubated in the presence of 3 M bisulfite for 20 hours at 20°C, excess reagent is removed by column chromatography, and the RNA further incubated in alkaline buffer to eliminate sulfonates. The kinetics of modification of 5S rRNA were monitored by base composition analysis after increasing incubation time as shown in the middle. On the right is a sequencing gel on which modified and unmodified spinach chloroplast 5S rRNA was loaded in parallel to detect "new" uridines arising from deamination of unpaired cytidines. C, control; L, nucleotide ladder; U, uridine-specific chemical sequencing reaction of unmodified RNA; UM, uridine sequencing reaction on modified RNA.

111

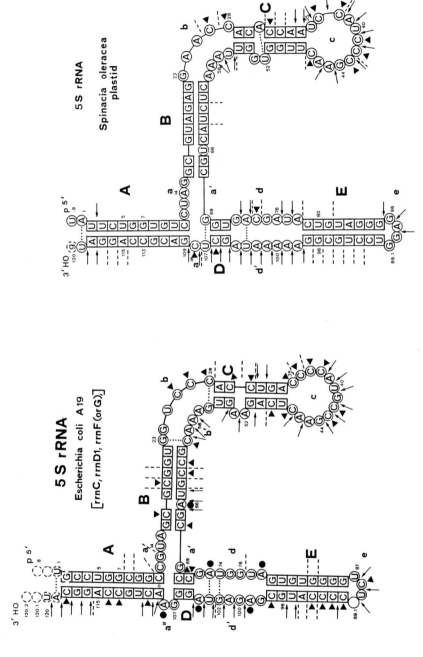

**FIGURE 7.** Structural analysis of eubacterial and chloroplast 5S rRNA. Results from enzymatic hydrolysis with the single-strand-specific nuclease S₁ arrows and the double-strand-specific ribonuclease CSV (dashed lines) as well as from base-specific chemical modification with Na-bisulfite (triangles) and MPPA (circles) are indicated.

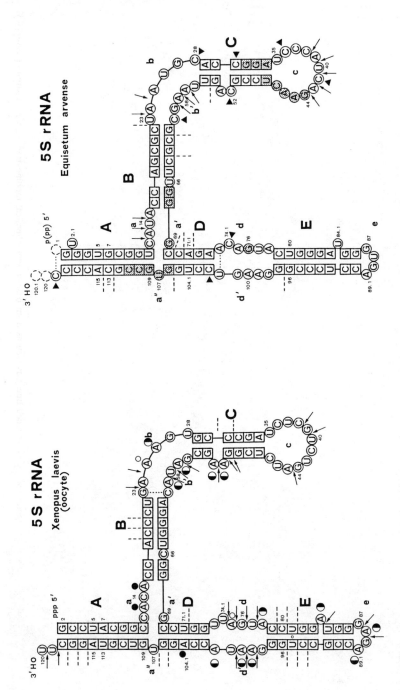

**FIGURE 8.** Structural analysis of eukaryotic 5S rRNA. Results from enzymatic hydrolysis with the single-strand-specific nuclease $S_1$ (arrows) and the double-strand-specific ribonuclease CSV (dashed line) as well as from basic specific chemical modification with MPPA (filled circles), diethylpyrocarbonate (half-filled circles), or Na-bisulfite (triangles) and phosphate modification studies with ethylnitrosourea (open circles) are indicated.

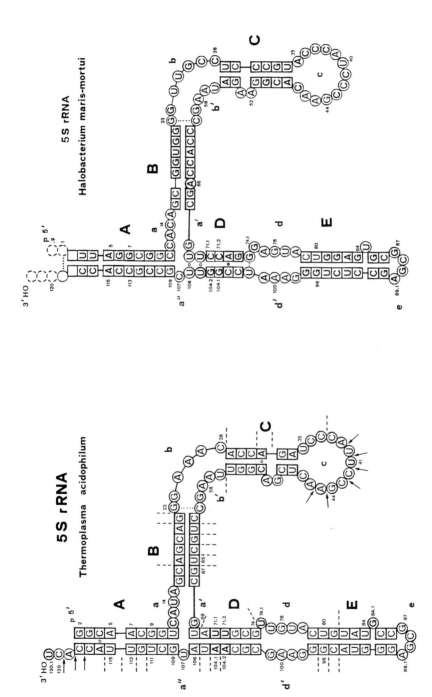

**FIGURE 9.** Structural analysis of archaebacterial 5S rRNA. Results from enzymatic hydrolysis with the single-strand-specific nuclease S$_1$ (arrows) and the double-strand-specific ribonuclease CSV (dashed lines) are indicated.

4) for helices A and E, as well as for helix D in eukaryotic and archaebacterial 5S rRNAs, suggest that these regular double-stranded elements are not involved in the formation of tertiary interactions or "buried" within the compact RNA molecule. Helices A and E might be coaxially stacked,[32] supported by stacking of the polypurine loop d'. Others propose the formation of a helix D equivalent structure in eubacterial 5S rRNA under the assumption of non-Watson-Crick base pairs (including the two conserved G-C base-pairs[33] (FIGURES 1, 6, and 7). Helices B and C as well as the eukaryotic helix E contain "looped-out" nucleotides; the exact nature of these structures, in particular whether they are actually "intercalating" rather than "looping out," remains to be determined. Comparative chemical modification studies[28] have led to the idea that there exists a dynamic equilibrium (influenced by the identity of the neighboring base pairs) between these two extreme conformations. Looped-out bases have also been discussed as protein recognition signals.[34]

Hairpin loop e as well as the internal loops a, b, b', and d' have unambiguously been identified as single stranded by chemical modification or digestion with the single-strand-specific nuclease S1. However, some of these structures are resistant to enzymatic hydrolysis, albeit modified with the base-specific reagents (i.e., loop a, part of loops b and d). As discussed above, these regions will probably be folded into the interior domains within the tertiary structure, rendering them invisible for their relatively large enzyme. Most interesting (because mainly unresolved) are the structures of the largest hairpin loop c and of the interior loop d. Major parts of loop c do not react with the single-strand-specific nucleases. Several laboratories, including our own,[32,35,36] have suggested different forms of base pairing between loop c and loop d. Hancock and Wagner have provided direct experimental evidence for a close proximity of these two domains by cross-linking G41 to G72 in *E. coli* 5S rRNA.[37] Thus, it is very likely that loop c and possibly also loop d are involved in the formation of tertiary interactions; these, however, must not necessarily be of a Watson-Crick base pair character and they might involve functional groups of the nucleobases as well as phosphates. In this respect it is interesting to note protection of phosphate groups located within the helix C/loop c as well as within the helix A/B/C domains of *Equisetum arvense* 5S rRNA (FIGURE 9).

## CLADISTIC ANALYSIS OF 5S rRNA SECONDARY AND PRIMARY STRUCTURE

Phylogenetic systematics *sensu* Hennig[38] (cladistics) is based on the distinction between plesiomorphic (primitive, ancient) and apomorphic (derived) characters. Monophyletic taxa (natural groups) are only defined by synapomorphic characters, i.e., derived characters that are present in all members of the taxon. The application of cladistics to nucleic acid (or protein) secondary and primary structure requires finding characteristics of low variability. One of the alternatives should easily be definable as plesiomorphic. Best suited are insertions and deletions, as already used in the past (TABLE 1). A second type of characteristic is odd base pairs versus Watson-Crick base pairs (TABLE 2). Even single bases of low variability (low number of parallel and back mutations) can be of phylogenetic significance (TABLE 3). So-called signature nucleotides have been defined for special taxa by Küntzel *et al.*[14] and Delihas and Andersen[11] and even "signature nucleotide combinations" by Walker.[39] But all

**TABLE 1.** Insertions and Deletions in 5S rRNA[a]

| Position | Plesiomorphic Character | Apomorphic Character | Variability | Monophyletic Taxons |
|---|---|---|---|---|
| 1 | N | DEL | 3 | 1. Thermoplasma<br>2. eukaryota |
| | | INS | | 3 orders of Zygomycota [8] (Mucorales, Entomophthorales, Harpellales) |
| 4.1 | — | INS | 2 | 1. Methanococcales and -bacteriales [3]<br>2. Thermococcus |
| 5.1/5.2 | — | INS | 1 | Octopus Spring Isolate 1 (arc) |
| 6/114 | NN | DEL | 2 | 1. eukaryota<br>2. Thermoplasma |
| 17.1 | — | INS | 1 | Octopus Spring Isolate 1 (arc) |
| 20.1 | — | INS | 2 | 1. Rhodophyta [3]<br>2. mt Angiospermae [4] |
| 30.1 | — | INS | 1 | Cyanobacteria/pt [15] |
| 36 | C | DEL | 1 | Synechococcus/pt [13] |
| 41 | Y | DEL | 3 | Octopus Spring Isolate 1 (arc) and eukaryota |
| | | INS | | 1. Rhodophyta [3]<br>2. Coemansia (Zygomycota) |
| 52.1 | — | INS | 1 | Endomycetidae and Ascomycetidae (Plecto-, Pyreno-, Disco-, and Hyphomycetes) [31] |
| 66 | A | DEL | 2 | 1. Sulfolobus<br>2. Thermoplasma |
| 74.1 | N | DEL | 2 | 1. Sulfolobus<br>2. eubacteria |
| 74.2 | — | INS | 2 | 1. Methanobrevibacter<br>2. Saccharomycetales (Ascomycota) [7] |
| 84.1 | U | DEL | 2 | 1. Methanococcales and -bacteriales<br>2. eubacteria |
| 104.2 | N | DEL | 1 | Octopus Spring Isolate 1 (arc) |
| 104.3 | — | INS | 1 | A group of ustomycetes not yet named comprising Microbotryum, Rhodosporidium, Aessosporon, Ustilago scabiosae, Sphacelotheca, Rhizoctonia crocorum and hiemalis, Pachnocybe and Agaricostilbum [10] |
| 107 | U | DEL | 1 | Sulfolobus |
| 108 | — | INS<br>DEL | 2 | eubacteria<br>Vibrio marinus (Rhodobacteria gamma) |
| 114.1 | — | INS | 1 | Sulfolobus |

[a] The number in parentheses following the taxon name refers to the number of different 5S rRNA sequences. Abbreviations: arc, archaebacteria; INS, insertion; DEL, deletion; N, variable position; pt, plastid; mt, mitochondrion.

of them lack the distinction between the primitive and the derived condition necessary for real significance, or they predefine groups that are clearly polyphyletic.

Eukaryotes as well as eubacteria clearly represent natural groups defined by a large number of derived characteristics, while the monophyletic nature of the archaebacteria has been recently questioned.[40–42] A Ur-5S rRNA (FIGURE 10) has been constructed by considering a characteristic to be plesiomorphic if it is present in most archaebacteria and either eukaryotes or eubacteria.[43] For this reason the Ur-5S rRNA resembles most an archaebacterial 5S rRNA, but only one archaebacterial species shares the exact number of positions, namely, *Methanospirillum*. The region of helix D is still uncertain. Eukaryotes are defined by three derived characteristics: (1) a deletion of base pair 6/114 in helix A (shared with the archaebacterium *Thermoplasma*); (2) the deletion of position 41 in loop c (shared with the archaebacterium Octopus Spring

TABLE 2. Odd Base Pairs in 5S rRNA[a]

| Position | Plesiomorphic Character | Apomorphic Character | Variability | Monophyletic Taxons |
|---|---|---|---|---|
| 3/117 | NN | UU | 2 | 1. A group of basidiomycota preliminarily named Doliporomycetes [28]<br>2. Nadsonia (Ascomycota) |
| 73/103 | NN | AC | 2 | 1. Chlorobiota [33]<br>2. Pyrenomycetes and Monilinia, Aureobasidium [12] |
| 81/95 | NN | UU | ? | 1. Metazoa [68]<br>2. Dicyema (Mesozoa) |
| | | UU/YY | | 1. Gram positives [18 of 25]<br>2. Rhodobacteria alpha [1 of 9]<br>3. Rhodobacteria beta [2 of 9]<br>4. Cyanobacteria and plastids [10 of 15] |
| | | AC | | A group of higher basidiomycetes including rusts, not yet named [13] |

[a] Odd base pairs are defined to be able to compensate for Watson-Crick base pairs not disturbing the helical conformation so that double-strand-specific nuclease will still work. This has been shown for AC and UU. For abbreviations see TABLE 1.

Isolate 1), which has been reinserted twice—in rhodophytes (red algae) and the zygomycete *Coemansia;* (3) deletion of a base pair (71.2./104.2) in helix D (a deletion in the 3′ segment of helix D is also apparent in the archaebacteria Octopus Spring Isolate 1, *Sulfolobus,* and *Thermococcus,* but these all adopt different base-pairing schemes).

Nevertheless, a sister group of eukaryotes cannot be determined by 5S rRNA analysis, yet. Eubacteria are defined by two derived characteristics: (1) deletion of position 84.1 (shared with the archaebacterial orders Methanococcales and -bacterials); (2) rearrangement of helix D and vicinity resulting in 2 Watson-Crick base pairs: (a) deletion of three bases in the 5′ segment (71.1, 71.2, 74.1); (b) net deletion of one base in the 3′ segment (104.2, 104.1, insertion of 108).

The structural difference between eukaryotic and eubacterial 5S rRNA has been discussed in detail in Reference 6. The structural variability of archaebacterial 5S rRNA is greatest, which is expected for the group believed to be most ancient.

Within the archaebacteria, the "thermoacidophiles" (represented by *Sulfolobus,* Octopus Spring Isolate 1, *Thermococcus,* and *Thermoplasma*) show the highest structural diversity in 5S rRNA, while the halophile-methanogens show similar features. The latter group must be monophyletic, whereas the thermoacidophilic condition is plesiomorphic. *Thermoplasma* clusters with the halophile-methanogens instead, when based on 16S oligonucleotide analysis. The rest, given the name eocytes by Lake,[41]

TABLE 3. Bases of Low Variability in 5S rRNA[a]

| Position | Plesiomorphic Character | Apomorphic Character | Variability | Monophyletic Taxons |
|---|---|---|---|---|
| 37 | U/C | A | 2 | 1. pt Euglena [3]<br>2. Streptomyces (Actinobacteria) |
| 39 | A | G | 3-5 | 1. Methanobrevibacter<br>2. Thermococcus<br>3.-5. eukaryotic group H [164] |
| 43 | C | A | 4 | 1. Octopus Spring Isolate 1 (arc)<br>2. Dipsacomyces/Linderina (Zygomycota) [2]<br>3. Artemia (Crustacea)<br>4. Chlorobiota [33] |
| 45 | A | C | 2 | 1. Chrysophyta [2] and Oomycetes [2]<br>2. Asomycetidae (Plecto-, Pyreno, Disco-, and Hyphomycetes) [22] |
| 46 | A | U | 2-4 | 1. Methanococcales and -bacteriales [3]<br>2.-4. eukaryotic group H [164] |
| 47 | C | U | 1 | Kinetoplastida [2] and Euglena [2] |
| 69 | G | U | 1 | pt Euglena [3] |
| 77 | U | C | 1 | Sulfolobus |
| 78 | A | C | 1 | Sulfolobus |
| 99 | A | G | 1 | Sulfolobus |

[a] Eukaryotic group H comprises: (a) Chrysophyta [2] and Oomycetes [2]; (b) Dictyostelium, Physarum, Amoebidium; (c) Cryptophyta; (d) Chytridio-[2], Zygo-[11], Asco-[33], Basidiomycota [39]; (e) Kinetoplastida [2] and Euglena [2]; and (f) Metazoa except Haliclona (Porifera) [67]. For abbreviations see TABLE 1.

have the characteristics of being sulfur dependent, but again this seems to be plesiomorphic.

We should be very cautious not to name groups by primitive characteristics. A cladistic analysis of 16S rRNA sequences reveals five positions in which eukaryotes share a nucleotide with the "eocytes/sulfur dependents," while at these positions eubacteria share another with the halophile-methanogens (TABLE 4). This would support the proposal of Zillig[40] and Lake[41] that eukaryotes are closer to eocytes/sulfur dependents and that eubacteria are closer to halophile-methanogens. But again,

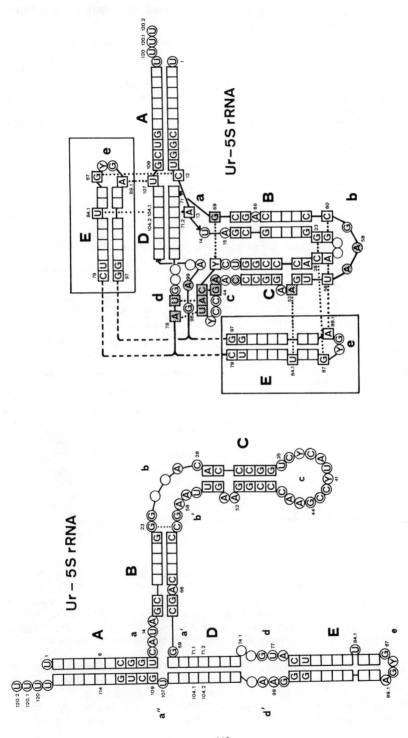

**FIGURE 10.** A Ur-5S rRNA has been constructed by considering a characteristic to be plesiomorphic if it is present in most archaebacteria and either eukaryotes or eubacteria.[43] For other details see text.

119

one has to decide which conditions one considers to be derived. At this point there is no help from sequence data—it is more or less a matter of opinion. 5S rRNA data show the halophilic-methanogenic condition to be derived, predicting a common ancestry with eubacteria. But this does not give any arguments for a relationship between eukaryotes and any of the sulfur-dependent archaebacteria. There is one additional base pair in 16S rRNA (positions 1413/2094 in the alignment of the 1986 compilation,[44] where a nucleotide is shared between eukaryotes and *Thermoproteus* but not with the other sulfur-dependent species, *Sulfolobus* (TABLE 4). This fact clearly proves that even the sulfur-dependent archaebacteria are no natural group. FIGURES 11 and 12 show phylogenies of eukaryotes and eubacteria based on the cladistic analysis. 5S rRNA data can be used to evaluate three far-reaching proposals concerning the phylogeny of eubacteria and eukaryotes:

TABLE 4. Bases of Low Variability in 16S rRNA that Show the Paraphyletic Character of Archaebacteria[a]

| Position | Plesiomorphic Character | Apomorphic Character | Variability | Monophyletic Taxon |
|---|---|---|---|---|
| 1413/2098 | AU | GC CG | 2 | Thermoproteus and eukaryotes eubacteria |
| 1440/1923 | CG | UA | 1 | halophile-methanogens and eubacteria |
| 2012 | G | C | 1 | halophile-methanogens and eubacteria |
| 2043 | G | C | 1 | halophile-methanogens and eubacteria |
| 2044 | G | C | ? | halophile-methanogens |
| | | Y | | Bacillus and Mycoplasma (Gram positives) Rhodobacteria (purple) and Desulfovibrio Bacteroides and Flavobacterium |
| | | G | | 1. Heliobacterium (Gram positives?) 2. Myxococcus (Myxobacteria) 3. Anacystis (Cyanobacteria) |

[a] The numbering is according to the annual compilation of 16S rRNA sequences.[44]

1. Since the discovery of the photosynthetic prokaryote *Prochloron,* it has been claimed to be the ancestor of chlorophyte plastids (chloroplasts) as both contain chlorophyll b instead of d as an accessory pigment. A cladistic analysis can rule out this possibility. Delihas found that the sister group of chloroplasts is comprised of cyanobacteria of the *Synechococcus* type, sharing a special deletion.[45] The occurrence of chlorophyll b in *Prochloron* must be a parallel mutation.
2. Demoulin has proposed a rhodophyte ancestry for asco- and basidiomycota.[46,47] A cluster analysis of 5S rRNA sequences favors an ancestry together with zygomycota from flagellated chytridiomycota as shown by Walker.[48]
3. Taylor[49] and Stewart and Mattox[50,51] postulate a polyphyletic origin for mitochondria; the three different types with lamellar, tubular, and vesicular cristae

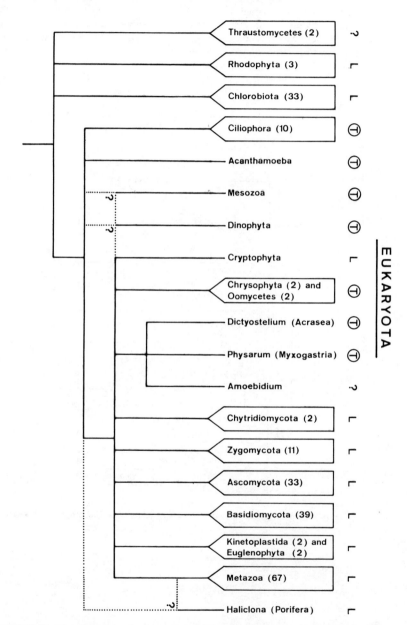

**FIGURE 11.** The phylogeny of eukaryotes based on cladistic analysis. For other details see text.

should have arisen independently from purple bacteria of the alpha subdivision with the respective cristae. This trichotomy would then be reflected in the eukaryote phylogeny. 5S rRNA data do not support this idea, but indicate that organisms with mitochondria having tubular cristae are rather derived from those with lammellar cristae, either once or twice: Heterokontophytes with their unpigmented relatives and ciliates.

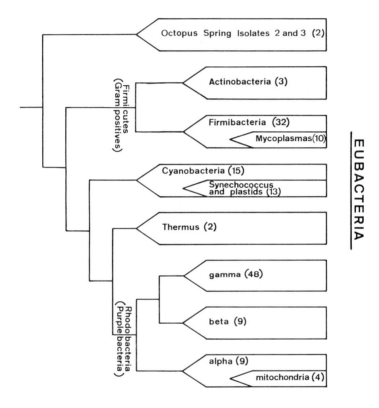

**FIGURE 12.** The phylogeny of eubacteria based on cladistic analysis. For other details see text.

## SUMMARY

The results documented in this publication demonstrate that for evolutionary studies the ribosomal 5S rRNA is a suitable object for such an investigation and that as many methods as possible should be consulted. In this study the results of biochemical and chemical experiments were combined with those of computer sequence analyses, and they revealed that these methods complement each other nicely. We are currently at a state at which we are able to well define the secondary structures of the 5S rRNAs for eubacteria, organelles, archaebacteria, and eukaryotes and we are even able to propose a secondary structure for a Ur-5S rRNA. It is also clear that in the future the present studies should be continued and extended in such a way that the tertiary structures of these molecules will become known.

## ACKNOWLEDGMENT

We would like to thank I. Brauer for typing the manuscript and A. Schreiber for the drawing of the figures.

## REFERENCES

1. ROSSET, R., R. MONIER & J. JULIEN. 1963. Bull. Soc. Chim. Biol. **46:** 87-109.
2. ERDMANN, V. A. 1976. Prog. Nucleic Acids Res. Mol. Biol. **18:** 45-90.
3. ERDMANN, V. A., B. APPEL, M. DIGWEED, D. KLUWE, S. LORENZ, A. LÜCK, A. SCHREIBER & L. SCHUSTER. 1980. *In* Genetics and Evolution of RNA Polymerase, tRNA and Ribosomes. S. Osawa, H. Ozeki, H. Uchida & T. Yura, Eds.: 553-568. Elsevier/North-Holland Biomedical Press. Amsterdam, Holland.
4. WITTMANN, H. G. 1983. Annu. Rev. Biochem. **52:** 35-65.
5. PIELER, T., M. DIGWEED & V. A. ERDMANN. 1984. *In* Gene Expression. B. F. C. Clark & H. F. Petersen, Eds. Alfred Benzon Symposium **19:** 353-376. Munksgaard. Copenhagen, Denmark.
6. ERDMANN, V. A., T. PIELER, J. WOLTERS, M. DIGWEED, D. VOGEL & R. HARTMANN. 1986. *In* Structure, Function and Genetics of Ribosomes. B. Hardesty & G. Kramer, Eds.: 164-183. Springer Publisher. New York, N.Y.
7. FOX, G. E. & C. R. WOESE. 1975. Nature **256:** 505-507.
8. HORI, H. 1975. J. Mol. Evol. **7:** 75-86.
9. HORI, H., T. ITOH & S. OSAWA. 1982. Bakt. Hyg. I. Abt. Orig. C **3:** 18-30.
10. STUDNICKA, G. M., F. A. EISERLING & J. A. LAKE. 1981. Nucleic Acids Res. **9:** 1885-1904.
11. DELIHAS, N. & J. ANDERSEN. 1982. Nucleic Acids Res. **10:** 7323-7344.
12. MACKAY, R. M., D. F. SPENCER, M. N. SCHNARE, W. F. DOOLITTLE & M. W. GRAY. 1982. Can. J. Biochem. **60:** 480-489.
13. DE WACHTER, R., M. W. CHEN & A. VANDENBERGHE. 1982. Biochemie **64:** 311-329.
14. KÜNTZEL, H., B. PIECHULLA & U. HAHN. 1983. Nucleic Acids Res. **11:** 893-900.
15. MACDONELL, M. T. & R. R. COLWELL. 1985. System. Appl. Microbiol. **6:** 171-182.
16. WOLTERS, J. & V. A. ERDMANN. 1985. *In* The Role of Data in Scientific Progress. P. S. Glaeser, Eds.: 103-109. Elsevier Science Publisher B. V., North-Holland. Amsterdam, Holland.
17. ERDMANN, V. A. & J. WOLTERS. 1986. Nucleic Acids Res. **14:** r1-r59.
18. WOLTERS, J. & V. A. ERDMANN. 1986. Berlin RNA Databank. (Data are available on request on magnetic tapes or diskettes.)
19. KRÜGER, M. & G. OSTERBURG. 1983. Comp. Prog. Biomed. **16:** 68-70.
20. SPRINZL, M., J. MOLL, F. MEISSNER & T. HARTMANN. 1985. Nucleic Acids Res. **13:** r1-r49.
21. VOGT, V. M. 1973. Eur. J. Biochem. **33:** 192-200.
22. VASSILENKO, S. & V. BABKINA. 1965. Biokhimiya **30:** 705-712.

23. DONNIS-KELLER, H., A. M. MAXAM & W. GILBERT. 1977. Nucleic Acids Res. **4:** 2527-2538.
24. PEATTIE. D. A. 1979. Proc. Nat. Acad. Sci. USA **76:** 1760-1764.
25. DIGWEED, M., T. PIELER, D. KLUWE, L. SCHUSTER, R. WALKER & V. A. ERDMANN. 1986. Eur. J. Biochem. **154:** 31-39.
26. PIELER, T., M. DIGWEED & V. A. ERDMANN. 1985. J. Biomol. Struct. Dynamics **3:** 495-514.
27. CRAMER, F. & V. A. ERDMANN. 1968. Nature **218:** 92-93.
28. PIELER, T., A. SCHREIBER & V. A. ERDMANN. 1984. Nucleic Acids Res. **12:** 3115-3126.
29. PEATTIE, D. A. & W. GILBERT. 1980. Proc. Nat. Acad. Sci. USA **77:** 4679-4682.
30. SHAPIRO, R., B. I. COHEN & R. E. SERVIS. 1970. Nature **227:** 1047-1048.
31. McDOUGAL, J. & R. N. NAZAR. 1983. J. Biol. Chem. **258:** 5256-5259.
32. PIELER, T. & V. A. ERDMANN. 1982. Proc. Nat. Acad. Sci. USA **79:** 4599-4603.
33. STAHL, D. A., K. R. LUEHRSEN, C. R. WOESE & N. R. PACE. 1981. Nucleic Acids Res. **9:** 6129-6137.
34. PEATTIE, D., S. DOUTHWAITE, R. A. GARRETT & H. F. NOLLER. 1981. Proc. Nat. Acad. Sci. USA **78:** 7331-7335.
35. JAGADEESWARAN, P. & J. D. CHERAYIL. 1980. J. Theor. biol. **83:** 369-375.
36. BÖHM. S., H. FABIAN & H. WELFLE. 1982. Acta Biol. Med. Germany **41:** 1-16.
37. HANCOCK, J. & R. WAGNER. 1982. Nucleic Acids Res. **10:** 1257-1269.
38. HENNIG. W. 1966. Phylogenetic Systematics. University of Illinois Press. Urbana, Chicago & London.
39. WALKER, W. F. 1985. Biosystems **18:** 269-278.
40. ZILLIG, W., R. SCHNABEL & J. TU. 1982. Naturwissenschaften **69:** 197-204.
41. LAKE, J. A., E. HENDERSON, M. OAKES & M. W. CLARK. 1984. Proc. Nat. Acad. Sci. USA **81:** 3786-3790.
42. LAKE, J. A., M. W. CLARK, E. HENDERSON, S. P. FAY, M. OAKES, A. SCHEINMAN, J. P. THORNBER & R. A .MAH. 1985. Proc. Nat. Acad. Sci. USA **82:** 3716-3720.
43. WOLTERS, J. & V. A. ERDMANN. 1986. J. Mol. Evol. **24:** 152-166.
44. HUYSMANS, E. & R. DEWACHTER. 1986. Nucleic Acids Res. **14:** r73-118.
45. DELIHAS, N., W. ANDRESINI, J. ANDERSEN & D. BERNS. 1982. J. Mol. Biol. **162:** 721-727.
46. DEMOULIN, V. 1975. Bot. Rev. **40:** 315-345.
47. DEMOULIN, V. 1985. Biosystems **18:** 347-356.
48. WALKER, W. F. 1984. System. Appl. Microbiol. **5:** 448-456.
49. TAYLOR, F. J. R. 1978. Biosystems **10:**67-89.
50. STEWART, C. D. & K. MATTOX. 1980. *In* Phytoflagellates. E. R. Cos, Ed. **2:** 433-462. Elsevier/North-Holland, Inc. New York, N. Y.
51. STEWART, K. D. & K. MATTOX. 1984. J. Mol. Evol. **21:** 54-57.

# Structural Diversity of Eukaryotic Small Subunit Ribosomal RNAs[a]

## Evolutionary Implications

MITCHELL L. SOGIN AND JOHN H. GUNDERSON

*Department of Molecular and Cellular Biology*
*National Jewish Center for Immunology and Respiratory Medicine*
*1400 Jackson Street*
*Denver, Colorado 80206-1997*

## INTRODUCTION

The RNA components of the ribosome are among the most evolutionarily conserved macromolecules in all living systems. Their functional roles in primitive information processing systems must have been well established in the earliest common ancestors of the eubacteria, archaebacteria, and eukaryotes. Ribosomal RNA genes in all contemporary organisms share a common ancestry, and they do not appear to undergo lateral gene transfer between species. Because of functional constraints, large portions of rRNA genes are well conserved and their sequences can be used to measure phylogenetic distances between even the most distantly related organisms.

The first homology measurements between ribosomal RNA genes were obtained from DNA/RNA hybridization experiments;[1,2] however the interspersion of conserved and nonconserved nucleotide sequences in ribosomal RNA genes make it difficult to interpret the hybridization data.[3–5] Furthermore, this methodology cannot be used to resolve complex phylogenetic relationships and experimental pairwise comparisons must be made for all organisms considered in a given analysis. These difficulties can be avoided by comparing complete ribosomal RNA sequences when estimating the extent of genetic similarity between organisms. An impressive number of 5S and 5.8S rRNA sequences have now been reported in the literature.[6–8] Their relatively small size (fewer than 120 nucleotides for 5S and 170 nucleotides for 5.8S rRNAs) has facilitated complete sequence determinations, but the utility of 5S and 5.8S rRNAs as phylogenetic markers is compromised by their limited number of independently variable positions. In comparisons among organisms separated by vast evolutionary distances, the majority of mutable sites in the 5S and 5.8S rRNAs may have changed multiple times. Moreover, for closely related species, the number of mutational differences may be too few to be statistically meaningful.

The limitations of comparative analyses between 5S and 5.8S rRNAs were recognized by Woese and his colleagues who turned their attention to characterizations

[a]This work was supported by Grant GM 32964 from the National Institutes of Health.

125

of small subunit rRNAs (SSU rRNAs). These RNAs have an increased number of independently variable positions (relative to 5S and 5.8S rRNAs), but because of their size, it has not been possible to determine full sequences from a large number of organisms. As an alternative, approximately 25% of the evolutionary information in SSU rRNAs can be obtained by determining the composition of oligonucleotides released after digesting radiolabeled SSU rRNAs with T₁ ribonuclease. In one remarkable study, well over 200 SSU rRNAs were analyzed and the comparisons revealed that instead of two major kingdoms, living systems could be divided into three major evolutionary lineages.[9] This viewpoint has now been substantiated on the basis of a large number of biochemical analyses.[10]

## CLONING AND SEQUENCING STRATEGIES

The first complete SSU rRNA gene sequence was determined for *Escherichia coli*.[11] A comparison of this sequence with the oligonucleotide catalogue data revealed that "universally conserved" elements (short sequences that appear to be conserved in all organisms) are distributed along the entire length of the *E. coli* SSU rRNA. Similar sequence analyses of SSU rRNA coding regions from *Saccharomyces cerevisiae*,[12] *Xenopus laevis*,[13] *Dictyostelium discoideum*,[3] *Halobacterium volcanii*,[14] and from a number of mitochondrial[15-18] and chloroplast genomes[19-21] confirmed this observation and identified the existence of "kingdom-specific" conserved elements (sequences that are conserved only in the eubacteria, the archaebacteria, or the eukaryotes, respectively). The universal and kingdom-specific conserved regions can be used for the rapid determination of complete SSU rRNA sequences. We have synthesized a collection of DNA oligomers (15-18 nucleotides in length) that are complementary to coding and noncoding strands of universally conserved and eukaryote-specific regions. These conserved sequence elements are strategically located in eukaryotic SSU rRNAs and can be used as primer sites in the base-specific, dideoxynucleotide chain termination sequencing method.[22]

Initially, clones containing ribosomal RNA genes were identified in genomic libraries constructed in pBR or Charon vectors. Radioactive SSU rRNA probes for "screening" the libraries were prepared by using bulk RNA populations from the cognate organisms as templates in primer extension syntheses. Reverse transcriptase reactions containing bulk RNA and radioactive dATP were "primed" with synthetic oligonucleotides that are complementary to 5' or 3' proximal "eukaryote-specific" sequence elements.[5] The primers only hybridize with specific sites on the cytoplasmic SSU rRNA templates;[23,24] thus, the synthesis of DNA probes that are specific for rRNAs coded by nuclear genes is assured. Clones that anneal to the probes in colony or plaque hybridization assays were selected and their SSU rRNA coding regions were then subcloned into one of the M13 single-stranded phages. Single-stranded templates containing either the coding or noncoding strands for SSU rRNAs were prepared and used to direct DNA synthesis in the primer extension sequencing protocols as described by Messing.[25]

The universal and eukaryote-specific oligonucleotide primers listed in TABLE 1 were used to prime DNA synthesis in the dideoxynucleotide chain terminated sequence analyses of M13 subclones containing SSU rRNA coding regions. The strategy for sequencing a typical eukaryotic SSU rRNA coding region (as represented by *Paramecium tetraurelia*) is outlined in FIGURE 1. Using M13 phage single-stranded DNA templates, 350-500 nucleotide positions can be determined from a given primer. As an alternative to the M13 system, duplex DNA templates from pBR or Charon recombinants can be used directly in modified sequencing reactions, but the number of positions that can be determined from a given primer is generally limited to 250 nucleotides. The use of universal and eukaryote-specific primers eliminates the requirement for subcloning multiple overlapping sequences into M13 vectors or custom synthesizing a large number of oligonucleotide primers for each gene. Our current collection of primers allows the rapid sequence determination of any eukaryotic SSU rRNA on both strands of the DNA. The limiting factor is the time required to grow the organism of interest and isolate the initial rDNA clone.

## Paramecium tetraurelia SSU rRNA

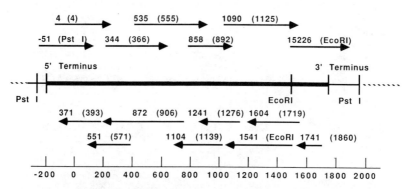

**FIGURE 1.** Restriction map and strategies used to determine the DNA sequence of the *Paramecium tetraurelia* small subunit ribosomal RNA gene. Synthetic DNA oligomers that are complementary to universally conserved and "eukaryote-specific" coding and noncoding regions were used as primers in dideoxynucleotide chain termination sequencing reactions. The direction and extent of sequence analysis from a given primer are shown by the arrows. The locations of the primers in the *P. tetraurelia* rRNA are indicated on the arrows, and the analogous positions in the *Dictyostelium discoideum* small subunit rRNA are given in parentheses.

## COMPARATIVE ANALYSIS OF EUKARYOTIC SSU rRNA SEQUENCES

Eukaryotic SSU rRNAs range in size from 1753 bases in *Tetrahymena thermophila*[26] to 2305 in *Euglena gracilis.*[27] Their coding regions are evolutionary mosaics in which highly and partially conserved sequences are interspersed among regions that display very high rates of genetic drift. The length variation and interspersion patterns for conserved and nonconserved regions in eukaryotic SSU rRNA genes are depicted by the sequence conservation histogram presented in FIGURE 2. For this analysis, complete SSU rRNA sequences from mouse,[28] rat,[29] *X. laevis,*[13]

*Artemia salina,*[30] rice,[31] *Zea mays,*[32] soy bean,[33] *Acanthamoeba castellanii,*[34] *S. cerevisiae,*[12] *Podospora anserina* (Sogin, Elwood, and Ingold, unpublished data), *T. thermophila,*[26] *P. tetraurelia,*[35] *Oxytricha nova* and *Stylonychia pustulata,*[5] *Euplotes aediculatus,*[36] *Plasmodium berghei* (Gunderson and Sogin, unpublished data), *Trypanosoma brucei* and *E. gracilis,*[27] and *Crithidia fasciculata*[37] were aligned by a procedure that considers the phylogenetic conservation of both primary and secondary structural features.[5] Initially the sequences were scanned for short subregions of similar or identical primary structure in approximately homologous positions. Alignment gaps

**TABLE 1.** Synthetic Primers Complementary to Conserved Regions in Eukaryotic Small Subunit Ribosomal RNA Sequences

| D. discoideum Location | Eukaryotic Location | Alignment Sequence[a] |
|---|---|---|
| 4- > 20 | 4- > 20 | CTGGTTGATCCTGCCAG[b] |
| 366- > 382 | 533- > 549 | AGGGTTCGATTCCGGAG[b] |
| 555- > 570 | 743- > 758 | GTGCCAGCMGCCGCGG[b] |
| 568- > 583 | 756- > 771 | CGGTAATTCCAGCTCC[b] |
| 892- > 906 | 1441- > 1455 | YAGAGGTGAAATTCT[b] |
| 962- > 976 | 1514- > 1529 | ATCAAGAACGAAAGT[b] |
| 1125- > 1141 | 1816- > 1832 | GAAACTTAAAKGAATTG[b] |
| 1180- > 1195 | 1872- > 1887 | TTTGACTCAACACGGG[b] |
| 1262- > 1276 | 1959- > 1973 | GGTGGTGCATGGCCG[c] |
| 1504- > 1519 | 2266- > 2281 | CAGGTCTGTGATGCTC[c] |
| 1704- > 1720 | 2542- > 2558 | TGYACACACCGCCCGTC[c] |
| 108- > 94 | 119- > 102 | CTGTTTTAATGAGCC[c] |
| 393- > 377 | 560- > 544 | TCAGGCTCCCTCTCCGG[c] |
| 556- > 541 | 744- > 728 | ACCAGACTTGCCCTCC[c] |
| 571- > 557 | 759- > 745 | ACCGCGGCKGCTGGC[c] |
| 576- > 559 | 764- > 747 | WATTACCGCGGCKGCTG[c] |
| 906- > 892 | 1455- > 1441 | AGAATTTCACCTCTG[c] |
| 1139- > 1125 | 1830- > 1816 | ATTCCTTTRAGTTTC[c] |
| 1276- > 1262 | 1973- > 1959 | CGGCCATGCACCACC[c] |
| 1519- > 1504 | 2281- > 2266 | GGGCATCACAGACCTG[c] |
| 1719- > 1705 | 2557- > 2543 | ACGGGCGGTGTGTRC[c] |
| 1860- > 1845 | 2731- > 2716 | YGCAGGTTCACCTAC[c] |

[a] The symbols used here are those recommended by the International Union of Biochemistry nomenclature committee; K represents G or T, R represents G or A, W represents A or T, and Y represents T or C.
[b] Synthetic DNA oligonucleotides complementary to the coding strand of the rRNA gene.
[c] DNA oligonucleotides complementary to the noncoding strand of the rRNA gene.

were introduced to juxtapose the regions of homology in the sequence collection. A Needleman-Wunsch-Sellers computerized algorithm[38] was then used to align sequence elements of weaker homology. The alignments within regions that displayed extreme length variation were refined by a consideration of phylogenetically conserved higher order structures. All SSU rRNAs can be folded into similar secondary structures which contain on the order of 50 helices.[39,40] Sequence elements that define phylogenetically conserved helical structures were juxtaposed by the appropriate placement of alignment gaps. (Helical structures are considered to be phylogenetically conserved

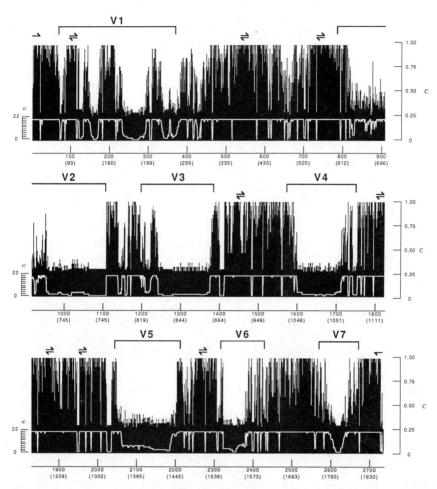

**FIGURE 2.** Sequence conservation histogram for eukaryotic small subunit rRNAs. The histogram bars represent the conservation of nucleotide usage at each of the 2744 aligned positions in a collection of 22 eukaryotic SSU rRNAs. The superimposed white line represents the number of sequences that contain a nucleotide at each position in the sequence alignment. We define the conservation of nucleotide usage, $c$, as $c = (p + (n - n')/4)/n$, where $n$ is the number of sequences considered in the alignment, $n'$ is the number of sequences represented by a residue at a given position, and $p$ is the number of times that the most conserved nucleotide is represented at a given position.

if they are found in the majority of SSU rRNAs and their formation is independent of absolute primary structure conservation; several compensating base changes must be found that maintain the helical structures.) This procedure is not rigorously defined, but the positions aligned by this method have a higher probability of being in homologous alignment than those aligned by maximizing the number of matching nucleotides. Sequence comparisons based upon this alignment technique are referred to as "structural" in order to distinguish this procedure from those using maximal matches.

The histogram bars in FIGURE 2 represent the conservation of nucleotide usage at each of the 2744 aligned positions in a collection of 22 eukaryotic SSU rRNAs. The superimposed white line represents the number of sequences that contain a nucleotide at each position in the sequence alignment. We define the conservation of nucleotide usage, $c$, as

$$c = (p + (n - n')/4)/n$$

where $n$ is the number of sequences considered in the alignment, $n'$ is the number of sequences represented by a residue at a given position, and $p$ is the number of times that the most conserved nucleotide is represented at a given position. The resulting plot displays the interspersion pattern for conserved, semiconserved, and hypervariable regions, as well as the relative locations of regions that display extreme length variation.

The highly conserved regions are nearly identical in all eukaryotic SSU rRNAs. Because of a lack of sequence variation, they contribute little information about evolutionary divergence but they are essential for establishing meaningful alignments between SSU rRNAs and, as described above, they are useful for rapidly sequencing the coding regions for SSU rRNAs. The locations of the conserved regions that are complementary to the sequencing primers in TABLE 1 are indicated in FIGURE 2. Presumably the eukaryote-specific and universal regions reflect the existence of critical functional domains that were established in the earliest eukaryotes or in common ancestor(s) of the eubacteria, archaebacteria, and eukaryotes.

The length variations displayed by eukaryotic SSU rRNA coding regions appear to be due to the insertion of large blocks of nucleotides that are highly variable in sequence. These blocks are specifically located in regions that display very high rates of genetic drift and presumably are not subject to strong functional constraints. FIGURE 3 shows the locations of the variable length regions (V1-V7) in secondary structure models of T. thermophila and A. castellanii SSU rRNAs. Despite the absence of phylogenetic proof, these extra nucleotides are generally represented in our structures as extensions of evolutionarily conserved helical regions because of their impressive capacity for extensive base pairing. The variable length regions in SSU rRNAs are not analogous to the "self-splicing" intron structures reported in large subunit rRNAs. Direct sequence analyses of the corresponding domains in the SSU rRNAs of A. castellanii and E. gracilis (which are 2303 and 2305 nucleotides in length respectively) demonstrate that the variable length regions in these SSU rRNA genes are fully represented in the mature rRNAs.[34] The evolutionary origin(s) for the variable length regions is unknown; they may have resulted from multiple insertion events or they may be the products of transposition events in which large numbers of nucleotides were inserted simultaneously into the genome.

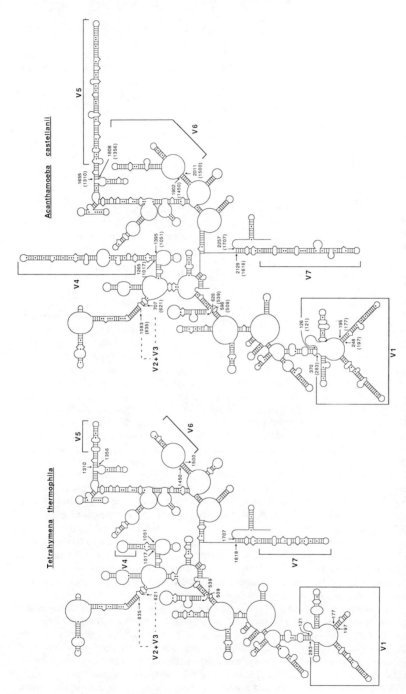

**FIGURE 3.** Secondary structure models of the SSU rRNAs from *T. thermophila* and *A. castellanii*. The secondary structures for the *T. thermophila* and *A. castellanii* SSU rRNAs were drawn according to the pairing schemes proposed by Gutell *et al.* [40] The brackets V1–V7 denote regions that display extreme length variation. The numbers indicate absolute sequence positions in each SSU rRNA, and the numbers in parentheses denote the corresponding positions in the *T. thermophila* SSU rRNA.

## PHYLOGENETIC ANALYSES

Comparing the similarities of functionally equivalent biopolymers is a proven method for inferring genealogical relationships.[41] Pairwise comparisons of all homologous nucleotide positions in the coding regions for SSU rRNAs can be used to calculate structural similarities. We define structural similarity as

$$s = m/(m + u + g/2)$$

where $m$ is the number of sequence positions with matching nucleotides, $u$ is the number of positions with nonmatching nucleotides, and $g$ is the number of sequence gaps. (Only the first 5 positions in a gap are considered in making the calculations. Large insertions or deletions probably reflect single rare events.) This similarity value was used in the formula of Jukes and Cantor[42] to approximate the "structural distance" (the number of evolutionary changes per 100 nucleotide positions) separating each pair of sequences.

The structural distance values were converted to phylogenetic trees by a variation of the method of Fitch and Margoliash.[43] The evaluation of alternative phylogenetic trees was based on the agreement of the structural distance data separating pairs of organisms and the sum of tree segment lengths joining the organisms in the tree. The computer-assisted algorithms for inferring phylogenetic trees from the structural distance data and for evaluating alternative tree topologies have been described in greater detail by Elwood et al.[5] A practical limit of 30 organisms is imposed by the computational time required for the analysis.

The phylogenetic tree presented in FIGURE 4 was based upon comparisons of all nucleotide positions that can be unambiguously aligned in selected SSU rRNAs from the archaebacteria (*Sulfolobus solfataricus*[44] and *H. volcanii*[14]), the eubacteria (*E. coli,*[11] *Proteus vulgaris,*[45] *Agrobacterium tumefaciens,*[46] *Bacillus subtilis,*[47] *Flavobacterium heparinum,*[48] and *Anacystis nidulans*[49]), and 16 of the eukaryotes represented in FIGURE 2 (approximately 1100 nucleotides were used for this analysis). In this "multikingdom" phylogeny the known primary lines of descent separate into three subtrees, each of which includes organisms that exemplify the known limits of SSU rRNA structural diversity for that subtree. Within the eukaryotic subtree, the extent of phylogenetic diversity appears to exceed the known depths of branching within the eubacteria or the archaebacteria. In a morphological sense, the diversity and complexity of eukaryotic organisms dwarf that seen in the prokaryotic world and yet eukaryotes appear to be metabolically far less diverse.[50] Since differences in biochemical characters are frequently used for inferring phylogenetic relationships,[51] the extreme divergence in eukaryotic SSU rRNAs is unexpected.

The euglenoids (as represented by *E. gracilis*) and kinetoplastids (as represented by *C. fasciculata* and *T. brucei*) display the earliest eukaryotic branching patterns yet characterized by comparisons of rRNA sequences. These flagellated protists do not appear to be closely related to one another; euglenoids and kinetoplastids diverged from one another soon after their separation from the other eukaryotes. It is likely that other organisms will be identified that diverged before *Euglena* and kinetoplastids. Organisms such as dinoflagellates, red algae, and *Pelomyxa* may have separated from other eukaryotes before *E. gracilis* did. Furthermore, preliminary sequence information indicates that organisms that gave rise to the microsporidians predated the euglenoids and kinetoplastids (Vossbrink and Woese, unpublished data).

The low homology between *E. gracilis* or *T. brucei* and the other eukaryotes is not an artifact resulting from unusually high rates of genetic drift (fast evolutionary clock speeds). The evolutionary distance between *E. coli* and *E. gracilis* or *T. brucei* is comparable to the distance between *X. laevis* and *E. coli,* indicating that these three eukaryotic SSU rRNAs are evolving at nearly equivalent rates. Similarly, the deep branching patterns exhibited by *E. gracilis* and *T. brucei* cannot be explained by postulating convergent evolution in other eukaryotic SSU rRNAs. Sequence convergence affecting the reliability of molecular phylogenies is a recurring question. There is little evidence that selection of a biochemical capability results in the selection of a particular nucleotide or amino acid sequence. Indeed there are numerous examples where functionally equivalent macromolecules show no statistically significant sequence homology.[52,53] One test for convergence in a molecular phylogeny is the comparison of phylogenetic trees that have been derived from functionally independent markers. If the trees are nearly identical, they must reflect divergent evolution from a common ancestor; it is improbable that multiple, functionally independent traits would converge at the same rate during evolution. Ribosomal RNAs can be considered to have multiple functional domains, and phylogenetic trees inferred from sequence comparisons for different regions of the SSU rRNA[3] have identical, or very nearly so, topologies.

The *T. brucei* and *E. gracilis* branchings were followed by the relatively early divergences of two independent lineages leading to the cellular slime molds *(D. discoideum)* and the apicomplexans *(P. berghei).* These groups do not appear to be closely related to any other branches in the tree. The absence of obvious relatives to the slime molds or the apicomplexans may be due to the small number of organisms represented in this phylogeny or to the actual absence of closely related forms among extant organisms. Independent lineages for the other eukaryotes including the fungi *(S. cerevisiae* and *P. anserina*), the plants (rice and *Z. mays*), the animals (*A. salina, X. laevis,* and rat), the Ciliophora *(P. aurelia* and *O. nova*) and the acanthamoebae *(A. castellanii)* diverged during a period of massive radiation. Branches representing the mitochondrial and chloroplast compartments of eukaryotic cells converge on different parts of the multikingdom tree. The mitochondrial SSU rRNAs affiliate with the purple bacterium *A. tumefaciens,* and chloroplast rRNAs affiliate with the cyanobacterium "*A.* " *nidulans.* If trees constructed from rRNA sequence comparisons have any validity, the eukaryotic cell must be a chimera.

According to the endosymbiotic theory for the origins of eukaryotic cells, mitochondria arose from a prokaryote invading an ancestral eukaryote. Our multikingdom tree is consistent with this theory, but the number of endosymbiotic events giving rise to mitochondria is still unresolved. The type of mitochondrial cristae an organism possesses is regarded as having phylogenetic significance. The possession of lamellar or tubular cristae is a dividing characteristic sometimes placed at the base of a protistan phylogeny.[54] The deeper branchings in our tree are distinctly segregated according to crista type. Unusual Ping-Pong paddle shaped cristae are peculiar to the euglenoids and kinetoplastids, two groups that shared a short period of common ancestry after their early divergence from other protists. *D. discoideum* and *P. berghei* have tubular cristae, while organisms that diverged during the radiative period have either lamellar or tubular cristae. Mitochondria with different types of cristae may have evolved from different bacteria with each type marking a separate endosymbiotic lineage.[55] If different bacteria gave rise to independent endosymbiotic lineages, the original endosymbionts probably arose from a single bacterial grouping. In a phylogenetic analysis based upon comparisons of eubacterial SSU rRNAs with mitochondrial encoded SSU

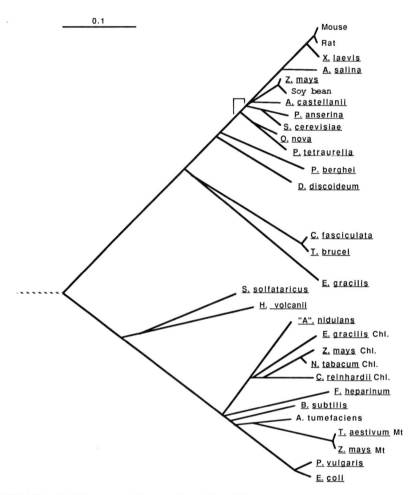

**FIGURE 4.** Multikingdom phylogeny inferred from SSU rRNA sequence homologies. A phylogenetic tree was inferred using structural distance data as described in the text. The analysis was limited to approximately 1100 positions which could be unambiguously aligned in all of the considered sequences. The evolutionary distances between nodes of the tree are represented by the horizontal component of their separation. The bracket in the tree corresponds to one change per 100 nucleotide positions.

rRNAs, all mitochondria appear to be related to the same purple bacterium *A. tumefaciens*. However, only a single mitochondrial representative containing tubular cristae was included in this analysis.[46] Additional mitochondrial SSU rRNA sequences will be required before firm conclusions can be drawn about multiple bacterial lineages giving rise to different types of mitochondria through multiple symbiotic events. In any case, our tree suggests that this difference in cristae structure does not extend to the root of the protist tree; acquisition of lamellar cristae by whatever route is a more recent event than the acquisitions of tubular cristae.

Chloroplasts almost certainly have been acquired independently in different lines.[56-58] As shown in the multikingdom tree, chloroplast SSU rRNA sequences from *E. gracilis, Chlamydomonas reinhardii,* and higher plants converge upon a single cyanobacterial lineage but this branching pattern does not agree with the relationships inferred from comparisons of nuclear encoded rRNAs. *E. gracilis* has often been classed with the green algae (which presumably gave rise to the higher plants) on the basis of chloroplast characteristics;[59] however our tree indicates that euglenoids are very distantly related to higher plants. It is likely that the two groups acquired similar chloroplasts independently. The independent acquisitions could be the results of multiple eukaryotic/prokaryotic or eukaryotic/eukaryotic endosymbiotic events.

In view of the small number of sequence determinations that have been made, it is premature to make major statements about the general patterns of eukaryote evolution. The little that is known supports Raikov's ideas concerning the evolution of mitosis.[60] He regarded closed mitosis with an intranuclear spindle as a central form from which other mitoses were derived. Indeed, kinetoplastids and euglenoids, which represent the earliest branches in the eukaryotic subtree, have an intranuclear spindle and a persistent nucleolus during nuclear division. Closed mitosis persists up through the fungal groups. Variants of closed mitosis appear independently in several lineages (e.g., in cellular slime molds and apicomplexans), while the types of mitoses he regards as most changed from an ancestral type are found in the most recently diverged organisms—multicellular plants and animals.

Despite the unexpected diversity in the eukaryotic subtree, lineages that gave rise to the three "higher" kingdoms diverged during a relatively short evolutionary period. In our multikingdom tree, the order of branching for organisms that arose during the massive period of radiation is not well resolved. The bracket in the tree corresponds to evolutionary distances of fewer than one nucleotide change per 100 nucleotide positions. Greater resolution in the distance matrix trees can be achieved by excluding archaebacterial and eubacterial SSU rRNAs from the analysis. Regions that define primary and secondary structures that are found only in the eukaryotic SSU rRNAs (and hence regions that cannot be juxtaposed with homologous sequence elements in prokaryotic SSU rRNAs) can now be included in structural similarity calculations. Such an analysis, which is based upon the comparisons of 1560 nucleotide positions, is represented by the phylogenetic tree in FIGURE 5. This expanded eukaryotic phylogeny includes the eukaryotes represented in FIGURE 4 as well as rice, *Blastocladiella emersonii* (Sogin and Elwood, unpublished data). *E. aediculatus, S. pustulata,* and *Tetrahymena hegewischi* (Sogin, unpublished data). The root of the tree is based upon the deep divergence of *E. gracilis* and the kinetoplastids in the multikingdom tree shown in FIGURE 4. In this phylogeny, protists such as *A castellanii* and *B. emersonii* diverge simultaneously with the fungal-plant-animal radiation while the ciliates diverged earlier.

The major ciliate classes seem to have a single common ancestor but the depth of branching within this phylogenetic grouping is striking. Two of the three ciliate classes proposed by Corliss[61] are included in our phylogeny. The Oligohymenophora are represented by two *Tetrahymena* species and *P. tetraurelia*. The Polyhymenophora are represented by the hypotrichous ciliates *E. aediculatus, S. pustulata,* and *O. nova.* The phylogenetic distance between these two ciliate classes approaches the depths of branching between *Z. mays, A. castellanii,* and *S. cerevisiae.* Other deep branching patterns can be noted within these ciliate classes. *Tetrahymena* and *Paramecium* are placed in the same order yet the phylogenetic distance separating these organisms approaches the distances between *P. tetraurelia* and *O. nova* or *S. pustulata.* By the criterion of rRNA relatedness, *T. thermophila* and *P. tetraurelia* might be placed in different subclasses. Similarly the depth of branching within the Polyhymenophora is

**FIGURE 5.** Eukaryote phylogeny inferred from SSU rRNA sequence homologies. A phylogenetic tree was inferred using structural distance data as described in the text. The analysis was limited to those positions that could be aligned in all of the eukaryotic SSU rRNAs (approximately 1560 positions). The evolutionary distances between nodes of the tree are represented by the horizontal component of their separation.

unexpected. In the phylogenies proposed by Corliss, *E. aediculatus, S. pustulata,* and *O. nova* are placed in the same suborder, Sporadotrichina. The *S. pustulata/O. nova* rRNA homology is consistent with this placement and is comparable to the divergence between rice and *Z. mays.* In contrast, the *E. aediculatus/S. pustulata* and the *E. aediculatus/O. nova* homology levels are similar to that found in comparisons between *P. tetraurelia* and *T. thermophila.* The deep branching pattern is consistent with the ciliates diverging into several independent lineages soon after their common ancestor separated from the other eukaryotes.

As more eukaryotic SSU rRNA sequence data become available, it should be possible to identify which protist groups gave rise to the three "higher" eukaryotic kingdoms. In our current phylogeny *B. emersonii,* recently removed from the fungal kingdom and placed with the "Protoctista," displays a slight affiliation with *P. anserina* and *S. cerevisiae.* The acanthamoebae unexpectedly branch very early in the lineage leading to the plant kingdom. Whether these are true affiliations or reflections of the inherent problems in ordering deep branches remains to be determined. The sequence analysis of SSU rRNAs from other chytrids, at least one oomycete, and several algal representatives will contribute to our understanding of this radiation.

We can only speculate about the reasons for this evolutionary radiation which we estimate to span a time frame of approximately 50-100 million years. Perhaps new methods of organizing and processing genetic information conferred upon an ancestral organism a new and expanded evolutionary potential. The radiation could correspond to the concurrent acquisition of mitochondria and an environmental shift to an oxidizing environment. A final scenario is that the radiation reflects a mass extinction. This cataclysmic event would have been followed by the adaptation of survivors into a wide spectrum of ecological niches.

Our comparison of available SSU rRNAs provides only a minimal phylogenetic framework for the eukaryotes. The observed depth of branching is consistent with the eukaryotes being a very old kingdom even though the enormous diversity of protists is only touched on by the few SSU rRNA sequences determined to date.

# SUMMARY

The phylogenetic diversity of the eukaryotic kingdom was assessed by comparing the structural and evolutionary diversity of 18-20S ribosomal RNA genes. The coding regions for cytoplasmic small subunit ribosomal RNA genes vary in length from 1753 to 2305 nucleotides, and they appear to be evolutionary mosaics in which highly and partially conserved sequences are interspersed among regions that display very high rates of genetic drift. Structural similarities between these gene sequences were used to establish a phylogenetic framework for the eukaryotes. The extent of sequence variation within the eukaryotes exceeds that displayed within the eubacterial or archaebacterial lines of descent. The kinetoplastids and euglenoids represent the earliest branchings among the eukaryotes. These branchings preceded the divergence of lineages leading to the slime molds and apicomplexans and far antedate a radiative period that gave rise to the plants, animals, fungi, and other protists.

# REFERENCES

1. MOORE, R. L. & B. J. MCCARTHY. 1967. J. Bacteriol. **94:** 1066-1074.
2. PACE, B. & L. L. CAMPBELL. 1971. J. Bacteriol. **107:** 543-547.
3. MCCARROLL, R., G. J. OLSEN, Y. D. STAHL, C. R. WOESE & M. L. SOGIN. 1983. Biochemistry **22:** 5858-5868.
4. STACKEBRANDT, E. & C. R. WOESE. 1981. *In* Molecular and Cellular Aspects of Microbial Evolution. M. J. Carlile, J. F. Collins & B. E. B. Moseley, Eds.: 1-31. Cambridge University Press. Cambridge, England.

5. ELWOOD, H. J., G. J. OLSEN & M. L. SOGIN. 1985. Mol. Biol. Evol. **2:** 399-410.
6. OLSEN, G. J. & M. L. SOGIN. 1982. Biochemistry **21:** 2335-2343.
7. HORI, H. & S. OSAWA. 1979. Proc. Nat. Acad. Sci. USA **76:** 381-385.
8. ERDMANN, V. A., E. HUYSMANS, A. VANDENBERGHE & R. DE WACHTER. 1983. Nucleic Acids Res. **11:** r105-r133.
9. FOX, G. E., E. STACKEBRANDT, R. B. HESPELL, J. GIBSON, J. MANILOFF, T. A. DYER, R. S. WOLFE, W. E. BALCH, R. S. TANNER, L. J. MAGRUM, L. B. ZABLEN. R. BLAKEMORE, R. GUPTA, L. BONEN, B. J. LEWIS, D. A. STAHL, K. R. LUEHRSEN, K. N. CHEN & C. R. WOESE. 1980. Science **209:** 457-463.
10. WOESE, C. R. & R. S. WOLFE. 1985. The Bacteria, **8.** Academic Press. New York, N.Y.
11. BROSIUS, J., M. L. PALMER, P. J. KENNEDY & H. F. NOLLER. 1978. Proc. Nat. Acad. Sci. USA **75:** 4801-4805.
12. RUBSTOV, P. M., M. M. MUSAKHANOV, V. M. ZAKHARYEV, A. S. KRAYEV, K. G. SKRYABIN & A. A. BAYEV. 1980. Nucleic Acids Res. **8:** 5779-5794.
13. SALIM, M. & B. E. H. MADEN. 1981. Nature **291:** 205-208.
14. GUPTA, R., J. M. LANTER & C. R. WOESE. 1983. Science **221:** 656-659.
15. EPERON, I. C., S. ANDERSON & D. P. NIERLICH. 1980. Nature **286:** 460-467.
16. VAN ETTEN, R. A., M. W. WALBERG & D. A. CLAYTON. 1980. Cell **22:** 157-170.
17. SOR, F. & H. FUKUHARA. 1980. C. R. Hebd. Seances Acad. Sci. Ser. D. **291:** 933-936.
18. KUNTZEL, H. & H. G. KOCHEL. 1981. Nature **293:** 751-755.
19. GRAF, L., E. ROUX, E. STUTZ & H. KÖSSEL. 1982. Nucleic Acids Res. **10:** 6369-6381.
20. DRON, M., M. RAHIRE & J.-D. ROCHAIX. 1982. Nucleic Acids Res. **10:** 7609-7620.
21. SCHWARZ, ZS. & H. KÖSSEL. 1980. Nature **283:** 739-742.
22. SANGER, F., S. NICKLEN & A. R. COULSON. 1977. Proc. Nat. Acad. Sci. USA **74:** 5463-5467.
23. QU, L. H., B. MICHOT & J.-P. BACHELLERIE. 1983. Nucleic Acids Res. **11:** 5903-5920.
24. LANE, D. J., B. PACE, G. J. OLSEN, D. A. STAHL, M. L. SOGIN & N. R. PACE. 1985. Proc. Nat. Acad. Sci. USA **82:** 6955-6959.
25. MESSING, J. 1983. Methods Enzymol. **101:** 20-78.
26. SPANGLER, E. A. & E. H. BLACKBURN. 1985. J. Biol. Chem. **260:** 6334-6340.
27. SOGIN, M. L., H. J. ELWOOD & J. H. GUNDERSON. 1986. Proc. Nat. Acad. Sci. USA **83:** 1383-1387.
28. RAYNAL, F., B. MICHOT & J.-P. BACHELLERIE. 1984. FEBS Lett. **167:** 263-268.
29. CHAN, Y. L., R. GUTELL, H. F. NOLLER & I. G. WOOL. 1984. J. Biol. Chem. **259:** 224-230.
30. NELLES, L., B. FANG, G. VOLCKAERT, A. VANDENBERGHE & R. DE WACHTER. 1984. Nucleic Acids Res. **12:** 8749-8768.
31. TAKAIWA, F., K. OONO & M. SUGIURA. 1984. Nucleic Acids Res. **12:** 5441-5448.
32. MESSING, J., G. CARLSON, G. HAGEN, I. RUBENSTEIN & A. OLESON, 1984. DNA **3:** 31-40.
33. ECKENRODE, V. K., J. ARNOLD & R. B. MEAGHER. 1985. J. Mol. Evol. **21:** 259-269.
34. GUNDERSON, J. H. & M. L. SOGIN. 1986. Gene **44:** 63-70.
35. SOGIN, M. L. & H. J. ELWOOD. 1986. J. Mol. Evol. **23:** 53-60.
36. SOGIN, M. L., M. T. SWANTON, J. H. GUNDERSON & H. J. ELWOOD. 1986. J. Protozool. **33:** 26-29.
37. SCHNARE, M. N., J. C. COLLINGS & M. W. GRAY. 1986. Curr. Genet. **10:** 405-410.
38. SMITH, T. F., M. S. WATERMAN & W. M. FITCH. 1981. J. Mol. Evol. **18:** 38-46.
39. NOLLER, H. F. & C. R. WOESE. 1981. Science **212:** 403-411.
40. GUTELL, R. R., B. WEISER, C. R. WOESE & H. F. NOLLER. 1985. Prog. Nucleic Acids Res. Mol. Biol. **32:** 155-216.
41. ZUCKERKANDL, E. & L. PAULING. 1965. J. Theor. Biol. **8:** 357-366.
42. JUKES, T. H. & C. R. CANTOR. 1969. *In* Mammalian Protein Metabolism. H. N. Munro, Ed.: 21-132. Academic Press. New York, N.Y.
43. FITCH, W. M. & E. MARGOLIASH. 1967. Science **155:** 279-284.
44. OLSEN. G. J., N. R. PACE, M. NUELL, B. P. KAINE, R. GUPTA & C. R. WOESE. 1985. J. Mol. Evol. **22:** 301-307.
45. CARBON, P., J. P. EBEL & C. EHRESMANN. 1981. Nucleic Acids Res. **9:** 2325-2333.
46. YANG, D., Y. OYAIZU, H. OYAIZE, G. J. OLSEN & C. R. WOESE. 1985. Proc. Nat. Acad. Sci. USA **82:** 4443-4447.

47. GREEN, C. S., G. C. STEWART, M. D. HOLLIS, B. S. VOLD & K. F. BOTT.1985. Gene 37: 261-266.
48. WEISBURG, W. G., Y. OYAIZU, H. OYAIZU & C. R. WOESE. 1985. J. Bacteriol. 164: 230-236.
49. TOMIOKA, N. & M. SUGIURA. 1983. Mol. Gen. Genet. 191: 46-50.
50. MARGULIS, L. & K. V. SCHWARTZ. 1982. Five Kingdoms: An Illustrated Guide to the Phyla of Life on Earth. Freeman. San Francisco, Calif.
51. RAGAN, M. A. & D. J. CHAPMAN. 1978. A Biochemical Phylogeny of the Protists. Academic Press. New York, N.Y.
52. DRATZ, E. A. & P. A. HARGRAVE. 1983. Trends Biol. Sci. 8: 128.
53. KRAUT, J. 1977. Annu. Rev. Biochem. 46: 331-358.
54. TAYLOR, F. J. R. 1978. Biosystems 10: 67-89.
55. STEWART, K. D. & K. R. MATTOX. 1984. J. Mol. Evol. 21: 54-57.
56. GIBBS, S. P. 1981. Int. Rev. Cytol. 72: 49-99.
57. WHATLEY, J. M. & F. R. WHATLEY. 1981. New Phytol. 87: 233-247.
58. WILCOX. L. W. & G. J. WEDEMAYER. 1985. Science 227: 192-194.
59. LEEDALE, G. F. 1978. Biosystems 10: 183-187.
60. RAIKOV, I. B. 1982. The Protozoan Nucleus: Morphology and Evolution. Cell Biol. Monog. 9. Springer-Verlag. Vienna & New York.
61. CORLISS, J. O. 1979. The Ciliated Protozoa: Characterization, Classification and Guide to Literature. 2nd edit. Pergamon Press. New York, N.Y.

# *Paracoccus* as a Free-Living Mitochondrion

PHILIP JOHN

*Department of Agricultural Botany*
*University of Reading*
*Reading RG6 2AS, United Kingdom*

The proposal that *Paracoccus denitrificans* is closely related to the mitochondrion[1,2] was made originally because of the large number of mitochondrial features shown by the respiratory chain and oxidative phosphorylation system in *Paracoccus*. These features either were characterized physiologically or were identified chemically or spectroscopically. Since that initial proposal was made, information at the structural and molecular level has become available on the protein complexes isolated from the respiratory chain of *Paracoccus*. In the present paper I shall take the opportunity to review this new information and to compare analogous components of the mitochondrial and *Paracoccus* respiratory chains at a deeper level than was possible some 10 years ago. The purpose of the comparison will be to determine what changes in the catalytic and regulatory functioning of the respiratory chain were likely to have accompanied the hypothetical evolutionary transition from a free-living protomitochondrion resembling *Paracoccus* to the endosymbiotically derived mitochondrion.

Before proceeding with that comparison, I shall summarize the current phylogenetic position of *Paracoccus* in relation to its photosynthetic relatives and I shall compare the mitochondrial affinities of *Paracoccus* with those of other eubacteria. In their original proposal, John and Whatley emphasized tht none of the mitochondrial features shown by *Paracoccus* was unique to *Paracoccus*, but that no other prokaryote possessed as many.[1] It was apparent then that the unique position of *Paracoccus* was challenged by the purple nonsulfur photosynthetic eubacterium *Rhodobacter* (formerly *Rhodopseudomonas*)[3] *sphaeroides*, which had a mitochondrial type of respiratory chain when it grew aerobically in the dark. It was predicted that future research would reveal that *Paracoccus* will be viewed as a "representative of a small group of aerobic bacteria (probably including *R. sphaeroides*) all having an obvious affinity with the mitochondrion."[1] That small group of aerobic bacteria has now acquired a taxonomic status, as subsequent comparative studies of cytochrome *c* amino acid sequences (combined with x-ray crystal structure) and ribosomal RNA (rRNA) nucleotide sequences have provided strong evidence that *Paracoccus* and *R. sphaeroides* are closely related to each other, and that they belong to a subdivision of eubacteria that contains all those bacteria having an obvious affinity with the mitochondrion.

# SEQUENCE DATA PLACE PARACOCCUS AMONG PURPLE NONSULFUR PHOTOSYNTHETIC EUBACTERIA IN A SUBDIVISION FROM WHICH MITOCHONDRIA PROBABLY AROSE

The *Paracoccus* cytochrome *c*-550, like the cytochromes $c_2$ of *Rhodobacter sphaeroides* and *Rhodobacter capsulatus*, belongs to the long class all of which have short sequences present as loops on the surface of the molecule.[4] Although there is extensive homology between the *Paracoccus* cytochrome *c*-550 and mitochondrial cytochrome *c*, the sequences of the cytochromes $c_2$ from the purple nonsulfur photosynthetic bacteria *Rhodomicrobium vannielii* and *Rhodopseudomonas viridis* are closer still to the sequence of mitochondrial cytochrome *c*.[5] These photosynthetic cytochromes $c_2$ lack the insertions apparent in *Paracoccus* cytochrome *c*-550 and are classed alongside mitochondrial cytochrome *c* in the medium class of *c*-type cytochrome, while the *Paracoccus* cytochrome *c*-550 appears to be most closely related to the cytochrome $c_2$ of *R. capsulatus* and *R. sphaeroides* when the additional loops are taken into consideration.[4]

The schematic phylogenetic tree proposed by Dickerson, based largely on the *c*-type cytochrome structure and sequence data, shows *Paracoccus* emerging by loss of photosynthesis from ancestral types represented today by *R. capsulatus* and *R. sphaeroides*, and mitochondria emerging with the loss of photosynthesis from ancestral types represented today by *Rm. vannielii*, *Rps. viridis*, and *Rhodopseudomonas acidophila*.[4] An essentially similar conclusion was reached by Dayhoff and Schwartz,[6] who used a least-squares matrix method to construct a calibrated evolutionary tree of soluble *c*-type cytochromes. In their tree *Paracoccus* cytochrome *c*-550 is again most closely related to *R. sphaeroides* and *R. capsulatus*. However their tree also indicates that an ancestral cytochrome *c* diverged from a structure represented now by the mitochondrial-like cytochromes $c_2$ of *Rm. vannielii*, *Rps. viridis*, and *Rps. acidophila* in one direction toward that of *Paracoccus*, and in another direction toward *Rhodopila* (formerly *Rhodopseudomonas*)[3] *globiformis*. Side branches diverging from the *R. globiformis* line lead (via presumed endosymbioses and loss of photosynthetic capacity) to the mitochondria of the major eukaryotic kingdoms.[6]

The construction of phylogenetic trees on the basis of sequence data obtained from molecules like *c*-type cytochromes that can show quite different physical and physiological properties in different organisms has recently been criticized[7] on the grounds that functional considerations set limits to structural change so as to distort phylogenetic relationships between widely divergent organisms. However it is conceded that where sequence variation is less than the limit set for that protein, valid relationships can be inferred from sequence data from *c*-type cytochromes.[7] One such relationship is that between *Paracoccus* and *R. capsulatus* and *R. sphaeroides*.[7]

The availability of nucleotide sequences of the 16S rRNA from a large number of eubacteria has allowed the construction of a phylogenetic tree of the eubacteria based on a comparative analysis of these sequences. Within this new phylogeny, *Paracoccus* is placed in the α subdivision of a major group of eubacteria provisionally termed the purple photosynthetic bacteria and their relatives.[8,9] Further division of the α subdivision distinguishes three subgroups all of which contain both photosynthetic and nonphotosynthetic types.[8] *Paracoccus* is closely associated with *R. capsulatus* and *R. sphaeroides* within the small α-3 subgroup,[8] in accordance with the affinities of their *c*-type cytochrome sequence.[4,6] In the α-2 subgroup are found the photosynthetic bacteria known to possess the medium-sized cytochrome $c_2$ molecules, which show the closest homology to mitochondrial cytochrome *c*: *Rm. vannielii, Rps. viridis,* and *Rps. acidophila.*[8] But the other possessor of a mitochondrial-like cytochrome $c_2$, *R. globiformis,* is placed some distance away in α subgroup-1.[8] Among the nonphotosynthetic members of the α-2 subgroups are the agrobacteria, rhizobia, and rickettsiae, all of which are capable of entering into intracellular associations with eukaryotic cells.[8,10]

In general, mitochondrial rRNA sequences show wide variation among mitochondria from different sources and their rRNA sequences differ greatly from those of prokaryotes. Therefore they have not proved useful in localizing the mitochondrial ancestor among the different bacterial subgroups. However since plant mitochondrial rRNA sequences do show some homology with those of eubacteria, Yang *et al.* chose to compare the equivalent of the 16S rRNA sequence of wheat mitochondria with the 16S rRNA sequences from a variety of eubacteria representative of some the main taxonomic groupings.[10] Of the eubacteria examined the closest mitochondrial relative was *Agrobacterium tumefaciens,*[10] the selected representative of the α subdivision of the major group—the purple photosynthetic bacteria and their relatives.

The grouping of *Paracoccus* with *R. sphaeroides* on the basis of the 16S rRNA sequence receives support from sequence analysis of the large-subunit rRNA of these bacteria. Thus McKay *et al.* showed that not only is there extensive sequence homology between the 14S fragments from *Paracoccus* and *R. sphaeroides* but that in both cases these fragments arise from a specific cleavage of the 23S rRNA[11]—an event that they note is rarely observed in other prokaryotes, but has been observed in *Agrobacterium tumefaciens* and *R. capsulatus.*[11]

Additionally, when the complete sequence of the *Paracoccus*[12] and *R. sphaeroides*[13] 5S rRNAs became available extensive homology between the two sequences became apparent. Moreover when the sequences of the 5S rRNA of plant mitochondria were included in the comparison, there was a greater homology between those sequences and the three members of the α subdivision of eubacteria, *Paracoccus, R. sphaeroides,* and *Rhodospirillum rubrum,* than for other bacteria for which comparable sequences were available.[13] Similarly the phylogenetic relationships revealed by 5S rRNA sequence analysis carried out by Walters and Erdmann have plant mitochondria originate from an ancestor in the α subdivision of the purple bacteria and their relatives.[14]

The concept of *Paracoccus* as a photosynthetically incompetent descendant of photosynthetic bacteria is supported by the discovery of respiratory enzymes of denitrification in purple nonsulfur photosynthetic bacteria. Thus strains of *R. sphaeroides*[15] and *Rps. palustris*[16] have been shown to be capable of denitrification; strains of *R. capsulatus* possess a respiratory nitrate reductase activity;[17] and $N_2O$ has been shown to be used as an electron acceptor by strains of *R. capsulatus* and *R. rubrum.*[18] In both *R. capsulatus* and *Paracoccus,* electron flow to $N_2O$ is linked to ATP synthesis and the $N_2O$ reductase is periplasmic, but the pathway of electron transport to $N_2O$ in *R. capsulatus* probably differs from that in *Paracoccus.*[18] In *R. capsulatus* there is evidence that electron flow via the nitrate and $N_2O$ reductase, like electron flow to $O_2$, can avoid overreduction during photosynthesis[19] and thus it may help to maintain the optimum redox potential for efficient photosynthetic electron transport.

The respiratory chains of *Paracoccus* and *R. sphaeroides* have many features in common,[1] and thus an evolution of the *Paracoccus* chain from that of *R. sphaeroides* is not difficult to explain. However evidence of other mitochondrial features among the photosynthetic bacteria that possess cytochrome $c_2$ resembling mitochondrial cytochrome *c* is more difficult to find. As a group they seem not to be fully aerobic, and perhaps for this reason little is known of their respiratory systems. Thus both *R.*

*globiformis*[20,21] and *Rm. vannielii*[21] are microaerophilic, and respiratory-dependent growth of *Rps. viridis* is limited.[22]

Taken together, sequence comparisons of *c*-type cytochromes, small-subunit rRNA, and large-subunit rRNAs all indicate an origin of the mitochondria from the α subdivision of the purple photosynthetic bacteria and their relatives. *Paracoccus*, classed in the α-3 subgroup, has probably evolved by a loss of photosynthesis from an ancestor resembling the present-day *R. sphaeroides* and *R. capsulatus.* On the basis of cytochrome *c* affinities, mitochondria are more likely to have evolved from a photosynthetic ancestor in the α-2 subgroup. Hence *Paracoccus* and mitochondria, as respiratory member of the α subdivision, may be viewed as evolutionary codescendants. It is in the context of this parallel evolutionary development that I shall now compare, at the structural and molecular level, the mitochondrial respiratory components with those of *Paracoccus* and its close relative *R. sphaeroides.*

## THE ENERGY-CONSERVING COMPLEXES OF THE PARACOCCUS RESPIRATORY CHAIN ARE RELATED TO THOSE OF THE MITOCHONDRIAL RESPIRATORY CHAIN BUT ARE OF SIMPLER CONSTRUCTION

It has been known for some time that the spectroscopically distinguishable components of the respiratory chain of aerobically grown *Paracoccus* closely resemble those of mammalian mitochondria.[1] It now seems likely that these redox components of the *Paracoccus* respiratory system are organized into counterparts of the four complexes of the mitochondrial respiratory chain (FIGURE 1).

The respiratory chain of *Paracoccus,* like that of other bacteria, has usually been represented as a branched chain of redox components with no reference to the organization of protein complexes, probably because bacterial respiratory chains were characterized until recently almost entirely from the prosthetic groups present. As noted by Ferguson, the usual representations of bacterial respiratory chains do not provide a useful framework for showing how the non-mitochondrial-type components relate to those that have mitochondrial characteristics.[23] The representation adopted here (FIGURE 1) shows both the mitochondrial and *Paracoccus* electron-transfer components organized into protein complexes. This provides a useful basis for comparing the respiratory complexes isolated from *Paracoccus* with the corresponding, better known mitochondrial complexes. In addition this new representation avoids inaccuracies (see Reference 23) in previously published[1] electron transport chains of *Paracoccus.* Mitochondria of fungi, protozoa, and plants possess a variety of dehydrogenases and oxidases in addition to those possessed by mammalian mitochondria.[2] However for the most part these additional components have not been characterized structurally in sufficient detail for a useful comparison to be made with components of bacterial respiratory chains (see for example Reference 24).

The reduced nicotinamide-adenine dinucleotide (NADH) dehydrogenases of both mammalian mitochondria and *Paracoccus* contain a similar array of iron-sulfur centers,[24] and both enzymes catalyze the rotenone-sensitive transfer of electrons from NADH to ubiquinone coupled to a transmembrane proton translocation which is associated with energy conservation at site I. However, while the NADH dehydrogenase complex isolated from mitochondria consists of at least 20 different polypeptide subunits,[26] the *Paracoccus* NADH dehydrogenase appears to consist of far fewer polypeptide subunits.[27,28] The 2 subunits of the NADH dehydrogenase isolated from detergent-solubilized plasma membranes of *Paracoccus* have relative molecular masses ($M_r$) of 48,000 and 25,000.[27] These correspond in $M_r$ to the two large subunits of the catalytic fragment, which is released from the mitochondrial NADH dehydrogenase by treatment with sodium perchlorate.[26] This fragment contains flavin mononucleotide (FMN) and iron, has NADH-ubiquinone oxidoreductase activity (although no longer rotenone sensitive), and consists of just 3 subunits of $M_r$ 51,000, 24,000, and 9000-10,000.[26] The largest of these subunits probably has structural features in common with either the large or small subunits of the *Paracoccus* NADH dehydrogenase, since an antibody raised against the *Paracoccus* enzyme recognizes the subunit of $M_r$ 51,000 from the mitochondrial enzyme.[27] Evidence for a third subunit of the *Paracoccus* NADH dehydrogenase has been obtained by George *et al.*, who found that an antibody

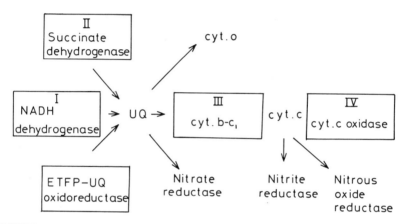

**FIGURE 1.** The respiratory chain of *Paracoccus denitrificans.* Respiratory complexes also present in the mitochondrial respiratory chain are shown enclosed. Reductases for nitrate, nitrite, and nitrous oxide are present in *Paracoccus* only under anaerobic conditions. Cytochrome *o* may be present in the mitochondria of fungi and protozoa but is absent from mammalian mitochondria.[2] Abbreviations: cyt., cytochrome; ETFP, electron-transfer flavoprotein; UQ, ubiquinone. (Modified from Reference 23.)

raised against the subunit of $M_r$ 49,000 from the mitochondrial NADH dehydrogenase cross-reacted with a polypeptide of $M_r$ 46,000 in the *Paracoccus* membrane.[28] George *et al.* have interpreted this observation to indicate that there may be a third subunit of the *Paracoccus* NADH dehydrogenase that was lost on partial purification of the 2-subunit enzyme.[28]

The two-subunit NADH dehydrogenase partially purified from *Paracoccus* has lost the rotenone sensitivity of the enzyme complex *in situ.* Moreover the isolated complex has not yet been shown to be capable of transmembrane $H^+$ translocation. Thus we cannot yet be sure that the isolated enzyme complex is functionally complete. However it already seems probable that the *Paracoccus* enzyme is constructed of fewer subunits than its mitochondrial counterpart. It is also apparent that the polypeptide subunits of the *Paracoccus* NADH dehydrogenase are structurally related to the flavoprotein and iron proteins that are at the catalytic heart of the mitochondrial enzyme. This conclusion receives independent support from the earlier work of Albracht *et al.,*[25] who had predicted from the characteristics of the electron paramagnetic resonance spectra obtained with *Paracoccus* membrane vesicles that the protein structure of the iron-containing molecules would be similar in *Paracoccus* and in mitochondria, but that their NADH dehydrogenases probably would have a different architecture arising from a different subunit composition.

The succinate dehydrogenase of *Paracoccus* has not yet been characterized in terms of its subunit composition and structure. However Barassi *et al.* have recently isolated the succinate dehydrogenase from aerobically grown *Rhodobacter sphaeroides* and shown it to contain two subunits of $M_r$ 68,000 and 30,000.[29] These resemble in relative molecular mass the large and small subunits of the succinate dehydrogenase isolated from beef heart mitochondria and *Rhodospirillum rubrum.*[30] The mitochondrial succinate dehydrogenase has already been shown to be related to the enzyme from *R. rubrum* not only in subunit composition, but also in the amino acid content and presence of prosthetic groups.[30] Presumably when the succinate dehydrogenase from *Paracoccus* comes to be examined it, like the enzyme from its photosynthetic relatives, will be seen to resemble the mitochondrial succinate dehydrogenase. This complex differs from the other three complexes of the mitochondrial respiratory chain in that the oxidoreductase reactions catalyzed are not associated with $H^+$ translocation and energy conservation. This difference may account for the succinate dehydrogenase having fewer subunits than the other complexes.

In both *Paracoccus* and mitochondria, electrons derived from the oxidation of fatty acids enter the membrane-bound respiratory chain via the electron transfer flavoprotein-ubiquinone complex[31] (FIGURE 1). This complex has the same iron-sulfur centers in both *Paracoccus* and mitochondria.[25,31] The complex has been partially purified from *Paracoccus,*[31] but its subunit composition has not yet been established.

The cytochrome $b$-$c_1$ complex of the mitochondrial respiratory chain catalyzes an antimycin-sensitive transfer of electrons from reduced ubiquinone to cytochrome $c$ coupled to a transmembrane proton translocation and energy conservation at site II. The isolated complex consists of two core proteins ($M_r$ 47,000), cytochrome $b$ ($M_r$ 44,000), cytochrome $c_1$ ($M_r$ 28,000), iron sulfur protein ($M_r$ 26,000), and smaller subunits, which are more variable in size between different preparations and which are more difficult to identify functionally.[32] The complex isolated from *Paracoccus* consists mainly of subunits of $M_r$ 62,000, 39,000, and 19,000 which were tentatively identified with cytochrome $c_1$, cytochrome $b$, and iron-sulfur protein respectively.[33] Cytochrome $c_1$ had previously been purified from *Paracoccus* and shown to have an $M_r$ of 65,000 with a single attached heme.[34] This value is twice that of mitochondrial cytochrome $c_1$, the $M_r$ of which varies from 29,000-31,000 for cytochrome $c_1$ from different sources. The structural relationship of the *Paracoccus* and mitochondrial cytochromes $c$ has been revealed by immunological cross-reactivity and by comparison of amino acid sequences in the heme-binding region.[34]

The cytochrome $b$-$c_1$ complex isolated from *Rhodobacter sphaeroides,* like that of *Paracoccus,* consists of just three major subunits of $M_r$ 40,000-48,000, 30,000-34,000, and 24,000.[35,36] One of these subunits is likely to be the cytochrome $c_1$ of *R. sphaeroides,* which resembles the mitochondrial cytochrome $c_1$ in size with an $M_r$ of 30,000.[37] There may also be smaller polypeptides present in the cytochrome $b$-$c_1$ complex of *R. sphaeroides,* but as for the mitochondrial complex these smaller subunits are more difficult to determine and identify functionally. The cytochrome $b$-$c_1$ complex of *Paracoccus* resembles the mitochondrial complex functionally in many respects: its reaction to the well-known inhibitor antimycin;[38] its sensitivity to the newly introduced inhibitor myxothiazol;[39] in the characteristics of the iron-sulfur centers as revealed by electron paramagnetic resonance spectra;[25] in the possession of two cytochrome $b$

hemes and a cytochrome $c_1$ distinguishable potentiometrically and spectrophotometrically.[1,2] However it appears that despite these functional similarities, the cytochrome $b$-$c_1$ complex of *Paracoccus,* like that of *R. sphaeroides,* is structurally simpler than the cytochrome $b$-$c_1$ complex of mitochondria.

The soluble cytochrome $c$-550 of *Paracoccus* shows structural affinities with mitochondrial cytochrome $c$ and the soluble cytochromes $c_2$ of purple nonsulfur photosynthetic bacteria as revealed by sequence homology,[40,41] immunological cross-reactivity,[42] and physiological cross-reactivity with reductases and oxidases of mitochondrial and bacterial origin.[43] Yet there is evidence that mitochondrial and bacterial $c$-type cytochromes may function in different ways. The function of the mitochondrial cytochrome $c$ appears to be a relatively simple one: it transfers electrons from the energy-conserving cytochrome $b$-$c_1$ complex to the next energy-conserving complex, the cytochrome $c$ oxidase. By contrast the soluble cytochrome $c$-550 of *Paracoccus* has a variety of roles: it transfers electrons from a methanol dehydrogenase, as well as from the cytochrome $b$-$c_1$ complex, and it transfers electrons to periplasmic nitrite reductase and nitrous oxide reductase as well as to the cytochrome $c$ oxidase (FIGURE 1). Similarly, the soluble cytochrome $c_2$ of *R. sphaeroides* can mediate a flow of electrons from the cytochrome $b$-$c_1$ complex both to the photosynthetic reaction center (in the light) and to the cytochrome $c$ oxidase (in the dark). The adaptation of the *Paracoccus* cytochrome $c$-550 toward one of its alternative oxidants can be seen when the relative cross-reactivities of horse cytochrome $c$ and *Paracoccus* cytochrome $c$-550 with the purified *Paracoccus* cytochrome $cd_1$ are compared.[44] With horse cytochrome $c$ the electron transport rate is 14% of the rate observed with *Paracoccus* cytochrome $c$-550.[44] Even in the aerobic respiratory chain the *Paracoccus* cytochrome $c$-550 may not function in the same way as the mitochondrial cytochrome $c$.[45,46] Thus all of the cytochrome $c$ is easily removed by washing mitochondria, and mitochondrial respiration is thereby abolished, but in *Paracoccus* a proportion of the cytochrome $c$-550 remains membrane bound on washing and aerobic respiration is relatively unaffected. This tightly bound cytochrome $c$-550 reacts rapidly with the cytochrome $c$ oxidase of membrane vesicles and may be closely associated with cytochrome $c$ oxidase on the membrane.[45] Furthermore the ubiquinol oxidase isolated from *Paracoccus* (a combination of cytochrome $b$-$c_1$ and cytochrome $c$ oxidase complexes) did not possess cytochrome $c$-550, yet oxidized ubiquinol efficiently.[33] It was suggested that electrons were passed from the cytochrome $b$-$c_1$ complex to the cytochrome $c$ oxidase via a tightly-bound $c$-type cytochrome, cytochrome $c$-552 ($M_r$ 22,000), that was detected.[33]

It has recently been reported that mutants of *Rhodobacter capsulatus* devoid of cytochrome $c_2$ can grow either aerobically in the dark (respiration dependent) or anaerobically in the light (photosynthesis dependent).[47] Thus the soluble cytochrome $c_2$ of *R. capsulatus,* like the cytochrome $c$-550 of *Paracoccus,* seems to be somewhat less important for electron transport than is the corresponding cytochrome $c$ in the mitochondrial respiratory chain.

Cytochrome $c$ oxidase is the terminal complex of respiratory chains of mitochondria and many bacteria.[48] It catalyzes the transfer of electrons from the 1e$^-$ donor cytochrome $c$ via four redox centers (two a-hemes and two Cu atoms) to the 4e$^-$ acceptor $O_2$. Cytochrome $c$ oxidase was the first of the respiratory complexes of *Paracoccus* to be isolated[49] and is the best characterized both functionally and structurally. The cytochrome $c$ oxidase purified from *Paracoccus* consists of just 2 polypeptide subunits of $M_r$ 45,000 and 28,000.[49] These 2 subunits correspond in $M_r$ to the 2 largest subunits of the mitochondrial cytochrome $c$ oxidase, which contains in addition 5-10 subunits apparently not represented in the *Paracoccus* enzyme.[50] The structural relationship of the *Paracoccus* enzyme with the 2 larger subunits of the mitochondrial enzyme has been revealed in two ways. First, by the homology apparent between partial sequences of the equivalent bacterial and mitochondrial subunits.[51] Second by immunological cross-reactivity observed between the smaller subunit of the *Paracoccus* enzyme with subunit II of the cytochrome *c* oxidase of yeast mitochondria.[48]

Despite the structural simplicity of the *Paracoccus* enzyme, functionally it closely resembles the mitochondrial enzyme.[48] Of particular significance is the demonstration that the purified *Paracoccus* enzyme incorporated into liposomes and supplied with reduced cytochrome *c* acts as an electrogenic proton pump.[52] This observation is significant because there is evidence that the mitochondrial cytochrome *c* oxidase subunit III, which is not apparently represented in the *Paracoccus* enzyme, is responsible for proton pumping. This subunit of the mitochondrial oxidase is specifically labeled by [$^{14}$C]dicyclohexyl-carbodiimide (DCCD), an inhibitor of proton pumping in cytochrome *c* oxidase and other mitochondrial $H^+$-translocating systems. However $H^+$ pumping by the *Paracoccus* cytochrome *c* oxidase is not only resistant to inhibition by concentrations of DCCD that inhibit the mitochondrial oxidase, but neither of the *Paracoccus* subunits is labeled by [$^{14}$C]DCCD.[53] As Solioz *et al.* note, it may be that the structure responsible for $H^+$ pumping is present on subunit III of the mitochondrial oxidase but on subunit II of the *Paracoccus* enzyme.[52] Now that the gene for subunit II of the *Paracoccus* enzyme has been cloned in *E. coli*,[54] a DNA sequence should become available for comparison with those of the mitochondrial subunits, and the degree of homology should become apparent.

In yeast the three large, catalytically important subunits, I, II, and III, are synthesized on mitochondrial ribosomes while the smaller subunits are of extramitochondrial origin.[50] These smaller subunits are variable in size and occurrence between different organisms and can differ between different tissues of the same organism. In some cases they may be necessary for the proper assembly of the respiratory complex.[55] It has also been suggested that they have a regulatory rather than a catalytic function.[50] Kadenbach has recently elaborated on this hypothesis proposing that both respiration and oxidative phosphorylation are regulated by allosteric modification of regulatory subunits of the cytochrome *c* oxidase.[56] The allosteric effectors would include substrates, other metabolites, and hormones. When these effectors bind to a regulatory subunit, the conformational change would be transmitted to the catalytic subunits resulting in an appropriate modulation either of the rate of electron transfer or of the $H^+/e^-$ ratio. Evidence to support this model[56] includes the variability in the $H^+/e^-$ ratio that has been observed with the mitochondrial cytochrome *c* oxidase, the stimulation of respiration by fatty acids, and the effect of thyroid hormones on mitochondrial ATP synthesis. An additional regulatory principle in higher organisms is suggested by the expression of isozymes of cytochrome *c* oxidase specific to particular tissues or development stages.[56]

TABLE 1 summarizes the polypeptide subunit composition of the respiratory complexes isolated from mitochondria and from *Paracoccus* and *R. sphaeroides*. The mitochondrial complexes have been subjected to far more extensive study than the eubacterial complexes, work on which has begun only in the last few years. Nevertheless the subunit composition of complexes from neither mitochondrial nor eubacterial sources is known with certainty. For the mitochondrial complexes uncertainty arises from the possibility that the smaller polypeptides that consistently copurify with the complex do so fortuitously and are not true subunits of the complex *in situ*. For the bacterial complexes uncertainty arises from the possibility that catalytically nonfunctional subunits have been lost during extraction and purification. This possibility may be more likely for the NADH dehydrogenase and cytochrome *b-c*$_1$ complexes than for the cytochrome *c* oxidase, but it is always difficult to demonstrate that a respiratory complex *in situ* does not contain regulatory subunits in addition to those catalytic subunits apparent in the isolated enzyme complex. However having noted these reservations the currently available evidence does indicate that the energy-conserving complexes of the *Paracoccus* respiratory chain are related to those of the mitochondrial chain but are of simpler construction with fewer polypeptide subunits. The function of the extra subunits present in the mitochondrial complexes and absent from the *Paracoccus* complexes is not yet clear, but it is possible that they are involved in the assembly of the mitochondrial complexes and in the regulation of their redox and energy-conserving activities.

## CONCLUDING REMARKS

From the present comparison of the structure of the respiratory complexes in *Paracoccus* and mitochondria it can be concluded that in the evolution of mitochondria

**TABLE 1.** Polypeptide Subunit Composition of the Respiratory Complexes Isolated from Mitochondria, *Paracoccus denitrificans,* and *Rhodobacter sphaeroides*

| Complex | Mitochondria | *Paracoccus* | *R. sphaeroides* |
|---|---|---|---|
| I   NADH dehydrogenase | > 20[26] | ≥ 2 (or 3?)[28] | NA[a] |
| II  Succinate dehydrogenase | 3[30] | NA[a] | 2 (or 3?)[29] |
| III Ubiquinol-cytochrome *c* oxidoreductase | 8-10[32] | 3 (or 4?)[33] | 3 (or 4?)[35, 36] |
| IV  Cytochrome *c* oxidase | 7-12[50] | 2[49] | 3 (or 2?)[57] |

[a] NA means not available.

from a protomitochondrion resembling *Paracoccus,* the protomitochondrial respiratory complexes were retained to constitute the catalytic subunits of the mitochondrial complexes. Additional subunits, probably regulatory in function, must have been imposed upon the catalytic subunits by the nucleocytoplasm in order to help integrate respiration and oxidative phosphorylation in the protomitochondria with the energy demands of the rest of the cell. The present comparison of respiratory complexes isolated from *Paracoccus* and mitochondria has been undertaken in order to identify those proteins that were likely to have been inherited by mitochondria from their free-living, eubacterial ancestor, and to determine how those complexes were subsequently modified in the evolution of the organelle. Other protein components of *Paracoccus* such as the $F_1 F_0$-ATPase of the plasma membrane,[58] the porins of the outer envelope,[59,60] and the oxidative enzymes of the cytosol[31] may be structurally related to corresponding proteins in mitochondria as well. When future research reveals the extent of this relatedness, it should become possible to identify more accurately than at present the eubacterial contribution to the construction of present-day mitochondria.

# ACKNOWLEDGMENTS

I am grateful to Christopher Darwen for an initial literature search, to Howard Gest and Tina George for useful comments, and to Stuart Ferguson for much helpful advice.

## REFERENCES

1. JOHN P. & F. R. WHATLEY. 1975. Nature **254:** 495-498.
2. JOHN, P. & F. R. WHATLEY. 1977. Adv. Bot. Res. **4:** 51-115.
3. IMHOFF, J. F., H. G. TRÜPER & N. PFENNIG. 1984. Int. J. Syst. Bacteriol. **34:** 340-343.
4. DICKERSON, R. E. 1980. Nature **283:** 210-212.
5. AMBLER, R. P., T. E. MEYER & M. D. KAMEN. 1976. Proc. Nat. Acad. Sci. USA **73:** 472-475.
6. DAYHOFF, M. O. & R. M. SCHWARTZ. 1981. Ann. N.Y. Acad. Sci. **361:** 92-104.
7. MEYER, T. E., M. A. CUSANOVICH & M. D. KAMEN. 1986. Proc. Nat. Acad. Sci. USA **83:** 217-220.
8. WOESE, C. R., E. STACKEBRANDT, W. G. WEISBURG, B. J. PASTER, M. T. MADIGAN, V. J. FOWLER, C. M. HAHN, P. BLANZ, R. GUPTA, K. H. NEALSON & G. E. FOX. 1984. Appl. Microbiol. **5:** 315-326.
9. WOESE, C. R., E. STACKEBRANDT, T. J. MACKE & G. E. FOX. 1985. Appl. Microbiol. **6:** 143-151.
10. YANG, D., Y. OYAIZU, H. OYAIZU, G. J. OLSEN & G. R. WOESE. 1985. Proc. Nat. Acad. Sci. USA **82:** 4443-4447.
11. MACKAY, R. M., L. B. ZABLEN, C. R. WOESE & W. F. DOOLITTLE. 1979. Arch. Microbiol. **123:** 165-172.
12. MACKAY, R. M., D. SALGADO, L. BONEN, E. STACKEBRANDT & W. F. DOOLITTLE. 1982. Nucleic Acids Res. **10:** 2963-2970.
13. VILLANUEVA, E., K. R. LUERSON, J. GIBSON, N. DELIHAS & G. E. FOX. 1985. J. Mol. Evol. **22:** 46-52.
14. WOLTERS, J. & V. A. ERDMANN. 1984. Endocytol. C. Res. **1:** 1-23.
15. SATOH, T., Y. HOSHINO & H. KITIMURA. 1976. Arch. Microbiol. **108:** 265-269.
16. KLEMME, J-H., I. CHYLA & M. PREUSS. 1980. FEMS. Lett. **9:** 137-140.
17. MCEWAN, A. G., C. L. GEORGE, S. J. FERGUSON & J. B. JACKSON. 1982. FEBS Lett. **150:** 277-280.
18. MCEWAN, A. G., A. J. GREENFIELD, H. G. WETZSTEIN, J. B. JACKSON & S. J. FERGUSON. 1985. J. Bacteriol. **164:** 823-830.
19. MCEWAN, A. G., N. P. J. COTTON, S. J. FERGUSON & J. B. JACKSON. 1985. Biochim. Biophys. Acta **810:** 140-147.
20. PFENNIG, N. 1974. Arch. Microbiol. **100:** 197-206.
21. TRÜPER, H. G. & N. PFENNIG. 1978. *In* The Photosynthetic Bacteria. R. K. Clayton & W. R. Sistrom, Eds.: 19-27. Plenum Press. New York, N.Y.
22. SAUNDERS, V. A. & O. T. G. JONES. 1973. Biochim. Biophys. Acta **305:** 581-589.
23. FERGUSON, S. J. 1982. Biochem. Soc. Trans. **10:** 198-200.
24. BONNER, W. D., JR., S. D. CLARKE & P. R. RICH. 1986. Plant Physiol. **80:** 838-842.
25. ALBRACHT, S. P. J., H. W. VAN VERSEVELD, W. R. HAGEN & M. L. KALKMAN. 1980. Biochim. Biophys. Acta **593:** 173-186.
26. RAGAN, C. I. 1985. Coenzyme Q.: 315-336.
27. GEORGE, C. L. & S. J. FERGUSON. 1984. Eur. J. Biochem. **143:** 567-573.
28. GEORGE, C. L., S. J. FERGUSON, M. W. J. CLEETER & C. I. RAGAN. 1986. FEBS Lett. **198:** 135-139.

29. BARASSI, C. A., R. G. KRANZ & R. B. GENNIS. J. Bacteriol. **163:** 778-782.
30. HEDERSTEDT, L. & L. RUTBERG. 1981. Microbiol. Rev. **45:** 542-555.
31. HUSAIN, M. & D. J. STEENKAMP. 1985. J. Bacteriol. **163:** 709-715.
32. RIESKE, J. S. & S. H. K. HO. 1985. Coenzyme Q.: 337-363.
33. BERRY, E. A. & B. L. TRUMPOWER. 1985. J. Biol. Chem. **260:** 2458-2467.
34. LUDWIG, B., K. SUDA & N. CERLETTI. 1983. Eur. J. Biochem. **137:** 597-602.
35. GABELLINI, N., J. R. BOWYER, E. HURT & B. A. MELANDRI. Eur. J. Biochem. **126:** 105-111.
36. YU, L. & C-A. YU. 1982. Biochem. Biophys. Res. Commun. **108:** 1285-1292.
37. WOOD, P. M. 1980. Biochem. J. **189:** 385-391.
38. JOHN, P. & S. PAPA. 1978. FEBS Lett. **85:** 179-182.
39. THIERBACH, G. & H. REICHENBACH. 1983. Arch. Microbiol. **134:** 104-107.
40. AMBLER, R. P., T. E. MEYER, M. D. KAMEN, S. A. SCHICHMAN & L. SAWYER. 1981. J. Mol. Biol. **147:** 351-356.
41. TIMKOVICH, R. & R. E. DICKERSON. 1973. J. Mol. Biol. **79:** 39-56.
42. KUO, L. M. & H. C. DAVIES. 1983. Mol. Immunol. **20:** 805-810.
43. ERREDE, B. & M. D. KAMEN. 1978. Biochemistry **17:** 1015-1027.
44. TIMKOVICH, R., R. DHESI, K. J. MARTINKUS, M. K. ROBINSON & T. M. REA. Arch. Biochem. Biophys. **215:** 47-58.
45. DAVIES, H. C., L. SMITH & M. E. NAVA. 1983. Biochim. Biophys. Acta **725:** 238-245.
46. KUO, L. M., H. C. DAVIES & L. SMITH. 1985. Biochim. Biophys. Acta **809:** 388-395.
47. DALDAL, F., S. CHENG, J. APPLEBAUM, E. DAVIDSON & R. C. PRINCE. 1986. Proc. Nat. Acad. Sci. USA **83:** 2012-2016.
48. LUDWIG, B. 1980. Biochim. Biophys. Acta **594:** 177-189.
49. LUDWIG, B. & G. SCHATZ. 1980. Proc. Nat. Acad. Sci. USA **77:** 196-200.
50. KADENBACH, B. & P. MERLE. 1981. FEBS Lett. **135:** 1-11.
51. STEFFENS, G. C. M., G. BUSE, W. OPPLIGER & B. LUDWIG. 1983. Biochem. Biophys. Res. Commun. **116:** 335-340.
52. SOLIOZ, M., E. CARAFOLI & B. LUDWIG. 1982. J. Biol. Chem **257:** 1579-1582.
53. PUTTNER, I., M. SOLIOZ, E. CARAFOLI & B. LUDWIG. 1983. Eur. J. Biochem. **134:** 33-37.
54. PAETOW, B., G. PANSKUS & B. Ludwig. 1985. J. Inorg. Biochem. **23:** 183-186.
55. DOWNHAM, W., C. R. BIBUS & G. SCHATZ. 1985. EMBO J. **4:** 179-184.
56. KADENBACH, B. 1986. J. Bioenerg. Biomembr. **18:** 39-54.
57. GENNIS, R. B., R. P. CASEY, A. AZZI & B. LUDWIG. 1982. Eur. J. Biochem. **125:** 189-195.
58. WALKER, J. E., I. M. FEARNLEY, N. J. GRAY, B. W. GIBSON, F. D. NORTHROP, S. J. POWELL, M. J. RUNSWICK, M. SABASTE & V. L. J. TYBULEWICZ. 1985. J. Mol. Biol. **184:** 677-701.
59. ZALMAN, L. S. & H. NIKAIDO. 1985. J. Bacteriol. **162:** 430-433.
60. NAUYALIS, P. A., M. S. HINDAHL & B. J. WILKINSON. 1985. Biochim. Biophys. Acta **840:** 297-308.

# Cyanelles[a]

## From Symbiont to Organelle

HAINFRIED E. A. SCHENK, MANFRED G. BAYER,
AND DOUGLAS ZOOK

*Institute of Biology I*
*University of Tübingen*
*Auf der Morgenstelle I*
*D-7400 Tübingen, Federal Republic of Germany*

## INTRODUCTION: GENETIC MECHANISMS IN THE ESTABLISHMENT OF SYMBIOSES

Following Darwin, Ernst Mayr indicated the fundamental dualistic character of evolutionary processes. The initial reactions, the so-called tandem dualisms, advance evolution. Such dualisms are, e.g., those of genotype and phenotype or mutation and selection or, as a special case, symbiosis and selection, symbiosis functioning as a large-step quasi mutation. Even in the beginning stages of a symbiosis, such mutations would tend to stabilize the new consortium that must follow. Natural selection factors would then protect any symbiotically derived advantages. Further, evolution of such a consortium leads to mutual adaptations and coevolved processes. These adaptations may allow the dominant partner to colonize new ecological niches and even evolve new morphological structures and metabolic pathways. It is then not such a great step to see that an established endocytobiosis can result in a new species. Clearly, on the genetic level, particular events must have occurred that then come forth on the metabolic level of the phenotype. These events were described with the terms gene, chromosome, genome, or ploidy mutation. Very important evolutionary processes (TABLE 1) for the endocytobiont genome are point mutations, intrachromosomal recombination, inversions, duplications, additions (insertions), deletions, gene transfer, polyploidy, and segregation.

For all intents and purposes, these potential mechanisms in endocytobiont evolution are not yet understood. With this in mind, regarding possible substitution or import processes for metabolites and proteins into the endocytobiont, we would like to focus on the special case of endocyanomes.

[a]This work was supported by a grant from the Deutsche Forschungsgemeinschaft. Research work by D. Zook was made possible by a Fulbright Fellowship grant and by Clark University, Worcester, Mass.

## ENDOCYANOMES

Symbioses of cyanobacteria or cyanobacterial descendants with bacteria or eukaryotic unicellular or multicellular hosts (flagellates, rhizopodes, algae, fungi, lower and higher plants, sponges) are defined as syncyanoses by Pascher.[1] The whole consortium is a cyanome,[2–6] and the symbiotic cyanobacterium is a cyanelle.[2] The symbiosis is an ectocyanosis if the cyanobacteria are attached outside the host cells, or an endocyanosis if they live inside the cytoplasm of the usually colorless eukaryotic host cell (TABLE 2). The first discovery of an endocyanosis was described for the thecamoeba *Paulinella chromatophora* by Lauterborn in 1895.[17] After that time, more endocyanomes were discovered.[6,7] Some of these organisms have been classified for a relatively long time as cyanomes, but some have been proposed more recently as endocyanomes, such as *Cyanidium caldarium*[25,26] or *Porphyridium aerugineum*.[31] Many of the de-

TABLE 1. Possible Basic Evolutionary Processes on the Genetic Level Establishing an Endocytobiont as Organelle

| Evolutionary Mechanisms | Phenotypic Processes |
|---|---|
| 1. Mutation | Adaptation of enzymes (pH, substrate, substrate concentration, temperature, partial pressure of oxygen or carbon dioxide, morphological variation, etc.) |
| 2. Deletion | Tolerable following a substitution of <br> (a) metabolites; <br> (b) host proteins (protein import); <br> (c) ? RNA (RNA-import) |
| 3. Gene transfer | In consequence with or without <br> (a) inactivation of the gene; <br> (b) substitution with host metabolites; <br> (c) import of host or own (endocytobiontic) proteins coded on host genome from host into the guest compartment |

scribed endocyanomes were found only one or two times, and, for the most part, they are not stored in algal collections. This is a really important field for limnologists and others to (re)discover these and other consortia anew, including isolation, purification, and cultivation of these microbes. The reviews can serve as good guides.[1–6,15,21,33–37]

## THE "CAROTENOID KEY" FOR FINDING ENDOCYANOMES

Surely there are endocyanomes with more or less adapted partners. In some cases, the adaptation is so far evolved between the partners that it is not easy to discriminate between cyanomes and representatives of the established primitive red algae.

As help toward better discrimination between the different chromatophoric forms of plastids and the probably phylogenetically younger forms of the endocyanelles, we have proposed, in addition to the morphological criteria, a biochemical estimation

TABLE 2. Syncyanosis: Syncyanomes and Endocyanomes[a]

| | | | References |
|---|---|---|---|
| Hypo-<br>thetical<br>phylo-<br>genetic<br>age of<br>symbiosis | 1 | Hepaticae: *Anthoceros vincentianus* | 7 |
| | | *Blasia pusilla* | 7 |
| | | Filicatae: *Azolla filiculoides* | 7-9 |
| | | Cycadatae: e.g., *Cycas revoluta* | 7, 10, 11 |
| | 1 + 2 | Rosidae: *Gunnera macrophylla* | 7, 12 |
| | 2a | Porifera: e.g., *Aplysina aerophoba* | 13 |
| | | Phycomycetes: *Geosiphon pyriforme* | 14 |
| | | Bacillariophyta: *Rhizosolenia* spp. | 15 |
| | | *Hemiaulas membranaceus* | 15 |
| | 2b | *Epithemia* spp. | 15 |
| | | *Rhopalodia gibba* | 16, 15 |
| | | Chlorophyta: *Cyanomastix morgani* | 6, 15 |
| | | Thekamoebaea: *Paulinella* | |
| | | *chromatophora* | 17-19 |
| | | Cryptophyta: *Chroomonas gemma* | 6 |
| | | *Cryptella cyanophora* | 6 |
| | | *Cyanomonas americana* | 6 |
| | | Glaucophyta: *Gloeochaete* | |
| | | *wittrockiana* | 20, 32 |
| | | *Peliaina cyanea* | 6, 15 |
| | | *Cyanophora* | |
| | | *tetracyanea* | 15, 20 |
| | | *Cyanophora paradoxa* | 15, 21 |
| | | *Glaucocystis* | |
| | | *nostochinearum* | 22, 15 |
| | | (Glaucocystaceae) | 15, 33 |
| | | *Chalarodora azurea* | 6, 15 |
| | | *Cyanoptyche* | |
| | | *gloeocystis* | 6, 15 |
| | | *Glaucosphaera* | |
| | | *vacuolata* | 23, 24, 15 |
| | 3* | Rhodophyta: *Cyanidium* | |
| | | *caldarium* | 25-29 |
| | 3* | *Porphyridium* | |
| | | *aeruginosa* | 30, 31 |
| | 3* | *Asterocytis* | |
| | | *ramosa* | 30 |
| | 3 | *Asterocytis ornata* | 6 |
| | 3 | *Asterocytis* | |
| | | *smaragdina* | 6 |
| | 3 | *Chroothece* | |
| | | *mobilis* | 6 |

———Endocytobiotic adaptation stage (morphological characteristics)——▶

[a] Comparison of the hypothetical phylogenetic age of symbiotic consortia with the morphological adaptation stage of the partners: (1) endosomatic endosymbiosis (only intercellular syncyanoses); (2) intracellular endosymbiosis (endocytobiosis); (2a) facultative; (2b) obligate; (3) hypothetical endocyanomes; (3*) estimated with the carotenoid key.

(FIGURE 1 and TABLE 3), the "carotenoid key."[37] This key seems to allow a discrimination between the characteristics of cyanobacterial carotenoids and plastidic carotenoids of all other photoautotrophs. It is based on the occurrence of two enzyme systems—the α-cyclase and/or the 5,6-epoxidase—which have been found neither in cyanobacteria nor in some primitive rhodophytes. The proposed hypothesis is as follows. The occurrence of one or both of these enzymes in a eukaryotic, photoassimilating compartment (plastid) demonstrates a protein import from the eukaryotic

TABLE 3. Pigments of Thylakoids, as Possible Markers for the Plastid Nature of an Anucleated, Intracellular Photoassimilator[a]

| Phylum ("algae") | α-Cyclase | 5,6-ep- or de-ep- Oxidase | α-Carotenoids | 5,6-Epoxides or Potential 5,6- Epoxide Marker |
|---|---|---|---|---|
| Cyanobacteria | − | − | − | − |
| Glaucocystophyta[32] | − | − | − | − |
| Rhodophyta | | | | |
|   Bangiophyceae | (−) | (−)? | (−) | (−) |
|   Florideophyceae | + | (+)? | α-c, lut | (anthera-x) |
| Cryptophyta | + | +? | α-c | (croco-x) |
| Dinophyta | − | + | − | peridinin, (fuco-x) |
| Xanthophyta | (+) | + | ε-c | neo-x, vaucheria-x |
| Bacillariophyta | (+) | + | ε-c, α-c | fuco-x |
| Phaeophyta | − | + | − | fuco-x |
| Chloromonadophyta | (+) | (+) | lut-epox | anthera-x |
| Chrysophyta | − | + | − | fuco-x |
| Haptophyta | − | + | − | fuco-x |
| Eustigmatophyta | − | + | − | viola-x, vaucheria-x |
| Euglenophyta | − | + | − | neo-x |
| Chlorophyta (with Prasinophyceae) | + | + | α-c, lut | viola-x, neo-x |

[a] "Carotenoid key:" presence of C-phycocyanin, but no findings of α-cyclase or 5,6-epoxidase or -de-epoxidase could be an indication for the cyanobacterial origin of the intracellular assimilator which then could named: endocyanelle. +, present in all investigated species of the given phylum; −, not detected in all investigated species of the given phylum; specification in parentheses, is not significant for each species of the investigated phylum, α-c, α-carotene; lut, lutein; lut-epox lutein-epoxide; x, xanthin; ?, finding of a marker xanthophyll regarding the epoxidase is no indication for a de-epoxidase (or cyclic changes within the xanthophyll-cycle). (For detail, see references in Reference 37).

host into the assimilator. Therefore, if a chromatophoric structure is found in a cell that contains C-phycocyanin and no α-carotene derivatives or only carotenoids that do not stand in biochemical relation to the xanthophyll cycle, then one can assume that this structure is derived from an original endocytobiotic cyanobacterium-like ancestor.[37] This statement is corroborated if the photoassimilator is still surrounded with a murein sacculus (peptidoglycan) (for comparison see primitive "rhodophytes" in TABLE 2).

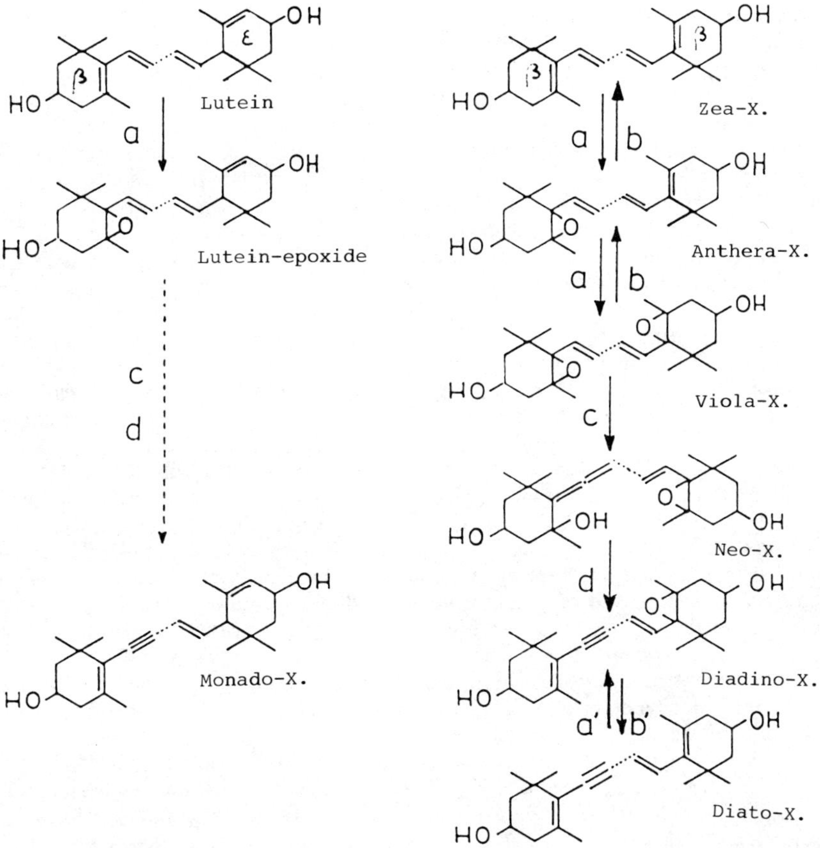

**FIGURE 1.** Postulated enzyme steps during biosynthesis of typical xanthophylls (for References see Reference 37). Left: α-carotene derivatives (ε: cyclization by α-cyclase); right: β-carotene derivatives (β: cyclization by β-cyclase); a or a′ = 5,6-epoxidase; b or b′ = 5,6-de-epoxidase; c = violaxanthin isomerase; d = allenic-dehydratase.

## *CYANOCYTA KORSCHIKOFFIANA*, THE CYANELLE OF *C. PARADOXA:* CRITERIA FOR CYANOBACTERIAL ORIGIN AND PLASTIDLIKE CHARACTERISTICS

The previously described characteristics for determination of an assimilator as an original endocytobiont apply to the endocyanelles in *C. paradoxa*[38] and in *Glaucocystis nostochinearum.*[22] In the following, we summarize observations that demonstrate the cyanobacterial origin of the *Cyanophora* endocyanelles. (1) Morphological criteria: (a) the *C. paradoxa* endocyanelles are coccoid, and they possess (b) a peripheral chromatoplasm, (c) a cyanobacterial-like inner envelope membrane,[39] and (d) a cyanobacterial-like dividing behavior.[35,39,40] (2) Biochemical characteristics: (a) the incomplete tricarboxylic acid cycle,[37,41] (b) presence and biosynthesis of C-phycocyanin,[30,42,43] (c) presence of the small subunit of the ribulose-bisphosphate-carboxylase,[44] (d) of the murein sacculus[38] with muraminic acid and diaminopimelinic acid,[36,45,46] (e) 5S rRNA sequence homology,[47] and (f) substructure of cyanelle genome.[48]

In contrast to the previous remarks, endocyanelles also possess some plastidlike characteristics. These endocyanelles have lost (or have never possessed) (a) respiration,[24,41] (b) nitrogen fixation,[49] (c) nitrate reductase, which is only found in the host cytoplasm,[50] (d) they possess nitrite reductase (like the chloroplasts of higher plants),[50,51] (e) a reduced circular 133.4 kilobase (kb) DNA genome[48,52,53] with considerable similarity to the higher plant plastid DNA regarding size and structure (e.g., two inverted repeat segments)[54] and (f) the lipid pattern (see next section). Most of these findings are reviewed by Trench[21] and Kies and Kremer[33] in detail.

Even if we apply very strong criteria to our hypotheses regarding the origin of phototrophic eukaryotic cells and their plastids (especially if we do not accept that the plastids have *endocytobiotic* prokaryotic ancestors), we must still deduce, given our current knowledge, that clearly the endocyanelles from *C. paradoxa* and *G. nostochinearum* derived originally from endocytobiotic cyanobacteria and, therefore, are not plastids *sensu stricto*. We are more uncertain concerning algae, like *C. caldarium, P. aerugineum,* and others, listed under the rhodopyhtes in TABLE 2, as to their status as endocyanelles. This in mind, it is interesting in the context of the serial endosymbiotic hypothesis (SEH) to investigate the adaptation steps of *Cyanocyta korschikoffiana,* the endocyanelle of *C. paradoxa.*

It has been demonstrated that these endocyanelles could be isolated, but not cultivated in an artificial medium.[55,56] It has also been shown that the photoassimilation of isolated cyanelles can be converted rapidly into a light-dependent oxygen consumption.[57] Additionally, it has been shown that cycloheximide has a very strong inhibitory effect on the biosynthesis of certain proteins and metabolites, also in the cyanelles of *C. paradoxa.*[21,58–61] In this paper we will refer to ongoing further experiments showing the high degree of adaptation of the cyanelles and their dependence on the host compartment. These experiments include (1) the estimation and identification of lipids and their respective fatty acid composition, and (2) protein synthesis in the endocyanelles of *C. paradoxa.*

## LIPIDS OF *CYANOPHORA PARADOXA*

The gas chromatographic identification of the fatty acids in *C. paradoxa* was very surprising[62–64] (FIGURE 2A). As main fatty acids, we found hexadecanoic (palmitic) acid, eicosatetraenoic (arachidonic) acid, and eicosapentaenoic acid.[63] In a lower concentration, we could still identify octadecanoic (stearic), octadecenoic (oleic), and octadecadienoic (α- and γ-linolenic) acids. The fatty acid pattern of the whole organism and of the isolated cyanelles (FIGURE 2B) is relatively similar. The pattern seems, in comparison with other classes of algae, rather like that of rhodophyta than that of other algal classes (TABLE 4). Despite several previous studies of fatty acids in cyanobacteria,[62,68] no one has found polyunsaturated eicosanoic acids in cyanobacteria. Therefore, we were at first unsure about our isolation method of cyanelles. To demonstrate the correctness of our findings regarding the presence of arachidonic acid and other polyunsaturated eicosanoic acids in the isolated cyanelles, we thought it would be revealing to isolate monogalactosyldiglyceride (MGDG). This lipid is known to be a main lipid of thylakoid membranes and should therefore be contained only

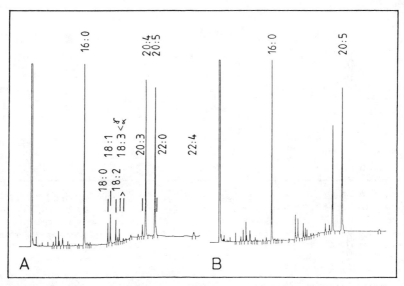

**FIGURE 2.** *C. paradoxa.* Gas chromatographic pattern of fatty acid methylesters. Transmethylation of fatty acids after extraction of the lipids: (A) from the whole organism; (B) from isolated endocyanelles. (Reproduced from Reference 63 with permission.)

or mainly in the cyanelles. The demonstration of arachidonic and the other C-20 acids in MGDG would indicate that our findings regarding the presence of these acids in the cyanelles could not result from undetected contamination of membranes of the eukaryotic compartment. Thin-layer chromatography of the lipids (FIGURE 3), identification, isolation, and rechromatography of MGDG—the main lipid in the endocyanelles—confirmed clearly the occurrence of the tetra-and pentaenoic C-20 acids[63] in the cyanelles (FIGURE 4: MGDG). It remains a very important question as to whether these acids are synthesized in the endocyanelles or are imported from the eukaryotic host compartment. Regarding the arachidonic acid, we believe the importation possibility more likely because (1) cyanobacteria have not been found with such acids and (2) cycloheximide shows an inhibitory effect on the biosynthesis of

**TABLE 4.** Fatty Acid Pattern of *C. korschikoffiana* in Comparison with Other Photosynthetic Organisms[a]

| Fatty Acid | Fatty Acid as Percent of Total Acid | | | | | | |
|---|---|---|---|---|---|---|---|
| | (1) | (2) | (3) | (4) | (5) | (6) | (7) |
| 10:0 | nd | nd | − | − | − | − | − |
| 12:0 | nd | nd | − | − | − | − | − |
| 14:0 | t | t | 2 | 8 | t | +, + + C | +, + + C |
| 15:0 | t | t | − | − | − | − | − |
| 16:0 | 31 | 39 | 23 | 30 | 18 | +, + + B | + + A |
| 16:1 | t | ? | 2 | 7 | t | + | + A |
| 18:0 | 1 | 1 | 2 | 11 | 7 | + | + C |
| 18:1 | 3 | 7 | 3 | 28 | 34 | + | +, + + C |
| 18:2 | 1 | 6 | 16 | 8 | 11 | + A, + + B | + B |
| 18:3 | 2 | 1 | t | 3 | 30 | + A | +, + + B |
| 18:4 | − | − | − | − | − | + C | + + A |
| 20:0 | − | − | − | − | − | − | + C |
| 20:1 | − | − | − | − | − | − | + + B |
| 20:2 | − | − | − | − | − | − | − |
| 20:3 | t-2 | t-2 | 2 | − | − | − | − |
| 20:4 | 8-19 | 15-22 | 36 | − | − | − | + C |
| 20:5 | 21-46 | 22-33 | 17 | − | − | − | +, + + B |
| 22:0 | t | 1-2 | − | − | − | − | − |
| 22:1 | − | − | − | − | − | − | − |
| 22:2 | − | − | − | − | − | − | − |
| 22:3 | − | ? | − | − | − | − | − |
| 22:4 | 6 | 6 | − | − | − | − | + B |
| 22:5 | − | − | − | − | − | − | + B |
| 22:6 | − | − | − | − | − | − | + B |

[a] (1) *C. paradoxa,* isolated endocyanelles;[62, 63] (2) *C. paradoxa,* whole organism;[62-64] (3) *Porphyridium cruentum;*[65] (4) *Porphyridium aeruginosum* (see references in Reference 62); (5) *Cyanidium caldarium* (20°C),[66] after Reference 67 *C. caldarium* has only traces of linolenic acid; (6) Cyanobacteria (see references in Reference 62);[68] (7) *Cryptomonades* 10 species (see references in Reference 62). Abbreviations: nd, not determined; −, not found; ?, too little for estimation; t, less than 1%; +, 1-10%; + +, 10%; A, found nearly in all investigated species; B, in half or most; C, only in a few.

arachidonic acid (Schenk and Zook, unpublished results). Also, our lipid identification investigations give more questions than answers.[68] In addition to a nearly cyanobacterial lipid pattern, including monogalactosyldiglyceride, digalactosyldiglyceride, phosphatidylglycerol, and sulfolipid, we find phosphatidylethanolamine (PE) and -choline (PC) in the cyanelles. The indication of PE seems to be due to contaminating mitochondria, although PC appears to be in both (cyanelle and host/mitochondrial) compartments. However, it is not totally clear whether PC is only synthesized in the host cytoplasm or also in the cyanelles.[64] Our recent investigations (Schenk and Zook, unpublished data, and Reference 68) show that PC is significantly inhibited by cycloheximide, thus indicating host cytoplasm synthesis. It is, therefore, possible that we have found—regarding the presence of arachidonic acid and PC in the cyanelles—the first demonstration of a host-dependent substitution process for important molecular structure elements in the membranes of an endocytobiont (2a in TABLE 1).

# PROTEIN IMPORT INTO THE ENDOCYANELLES OF
## *CYANOPHORA PARADOXA*

As we have seen in earlier experiments (FIGURE 5 summarizes our inhibition experiments regarding the biosyntheses of different pigments),[60,61] the biosynthesis of the phycobiliproteins is not only inhibited by chloramphenicol but also very strongly by cycloheximide (phycocyanin more than allophycocyanin!), the most commonly used eukaryotic translation inhibitor. Many experiments without radioactive labeling

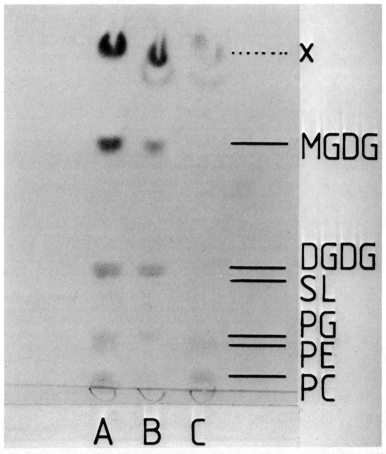

**FIGURE 3.** *C. paradoxa.* One-dimensional silica gel thin-layer chromatography. Solvent (Pohl *et al.,* Reference 68): benzene/acetone/water (30:91:8). (A) Whole organism. (B) Isolated cyanelles. (C) Mitochondria/host membranes fraction. DGDG = digalactosyldiglyceride; MGDG = monogalactosyldiglyceride; PC = phosphatidylcholine; PE = phosphatidylethanolamine; PG = phosphatidylglycerol; SL = sulfoquinovosyldiglyceride; x = position of pigments and neutral lipids.

of the proteins have confirmed these results. But the reason for this inhibition (metabolic or genetic dependency of the cyanelles from the host environment) remained unclear. The thorough studies of Löffelhardt and Bohnert[53,54] and others[48,69] on the cyanelle DNA has demonstrated a considerable similarity with the plastid DNA of higher plants regarding the size and structure (e.g., two inverted repeat segments) of the circular cyDNA as well as the organization of the tRNA genes on the cyDNA and the high degree of homology (received by hybridization experiments). Following these data, one could assume that the cyanelle chromosome is able to code for only approximately 120-150 proteins in the cyanelles (126, see Reference 21). But it was unclear how many proteins are necessary to assemble the cyanelles. Therefore, the

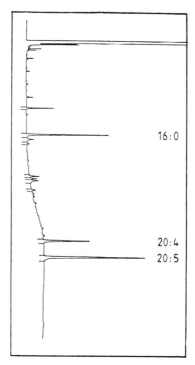

**FIGURE 4.** *C. Paradoxa.* MGDG, twice purified with TLC: gas chromatogram of transmethylated fatty acids. (Reproduced from Reference 63 with permission.)

possibility remained that all established deletions of original cyanelle enzymes were compensated only by import of obligate metabolites from host compartment and not by protein import.

But a two-dimensional electrophoretic separation of water-soluble cyanelle proteins shows without consideration of the membrane proteins the presence of more then 200 polypeptides in the endocyanelles (FIGURE 6). With this result in mind we could argue that at least 50-70 cyanelle proteins must be synthesized in the host cytoplasm. To clarify this question regarding the number of cyanelle proteins synthesized in the cyanelle or host compartment, we have carried out an inhibition experiment in which

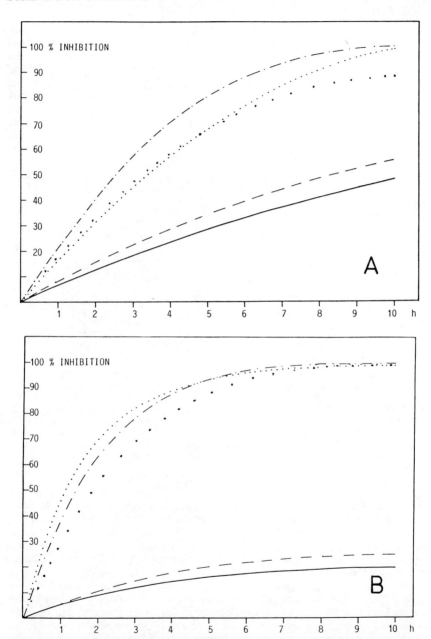

**FIGURE 5.** *C. paradoxa.* Summary[60,61] of inhibitory effects of cycloheximide (A) and chloramphenicol (B) on the biosynthesis of pigments (*in vivo* VIS measurements; inhibition as percent of control): solid line, inhibition of photosynthetic oxygen evolution; dashed line, inhibition of carotenoid synthesis; alternate dots and dashes, inhibition of chlorophyll synthesis; small-dotted line, inhibition of C-phycocyanin synthesis; large-dotted line, inhibition of allophycocyanin synthesis.

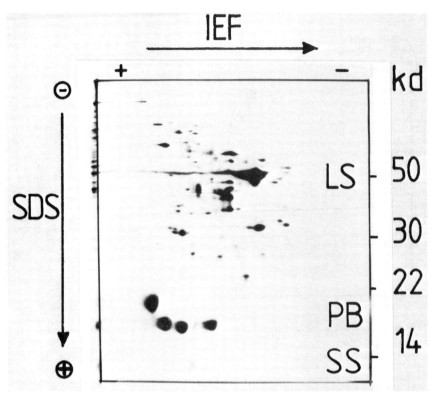

**FIGURE 6.** *C. paradoxa.* Autoradiography of a two-dimensional polyacrylamide gel electrophoretic (PAGE) separation of water-soluble proteins, extracted from isolated endocyanelles after photosynthetic *in vivo* protein synthesis of the whole organism in presence of (C-14) sodium bicarbonate. First dimension, IEF (isoelectric focusing); second dimension, SDS-PAGE; LS and SS, large and small subunit of ribulose-1,5-bisphosphate carboxylase; PB, subunits of phycobiliproteins.

we used C-14 to label the new synthesized proteins.[70] Different cultures of *C. paradoxa* growing in exponential phase under continuous light (3000 lx) were incubated with the translation inhibitors chloramphenicol (CA) or cycloheximide (CHI) and 10 minutes later with radioactive $NaHCO_3$. After 4 hours, the cultures were harvested and the cyanelles were separated from the cytoplasm by osmotic shock. A further fractionation gives finally three extracts: host cytosol, water-soluble proteins of cyanelles, and membrane-bound proteins of cyanelles (solubilized with a neutral detergent). The proteins were separated in a two-dimensional ultrathin electrophoresis (using isoelectric focusing and sodium dodecyl sulfate electrophoresis) and visualized with silver staining and autoradiography. We show here the results regarding the water-soluble proteins of the cyanelle compartment (FIGURES 6, 7, and 8). The biosynthesis of the most-water-soluble proteins in the cyanelles is inhibited by CHI (FIGURE 7) (It seems that Wasman and Wolk have gotten similar results.)[71] CA inhibits the biosynthesis of only a few peptides (FIGURE 8): the 4 subunits of the phycobiliproteins, the large and small subunit of ribulose-bisphosphate-carboxylase, and approximately 20 other as yet unidentified polypeptides. These are only about 10% of the total amount of the approximately 220 (until now visualized) soluble proteins. The biosynthesis of the remaining peptides (until now approximately more than 200) should take place on the 80S ribosomes of the host cell. As a proof of these results, we superimposed the two (CHI and CA) polypeptide patterns (FIGURES 7 and 8). If the experiments were correct, then we should receive the pattern of the control (FIGURE 6). This is nearly the case.[70]

We interpret this fact as giving biochemical support to the SEH: namely, we have now shown that, obviously, many proteins of the original endocytobiont are synthesized on 80S ribosomes and therefore they must be imported from the eukaryotic into the prokaryotic compartment. The mechanisms appear totally unclear regarding movement through the remnant cell wall. There is perhaps[72] analogous movement with chloroplast envelopes. If so many cyanelle proteins are synthesized on 80S ribosomes, then it is not only probable that they are coded in the eukaryotic genome, but also that most of them have been integrated in the eukaryotic genome by gene transfer from endocytobiont to host genome, during the coevolution of both the host flagellate and the endocyanelle.

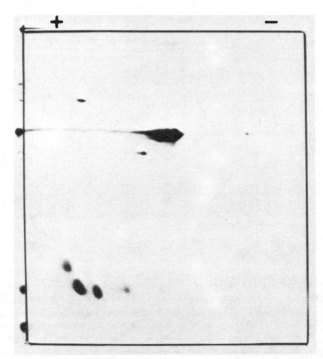

**FIGURE 7.** *C. paradoxa.* As in FIGURE 6, but during *in vivo* protein synthesis, additionally, with cycloheximide (3 µg/ml culture).

## CONCLUSIONS

Regarding firstly the fatty acid substitution in the cyanelle lipids and secondly the synthesis of so many cyanelle proteins on 80S ribosomes, it seems to us possible, in spite of the outstanding proof of the postulated gene transfer, to make the statement that some endocyanomes are more than just model organisms for the evolution of the

**FIGURE 8.** *C. paradoxa.* As in FIGURE 6, but during *in vivo* protein synthesis, additionally, with chloramphenicol (300 μg/ml culture).

plastid-containing cells, but rather are living fossils and perhaps—in a parallel lineage with the evolution of rhodoplasts (plastids of rhodophytes)—missing links in the evolution to "cyanoplasts,"[36,37] in which, at least in *C. paradoxa,* the original endocytobiont has clearly reached physiologically[73] as well as genetically the status of an organelle.[70] These results combined with those of the structure of cyanelle DNA[52,53] and compared with the characteristics of plastids seem to be direct proof along with the discoveries from Gibbs[74] for the SEH and elevate this hypothesis regarding the plastid evolution to a relatively well-founded theory (SET).

# SUMMARY

Cyanomes are known as model organisms for study in the mechanisms of plastid evolution with reference to the serial endosymbiotic hypothesis (SEH). Following a brief introduction of history of cyanelle research, including a definition and phenomenological description of endocyanomes, it is proposed that the "carotenoid key" can be used for biochemical detection of additional endocyanomes. Important adaptation mechanisms for evolution of an endocytobiont to a plastidlike organelle are deletions as well as metabolic substitutions and protein import from host into the endocytobiont. These are demonstrated for the cyanelles of *C. paradoxa,* with particular reference to (a) their genome size and structure, (b) the donation of arachidonic acid, possibly also of the transfer of phosphatidylcholine, from host to cyanelle, and (c) the import of about 200 proteins, synthesized on 80S ribosomes. These results give the SEH the status of a theory.

# ACKNOWLEDGMENTS

We thank our colleagues Prof. Dr. K. Poralla, Dr. H. G. Frank, and Prof. Dr. U. Schwarz (all Tübingen) without whose technical and financial assistance this work would not be possible, and Prof. Dr. L. Kies and Dr. B. P. Kremer for allowing us to read their very interesting, but still unpublished article about the Glaucophyta.[33]

## REFERENCES

1. PASCHER, A. 1914. Ber. Dtsch. Bot. Ges. **32:** 339-352.
2. PASCHER, A. 1929. Jahrb. Wiss. Bot. **71:** 386-462.
3. GEITLER, L. 1936. Schizophyzeen. *In* Handbuch der Pflanzenanatomie. G. Tischler & A. Pascher, Eds. **6**(1B): 106-117. Borntraeger. Berlin, Germany.
4. GEITLER, L. 1959. Syncyanosen. *In* Handbuch der Pflanzenphysiologie. W. Ruhland, Ed. **11:** 530-545. Springer. Berlin, FRG.
5. TAYLOR, D. L. 1970. Int. Rev. Cytol. **27:** 29-64.
6. PRINGSHEIM, E. G. 1958. Organismen mit blaugrünen Assimilatoren. Stud. Plant. Physiol. Praha: 165-184.
7. SCHAEDE, F. 1962. *In* Die pflanzlichen Symbiosen. F. H. Meyer, Ed. Fischer Verlag. Stuttgart, FRG.
8. TYAGI, V. V. S., B. C. MAYNE & G. A. PETERS. 1980. Arch. Microbiol. **128:** 41-44.
9. GATES, J. E., R. W. FISHER, T. W. GOGGIN & N. I. AZROLAN. 1980. Arch. Microbiol. **128:** 126-129.
10. CAIOLA, M. G. 1980. New Phytol. **85:** 537-544.
11. PERRAJU, B. T. V., A. N. RAI, A. P. KUMAR & H. N. SINGH. 1986. Symbiosis **1:** 239-250.
12. NEUMANN, D., M. ACKERMANN & F. JACOB. 1970. Biochem. Physiol. Pflanz. **161:** 483-498.
13. WILKINSON, C. R. 1983. *In* Endocytobiology II. Intracellular Space as Oligogenetic Ecosystem. H. E. A. Schenk & W. Schwemmler, Eds.: 993-1002. De Gruyter. Berlin, FRG.
14. MOLLENHAUER, D. 1970. Nat. Mus. **100:** 213-223.
15. KIES, L. 1984. Biol. Rdsch. **22:** 145-157.

16. FLOENER, L. & H. BOTHE. 1980. *In* Endocytobiology. Endosymbiosis and Cell Biology. W. Schwemmler & H. E. A. Schenk, Eds.: 541-552. De Gruyter. Berlin, FRG.
17. LAUTERBORN, R. 1895. Zeitschr. wiss. Zool. **59:** 537-544.
18. KIES, L. 1974. Protoplasma **80:** 69-89.
19. KIES, L. & B. P. KREMER. 1979. Naturwissenschaften **66:** 578.
20. KIES, L. 1976. Protoplasma **87:** 419-446.
21. TRENCH, R. K. 1982. *In* Progress in Phycology Research F. E. Round & D. J. Chapman, Eds. **1:** 257-288. Elsevier Biomedical Press. Amsterdam, Holland.
22. SCOTT, O. T., R. W. CASTENHOLZ & H. T. BONNETT. 1984. Arch. Microbiol. **139:** 130-138.
23. RICHARDSON, F. L. & T. E. BROWN. 1970. J. Phycol. **6:** 165-171.
24. KREMER, B. P., L. KIES & A. ROSTAMI-RABET. 1979. Z. Pflanzenphysiol. **92:** 303-317.
25. KREMER, B. P., G. B. FEIGE & HJ. A. W. SCHNEIDER. 1978. Naturwissenschaften **65:** 157-158.
26. KREMER, B. P. 1983. *In* Endocytobiology II. Intracellular Space as Oligogenetic Ecosystem. H. E. A. Schenk & W. Schwemmler, Eds.: 963-970. De Gruyter. Berlin, FRG.
27. FUKUDA, I. 1981. Literature survey of *Cyanidium caldarium* III. *In* Science: 11-15. University of Tokyo. Tokyo, Japan.
28. NAGASHIMA, H. & I. FUKUDA. 1983. Phytochemistry **22:** 1949-1951.
29. SECKBACH, J., J. F. FREDRICK & D. J. GARBARY. 1983. *In* Endocytobiology II. Intracellular Space as Oligogenetic Ecosystem. H. E. A. Schenk & W. Schwemmler, Eds.: 947-962. De Gruyter. Berlin, FRG.
30. CHAPMAN, D. J. 1966. Arch. Microbiol. **55:** 17-25.
31. ZETSCHE, K., M. KALING & K. STEINMÜLLER. 1983. *In* Endocytobiology II. Intracellular Space as Oligogenetic Ecosystem. H. E. A. Schenk & W. Schwemmler, Eds.: 971-978. De Gruyter. Berlin, FRG.
32. KIES, L. & B. P. KREMER. 1986. Taxon **35:** 128-133.
33. KIES, L. & B. P. KREMER. 1987. *In* Handbook of Protoctists. L. Margulis, D. J. Chapman & J. Corliss, Eds. Jones & Bartlett. Boston, Mass.
34. SCHILLER, J. 1954/55. Arch. Protistenkd. **100:** 116-126.
35. KIES, L. 1984. *In* Compartments in Algal Cells and Their Interaction. W. Wiessner, D. Robinson & R. C. Starr, Eds.: 191-199. Springer. Berlin, FRG.
36. SCHENK, H. E. A. 1973. Chloroplasten Evolution in biochemischer Sicht. Habilitation: 120. University of Tübingen. Tübingen, FRG.
37. SCHENK, H. E. A. 1977. Arch. Protistenkd. **119:** 274-300.
38. SCHENK, H. E. A. 1970. Z. Naturforschung **25b:** 640, 656.
39. GIDDINGS, T. H., C. WASMAN & L. A. STAEHELIN. 1983. Plant Physiol. **71:** 409-419.
40. PICKETT-HEAPS, J. 1972. New Phytol. **71:** 561-567.
41. ROSTAMI-RABET, A. 1980. Untersuchungen zur Enzymausstattung der blaugrünen Endosymbionten (Cyanellen) von Cyanophora paradoxa Korsch. Ph.D. Dissertation. University of Hamburg. Hamburg, FRG.
42. LEMAUX, P. G. & A. GROSSMAN. 1984. Proc. Nat. Acad. Sci. USA **81:** 4100-4104.
43. BRYANT, D. A., *et al.* 1985. Proc. Nat. Acad. Sci. USA **82:** 3242-3246.
44. HEINHORST, S. & J. M. SHIVELY. 1983. Nature **304:** 373-374.
45. HEINZ, G. 1973. Versuche zur Isolierung der lysozymempfindlichen Stützmembran von Cyanocyta korschikoffiana, der Endocyanelle aus Cyanophora paradoxa KORSCH. Dipl. Arb. University of Tübingen. Tübingen, FRG.
46. AITKEN, A. & R. Y. STANIER. 1979. J. Gen. Microbiol. **112:** 219-223.
47. DELIHAS, N. & G. E. FOX. 1987. Ann. N.Y. Acad. Sci. (This volume.)
48. LAMBERT, D. H., *et al.* 1985. J. Bacteriol. **164:** 659-664.
49. BOTHE, H. & L. FLOENER. 1978. Z. Naturforsch. **33c:** 981-987.
50. BÖTTCHER, U., *et al.* 1982. Z. Pflanzenphysiol. **106:** 167-172.
51. FLOENER, L., G. DANNEBERG & H. BOTHE. 1982. Planta **156:** 70-77.
52. HERDMAN, M. & R. Y. STANIER. 1977. FEMS Microbiol. Lett. **1:** 7-11.
53. LÖFFELHARDT, W., H. MUCKE & H. J. BOHNERT. 1980. *In* Endocytobiology. Endosymbiosis and Cell Biology. W. Schwemmler & H. E. A. Schenk, Eds.: 523-530. De Gruyter. Berlin, FRG.

54. BOHNERT, H. J. & W. LÖFFELHARDT. 1984. *In* Compartments in Algal Cells and Their Interaction. W. Wiessner, D. Robinson & R. C. Starr, Eds.: 58-68. Springer. Berlin, FRG.
55. EDELMAN, M., *et al.* 1967. Bacteriol. Rev. **31:** 315-331.
56. TRENCH, R. K., *et al.* 1978. Proc. R. Soc. London Ser. B **202:** 423-443.
57. HANF, J. & H. E. A. SCHENK. 1980. *In* Endocytobiology. Endosymbiosis and Cell Biology. W. Schwemmler & H. E. A. Schenk, Eds.: 531-539. De Gruyter. Berlin, FRG.
58. TRENCH, R. K. & H. C. SIEBENS. 1978. Proc. R. Soc. London Ser. B **202:** 473-482.
59. MARTEN, S., P. BRANDT & W. WIESSNER. 1982. Planta **155:** 190-192.
60. NEUMÜLLER, M. & H. E. A. SCHENK. 1983. *In* Endocytobiology II. Intracellular Space as Oligogenetic Ecosystem. H. E. A. Schenk & W. Schwemmler, Eds.: 451-463. De Gruyter. Berlin, FRG.
61. NEUMÜLLER, M. & H. E. A. SCHENK. 1983. Z. Naturforsch. **38c:** 972-977.
62. SCHENK, H. E. A., *et al.* 1985. Endocyt. Cell Res. **2:** 233-238.
63. ZOOK, D., *et al.* 1986. Endocyt. Cell Res. **3:** 99-103.
64. KLEINIG, H., *et al.* 1986. Z. Naturforsch. **41c:** 169-171.
65. NICHOLS, B. W. & R. S. APPLEBY. 1969. Phytochemistry **8:** 1907-1915.
66. KLEINSCHMIDT, M. G. & V. A. MCMAHON. 1970. Plant Physiol. Lancaster **46:** 286-289.
67. IKAN, R. & J. SECKBACH. 1972. Phytochemistry **11:** 1077-1082.
68. ZOOK, D. & H. E. A. SCHENK. 1986. Endocyt. Cell Res. **3:** 203-211.
69. KO, K., J. M. JAYNES & N. A. STRAUS. 1985. Plant Sci. **42:** 115-123.
70. BAYER, M. & H. E. A. SCHENK. 1986. Endocyt. Cell Res. **3:** 197-202.
71. WASMAN, C. & C. P. WOLK. 1980. Annual Report **15:** 96-97. MSU-DOE Plant Research Laboratory. Michigan State University. East Lansing, Mich.
72. HARINGTON, A. & A. L. THORNLEY. 1982. J. Mol. Evol. **18:** 287-292.
73. MARTEN, S. & P. BRANDT. 1984. *In* Compartments in Algal Cells and Their Interaction. W. Wiessner, D. Robinson & R. C. Starr, Eds.: 69-75. Springer. Berlin, FRG.
74. LUDWIG, M. & S. P. GIBBS. 1985. Protoplasma **127:** 9-20.

# Phylogenetic and Phenotypic Relationships between the Eukaryotic Nucleocytoplasm and Thermophilic Archaebacteria[a]

DENNIS G. SEARCY

*Zoology Department*
*University of Massachusetts*
*348 Morrill Science Center*
*Amherst, Massachusetts 01003*

## INTRODUCTION

The evolutionary origin of the eukaryotic nucleus and cytoplasm has been a subject of considerable debate. It has been difficult to formulate a logical "story" for why mitochondria might have originated from intracellular symbionts rather than by compartmentation of a single ancestral cell. This point was clearly made by Raff and Mahler, who in 1972 made a plausible case for such a nonsymbiotic origin of mitochondria.[3]

However, more recent sequence-homology studies have shown that chloroplasts and mitochondria are unmistakably related to certain groups of free-living eubacteria.[2,4,5] In addition, both chloroplasts and mitochondria have numerous phenotypic characters consistent with their bacterial origins.[6-8]

Similarly, sequence-homology studies suggest that the eukaryotic cytoplasm is most nearly related to the thermophilic Archaebacteria (see later). In addition, the thermophilic Archaebacteria have phenotypic features that resemble those of eukaryotic cells. These features undoubtedly function in the Archaebacteria as adaptations to their extreme environments. However, they may have been preadaptations for the evolution of the eukaryotic cells, as will be discussed below.

Thus, an evolutionary story is becoming increasingly evident for how and why eukaryotic cells might have originated from an Archaebacterial ancestor. This will be described with particular reference to the organism most nearly resembling the hypothetical ancestor of the nucleus and cytoplasm, *Thermoplasma acidophilum.*

But first we should review some of the sequence-homology studies that have examined the question of eukaryotic origins. Woese has argued forcefully that the

---

[a] "Thermophilic Archaebacteria" is a term synonymous, more or less, with the following taxa: Thermo-acidophilic Archaebacteria, Thermoproteales, Sulfolobales, *Caldariella*-group, Sulfur-dependent Archaebacteria, and Eocytes. *Thermoplasma acidophilum* is phenotypically within this group, but is not closely related to any other known species according to the criterion of nucleic acid homology.[1,2]

three Urkingdoms (Eubacteria, Archaebacteria, and Eukaryotes) all diverged more or less simultaneously from a single ancestor: the "Progenote."[2,9] However, if one examines his recent phyletic tree, the estimated genetic distance from the Eukaryotes to the Archaebacteria is shorter than the distance from the Eukaryotes to the Eubacteria.[2] The authors infer this to indicate that the Archaebacteria have evolved at a slower rate than have the other two main branches of evolution. They suggest that the "Progenote," which was the ancestor also of the eukaryotic branch, probably resembled most nearly a thermophilic Archaebacterium.

Hunt et al. analyzed similar data, but came to a somewhat different conclusion: that the Eubacteria and Archaebacteria diverged first, and then more recently the eukaryotic cytoplasm separated from the Archaebacteria.[5] Their analyses included sequences of 16S rRNAs, 5S rRNAs, and proteins. Although less prominently published than the study of Woese et al., this study is noteworthy because it comes from the group founded by M. O. Dayhoff, and which has considerable expertise in the analysis of evolutionary sequence data. They established the ancestral "root" of their tree from ferridoxin data. Then, the most complete set of data available was that for 5S rRNAs, from which they concluded that the closest relatives to the eukaryotic nucleocytoplasm are the thermophilic Archaebacteria. T. acidophilum was found to be the prokaryote highest on the branch leading to the eukaryotic cytoplasm.

However, since no modern eukaryotes are thermophilic, on first thought it should appear unlikely that the ancestor of eukaryotic cells could have been thermophilic. Nevertheless, there are a number of eukaryotic proteins that are exceptionally heat stable and might be vestiges of a thermophilic ancestor. These include actin, calmodulin, and the Cu,Zn-superoxide dismutase. Eukaryotic histones are heat stable by themselves, and in addition are highly effective in protecting DNA against thermal denaturation. Remarkably, all of these proteins are thought to be diagnostic of eukaryotic cells, since they are found in all groups of eukaryotes, but classically are assumed to be absent from prokaryotes. Nevertheless, each one is similar to a protein that is present in T. acidophilum.

In the thermophilic Archaebacteria these proteins evolved in response to the harsh environment. Yet certain of the features may have served also as preadaptations for the origin of eukaryotes. The most obvious of these features will be discussed below, including the presence of a histonelike protein, a cytoskeleton, calcium-binding protein, and fermentative energy metabolism.

## HISTONELIKE PROTEIN

T. acidophilum has a small basic protein ubiquitously associated with its DNA.[10] This protein contains 22% basic amino acid residues and is soluble in 0.2 M $H_2SO_4$, which are properties identical to those of eukaryotic histones. The amino acid sequence of the protein is known.[11] In the N-terminal 22 residues, there is a small but significant homology with calf histone H2A. There also is a trace of homology with all of the other eukaryotic histones except for histone H1.[12,13]

The physiological function of the protein in T. acidophilum appears to be in stabilizing the DNA against thermal denaturation.[14] As shown in FIGURE 1, in low ionic concentrations the protein can increase the melting temperature of DNA by about 40°C. T. acidophilum normally grows at 59°C, but can withstand exposure to temperatures as high as 80°C. For osmotic reasons, the cells have an exceptionally

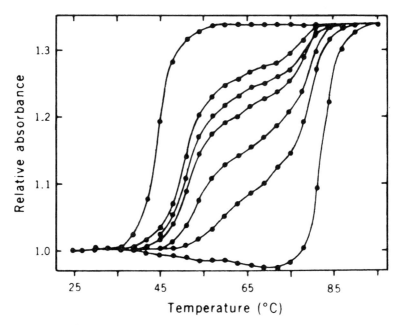

**FIGURE 1.** Stabilization of DNA against thermal denaturation by the *T. acidophilum* histonelike protein. Thermal stabilization is assumed to be the physiological function of this protein. Calf thymus DNA was combined with increasing amounts of the histonelike protein in the following ratios of protein: DNA (w/w): 0, 0.61, 0.80, 0.97, 1.34, 1.82, and 2.43, left to right. The nucleoprotein was dissolved in 0.25 mM sodium ethelenediaminetetraacetate (EDTA), pH 8. Denaturation of the DNA was detected by the increase in absorbance at 260 nm, which has been normalized to the same end points. (Reproduced from Reference 14 with permission. ©1978 American Association for the Advancement of Science.)

low intracellular ionic concentration, which has the effect of destabilizing DNA. Without the chromosomal protein, it appears that the chromosomal DNA would indeed undergo thermal denaturation.[14]

An additional property of the protein is that it condenses DNA into globular subunits (FIGURE 2). These subunits resemble the structure of eukaryotic nucleosomes, except that they are smaller and contain only a single coil of DNA. The physiological significance of the subunit structure is not known. However, the DNA is condensed, and so this may have been a preadaptation for the evolution of condensed eukaryotic chromatin, and of larger and more complex genomes.

Homologous histonelike proteins occur also in eubacterial cells, such as *Escherichia coli.*[15,16] However, the *E. coli* protein is not highly basic, and not acid soluble. It was effective in condensing the DNA only in ionic concentrations that were significantly below physiological. There have been reports suggesting that the *E. coli* histonelike protein HU is not associated with the chromosomal DNA,[17] that it has a higher affinity for RNA than for DNA,[18] and that it cannot be detected immunologically within the nucleoid of frozen and sectioned bacterial cells (E. Kellenberger, personal communication). Thus, the functions of the eubacterial histonelike proteins remain controversial.

In contrast, the *T. acidophilum* histonelike protein is bound tightly and unambiguously to the chromosome.[10] In ionic concentrations up to 0.5 M NaCl, which is about 10 times higher than physiological, it condenses the DNA into nucleosome-like particles. The deformation of the DNA can be detected by changes in circular dichroism (J. Cook and D. Searcy, in preparation). In conclusion, the properties of the nucleoprotein complex in *T. acidophilum* are different from those in any other prokaryote yet described.

## CYTOSKELETON

An actin-based cytoskeleton is apparently a universal feature of eukaryotic cells, but so far has not been shown convincingly in any prokaryote. There had been reports earlier of actinlike proteins in Eubacteria, but these were found later to be the ribosomal elongation factor EF-Tu. The Eubacterial mycoplasmas may have some type of cytoskeleton, but it is apparently not homologous with that of eukaryotes. For a review see Maniloff.[19]

However, the thermophilic Archaebacteria are often without rigid cell walls, and have a plastic cellular shape. Some have segmented cell walls resembling medieval chain mail. *T. acidophilum* is without any cell wall whatever. Elongated pleomorphic cells of *T. acidophilum* can be seen in living cultures at 59°C. At room temperature these cells quickly contract and become spherical, suggesting the presence of a labile and possibly dynamic cytoskeleton.[20]

There are several indications that a myosinlike protein may be present in *T. acidophilum*. For example, when a clear cytoplasmic extract was incubated, a precipitate spontaneously formed.[20,21] The precipitate could be redissolved by the addition of adenosine triphosphate (ATP). Upon further incubation the precipitate formed again, but could be dissolved several times sequentially by additions of more ATP,

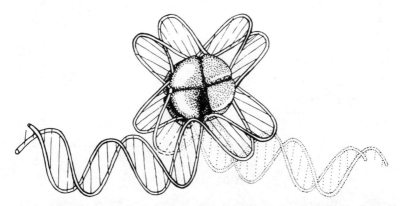

**FIGURE 2.** Nucleosome-like structure of the *T. acidophilum* nucleoprotein complex. Each particle consists of a tetramer of protein subunits around which are coiled 40 base pairs of DNA. This model was established by studies that utilized nuclease digestion, protein cross-linking, and electron microscopy. (Reproduced from Reference 48 with permission. ©1982 Elsevier Biomedical Press.)

as shown in FIGURE 3. This behavior is suggestive of eukaryotic actomyosin, particularly from non-muscle cells, where ATP is required for separation of the actin-myosin complex.[22,23] In addition, in *T. acidophilum* whole cell extracts there is an ATPase activity. When $10^{-6}$ M $Ca^{2+}$ was added, the ATPase activity increased up to threefold (preliminary results). This observation suggests that micromolar concentrations of $Ca^{2+}$ may indeed have a regulatory function in *T. acidophilum*, as in eukaryotic cells. Finally, the ATPase is active in 0.5 M KCl plus 10 mM ethylene-diaminetetraacetate; in this buffer only myosin-type ATPases are reported to be active.[23]

In cellular extracts that were negatively stained and then examined by electron microscopy, short filaments of about 6 nm were visible. This diameter is characteristic of a eukaryotic actin filament. However, rabbit heavy meromyosin did not bind to these filaments,[20] and so a close homology with eukaryotic actin is not indicated. Nevertheless, the possibility of distant homology remains, and will be tested by isolating and sequencing the actinlike protein (work in progress).

Regarding the plastic cell shape of the thermophilic Archaebacteria, one possible reason for this unusual adaptation can be seen in the accompanying scanning electron micrograph of *Sulfolobus brierleyi*, which is shown growing on iron pyrites (FIGURE 4). The cells appear to flatten out on the surface of the pyrites, which is oxidized for metabolic energy. Contact with the substratum may be advantageous for this type of metabolism,[24,25] and the flexible cell shape may be an adaptation for greater surface

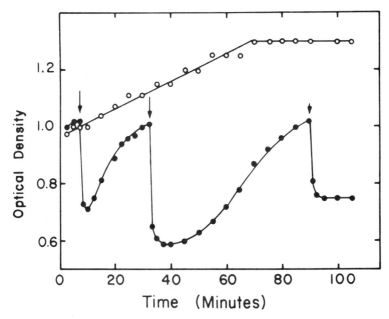

**FIGURE 3.** Dissociation of precipitated *T. acidophilum* cytoplasmic extract by repeated additions of adenosine triphosphate (ATP). Such ATP-dependent dissociation is typical of actin-myosin interactions: open circles, no ATP added; filled circles, ATP added, where arrows indicate successive additions of 0.25 mM, 0.5 mM, and 1.0 mM ATP. The optical density due to light scattering was measured at 540 nm. (Reproduced from Reference 21 with permission. ©1978 Elsevier/North-Holland Scientific Publishers.)

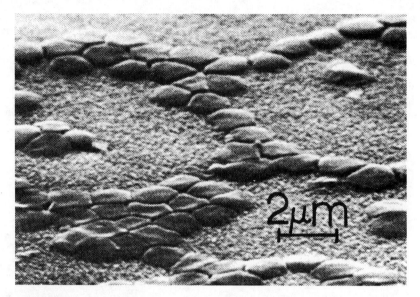

**FIGURE 4.** Scanning electron micrograph of *Sulfolobus acidocaldarius* adhering to the surface of pyrites ($FeS_2$). The ability of certain sulfur-dependent Archaebacteria to flatten onto their substrates may account for the evolutionary origin of the cytoskeleton. (Photograph by Vinod K. Berry.[25] Reproduced with permission.)

contact. However, certain mesophilic Eubacteria with rigid cell walls, such as *Thiobacillus ferrooxidans,* can also oxidize pyrites, but at room temperature. The hot acid environment of these Archaebacteria may be part of the explanation also for the absence of rigid cell walls.

One can speculate that an active ameboid cytoskeleton first evolved in thermophilic Archaebacteria for reasons related to lithotrophic respiration. Subsequently, the cytoskeletal proteins and flexible cell shape might have been preadaptations for the evolution of endocytosis, compartmentalized cells, and other functions now identified with eukaryotes.

## CALCIUM-BINDING PROTEINS

The protein calmodulin has been identified in all groups of eukaryotes.[26] Thus, adoption of $Ca^{2+}$ as an intracellular messenger and the evolution of the $Ca^{2+}$-receptor protein "calmodulin" occurred early in evolution. Nevertheless, calmodulin has not yet been reported in any prokaryotic cell. However, we recently have found evidence of a probable homologue of calmodulin in *T. acidophilum,* as described below.

The first observations were of a pharmacological nature. The growth of *T. acidophilum* was inhibited by $10^{-7}$ M Calmidazolium (a calmodulin inhibitor) and by $10^{-8}$ M A23187 (a calcium ionophore).[27] In contrast, in similar conditions the growth of

*Bacillus acidocaldarius* was not affected. These observations suggested the presence in *T. acidophilum* of an intracellular messenger system utilizing $Ca^{2+}$.

Additional evidence was obtained when soluble extracts of *T. acidophilum* were analyzed by two-dimensional polyacrylamide gel electrophoresis. The first dimension was run without $Ca^{2+}$, and the second dimension was run with 5 mM $CaCl_2$ present. Since most proteins are indifferent to the presence of $Ca^{2+}$, they fall into a diagonal line in the two-dimensional gel (FIGURE 5). However, the mobility of one protein increased significantly in the $CaCl_2$ dimension, thus falling below the diagonal line. This behavior is characteristic of eukaryotic calmodulin.[26]

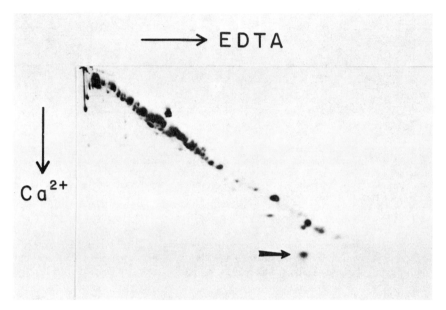

**FIGURE 5.** Two-dimensional polyacrylamide gel electrophoresis of a *T. acidophilum* cytoplasmic extract. The arrow indicates a protein specifically affected by calcium, similar to eukaryotic calmodulin. An extract was made in 10 mM sodium etheleneglycol-bis-(b-aminoethyl ether) *N,N'*-tetraacetate (EGTA), pH 7. After centrifugation at 100,000 × *g*, the supernatant solution was analyzed by electrophoresis in an 18% polyacrylamide gel containing 0.1% sodium dodecylsulfate. The first dimension contained 0.5 mM EDTA. The second dimension contained 5 mM $CaCl_2$. (From M. Colón and D. Searcy, unpublished results.)

For controls, the same experiment was performed with extracts of *E. coli* and of rabbit brain. With *E. coli* no proteins departed significantly from the diagonal line. With rabbit brain there were two $Ca^{2+}$-affected proteins detected: these were calmodulin and (presumably) parvalbumin. (Parvalbumins are $Ca^{2+}$-binding proteins that occur in certain eukaryotic tissues, homologous to calmodulin, but with no known signaling function.)[28]

In sodium dodecylsulfate and in the absence of $Ca^{2+}$, the *T. acidophilum* protein migrated at a velocity corresponding to a molecular weight of about 16 kilodaltons (kD). This is similar in size to calmodulins that have been described previously from lower eukaryotes.[29]

However, we have not yet determined if the $Ca^{2+}$-affected protein from *T. acidophilum* has any regulatory function. Therefore, it is premature to identify it as calmodulin. Further studies are under way.

## ENERGY METABOLISM

In modern eukaryotic cells the metabolism of glucose is partitioned between the cytoplasm and the mitochondrion. The cytoplasm is fermentative and secretes pyruvate, which then is oxidized by mitochondria to $CO_2$ and $H_2O$. One can speculate that this distribution of functions is vestigial of the original arrangement between the cytoplasm and the bacterial symbiont ancestral to the mitochondrion.

Metabolically, *T. acidophilum* fits the anticipated features of the cytoplasmic host cell fairly well. When grown on [$^{14}$C]-glucose, the excretory products formed were $CO_2$ and acetic acid. Acetic acid accounted for up to one-third of the carbon excreted.[30] Since $O_2$ was abundantly available, these observations suggest that oxidative phosphorylation is probably not functional in this organism.

The excreted acetic acid was found to be toxic to *T. acidophilum* at concentrations similiar to those that accumulated in the culture medium. The apparent reason for this toxicity can be explained by "ion trapping." At pH 1.7, which is the pH of the external medium, acetic acid is not ionized, and is lipid soluble. It can diffuse into the cell through the membrane. Inside the cell it ionizes, becoming nonpermeant, and is trapped inside. The net result is both cytoplasmic acidification and osmotic stress.[30]

*T. acidophilum* would benefit by association with another organism that could remove the acetate. Such an association can be envisioned as the initial step toward the evolutionary acquisition of a mitochondrion. This association would place no energetic burden upon *T. acidophilum,* since the acetic acid is excreted anyway. Instead, the symbiosis would be of immediate benefit to both species. The host would be relieved of a toxic waste product, and the bacterial symbiont would be given a nutrient plus be sheltered from the acidic environment.

An additional benefit would accrue if the bacterial symbiont could not utilize glucose directly. Then it could not compete with the host for this nutrient. Glucose would be converted into acetate by the fermentative host cell, and the acetate would be metabolized by the bacterial symbiont via the Krebs cycle to $CO_2$. Such an arrangement would be particularly efficient because the two partners' activities would complement, and not be competitive.

In fact, cultures of *T. acidophilum* have become contaminated by eubacteria. In some of these situations we have observed that the bacterial contaminants increased the growth of *T. acidophilum*. Unfortunately, these mixed cultures have not been studied further.

In addition to energy metabolism, there are other significant ways in which *T. acidophilum* resembles the eukaryotic cytoplasm. There is a membrane-bound cytochrome present which is of the *b*-type.[31] This class of cytochrome occurs also in eukaryotic cellular membranes, such as cytochrome P-450. Other cytochrome pigments appear to be absent from both *T. acidophilum* and the eukaryotic cytoplasm. An $H^+$-translocating ATPase appears to be absent from both types of cells, or if present is at least greatly different from the $H^+$-ATPase of mitochondria and eubacteria. And finally, there is present in *T. acidophilum* a superoxide dismutase that resembles the Cu,Zn-superoxide dismutase of eukaryotic cells.[32]

It has been correctly pointed out that the cytoplasm of eukaryotic cells is not anaerobic, regardless of its apparent fermentative energy metabolism.[3,33,34] Numerous oxygen-dependent reactions are indeed present, including sterol synthesis, cytoplasmic mono- and dioxygenases, and peroxisomal-type respiration. Similarly, we found that *T. acidophilum* is obligately aerobic, even though oxidative phosphorylation appears to be absent.[30] Much of the oxygen consumption occurs via soluble flavin-linked oxidases, which are apparently similar to enzymes found in eukaryotic microbodies.[30] Interestingly, de Duve had speculated previously that the microbody enzymes might be vestiges of the premitochondrial respiratory metabolism of the cytoplasm.[33] Thus, the metabolic properties of *T. acidophilum* conform in several ways to the postulated features of the ancient host cell that became the cytoplasm of eukaryotes.

## OTHER PHENOTYPIC SIMILARITIES

There are a number of additional observations that also are suggestive of a relationship between eukaryotes and Archaebacteria. For example, compared to eubacterial tRNAs, halobacterial tRNAs are preferentially charged by eukaryotic aminoacyl-tRNA synthetases.[35] Both the Eukaryotic and the Archaebacterial elongation factors EF-2 can be ADP-ribosylated by diptheria toxin.[36] Archaebacterial ribosomes are generally inhibited by neither chloramphenicol nor streptomycin, but are inhibited by anisomycin, and otherwise respond generally to a variety of drugs in a pattern that is similar to that of eukaryotes.[37–39] Archaebacterial RNA polymerases cross-react immunologically with eukaryotic RNA polymerase, which they resemble also in their response to certain inhibitors (e.g., not affected by rifampicin, but stimulated by silybin).[40] The RNA polymerases resemble those of eukaryotes in the complexity of their subunit structures, particularly among the thermophiles.[40] The shape of the ribosomes of thermophilic Archaebacteria is similar to that of eukaryotes.[41] Certain Archaebacterial genes contain introns.[42,43] Long poly(A) tails of about 120 nucleotides are found among the mRNAs (T. Oshima, personal communication). There are eukaryotic-like membrane glycoproteins.[44] The carbohydrate component of the glycoproteins is assembled while linked to dolichol phosphates,[45] and so on.

However, it must be admitted that there are other observations that are not consistent with a close relationship between the Archaebacteria and Eukaryotes. The membrane lipids of Archaebacteria are unique,[46] a eubacterial-type Shine-Dalgarno sequence is present,[47] and the ribosomes lack a 5.8S rRNA.[37] Nevertheless, in balance it appears that the majority of features are most consistent with an Archaebacterial-eukaryotic relationship.

## CONCLUSION

Several distinctively eukaryotic features, including histones, cytoskeleton, calcium modulation, and acetogenic metabolism, have evolved in the thermophilic Archaebacteria in response to their harsh environment. Their appearance in the thermophilic Archaebacteria might be examples of preadaptation, as summarized in FIGURE 6. Thus, the conjecture is that histonelike proteins evolved first to protect DNA from thermal denaturation, but also condensed the DNA, and therefore were a preadaptation for the accumulation of more DNA and for the evolution of eukaryotic chromosomes. The ameboid shape and internal cytoskeleton of certain Archaebacteria may have first evolved as an adaptation to thermophily and lithotrophy, but were preadaptive for the evolution of endocytosis, ameboid movement, and other functions common to

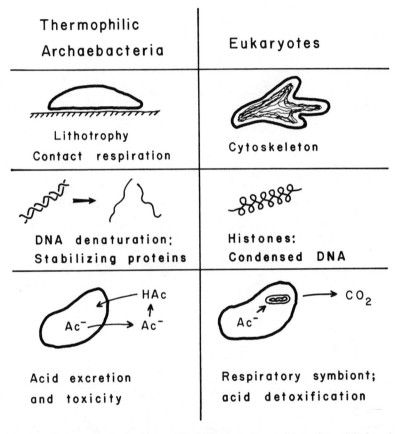

| Thermophilic Archaebacteria | Eukaryotes |
|---|---|
| Lithotrophy<br>Contact respiration | Cytoskeleton |
| DNA denaturation:<br>Stabilizing proteins | Histones:<br>Condensed DNA |
| Acid excretion<br>and toxicity | Respiratory symbiont;<br>acid detoxification |

**FIGURE 6.** Overview of certain thermoacidophilic adaptations and how they might have been preadaptive for the evolution of eukaryotic cells. (1) Thermophilic Archaebacteria often lack rigid cells, which may facilitate mineral respiration. (2) Thermophilic Archaebacteria have histonelike proteins that protect DNA from thermal denaturation. (3) *T. acidophilum* excretes acetic acid, which is toxic in its acidic environment (see text).

eukaryotes. In the extreme environment of *T. acidophilum* the $H^+$-translocating ATPase was lost, resulting in a fermentative metabolism, which was preadaptive for the acquisition of an aerobic bacterial symbiont.

Thus, a plausible story can indeed be made for why thermophilic Archaebacteria might have been particularly preadapted to give rise to the nucleus and cytoplasm of eukaryotic cells.

## REFERENCES

1. Tu, J., D. Prangishvilli, H. Huber, G. Wildgruber, W. Zillig & K. O. Stetter. 1982. Taxonomic relations between Archaebacteria including 6 novel genera examined by cross hybridization of DNAs and 16S rRNAs. J. Mol. Evol. **18:** 109-114.
2. Pace, N. R., G. J. Olson & C. R. Woese. 1986. Ribosomal RNA phylogeny and the primary lines of evolutionary descent. Cell **45:** 325-326.
3. Raff, R. A. & H. R. Mahler. 1972. The non symbiotic origin of mitochondria. Science **177:** 575-582.
4. Dayhoff, M. O. & R. M. Schwartz. 1981. Evidence on the origin of eukaryotic mitochondria from protein and nucleic acid sequences. Ann. N.Y. Acad. Sci. **361:** 92-104. (See also **361:** 260-272.)
5. Hunt, L. T., D. G. George & W. C. Barker. 1985. The prokaryote-eukaryote inferface. BioSystems **18:** 223-240.
6. Raven, P. H. 1970. A multiple origin for plastids and mitochondria. Science **170:** 641-646.
7. Whatley, F. R. 1981. The establishment of mitochrondria: *Paracoccus* and *Rhodopseudomonas.* Ann. N.Y. Acad. Sci. **361:** 330-340.
8. Margulis, L. 1981. Symbiosis in Cell Evolution. W. H. Freeman & Company. San Francisco, Calif.
9. Woese, C. R. 1981. Archaebacteria. Sci. Am. **244**(6): 98-122.
10. Searcy, D. G. & D. B. Stein. 1980. Nucleoprotein subunit structure in an unusual prokaryotic organism: *Thermoplasma acidophilum.* Biochim. Biophys. Acta **609:** 180-195.
11. DeLange, R. J., L. C. Williams & D. G. Searcy. 1981. A histone-like protein (HTa) from *Thermoplasma acidophilum.* II. Complete amino acid sequence. J. Biol. Chem. **256:** 905-911.
12. Searcy, D. G. & R. J. DeLange. 1970. *Thermoplasma acidophilum* histone-like protein. Partial amino acid sequence suggestive of homology to eukaryotic histones. Biochim. Biophys. Acta **609:** 197-200.
13. Searcy, D. G. 1986. The Archaebacterial histone "HTa". *In* Bacterial Chromatin. C. O. Gualerzi & C. L. Pon, Eds.: 175-184. Springer-Verlag. Berlin, FRG.
14. Stein, D. B. & D. G. Searcy. 1978. Physiologically important stabilization of DNA by a prokaryotic histone-like protein. Science **202:** 219-221.
15. Rouvière-Yaniv, J., M. Yaniv & J.-E. Germond. 1979. E. coli DNA binding protein HU forms nucleosome-like structure with circular double-stranded DNA. Cell **17:** 265-274.
16. Broyles, S. S. & D. E. Pettijohn. 1986. Interaction of the *Escherichia coli* HU Protein with DNA. Evidence for formation of nucleosome-like structures with altered DNA helical pitch. J. Mol. Biol. **187:** 47-60.
17. Shouten, J. P. 1985. Detection and identification of protein cross-linked *in vivo* to specific nucleic acid sequences. Doctoral Thesis. University of Amsterdam. Amsterdam, Holland.
18. Holck, A. & K. Kleppe. 1985. Affinity of protein HU for different nucleic acids. FEBS Lett. **185:** 121-124.
19. Maniloff, J. 1981. Cytoskeletal elements in mycoplasmas and other prokaryotes. BioSystems **14:** 305-312.
20. Searcy, D. G., D. B. Stein & K. B. Searcy. 1981. A mycoplasma-like Archaebacterium possibly related to the nucleus and cytoplasm of eukaryotic cells. Ann. N.Y. Acad. Sci. **361:** 312-324.
21. Searcy, D. G., D. B. Stein & G. R. Green. 1978. Phylogenetic affinities between eukaryotic cells and a thermophilic mycoplasma. BioSystems **10:** 19-28.
22. Matsumura, S., A. Kumon & T. Chiba. 1985. Proteolytic substructure of brain myosin. J. Biol. Chem. **260:** 1959-1966.
23. Côté, G. P., J. P. Albanesi, T. Ueno, J. A. Hammer III & E. D. Korn. 1985. Purification from *Dictyostelium discoideum* of a low-molecular-weight myosin that resembles Myosin I from *Acanthamoeba castellanii.* J. Biol. Chem. **260:** 4543-4546.
24. Silverman, M. P. 1967. Mechanism of bacterial pyrite oxidation. J. Bacteriol. **94:** 1046-1051.

25. BERRY, V. K. & L. E. MURR. 1978. Direct observations of bacteria and quantitative studies of their catalytic role in the leaching of low-grade, copper-bearing waste. *In* Metallurgical Applications of Bacterial Leaching and Related Microbiological Phenomena. L. E. Murr, A. E. Torma & J. A. Brierley, Eds.: 103-136. Academic Press. New York, N.Y.

26. KLEE, C. B. & T. C. VANAMAN. 1982. Calmodulin. Adv. Protein Chem. **35**: 213-321.

27. SEARCY, D. G. 1986. Some features of thermo-acidophilic Archaebacteria preadaptive for the evolution of eukaryotic cells. System. Appl. Microbiol. **7**: 198-201.

28. PECHÈRE, J.-F., J.-P. CAPONY & J. DEMAILLE. 1973. Evolutionary aspects of the structure of muscular parvalbumins. System. Zool. **22**: 533-548.

29. FULTON, C., K.-L. CHENG & E. Y. LAI. 1986. Two calmodulins in *Naegleria* flagellates: characterization, intracellular segregation, and programmed regulation of mRNA abundance during differentiation. J. Cell Biol. **102**: 1671-1678.

30. SEARCY, D. G. & F. R. WHATLEY. 1984. *Thermoplasma acidophilum:* glucose degradative pathways and respiratory activities. System. Appl. Microbiol. **5**: 30-40.

31. SEARCY, D. G. & F. R. WHATLEY. 1982. *Thermoplasma acidophilum* cell membrane: cytochrome *b* and sulfate-stimulated ATPase. Zbl. Bakt. Hyg. I. Abt. Orig. C **3**: 245-257.

32. SEARCY, K. B. & D. G. SEARCY. 1981. Superoxide dismutase from the Archaebacterium *Thermoplasma acidophilum.* Biochim. Biophys. Acta **670**: 39-46.

33. DEDUVE, C. 1973. Origin of mitochondria. Science **182**: 85.

34. HALL, J. B. 1973. The nature of the host in the origin of the eukaryotic cell. J. Theor. Biol. **38**: 413-418.

35. KWOK, Y. & J. T.-F. WONG. 1980. Evolutionary relationship between *Halobacterium cutirubrum* and eukaryotes determined by use of aminoacyl-tRNA synthetases as phylogenetic probes. Can. J. Biochem. **58**: 213-218.

36. KLINK, F. 1985. Elongation Factors. *In* The Bacteria. C. R. Woese & R. S. Wolfe, Eds. **8**: 379-410. Academic Press. Orlando, Fla.

37. MATHESON, A. T. 1985. Ribosomes of Archaebacteria. *In* The Bacteria. C. R. Woese & R. S. Wolfe, Eds. **8**: 345-377.

38. BÖCK, A. & O. KANDLER. 1985. Antibiotic sensitivity of Archaebacteria. *In* The Bacteria. C. R. Woese & R. S. Wolfe, Eds. **8**: 525-544. Academic Press. Orlando, Fla.

39. CAMMARANO, P., A. TEICHNER, P. LONDEI, M. ACCA, B. NICOLAUS, J. L. SANZ & R. AMILS. 1985. Insensitivity of archaebacterial ribosomes to protein synthesis inhibitors. Evolutionary implications. EMBO J. **4**: 811-816.

40. ZILLIG, W., K. O. STETTER, R. SCHNABEL & M. THOMM. 1985. DNA-dependent RNA polymerases of the Archaebacteria. *In* The Bacteria. C. R. Woese & R. S. Wolfe, Eds. **8**: 499-524. Academic Press. Orlando, Fla.

41. LAKE, J. A., E. HENDERSON, M. OAKES & M. W. CLARK. 1984. Eocytes: a new ribosome structure indicates a kingdom with a close relationship to eukaryotes. Proc. Nat. Acad. Sci. USA **81**: 3786-3790.

42. KAINE, B. P., R. GUPTA & C. R. WOESE. 1983. Putative introns in tRNA genes of prokaryotes. Proc. Nat. Acad. Sci. USA **80**: 3309-3312.

43. KJEMS, J. & R. A. GARETT. 1985. An intron in the 23S ribosomal RNA gene of the archaebacterium *Desulfurococcus mobilis.* Nature **318**: 675-677.

44. MESCHER, M. F. 1981. Glycoproteins as cell-surface structural components. Trends Biochem. Sci. **6**: 97-99.

45. LECHNER, J., F. WIELAND & M. SUMPER. 1985. Biosynthesis of sulfated saccharides *N*-glycosidically linked to the protein via glucose. Purification and identification of sulfated dolichyl monophosphoryl tetrasaccharides from Halobacteria. J. Biol. Chem. **260**: 860-866.

46. LANGWORTHY, T. A. 1985. Lipids of Archaebacteria. *In* The Bacteria. C. R. Woese & R. S. Wolfe, Eds. **8**: 459-497. Academic Press. Orlando, Fla.

47. STEITZ, J. A. 1978. Methanogenic bacteria. Nature **273**: 10.

48. SEARCY, D. G. 1982. *Thermoplasma:* a primordial cell from a refuse pile. Trends Biochem. Sci. **7**: 183-185.

# Photosomes and Scintillons

## Intracellular Localization and Control of Luminescent Emissions

J. WOODLAND HASTINGS,[a] JEAN-MARIE BASSOT,[b]
AND MARIE-THÉRÈSE NICOLAS[b]

[a]Department of Cellular and Developmental Biology
Harvard University
16 Divinity Avenue
Cambridge, Massachusetts 02138

[b]Bioluminescence Laboratory
CNRS
91190 Gif-sur-Yvette, France

Phylogenetically, bioluminescence occurs sporadically, from bacteria and unicellular eucaryotes to very diverse invertebrates and many fishes.[1,2] The character and chemistry of the luminescent systems differ markedly in different groups: biochemical, physiological, and structural features exhibit few homologies. Although sophisticated light organs or photophores show striking convergences in their eyelike shapes and in the configuration of dioptric structures such as lenses, interference reflectors, or screens, the basic property of light emission must have evolved independently many different times.[3]

Nevertheless, systems that are able to flash have evidently utilized similar strategies such that (a) the luminous reagents, whatever they may be, are compartmentalized and often localized within specialized cells (the photocytes), which (b) respond to a specific stimulation, inasmuch as bioluminescence is integrated as a behavioral response, and (c) the reaction is controlled so that it is turned off promptly, giving a flash rather than a longer lasting glow.

By contrast, luminous bacteria do not flash; they glow continuously and no compartmentation is known to occur. The luciferase reaction appears to be cytoplasmic, albeit linked to the electron transport pathway, thus suggesting the existence of some membrane-associated steps. Luminous bacteria are also cultivated as symbionts in several types of light organs in fishes and squids.[4] In such cases, the host is able to switch on or off or to modulate the continuous light of its symbionts (thus a flash or a blink) by quickly rotating the photophore or covering it with a lid (*Anomalops, Photoblepharon*).

If bacteria are an exception to the general rule of compartmentation, they also prove the rule. Tunicates (pyrosomes and salps) exhibit bright flashes which arise from photocytes containing bacteroid organelles.[5] This system responds to a stimulation and is under host control, but it possesses luciferase of the bacterial type,[6] suggesting that these organelles may very well represent or have originated as luminous bacterial endosymbionts. The excitation-bioluminescence coupling very likely involves an epithelial conducted action potential, but its triggering mechanism remains unknown.

In most cases, however, bioluminescence does not appear to involve symbionts but is an intrinsic property, generally restricted to specialized—often sophisticated—cells or organs. In fact, there are few systems in which the correlations have been made between first the microsources, as observed *in vivo* through the microscope (usually with image intensifier), second a definite organelle visualized and characterized at the level of the electron microscope, and third a biochemical fraction with luminous activity *in vitro*. While subcellular compartmentation appears to be a common feature, the organelles and/or membrane associations are very different in photocytes from different groups. The situation is thus markedly different from that in photoreception with its clear ciliary or rhabdomeric phyletic lines.[7]

In many coelenterates luminescence involves a $Ca^{++}$-activated photoprotein.[8,9] In sea pens and hydroids luminescence and fluorescence are localized within photocytes.[10] Flashing is triggered by calcium entry into neighboring support cells via chemical signaling through intercellular gap junctions.[11] In siphonophores the reported spreading and facilitation of responses to repeated epithelial $Ca^{++}$ action potentials[12] may have a similar explanation. Reports that emission in certain coelenterates is associated with an intracellular "lumisome" or "luminelle"[13,14] have not been pursued.

Luminescence in the firefly (coleopterid) system is very different biochemically, involving ATP and the oxidation of a unique luciferin, a highly fluorescent molecule whose structure involves a benzothiazole nucleus.[15] All luminous coleopterids possess photocytes with similar cytoplasmic granules, which are the origin of fluorescence and bioluminescence. But the control mechanisms differ. In the more primitive groups (*Phengodes*), the photocytes are huge isolated cells, and light is emitted continuously.[16] In glowworms and larval fireflies luminescence is turned on and off slowly (minutes); photocytes form a compact tissue irregularly penetrated by trachea. In adult fireflies, which communicate by flashing signals during courtship, the ventral light organs contain photocytes grouped in rosettes around precisely spaced trachea.[17] In these cells the luciferin and luciferase (and light emission) are localized to special granules, reported to possess also some peroxisome reactivites.[18,19] However, the granules (1-2 $\mu$m) are of at least three types, characterized differently but comprising a matrix with microtubules and sometimes crystalline inclusions.[20]

Scale worm (marine annelid) luminescence is less well characterized biochemically; it appears to involve a membrane photoprotein triggered by oxygen radicals.[21] The system has been described in considerable detail at the structural and physiological levels.[22,23] The photocytes contain paracrystallinelike structures termed photosomes, from whence the luminescence is emitted.[24]

The photogenic epithelium is flat and single layered, so that it is essentially a two-dimensional preparation. The photosomes (FIGURE 1) are the microsources of bioluminescence, as visualized by image-intensified light microscopy, and emission results in the formation of a fluorescent compound. Thus the fluorescence of the photosomes is a function of the previous luminescent activity. Facilitation, whereby repeated stimuli result in progressively brighter flashes, involves recruitment of an increasing number of active photosomes.

At the electron microscopic (EM) level, photosomes are exceptionally well characterized organelles. They are formed of membrane tubules, 20 nm in diameter, folded in a regular serpentine fashion (FIGURE 1). The disposition of such tubules in parallel rows, then in parallel or intercrossing planes, results in a paracrystalline network which develops a considerable membrane surface and separates two compartments. The tubules of the photosomes are continuous with the endoplasmic reticulum membranes,[25] so that one compartment is intracisternal, the other one being cytosolic.

Upon stimulation, the photogenic epithelium, whose cells are linked by gap junctions, conducts a triggering action potential.[26,27] But how is the excitation transmitted to the photosomes? We used fast-freeze fixation,[28] immobilizing the cytophysiological state in a matter of milliseconds, to compare stimulated and unstimulated preparations. In preparations fixed during stimulation (FIGURE 1), we observed sophisticated junc-

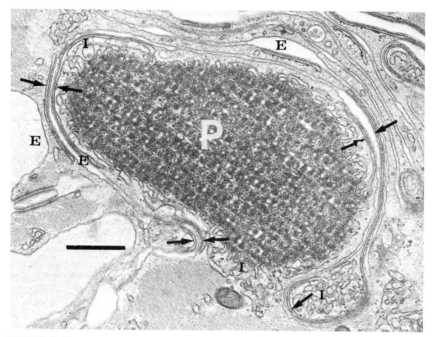

**FIGURE 1.** Photosome in a luminous cell of the scale worm *Harmothoe lunulata,* after fast-freeze fixation and freeze substitutions in $OsO_4$-acetone during repetitive stimulation. A photosome (P) is enclosed in a pouch wrapped by several slender digitations, separated by narrow extracellular spaces (E). A photosome is surrounded by arrays of intermediate reticulum (I) which connect terminal saccules forming dyad junctions with the plasma membrane (arrows). This coupling arrangement is characteristic of the photosomes during bioluminescent emission. Bar = 1 $\mu$m.

tional complexes, strikingly reminiscent of those made by the sarcoplasmic reticulum in striated muscle. Terminal saccules of endoplasmic reticulum were seen to be coupled with the plasma membrane by dyad junctions, sometimes extending over several microns.[29] The terminal saccule opens in turn in an irregular but continuous array of intermediate reticulum, which ultimately connects to the paracrystalline tubules of the photosomes. This continuous membranous conduction path, establishing a dyad junction with the plasma membrane, occurred only rarely in unstimulated preparations which, however, accounts for the fact that only a few photosomes are able to respond to the first stimulation. During facilitation, more and more photosomes become coupled. But coupling is a transient state, with a lifetime of the order of one second, as shown by the facts that facilitation is a function of the stimulation frequency and that after interruption of repetitive stimuli, the next response is weaker, and in proportion to the duration of the interruption. The rapidity with which this coupling differentiates and dedifferentiates, if not sustained by continued stimulation, is extraordinary.

The luminescent system of dinoflagellates is again different from those already described, both biochemically, physiologically, and with regard to subcellular localization.[30-32] In *Gonyaulax polyedra* the biochemical system is highly pH dependent and involves a tetrapyrrole substrate (*Gonyaulax* luciferin), with a fluorescence emission that corresponds spectrally to the bioluminescence emission.[30,33] The reason for this identity is not known, but it has allowed us to demonstrate that the subcellular distribution of the fluorescence, which occurs on cortically located spherical granules of dimensions less than 0.5 μm, colocalizes with the bioluminescent flashing emission.[34]

To visualize the emitting structures, we used immunocytochemical staining with a polyclonal antibody[35] raised in rabbits against purified *G. polyedra* luciferase. This was combined with fast-freeze fixation[28] and freeze substitution with acetone or acetone/osmium tetroxide. The ultrastructural preservation was greatly improved compared to the liquid chemical fixation previously used;[36] the results were confirmed and extended. In both cases sections were treated with antiluciferase followed by a goat-antirabbit antibody labeled with colloidal gold particles, following the technique of De Mey.[37] Two structures are labeled (FIGURE 2): dense bodies, which we identify as the luminous organelles, and the trichocyst sheaths. Labeling of both organelles still occurs with antibody after its affinity purification with pure luciferase. Thus, while the trichocysts are evidently nonluminous (nor are they fluorescent), they appear to contain either luciferase (or an antigenically reactive fragment thereof), or a different (but presumably related) protein carrying similar epitopes.

In *G. polyedra* the luminous organelles, which we refer to as scintillons (flashing units),[38] are made of a dense heterogeneous matrix enveloped by the vacuolar membrane (the tonoplast); only the matrix is labeled with antiluciferase. The dense light-emitting bodies actually protrude into the vacuole and may often be seen as hanging drops. But they are linked to the cytoplasm by narrow bridges; a tonoplast action potential could thus trigger the pH-sensitive flashing by opening voltage-gated $H^+$ channels.[31]

In cell extracts made at pH 8, we postulate that particulate scintillons are formed during cell extraction by pinching off of dense bodies to form vesicles. Scintillons flash *in vitro* when subjected to a rapid decrease in pH, and in purified fractions there are dense vesicles resembling the *in vivo* bodies that label with antiluciferase. After flashing, scintillons may be "recharged" by incubating with fresh luciferin at pH 8.

The results of these experiments, as well as the observations that the size, shape, and distribution of the dense bodies correspond closely with the microsources seen in the light microscope, suggest that they are the principal bioluminescent organelles in this alga, and that flashes are triggered *in vivo* by the entry of $H^+$ ions from the vacuole.

How did the ability of a cell to emit light originate in evolution? Do these several different systems, exemplifying, as they do, the diversity of bioluminescent mechanisms, have any common origin or evolutionary relationships?[3] An endosymbiotic origin of at least one system seems evident,[6] but for the others—coelenterates, fireflies, annelids, and dinoflagellates—there is no indication that this might be the case. Structures with other functions in nonluminous cells seem—perhaps surprisingly—to be called upon to serve as the structural site for light emission. Links are made—some very rapidly—to establish flash-triggering membrane events, involving $Ca^{++}$, $H^+$, and other still-to-

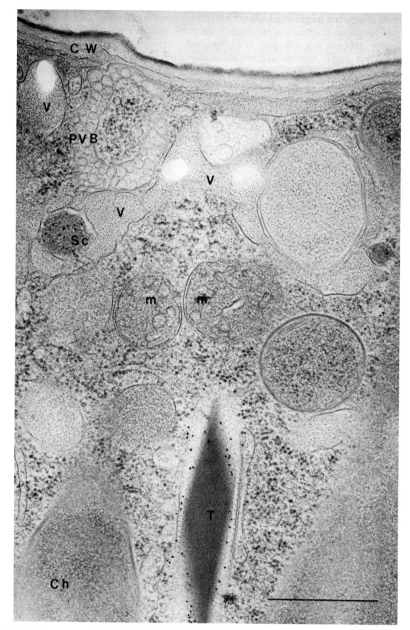

**FIGURE 2.** Immunogold staining of a *Gonyaulax polyedra* section after fast-freeze fixation and freeze substitution in $OsO_4$-acetone: the labeling was done by applying first the antibody raised in rabbits directed against luciferase followed by a second antibody, goat antirabbit with 10 nm gold particles attached (GAR-G$_{10}$ from Janssen Pharmaceutica). Two organelles are specifically labeled: the scintillons (Sc) protruding in the vacuole, and the space between the shaft and the membrane of the trichocyst (T). CW, cell wall; Ch, chloroplast; m, mitochondrion; PVB, polyvesicular body; Sc, scintillon; T, trichocyst; V, vacuole. Bar = 1 $\mu$m.

be-defined factors. Coevolution is clearly a theme, but the actors seem to enter from all sides, and with diverse capabilities.

## REFERENCES

1. HARVEY, E. N. 1952. Bioluminescence. Academic Press. New York, N.Y.
2. HERRING, P. J., Ed. 1978. Bioluminescence in Action. Academic Press. New York, N.Y.
3. HASTINGS, J. W. 1983. Biological diversity, chemical mechanisms and evolutionary origins of bioluminescent systems. J. Mol. Evol. **19:** 309-321.
4. HASTINGS, J. W. & K. H. NEALSON. 1981. The symbiotic luminous bacteria. *In* The Procaryotes. M. P. Starr, H. Stolp, H. G. Trüper, A. Balows, & H. G. Schlegel, Eds.: 1332-1345. Springer-Verlag. Berlin, Heidelberg & New York.
5. MACKIE, G. O. & Q. BONE. 1978. Luminescence and associated effector activity in pyrosoma (tunicata: pyrosomida). Proc. R. Soc. London Ser. B **202:** 483-495.
6. NEALSON, K. & J. W. HASTINGS. 1980. Luminescent bacterial endosymbionts in bioluminescent tunicates. In *Endocytobiology*. W. Schwemmler & H. E. A. Schenk, Eds.: 461-466. Walter de Gruyter & Co. Berlin, FRG.
7. EAKIN, R. M. 1968. Evolution of photoreceptors. *In* Evolutionary Biology. T. Dobzhansky, M. K. Hecht & W. C. Steere, Eds. **2:** 194-242. Appleton-Century-Crofts. New York, N.Y.
8. SHIMOMURA, O. & F. H. JOHNSON. 1979. Chemistry of the calcium-sensitive photoprotein, Aequorin. *In* Detection and Measurement of Free Calcium Ions in Cells. C. C. Ashley & L. K. Campbell, Eds.: 73-83. Elsevier/North Holland. Amsterdam, Holland.
9. MORIN, J. G. & J. W. HASTINGS. 1971. Biochemistry of the bioluminescence of colonial hydroids and other coelenterates. J. Cell Physiol. **77:** 305-312.
10. MORIN, J. G. & G. T. REYNOLDS. 1974. The cellular origin of bioluminescence in the colonial hydroid *Obelia*. Biol. Bull. **147:** 397-410.
11. DUNLAP, K., K. TAKEDA & P. BREHM. 1987. Activation of a calcium-dependent photoprotein by chemical signalling through gap junctions. Nature **325:** 60-62.
12. BASSOT, J.-M., A. BILBAUT, G. O. MACKIE, L. M. PASSANO & M. PAVANS DE CECCATTY. 1978. Bioluminescence and other responses spread by epithelial conduction in the siphonophore *Hippopodius*. Biol. Bull. **155:** 473-479.
13. ANDERSON, J. M. & M. J. CORMIER. 1973. Lumisomes, the cellular site of bioluminescence in coelenterates. J. Biol. Chem. **248:** 2937-2943.
14. SPURLOCK, B. O. & M. J. CORMIER. 1975. A fine structure study of the anthocodium in *Renilla mulleri*. Evidence for the existence of a bioluminescent organelle, the luminelle. J. Cell Biol. **64:** 15-28.
15. MCELROY, W. D. & M. DELUCA. 1978. Chemistry of firefly luminescence. *In* Bioluminescence in Action. P. J. Herring, Ed.: 109-127. Academic Press. New York, N.Y.
16. BASSOT, J. M. 1974. Des cellules lumineuses du Coleoptere *Phengodes*. *In* Recherches Biologiques Contemporaines. L. Arvy, Ed.: 79-96. Vagner. Nancy, France.
17. BEAMS, H. W. & E. ANDERSON. 1955. Light and electron microscope studies of the light organ of the firefly (*Photinus pyralis*). Biol. Bull. **109:** 375-393.
18. HANNA, C. H., T. A. HOPKINS & J. BUCK. 1976. Peroxisomes of the firefly lantern. J. Ultrastruct. Res. **57:** 150-162.
19. SMALLEY, K. N., D. E. TARWATER & T. L. DAVIDSON. 1980. Localization of fluorescent compounds in the firefly light organ. J. Histochem. Cytochem. **28**(4): 323-329.
20. NEUWIRTH, M. 1981. Ultrastructure of granules and immunocytochemical localization of luciferase in photocytes of fireflies. Tissue Cell **13:** 599-607.
21. NICOLAS, M.-T., J.-M. BASSOT & O. SHIMOMURA. 1982. Polynoidin: a membrane photoprotein isolated from the bioluminescent system of scale-worms. Photochem. Photobiol. **35:** 201-207.
22. PAVANS DE CECCATTY, M., J.-M. BASSOT, A. BILBAUT & M.-T. NICOLAS. 1977. Bioluminescence des elytres de'Acholoe. I. Morphologie des supports structuraux. Biol. Cell. **28:** 57-64.

23. BILBAUT, A. & J.-M. BASSOT. 1977. Bioluminescence des elytres d'Acholoe. II. Donnees photometriques. Biol. Cell. **28:** 145-154.
24. BASSOT, J.-M. & A. BILBAUT. 1977. Bioluminescence des elytres d'Acholoe. III. Deplacement des sites d'origine au cours des emissions. IV. Luminescence et fluorescence des photosomes. Biol. Cell. **28:** 155-162; 163-168.
25. NICOLAS, M.-T. 1977. Bioluminescence des elytres d'Acholoe. V. Les principales etapes de la regeneration. Arch. Zool. Exp. Gen. **118**(1): 103-120.
26. HERRERA, A. A. 1979. Electrophysiology of bioluminescent excitable epithelial cells in a polynoid poly-chaete worm. J. Comp. Physiol. A **129:** 67-78.
27. BILBAUT, A. 1980. Cell junctions in the excitable epithelium of bioluminescent scales on a polynoid worm: a freeze fracture and electrophysiological study. J. Cell Sci. **41:** 341-368.
28. ESCAIG, J. 1982. New instruments which facilitate rapid freezing at 83K and 6K. J. Micros. **126:** 221-229.
29. BASSOT, J.-M. 1985. Fast membrane transformation in a flashing endoplasmic reticulum. *In* Recent Advances in Biological Membranes Studies. L. Packer, Ed.: 259-284. Plenum Press. New York, N.Y.
30. HASTINGS, J. W. 1986. Bioluminescence in bacteria and dinoflagellates. *In* Light Emission in Plants and Bacteria. Govindjee, J. Amesz & D. C. Fork, Eds.: 363-398. Academic Press. New York, N.Y.
31. ECKERT, R. & T. SIBAOKA. 1968. The flash-triggering action potential of the luminescent dinoflagellate, *Noctiluca.* J. Gen. Physiol. **52:** 258-282.
32. WIDDER, E. A. & J. F. CASE. 1982. Distribution of subcellular bioluminescent sources in a dinoflagellate *Pyrocystis fusiformis.* Biol. Bull. **162:** 423-448.
33. DUNLAP, J., J. W. HASTINGS & O. SHIMOMURA. 1981. Dinoflagellate luciferin is structurally related to chlorophyll. FEBS Lett. **135:** 273-276.
34. JOHNSON, C. H., S. INOUE, A. FLINT & J. W. HASTINGS. 1985. Compartmentation of algal bioluminescence: autofluorescence of bioluminescent particles in the dinoflagellate *Gonyaulax* as studied with image intensified video microscopy and flow cytometry. J. Cell Biol. **300:** 1435-1446.
35. DUNLAP, J. & J. W. HASTINGS. 1981. The biological clock in *Gonyaulax* controls luciferase activity by regulating turnover. J. Biol. Chem. **256:** 10509-10518.
36. NICOLAS, M.-T., C. H. JOHNSON, J.-M. BASSOT & J. W. HASTINGS. 1985. Specific labeling of two organelles in the bioluminescent dinoflagellate *Gonyaulax polyedra* with antiluciferase antibody. Cell Biol. Int. Rep. **9:** 797-802.
37. DE MEY, J. 1983. Colloidal gold probes in immunocytochemistry. *In* Immunocytochemistry. J. M. Polack & S. Van Noorden, Eds.: 82-112. Wright-PSG. Bristol, London, & Boston.
38. FOGEL, M., R. SCHMITTER & J. W. HASTINGS. 1972. On the physical identity of scintillons: bioluminescent particles in *Gonyaulax polyedra.* J. Cell Sci. **11:** 305-317.

# Prokaryotic Origin of Undulipodia[a]

## Application of the Panda Principle to the Centriole Enigma

DAVID BERMUDES,[b] LYNN MARGULIS,[b,d] AND
GEORGE TZERTZINIS[c]

[b]*Department of Biology*
*Boston University*
*2 Cummington Street*
*Boston, Massachusetts 02215*

[c]*Department of Chemistry*
*Boston University*
*590 Commonwealth Avenue*
*Boston, Massachusetts 02215*

## INTRODUCTION

Darwin answers that we must look for imperfections and oddities, because any perfection in organic design or ecology obliterates the paths of history and might have been created as we find it. This principle of imperfections became Darwin's most common guide. . . . I like to call it the "panda principle."[1]

S. J. Gould, 1986

The origin of the eukaryotic motility organelle (undulipodium, cilium, or 9+2 flagellum, henceforth referred to as the undulipodium),[2] and its underlying structure the kinetosome (homologous to the centriole) is considered an enigma.[3] Although bacteria possess what are termed flagella, there is neither structural nor physiological homology of prokaryotic flagella with eukaryotic undulipodia (FIGURE 1). Prokaryotic flagella are much smaller and more simple than undulipodia.[4] The extracellular flagellar shaft, which is solid and intrinsically nonmotile, is composed of subunits of a single structural globular protein belonging to a class of proteins called "flagellins." Each flagellar base is embedded in the bacterial cell wall, and motility is generated in a basal structure (i.e., the "rotary motor") which rotates in response to proton

[a]This work is based on work supported by a National Science Foundation (NSF) Graduate Fellowship to David Bermudes, the Lounsbery Foundation, National Aeronautics and Space Administration Grant NGR 004-025 to Lynn Margulis, NSF Grant DMB-8503940 to Richard Laursen (for George Tzertzinis), and the Boston University Graduate School.

[d]To whom correspondence should be addressed.

gradients across the cell membrane. In contrast, undulipodia are far larger and more complex. The shaft of the eukaryotic structure, called the axoneme, is always inside the cell. The intracellular axoneme is covered by the undulipodial membrane which is an extension of the plasma membrane. Axonemes from protoctists, animals, and plants display a remarkable structural and biochemical uniformity. Axonemes are composed of nine doublet peripheral and one pair of singlet central microtubules. The tubules are made of dimeric proteins, called alpha and beta tubulins. Although tubulins are the most abundant axonemal proteins, gel electrophoresis of axonemal preparations reveal an additional 200 or so nontubulin proteins. The molecular weight of each tubulin molecule is approximately 50,000 daltons. Alpha and beta subunits are closely related to each other in amino acid sequence. Furthermore, they are homologous to

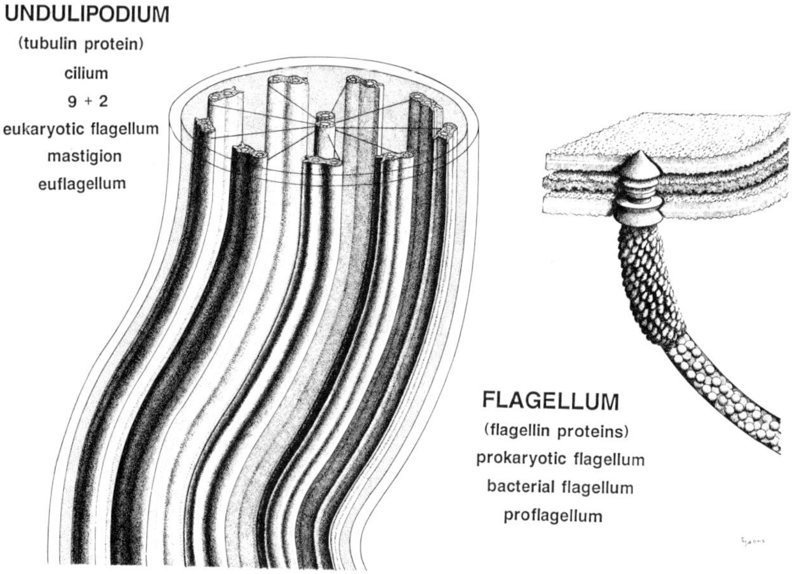

**UNDULIPODIUM**
(tubulin protein)
cilium
9 + 2
eukaryotic flagellum
mastigion
euflagellum

**FLAGELLUM**
(flagellin proteins)
prokaryotic flagellum
bacterial flagellum
proflagellum

**FIGURE 1.** Comparison between the flagellum of prokaryotes (bacteria) (based on Shapiro *et al.*)[4] and the undulipodium of eukaryotes (protoctists, animal cells, and plant sperm). Various names for these structures are employed. Drawing by Christie Lyons.

other microtubule proteins of eukaryotes such as those from brain tissue, the mitotic apparatus, and other nonaxonemal microtubules.[5,6] Tubulins have no known relation with bacterial flagellins. Tubulin dimeric proteins are linearly aligned to form 13 protofilaments arranged around a lumen to comprise the walls of hollow microtubules. Axonemal and nearly all other cytoplasmic microtubules measure 24-25 nm in diameter. The 9+2 microtubular axonemes are intrinsically motile along their length and move in response to ATP in aqueous solutions containing controlled concentrations of calcium and magnesium. Because of their profound differences no biologist has argued that the bacterial flagellum evolved directly into the undulipodium. Rather two classes of hypotheses (the endogenous or direct filiation and the exogenous or symbiotic) have been proposed to explain the origin of undulipodia.

Common to all versions of the endogenous hypothesis[7-14] is the differentiation of the undulipodium within mitotic eukaryotes. Proponents of the endogenous hypothesis claim that nonundulipodiated cells (already containing mitotic microtubules and microtubule organizing centers) evolved into undulipodiated cells. In contrast, the exogenous hypothesis states that motility symbioses gave rise to undulipodiated mitotic cells. These symbioses formed between a host that lacked microtubules and motile prokaryotes that contained them. Such motility symbioses—complexes of a nonmotile host rendered motile by attached symbionts—were predecessors to all motile protists. Permanent establishment of symbiotic motile prokaryotes led to the origin of undulipodia as organelles and subsequently to mitosis.[15-20] We review here the major details of these two classes of hypotheses by comparing their a priori assumptions. We suggest testable consequences for the symbiotic hypothesis.

# UNDULIPODIA BY DIRECT FILIATION (THE ENDOGENOUS HYPOTHESIS)

## *Statement of the Hypothesis*

The endogenous hypothesis, as stated by its current major proponent Cavalier-Smith, postulates the origin of primitively "nonflagellate" fungi from bacteria via direct filiation.[7-11] In this hypothesis, microtubules utilized in mitosis evolved inside the eukaryotic cell. The subsequent origin of the "protocilium," i.e., the protoundulipodium, occurred by endogenous differentiation through selection for an elongation of microtubules from a "centrosome" associated with the plasma membrane (Reference 11, p. 475). Primitive characters of eukaryotes are assumed to be a Golgi apparatus with separate scattered cisternae, flattened mitochondrial cristae, mitotic nuclear division, and lack of undulipodia at all stages. The first organisms to possess undulipodia are hypothesized to have been similar to the hemiascomycetes.

## *Problems with the Direct Filiation Argument*

In presenting the direct filiation hypothesis, Cavalier-Smith criticizes the symbiotic hypothesis because of its lack of parsimony: loss of undulipodia must be used to explain the thousands of protists which lack them (Reference 11, p. 467). Yet later in the same paper he argues that undulipodia were lost in at least cnidosporidians and acrasid cellular slime molds (Reference 11, p. 470). He also admits (assuming the monophyletic origin of vascular plants)[21,22] that hundreds of thousands of angiosperms lacking undulipodia must have descended from plant ancestors that lost them. If he can argue that these organisms lost their undulipodia, we find no reason that any or all groups of eukaryotes with mitosis but lacking undulipodia may not have secondarily lost their undulipodia. Thus we believe the statement that "the limited phylogenetic distribution of cilia and the universality of spindle microtubules makes implausible the idea that cilia evolved before mitosis [pp. 467-468]"[11] is invalid. We find plausible the idea that undulipodia evolved before mitosis.

The argument for direct filiation in the origin of undulipodia has been tied to claims of direct filiation for mitochondria and their membrane structure (flattened and tubular cristae) (Reference 11, pp. 469-470). It is claimed that flattened mitochondrial cristae represent the "primitive state" and that tubular cristae evolved from them. This notion is then used as a constraint for the phylogeny of the undulipodia. However, there is ever-increasing evidence that mitochondria are polyphyletic, that is, they arose through multiple symbioses.[23,24] Systematic differences in the 16S rRNA nucleotide sequence of mitochondria from different sources can be correlated with different (alpha, gamma) lineages of free-living phototrophic and oxygen-respiring bacteria.[25] Morphological arguments have also been presented which show fundamentally different mitochondrial types.[26] In addition, several types of endosymbiotic bacteria that, at least in some cases, are hydrogen acceptors have been found in the cytoplasm of certain protists living in less than fully aerobic environments (e.g., in *Pelomyxa*,[27] *Mastigamoeba*,[28] *Mixotricha paradoxa*,[29] plagiopylid and other ciliates[30]). These occurrences support the idea that several different symbioses were involved in the origin of mitochondrialike intracellular organelles, and mitochondria themselves. The origin of mitochondria most likely required a complex series of steps involving metabolic and genetic integration of intracellular symbionts.[31] Morphological and physiological differences in the mitochondria of protists are evidence that the integration of symbiotic bacteria into those complexes was unlikely to have been linear and monophyletic.

If mitochondria are polyphyletic, *neither* flattened vesicular nor tubular mitochondrial morphologies are necessarily representative of "the primitive states," where one type is derived from the other. Recognition of the multiple origins of mitochondria (and the possible origin of the undulipodia before the mitochondria—see below) releases from arbitrary constraint any claims of necessary correlation between mitochondrial morphology and undulipodial-microtubular systems.

Furthermore, the statement that the "strongest argument that the direction of evolution was from microtubules to flagella and not vice versa is that a flagellum is about 100 times more complicated as a microtubule [p. 96]"[9] does not distinguish the direct filiation from the symbiotic hypothesis since *both* hypotheses postulate the origin of microtubules prior to the origin of the 9+2 axoneme. The exogenous hypothesis merely assumes that the transition from tubules to complex axoneme occurred in bacteria (prior to or during the establishment of their motility associations with microbes lacking microtubules) whereas the endogenous hypothesis postulates the origin of axonemes in fungal-like mitotic eukaryotes making the transition from immotile mitotic cell to mastigote.

## WHICH CAME FIRST: UNDULIPODIA OR MITOCHONDRIA?

Some, if not all, mitochondria were probably acquired subsequent to the origin of the undulipodia.[32] Taxa of protists that lack mitochondria (e.g., pyrsonymphids, hypermastigotes, etc.) but bear large numbers of perfectly standard undulipodia support the "undulipodia first" idea. On the other hand, the existence of eukaryotic organisms that entirely lack microtubules suggests that in some lineages, mitochondria and even photosynthetic prokaryotes that became plastids preceded the appearance of microtubule-based mitosis and undulipodia. Amebae such as *Entamoeba histolytica*[33] and tiny green algae such as *Nanochlorum eucaryoticum*[34] appear to lack all microtubules, including undulipodial and mitotic. If the multiple independent acquisition of mitochondria (and of course plastids)[35-37] is accepted, the existence of all of these organismic types can be understood in one or more ways: in different lineages of protists mitochondria may have been acquired symbiotically before and/or after the appearance of undulipodia.


# UNDULIPODIA BY SYMBIOTIC ACQUISITION (THE EXOGENOUS HYPOTHESIS)


## *Historical Analysis and the Symbiotic Hypothesis*


The peculiar methods of historical sciences, including those of evolution, have been described by Gould.[1] Ultimately, in evolutionary studies, history is inferred from its results. This is done by accounting for peculiarities and reconstructing possible scenarios from clues taken to be representative. The panda principle, using imperfections and oddities, is a guide to this reconstruction. Recognizing the nature of historical reconstruction we infer the prokaryotic origin of the undulipodium from the biology of the organelle itself, its distribution and the occurrence of related and analogous structures.

There are four major reasons that together suggest undulipodia may be derived from bacteria through motility symbioses. The data, although not well understood in all cases, are well enough supported by so many studies that they cannot be discounted. Their unification in papers such as this one may further our understanding.

Firstly, non-Mendelian inheritance of a given trait has proved to be a good indicator of possible endosymbiotic origin of that trait. In cases where it is known, "non-Mendelian" genes are strictly correlated with the presence of bacteria, plasmids, viruses, or cell organelles such as mitochondria and plastids. The classical "cytoplasmic genes" of ciliates (e.g., kappa, lambda, mate-killer, etc., in *Paramecium*) are now known to be associated with gram-negative symbiotic bacteria called *Caedobacter*.[38] Genes associated with organelles such as mitochondria or plastids are acknowledged to have been derived from symbiotic bacteria.[23] Other genes are associated with small replicons such as plasmids or viruses.[39,40] Many aspects of the pattern and distribution of kinetosomes and undulipodia in the formation of kineties in the ciliate cortex also display a non-Mendelian form of inheritance.[41-45] In addition to this non-Mendelian inheritance of undulipodia in ciliates, mapping of undulipodial mutants of *Chlamydamonas* has shown that the mutations are linked to each other but do not map to any of the known nuclear linkage groups.[46,47] Inductively, the most likely explanation of these non-Mendelian genetic systems is the fixed presence, in the ciliate cortex, of originally exogenous bacterial[17] or viral[48] genomes.

Secondly, the development of undulipodia is analogous to that of the other organelles known to be of bacterial origin: mitochondria and plastids. In certain cells mitochondria and plastids lose their prominence. For example, in facultatively respiring yeast under fermentative conditions mitochondria dedifferentiate. Likewise, in photosynthetic mastigotes and plant tissue under heterotrophic conditions, plastids dedifferentiate. These organelles are capable of cyclical morphological changes. They develop from inconspicuous precursors called promitochondria or protoplastids into mature differentiated mitochondria or chloroplasts. Although not necessarily apparent, even with high-resolution electron microscopy, the dedifferentiated organelles, or at least the cytoplasm in which they are capable of developing, contain organellar nucleic acids.[49] In an analogous fashion undulipodial precursors are visually undetectable, but genetically present, in many types of cells in which they are known to arise, apparently "de novo" from amorphous microtubule organizing centers (MTOC). Cells of chytrids, amebomastigotes, myxomycetes, brown algae, animals, and some plants all have the ability to produce undulipodia under appropriate conditions even if no centrioles or kinetosomes can be seen by transmission electron microscopy in thin section prior to their development.

In well-studied cases, cells capable of making undulipodia contain microtubule organizing centers harboring ribonucleoprotein.[50–53] Furthermore, in *Stentor* there is an absolute requisite for RNA synthesis to accompany the production of new kinetosomes while no such DNA requirement has been demonstrated.[54] In addition to their "de novo" appearance in certain cell lineages, the production of offspring undulipodia, at least in many ciliates and mastigotes, appears to occur indirectly on a parental kinetosome template at a 90° angle. Although all details are not yet clear, these phenomena suggest that nucleic acid replicative and developmental processes analogous to dedifferentiation and reproduction of plastids involve RNA instead of DNA.

A third factor in favor of the symbiotic hypothesis of the origin of undulipodia, relative to the concept of direct filiation, is the explanatory power of the former. Furthermore, many questions generated by the symbiotic hypothesis for the origin of undulipodia have led to explanations of otherwise bizarre phenomena, as well as to explicit assertions testable by experimentation (Reference 55, pp. 108-110). For instance, the answer to one of the "enigmas" posed by Wheatley, the dual nature of the centriole in motility and cell division,[3] is that the centriole is a legacy of a motility symbiosis; all undulipodiated organisms require the 9+0 structure to nucleate the development of axonemes. In many protoctists and descendants of these protoctists, including all animals and plants, some components of the undulipodia must be used directly in mitosis, as is the case in *Chlorogonium*.[56] In some lineages, including those ancestral to animals and plants, the processes of undulipodial motility and mitotic karyokinesis are mutually exclusive: the two processes are never seen in the same cell at the same time. The mutual exclusivity of mitosis and undulipodial motility is explained by the symbiotic hypothesis in that the undulipodium-kinetosome-centriole complex is a single *genetic* entity, not simply a collection of gene products. This genetic entity undergoes differentiation resulting in the ability to perform one act or another. The fact that many undulipodiated cells are incapable of cell division by mitosis, part of the symbiotic-evolutionary scenario we continue to develop, is extremely difficult to explain by the direct filiation alternative. If mitotic spindle tubules were well developed before they differentiated into undulipodia, as claimed by Pickett-Heaps and Cavalier-Smith, why are mitosis and undulipodial motility mutually exclusive in all animal and plant cells? Many other aspects of the explanatory power of the symbiosis concept are explored in Margulis and Sagan.[55]

Our fourth argument stems from the application of Gould's "panda principle." If the evolution of a seme[24,57] as critical and apparently monophyletic as the undulipodium had evolved in fungi or other mitotic, nonmotile lineages of eukaryotes (e.g., red algae, conjugating green algae, amebae, etc.), one would expect to see significant variation associated with the appearance of this organelle of motility. Fundamental variation, "panda thumbs" or imperfections and oddities are expected to occur in the group of organisms in which the seme evolves. On the contrary, the cirri of ciliates, the kinetosomes and axonemes of the light-sensitive rods and cones, olfactory and auditory cells of animals, the mechanoreceptors of roaches, and the 6+0 haptonemes that develop from (9+2) kinetosomes are among many examples of structures derived directly from undulipodia. As such, they provide evidence of subsequent diversification of entire undulipodia that had already completely evolved in protoctist ancestors. Indeed, using the same panda principle it is the seme of *mitosis,* with its "imperfections and oddities," that clearly has originated and diversified in protoctists.[18] The same principle that tells us that the process of mitosis evolved in eukaryotic microorganisms and their descendants denies that undulipodia evolved in these same organisms because of the absence of undulipodial "imperfections and oddities" in the group. The panda principle directs us to prokaryotes and prokaryote motility symbioses where we have seen variations on the theme of microtubules and attachment organelles in motility associations.[18]

Historical "imperfections and oddities" ought not be ignored as exceptions, rather they should be comprehended as products of opportunistic, not foresighted, evolution. Symbiotic spirochetes and other bacteria that propel protoctists in well-developed and different kinds of motility symbioses[29,58] inside the hindguts of termites are one such oddity that may shed light on the origin of the undulipodia.

Taken together, the circumstantial evidence suggests a symbiotic origin of undulipodia. This contrasts with the direct filiation hypothesis which offers no such circumstantial evidence but rather makes an argument simply by the assertion that direct filiation is a more simple mechanism. We now discuss the consequences of assuming the symbiotic origin of undulipodia.

### Which Bacteria Were Ancestral to Undulipodia?

On the assumption that precursors to undulipodia were free-living bacteria, the biology of these motile prokaryotes and their interaction with their hosts become matters of primary concern. If we accept monophyly of the 9+2 undulipodia, then two or more mutually exclusive possibilities for their bacterial ancestry exist. In one case a fully developed seme might be acquired by symbiosis. This would be comparable to the origins of plastids, i.e., the acquisition of the seme of oxygenic photosynthesis. Pigmented thylakoids, already developed in phototrophic bacteria, were acquired in toto by nonphototrophic hosts in the origins of plastids. Thus perhaps the original motile prokaryotes that became undulipodia already had the 9+0/9+2 microtubular motile axoneme which used an ATP motility system. A second possibility is that the appearance of complex structures occurred only after the formation of well-established functioning symbioses. This would be analogous to the appearance of light-focusing, guanine-crystalline reflective light organs in marine fish as the product of coevolution with luminescent bacteria.[59] The symbiosis of the two organisms allows for selection of the new (emergent) feature, i.e., the reflective guanine-crystalline light-reflecting organ. By this analogy the original motile prokaryotes may have contained singlet, doublet, or other arrangements of microtubules, but their 9+0/9+2 axonemal arrangement and ATP sensitivity evolved as an emergent property of the symbiosis. Two morphological and physiological possibilities for the bacterial ancestry can be derived from these hypotheses:

1. The original prokaryote was an undulating motile heterotroph that contained microtubules in 9+0/9+2 array. No such 9+0/9+2 prokaryotes with ATP-based motility are yet known.

2. The original motile prokaryote that became the undulipodium was a member of the genus *Spirochaeta* or related to this genus of free-living spirochetes. If so, these bacteria must have had periplasmic flagella (these are equivalent to "bacterial flagella," "axial filaments," "axial fibrils," or "endoflagella"). Since all spirochetes have motility mediated by rotary motors run by the proton motive force, they too must have had this typical flagellar motility system and must have lacked motility sensitive to external concentrations of ATP. If they began as members of the genus *Spirochaeta,* these prokaryotes would not only have had to lose their flagella but they would also have had to gain an ATP-sensitive motility system in the course of becoming undulipodia.

## CONCLUSIONS: TESTABLE CONSEQUENCES OF THE SYMBIOSIS HYPOTHESIS

The hypothesis of a symbiotic origin of the undulipodium from prokaryotes leads to many predictions concerning the nature of certain prokaryotes. The following assertions are testable consequences of the symbiotic prokaryotic origin of undulipodia which contradict or at least do not follow from the endogenous hypothesis.

1. Proteins homologous to tubulins and their molecular associates (high-molecular-weight microtubule-associated proteins, called MAPS, tau, dynein, and so forth) and tubule structures universal in undulipodiated cells will be found in prokaryotes dissimilar to the codescendants of the nonundulipodiated components of eukaryotic cells. That is, tubulin proteins will not be found as an intrinsic component of plastids (or oxygenic phototropic bacteria), mitochondria (or oxygen-respiring bacteria), or presumed ancestors to the nucleocytoplasm such as *Thermoplasma*-like archaebacteria. Accordingly, certain motile prokaryotes that form motility symbioses such as spirochetes will contain tubulin proteins. These prokaryotic tubulins, possibly restricted to such spirochetes, will be more closely allied by amino acid sequence to eukaryotic tubulins than to cytoplasmic proteins of other arbitrarily chosen prokaryotes. (For the current status of this assertion see References 24, 60, 61, and 62.)
2. Fundamental variations on the theme of microtubule morphology, arrangement, and tubulin protein chemistry will be found in spirochetes or related motile prokaryotes. Such variations will not be found in fungi (nonmotile nonundulipodiated eukaryotes) like hemiascomycetes, which by the symbiotic hypothesis ultimately derived their mitotic tubulin by symbiotic acquisition of undulipodia. (Undulipodia were subsequently lost during the evolution of mitosis and meiosis in common ancestors to fungi and other eukaryotic lineages.)
3. Transitions from proton pump to ATP-sensitive motility systems occurred in symbiotic or free-living motile prokaryotes. Proton pump and membrane ATPases of undulipodiated cells will show sequence homologies to the spirochetes or other motile prokaryotic codescendant ancestors of the undulipodia.

There is at present insufficient evidence to prove either direct filiation or the symbiotic hypothesis for the origin of the undulipodia.[24] We are currently attempting to test aspects of these predictions with the ultimate goal of distinguishing definitively between these two hypotheses.

## ACKNOWLEDGMENTS

We wish to thank Christie Lyons for FIGURE 1; B. Dyer, E. P. Greenberg, R. Obar, and L. Shapiro for helpful communication; B. Dorritie for manuscript proofreading; and J. Kearney for manuscript preparation.

## REFERENCES

1. GOULD, S. J. 1986. Evolution and the triumph of homology, or why history matters. Am. Sci. **74:** 60-69.
2. MARGULIS, L. & D. SAGAN. 1985. Order amidst animalcules: the protoctista kingdom and its undulipodiated cells. BioSystems **18:** 141-147.
3. WHEATLY, D. N. 1982. The Centriole: A Central Enigma of Cell Biology. Elsevier Biomedical Press. New York, N. Y.
4. SHAPIRO, L., W. ALEXANDER, R. BRYAN, R. CHAMPER, P. FREDERISKE, S. L. GOMES, K. HAHNENBERGER & B. FLY. 1985. Biogenesis of a polar flagellum and a chemosensory system during *Caulobacter* cell differentiation. *In* Sensing and Response in Microorganisms. M. Eisenbach & M. Balaban, Eds.: 93-106. Elsevier Science Publishers B. V. Amsterdam, Holland.
5. LITTLE, M., R. F. PONSTINGLE & E. KRAUHS. 1981. Tubulin sequence conservation. BioSystems **14:** 239-246.
6. LITTLE, M. 1985. An evaluation of tubulin as a molecular clock. BioSystems **18:** 241-247.
7. CAVAILER-SMITH, T. 1975. The origin of nuclei and of eukaryotic cells. Nature **256:** 463-468.
8. CAVALIER-SMITH, T. 1978. The evolutionary origin and phylogeny of microtubules, mitotic spindles and eukaryote flagella. BioSystems **10:** 93-114.
9. CAVALIER-SMITH, T. 1980. Cell compartmentation and the origin of eukaryotic membranous organelles. *In* Endocytobiology: Endosymbiosis and Cell Biology, a Synthesis of Recent Research. W. Schwemmler & H. E. A. Schenk, Eds.: 893-916. Walter de Gruyter. Berlin, FRG.
10. CAVALIER-SMITH, T. 1981. The origin and early evolution of the eukaryotic cell. Symp. Soc. Gen. Microbiol. **32:** 33-84.
11. CAVALIER-SMITH, T. 1982. The evolutionary origin and phylogeny of eukaryote flagella. Symp. Soc. Exp. Biol. **35:** 465-493.
12. PICKETT-HEAPS, J. D. 1974. Evolution of mitosis and the eukaryotic condition. BioSystems **6:** 37-48.
13. SLEIGH, M. A. 1985. The origin of flagella—autogenous or symbiotic? Abstr. Cell Motility Symposium, Villefranche-sur-Mer, France, September 8-12, 1985.
14. TAYLOR, F. J. R. 1976. Autogenous theories for the origin of eukaryotes. Taxon **25:** 377-390.
15. MARGULIS, L. 1970. The Origin of Eukaryotic Cells. Yale University Press. New Haven, Conn.
16. MARGULIS, L. 1975. Microtubules and evolution. *In* Microtubules and Microtubule Inhibitors. M. Borgers & M. de Brabander, Eds.: 3-18. North-Holland Publishing Company. Amsterdam, Holland.
17. MARGULIS, L. 1975. Symbiotic theory of the origin of eukaryotic organelles: criteria for proof. Symp. Soc. Exp. Biol. **24:** 21-38.
18. MARGULIS, L. 1981. Symbiosis in Cell Evolution: Life and Its Environment on the Early Earth. W. H. Freeman and Co. San Francisco, Calif.
19. MARGULIS, L., D. CHASE & L. P. To. 1979. Possible evolutionary significance of spirochetes. Proc. R. Soc. London Ser. B. **204:** 189-198.
20. MARGULIS, L., L. P. To & D. CHASE. 1981. Microtubules, undulipodia and *Pillotina* spirochetes. Ann. N. Y. Acad. Sci. **361:** 356-367.
21. BANK, H. P. 1972. Evolutionary History of Plants. Wadsworth. Belmont, Calif.

22. DELEVORYAS, T. 1966. Morphology and Evolution of Fossil Plants. Holt, Rinehart and Winston. New York, N. Y.
23. GRAY, M. W. 1983. The bacterial ancestry of plastids and mitochondria. BioScience 33: 663-699.
24. MARGULIS, L. & D. BERMUDES. 1985. Symbiosis as a mechanism of evolution: status of cell symbiosis theory. Symbiosis 1: 101-124.
25. FOX, G. E. 1985. Insights into the phylogenetic positions of photosynthetic bacteria obtained form 5s rRNA and 16s rRNA sequence data. In The Global Sulphur Cycle. D. Sagan, Ed.: 30-39. NASA Technical Memorandum 87570. Life Sciences Division, NASA Office of Space Science and Applications. Washington, D. C.
26. ROBERTS, K. R., K. D. STEWART & D. W. MALLOCH. 1981. The flagellar apparatus of Chilomonas paramecium (Cryptocphyceae) and its comparison with certain zooflagellates. J. Phycol. 177: 159-167.
27. VAN BRUGGEN, J. J. A., C. K. STUMM & G. D. VOGELS. 1983. Symbiosis of methanogenic bacteria and sapropelic protozoa. Arch. Microbiol. 136: 89-95.
28. VAN BRUGGEN, J. J. A. 1986. Methanogenic bacteria as endosymbionts of sapropelic protozoa. Ph.D. Thesis. University of Nijmegen. Nijmegen, Holland.
29. CLEVELAND, L. R. & A. V. GRIMSTONE. 1964. The fine structure of the flagellate Mixotricha paradoxa and its associated microorganisms. Proc. R. Soc. London Ser. B. 157: 668-683.
30. DYER, B. D. 1984. Protoctists from the microbial communities of Baja California, Mexico. Ph.D. Thesis. Boston University. Boston, Mass.
31. OBAR, R. & J. GREEN. 1985. Molecular archaeology of the mitochondrial genome. J. Mol. Evol. 22: 243-251.
32. SCHWEMMLER, W. 1984. Reconstruction of Cell Evolution: a Periodic System. CRC Press, Inc. Boca Raton, Fla.
33. MARTINEZ-PALOMO, A. 1982. The Biology of Entamoeba hystolitica. Research Studies Press. New York, N. Y.
34. ZAHN, R. K. 1984. A green alga with minimal eukaryotic features: Nanochlorum eucaryotum. Origins Life 13: 289-303.
35. DOOLITTLE, W. F. & L. BONEN. 1981. Molecular sequence data indicating an endosymbiotic origin for plastids. Ann. N. Y. Acad. Sci. 361: 248-259.
36. RAVEN, P. H. 1970. A multiple origin for plastids and mitochondria. Science 169: 641-646.
37. SCHWARTZ, R. M. & M. O. DAYHOFF. 1981. Chloroplast origins: inferences from proteins and nucleic acid sequences. Ann. N. Y. Acad. Sci. 361: 260-272.
38. PREER, L. B. 1981. Prokaryotic symbionts of Paramecium. In The Prokaryotes. A Handbook on Habitats, Isolation, and Identification of Bacteria. M. P. Starr, H. Stolp, H. Truper, A. Balows & H. G. Schlegel, Eds. 2: 2127-2136. Springer Verlag. Berlin, Heidelberg & New York.
39. SONEA, S. & M. PANISSET. 1980. A New Bacteriology. Jones and Bartlett Pub., Inc. Boston, Mass.
40. LEMKE, P. A., Ed. 1979 Viruses and Plasmids in Fungi. Marcel Dekker, Inc. New York, N. Y.
41. JINKS, J. L. 1964. Extrachromosomal Inheritance. Prentice-Hall. Englewood Cliffs, N. J.
42. BEISSON, J. & T. M. SONNEBORN. 1965. Cytoplasmic inheritance of the organization of the cell cortex in Paramecium aurelia. Proc. Nat. Acad. Sci. USA 53: 265-282.
43. TARTAR, V. 1967. Morphogenesis in Protozoa. In Research in Protozoology. Tze-Tuan Chen, Ed. 2: 1-116. Pergamon Press. Oxford, England.
44. SONNEBORN, T. M. 1974. Ciliate morphogenesis and its bearing on general cellular morphogenesis. Actual. Protozool. 1: 337-355.
45. GRIMES, G. 1982. Nongenic inheritance: a determinant of cellular architecture. BioScience 32: 279-280
46. HUANG, B., Z. RAMANIS, S. K. DUTCHER & D. J. L. LUCK. 1982. Uniflagellar mutants of chlamydomonas: evidence for the role of basal bodies in transmission of positional information. Cell 29: 745-753.
47. RAMANIS, Z. & D. J. L. LUCK. 1986. Loci affecting flagellar assembly and function map to an unusual linkage group in Chlamydomonas reinhardtii. Proc. Nat. Acad. Sci. USA 83: 423-426.

48. TAYLOR, F. J. R. 1979. Symbioticism revisited: a discussion of the evolutionary impact of intracellular symbioses. Proc. R. Soc. London Ser. B **204**: 267-286.
49. GILLHAM, N. W. 1978. Organelle Heredity. Raven Press. New York, N. Y.
50. DIPPELL, R. V. 1976. Effects of nuclease and protease on the ultrastructure of *Paramecium* basal bodies. J. Cell Biol. **69**: 622-637.
51. HARTMAN, H., J. P. PUMA & T. GURNEY. 1974. Evidence for the association of RNA with ciliary basal bodies of *Tetrahymena*. J. Cell Sci. **161**: 241-260.
52. HEIDEMANN, S. R., G. SANDER & M. W. KIRSCHNER. 1977. Evidence for a functional role of RNA in centrioles. Cell **10**: 337-350.
53. REIDER, C. L. 1979. Ribonucleoprotein staining of centrioles and kinetochores in newt lung cell spindles. J. Cell Biol. **80**: 1-9.
54. YOUNGER, K. B., S. BANERJEE, J. K. KELLEHER, M. WINSTON & L. MARGULIS. 1972. Evidence that synchronized production of new basal bodies is not associated with DNA synthesis in *Stentor coeruleus*. J. Cell Sci. **11**: 621-637.
55. MARGULIS, L. & D. SAGAN. 1986. Origins of Sex: Three Billion Years of Genetic Recombination. Yale University Press. New Haven, Conn.
56. HOOPS, H. J. & G. B. WHITMAN. 1985. Basal bodies and associated structures are not required for normal flagellar motion or phototaxis in the green alga *Chlorogonium elongatum*. J. Cell Biol. **100**: 297-309.
57. HANSON, E. D. 1977. The Origin and Early Evolution of Animals. Wesleyan University Press. Middleton, Conn.
58. TAMM, S. L. Flagellated ectosymbiotic bacteria propel a eukaryotic cell. J. Cell Biol. **94**: 697-709.
59. NEALSON, K., D. COHN, G. LEISMAN & B. TEBO. 1981. Co-evolution of luminous bacteria and their hosts. Ann. N. Y. Acad. Sci. **361**: 76-91.
60. OBAR, R. 1985. Purification of tubulin-like proteins from a spirochete. Ph.D. Thesis. Department of Chemistry. Boston University. Boston. Mass.
61. BERMUDES, D., R. OBAR & G. TZERTZINIS. 1986. Tubulin-like proteins in prokaryotes. American Society of Microbiology Annual Meeting, Washington, D. C. (Abstract I-85, p. 79).
62. BERMUDES, D., S. FRACEK, R. A. LAURSEN, L. MARGULIS, R. OBAR & G. TZERTZINIS. 1987. Tubulinlike proteins from *Spirochaeta* bajacaliforniensis. Ann. N.Y. Acad. Sci. (This volume.)

# Are the Nucleomorphs of Cryptomonads and *Chlorarachnion* the Vestigial Nuclei of Eukaryotic Endosymbionts?[a]

MARTHA LUDWIG AND SARAH P. GIBBS

*Department of Biology*
*McGill University*
*1205 Avenue Docteur Penfield*
*Montreal, Quebec, Canada H3A 1B1*

Originally thought to be an invagination of the cytoplasm into the periplastidal compartment,[1] the cryptomonad nucleomorph was first recognized as a distinct, double-membrane-bound organelle by Greenwood in 1974.[2] Because of its morphology and location between the two pairs of membranes limiting the cryptomonad chloroplast, Greenwood hypothesized that the nucleomorph contained a subsidiary genome which was involved in the maintenance of this periplastidal compartment.[2] Since then other investigators have postulated that the nucleomorph is the vestigial nucleus of a eukaryotic endosymbiont which gave rise to the cryptomonad chloroplast.[3–8] According to this hypothesis, the two membranes of the cryptomonad chloroplast envelope would originally have been the plasma membrane of a photosynthetic prokaryote and the phagocytic vacuole membrane of a colorless eukaryotic host cell. The resulting photosynthetic eukaryote, subsequently, became the endosymbiont of a colorless, biflagellated cryptomonad cell. Thus the outer pair of membranes surrounding the cryptomonad chloroplast, the chloroplast endoplasmic reticulum (CER), would originally have been the plasma membrane of the photosynthetic eukaryote and the phagocytic vacuole membrane of the cryptomonad host cell.

Recently, nucleomorphlike organelles have been discovered in the periplastidal compartments of the green ameba, *Chlorarachnion reptans*.[9] The several chloroplasts of this unique ameba are delimited by an extra pair of membranes which, unlike CER, does not have ribosomes on its outer membrane. A single nucleomorphlike organelle is present in the periplastidal compartment of each chloroplast. The discovery of another organism containing a reduced, nucleuslike organelle associated with the chloroplast further supports the hypothesis that the chloroplasts of some algal groups were acquired through eukaryotic endosymbionts. Hibberd and Norris have hypothesized that the chloroplasts of *C. reptans* were acquired in this manner and that the nucleomorphlike organelle was originally the nucleus of the eukaryotic endosymbiont.[9]

We have been studying the cryptomonad nucleomorph and the nucleomorphlike organelle of *Chlorarachnion,* and, based on morphological and cytochemical data, we

[a] This research was supported by the Natural Sciences and Engineering Research Council of Canada (Grant No. A-2921).

198

believe they are analogous organelles. Our data indicate that the nucleomorphs of both organisms contain independent genomes and lend strong support to the hypothesis that these organelles are vestigial nuclei.

A longitudinal section through *Cryptomonas abbreviata* shows the two lateral lobes of the cell's single chloroplast (FIGURE 1). A pyrenoid bridge connects the two lobes of the chloroplast, and in favorable longitudinal sections (FIGURE 5b), a groove, which bisects the inner face of the pyrenoid, can be seen. The single nucleomorph of *C. abbreviata* is located in this groove. The cell in FIGURE 1 has been sectioned such that the nucleomorph is seen at the edge of the pyrenoid (arrow).

The ameboid cells of *Chlorarachnion reptans* contain several chloroplasts, each with a projecting pyrenoid (FIGURE 2). The chloroplasts are typically located along the periphery of the cell with the pyrenoids projecting toward the cell's interior. At the apex of each pyrenoid is a groove, and located in each of these grooves is a nucleomorph (arrows, FIGURE 2).

The nucleomorphs of cryptomonads and *Chlorarachnion* are surprisingly similar. As previously described and shown again at higher magnification in FIGURES 3a and 3b, the nucleomorphs of *Cryptomonas abbreviata* and *Chlorarachnion reptans* are located in the pyrenoid regions of the chloroplasts. However, in some cryptomonads[6,10,11] and in *Chlorarachnion* sp., isolated by Dr. P. W. Cook, the nucleomorphs are not found in the pyrenoids but, instead, lie along the inner face of the chloroplasts (FIGURES 4a and 4b). Regardless of the location with respect to the pyrenoid, the nucleomorphs of cryptomonads and *Chlorarachnion* are always located in the periplastidal compartment of the chloroplasts.

A double membrane surrounds the nucleomorphs of both organisms (FIGURES 3a, 3b, 4a, and 4b). Pores or larger gaps interrupt the nucleomorph envelope at several places (arrows, FIGURES 3a, 3b, 4a, and 4b). Although the nucleomorphs of both organisms contain a fibrillogranular body that has been compared to a small nucleolus,[6] this region is much less prominent in the nucleomorph of *Chlorarachnion* than in the cryptomonad organelle (cf. FIGURES 3a and 4a with 4b). The other inclusion invariably present in the nucleomorphs are rod-shaped dense globules. These structures are typically clustered (FIGURE 4a) or, in *Chlorarachnion,* they are often arranged along the inner nucleomorph membrane (FIGURES 3b and 4b). The function of these globules is not known, although it has been suggested that they may be an RNA-containing virus.[12]

Starch grains and putative eukaryotic-sized ribosomes are also present in the periplastidal compartment of cryptomonads (FIGURES 3a and 4a). Ribosomelike particles are seen in the periplastidal compartment of *Chlorarachnion* (FIGURES 3b and 4b). However, the storage carbohydrate of *Chlorarachnion,* reportedly not starch,[9] is located in the cytoplasm in a membrane-limited vesicle which caps the pyrenoid (FIGURES 3b and 4b).

We have recently obtained evidence that the nucleomorph of *Cryptomonas abbreviata* contains DNA.[12] Sections of cells fixed in 3:1 ethanol-acetic acid and embedded in JB-4 were stained with 4'-6-diamidino-2-phenylindole (DAPI), which fluoresces blue white when bound to double-stranded DNA.[13]

The nucleomorph can be identified in favorable sections by its characteristic location in the pyrenoid (arrows, FIGURES 5b and 5e). Following DAPI staining, bright blue white fluorescence is exhibited by the nucleomorphs of *C. abbreviata* cells (arrows, FIGURES 5c and 5f). In some cell sections the fluorescence is observed to be confined to a particular region of the nucleomorph. For example, only the anterior tip of the nucleomorph of the cell shown in FIGURE 5c fluoresces following DAPI staining.

DNA-DAPI fluorescence is also demonstrated by the nuclei of *C. abbreviata* cells (FIGURES 5c and 5f). The chloroplast DNA of these cells is identified as discrete

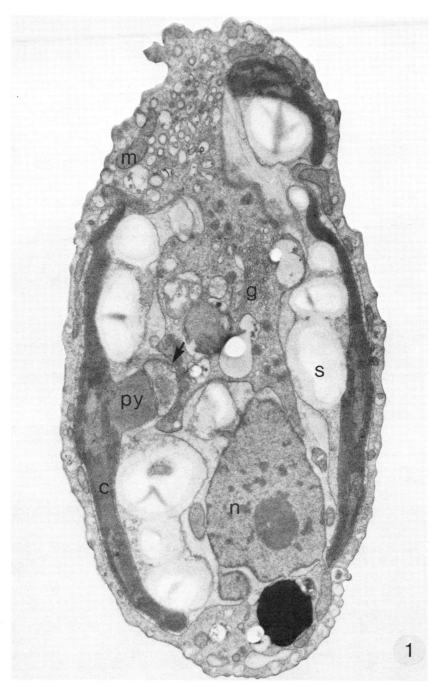

**FIGURE 1.** Longitudinal section through *Cryptomonas abbreviata*. The nucleomorph (arrow) is seen at the edge of the pyrenoid (py). Numerous starch grains (s) lie along the inner surface of the single, bilobed chloroplast (c). The nucleus (n) contains a prominent nucleolus and is located in the posterior region of the cell. g, Golgi body; m, mitochondrion. ×13,440.

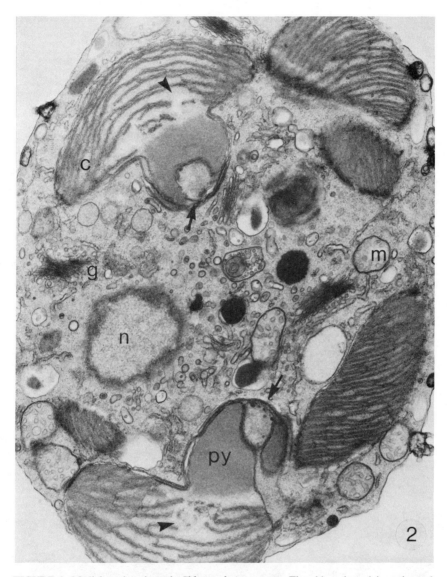

**FIGURE 2.** Medial section through *Chlorarachnion reptans*. The chloroplasts (c) are located along the periphery of the cell and the pyrenoids (py) project toward the cell's interior. Located in the groove of each pyrenoid is a nucleomorph (arrows). A less dense area in the chloroplast stroma (arrowheads), probably containing chloroplast DNA, is seen at the base of each pyrenoid. Also present in the cell are mitochondria (m) with tubular cristae, several Golgi bodies (g), numerous vesicles, and an extensive network of smooth and rough endoplasmic reticulum. n, nucleus ×18,200.

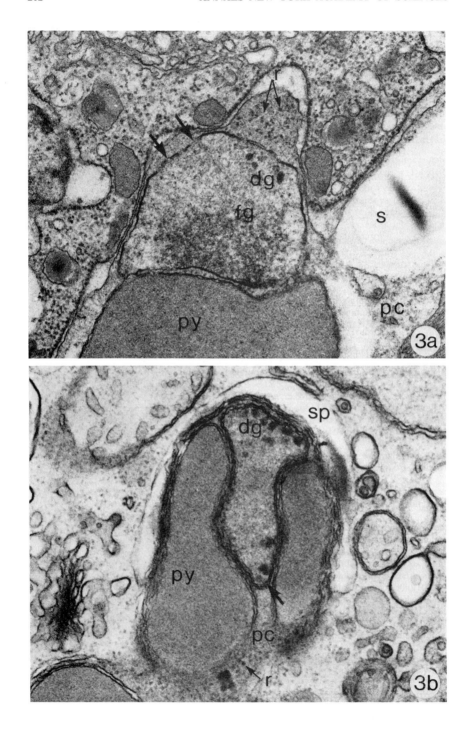

fluorescing dots arranged either along the cell periphery (FIGURE 5c) or as a cluster in a lobe of the chloroplast (FIGURE 5f). *C. abbreviata* cells also contain polyphosphate granules which fluoresce after DAPI staining (FIGURE 5f). However, since polyphosphate-DAPI complexes fluoresce yellow gold, these structures are easily distinguished from DNA-containing organelles.[14,15]

The DAPI fluorescence exhibited by the nucleomorphs of *C. abbreviata* cells, as well as that demonstrated by the nuclei and chloroplast nucleoids, proved to be susceptible to DNase digestion.[12] RNase treatment had no effect on the fluorescence exhibited by the DNA-containing organelles.[12]

Preliminary results obtained from DAPI-stained sections of *Chlorarachnion* cells fixed in 2% glutaraldehyde and embedded in JB-4 indicate that the nucleomorphs of these organisms also contain DNA. The location of the nucleomorphs in the pyrenoid grooves of *C. reptans* also proved advantageous since in favorable cell sections nucleomorphs are identified as less dense areas at the distal end of the pyrenoids (arrows, FIGURES 6b and 6e). When cell sections were stained with DAPI, blue white fluorescing, sometimes slightly elongate, dots corresponding exactly to the location of the nucleomorphs are present in the pyrenoids of *C. reptans* (arrows, FIGURES 6c and 6f). Chloroplast nucleoids, typically located at the cell periphery (FIGURE 6c) or at the base of the projecting pyrenoid (FIGURE 6f), also display DNA-DAPI fluorescence. The nuclei of *C. reptans* cells also fluoresce after cell sections are stained with DAPI (FIGURES 6c and 6f); however, fluorescence is much reduced in the region of the nucleolus (FIGURE 6c).

To visualize DNA fibrils, cells of *Cryptomonas abbreviata* and *Chlorarachnion reptans* were fixed according to the method of Ryter and Kellenberger.[16] As a result of this fixation technique DNA appears as a network of fine fibrils in electron micrographs.[16,17] Such a network of DNA-like fibrils is present in the nucleomorphs of both species. The cryptomonad nucleomorph shown in FIGURE 7a displays a single region of DNA-like fibrils (arrow). The large convoluted nucleomorph of *C. reptans* shown in FIGURE 7b displays three areas of fine fibrils (arrows). In both organisms the fibrils are located in the part of the nucleomorph matrix not occupied by the fibrillogranular body or the dense globules. In addition, in similarly sectioned *Cryptomonas abbreviata* cells, the region of the nucleomorph containing the network of DNA-like fibrils was shown to correspond exactly to the area of the organelle exhibiting DNA-DAPI fluorescence.[12]

Since the nucleomorphs of *Chlorarachnion* were discovered only recently, all previous data on the possible nuclear origin of the nucleomorph were obtained by studying the cryptomonad organelle. The pores interrupting the nucleomorph envelope have

---

FIGURE 3. Sections through the nucleomorphs of *Cryptomonas abbreviata* (3a) and *Chlorarachnion reptans* (3b). The nucleomorphs of both organisms are located in the periplastidal compartment (pc), associated with the pyrenoid (py). The nucleomorphs are limited by a double membrane which is interrupted by pores (arrows). Dense globules (dg) are seen in the nucleomorphs of both organisms. A fibrillogranular body (fg) is seen in the cryptomonad nucleomorph. The periplastidal compartments contain ribosomelike particles (r). Note starch (s) is stored in the periplastidal compartment of *Cryptomonas abbreviata* while the storage product (sp) of *Chlorarachnion reptans* is located in the cytoplasm. (3a) ×42,900; (3b) ×51,100

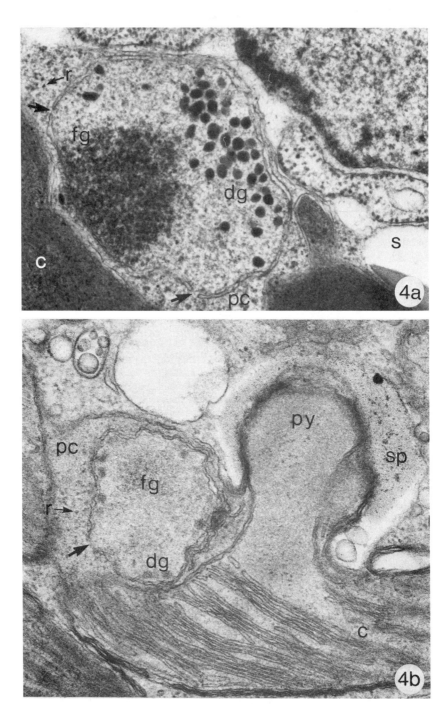

been looked at in several cryptomonads. Circular pores, 70 nm in diameter, have been observed in three freshwater cryptomonads.[10,18] Whereas these pores and the pores of the nuclear envelope are similar in appearance, the nuclear pores are slightly larger, with a diameter of about 100 nm. The pores interrupting the nucleomorph membranes of two marine species of cryptomonads were found to be elliptical.[6] This is believed to be a degenerative feature. A dense central granule or core is observed in both the circular[10,18] and elliptical[6] pores.

The cryptomonad nucleomorph is a self-replicating organelle. Microtubules do not appear to be involved in its division. Instead, during preprophase, the envelope of the nucleomorph infolds, a dumbbell shape is formed, and the organelle then simply pinches in two.[19] This mode of division is also thought to be a degenerative feature since the nucleus of the putative chrysophyte endosymbiont in *Peridinium balticum* also divides amitotically by pinching in two.[20]

Even the earliest descriptions of the cryptomonad nucleomorph suggested that the fibrillogranular body was morphologically similar to a nucleolus.[2,6] Cytochemically, by employing a modification of Bernhard's regressive uranyl acetate-EDTA-lead staining technique,[21] the fibrillogranular body was shown to bleach at about the same time as the cell's nucleolus, e.g., after a relatively long EDTA treatment, thus suggesting that it contains ribonucleoprotein.[6]

These data indicate that a nuclear origin for the cryptomonad nucleomorph is very likely. Similarly, the presence of DNA-DAPI fluorescence and DNA-like fibrils in the *Chlorarachnion* nucleomorph strongly supports the hypothesis that this organelle is also a vestigial nucleus. However, further data on the structure of the pores interrupting the envelope of this organelle (e.g., tangential sections) should be obtained. And it will be very interesting to determine if the nucleomorph of *Chlorarachnion*, like its cryptomonad counterpart, divides amitotically.

The presence of a nuclear protein in the nucleomorphs would be still further evidence that they are remnants of eukaryotic nuclei. We have obtained preliminary evidence, using antibodies directed against histone H1, H2A, H2B, and H3, combined with the protein A-gold technique,[22] that indicates that the nucleomorph of *Cryptomonas abbreviata* probably does not contain histones. The absence of histones in the nucleomorph, however, is not at all unexpected since condensed chromatin has never been seen in the organelle. Similar experiments with antibodies directed against other nuclear proteins are currently in progress.

Since it appears that the nucleomorphs of cryptomonads and *Chlorarachnion* are vestigial nuclei, two further questions of interest are the identities of the endosymbionts that gave rise to the chloroplasts of these organisms and the coding functions of the nucleomorph DNAs.

---

**FIGURE 4.** Sections through the nucleomorphs of *Cryptomonas* sp. (φ) (4a) and *Chlorarachnion* sp. (4b). The nucleomorphs of both organisms lie along the inner face of the chloroplast (c) in the periplastidal compartment (pc). A double membrane, interrupted by pores or larger gaps (arrows), delimits each nucleomorph. Located in the nucleomorph matrix of both organisms are a fibrillogranular body (fg) and dense globules (dg). Putative ribosomes (r) are also present in the periplastidal compartments. py, pyrenoid; s, starch; sp, storage product. (4a) × 49,000; (4b) × 66,750.

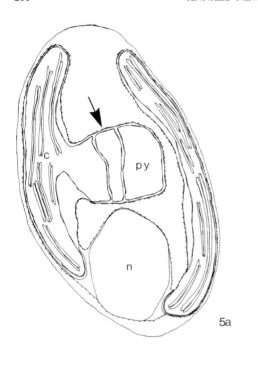

5a

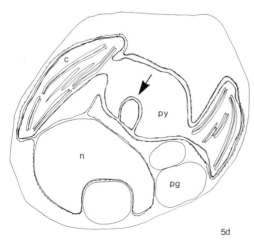

5d

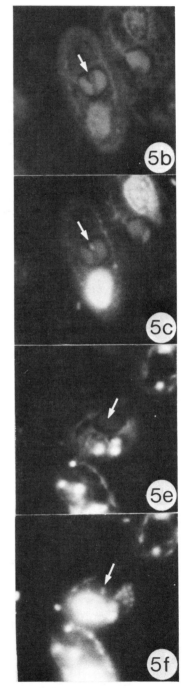

It has been suggested previously that the cryptomonad chloroplast evolved from a redlike alga that had evolved chlorophyll $c$ while retaining phycobiliprotein pigments.[5,12,23] However, no known red alga containing chlorophyll $c$ and phycobiliproteins exists today. Alternatively, the ancestral red alga could have evolved chlorophyll $c$ after it had been taken up by the cryptomonad host cell. A morphological observation supporting a redlike algal origin of the cryptomonad chloroplast is the presence of starch grains in the cryptomonad periplastidal compartment (in the cytoplasm of the endosymbiont); red algae store starch outside the chloroplast in the cytoplasm. A second line of evidence that supports a direct evolutionary relationship between a redlike alga and the cryptomonad chloroplast is that the cryptomonads, red algae, and cyanobacteria contain phycobiliproteins. Cyanobacterial and red algal biliproteins are evolutionarily very closely related,[24-26] thus supporting the hypothesis that cyanobacteria gave rise to the chloroplasts of red algae. Recently, several studies have shown that amino acid sequence homology exists among the subunits of several cryptomonad biliproteins and the corresponding biliprotein subunits of cyanobacteria and red algae.[27-29] In addition, cryptomonad phycobiliproteins also demonstrate immunological cross-reactivity with red algal phycobilins.[30,31]

Pigment analysis has shown that *Chlorarachnion reptans* contains chlorophylls $a$ and $b$; thus it has been proposed that the chloroplasts of these organisms have evolved from a symbiotic green alga.[9] Thylakoid arrangement in *C. reptans* is also similar to that found in some green algae. The lamellae contain paired thylakoids, although deeper stacks are also found[9] and interchange of thylakoids between adjacent lamellae is often seen. Although the chloroplasts of euglenoids also demonstrate many of these characteristics, it is unlikely that a euglenoid gave rise to the *Chlorarachnion* chloroplast. If a euglenoid, which contains a chloroplast limited by three membranes, were taken up by *Chlorarachnion,* then five membranes, not four, would separate the cytoplasm of *Chlorarachnion* from the chloroplast stroma.

The endosymbionts that gave rise to the chloroplasts of cryptomonads and *Chlorarachnion* were, obviously, morphologically and phylogenetically diverse. Because of these differences and as a result of the endosymbioses and the subsequent transfer of genes from one organelle to another, the nucleomorph DNAs of cryptomonads and *Chlorarachnion* probably code for different sets of proteins. The nucleomorph DNA of both *Chlorarachnion* and the cryptomonads, however, may contain cytoplasmic ribosomal RNA genes since, as shown above, putative eukaryotic-sized ribosomes are present in the periplastidal compartments (the cytoplasm of the eukaryotic endosymbiont) of these organisms. Clearly, future experiments employing molecular analyses are needed to determine the coding functions maintained by the DNA of cryptomonad and *Chlorarachnion* nucleomorphs.

**FIGURE 5.** Autofluorescence (excitation 445 nm, barrier filter 515 nm; 5b and 5e) and DAPI-associated fluorescence (excitation 365 nm, barrier filter 430 nm; 5c and 5f) of DAPI-stained sections of *Cryptomonas abbreviata.* Line drawings (5a and 5d) are schematic representations of the same cells. Blue white fluorescence, characteristic of DNA-DAPI complexes, is exhibited by the nucleomorph (arrows, 5c and 5f). The chloroplast nucleoids and the nucleus (n) also fluoresce. Polyphosphate granules (pg) demonstrate yellow fluorescence. c, chloroplast; py, pyrenoid. $\times 2500$.

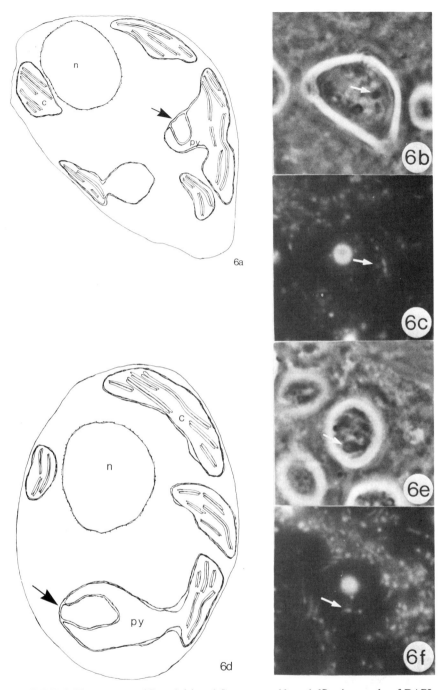

**FIGURE 6.** Phase contrast (6b and 6e) and fluorescence (6c and 6f) micrographs of DAPI-stained sections of *Chlorarachnion reptans*. Line drawings (6a and 6d) are schematic representations of the same cells. The nucleomorph demonstrates DNA-DAPI fluorescence (arrows, 6c and 6f). Fluorescence is also exhibited by the nucleus (n) and the chloroplast nucleoids of the cell. c, chloroplast; py, pyrenoid. ×2500.

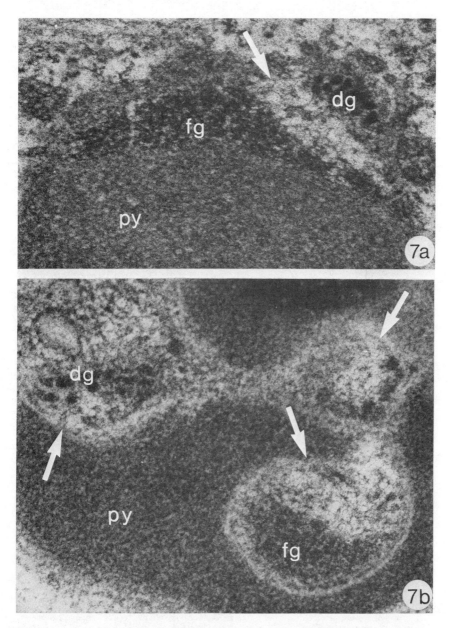

**FIGURE 7.** Sections through the nucleomorphs of *Cryptomonas abbreviata* (7a) and *Chlorarachnion reptans* (7b) following Ryter and Kellenberger fixation. In both organisms DNA-like fibrils (arrows) are located in the part of the nucleomorph matrix not occupied by the fibrillogranular body (fg) or the dense globules (dg). py, pyrenoid. (7a) × 61,420; (7b) × 107,250.

## REFERENCES

1. GIBBS, S. P. 1962. Nuclear envelope-chloroplast relationships in algae. J. Cell Biol. **14:** 433-444.
2. GREENWOOD, A. D. 1974. The Cryptophyta in relation to phylogeny and photosynthesis. Abstr., Eighth Int. Congr. Electron Microsc. **2:** 566-567.
3. GREENWOOD, A. D., H. B. GRIFFITHS & U. J. SANTORE. 1977. Chloroplasts and cell compartments in Cryptophyceae. Br. Phycol. J. **12:** 119.
4. GIBBS, S. P. 1978. The chloroplasts of *Euglena* may have evolved from symbiotic green algae. Can. J. Bot. **22:** 2883-2889.
5. WHATLEY, J. M., P. JOHN & F. R. WHATLEY. 1979. From extracellular to intracellular: the establishment of mitochondria and chloroplasts. Proc. R. Soc. London Ser. B **204:** 165-187.
6. GILLOTT, M. A. & S. P. GIBBS. 1980. The cryptomonad nucleomorph: its ultrastructure and evolutionary significance. J. Phycol. **16:** 558-568.
7. WHATLEY, J. M. 1981. Chloroplast evolution—ancient and modern. Ann. N.Y. Acad. Sci. **361:** 154-165.
8. GIBBS, S. P. 1981. The chloroplasts of some algal groups may have evolved from endosymbiotic eukaryotic algae. Ann. N.Y. Acad. Sci. **361:** 193-208.
9. HIBBERD, D. J. & R. E. NORRIS. 1984. Cytology and ultrastructure of *Chlorarachnion reptans* (Chlorarachniophyta divisio nova, Chlorarachniophyceae classis nova). J. Phycol. **20:** 310-330.
10. MORRALL, S. & A. D. GREENWOOD. 1982. Ultrastructure of nucleomorph division in species of Cryptophyceae and its evolutionary implications. J. Cell Sci. **54:** 311-328.
11. SANTORE, U. J. 1984. Some aspects of taxonomy in the Cryptophyceae. New Phytol. **98:** 627-646.
12. LUDWIG, M. & S. P. GIBBS. 1985. DNA is present in the nucleomorph of cryptomonads: further evidence that the chloroplast evolved from a eukaryotic endosymbiont. Protoplasma **127:** 9-20.
13. WILLIAMSON, D. H. & D. J. FENNELL. 1975. The use of fluorescent DNA-binding agent for detecting and separating yeast mitochondrial DNA. Methods Cell Biol. **12:** 335-351.
14. COLEMAN, A. W. 1978. Visualization of chloroplast DNA with two fluorochromes. Exp. Cell Res. **114:** 95-100.
15. COLEMAN, A. W., M. J. MAGUIRE & J. R. COLEMAN. 1981. Mithramycin- and 4'-6-diamidino-2-phenylindole (DAPI)-DNA staining for fluorescence microspectrophotometric measurement of DNA in nuclei, plastids, and virus particles. J. Histochem. Cytochem. **29:** 959-968.
16. SCHREIL, W. H. 1964. Studies on the fixation of artificial and bacterial DNA plasms for the electron microscopy of thin sections. J. Cell Biol. **22:** 1-20.
17. NASS, M. M. K. & S. NASS. 1963. Intramitochondrial fibers with DNA characteristics. I. Fixation and electron staining reactions. J. Cell Biol. **19:** 593-611.
18. SANTORE, U. J. 1982. The distribution of the nucleomorph in the Cryptophyceae. Cell Biol. Int. Rep. **6:** 1055-1063.
19. MCKERRACHER, L. & S. P. GIBBS. 1982. Cell and nucleomorph division in the alga *Cryptomonas.* Can. J. Bot. **60:** 2440-2452.
20. TIPPIT, D. H. & J. D. PICKETT-HEAPS. 1976. Apparent amitosis in the binucleate dinoflagellate *Peridinium balticum.* J. Cell Sci. **21:** 273-289.
21. BERNHARD, W. 1969. A new staining procedure for electron microscopical cytology. J. Ultrastruct. Res. **27:** 250-265.
22. ROTH, J. 1982. The protein A-gold (pAg) technique—a qualitative and quantitative approach for antigen localization on thin sections. *In* Techniques in Immunocytochemistry. G. R. Bullock & P. Petrusz, Eds. **1:** 107-133. Academic Press. New York, N.Y.
23. WHATLEY, J. M. & F. R. WHATLEY. 1981. Chloroplast evolution. New Phytol. **87:** 233-247.
24. STANIER, R. Y. 1974. The origins of photosynthesis in eukaryotes. Symp. Soc. Gen. Microbiol. **24:** 219-240.
25. MACCOLL, R. & D. S. BERNS. 1979. Evolution of the biliproteins. Trends Biochem. Sci. **4:** 44-47.

26. GLAZER, A. N. 1980. Structure and evolution of photosynthetic accessory pigment systems with special reference to phycobiliproteins. *In* The Evolution of Protein Structure and Function. D. S. Sigman & M. A. B. Brazier, Eds.: 221-243. Academic Press. New York, N.Y.

27. GLAZER, A. N. & G. S. APELL. 1977. A common evolutionary origin for the biliproteins of cyanobacteria, Rhodophyta and Cryptophyta. FEMS Lett. **1:** 113-116.

28. SIDLER, W., B. KUMPF, F. SUTER, W. MORISSET, W. WEHRMEYER & H. ZUBER. 1985. Structural studies on cryptomonad biliprotein subunits. Two different α-subunits in *Chroomonas* phycocyanin-645 and *Cryptomonas* phycoerythrin-545. Biol. Chem. Hoppe-Seyler **366:** 233-244.

29. TROXLER, R. F., G. D. OFFNER, F. G. OPPENHEIM, R. MACCOLL & D. S. BERNS. 1985. Phycoerythrin from the cryptophyte alga, *Rhodomonas lens.* Fed. Proc. **44:** 1227.

30. MACCOLL, R., D. S. BERNS & O. GIBBONS. 1976. Characterization of cryptomonad phycoerythrin and phycocyanin. Arch. Biochem. Biophys. **177:** 265-275.

31. WEHRMEYER, W. 1983. Phycobiliproteins and phycobiliprotein organization in the photosynthetic apparatus of cyanobacteria, red algae, and cryptophytes. *In* Proteins and Nucleic Acids in Plant Systematics. U. Jensen & D. E. Fairbrothers, Eds.: 143-167. Springer Verlag. Berlin, FRG.

# Viroids[a]

## Subcellular Location and Structure of
## Replicative Intermediates

DETLEV RIESNER, PETRA KLAFF, GERHARD
STEGER, AND ROLF HECKER

*Institute for Physical Biology*
*University of Düsseldorf*
*Universitätsstrasse 1*
*D-4000 Düsseldorf 1, Federal Republic of Germany*

### INTRODUCTION

Viroids are an independent class of plant pathogens which are distinguished from viruses by the absence of a protein coat and their unusually small size. They consist merely of a single-stranded, circular ribonucleic acid of a few hundred nucleotides. Until now, about 15 different viroids have been detected, the majority of them causing diseases of economically important crop plants such as potato, tomato, cucumber, avocado, citrus, coconut palms, etc.

The cadang-cadang viroid of the coconut palm, for example, causes severe losses in the vast palm plantations and hence seriously threatens the economy of a whole country, the Phillipines. It is hoped that a detailed knowledge of viroid structure, replication, and the interaction of the viroid with the host cell might finally contribute to the control of their propagation and thus to a decrease in the enormous losses they cause in agriculture.

Sequence analysis and physicochemical studies of several viroids have shown that, as a result of intramolecular base pairing, viroids form a unique rodlike secondary structure which is characterized by a serial arrangement of double-helical sections and internal loops. It could be deduced from several lines of experimental evidence that there is no indication for an additional tertiary structure. In FIGURE 1A the structure of the viroid of the potato spindle tuber (PSTV) disease is depicted in a two-dimensional scheme and in the form of a three-dimensional model. During denaturation, all of the native base pairs of viroids are dissociated in one highly cooperative transition, and in the same process very stable hairpins which are not present

[a]This work was supported by grants from the Deutsche Forschungsgemeinschaft, the Minister für Wissenschaft und Forschung des Landes Nordrhein-Westfalen, and the Fonds der Chemischen Industrie.

in the native structure are newly formed. This mechanism is shown in FIGURE 1B. The close similarity between different viroids is more expressed in the overall structure and in thermodynamic and functional domains than in the primary sequence. The biochemical and physicochemical properties of viroids have been summarized in several recent reviews. [1,4,5]

Because of their infectivity as naked RNA, one may speculate that viroids do not underlie all topical barriers inside the cell and that they may be present as dissolved molecules and not as part of a tight complex. Such a simplified view would actually be in accordance with the very early experiments that led to the concept of viroids as a new class of infectious agents (for review, see Reference 6.) In those experiments it was shown that viroid infectivity in a tissue homogenate was unprotected against RNase digestion, and that viruslike particles were not visible in infected tissue. From the mechanistic point of view, one has to assume specific complexes of viroids with components of the host cell. Those complexes are required for protection against degradation, for replication, for the expression of symptoms, and possibly for the spreading of the systemic infection.

Another challenging problem deals with the cellular structure of the replicative intermediates of viroids. In contrast to our knowledge of the structure of mature viroids, there has been only speculation about the secondary or tertiary structure of the replicative intermediates. In the infected plant cells there are found monomeric and oligomeric linear viroids of both polarities in addition to the mature circular viroids (for review, see Reference 1). Therefore, a rolling-circle type of replication was proposed. [7,8] These oligomeric replication intermediates require at least specific cleavage of multimeric ( + ) intermediates to monomeric units, followed by ligation of the monomers to covalently closed circular molecules. Because of the secondary structure of circular viroids, the multimeric molecules may form very peculiar structures, allowing the cleavage and ligation enzymes to recognize a monomeric unit. These highly specialized structures do not have to be saved in the circular molecules or be the same for ( + ) and ( − ) intermediates.

In this article we will concentrate on the two problems outlined above: (1) the intracellular complexes of mature viroids, and (2) the secondary structure of the replicative intermediates of viroids.

# MATERIALS AND METHODS

## Chemicals

DNase I was purchased from Boehringer (Mannheim, Federal Republic of Germany), $N$-cetyl-$N,N,N$,trimethylammoniumbromide (CTAB) from Merck (Darmstadt, Federal Republic of Germany); Hydroxylapatite Fast-Flow from Calbiochem-Bering (LaJolla, Calif.), and polyethyleneglycole, $M_r$ 6000 (PEG 6000), was from Roth (Karlsruhe, Federal Republic of Germany). All buffers were made from high purity water (Milli-Q-System, Millipore) and filtered through a 1.0 $\mu$m membrane (RC 60, Schleicher und Schüll).

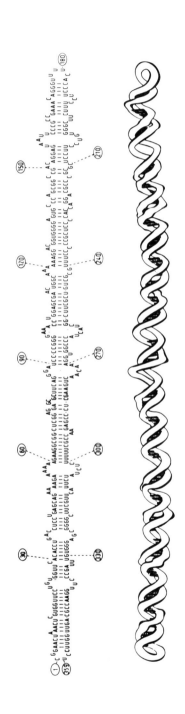

**FIGURE 1.** Structure and structure formation of viroids. FIGURE 1A (above). A two-and a three-dimensional representation of the structure is given according to References 2 and 3.

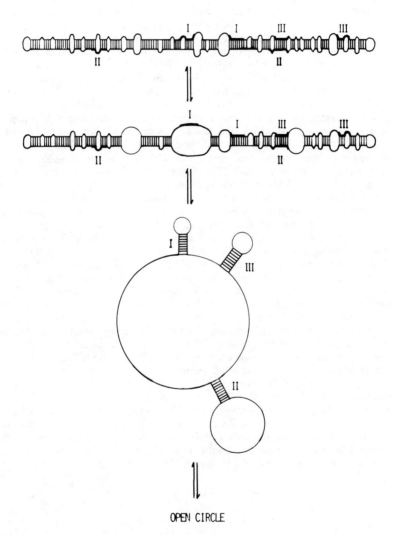

**FIGURE 1B.** The different stages during thermal denaturation are shown schematically.[21]

## *Preparation of PSTV Containing RNA Extract*

For viroid isolation, tomato plants (*Lycopersicon esculentum Mill.,* variety Rutgers) were infected with the potato spindle tuber viroid (PSTV) originally supplied by Dr. T. O. Diener, Beltsville, Md. The host plants were grown in a green house at 30°C.

Five hundred grams of infected tomato plants frozen in liquid nitrogen were homogenized in a Waring Blendor for 10 minutes in 750 ml TES-SDS buffer (100 mM Tris-HCl, pH 8.9; 100 mM NaCl; 10 mM EDTA; 0.5% sodium dodecyl sulfate; 0.5% sodium lauryl sarcosinate), 10 g bentonite, and 2.5 ml 2-mercaptoethanol. After centrifugation, the nucleic acids of the clear supernatant were precipitated with one volume of ethanol for 1 hour at −20°C. The precipitate was dissolved in 100-150 ml TES-SDS and extracted with the same volume of TES-saturated phenol. The phenolic phase was reextracted with TES-SDS. Both aqueous phases were combined and extracted first with phenol-chloroform (1:1) and second with chloroform.

An effective fractionation was carried out with an aqueous two-phase system.[9] The aqueous phase is adjusted with PEG 6000 and 30% (w/w) potassium phosphate (56.04 g $KH_2PO_4$ and 93.96 g $K_2HPO_4$ dissolved in 350 ml $H_2O$) to a final concentration of 8.2% PEG and 52.6% (v/v) of the potassium phosphate solution. After phase separation of the aqueous two-phase system, the potassium phosphate phase (bottom phase with larger volume) was diluted with the same volume of $H_2O$, and the nucleic acids were precipitated for 2 hours on ice by adding 0.1 volume of a solution containing 1% CTAB, 0.1% NaCl, and 10 mM EDTA. The precipitate was washed two times with 70% ethanol, 200 mM NaOAc, pH 7.5. The precipitate was dissolved in 20 ml STE buffer (100 mM NaCl, 10 mM Tris-HCl, pH 7.0; 1 mM EDTA), and 5 ml 10 M LiCl were added to reach a final concentration of 2 M LiCl. Precipitation of large single-stranded RNA was allowed for 1 hour on ice. The pellet was dissolved in 20 ml STE and precipitated again with 0.25 volume of 10 M LiCl. The supernatants of both LiCl precipitations containing the viroid were combined, and the nucleic acids were precipitated with 3 volumes of ethanol for 1 hour at −70°C. The pellet was dissolved in DNase I buffer (10 mM Tris-HCl, pH 8.0, 15 mM NaCl, 6.25 mM $MgCl_2$) containing also 0.1 volume of 200 mM ribonucleoside-vanadyl complex.[10] DNase I (200 U) was added. After 15 minutes at 37°C the solution was extracted twice with phenol-chloroform and twice with chloroform. To the aqueous phase 0.5 volume of 30% PEG 6000, 1.5 M NaCl was added and precipitation was allowed on ice for 1 hour. From the supernatant, the viroid-enriched nucleic acids were precipitated with 3 volumes of ethanol. After 1 hour at −70°C and centrifugation the precipitate was dissolved in 10 mM Tris-HCl, pH 7.0, 0.1 mM EDTA. The concentration of the RNA was determined by UV spectroscopy.

## *High Performance Liquid Chromatography of PSTV*

High performance liquid chromatography (HPLC) was performed on an LC 850 liquid chromatograph (Du Pont, Bad Nauheim, Federal Republic of Germany) equipped with a UV detector type 87.00 (Knauer, Bad Homburg, Federal Republic of Germany). Samples were injected with the loop-sample injector 7125 (Rheodyne Inc., Berkeley, Calif.) and a 2.5 ml sample loop.

For HPLC, the weak anion exchanger Nucleogen-DEAE 500-IWC (Macherey-Nagel, Düren, Federal Republic of Germany) in columns of 10 mm internal diameter and 125 mm length was used.[11] The best results are obtained with linear salt gradients from 0.4 M KCl to 0.9 M KCl in a buffer containing 6 M Urea, 30 mM potassium phosphate, pH 6.5, 0.1 mM EDTA at ambient temperature. For analytical chromatograms, the salt gradient was 10 mM KCl/minute at a flow rate of 1 to 3 ml/minute. For preparative chromatograms, the gradient was lowered to 2 mM KCl/minute at the same flow rate. In analytical runs, 100 to 200 μg total RNA of the crude extract were analyzed. In the preparative runs, up to 20 mg RNA of the crude extract were separated in 200 minutes yielding about 100 μg pure PSTV. The chromatograms were similar as in References 11 and 12. The different fractions after HPLC were analyzed by 5% polyacrylamide gel electrophoresis (PAGE), as described elsewhere.[11] For desalting, the desired fractions were pooled and adsorbed on hydroxylapatite, washed with 20 mM potassium phosphate, pH 6.5, and eluted with 0.5 M potassium phosphate, pH 6.5. The RNA was precipitated with 0.1 volume of 1% CTAB, 1% NaCl, 0.1 mM EDTA for 1 hour on ice. The pellet was dissolved in 7.5 M ammonium acetate, pH 7.4, afterwards diluted with 2 volumes $H_2O$ to 2.5 M, and then precipitated with 2.5 volumes of ethanol. The precipitate was dissolved in STE and precipitated again with 2.5 volumes of ethanol. The final precipitate contains the sodium salt of the viroid RNA and may be dissolved in any buffer wanted for further experiments.

## SP6 Transcripts of PSTV

Synthetic PSTV replication intermediates used for thermodynamic investigations were provided by M. Tabler and H. L. Sänger. As described earlier,[13] they were constructed by insertion of one to six head-to-tail connected DNA copies of PSTV RNA into plasmid pSP62-PL downstream of the promoter for SP6 RNA polymerase. *In vitro* transcription yielded the corresponding monomeric or oligomeric single-stranded linear PSTV RNA molecules of (+) and (−) polarity. Short terminal sequences are vector derived: 47 nucleotides at the 5′ terminus and 8 nucleotides at the 3′ terminus. The PSTV sequence is, in the case of the (+) species, nucleotide 282-359/1-281 according to the numbering of circular PSTV (cf. FIGURE 1A) and, in case of the (−) species, 281-1/359-282.

## Fractionation of Subcellular Fractions

Nuclei were prepared from PSTV-infected tomato leaves by the method of Schumacher *et al.*[14] The fractionation of nuclei into nucleoli-containing and chromatin-containing fractions was carried out according to Pederson,[15] and the subfractionation of these components according to Wolff *et al.*[16] Proteins were isolated by acid extraction as published by Prestayko and Busch.[17] Western blots were carried out in an LKB 2005 Transphor Electroblotting Unit.

### Ultraviolet Cross-Linking

For UV cross-linking of cellular protein-nucleic acid complexes, the procedure of Schoemaker and Schimmel was applied.[26] For UV cross-links, 10 ml of a solution containing $10^7$ nuclei/ml were irradiated with 254 nm light (Fluotest 406 AC lamp, Heraeus, Hanau, Federal Republic of Germany) at 4°C for 1 hour. The solution was gently shaken in a petri dish at a distance of 5 cm from the lamp.

### Analytical Ultracentrifugation

Sedimentation coefficients were determined by analytical ultracentrifugation in a Spinco model E equipped with a fluorescent detection system.[18] About 0.05 $A_{260}$/ml RNA and $4.7 \times 10^{-7}$ M ethidiumbromide in 500 mM NaCl, 1 mM EDTA, 10 mM sodium cacodylate, pH 6.8 were used for the measurements. Molecular weights were determined in a Spinco model E equipped with a UV detection system. About 0.15 $A_{260}$/ml RNA in 4 M urea, 500 mM NaCl, 0.1 mM EDTA, 1 mM sodium cacodylate, pH 6.8 was used.

### Thermal Transition Curves

The equilibrium thermal denaturation curves were measured as described earlier.[19] The experimental solution conditions were 0.2 $A_{260}$/ml RNA in 4 M urea, 500 mM NaCl, 0.1 mM EDTA, 1 mM sodium cacodylate, pH 6.8. The experimental curves had to be extrapolated to reference conditions, i.e., 1 M ionic strength without urea; all theoretical curves refer to these reference conditions. For extrapolation the whole curve was shifted by the difference in $T_m$ between the reference and experimental conditions, which was experimentally determined for the main transition of circular viroids ($\Delta T_m = 15°C$)[20] and for the main transition of SP6 PSTV 1.8 (+) ($\Delta T_m = 15°C$).

The most stable secondary structures of the different RNAs were calculated with a slightly modified[21] Zuker-Nussinov[22,23] program on a VAX 11/750. The calculation of "thermal denaturation curves" from secondary structures at different temperatures was done as described earlier.[21]

# RESULTS

## Subcellular Location of Viroids

### Distribution of Viroids in the Cell

The problem of the *in vivo* structure of viroids was approached by asking the question, At which site and in which organelle of the cell are viroids located?[14] Isolated nuclei from green leaf tissue of infected tomato plants were analyzed by bidirectional gel electrophoresis[24] for the presence of viroids. Depending upon the progress of the systemic infection and the corresponding increase in the intensity of disease symptoms, between 200 and 10,000 viroid molecules per nucleus were observed. During the isolation of chloroplasts the nuclei were lysed and sedimented in the form of an aggregated network of chromatin and other nuclear components. One or two viroid molecules were observed per chloroplast organelle; in contrast, the chromatin-rich networks contained up to 10,000 viroid molecules per nucleus. If viroids were isolated on a preparative scale[11] from leaf tissue without prior isolation of nuclei, 1 mg ($5 \times 10^{15}$ molecules) purified viroid was obtained from 1 kg highly diseased tissue. About $5 \times 10^8$ cells/g tissue were counted from leaf sections, i.e., an average of 10,000 viroid molecules are present per cell. The numbers of viroids in the cell and in the different subcellular fractions are summarized in TABLE 1. Because they represent average numbers, the actual copy number may vary between individual cells as well as between different cell types.

Subfractionation of the nuclei showed that $\approx 90\%$ of the viroid RNA is present in the nucleolar fraction. The presence of viroids in the nucleus was expected on the basis of earlier experiments; not only was this expectation verified, beyond that it could now be shown that the location of viroids inside the nucleus is specifically the nucleolus. Cosedimentation of the viroid with the chromatin network as mentioned above is not due to a molecular association with the bulk part of the chromatin but due to mechanical inclusion of nucleoli. The association of the viroid RNA with the nucleoli was studied by the ionic strength dependence of the presumed complex. After treatment with "high salt buffer" (0.5 M NaCl, 50 mM $MgCl_2$, 10 mM Tris-HCl, pH 7.0), the viroid RNA was released from its association with the nucleoli. Because such high salt treatment has been shown to abolish most of the nuclear protein-nucleic acid interactions,[25] it was concluded that viroid RNA is present in the nucleoli in a protein-nucleic acid complex.

### 10S Particles

For further characterization, viroid-protein complexes were prepared from the nucleus without using dissociating conditions.[16] Nucleoli from infected plants were digested with DNase I and sonicated. After centrifugation through a 30% sucrose cushion, viroids were found in the supernatant. Normally the supernatant contains the fraction of the chromatin that is organized in the nucleolus in the form of nucleosomes; it is mainly the ribosomal DNA. The components of the supernatant were further characterized by their sedimentation behavior. After centrifugation through a gradient from 5 to 30% sucrose, fractions were collected and tested for the presence of viroids by bidirectional gel electrophoresis. FIGURE 2 shows the analysis of the fractions harvested from the gradient. PSTV RNA was detected mainly in fractions 4, 5, 6, 7, and 8. The position of the viroid RNA in the gradient allows estimation of the $s_{20,w}$ values of the viroid-containing complexes. The fractions of the gradient containing viroid RNA correspond to the following $s_{20,w}$ values: 5.8S, 8.0S, 11.1S, 13.5S, and 17S. The average sedimentation coefficient is around 10S, slightly different from a value of 12S published earlier.[16]

## In vitro *Reconstitution*

In order to identify proteins that form complexes with viroids, the proteins of the chromatin-containing fraction and of the nucleolar fraction were isolated and tested for viroid binding. A summary of the procedure used is shown in FIGURE 3 (cf. also

TABLE 1. Cellular Distribution of Viroids

| | |
|---|---|
| Intact cells | Up to 10,000 viroids (depending upon the progress of the infection) |
| Nuclei | Up to 10,000 viroids |
| Lysed nuclei | All viroids cosediment with the chromosomal network |
| Chloroplasts | 1-2 viroids (?) |
| Nucleoli | 90% of all viroids from the nucleus |

Reference 16). The method is based on the reconstitution of non-covalently bound complexes. These are covalently fixed in a second step, because they would dissociate otherwise during the subsequent detection of the viroid RNA by molecular hybridization.

In each experiment about 20 $\mu$g of the protein preparations obtained from chromatin or nucleoli-containing fraction were analyzed by the procedure of FIGURE 3. From the left two slots of FIGURE 4 it is seen that a whole series of proteins are able to bind viroids.

In order to obtain information about the specificity of the reconstituted complexes, two types of experiments were carried out. First, after reconstitution but before the cross-linking reaction the filters were washed with buffers of increasing ionic strength. As shown in FIGURE 4, washing with an ionic strength as high as 0.6 M NaCl did not completely dissociate the complex of PSTV and a protein with a molecular weight of about 41,000. The binding to histones is weakened abruptly by washing with 0.3 M NaCl. Second, the reconstitution of the viroid complexes was tested for specificity by increasing competition with other nucleic acids such as 5S RNA or calf thymus DNA. The binding of PSTV to the proteins with $M_r$ of 41,000, 39,000, and 29,000 and to the histones is still possible in the presence of a 1000-fold excess of 5S RNA or calf thymus DNA, as shown in FIGURE 5.

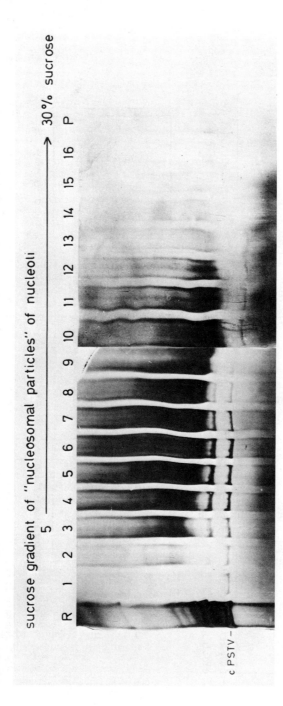

**FIGURE 2.** Distribution of viroids after a sucrose gradient centrifugation of the nucleosomal fraction from nucleoli, analyzed by bidirectional gel electrophoresis: lane R, PSTV-containing crude extract as reference; lanes 1-16, fractions 1-16; lane P, pellet.

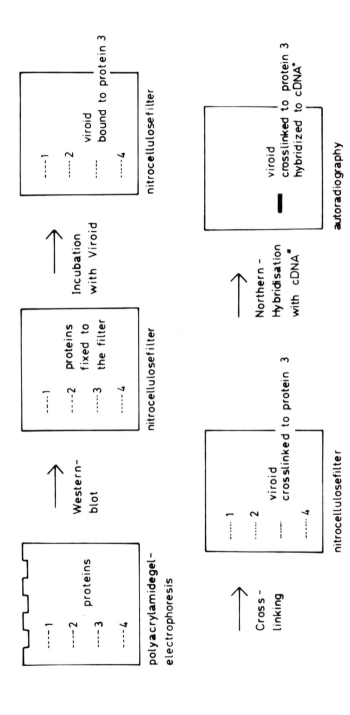

**FIGURE 3.** Procedure to detect *in vitro* reconstituted PSTV-protein complexes according to Reference 16. Step 1, separation of the proteins by SDS-PAGE; step 2, transfer of the proteins to nitrocellulose filters (Western blot); step 3, incubation with viroid RNA to establish the binding; step 4, fixation of the complexes by cross-linking with glutaraldehyde; step 5, detection of the viroid RNA by molecular hybridization with radioactive viroid-specific RNA.

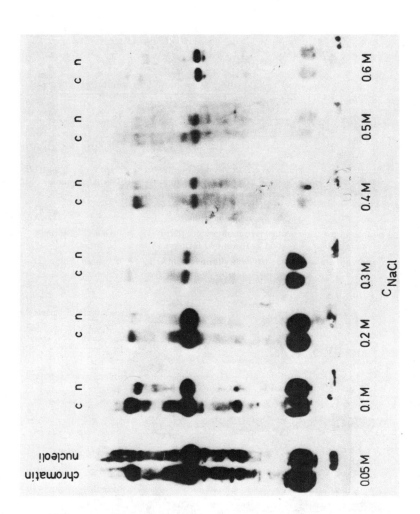

**FIGURE 4.** Viroid-binding proteins and the dependence of the binding upon the ionic strength. After incubation with PSTV the filters were washed with buffers of different NaCl concentrations; otherwise the procedure was as in FIGURE 3. C, proteins from chromatin; N, proteins from nucleoli.

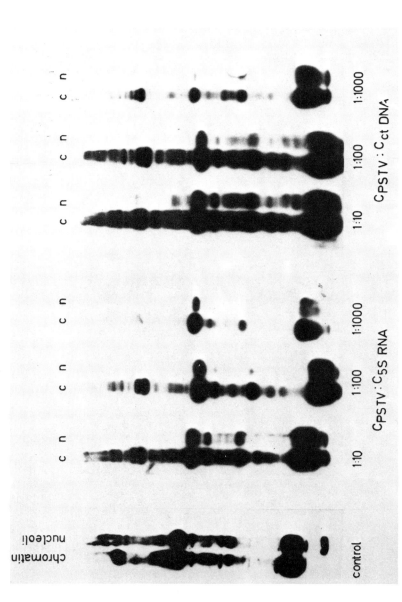

**FIGURE 5.** Viroid-binding proteins in the presence of competing 5S RNA and calf thymus (ct) DNA. The binding reaction of PSTV to the proteins was carried out in the presence of excess concentration of 5S RNA and ct DNA; otherwise the procedure was as in FIGURE 3. C, proteins from chromatin; N, proteins from nucleoli; panel 1, control, no 5S RNA or ct DNA present; panel 2, 10-fold excess of 5S RNA; panel 3, 100-fold excess of 5S RNA; panel 4, 1000-fold excess of 5S RNA; panel 5, 10-fold excess of ct DNA; panel 6, 100-fold excess of ct DNA; panel 7, 1000-fold excess of ct DNA. The salt concentration was 0.05 M NaCl.

*Ultraviolet Cross-Links*

Specific viroid-protein complexes may be identified also *in vivo*. Nuclei were irradiated with UV light to introduce covalent cross-links between proteins and nucleic acids. After irradiation, the nuclei were fractionated according to the protocol for the isolation of the 10S particles. The fractions of the gradient, similar to those in FIGURE 2, were extracted with phenol in order to test for covalent protein-nucleic acid complexes.[26] From the phenol phases the proteins and the potential protein-nucleic acid complexes were precipitated with ethanol, separated by polyacrylamide gel electrophoresis, and analyzed for their content of viroid after Western blotting and molecular hybridization with radioactive SP6 RNA. The result is shown in FIGURE 6. In fractions, 5 and 6, clear bands over a background are detectable. As expected free viroid is not detectable on the filter. The PSTV-containing bands are mediated by covalently cross-linked proteins. The result shows that viroids exist inside the cell in well-defined protein-viroid complexes.

## Structure of Replicative Intermediates

Due to the availability of viroid clones[27–29] and the possibility of transcribing the clones into RNA by the SP6 system, oligomeric viroid RNA of both polarities could be synthesized in amounts sufficient for physicochemical experiments.[13]

*Calculation of Secondary Structures*

The most stable native structure of circular viroids is a serial arrangement of short helices and small internal loops as depicted in FIGURE 1. This type of structure is true also for multimeric linear viroids, as was experimentally proven for the dimeric forms of coconut cadang-cadang viroid with (+) polarity.[30] For the oligomeric viroid RNA of PSTV, it was calculated now that star-shaped secondary structure arrangements are only insignificantly less stable than the complete linear arrangement (cf. FIGURE 7b). Because of kinetic reasons, the formation of the star-shaped structures is favored during the replication; after synthesis of each unit of an oligomer the most stable structure may be formed for this unit alone. The most stable structure of the whole molecule, however, may only be formed after synthesis of the complete oligomer by denaturation of the existing structure and rearrangement into the most stable structure.

As may be seen from FIGURE 7a the most stable structure of SP6 PSTV 1.0 (+) is nearly identical to that of native circular PSTV. The additional hairpin and two small helices in the middle of the secondary structure are due to the 55 vector-derived nucleotides. Because the structure of PSTV is self-complementary except the G:U base pairs, the structure of SP6 PSTV 1.0 (−) is very similar to that of SP6 PSTV 1.0 (+).

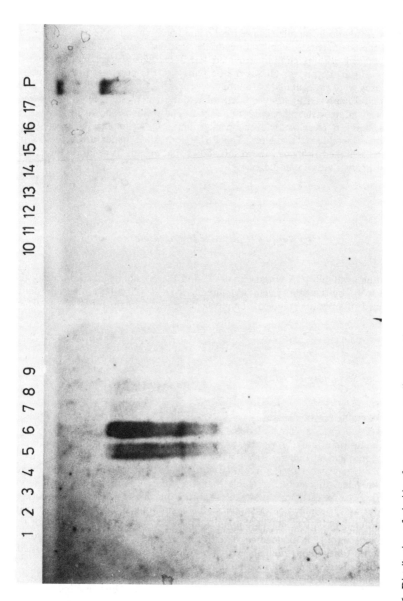

**FIGURE 6.** Distribution of viroids after a sucrose gradient centrifugation of DNase I-digested nucleoli obtained from UV-irradiated nuclei. The precipitates from the phenolic phases of the fractions were tested for the presence of viroids by blotting and molecular hybridization with radioactive viroid-specific RNA. Lanes 1-17, fractions 1-17; lane P, pellet.

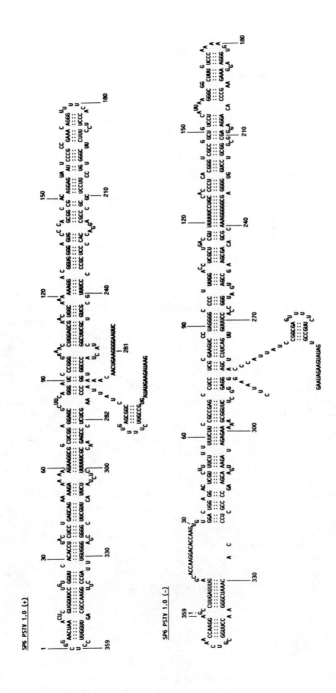

**FIGURE 7A.** Secondary structures of SP6 PSTV 1.0 (+) and 1.0 (−) at 60°C as calculated with the modified Zuker program.

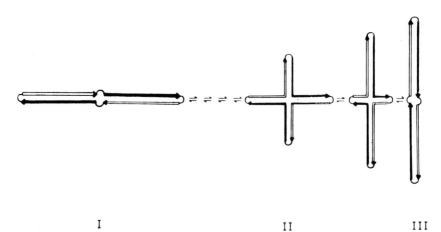

I                                    II                      III

**FIGURE 7B.** Possible secondary structures of a linear dimer of PSTV which show only minor differences in thermodynamic stability.

*Optical Denaturation Curves*

Experimental optical melting curves of SP6 PSTV 1.0 (+) and 1.0 (−) are shown in FIGURE 8. For comparison the figure contains also the theoretical curves calculated from the secondary structures at different temperatures. The experimental curves were measured in 4 M urea, 500 mM NaCl and extrapolated to 1 M NaCl for better comparison with the theoretical curves.

The main transition found earlier in circular PSTV is split into two transitions in the 1.0 (−) transcript due to the inability to form the stable hairpins I and II of PSTV (cf. FIGURE 1B). Both transitions of 1.0 (−) are similar in hypochromicity and cooperativity. The first transition of these is at the denaturation temperature of circular PSTV. An additional transition with very low hypochromicity which may be seen at lower temperatures is due to the vector-derived sequences.

Whereas for PSTV 1.0 (−) the measured and calculated transition curves are in accordance, for the 1.0 (+) transcript they differ significantly from each other. The low-temperature transition is again due to the vector-derived sequences. The main transition is at the same temperature as in circular PSTV but less cooperative. At higher temperatures the calculated curve shows only one additional transition whereas the experimental curve shows clearly two transitions. The last experimental transition is near 90°C and contains about 25 base pairs with about 80% G:C. A structure that may be responsible for this transition does not exist in a monomeric 1.0 (+) transcript.

The transition curves measured with oligomeric linear PSTVs are very similar to those of the monomers. Only the two transitions between 70 and 80°C become less cooperative and may not be seen as separate transitions.

*Isolation of PSTV Complexes with HPLC*

If a mixture of pure circular PSTV and PSTV with a single nick is injected on an HPLC column under denaturing conditions (300 mM KCl, 50% formamide, 60°C), two RNA peaks are eluted. The chromatography was carried out on the anion exchanger Nucleogen-DEAE 500[11,12] with a linear gradient from 400 to 1000 mM KCl, 50% purified formamide, 25 mM potassium phosphate, pH 6.7, 60°C.[31] Native 5% polyacrylamide gel electrophoresis showed for both peaks the normal PSTV band (FIGURE 9). Peak II contained an additional RNA species migrating at higher molecular weights. If the RNA of peak II was preincubated at temperatures higher than 50°C and run on a native 5% polyacrylamide gel, the high molecular weight band vanished and the RNA migrated as normal PSTV (FIGURE 9). Under denaturing conditions (8 M urea, 60°C) a large band of linear PSTV was shown in addition to small amounts of circular PSTV. The electron microscopic pictures of the RNA from peak II showed dimeric complexes of PSTV molecules under native conditions (FIGURE 10) and circular and linear PSTV under denaturing conditions.

*Determination of Sedimentation Coefficients and Molecular Weights*

The sedimentation coefficients of circular mature viroids are proportional to the logarithm of their molecular weights (see FIGURE 11), because the rod-shaped mol-

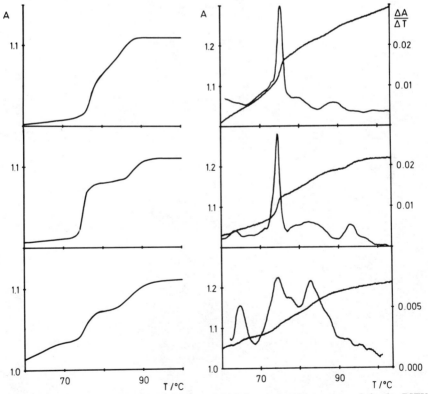

**FIGURE 8.** Calculated (left) and experimental (right) optical melting curves of circular PSTV (top), SP6 PSTV 1.0 (+) (middle), and SP6 PSTV 1.0 (−) (bottom). The experimental curves are extrapolated to 1 M ionic strength.

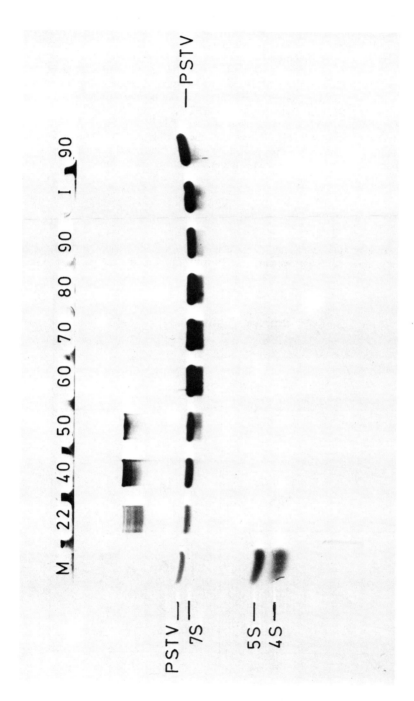

**FIGURE 9.** Analysis of an HPLC of a mixture of circular and nicked PSTV. FIGURE 9A (above). Native 5% PAGE of 0.05 μg RNA of peak I (I) of the HPLC and 0.05 μg of peak II preincubated at given temperatures for 10 minutes.

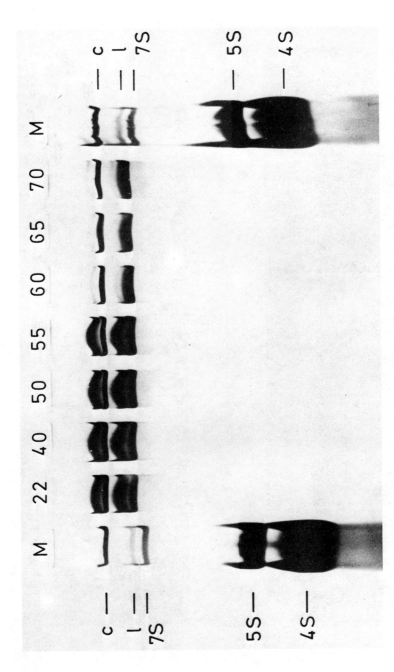

**FIGURE 9B.** Denaturing 5% PAGE of 0.05 μg RNA of peak II preincubated at given temperatures for 10 minutes. The RNA peaks were isolated with a linear gradient of 400 to 1000 mM KCl, 50% purified formamide, 25 mM potassium phosphate, pH 6.7, 60°C with HPLC on Nucleogen DMA-500 and dialyzed to 10 mM NaCl, 0.1 mM EDTA, 1 mM sodium cacodylate, pH 6.9 as described in Reference 31. M, 4 μg crude extract of PSTV containing RNA; c, circular PSTV; l, linear PSTV.

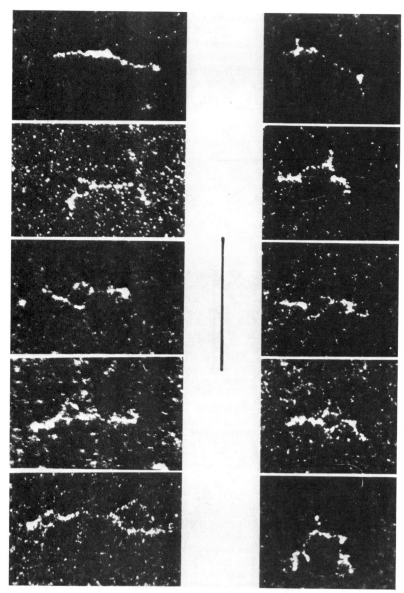

**FIGURE 10.** Electron microscopic pictures of the RNA of peak II of the same HPLC as in FIGURE 9. Dimeric complexes of PSTV under native conditions (24°C, no urea). The scale bar represents 100 nm. The figure is a courtesy from Prof. G. Klotz of Ulm University. Under the same conditions, PSTV is a rod with an average length of 37 nm.[2,3]

ecules behave like rigid cylinders over the range previously investigated.[30] The s values of viroid intermediates with ( + ) polarity were found drastically higher in comparison to that of mature viroids. This indicates a deviation from the rod-shaped structure, and/or that the intermediates do not sediment as a single molecules but as complexes.

The molecular weights of PSTV and intermediates determined by sedimentation diffusion equilibrium were the expected ones with one exception. If SP6 PSTV 1.0 ( + ) was denatured up to a temperature below the last transition and then renatured slowly (0.2°C/minute), a doubling of the molecular weight was observed (cf. TABLE 2).

*Model for the Secondary Structure of ( + ) Strand Intermediates*

A secondary structure model for replicative intermediates of viroids was derived mainly from three lines of experimental evidence described above: (1) in ( + ) strand

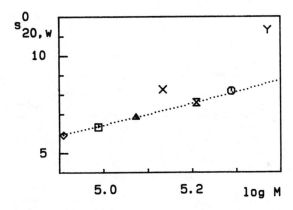

**FIGURE 11.** Dependence of $s^\circ_{20,w}$ upon log M of circular mature viroids[30] and viroid intermediates. The straight line is according to theory.[36,30] x, SP6 PSTV 1.0 (+); y, SP6 PSTV 1.8 (+). Buffer conditions are 500 mM NaCl, 0.1 mM EDTA, 1 mM sodium cacodylate, pH 6.9.

intermediates a transition was observed that was more stable than in mature viroids; (2) the appearance of this transition may clearly differentiate between ( + ) and ( − ) strand intermediates; and (3) HPLC, electron microscopy, and analytical ultracentrifugation demonstrate that ( + ) strand intermediates form dimeric complexes. The structure includes three successive helices out of nucleotides 79-110 involving hairpin I and the central conserved region with 28 base pairs and a G:C content of 71% (cf. FIGURE 12). This structure is formed by intermolecular base pairing from two monomers; in the case of multimeric viroids, the same structure may be formed by intramolecular base pairing. Due to higher electrostatic repulsion of the different molecules the intermolecular complex formation is not possible in low salt (i.e., 10 mM, data not shown). The complex is stable, however, in low salt if it has been formed in high salt. One has to conclude that the activation energy for the rearrangement of the metastable structure involving the mentioned 28 base pair structure into the equilibrium structure is too large. The metastable and the equilibrium structures of a trimeric (+) PSTV intermediate are depicted in FIGURE 12. The monomeric units are easily seen in this scheme. The difference in energy between the metastable structure and the completely extended equilibrium structure of a dimer is 34 KJ/mol or 7% at 60°C.

The formation of a corresponding intermediate structure is not possible for viroids of (−) polarity. The two G:U base pairs in the (+) strand structure (FIGURE 12) would result in formation of two additional internal loops in the (−) strand secondary structure. These loops in the minus strand may lead to such a destabilization that the formation of the structure is no longer possible. In accordance, a complex formation of (−) intermediates was not observed under the same conditions as for complex formation of (+) strands.

TABLE 2. Molecular Weights of PSTV and PSTV Transcripts[a]

| RNA | Experimental $M_r$ | Theoretical $M_r$ |
|---|---|---|
| PSTV | 120,000 | 118,000 |
| SP6 PSTV 1.0 (−) | 150,000 | 137,000 |
| SP6 PSTV 1.0 (+) | 140,000 | 137,000 |
| SP6 PSTV 1.0 (+)* | 270,000 | 137,000 |

[a] Molecular weights were determined by sedimentation diffusion equilibrium in 4 M urea, 500 mM NaCl, 0.1 mM EDTA, 1 mM sodium cacodylate, pH 6.9. (*): SP6 PSTV 1.0 (+) was denatured up to a temperature just below the last transition (90° C in FIGURE 8) and then renatured slowly.

## CONCLUSIONS

From the fact that viroids are located at specific sites, it may be expected that these sites are relevant for viroid function. This functional relevance together with the structure of the intermediate forms of viroids will briefly be discussed under the particular aspect of viroid replication.

According to current models the invading viroid is transcribed into multimeric (−) strands by DNA-dependent RNA polymerase II.[33,35] This enzyme is located in the nucleoplasm, whereas the bulk part of circular viroids was found in the nucleolus. One may assume that the main template for polymerase II is the viroid originally invading the cell during infection. It has been published elsewhere that viroids when incubated with nuclei do enter the nucleoplasm but not the nucleolus.[5] The viroids found in the nucleolus have to be regarded as products of intranuclear replication. According to studies with specific replication inhibitors the oligomeric (−) strand is transcribed into the oligomeric (+) strand by the DNA-dependent RNA polymerase I.[34] Because polymerase I normally transcribes ribosomal RNA from the ribosomal DNA genes inside the nucleolus, the studies on the replication and on the location of viroids fit very well together showing independently that the nucleolus is the site of viroid (+) strand synthesis and of storage of mature viroids. One should assume

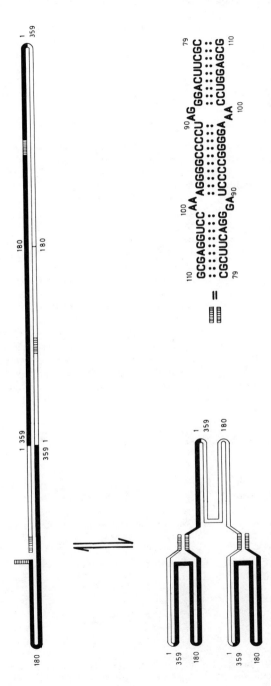

**FIGURE 12.** Secondary structure models of (+) strand viroid intermediates. Top, extended secondary structure of oligomeric (+) stranded PSTV. Bottom left, stable secondary structure of a trimeric PSTV intermediate including the secondary structure out of hairpin I and the central conserved region. Bottom right, secondary structure of PSTV from nucleotides 79–110 including hairpin I and the central conserved region. This structure may be formed by intermolecular complex formation of two monomeric PSTV molecules or by intramolecular base pairing in multimeric PSTV.

that splitting and ligation of oligomeric viroids to mature viroid circles occur also in the nucleolus.

From this brief discussion of the replication cycle, it is evident that viroids in three intracellular forms—mature circles, oligomeric (−) strands, and oligomeric (+) strands—are involved in quite different activities. By our physicochemical studies it could be shown that the three intracellular forms assume different secondary structures. It is not known whether the different structures or different intracellular locations are the reason for their different template activities, i.e., viroid circles for polymerase II and (−) strand oligomers for polymerase I.

At present, a more detailed model exists for the mechanism of forming monomeric viroid, linear or circular, from (+) strand oligomers. The stable region of 28 base pairs as found experimentally in this work (cf. FIGURE 12) was discussed by Diener as the site for splitting monomeric viroids from oligomers.[32] An adequate cut in the highly base paired region releases a unit PSTV molecule. Ligation of both remaining molecules results in a circular PSTV and an intact multimeric PSTV from which further circles may be processed. This hypothesis is supported by the minimum size of PSTV intermediates that are still highly infectious[27] (for review of infectious sequences see Reference 32). Such a PSTV intermediate contains a monomeric PSTV and additionally 11 vector-derived nucleotides doubling part of the central conserved region of PSTV. A sequence containing only 5 nucleotides of the central conserved sequence in addition to the monomer is only marginally infectious. It fits very well into the model that formation of this region was proven experimentally for (+) strand oligomers, whereas under the same conditions (−) strands do not form the region. Indeed the stability of the corresponding region in (−) strands is expected to be lower because of two A:C oppositions instead of G:U wobble pairs.

In summary, the subcellular location of viroids and the structural results on the replicative intermediates are able to explain some mechanistic details of the replication cycle. They also stimulate further studies on the interaction of the polymerases with the replicative intermediates, on the subnuclear location of the intermediates, and on the influence of mutations on the structure and infectivity of viroid RNA.

## ACKNOWLEDGMENTS

Part of the work that is summarized in this article was due to intensive cooperation with Prof. H. L. Sänger and Mr. M. Tabler (MPI für Biochemie/München) and Prof. G. Klotz (Universität Ulm). Furthermore, the stimulating discussions with Dr. M. Colpan (Diagen GmbH) and Mr. W. Rappold and R. Kapahnke (this laboratory) are gratefully acknowledged. The computer division of the Stadtwerke AG Düsseldorf supported the work by providing the computer time on the VAX 11/750. We thank Ms. E. Hupperts and Ch. Triebel for skillful technical assistance and Ms. H. Gruber for her help in preparing the manuscript.

## REFERENCES

1. RIESNER, D. & H. J. GROSS. 1985. Viroids. Annu. Rev. Biochem. **54:** 531-564.
2. GROSS H. J. & D. RIESNER. 1980. Angew. Chem. Int. Ed Engl. **19:** 231-243.

3. GROSS, H. J. & D. RIESNER. 1980. Angew. Chem **92:** 233-245.
4. SÄNGER, H. L. 1982. *In* Encyclopedia of Plant Physiology (NS). B. Parthier & D. Boulter, Eds. **14B:** 368-454. Springer-Verlag Berlin & Heidelberg.
5. RIESNER, D. *In* The Viroids T. O. Diener, Ed. Plenum Press, New York, N.Y. (In press.)
6. DIENER, T. O. 1979. Viroids and Viroid Diseases. Wiley, New York, N.Y.
7. BRANCH, A. D., H. D. ROBERTSON, & E. DICKSON. 1981. Proc. Nat. Acad. Sci. USA **78:** 6381-6385.
8. OWENS, R. A. & T. O. DIENER. 1982. Proc. Nat. Acad. Sci USA **79:** 113-117.
9. BEUTHER, E., J. SCHUMACHER, P. LOSS, S. KÖSTER & D. RIESNER. Manuscript submitted.
10. BERGER S. L. & L. S. BIRKENMEIER. 1979. Biochemistry **18:** 5143-5149.
11. COLPAN, M., J. SCHUMACHER, W. BRÜGGEMANN, H. L. SÄNGER & D. RIESNER. 1983. Anal. Biochem. **131:** 257-265.
12. COLPAN, M. & D. RIESNER. 1984. J. Chromatogr. **296:** 339-353.
13. TABLER, M. & H. L. SÄNGER. 1985. EMBO J. **4:** 2191-2199.
14. SCHUMACHER, J., H. L. SÄNGER & D. RIESNER. 1983. EMBO J. **2:** 1549-1555.
15. PEDERSON, T. 1974. Proc Nat. Acad. Sci. USA **71:** 617-621.
16. WOLFF, P., R. GILZ, J. SCHUMACHER & D. RIESNER. 1985. Nucleic Acids Res. **13:** 355-367.
17. PRESTAYKO, A. W. & H. BUSCH. 1973. Methods Cancer Res. **9:** 155-191.
18. KAPAHNKE, R., W. RAPPOLD, U. DESSELBERGER & D. RIESNER. 1986. Nucleic Acids Res. **14:** 3215-3228.
19. WERNTGES, H., G. STEGER, D. RIESNER & H. J. FRITZ. 1986. Nucleic Acids Res. **14:** 3773-3790.
20. LANGOWSKI, J., K. HENCO, D. RIESNER & H. L. SÄNGER. 1978. Nucleic Acids Res. **5:** 1589-1610.
21. STEGER, G., H. HOFMANN, J. FÖRTSCH, H. J. GROSS, J. W. RANDLES, H. L. SÄNGER & D. RIESNER. 1984. J. Biomol. Struct. Dyn. **2:** 543-571.
22. ZUKER, M. & P. STIEGLER. 1981. Nucleic Acids Res. **9:** 133-148.
23. NUSSINOV, R. & A. JACOBSON. 1980. Proc. Nat. Acad. Sci. USA **77:** 6309-6313.
24. SCHUMACHER, J., J. W. RANDLES & D. RIESNER. 1983. Anal. Biochem. **135:** 288-295.
25. ZIEVE, G. W., & S. PENMAN. 1981. J. Mol. Biol. **145:** 501-523.
26. SCHOEMAKER, H. J. P. & P. R. SCHIMMEL. 1974. J. Mol. Biol. **84:** 503-513.
27. TABLER, M. & H. L. SÄNGER. 1984. EMBO J. **3:** 3055-3062.
28. CRESS, D. E., M. C. KIEFER & R. A. OWENS. 1983. Nucleic Acids Res. **11:** 6821-6835.
29. VAN WEZENBEEK, P., P. VOS, J. VAN BOOM & A. VAN KAMMEN. 1982. Nucleic Acids Res. **10:** 7947-7957.
30. RIESNER, D., J. M. KAPER & J. W. RANDLES. 1982. Nucleic Acids Res. **10:** 5569-5586.
31. COLPAN, M. 1983. Thesis. Technische Hochschule. Darmstadt, Federal Republic of Germany.
32. DIENER, T. O. 1986. Proc. Nat. Acad. Sci. USA **83:** 58-62.
33. RACKWITZ, H. R., W. ROHDE & H. L. SÄNGER. 1981. Nature **291:** 297-301.
34. SPIESMACHER, E., H.-P. MÜHLBACH, M. TABLER & H. L. SÄNGER. 1985. Biosci. Rep. **5:** 251-265.
35. GOODMAN, T. C., L. NAGEL, W. RAPPOLD, G. KLOTZ & D. RIESNER. 1984. Nucleic Acids Res. **12:** 6231-6246.
36. YAMAKAWA, H. & M. FUJII. 1973. Macromolecules. **6:** 407-415.

# Predatory Bacteria in Prokaryotic Communities[a]

## The Earliest Trophic Relationships

RICARDO GUERRERO, ISABEL ESTEVE, CARLOS PEDRÓS-ALIÓ, AND NÚRIA GAJU

*Department of Genetics and Microbiology*
*Autonomous University of Barcelona*
*Bellaterra (Barcelona), Spain*

So, naturalists observe, a flea
Hath smaller fleas that on him prey;
And these have smaller still to bite 'em;
And so proceed *ad infinitum.*
Jonathan Swift, *On Poetry. A Rhapsody* (1733)

## INTRODUCTION

For more than a century, microbiologists have tried to work with axenic, or "pure", cultures. The results obtained for medical and industrial microbiology, microbial genetics, and physiology have made the effort worthwhile. Axenic technology is required to obtain clear-cut conclusions about cause-effect relationships in all these kinds of studies. Nevertheless, the axenic culture technique hides relationships among microorganisms, such as the ones described in this article, that are common in nature.

Ecological relationships among bacteria occasionally have been reported without the recognition of their meaning. In 1924, Bavendamm described several small, elongated bodies attached to the larger cells of the purple sulfur bacterium *Chromatium warmingii* which he thought were buds or bridges between cells. In his own words: "On a very hot day in June 1922..., shortly before a thunderstorm ... I saw an extraordinary surprising wealth of such buds and bridges.... The phenomenon could be seen for several days. After that most Chromatia had settled to the glass walls of the culture vessel."[1] Bavendamm pointed out similar observations in *Chromatium okenii, Chromatium weissei,* and *Thiospirillum* sp., and proposed that such bodies

[a]The experimental part of this study was supported by various Spanish grants to R.G. from the Comisión Asesora de Investigación Científica y Técnica (Ministerio de Educación y Ciencia), the Fondo de Investigaciones Sanitarias de la Seguridad Social (Ministerio de Sanidad y Consumo), and by the Lounsbery Foundation (to L. Margulis).

could be buds or even parasites. Doubtful as the correlation between parasites and thunderstorms must be, and despite the "Leeuwenhoekian" style of the statement, we have confirmed that the settling of Chromatiaceae and the abundance of those "buds" are related.

Bacteria play an essential role in the recycling of matter in the biosphere; they are the only organisms that have a full complement of enzymes capable of completing all the biogeochemical cycles. But how was organic matter recycled during the 2.5 billion years (more than half the whole history of the earth) before eukaryotes evolved? Furthermore, it is well known that many protists make their living grazing or preying on bacteria, and that predation is a system of feeding that successfully and repeatedly evolved afterwards in animals. But, is feeding by predation a discovery of the eukaryotes? Following Swift's naturalists, if any flea has smaller fleas, does it proceed *ad infinitum?*

This article describes several strictly aerobic bacteria shown to feed on a wide range of other prokaryotes, either dissolving the cytoplasm from the outside or penetrating in the periplasmic space, and details the ecology and physiology of two recently discovered bacteria[2] which feed only on members of the family Chromatiaceae, a group of phototrophic, anaerobic prokaryotes. The first of them, named *Vampirococcus,* is a strict anaerobe. It attaches to the surface of the prey cell, and divides epibiotically while degrading the cytoplasm of the prey. The other bacterium, *Daptobacter,* is facultatively anaerobic; it penetrates both the cell wall and cytoplasmic membrane, and divides in the cytoplasm of the prey by binary fission. The existence of these two anaerobic bacteria suggests that predation first appeared in prokaryotes in anoxic ecosystems in the early Archean and allows us to envision some major relationships of preeukaryotic ecology.

# EXTRACELLULAR PREDATION

In most cases bacterial cell lysis and digestion requires contact between the bacterium and its prey, while in a few cases it is caused by extracellular lytic enzymes. Although the ability of excreting enzymes outside the cell is a property widely spread in different bacterial groups, those enzymes capable of dissolving and digesting live bacterial cells are only produced by a small number of groups. The best known examples are those of the gliding bacteria, especially the myxobacteria.

### *Lysis by Means of Exoenzymes*

The heterogeneous group of the organotrophic gliding bacteria[3] are usually able to decompose polymeric compounds such as cellulose, agar, chitin, pectin, and proteins, as well as whole bacterial or yeast cells. In aquatic habitats some cytophaga lyse cyanobacteria and perhaps contribute to controlling water blooms of these phototrophs. Gliding motility moves them to places where nondiffusible substrates (highly polymeric molecules and dense microbial communities) are deposited, and where they come into close contact with these particulate structures. Such a close contact is essential because it reduces loss of both the substrates and the enzymes by diffusion, and brings enzymes that remain attached to the cell surface (like cellulases) into contact with the substrate. Most gliding bacteria are strictly aerobic. No gliding bacterium is known to have flagella at any stage of its life cycle.

Among the gliding bacteria, the phylogenetically uniform group of the myxobacteria[4] are the best known examples of microorganisms that obtain their nutrients by solubilizing extracellular macromolecules with excreted enzymes. All myxobacteria are strictly aerobic organotrophs. They are with the cyanobacteria the most morphologically sophisticated prokaryotes known, and serve as model systems for the study of developmental processes, interestingly still in the domain of the bacteria. They form typical multicellular structures of complex shape by the coordinated action of a large number of "swarm" cells (on the order of $10^5$-$10^7$), which retain their physical individuality. This process of cooperative morphogenesis offers a clear ecological advantage. A population, in contrast to an individual cell, can maximize the efficiency of extracellular enzymes. Thus, the high population density will result in increased enzymatic levels, and the overlapping zones of hydrolysis that surround the cells will inhibit diffusion of nutrients and increase the efficiency of growth. In nature, many myxobacteria seem to subsist mainly by feeding on different microorganisms (which they degrade by means of a variety of hydrolytic exoenzymes: cell wall-lytic enzymes, proteases, nucleases, lipases), and are especially efficient in the destruction of other bacteria and yeasts. Although they do not depend exclusively on living organisms, in the laboratory living bacteria are more efficient for the growth of myxobacteria than are dead ones. The two most frequently employed species for the preferential isolation of myxobacteria are either the gram-negative *Escherichia coli* or the gram-positive *Micrococcus luteus*, which gives us an idea of the wide range of prey they have.

### Attachment to the Prey by Flagellated Aerobic Predators

The morphology and growth cycle of an interesting bacterial predator of other bacteria in soil was described by Casida in 1982.[5] It was named *Ensifer adhaerens* ("ensifer," sword bearing). The cells are rounded rods (0.7-1.1 by 1.0-1.9 μm), occurring singly or in pairs. They are motile by means of a tuft of three to five flagella that are attached subterminally. This aerobic, gram-negative bacterium is not an obligate predator, for it grows well in the absence of prey cells on a variety of media, including some very poor in nutrients. During growth in the laboratory in the presence of, for example, *Micrococcus luteus*, *Ensifer* attaches to each prey cell in large numbers in a picket fence arrangement. A lytic factor which diffuses through agar was detected emanating from *Ensifer* during growth on a nutritionally dilute medium. *Ensifer* can attach to various living gram-positive and gram-negative bacteria, though it has not any apparent specialized terminal structure for anchoring purposes. *Ensifer* is not an obligate predator and it is not nutritionally fastidious in the absence of prey cells. In axenic cultures it demonstrates the same gyrating and tumbling motility that is observed in soil.

There are also recent descriptions of other bacteria whose relationship to their prey is more strict. Two obligate predators of two species of *Pseudomonas* have been isolated from wastewater by Lambina *et al.*[6,7] Very similar in morphology and physiology, they have been put together in the new genus *Micavibrio* ("mica", a crumb, viz. very small). Out of 55 bacteria from different taxonomic groups tested, the first one, *M. admirandus,* exhibits specificity exclusively for nearly all strains of *Pseudomonas maltophilia,* while the second one, *M. aeruginosavorus,* exclusively for those of *P. aeruginosa,* either in solid (forming plaques) or in liquid cultures (clearing a suspension of the suitable bacterium). *Micavibrio* is a gram-negative curved rod, small (0.25-0.4 by 0.6-1.0 μm), with a single polar unsheathed flagellum 15 nm in diameter. The mol% G + C of the DNA is 57.1 and 53.8, respectively.

*Micavibrio* attaches to the surface of the prey cells and destroys them without penetration. A single electron-lucent zone, which is adjacent to the nucleoid, is clearly visible in the cytoplasm. While dissolving the contents of the prey the nucleoid of the growing cell increases in size and divides, after which the cell itself divides. It is possible to observe the appearance of a second electron-lucent zone during growth or in the stage preceding division. The growth cycle and mode of attack of *Micavibrio* consist of the following stages: freely swimming of the polarly flagellated predator until it collides with a prey; perpendicular attachment to the cell wall of the prey by the end opposite to the flagellum; loss of the flagellum and of motility; taking of a position parallel to the longitudinal axis of the prey cell so that the area of contact is increased; development on the surface of the prey cell without penetration; and, finally, binary fission accompanied by lysis of the affected cell. Degenerative changes in the prey cell are observed even in early stages of the predator attachment. *Micavibrio* cannot grow in axenic cultures, even in complex nutrient media. Therefore, a necessary condition for its growth and multiplication is the presence in the incubation medium of the living prey cells, which indicates the obligate nature of its predatory manner of feeding.

Finally, it seems convenient to mention another predatory bacterium that has a specific eukaryote as its only prey. This bacterium resembles *Bdellovibrio* but is able to prey on the alga *Chlorella vulgaris.* It was described in 1972 by Gromov and Mamkaeva.[8] Although there are certain similarities (morphology and size, requirement for susceptible live cells), important differences and the intention to precisely define the already broad genus *Bdellovibrio* indicated the creation of a new genus for this bacterium. The important differences are (a) the prey is eukaryotic; (b) no elongated spirillar forms occur; (c) growth occurs outside the prey cell and the latter is not penetrated; and (d) the flagellum lacks a sheath. The provocative name of the predator later given by the two discoverers is *Vampirovibrio chlorellavorus.*[9]

## *Attachment to the Prey by an Anaerobic Nonmotile Predator:* Vampirococcus

Although other bacteria with similar characteristics to the ones just described have been cited in the literature, until now only one nonmotile anaerobic predator has been found and reported.[10] This new kind of predator has been observed in natural samples (coming from lakes with layers of highly dense populations of purple sulfur bacteria, i.e., the family Chromatiaceae). Unfortunately it has resisted attempts to grow it either in axenic culture or in two-membered cultures with its prey, irrespective of the media employed.

In the course of ecological studies on the phototrophic bacterial communities of sulfurous karstic lakes, we observed and characterized two kinds of predatory bacteria by light and electron microscopy.[2] The first bacterium, named *Vampirococcus,* is gram negative and ovoidal (0.6 μm wide), does not have any flagellum, and apparently is strictly anaerobic. The second one, named *Daptobacter,* is a gram-negative straight rod, with a single polar flagellum, and facultatively anaerobic; it will be described in more detail later in this article. Both of them are strikingly reminiscent of the "buds" described by Bavendamm more than 60 years ago.[1]

*Vampirococcus* and *Daptobacter* were first observed in Lake Estanya (near the town of Benabarre, in the province of Huesca) and Lake Cisó (near the town of Banyoles, in the province of Girona), in northeastern Spain. Afterwards, they were also detected in Lake Vilar (same location as Cisó). The three lakes are sinkholes of different size formed in karstic areas also rich in gypsum ($CaSO_4 \cdot 2H_2O$). Bacterial reduction of sulfate in the sediment produces considerable amounts of hydrogen sulfide, which ascends through the water.[11]

Phototrophic sulfur bacteria (both purple and green), which are anaerobic and anoxigenic, reach large population densities (up to $10^6$ to $10^7$ cells per $cm^3$) in horizontal layers where adequate amounts of light and sulfide are simultaneously present. In the highest layers where the phototrophic sulfur bacteria live, the dominant populations are those formed by several species of the genus *Chromatium*, a very actively motile purple sulfur bacterium, which possesses a tuft of polar flagella and typically accumulates conspicuous sulfur globules in its cytoplasm.

Samples for microscopic observation were taken from various depths in the layers of *Chromatium* spp. Many of the *Chromatium* cells were observed by light microscopy to have smaller ovoidal bacteria attached to them. Electron micrography (thin sections) showed that the different *Chromatium* cells to which the smaller bacteria were attached presented different phases of cytoplasmic lysis. The percentage of *Chromatium* individuals with attached cells increased with depth, in parallel with diminishing viability of the purple bacteria (lower viability is due to decreasing amounts of available light).[12]

Water from the bacterial layer was taken from the lake, kept under anaerobic conditions, and brought to the laboratory, where the purple bacteria sedimented at the bottom of flat vessels, previously filled up completely with water. After a few days, lytic plaques appeared. Observations of samples taken from these plaques were similar to those made in water taken directly from the lake. Therefore, it was possible to assign a predator-prey relationship to the observed association. This predatory bacterium has been named *Vampirococcus* (following the style of *Vampirovibrio*), from "vampire" (Serbian: vampir, bloodsucker) and "coccus" (Greek: coccus, a grain or berry).

A characterization of the relationship between *Vampirococcus* and its *Chromatium* prey was done by transmission electron microscopy, and is illustrated in FIGURE 1. Thin sectioning of the bacteria in natural samples allows clear observation of the morphology, attachment and mode of division of *Vampirococcus*, as well as the progressive degradation of the cytoplasmic content of its prey. In FIGURE 1a can be seen the conspicuous attachment structure that binds *Vampirococcus* to its prey. The breach in the outer membrane of *Vampirococcus*, the plaque of dense material at the attachment site, the bulging of the zone in *Chromatium* near the attachment site, as well as the beginning of the degradation under the cytoplasmic membrane can be observed. Note the cross-wall formed in the process of cell division of *Vampirococcus*. In FIGURE 1b two ovoidal offspring can be seen just after separation but still united at one point; the other attached *Vampirococcus* has not started the process of forming the cross-wall yet. From one to six *Vampirococcus* cells can simultaneously prey on a single *Chromatium* (see Table 1 of Reference 2). *Vampirococcus* is seen to multiply only when attached to its prey. The normal structure of the cytoplasm of *Chromatium* can also be observed (FIGURE 1b): the small granules that cover the cytoplasm are the photosynthetic membranes; the large, dark granules are inclusions of the reserve material poly-$\beta$-hydroxybutyrate (PHB); the clear zone almost at the center is a sulfur globule. Finally, in FIGURE 1c and d different stages of dissolution of the cytoplasmic

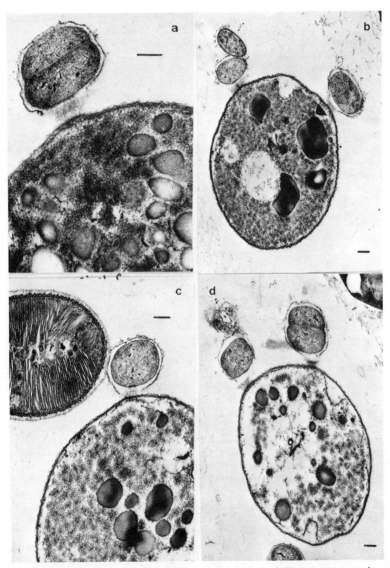

**FIGURE 1.** Morphology, mode of division, and attachment of *Vampirococcus* to its prey, *Chromatium* spp., as observed by transmission electron microscopy in natural samples. (a) The attachment structure (see text for details). (b) Two offspring after division but still united at one point. Other predatory cell is attached to the same prey. (c) Attachment to and degradation of a *Chromatium* cell by *Vampirococcus* while another species remains unaltered. (d) Lysis of the cytoplasm of the prey. While the cell wall remains essentially unaltered, the cytoplasmic membrane is invaginated in several places. The marker bar represents 250 nm.

matrix of *Chromatium* are shown: many photosynthetic vesicles are degraded, zones of lysis are abundant, and invaginations of the cytoplasmic membrane can be seen at different points. The only things that remain after lysis are the ruptured cell envelope and the PHB inclusions (unpublished results). By contrast, a cell of another species of phototrophic bacteria, which does not seem suitable for *Vampirococcus* to grow on (observe the absence of attachment structure between them, despite their touching each other), presents a totally normal cytoplasm (for that given species).

Many unanswered questions are left in considering the life cycle of *Vampirococcus.* How does it survive in the natural habitat after division and before entering in contact with a new prey? How does it locate its prey? Is any kind of chemotactic mechanism involved? Does *Vampirococcus* have any nonflagellar motility system? What kind of enzyme(s) does it produce? What is the mechanism of dissolution of the cytoplasm of the prey? What is the basis for its apparent specificity for purple sulfur bacteria? How much energy can *Vampirococcus* obtain through its strictly anaerobic metabolism? The resolution of at least a few of these problems will doubtless lead to a deeper understanding of the establishment of one of the simplest trophic chains among anaerobic bacteria.

# INTRACELLULAR PREDATION

## *Periplasmic Predators:* Bdellovibrio

Since it was isolated by Stolp and first described by Stolp and Petzold in 1962, only the polarly flagellated *Bdellovibrio* has been well characterized as able to attack a wide range of gram-negative bacteria, penetrate their cell wall, and henceforth generate several progeny swarmers by multiple fission in the periplasmic space of the prey cell.[9,13]

*Bdellovibrio* ("bdella," leech, sucker) has been the subject of intensive investigation, and until now more than 300 articles have been published on it. This section summarizes the general properties relevant to the characterization of its morphology and mode of feeding as well as the range of prey and taxonomical diversity.

The strains isolated from nature depend on their prey for growth. They are vibrioid to rod-shaped gram-negative bacteria of relatively small size (0.25-0.40 $\mu$m wide; 1-2 $\mu$m long). They are extremely motile, by means of one thick (28 nm in diameter) sheathed flagellum, which protudes at one pole of the cell. From populations of prey-dependent *Bdellovibrio* it is not difficult to select strains that can grow axenically in nutrient-rich media. There is also evidence that many wild-type *Bdellovibrio* are capable of multiplying in suspensions of heat- or UV-killed prey cells.[13] *Bdellovibrio* is characterized by a biphasic life cycle, in which freely motile cells of the attack-phase and intraperiplasmic, multiplication-phase cells alternate. The attack-phase cell (vibrio shaped, flagellated) is incapable of DNA synthesis, whereas the multiplication-phase cell (spirillum shaped, without flagellum) shows DNA replication, elongation, and fission, and produces within the periplasmic space of the prey cell motile, attack-phase progeny which swarm out and are capable of attacking new prey.[14] The complex life cycle of *Bdellovibrio* and its various alternatives under different conditions of growth have been clearly depicted by Shilo (see Figure 1 of Reference 14).

*Bdellovibrio* is widespread in nature and has been isolated from a variety of habitats. In spite of different claims, it is now agreed that the predatory activity of *Bdellovibrio* is limited to gram-negative bacteria. No strains have yet been isolated capable of lysing gram-positive bacteria or cyanobacteria. *Bdellovibrio* is also unable to attack eukaryotic organisms. Some *Bdellovibrio* strains have a broad range of prey; others, very limited. The bases of the specificity for the prey are still unknown. Taxonomically, all strains of *Bdellovibrio* are presently considered to belong to a single genus containing several species, *B. bacteriovorus, B. stolpii,* and *B. starrii,* plus an unnamed marine species.[9] This classification was based on the morphological, predatory, and unique life-cycle features just described. The study of a variety of *Bdellovibrio* cultures in terms of DNA analysis (mol% G + C, genome size, hybridization), cytochrome spectra, and prey range shows a high level of heterogeneity.[15] Therefore, *Bdellovibrio* should not be considered a single genus (the mol% of G + C spans from 33.4 to 51, which is a broad range, even for bacteria) but probably two or three genera that result from the convergent evolution of bacteria sharing properties inextricably connected with the predatory mode of life. Its reclassification is expected in the future.[14]

Several bacteria that resemble *Bdellovibrio* in one or more characteristics have been described. The best known is *Vampirovibrio* (see above). Another example is "Microvibrio," which presents a biphasic morphology (a small motile, vibrio-shaped form, and elongated and spirillar forms which are observed preceding the appearance of the former). They do not penetrate the cell wall of the prey (several gram-negative bacteria). After being attacked, the prey cells lose their viability as a result of the interaction but they are not lysed. Finally, a strain of spirillumlike bacteria which attack, penetrate, and feed within the alga *Scenedesmus acutus* has been described. The bacteria appear to enter the cytoplasm by lysing the cell membrane, develop into a long thin spirillar form, and digest the content of the algal cell.[9]

### Cytoplasmic Predators: Daptobacter

The study of natural samples from the bacterial layers of the lakes where *Vampirococcus* was first observed revealed the presence of another type of predatory bacteria. Between the large phototrophic bacteria were seen many smaller, actively swimming rods which frequently collided and attached to the *Chromatium* cells. At first glance they greatly resembled *Bdellovibrio* but further study showed them to be entirely different. The new isolate was named *Daptobacter* (from "dapto," Greek: to gnaw, to devour; "bacter," Latin from Greek: rod, shaft). Transmission electron micrographs of thin sections demonstrated that, unlike *Bdellovibrio, Daptobacter* penetrates both the cell wall and cytoplasmic membrane of its prey. Once inside, it digests the cytoplasm and subsequently divides by binary fission to form two offspring cells.

The morphology and life cycle of *Daptobacter* feeding on *Chromatium minus,* mainly from laboratory cultures, were studied by light and transmission electron microscopy, as shown in FIGURE 2. Water samples from the bottom of the layer of phototrophic bacteria, where different *Chromatium* species comprise 90% of the community biomass,[2,10] were used to establish enrichment cultures. When these cultures were passed through glass-fiber filters that had a layer of *Chromatium minus* on them, lytic plaques appeared. Such plaques are typically seen as pale pink circles on the deep purple opaque lawn of *Chromatium.* They reach 0.5 cm in diameter after seven days of incubation under microaerophilic conditions at 30°C (FIGURE 2a). When

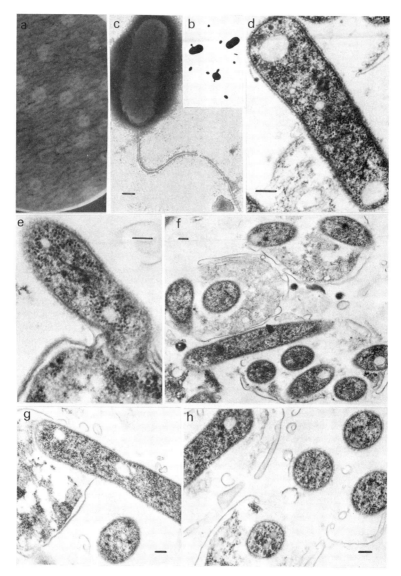

**FIGURE 2.** Morphology and life cycle of the predatory bacterium *Daptobacter* feeding on *Chromatium minus* as observed by light and transmission electron microscopy in laboratory cultures. (a) Plaques originated by the growth of *Daptobacter* on lawns of *C. minus* cells retained by a glass-fiber filter. (Actual size of each plaque is about 0.5 cm in diameter.) (b) Phase-contrast micrograph of the prey, *C. minus*, with one or two attached *Daptobacter* cells. (For scale, the vertical stroke of the "b" represents 4 μm.) (c) Electron micrograph, negative stain, of a single *Daptobacter* cell showing its unsheathed polar flagellum. (d) Thin section of *Daptobacter* showing electron-lucent structures at the poles of the cell. (e) Thin section of *Daptobacter* apparently penetrating the bulging wall of its prey. (f) Thin section of cells of *C. minus* preyed upon by several *Daptobacter* individuals; note lysis zones around the latter. (g) Thin section of a cell of *Daptobacter* multiplying by binary fission. (h) Young cells of *Daptobacter* liberated into the medium while *C. minus* is destroyed; only the fragmented cell wall and some cytoplasmic inclusions remain. The marker bar represents 200 nm.

samples were taken from the plaques and deposited on a thin dry agar film on a microscope slide, small motile rods attacking *Chromatium* cells were easily observed by phase-contrast microscopy (FIGURE 2b). Those small bacteria were *Daptobacter*, which gyrated rapidly in a clearly recognizable fashion and attached perpendicularly to much larger cells of *Chromatium*. After the corresponding staining *Daptobacter* was shown to be gram negative.

Material from a plaque was suspended into pure cultures of *Chromatium minus*. After incubation (three days at 30°C) the suspension was filtered again to form a lawn of the prey bacterium, on which *Daptobacter* plaques formed. This process was repeated several times to increase the number of *Daptobacter* cells. Throughout the successive passages, the shape and motility of the predatory cells were observed to be maintained.

## Morphology and Life Cycle of Daptobacter

The morphology, penetration into the prey cell, and mode of division of *Daptobacter* were studied in enriched cultures of natural samples as well as in cocultures with *Chromatium minus*. The electron microscope images obtained by negative stain and thin sections showed that individual cells of *Daptobacter* were rod-shaped bacteria (0.5-0.6 by 1.5-2.0 $\mu$m) bearing one to three polar flagella 15 nm wide. Flagella were also observed in axenically grown cells of *Daptobacter* (FIGURE 2c). In thin sections the morphology of *Daptobacter* cells was shown to be similar to that of other gram-negative rods. However, distinctive electron-lucent structures (0.1-0.3 $\mu$m in diameter) were observed near the poles of the cells (FIGURE 2d).

Penetration involved the attachment of *Daptobacter* to the *Chromatium* cell wall, resulting in a small bulge on the outer membrane of the cell wall of the prey in the region of contact. *Daptobacter* penetrated the cytoplasm of *Chromatium* by rupturing both its cell wall and cytoplasmic membrane (FIGURE 2e). *Daptobacter* located inside the prey reproduced by binary fission (FIGURE 2g). Each *Daptobacter* produced two offspring cells as *Chromatium* disintegrated (FIGURE 2g). During this process the cytoplasm of *Chromatium minus* cells was dramatically altered. Lytic zones appeared, especially in the immediate vicinity of the predator, and the photosynthetic vesicles were destroyed. Eventually only the fragmented cell wall and cytoplasmic inclusions remained as the young *Daptobacter* were liberated into the medium (FIGURE 2h). The already mentioned clear zones near the poles of *Daptobacter* cells are especially obvious in FIGURE 2d and 2g, but they can also be seen in 2e, and in some of the cells in 2f and 2h. They are reminiscent of the clear zones already mentioned for *Micavibrio*[7] but in the case of *Daptobacter* they seem associated with the septum during division.

To determine the range of bacterial prey for *Daptobacter*, various species of phototrophic and heterotrophic bacteria were tested. *Daptobacter* formed no plaques at all on lawns of any of the purple "nonsulfur" bacteria (Rhodospirillaceae) tested. The assayed bacteria were *Rhodobacter capsulatus, Rhodopseudomonas palustris,* and *Rhodocyclus gelatinosus*. In contrast, *Daptobacter* formed clearly visible plaques on different species of Chromatiaceae. The bacteria tested were *Chromatium minus, C. vinosum, C. minutissimum, Thiocapsa roseopersicina* and *Lamprocystis* sp. Generally, the times required to produce plaques in sheathed genera (*Thiocapsa* and *Lamprocystis*) were longer than in the unsheathed genera (7 to 10 days vs. 3 to 7 days). None of the heterotrophic gram-negative or gram-positive bacteria tested (*Escherichia coli, Pseudomonas aeruginosa, Bacillus subtilis,* and *Staphylococcus aureus*) were able to support the growth of *Daptobacter,* yielding visible plaques. Plaques were not formed on any of the Chlorobiaceae tested.

Axenic cultures of *Daptobacter* were obtained by plating samples from cultures with the prey on rich media, in aerobic and dark conditions. Colonies presented a slight orange pigmentation. The ability to form lytic plaques on *Chromatium minus* lawns was tested for the isolated bacteria. Further study of these isolates, including optical, scanning, and transmission electron microscopy, showing the bacteria were identical to *Daptobacter* confirmed the identification.

### *Differentiation between* Vampirococcus, Daptobacter, *and* Bdellovibrio

In spite of the coincidence in habitat, prey, and some physiological characteristics (anaerobiosis), the different morphology and mode of attack easily distinguish *Vampirococcus* from *Daptobacter* (and of course *Bdellovibrio*), by electron microscopy. The problem arises when observing wet mounts of natural samples, or plaques formed on lawns of sedimented prey cells in laboratory flasks. In the first case, during the initial stage of the attack, the attached cells of both *Vampirococcus* and *Daptobacter* are nonmotile, and difficult to distinguish with the light microscope, in spite of their different sizes. In the second case, the plaques formed in anaerobic conditions seem to have similar morphology in both bacteria, implying the necessity of preparation of samples for electron microscopy. The morphological changes effected on the prey are also clearly different in both cases, as shown by thin-section electron micrographs. *Daptobacter* breaks down both the cell wall and the cytoplasmic membrane, and destroys the cytoplasm from inside; *Vampirococcus* "sucking" from the outside causes invagination of the cytoplasmic membrane at many points.

The main morphological and biochemical differences between *Daptobacter* and *Bdellovibrio* are as follows. The cellular shape (rod vs. vibrio) and type of flagellum (slender and unsheathed vs. wider and ensheathed) differ. *Daptobacter* is also slightly larger. Differences in life cycle have been already pointed out in this article. Physiologically, the most important distinction is in the metabolic relation to oxygen: while *Daptobacter* is a facultative anaerobe, *Bdellovibrio* is strictly aerobic. Furthermore, *Daptobacter* is able to ferment several sugars, organic alcohols, and acids, as well as to respire them. *Bdellovibrio* is unable to ferment any substrate.

*Daptobacter* is versatile in its metabolic capabilities: it can ferment or respire, it lives in either aerobic or anaerobic conditions, and it grows in pure culture on rich media or in a two-membered predator-prey system (really as a "primary consumer" because it feeds on a phototrophic organism). The optimal temperature for growth of *Daptobacter* is 33°C but it grows over a wide range of temperatures, from 5°C to 42°C. In its natural habitat temperature ranges between 5°C in the winter and 23°C in the summer. *Daptobacter* can also tolerate high concentrations of sulfide, up to 1.75 mM, an important feature for an organism attacking purple sulfur bacteria in habitats where sulfide may range between 0.1 mM and 2 mM.

Both the mode of reproduction and the range of prey distinguish these two predatory bacteria. *Daptobacter,* capable of division inside the cytoplasm of phototrophic bacteria, so far has a rather narrow range of prey. A defining characteristic of *Bdellovibrio* is division in the periplasm of its prey, which can be a wide range of gram-negative heterotrophic bacteria, including by some accounts two phototrophic ones:

*Rhodospirillum rubrum* and the widely used strain D of *Chromatium vinosum*. *Bdellovibrio* develops into a spiral-shaped cell and has multiple division. *Daptobacter* undergoes binary fission involving the formation of a single septum separating the cells, such that no more than two offspring result. However, up to six (or maybe even more) *Daptobacter* (FIGURE 2f), all capable of binary fission, may penetrate a single prey, such that the total number of *Daptobacter* released may also depend on the size of the prey cell.

Thus, *Daptobacter* and *Bdellovibrio* are different bacteria with common properties determined by the predatory mode of life. The fermentation capacity of *Daptobacter* makes it more related to the genus *Aeromonas*, while *Bdellovibrio* presents a physiology more related to the strictly aerobic genus *Pseudomonas*, both of them belonging to two unrelated bacterial families. Further research is necessary to completely understand the life cycle and mechanism of predation of *Daptobacter* in nature.

# CONCLUDING REMARKS AND RETROSPECTIVE

Evolutionary and ecological implications of three different types of predatory prokaryotes have been recently published.[2] Attempts have been made to reconstruct the ecology of prokaryotic cell communities in the early Archaen eon, in which, although cyanobacterial-like prokaryotes were present, ecosystems were primarily anoxic. The discovery of these new anaerobic and facultatively anaerobic predatory bacteria implies that neither oxygen-rich environments nor eukaryotic organisms were prerequisite to the development of such ecosystems.

In recognition of the biotrophic modes of these new bacteria a short discussion on ecological terminology is warranted. Although complete consistency with usage of terms is desirable, widely used terminology is difficult to avoid. In this article, as we have discussed these necrotrophic (i.e., leading to the death of one of the partners) relationships, we have referred to the "victim" as "prey" rather than "host" and to the "consumer" as "predator" rather than "parasite." The distinction between predator and parasite is usually based on the relative sizes of the partners, and it appears inappropriate. We have tried not only to be consistent with more general ecological concepts but also to follow the recent usage in the last edition of Bergey's Manual[9] and by Shilo,[14] which call *Bdellovibrio* consistently a predator, and the bacteria in which it feeds a prey (thus avoiding the confusion of previous literature, in which, in dealing with *Bdellovibrio*, the terms parasite and host are extensively used).

Furthermore, *Daptobacter* presents a new mechanism of feeding among prokaryotes. Because the Chromatiaceae are primary producers, *Daptobacter* must be considered a primary consumer. This terminology applies when using the most general ecological relationships. Finally, if *Daptobacter* and *Vampirococcus* have an impact on natural populations as suggested,[2] trophic chains in the microbial world similar to those in the "visible" environment could be envisaged during the Archean eon. Thus, the possible existence of similar trophic relationships among former prokaryotes suggests that in the early history of earth, when phototrophic bacteria were the only primary producers and heterotrophic bacteria the only consumers, ecological relationships were entirely analogous to those extant among eukaryotes.

## ACKNOWLEDGMENTS

We are indebted to Prof. Hans G. Trüper for drawing our attention to the pertinent passages of the book by Bavendamm and for so kindly translating them from German. The invaluable assistance of Lynn Margulis and Gregory J. Hinkle, who provided the necessary peaceful environment and aided in the preparation of the final form of the manuscript at Boston University, is also acknowledged. We are grateful to David Chase (1933–1987) for the micrographs in FIGURE 1. This article is dedicated to his memory.

## REFERENCES

1. BAVENDAMM, W. 1924. Die farblosen und roten Schwefelbakterien de Süss und Salzwassers. Gustav Fisher Verlag. Jena, Germany.
2. GUERRERO, R., C. PEDRÓS-ALIÓ, I. ESTEVE, J. MAS, D. CHASE & L. MARGULIS. 1986. Predatory prokaryotes: predation and primary consumption evolved in bacteria. Proc. Nat. Acad. Sci USA **83:** 2138-2142.
3. REICHENBACH, H. & M. DWORKIN. 1981. Introduction to the gliding bacteria. *In* The Prokaryotes. A Handbook on Habitats, Isolation, and Identification of Bacteria. M. P. Starr, H. Stolp, H. G. Trüper, A. Balows & H. G. Schlegel, Eds.: 315-327. Springer-Verlag. New York, N.Y.
4. REICHENBACH, H. & M. DWORKIN. 1981. The order Myxobacterales. *In* The Prokaryotes. M. P. Starr *et al.* Eds.: 328-355. Springer-Verlag. New York, N.Y.
5. CASIDA, L. E. 1982. *Ensifer adhaerens* gen. nov., sp. nov.: a bacterial predator of bacteria in soil. Internat. J. Syst. Bacteriol. **32:** 339-345.
6. LAMBINA, V. A., A. V. AFINOGENOVA, Z. ROMAY, S. M. KONOVALOVA & A. P. PUSH-KAREVA. 1982. *Micavibrio admirandus* gen. et sp. nov. Mikrobiologiya **51:** 114-117.
7. LAMBINA, V. A., A. V. AFINOGENOVA, Z. ROMAY, S. M. KONOVALOVA & L. V. ANDREEV. 1983. A new species of exoparasitic bacteria from the genus *Micavibrio* destroying gram-negative bacteria. Mikrobiologiya **52:** 777-780.
8. GROMOV, B. V. & K. A. MAMKAEVA. 1972. Electron microscope examination of *Bdellovibrio chlorellavorus* parasitism on cells of the green alga *Chlorella vulgaris.* Tsitologiya **14:** 256-260.
9. KRIEG, N. R. 1984. Aerobic/microaerophilic, motile, helical/vibroid gram-negative bacteria. *In* Bergey's Manual of Systematic Bacteriology. N. R. Krieg & J. G. Holt, Eds.: 71-124. Williams & Wilkins. Baltimore, Md.
10. ESTEVE, I., R. GUERRERO, E. MONTESINOS & C. ABELLÀ. 1983. Electron microscopy study of the interaction of epibiontic bacteria with *Chromatium minus* in natural habitats. Microbiol. Ecol. **9:** 57-64.
11. GUERRERO, R., E. MONTESINOS, C. PEDRÓS-ALIÓ, I. ESTEVE, J. MAS, H. VAN GEM-ERDEN, P. A. G. HOFMAN & J. F. BAKKER. 1985. Phototrophic sulfur bacteria in two Spanish lakes: vertical distribution and limiting factors. Limnol. Oceanogr. **30:** 919-931.
12. VAN GEMERDEN, H., E. MONTESINOS, J. MAS & R. GUERRERO. 1985. Diel cycle of metabolism of phototrophic purple sulfur bacteria in Lake Cisó (Spain). Limnol. Oceanogr. **30:** 932-943.
13. STOLP, H. 1981. The genus *Bdellovibrio. In* The Prokaryotes. M. P. Starr *et al.,* Eds.: 618-629. Springer-Verlag. New York, N.Y.
14. SHILO, M. 1984. *Bdellovibrio* as a predator. *In* Current Perspectives in Microbial Ecology. M. J. Klug & C. A. Reddy, Eds.: 334-339. American Society for Microbiology. Washington, D.C.
15. TORRELLA, F., R. GUERRERO & R. SEIDLER. 1978. Further taxonomic characterization of the genus *Bdellovibrio.* Can. J. Microbiol. **24:** 1387-1394.

# Bacterial Viruses, Prophages, and Plasmids, Reconsidered

SORIN SONEA

*Department of Microbiology and Immunology*
*University of Montreal*
*C.P. 6128, Succursale A*
*Montreal, Canada H3C 3J7*

## INTRODUCTION

In this reappraisal of the small self-replicating DNA molecules found in bacteria, the accent will be on their usually neglected general properties—particularly their essential role in nature, including their influence on evolution.

The term bacterial virus covers very different entities, to the point of becoming a confusing terminology, with lasting detrimental consequences. This results in an incoherent approach to bacteria in general. Viruses cause illness or death and no favorable changes in the cells where they are multiplied. In bacteria they are called *virulent* bacteriophages (*phages*) and they will not be discussed in length here; being parasites their role in nature and their probable destiny are restricted. More controversial are the *temperate* phages, which are among the best known bacterial "viruses." Far from being harmful, they represent the extracellular, resistant, gene-broadcasting forms of *prophages.* The latter are the most advanced systems for exchanging genes between different types of bacteria. Prophages are closely related in structure, function, and probable origin to the other group of favorable "visiting" genes, the plasmids. Both categories, *the small replicons,* consist of a self-replicating double-stranded DNA, at least a hundred times smaller than the "chromosome," as the large bacterial replicon is often called (FIGURE 1). All replicons have their own replicator genes responsible for their autonomous division. The large replicon carries essential, stable, genes for the respective bacterium. The small replicons carry *conversion* genes which are always derepressed and, when present in a new cell, instruct it to synthesize new substances, usually enzymes. When the new substances constitute an asset for the recipient bacterium, the bacterium will function better, and consequently will multiply abundantly, broadcasting the corresponding small replicon. When the converting genes are useless or even harmful, the plasmid or prophage will be eliminated from the strain (*curing*). Thus, the small replicons are the available and disposable genetic tools of the bacteria. In nature, each bacterium contains one or more small replicons. Other small replicons are readily available for each bacterium, yet temporarily present in different bacteria. Some of them may be obtained only through intermediate hosts. This facility of access to genes from other, metabolically different strains increases tremendously the capacities of the bacterial world. Prophages and plasmids are active and essential participants in the life of bacteria; they seem to have played a role in the origin of eukaryotes and of viruses and have left many vestiges in the cells of animals and plants. These neglected aspects should be approached in a more realistic and coherent way by the biologists.

251

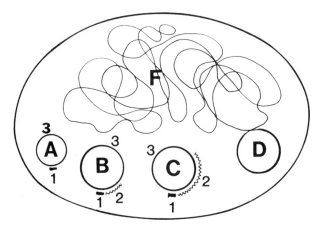

**FIGURE 1.** Prophages and plasmids in a bacterium (the quantitative and the sequential location of different types of genes is schematic and general): **A,** non-self-transmissible plasmid; **B,** conjugation plasmid; **C,** prophage; **D,** integrated prophage or conjugating plasmid in the length of the large replicon ("chromosome") **F;** 1, self-replicating genes; 2, transfer operons; 3, converting genes.

## PROPHAGES AND CONJUGATION PLASMIDS (SELF-TRANSMISSIBLE SMALL REPLICONS)

Both these categories of small replicons possess groups of genes (usually repressed) organized in one to three transfer operons and coding for an active transfer of their copies to different bacteria and often for the transfer of some genes from the host cell genome as well. Moreover, genes of these kinds of operons code for tubular protein structures, implied in the introduction into a new bacterium of copies of the genes carried by prophages or by conjugation plasmids (FIGURE 2). Specific receptors at the surface of a sensitive cell are needed to attract and attach the distal end of different types of tubular structures. Temperate phages and conjugation plasmids may benefit from the same receptors. The repressor of some transfer operons in prophages is similar to the one active in some conjugation plasmids. Moreover, there are a few DNA sequences common to prophages and conjugation plasmids. All these facts indicate a probable common origin for the two categories of self-transmissible small replicons.

### Prophages

About a fifth of the bacterial strains are *lysogenic,* in nature, which means that they contain one or more types of prophages. In a minority of bacteria of a lysogenic strain a spontaneous *induction* occurs. That is, the usually repressed transfer genes of the prophage are derepressed and start its rapid multiplication. This is followed by the synthesis of protein coatings, called the *heads,* that surround these small DNA molecules, then by the synthesis of the tubular structure for gene transfer, called the *tail*, and then by an assemblage from these separate parts of the complete *temperate* phages. It is followed by the production of enzymes to lyse the bacterial cell wall, resulting in the liberation of the temperate phages and the death of the cell. This phenomenon is named the lytic cycle of the prophage. It brings to mind the formation of flowers in plants and the subsequent dissemination of pollen and of seeds. In nature, the temperate phages keep the genes they carry protected against DNases, pH variations, etc., and therefore they may persist for weeks or even months. They may travel great distances, carried by water curents, winds, and animals. When they find a cell with appropriate receptors they attach themselves to it with the free end of their tails and inject their DNA into its cytoplasm (FIGURE 2F). Usually it is the genome of the prophage that is thus transferred. It transforms the receiving cell and its offspring into a lysogenic strain (*lysogenization*). There are new lysogenic strains formed in nature all the time. As previously mentioned, other lysogenic strains may lose their

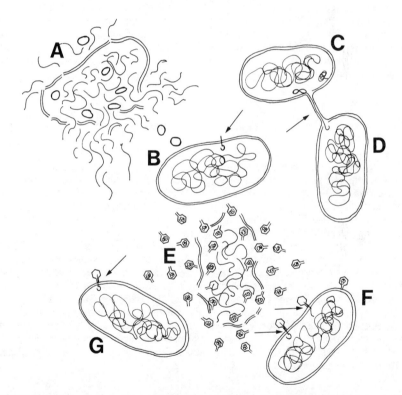

**FIGURE 2.** Examples of gene exchanges by prophages and plasmids, in bacteria: **A,** dead bacterium liberating, in solution, DNA fragments including circular small replicons (prophages or plasmids); **B,** bacterium receiving such a soluble small replicon (transfection); **C,** cell possessing conjugation plasmids and actively transferring a copy of them to cell **D** with the help of a tubular pilus (conjugation); **E,** lysed cell liberating temperate phages; **F,** cell injected through the tails of temperate phages with their prophage genome, which will settle in the new cell, with repressed transfer genes (lysogenization); **G,** cell injected by a phage from cell **E** carrying a small replicon usually a non-self-transmissible plasmid (transduction).

prophages. Very often, after the injection of the genome of the prophage into the cell of a new strain, its transfer operons may continue to instruct the new cell to produce copies of the prophage's heads and tails and assemble them. Then the cell is lysed and it liberates about a hundred completed temperate phages. This helps an isolated prophage genome to amplify rapidly its population in any new environment. It was this second type of lytic cycle that was first discovered and studied. It is thus understandable why in the early 1920s the temperate phages were classified with the viruses. The newer concept of prophages as small replicons, as source of temperate phages, and as responsible for lysogenic conversion, was proposed only around 1950.[1,2] The old simplistic opinion about temperate phages as being typical bacterial viruses lingered on for nearly 70 years. It was encouraged by most microbiologists, mostly due to entrenched but outdated views. This attitude inhibited or at least delayed the evolving of a coherent concept on the bacterial world and its decisive role in the biosphere. We also know that another essential role of prophages is to transfer, to other kinds of bacteria, genes from the large replicon or entire small replicons from the host bacterium, by incorporating them into the "heads" formed in the bacteria going through a lytic cycle. By this mechanism, called *transduction,* numerous, very peculiar phages will transfer other genes than those of their prophages (FIGURE 2G). This throws a new light on the role of these small replicons as general gene exchangers in the bacterial world, thereby singling them out even more from the viruses. All types of bacteria, including Archaebacteria, possess prophages.[3,4]

### *Conjugation Plasmids*

These small replicons transfer one of their copies (DNA) into a new bacterium with the help of a *pilus* (FIGURE 2C and D). This is their tubular structure with one end growing out through the bacterial wall and the other end attached in the cytoplasm of its bacterium. The free end of the pilus fixes itself on a receptor at the surface of a different type of cell. This physical contact is needed for the active transfer of a copy of the plasmid. As in the case of prophages, in nature conjugation plasmids may transfer from the other genes present in the donor cell: fragments of the large replicon or entire small replicons, particularly non-self-transmissible plasmids. Practically all gram-negative, heterotrophe bacteria may exchange conjugation plasmids. Conjugation plasmids meet a greater range of various metabolisms in the strains they "visit" than do prophages, whose range is restricted to metabolically similar strains. There is less conjugation among gram-positive strains. By contrast, the prophages are present in all types of bacteria. The majority of the prophages are integrated in the large replicon whereas only a minority of conjugation plasmids display this tendency. Thus, the usual term "extrachromosomal" genetic elements is inappropriate when applied generally to prophages or plasmids.

## NON-SELF-TRANSMISSIBLE PLASMIDS

These small replicons usually need the help of prophages or of conjugation plasmids for the transfer. They may also be accepted in many kinds of bacteria by *transfection.*

This is a form of gene transfer involving small replicons or virulent bacterial phages that cross the external membranes of a *competent* bacterium, which presents actively open receptor sites for soluble foreign DNA (FIGURE 2A and B). This method works only in the immediate vicinity of the involved bacteria and usually, as in the case of conjugation, it occurs in concentrated, mixed bacterial communities. Non-self-transmissible plasmids are present in all types of bacteria, including Archaebacteria.[5] Devoid of transfer genes, these smaller plasmids are more numerous than prophages or conjugation plasmids. Non-self-transmissible plasmids do not usually integrate in the large replicon. Therefore they are more easily eliminated if useless (curing) than are integrated prophages or conjugation plasmids.

## CONSEQUENCES OF THE PRESENCE AND ACTIVITIES OF SMALL REPLICONS

Bacterial genes are not permanently attached to one type of replicon. With the help of insertion sequences they may "jump" from one replicon to another in the same cell. Then, by selective pressure, the best intracellular location for any particular situation will finally prevail for each gene. Therefore, a conversion gene on a small replicon may have belonged, a few generations ago, to the large replicon of the cell. The reverse is equally valid. Moreover, we cannot determine which genes are old and which are recent arrivals in a bacterial strain. Genes may be playing the game of "musical chairs" within a cell, as well as between the different types of bacteria.

This stimulates the functioning of the bacterial world, conferring upon it an exceptional adaptability. This life-style lasted for a long period of time and became widespread among bacteria. Their cells had to remain small and to contain a minimum of genes, in order to multiply faster and to resist better in the competitive surroundings they inhabit. Thus, the survivors among the bacteria have always remained overspecialized, containing a rather incomplete set of genes and therefore only capable of an incomplete metabolism. By enjoying accessibility to the genes of other bacteria, the challenge was more than met. There is a dynamic, extremely adaptable, and optional "symbiosis" between bacteria and the small replicons. Its result is the global bacterial world, possessing and using a common potential genome. Thus, all bacteria, including Archaebacteria (in spite of a different opinion),[6] behave as a unified and partially coordinated entity. This entity, a clonal development from a unique ancestor, may be considered to be a single, planetary, noncontiguous, pluricellular superorganism.[7,9] Naturally there cannot be exact equivalents between the categories established for eukaryotes when applied to bacteria.

During our lifetime, the bacteria responsible for infectious diseases have been massively challenged with new antibacterial drugs, particularly antibiotics. It is a well-known fact that a few years of intense usage of any one of these drugs created without exception a selective pressure among the involved bacteria which led them to obtain from their faraway "cousins" (from the soil) the necessary genes of resistance. These genes arrived mostly by non-self-transmissible plasmids, transduced by prophages for the gram-positive bacteria and mobilized by conjugation for the gram-negative ones. Medical microbiology confirmed all these events, and is studying ways to circumvent the health problems they cause. A similar situation has been occurring since more and more active biological substances have been sprayed on cultivated soils. Each time that these products displayed an antibacterial activity, the "provoked" bacteria reacted with success. Resistance genes already present in some bacteria, multiplied and directed by selective pressures, found their way into local bacteria, thus protecting them. This was done by conjugation, by prophage action, or by a combined action which led, usually, to the digestion of the new chemical aggressor of the bacteria.[8]

We have witnessed these events because they were started by people and because the efficient bacterial reactions have upset our projects. Although we are not aware of most of them, such efficient problem-solving episodes by gene exchanges seem to be going on in nature on a permanent basis. All the bacterial world is kept unified and very efficient by the open access of its cells to genes of practically all the other metabolically different strains, and this situation should be considered as the central fact of the bacterial world.[7,9] Bacterial cells finally acquired more than a tolerance for genes arriving from different types of cells. These genes are everywhere "at home" in the same planetery bacterial entity and are highly welcomed (FIGURE 3). This is also proven by the hundreds of receptors present at the surface of each bacterial cell, waiting for different temperate phages or for conjugation plasmids to bring new genes. A very small number of them finally become established in the cell, after a period of trial and error. By countless such events, over a period of three billion years, the mechanisms that facilitate DNA adaptability evolved in the bacteria.

This amazing, newly discovered bacterial world has far-reaching meaning for biology. Unfortunately, the consequences are too often not yet accepted, or are simply ignored by scientists and, ironically, mostly by microbiologists. It is one more example where scientific progress contradicts entrenched ways of structured knowledge. However there is no more choice: once the role of small replicons was understood, new concepts became necessary in order to explain the bacterial world and to better understand their descendants, the eukaryotes, as well.[7,9]

# A NEW LOOK AT EVOLUTION IN THE LIGHT OF THE ROLE OF PROPHAGES AND OF PLASMIDS

## *Evolution in Bacteria and in Small Replicons*

Bacteria had an accelerated evolution since they could always use new, already "discovered" genes from other strains. This explains the impressive variety in bacteral metabolism, incomparably richer than the metabolism of eukaryotes. All the evolutionary episodes, including those implying the small replicons, ultimately occurred *inside* the common genetic heritage of all bacteria and seem to have evolved towards more cooperative cells with more efficient small replicons to keep most of the genes available. It was suggested and tacitly agreed that most of the present bacterial genes were already ''discovered'' about a billion years ago. Since then they were probably shuffled around in countless trial-and-error situations in different bacteria. By successive selections they are now in the best position in the different replicons of all bacteria. A large number of methods for modifying and recombining DNA were also discovered. Large replicons are rather stable today as they represent a near perfect gene arrangement for the present planetary conditions, kept stable by life-promoting Gaia mechanisms,[10] a good part of which are bacterial. Small replicons specialize in resistance genes and in those that may be temporarily useful. We know too that all the potential for other important change is there, ready to solve bacterial population problems.

All small replicons probably originated from non-self-transmissible plasmids. The latter were probably a normal constituent of the earliest cells, representing their genome. By the presence of many copies of each small replicon in each cell they corrected the probably not so efficient cell divisions. It is even possible that there was no large replicon in these cells. The self-transmissible small replicons derived from these simpler non-self-transmissible ones. Prophages and conjugation plasmids seem to have a common intermediate ancestor, able to transfer itself actively to other cells with the help of a tubular protein and of receptors for it in the sensitive cells. Later they branched into typical prophages on one hand and conjugation plasmids on the other.

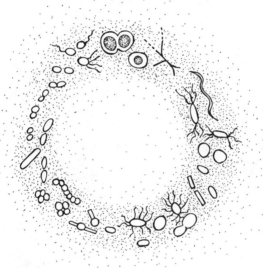

**FIGURE 3.** Simplified view of the unified bacterial world. Small dots represent available exchange genes, mostly small replicons, always present around and in all bacteria.

### Origin of Eukaryotes

The symbiotic theory of the origin of eukaryotes from two or more different bacteria[11,12] seems even more convincing in view of the previous facility for ectosymbiosis already existing in bacteria. Moreover, the tolerance for more than one genetic center in the same cell existed in all bacteria, with their large replicon being accompanied, when necessary, by a choice of different small replicons. The first nucleus in the ancestor of the eukaryotes had its number of genes increased nearly a thousand times compared to those in a bacterium, probably by integrating and keeping captive most of the "visiting" small replicons, which in bacteria usually bring help only on a temporary basis.[13] When no more small replicons were needed in the outer symbiont containing the nucleus, the ancestors of eukaryotes lost the capacity (receptors, etc.) to accept prophages or plasmids. This started the genetic isolation common today to all eukaryotes. The only known genetic transfer now occurring in nature from bacteria to eukaryotes is that of genes from Ti conjugation plasmids of *Agrobacterium tumefaciens,* conferring the "crown galls" illness to many dicotyledon plants, characterized by tumorlike deformations. This small replicon is used extensively as a means to introduce desired genes in dicotyledon cultivated plants.[14]

Neither morphology nor special cell arangements were transmitted from bacteria to eukaryotes. The rich possibilities of modifying DNA were their main inheritance. The facility for ectosymbiosis, which bacteria realized before the eukaryotes appeared, and which is seen everywhere in improvised bacterial communities, has also been extended to associations with eukaryotes producing everywhere successful ectosymbiosis (rumen, etc.) and many cases of endosymbiosis in arthropods, protozoa, etc.[15]

## *Origin and Evolution of Viruses*

The virulent phages have most probably originated from previous small replicons, which as we have seen present some viruslike properties. The tailed, double-stranded-DNA virulent phages seem to be direct descendants by retrograde evolution from prophages which lost the repressor genes necessary for lysogenization. Later, they lost the then useless conversion genes. They are by far the most frequent DNA virulent phages. They are also indicants of prophage presence in the group of bacteria where they multiply. The tubular, "filamentous" DNA virulent phages are rather heterogeneous. They might have a common ancestry with the conjugation plasmids. Their transfer genes code for a tubular protein in which the viral DNA is protected for weeks. Moreover, these phages do not kill the infected cell at once, only much later, after the successive liberation of many phages into the surroundings. We know much less about the few tailless, nonfilamentous DNA phages.

The RNA phages are the smallest and might be a vestige from the beginning of life on earth, when the genetic information was probably carried exclusively by RNA replicons which probably produced some parasitic forms, as did later DNA small replicons. When the bacteria swiched to DNA genes, better able to perform the different forms of recombination, only the already parasitic RNA genes (viral) probably kept their old structures.

The RNA viruses of eukaryotes probably originated directly from virulent RNA phages which, probably, continued to infect the outer symbiont of the emerging eukaryotes.

No DNA virus in eukaryotes resembles tailed DNA phages, virulent or temperate. Probably, from the beginning of their genetic isolation, the eukaryotes have closed the door to this large group of phages.

If small replicons accumulated at the origin of the first nucleus,[13] then the theory of the endogenous origin of eukaryotic DNA viruses stands a good chance of being true. It suggests that a part of the genome of a eukaryote may contain arrangements of genes that, with very few mutations, could produce an infectious and autonomous DNA particle. Hence, new DNA viruses of eukaryotes could be produced at any time. Nuclear vestiges of small replicons seem to be the most probable candidates for such a theory, but for the moment, we have no observations or experimental results to support it.

Viruses, in contrast with the small replicons, seem to constitute evolutive dead-end entities, as are most parasites. The only well-documented "conversions" by natural infection with eukaryote viruses are tumor (often cancer) producing ones. Their genes, or the DNA equivalent of RNA genes after reverse transcriptase action, integrate in the chromosomes of the infected eukaryote. This integrated viral DNA, the eukaryotic equivalent of a prophage, is called *provirus,* and the type of genetic action is called a viral ''transformation.'' These facts, added to the knowledge of the bacterial resourceful genetic mechanisms, have brought hopes for correcting genetic disabilities in humans and other eukaryotes. This will probably be achieved by using prophages, plasmids, and mild or avirulent stable mutants of DNA viruses of eukaryotes as the main tools of such genetic engineering. It is a positive manner of interrupting a billion and a half years of genetic separation between prokaryotes and eukaryotes.

## CONCLUSIONS

There is a need to reassess the general role in nature of the bacterial small replicons. Some of them have been classified with the viruses, since both these categories are "visiting" self-replicating groups of genes. Studies on plasmids[16] have left out the very similar prophages, and therefore their impact has been limited. It is not yet realized by most biologists that prophages and plasmids play a similar essential part in bacterial life. As autonomous groups of genes, they are adapted to be exchanged and to be available to different bacteria, helping them to adjust to changes and to realize planetary-wide complex functions. If we understand the unifying and stimulating role of the small replicons, our concept of the bacterial world changes from one of anarchic, primitive, and incomplete strains to one of a single entity, a diversified and efficient planetary clone. In bacterial evolution these small replicons were the main catalysts, along with the permanent fierce competition among different types of bacteria. Prophages and plasmids have, thus, accelerated the evolution of prokaryotes and increased immensely the variety of genetic solutions and therefore of cell metabolisms. The eukaryotes inherited ready-made formulas and seem to carry many vestiges from their bacterial ancestors. All the viruses, bacterial or eukaryotic, originated directly or indirectly from prophages or plasmids. All these facts and probabilities need to be structured in a new coherent concept.[7,9] This new approach may also fill the gap for an essential missing link in general biology.

## SUMMARY

Prophages and plasmids offer to the bacterial cells generalized access to each other's genes. The result is an extremely rich, available gene bank. It has successfully supported the original bacterial life since its beginnings and therefore it has conditioned all bacterial cells. Thus, most of the basic mechanisms for the living world, the richest variety of new genes, and particularly the improved ways of using DNA as an extremely adaptable genetic material happened in bacteria with the help of prophages and plasmids. This fact has profoundly marked all the biosphere. The ancestor of the nucleus probably started as an accumulation of prophages and plasmids integrated in the growing "chromosome" of the outer symbiont of the first eukaryotes. Many bacterial vestiges were probably retained in eukaryotes, mostly those related to the dominant and lasting role of small replicons in all their bacterial precursors. These vestiges may, for example, serve as an endogenic source for some DNA viruses in eukaryotes. The other animal and plant viruses seem to derive directly or indirectly from prophages or plasmids. In the case of RNA viruses they may have originated from probable RNA small replicons present in the first forms of life on earth. Some confusion arose in biology, as viruses were discovered first and therefore their most probable ancestors, the plasmids and the prophages which were discovered later, were thought to be viruslike, or viruses, as is the case with prophages.

## REFERENCES

1. FREEMAN, V. J. 1951. Studies of virulence of bacteriophage infected strains of *Corynebacterium diphtheriae*. J. Bacteriol. **61**: 675-688.
2. LWOFF, A. 1953. Lysogeny. Bacteriol. Rev. **17**: 209-337.
3. SCHNABEL, H. 1984. An immune strain of *Halobacterium halobium* carries the invertible L segment of phage H as a plasmid. Proc. Nat. Acad. Sci. USA **81**: 1017-1070.
4. ZILLIG, W., F. GRAPP, A. HENSCHEN, H. NEWMANN, P. PALM, W. D. REITER, M. RETTENBERGER, H. SCHNABEL & S. YEATS. 1986. Archaebacterial virus host systems. Syst. Appl. Microbiol. **7**: 58-66.
5. ZILLIG, W., S. YEATS, I. HOLZ, A. BOCK & S. LUTZ. 1985. Plasmid-related anaerobic autotrophy of the novel Archaebacterium, *Sulfobolus ambivaleus*. Nature **313**: 789-791.
6. WOESE, C. R. 1982. Archaebacteria and cellular origins: an overview. Zbl. Bakt. Hyg. I. Abt. Orig. C **3**: 1-17.
7. SONEA, S. 1971. A tentative unifying view of bacteria. Rev. Can. Biol. **30**: 239-244.
8. FISHER, P. R., J. APPLETON & J. M. PEMBERTON. 1978. Isolation and characterization of the pesticide-degrading plasmid pJP1 from *Alcaligenes paradoxus*. J. Bacteriol. **135**: 798-804.
9. SONEA, S. & M. PANISSET. 1983. A New Bacteriology. Jones and Bartlett Publishers. Boston, Mass.
10. LOVELOCK, J. E. & L. MARGULIS. 1974. Atmospheric homeostasis by and for the biosphere: the Gaia hypothesis. Tellus **26**: 2-9.
11. MARGULIS, L. 1981. Early Life. Science Books International. Boston, Mass.
12. DYER, B. D. & R. OBAR. 1985. The origin of eukaryotic cells. Van Nostrand Reinhold. New York, N.Y.
13. SONEA, S. 1972. Bacterial plasmids instrumental in the origin of eucaryotes. Rev. Can. Biol. **31**: 61-63.
14. MARX, J. L. 1985. Plant gene transfer becomes a fertile field. Science **230**: 1148-1150.
15. MARGULIS, L. 1981. Symbiosis in Cell Evolution. W. H. Freeman and Co. San Francisco, Calif.
16. BRODA, P. 1979. Plasmids. W. H. Freeman and Co. Oxford, England.

# Different Endocytobionts Simultaneously Colonizing Ciliate Cells

HANS-DIETER GÖRTZ

*Zoological Institute*
*University of Münster*
*Schlossplatz 5*
*D-4400 Münster, Federal Republic of Germany*

## INTRODUCTION

Ciliates are often colonized by microorganisms, many of which are pathogenic and kill their host cells, while others have become well-adapted parasites. A few have even developed into mutualistic, sometimes essential symbionts.[1-6] In most cases only one species of endocytobionts is present within the same host cell, and the simultaneous colonization of the same or different compartments of one cell by different microorganisms has been regarded as exceptional.[2,4,7,8] However, occasionally different microorganisms live in the same host population and sometimes even coexist within an individual cell. The aim of this article is to present observations on the coexistence of different endocytobionts (= multiple endocytobiosis) in ciliates. The observations will be discussed together with earlier reports.

## MULTIPLE ENDOCYTOBIOSIS IN CILIATES ISOLATED FROM NATURE

Several observations of different endocytobionts within individual host cells are reported for ciliates isolated from nature (TABLE 1). *Paramecium biaurelia* was found with *Caedibacter taeniospiralis* (formerly kappa particles) in the cytoplasm and *Holospora caryophila* (formerly alpha particles) in the macronucleus.[5,9] The endonucleobiotic *H. caryophila* were only infective to strains of *P. biaurelia* that simultaneously contained *Caedibacter* or had derived from such strains. This raised the question whether certain host genes were necessary for the maintenance of *H. caryophila*, since a dependence on host genes was proven for *Caedibacter*.[10,11]

In an investigation by Heckmann *et al.* it was shown that *Euplotes* species of the 9 type 1 cirrus pattern depend upon endosymbionts.[12] Most of these species harbor bacteria very similar to the omikron particles observed in *E. aediculatus*.[3] Particles that are similar to omikron are called omikronlike. Some *Euplotes* strains contained a second bacterial endocytobiont, and in *E. daidaleos* a symbiotic *Chlorella* (zoochlorella) was found in addition to omikronlike bacteria.[12]

*P. bursaria,* known as the green paramecium, contains a symbiotic *Chlorella* (zoochlorellae), and host and endocytobiont are well adapted to one another.[6,13] This seems to be the reason why the occurrence of other microorganisms in the cytoplasm of *P. bursaria* is rarely observed. It could be shown that the algal endocytobionts can indeed prevent infections of the cytoplasm of their host cells by other microorganisms (see Interaction of Different Endocytobionts). However, the nuclei of *P. bursaria* may be infected by bacteria even when symbiotic algae are present in the host cytoplasm, and the micronucleus occasionally is infected by *Holospora acuminata* or by other bacteria.[14-16] Also, in *Stentor polymorphus,* another ciliate harboring zoochlorellae, like in *P. bursaria* occasionally endonucleobionts are found in addition to the algae. The endonucleobionts, which obviously belong to the genus *Holospora,* colonize the macronucleus.

**TABLE 1.** Multiple Endocytobioses in Ciliates

| Host | 1. Endocytobiont[a] | 2. Endocytobiont[a] | 3. Endocytobiont[a] | Reference |
|---|---|---|---|---|
| *P. bursaria* | *Chlorella* (cy) | *H. acuminata* (mi) | — | 16, 22 |
| *P. bursaria* | *Chlorella* (cy) | *H. acuminata* (mi) | Small bacteria (mi) | 16 |
| *P. biaurelia* | *C. taeniospiralis* | *H. caryophila* (ma) | — | 9 |
| *P. caudatum* | *H. obtusa* (ma) | *H. elegans* (mi) | — | 20 |
| *P. caudatum* | *H. obtusa* (ma) | *Caedibacter* (ma) | — | |
| *P. caudatum* | *H. caryophila* (ma) | *Caedibacter* (ma) | — | |
| *P. caudatum* | *Caedibacter* (ma) | *H. elegans* (mi) | — | |
| *E. daidaleos* | Omikronlike (cy) | *Chlorella* (cy) | — | 12 |
| *E. eurystomus* | Omikronlike (cy) | Small bacteria (cy) | — | 12 |
| *E. octocarinatus* | Omikronlike (cy) | Small bacteria (cy) | — | 12 |
| *Euplotes* sp. | Omikronlike (cy) | *Leptomonas* sp. (ma) | — | 21 |
| *S. polymorphus* | *Chlorella* (cy) | Bacteria (ma) | — | |
| *S. polymorphus* | *Chlorella* (cy) | Microsporidia (cy) | — | |

[a] cy = cytoplasm; mi = micronucleus; ma = macronucleus.

The nuclei of *P. caudatum* are often infected by various *Holospora* species, while an infection of the cytoplasm of this ciliate has only been reported once.[17] However, the simultaneous infection of individual cells with more than one Holospora species (FIGURE 1) is rarely observed in cells isolated from nature.

Macronuclei of *P. caudatum* are also found infected with bacteria other than *Holospora.*[18-20] A local population of this ciliate quantitatively bearing a *Caedibacter* species in the macronucleus also contained different *Holospora* species. At a low percentage the cells were infected either by *H. obtusa* or *H. caryophila* in the macronucleus or *H. undulata* in the micronucleus. When taken into laboratory culture these lines lost the *Holospora* species and maintained *Caedibacter.* In other cases where different bacterial endonucleobionts are present in the macronuclei of *P. caudatum* in addition to *H. obtusa* or *H. caryophila,* only the *Holospora* species are maintained in culture.

*P. caudatum* appears to be preadapted for the reception of endonucleobionts. *P. aurelia* species on the other hand are regularly colonized by endocytobionts in the

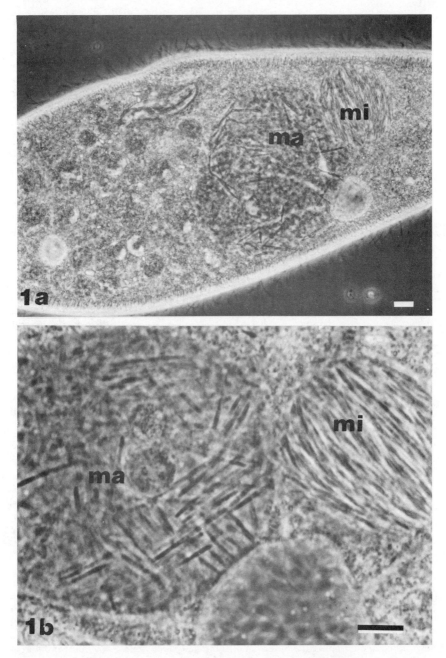

**FIGURE 1.** *Paramecium caudatum* with *Holospora elegans* in the micronucleus (mi) and *Holospora obtusa* in the macronucleus (ma). Bar, 10 μm.

cytoplasm but less frequently in the nuclei.[2] The reason for this difference between *P. caudatum* and the *P. aurelia* species is unknown.

Only few observations are reported that endocytobiont-bearing ciliates are infected by eukaryotic parasites. *Stentor polymorphus* is occasionally the host for microsporidia in the cytoplasm in addition to the zoochlorellae (FIGURE 2). Another example is the infection of a Euplotes species with a leptomonad flagellate reported by Wille *et al.;*[21] Figure 3 in their paper shows that omikronlike endocytobionts are present as well.


## INDUCED MULTIPLE ENDOCYTOBIOSIS IN CILIATES


Since the *Holospora* species all are highly infectious it is tempting to induce different types of multiple endonucleobiosis. Infection of *Caedibacter*-bearing *P. caudatum* with *H. obtusa* or *H. caryophila* (infecting the macronucleus) or *H. elegans* (infecting the micronucleus) is possible (TABLE 1), but these double infections are not stable and the *Holospora* species are gradually lost within some weeks first from individual host cells and finally from the whole culture. Infections of *Holospora*-bearing *P. caudatum* with other *Holospora* species are not as easily achieved and the double infections with different combinations of *Holospora* species under normal culture conditions can be maintained for a limited time only.[7,8,22] However, Dr. P. Daggett (personal communication) observed the simultaneous infection of an ATCC strain of *P. caudatum* with *H. obtusa* and *H. elegans*. A portion of the cells of this strain was infected with *H. obtusa* (in the macronucleus), a portion with *H. elegans* (in the micronucleus), a portion with both, and a portion with none, and the double infection turned out to be stable; the strain, which was kindly sent to me by P. Daggett, still contains both endonucleobionts (FIGURE 1).


## INTERACTION OF DIFFERENT ENDOCYTOBIONTS


Endocytobionts simultaneously living in one host cell may either depend on different metabolites supplied by their host, or compete at least partly for identical nutrients. A third possibility is that different endocytobionts even favor one another. This interdependence is evident between xenosomes (for the definition of the term xenosome see Corliss)[23] such as mitochondria and chloroplasts or omikron particles and mitochondria.

Also, a mutualistic interdependence can be concluded between endocytobiotic chlorellae and mitochondria, e.g., in *P. bursaria*. However, in *E. daidaleos* Heckmann *et al.* observed that the number of omikronlike endocytobionts was lower in cells harboring zoochlorellae than in cells without algae,[12] suggesting some competition between the two different endocytobionts. Nevertheless, *E. daidaleos* regularly harbors zoochlorellae in addition to omikronlike bacteria, and there appears to be only a limited competition. On the other hand, some mutual advantage for the two different endocytobionts might be assumed. *E. daidaleos* depends on omikronlike bacteria even when it contains zoochlorellae,[12] and under starvation conditions *E. daidaleos* like other ciliates harboring symbiotic algae can survive for a longer time than alga-free

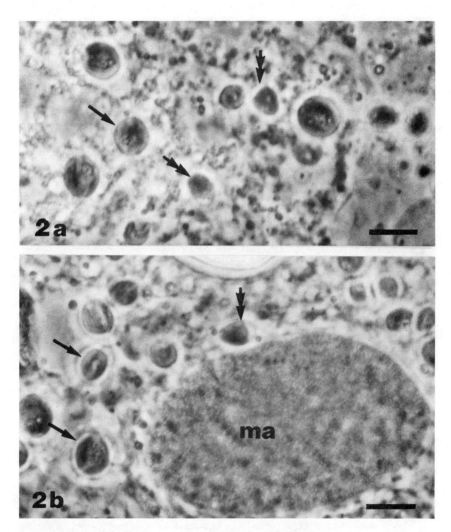

**FIGURE 2.** Symbiotic chlorellae (arrow) and microsporidia (double arrow) in *S. polymorphus*. ma, macronucleus. Bar, 10 μm.

*Euplotes.* So this ciliate might show an early step in a course of acquisition of different organelles (e.g., mitochondria and chloroplasts) by sequential endocytobiosis and could help us to understand the development of the eukaryotic cell. This example is especially interesting because these endocytobionts are not yet true organelles and the ciliates can be cured from either of these endocytobionts in order to test their significance.

In *P. bursaria* the endocytobiotic algae can prevent infections of the host cytoplasm with bacteria or yeast, both of which can easily infect aposymbiotic paramecia.[24] Also, bacterial endocytobionts could prevent an additional infection of this host with yeasts and vice versa. Possibly the establishment of an additional endocytobiosis is difficult because the endocytobiont that is already present is entirely using up certain essential metabolites. However, neither endocytobiotic bacteria nor yeasts could prevent an infection of *P. bursaria* with *Chlorella*. Even *P. bursaria* containing bacteria or yeasts in the cytoplasm are cured of these endocytobionts after an infection with symbiotic algae.

On the other hand, in *P. bursaria* with *Chlorella* in the cytoplasm and *H. acuminata* in the micronucleus no sign of competition could be observed, and similarly *C. taeniospiralis* in the cytoplasm and *H. caryophila* in the macronucleus of *P. biaurelia* do not exclude one another nor do they depend on each other.[2,9] In these cases different compartments of the hot cell are occupied, and possibly host metabolites limiting the growth of endocytobionts are different for the coexisting species.

In cases of multiple endonucleobiosis in *P. caudatum* and *P. bursaria,* competitive behavior of different *Holospora* species has been observed and discussed,[7,20,22,25] and it has been suggested that these endonucleobionts could also provide a resistance against pathogenic germs; such a resistance could be an advantage for *Holospora*-bearing paramecia.[7,25] A possible explanation for the competition again might be that the different *Holospora* species use identical metabolites.

However, in the strain where P. M. Daggett observed a stable double infection with *H. obtusa* in the macronucleus and *H. elegans* in the micronucleus (see above), the percentage of doubly infected paramecia is sometimes even higher than expected from the percentages of single infections with each of the bacteria. It can therefore be assumed that the two endonucleobionts are not competitive in this host strain, and this exception from the normal behavior still cannot be explained. The reason could either be a mutation of one of the bacteria or a mutation or adaptation of the host strain.

In a laboratory culture of *P. bursaria* bearing *H. acuminata* an infection of the micronucleus with an additional bacterium was observed and it had been suggested by Görtz and Freiburg that this additional endonucleobiont had been transported into the micronucleus together with *H. acuminata.*[16] Later *H. acuminata* was lost from this strain, whereas the small additional bacteria were maintained. The loss could be due to damage of *H. acuminata* caused by the small bacteria or to a competition of the two endonucleobionts. The latter seems more probable because *H. acuminata* was lost gradually and not rapidly.[16]

This is similar to when *P. caudatum* bearing a *Caedibacter* species in the macronucleus are infected with *H. obtusa* in addition, where the initially high rate of *Holospora*-infected cells decreases until in most cases only *Caedibacter* remains in the culture. Also, in *P. caudatum* isolated from nature which contain *H. obtusa* sometimes other additional endonucleobionts are observed, most of which are lost when cultured in the laboratory. So the various microorganisms are obviously adapted to the endonucleobiotic way of life to varying degrees. The reasons for the low stability of multiple endocytobiosis will be the different degrees of adaptation and the reciprocal interactions, since infections with a single type of these endocytobionts are normally maintained without difficulties.

Parasitic microsporidia infecting *Stentor* appear poorly adapted, and the ciliates are mostly killed under starvation conditions. The fatal effect of a microsporidia infection seems to be stronger in *Stentor roeseli,* which does not contain zoochlorellae, than in *S. polymorphus.* It may be suggested that the symbiotic algae can improve the prognosis for a *Stentor* infected with microsporidia.

## CONCLUSIONS

Endocytobionts being parasitic, mutualistic, or other use the host cell as a habitat, and the unit of host cell with its various organelles and one or more endocytobiont species represents a small but intricate ecosystem. Environmental factors, especially chemical ones affecting this ecosystem, operate on the endocytobionts only indirectly and to some extent are filtered or moderated by the host metabolism. So the environment of an endocytobiont will be more balanced than the environment of free-living organisms. However, the host cell may react to the presence of endocytobionts (see above), and Taylor concluded that the cell as a cytocosm with its endocytobionts can be regarded as a cybernetic (feedback control) ecosystem,[26] although we do not know the control systems involved in the regulation of endocytobiont growth. Even more obscure are the interactions between different endocytobionts coexisting in the same host cell. The observations that ciliates isolated from nature mostly maintain only one of the two or more initial endocytobionts indicate that environmental factors that are not yet recognized may eventually stabilize multiple endocytobiosis

### REFERENCES

1. BALL, G. H. 1969. *In* Research in Protozoology. T. T. Chen, Ed. **3:** 565-718. Pergamon Press. London & New York.
2. PREER, J. R., JR., L. B. PREER & A. JURAND. 1974. Bacteriol. Rev. **38:** 383-390.
3. HECKMANN, K. J. 1975. Protozool. **22:** 97-104.
4. HECKMANN, K. 1983. *In* Intracellular Symbiosis. K. W. Jeon, Ed.: 111-144. Academic Press. New York, N.Y.
5. PREER, J. R., JR. & L. B. PREER. 1984. *In* Bergey's Manual of Systematic Bacteriology. N. R. Krieg, Ed. **1:** 795-811. Williams and Wilkins. Baltimore & London.
6. GÖRTZ, H.-D. & J. DIECKMANN. 1977. J. Verh. Dtsch. Zool. Ges. **1977:** 267.
7. LEE, J. J., A. T. SOLDO, W. REISSER, M. J. LEE, K. W. JEON & H.-D. GÖRTZ. 1985. J. Protozool. **32:** 391-403.
8. GÖRTZ, H.-D. 1983. *In* Intracellular Symbiosis. K. W. Jeon, Ed.: 145-176. Academic Press. New York, N.Y.
9. PREER, L. B. 1969. J. Protozool. **16:** 570-578.
10. SONNEBORN, T. M. 1943. Proc. Nat. Acad. Sci. USA **29:** 329-343.
11. CHAO, P. K. 1953. Proc. Nat. Acad. Sci. USA **39:** 103-113.
12. HECKMANN, K., R. TEN HAGEN & H.-D. GÖRTZ. 1983. J. Protozool. **30:** 284-289.
13. REISSER, W., R. MEIER, H.-D. GÖRTZ & K. W. JEON. 1985. J. Protozool. **32:** 383-390.
14. OSSIPOV, D. V., O. N. BORCHSENIUS & S. A. PODLIPAEV. 1980. Acta Protozool. **19:** 315-326.
15. OSSIPOV, D. V., I. I. SKOBLO, O. N. BORCHSENIUS, M. S. RAUTIAN & S. A. PODLIPAEV. 1980. Tsitologiya **22:** 922-929.

16. GÖRTZ, H.-D. & M. FREIBURG. 1984. Endocytol. C. Res. **1:** 37-46.
17. DIECKMANN, J. 1981. Abstr. Int. Congr. Protozool., 6th, Warsaw: 77.
18. ESTEVE, J.-C. 1978. Protistologica **14:** 201-207.
19. GÖRTZ, H.-D. 1980. *In* Endocytobiology. W. Schwemmler & H. E. A. Schenk, Eds. **1:** 381-392. de Gruyter. Berlin, FRG.
20. GÖRTZ, H.-D. 1986. Int. Rev. Cytol. **102:** 169-213.
21. WILLE, J. J., JR., E. WEIDNER & W. L. STEFFENS. 1981. J. Protozool. **28:** 223-227.
22. OSSIPOV, D. V., I. I. SKOBLO & M. S. RAUTIAN. 1975. Acta Protozool. **14:** 263-280.
23. CORLISS, J. O. 1985. J. Protozool. **32:** 373-376.
24. GÖRTZ, H.-D. 1983. J. Cell Sci. **58:** 445-453.
25. HECKMANN, K. 1977. Proc. Int. Congr. Protozool., 5th.: 160-164.
26. TAYLOR, F. J. R. 1983. *In* Intracellular Symbiosis. K. W. Jeon, Ed.: 1-28. Academic Press. New York, N.Y.

# Initiation and Control of the Bioluminescent Symbiosis between *Photobacterium leiognathi* and Leiognathid Fish

P. V. DUNLAP[a,c] AND M. J. McFALL-NGAI[b]

[a] Department of Microbiology
Cornell University
Ithaca, New York 14853

[b] Department of Ophthalmology
Jules Stein Eye Institute
University of California at Los Angeles
Los Angeles, California 90024

## INTRODUCTION

Bioluminescent bacteria, common "free-living" members of planktonic and benthic microbial communities, can also be found in symbiotic associations with marine animals. These include nonspecific associations with invertebrates and vertebrates (saprophytic, commensal, and parasitic), in which bacterial luminescence, when produced, may have no benefit for the animal host, and species-specific, apparently mutualistic light-organ symbioses with certain fishes and squids, in which light production by the bacteria is central to the symbiosis. In the light-organ symbioses, the bacteria are housed as dense, pure cultures in highly structured, glandlike light organs. They presumably obtain nutrients and a protected growth environment from the animal host, which in turn utilizes the bacterial light in various bioluminescence displays. The light-organ symbioses, because of the pure-culture nature of the symbiotic bacteria, offer a great potential for revealing specific symbiotic interactions between prokaryotes and their eukaryote hosts.[1-4] However, this potential has not been realized, and at present very little is known about the ecological and physiological interactions between light-organ bacteria and the fishes and squids that harbor them. This lack of knowledge is due, at least in part, to a general difficulty in collecting and maintaining the animal hosts, and therefore also to a lack of detailed descriptive information on the symbiosis on which to base specific research questions.

One of these associations, however, the light-organ symbiosis between the bioluminescent bacterium *Photobacterium leiognathi* and fish of the family Leiognathidae, differs in this regard in that the host fish are particularly abundant and accessible

---

[c] Present affiliation: Department of Biology, Box 3AF, New Mexico State University, Las Cruces, New Mexico 88003.

and, when transported and handled properly, can be maintained in laboratory aquaria for extended periods (e.g., more than one year, McFall-Ngai, personal observation). Leiognathids are schooling, shallow-dwelling fish common in coastal and bay environments throughout the Indo-West Pacific region.[1,5-7] The family is divided into three genera (*Gazza, Leiognathus,* and *Secutor*) with approximately 30 species.[5,8,9] Bioluminescence in leiognathids is produced from a single, circumesophageal light organ by a pure, extracellular culture of *P. leiognathi,* a bacterium that can be readily cultured free of the host fish.[10-13] Leiognathids use the light produced by *P. leiognathi* in a versatile and complex array of bioluminescence displays.[14] Moreover, leiognathids exhibit species- and sex-based morphological variation in their light organs,[14-16] and strain variation occurs in *P. leiognathi* that may be host-species related.[17-19]

For these reasons, the light-organ symbiosis between *P. leiognathi* and leiognathid fish presents unique opportunities to examine the ecological and physiological interactions between a specific symbiotic bacterium and its animal host. In this report we address these opportunities for research in the context of recent work on the initiation of the association and on physiological control of light production in *P. leiognathi.*

## ANATOMY AND FUNCTION OF THE LEIOGNATHID LIGHT-ORGAN SYSTEM

The leiognathid light organ (FIGURE 1) is an internal, circumesophageal ring of tissue located just anterior to the stomach.[20-24] In general, light is released from the light organ (1) posteriorly into the closely abutted gas bladder, where it is reflected, scattered, and then emitted ventroposteriorly from the gas bladder into hypaxial musculature, which serves as a light guide and diffuser;[25,26] and (2) anteroventrally directly into the opercular region.[24] Proximal control of light emission is given by muscular shutters and chromatophores directly covering the light organ.[23] More distal control over the intensity and directionality of the emitted light is provided by secondary tissues of the light-organ system, including the differentially reflective and transparent lining of the gas bladder, the translucent hypaxial musculature, chromatophores, and transparent regions of the otherwise highly silvered skin,[15,25,26] and, for certain species (e.g., *G. minuta*), the anterior and posterior sliding of the opercular flap margin to expose a patch of unsilvered skin at the posterior margin of the opercular cavity.[14] The internal location of the light organ permits the incorporation of these additional tissues in the control of light emission, and differential employment of the various tissues of the light-organ system results in an array of bioluminescent displays (TABLE 1), one that is perhaps more versatile than that of other bioluminescent organisms.[14,16] In addition, the light organs of most examined leiognathid species are sexually dimorphic, with the light organ of the male fish larger and more highly pigmented than that of the female.[16] The hypertrophied lobes of the male light organ are associated with patches of transparent skin on the male's flanks and in some species also with lateral transparent patches in the otherwise highly silvered gas bladder lining.[16] These sex-based morphological differences in the light-organ systems of leiognathids indicate the potential for sex-specific bioluminescent signaling[16] (TABLE 1). Evidence for other types of displays and their possible functions are likely to be discovered as more leiognathid species are examined.

# INITIATION OF THE SYMBIOSIS

Information on the initiation of light-organ symbiosis has a central importance for our understanding of the specificity of the association (i.e., how a pure culture of *P. leiognathi,* and not another bacterium, becomes established in the leiognathid light organ) and for our understanding of the interactions between a symbiotic bacterium and host fish that lead to development of the light organ and maintenance of the bacteria. Information on this topic in just beginning to be obtained. In examining

**TABLE 1.** Summary of Luminescence Displays in Leiognathid Fish[a]

| Display | General Description | Proposed Function |
|---|---|---|
| Counterillumination[1,25] | Diffuse, sustained glow over entire ventrum | Antipredation (silhouette camouflage) |
| Discrete projected luminescence (DPL) | Anteroventrally directed beam from clear patch of skin at posterior margin of each opercular cavity; short or sustained emission | Communication (intra- and interspecific signaling), predation (foraging) |
| Ventral body flash | Brief, intense flash over entire ventral two-thirds of body | Antipredation (startle display or alarm signal) |
| Opercular flash[2,21] | Anteroventrally directed flashes of variable intensity and duration over entire opercular region | Antipredation (startle display?) |
| Buccal luminescence | Glow of variable brightness and duration from mouth, through transparent membrane separating light organ and buccal cavity | Predation (attractant?) |
| Sex-specific signaling | Possibly similar to DPL, with light emitted from clear patches at flanks of males having sexually dimorphic light organs | Mating (intraspecific sexual recognition), schooling |

[a] See References 14 and 16.

larvae that developed from artificially fertilized eggs of the monocentrid fish *Monocentris japonicus* (which harbors *Vibrio fischeri* in its mandibular light organs),[27] Yamada *et al.* observed neither light organs nor luminescence in specimens up to 6 mm in length.[28] Recently, McFall-Ngai and Cabanban (manuscript in preparation), using histology and scanning electron microscopy (SEM), observed melanophore-shrouded light organs containing packets of bacteria in larval leiognathids as small as 5 mm in length. The light organ is derived from outpocketings of the esophageal tissue (see Figure 4 of Harms).[20] Thus, infection by *P. leiognathi,* a bacterium present in tropical waters where leiognathids are found,[11] could occur through ingestion of food or water very early in the fish's development, but the precise timing of infection

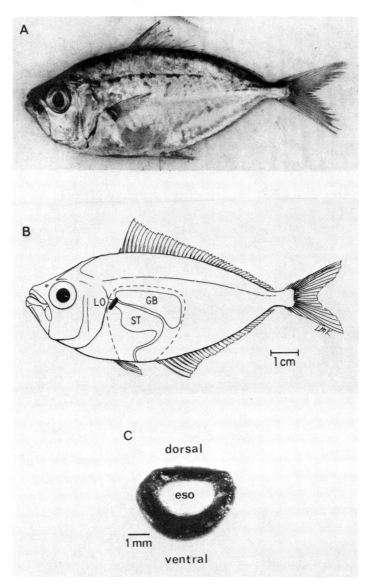

**FIGURE 1.** The leiognathid light organ. (A) *Gazza minuta,* a typical leiognathid from Manila Bay, Luzon, Phillippine Islands; (B) sketch of *G. minuta* with body wall cut away to show the position of the internal, circumesophageal light organ (LO) and its relationship to the gas bladder (GB) and stomach (ST); (C) ring-shaped light organ of *G. minuta* encircling the esophagus (eso), as viewed from the posterior.

is not known. Further, it remains to be determined whether the bacteria observed in the developing leiognathid light organs are a pure culture of *P. leiognathi* or are perhaps a mixed culture of, possibly, marine enterobacteria (i.e., *Photobacterium* and luminous and nonluminous *Vibrio* species).[13] Also, it is not clear if early development of the light organ requires the presence of the species-specific symbiotic bacterium. In this regard, Leis and Bullock observed developing light organs in the larvae of the luminous apogonid fish *Siphamia versicolor*[29] (which may also harbor *P. leiognathi* as its symbiotic light-organ bacterium.[1,29-31] In a larva 2.8 mm in length, the developing light organ lacked bacteria, whereas in larvae 3.5 mm and 10.4 mm in length, unidentified bacteria were present.[29] In this species, and thus perhaps in leiognathids and other bacterially bioluminescent fishes, the presence of the species-specific symbiotic bacteria (or other bacteria) may not be necessary for the early stages of light-organ morphogenesis.

## MORPHOLOGICAL STATE OF THE LEIOGNATHID LIGHT-ORGAN BACTERIA

Cells of *P. leiognathi* in the leiognathid light organ exhibit several morphological differences with those grown in laboratory culture,[24,32] and comparisons between the two states might provide some insight into conditions within the light organ. Within most tubules of the light organ (FIGURE 2), *P. leiognathi* cells are characteristically uniformly shaped coccobacilli approximately 1.6 $\mu$m wide by 3.2 $\mu$m long. They lack visible granules of the fatty acid storage product poly-$\beta$-hydroxybutyrate (PHB),[24,33] and, like *V. fischeri* in light organs of the monocentrid fish *M. japonicus,*[34] they lack flagella (and motility).[32] In culture, the cells are generally slightly more elongate (in exponential phase of growth), exhibit large PHB granules,[24,33] and develop flagella (and motility) within a few to several hours after transfer to culture medium.[32] These gross morphological differences presumably reflect responses to the physical and physiological constraints imposed on *P. leiognathi* by the light-organ environment. Identification of factors in culture that lead to repression of flagellar synthesis in *P. leiognathi* might provide information on the light-organ conditions that evoke this response.

Of importance in this regard, the light organ of an average leiognathid from the Philippine Islands is 0.036 ml in volume and contains approximately $2.5 \times 10^8$ bacteria, thus giving an apparent bacterial density of 6 to $7 \times 10^9$ cells $ml^{-1}$.[32] However, the bacteria are not freely dispersed within light-organ tubules, but are tightly packed at very high cell density ($10^{11}$ cells $ml^{-1}$) inside elongate, baglike structures, called light-organ saccules (FIGURE 3), that are themselves held within the light-organ tubules.[32] The composition, origin, and function of the saccules are unknown, but possibly similar structures have been described for the light organs of certain macrourid fish.[35-37] The high density of bacterial cells, due to their retention within the saccules, could lead to competition for nutrients and oxygen, resulting in slower growth; a higher concentration of the *P. leiognathi* autoinducer for luminescence, resulting in a sustained potential for a high level of light production; and possibly the observed repression of flagellar synthesis.[32] Distortion of cell shape occurs, as noted by Bassot,[24] apparently as a result of the high cell density in the light organs. In addition, apparently partially digested bacteria ("ghost" cells) have been observed in some cases,[24,32] and the bacteria sometimes occur in chains of up to 15 cells (Dunlap, unpublished observation).

The light-organ bacteria of *L. elongatus*, the species with the most highly pronounced light-organ sexual dimorphism (i.e., male light organ 100 times larger in volume than that of female for fish 100 mm in standard length),[16] exhibit a distinctively different cell morphology[15] from that seen for the bacteria of most leiognathids. Bacteria from the light organs of both male and female specimens are uncharacter-

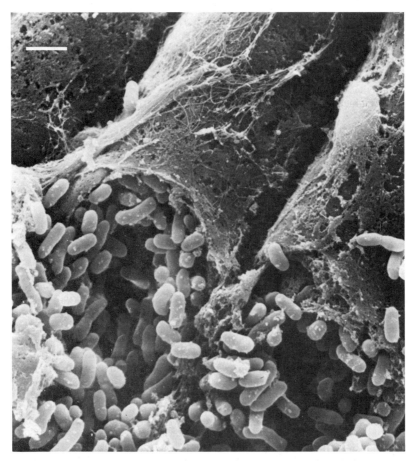

**FIGURE 2.** SEM appearance of the characteristic coccobacilloid cells of *Photobacterium leiognathi* seen in light-organ tubules of male and female specimens of *Gazza minuta* and most other leiognathid species. Bar = 5 μm.

istically elongate and have a somewhat irregular surface (FIGURES 4 and 5). Initial attempts to culture the light-organ bacteria from *L. elongatus* (possibly from male specimens) were unsuccessful,[15] which led to speculation that this species represents a transition, within the Leiognathidae, from harboring facultative symbionts to harboring obligately symbiotic bacteria.

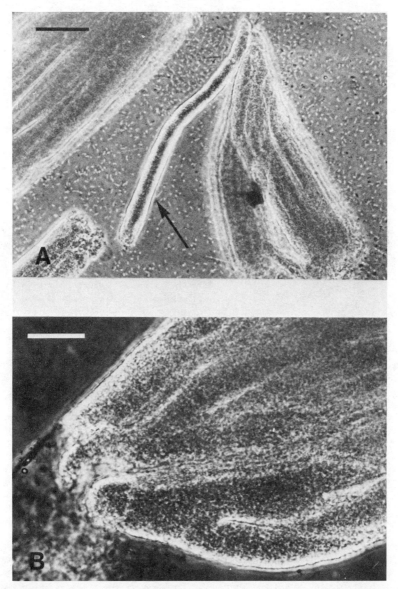

**FIGURE 3.** (A) Phase contrast micrograph of light-organ saccules (arrow) containing *Photobacterium leiognathi,* from one light-organ tubule of *Gazza minuta.* Saccules are fragile and are apparently destroyed in SEM preparation. Bar = 75 μm. (B) Close-up of end of a cluster of light-organ saccules from one tubule. Bar = 20 μm. (Reproduced from Reference 32 with permission.)

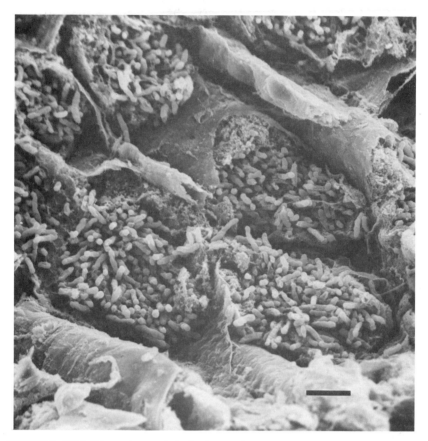

**FIGURE 4.** SEM of the uncharacteristically elongate, irregularly shaped cells of *Photobacterium leiognathi* occurring in light-organ tubules of male specimens of *Leiognathus elongatus.* Bar = 10 μm.

However, more recent attempts to culture the bacteria from light organs of male and female specimens of *L. elongatus* were successful,[9,17] indicating that this species, like all other examined leiognathids, harbors facultatively symbiotic, not obligately symbiotic, light-organ bacteria. Nonetheless, distinct differences are present between bacteria from light organs of male and female *L. elongatus.* Bacteria from the male fish grow very poorly on primary isolation and die readily between transfers, whereas bacteria from the female fish grow readily on primary isolation and survive as well in culture as bacteria from most other leiognathid species.[17] A similar case was observed for bacteria from the light organs of male and female specimens of *L. rivulatus,* another species that bears very sexually dimorphic light organs,[16] except that after primary isolation, bacteria from both male and female fish grew readily and survived well in culture.[17]

The relationship between light-organ dimorphism and host species- and sex-specific strain variation in bacterial morphology and cultivability is not clear, but the presence of distinct strain variation in leiognathid light-organ bacteria suggests that profound physiological (or other) differences exist in the light organs of different leiognathid species. Strain variation in the expression and activity of the bacteriocuprein (copper-zinc superoxide dismutase) of *P. leiognathi* also occurs and may be related to the species, sex, and geographical source of the host fish.[18,19] How such strain variation arises, perhaps through selection of variants among the population of *P. leiognathi* or by adaptation of *P. leiognathi* to the specific conditions of a given species' light organ, is an intriguing question. Further studies on *L. elongatus,* with its pronounced sexual dimorphism and distinct bacterial strain variation, may provide answers.

## PHYSIOLOGICAL STATE OF THE LEIOGNATHID LIGHT-ORGAN BACTERIA

Very little is presently known about the physiological activity of *P. leiognathi* in leiognathid light organs, or that of any light-organ bacterium in association with its

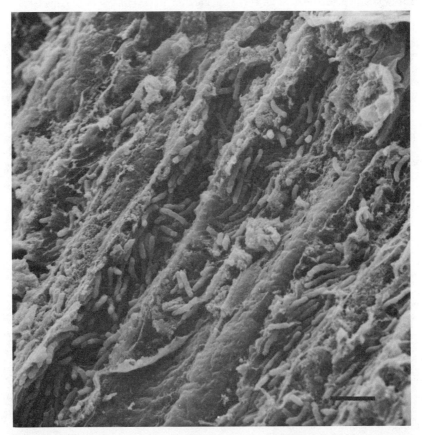

**FIGURE 5.** SEM of the uncharacteristically elongate, irregularly shaped cells of *Photobacterium leiognathi* occurring in light-organ tubules of female specimens of *Leiognathus elongatus.* Bar = 10 μm.

host. Such information, however, is crucial to understanding what controls two primary activities of *P. leiognathi* in the symbiosis, growth and light production. From the point of view of the fish, which supplies the bacteria with nutrients for growth and which requires bacterial light for its bioluminescence displays, a low rate of bacterial growth and a high level of bacterial light production would be adaptive. In this regard, Nealson proposed two alternative models for this symbiosis which are consistent with high levels of light production.[33] The first involves rapid growth of the bacteria in the light organ and their release into the gut tract of the fish where they are digested so that the fish regains nutrients expended in their growth. The second model involves restriction of the bacterial growth rate by limitation of nutrients supplied to the bacteria. Using the frequency-of-dividing-cells approach of Newell and Christian,[38] Dunlap estimated the doubling time of light-organ bacteria of one leiognathid specimen to be 23 hours.[32] This slow rate of growth is roughly consistent with that estimated, by another method, for *V. fischeri* in monocentrid fish light organs (7.5 to 135 hours per doubling),[39] and for the unidentified (not yet cultured) bacteria from anomalopid fish light organs (8 to 23 hours).[39] A more detailed analysis of the growth rate of leiognathid light-organ bacteria, and of their fate within and after release from the light organ, is needed before an energy budget for the symbiosis can be made.

To obtain data on the luminescence activity of *P. leiognathi* in the symbiosis, Dunlap examined bacteria taken directly from the light organs of several freshly captured leiognathids.[32] The level of light production was consistently high, approximately $10^4$ quanta sec$^{-1}$ cell$^{-1}$, a level 10 to 100 times higher than maximum values of luminescence produced by *P. leiognathi* grown under typical culture conditions[32] and 10 times higher than that of bacteria (also identified as *P. leiognathi*)[30,31] taken directly from the light organ of a specimen of the apogonid fish *Siphamia cephalotes* ($10^3$ quanta second$^{-1}$ cell$^{-1}$).[31] For the leiognathid bacteria, luminescence accounted for 1.7 to17% of the oxygen taken up[32] (the range is given because the *in vivo* quantum yield for oxygen in the luminescence reaction is not known but probably falls between 0.1 and 1.0).[40,41] Thus, the physiological state of *P. leiognathi* in the leiognathid light organs can be partially characterized as one of slow growth and high light production.

## PHYSIOLOGICAL CONTROL OF LIGHT PRODUCTION IN *P. LEIOGNATHI*

That symbiotic light-organ bacteria are actively producing light in the symbiosis indicates they are consuming oxygen, one of the substrates in the luciferase-mediated bioluminescence reaction.[2] For the monocentrid-*V. fischeri* symbiosis, a model for oxygen limitation of bacterial growth by the host fish has been proposed,[33] based in part on the fact that for *V. fischeri* in culture, growth limitation by low oxygen leads to an enhancement of luminescence.[42] In contrast for *P. leiognathi* in culture, a low oxygen tension results in a low level of light production,[42] so in the symbiosis a high level of luminescence most likely requires a high level of oxygen in the light organ.[33] However, compared to the relatively well-vascularized monocentrid light organ,[34] few capillaries are seen for the leiognathid light organ,[24] which suggests that direct delivery of oxygen from the circulatory system to the light organ would not be adequate to support a high level of luminescence in *P. leiognathi*.

Recently, McFall-Ngai resolved this anomaly by demonstrating that the bacteria in the leiognathid light organ could obtain a substantial amount of the oxygen for luminescence from the fish gas bladder.[43] The gas bladder contains a relatively high level of oxygen, approximately 20 to 25% by volume (McFall-Ngai, unpublished data). Further, the gas bladder bears a rete system for oxygen delivery from the circulatory system, and the gas bladder interfaces with the light organ across a thin, oxygen-permeable membrane.[26,43] Replacement of gases in the gas bladder with dinitrogen or oxygen results in immediate cessation or resumption, respectively, of luminescence by bacteria in the light organ (McFall-Ngai, unpublished data).

The gas bladder is therefore postulated to be a central component of the leiognathid light-organ symbiosis[43] (see FIGURE 6). Most areas of the gas bladder lining have an

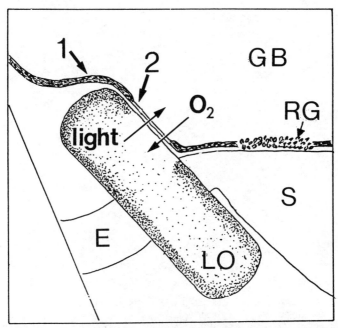

**FIGURE 6.** The gas bladder as a central component of the leiognathid light-organ symbiosis. The interface between light organ and gas bladder: (1) purine-rich gas bladder lining; (2) thin, oxygen-permeable membrane separating light organ (LO) from the gas bladder (GB) red gland (RG), a counter-current exchange system for the secretion of oxygen into the gas bladder; stomach (S), and esophagus (E). See text for details.

unusually high concentration of purine which serves both for the reflection of bacterial light and for the prevention of diffusion of gases.[26] Purine is lacking, however, in those areas of the lining where light is transmitted: (1) at the anterior end of the gas bladder where light enters from the light organ, and (2) at the posterior end where the reflected and scattered light exits the gas bladder to enter the hypaxial musculature.[26] The posterior clear area of the gas bladder is cupped by transparent bone that has a low coeffient of diffusion,[26] whereas at the anterior end, the oxygen-rich gas bladder space is separated from the light organ by only a thin, transparent membrane (FIGURE 6). Oxygen diffuses across the membrane from the gas bladder into the light organ, and consumption of oxygen by the light-organ bacteria is thought to create a steep diffusion gradient. Further, the dorsal purine-rich muscular shutter of the light organ, which controls release of light into the gas bladder, may also regulate oxygen delivery to the light organ from the gas bladder. Thus, the fish may regulate oxygen delivery from the gas bladder to the bacteria metabolically (rete system) and dynamically (muscular shutter). A high rate of oxygen delivery would be consistent with the high level of luminescence ($10^4$ quanta second$^{-1}$ cell$^{-1}$) of bacteria taken directly from the leiognathid light organ.[32]

Since luminescence levels reflect the physiological conditions (e.g., oxygen tension)[42] under which *P. leiognathi* is grown, one approach to understanding how the host fish could control the production of bacterial light is to vary the conditions under which *P. leiognathi* is grown in culture. Those conditions that lead to high levels of luminescence ($10^4$ quanta second$^{-1}$ cell$^{-1}$) might be indicative of or mimic the actual

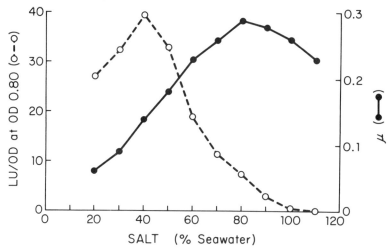

**FIGURE 7.** The luminescence (LU/OD at OD 0.8, light units per cell density) at peak luminescence and growth rate ($\mu$) of *Photobacterium leiognathi* grown in medium of increasing osmotic concentration (expressed as salt concentration relative to that of seawater). (Reproduced from Reference 44 with permission.)

physiological conditions of the light organ. With this approach, Dunlap identified osmolarity as a primary factor of the growth environment that differentially controls the luminescence and growth rate of *P. leiognathi*[44] (FIGURE 7). Lower osmolarity stimulates the luminescence and restricts the growth rate of this bacterium, whereas higher osmolarity decreases luminescence sharply but stimulates growth.[44] Peak light production occurs at an osmotic concentration very similar to that of fish tissue (400 mOsm), and luminescence at this osmolarity is equal to that of bacteria taken directly from the fish light organ ($10^4$ quanta second$^{-1}$ cell$^{-1}$).[32,44] In the medium (400 mOsm) that gave optimal luminescence, the level of extractable luciferase was also increased, by 8- to 10-fold, whereas rates of oxygen uptake and growth were 2- to 4-fold less than in the medium (800 mOsm) that gave optimal growth.[44] Also, the percentage of oxygen consumed due to luminescence, 2 to 20%, in the medium of lower osmotic concentration was very similar to that of bacteria taken directly from the light organ

(1.7 to 17%).[44] All tested strains of *P. leiognathi* from leiognathid light organs responded to osmolarity of the growth medium in a very similar way, whereas other species of luminous bacteria and a saprophytic strain of *P. leiognathi* gave distinctly different, and characteristic, luminescence and growth responses to osmolarity.[17,44] Thus, low osmolarity may be a host control factor, specific for the leiognathid-*P. leiognathi* symbiosis, that serves to stimulate the luminescence and restrict the growth of *P. leiognathi* in the leiognathid light organ. As such, *P. leiognathi* can be envisioned as inhabiting two ecologically different, osmotically distinct environments: (1) the leiognathid light organ (low osmolarity), where growth is restricted, and consequently where luminescence is physiologically adaptive for survival; and (2) seawater and the gut tracts of fish (high osmolarity) where growth is more rapid, and luminescence is less important for survival.[44]

The mechanism by which osmolarity differentially controls luminescence and luciferase levels on the one hand and oxygen uptake and growth rate on the other remains unclear at present, but could involve both physiological regulation of respiratory and luminescence enzyme activity and regulation of luminescence gene expression.[44] In this regard, iron limitation has been proposed as a possible host factor restricting the growth and optimizing the luminescence of *V. fischeri* in monocentrid light organs;[45] iron limitation could be functioning through inhibition of the respiratory system. In *P. leiognathi* from leiognathid light organs, iron limitation restricts the growth rate in culture and stimulates luminescence, but the effect is much less pronounced than that of low osmolarity.[17] A detailed understanding of the osmotic response in *P. leiognathi* will require much further work on the osmoregulation and respiratory enzyme activity in this bacterium, but such work may reveal the adaptive value for *P. leiognathi* of the production of high levels of luminescence under low osmolarity conditions. Further attempts to mimic in culture the slow growth rate and high level of light production of *P. leiognathi* in the symbiosis, perhaps through the use of continuous (chemostat) cultures, may lead to the identification of other possible host control factors.

## CONCLUSION

The bioluminescent light-organ symbiosis between *P. leiognathi* and leiognathid fish presents several opportunities for research on the ecological and physiological interactions between a specific symbiotic bacterium and its animal host. With the studies highlighted here on the initiation of the association and on the physiological control of light production in *P. leiognathi*, we have begun to address questions on the specificity and development of the association and on the regulation of bacterial activities in the symbiosis.

### REFERENCES

1. HERRING, P. J. & J. G. MORIN. 1978. Bioluminescence in fishes. *In* Bioluminescence in Action. P. J. Herring, Ed.: 273-329. Academic Press. New York, N.Y.
2. NEALSON, K. H. & J. W. HASTINGS. 1979. Bacterial bioluminescence: its control and ecological significance. Microbiol. Rev. **43**: 496-518.

3. HASTINGS, J. W. & K. H. NEALSON. 1981. The symbiotic luminous bacteria. *In* The Prokaryotes. M. P. Starr, H. Stolp, H. G. Truper, A. Balows & H. G. Schlegel, Eds.: 1332-1345. Springer. New York, N.Y.

4. FUKASAWA, S. & P. V. DUNLAP. 1986. Identification of luminous bacteria isolated from the light organ of the squid, *Doryteuthis kensaki.* Agric. Biol. Chem. **50:** 1645-1646.

5. TIEWS, K. & P. CACES-BORJA. 1965. On the availability of fish of the family Leiognathidae Lacepede in Manila Bay and San Miguel Bay and on their accessibility to controversial fishing gears. Philippine J. Fish. **7:** 59-83.

6. KUHLMORGEN-HILLE, G. 1974. Leiognathidae. *In* FAO Species Identification Sheets for Fishery Purposes, Eastern Indian Ocean (Fishing Area 57) and Western Central Pacific (Fishing Area 71). W. Fischer & P. J. P. Whitehead, Eds. **2.** Food and Agriculture Organization of the United Nations. Rome, Italy.

7. PAULY, D. & S. WADE-PAULY. 1981. An annotated bibliography of slipmouths (Pisces: Leiognathidae). ICLARM Bibliographies **2.** International Center for Living Aquatic Resources Management. Manila, Philippine Islands.

8. JAMES, P. S. B. R. 1975. A systematic review of the fishes of the family Leiognathidae. J. Mar. Biol. Assoc. India **17:** 138-172.

9. DUNLAP, P. V. & M. J. MCFALL-NGAI. 1984. *Leiognathus elongatus* (Perciformes: Leiognathidae): two distinct species based on morphological and light organ characters. Copeia **1984(4):** 884-892.

10. BOISVERT, H., R. CHATELAIN & J.-M. BASSOT. 1967. Etude d'un *Photobacterium* isole de l'organe lumineux des poissons Leiognathidae. Ann. Inst. Pasteur Paris **112:** 520-524.

11. HASTINGS, J. W. & G. MITCHELL. 1971. Endosymbiotic bioluminescent bacteria from the light organ of pony fish. Biol. Bull. **141:** 261-268.

12. REICHELT, J. L., K. NEALSON & J. W. HASTINGS. 1977. The specificity of symbiosis: pony fish and luminescent bacteria. Arch. Microbiol. **112:** 157-161.

13. BAUMANN, P. & L. BAUMANN. 1981. The marine gram-negative eubacteria. *In* The Prokaryotes. M. P. Starr, H. Stolp, H. G. Truper, A. Balows & H. G. Schlegel, Eds.: 1302-1311. Springer. New York, N.Y.

14. MCFALL-NGAI, M. J. & P. V. DUNLAP. 1983. Three new modes of luminescence in the leiognathid fish *Gazza minuta* (Perciformes: Leiognathidae): discrete projected luminescence, ventral body flash and buccal luminescence. Mar. Biol. **73:** 227-237.

15. HANEDA, Y. & F. I. TSUJI. 1976. The luminescent system of pony fishes. J. Morphol. **150:** 539-552.

16. MCFALL-NGAI, M. J. & P. V. DUNLAP. 1984. External and internal sexual dimorphism in leiognathid fishes: morphological correlates of sex-specific bioluminescent signalling. J. Morphol. **182:** 71-83.

17. DUNLAP, P. V. 1984. The ecology and physiology of the light organ symbiosis between *Photobacterium leiognathi* and ponyfishes. Ph.D. Dissertation. University of California. Los Angeles, Calif.

18. DUNLAP, P. V. & H. M. STEINMAN. 1986. Strain variation bacteriocuprein superoxide dismutase from symbiotic *Photobacterium leiognathi.* J. Bacteriol. **165:** 393-398.

19. STEINMAN, H. M. & P. V. DUNLAP. 1986. Distribution of bacteriocuprein superoxide dismutases in prokaryotes. *In* Superoxide and Superoxide Dismutase in Chemistry, Biology and Medicine. G. Rotilio, Ed.: 270-273. Elsevier. Amsterdam, Holland.

20. HARMS, J. W. 1928. Bau und entwicklung eines eigenartigen leuchtorgans bei *Equula* spec. (Teleost.). Z. Wiss. Zool. **131:** 157-179.

21. HANEDA, Y. 1940. On the luminescence of the fishes belonging to the family Leiognathidae of the tropical Pacific. Palao Trop. Biol. Stn. Studies **2:** 29-39.

22. HANEDA, Y. 1950. Luminous organs of fish which emit light indirectly. Pacif. Sci. **4:** 214-277.

23. AHRENS, G. 1965. Untersuchungen am leuchtorgan vom *Leiognathus klungzingeri* (Steindachner). Z. Wiss. Zool. **173:** 90-113.

24. BASSOT, J.-M. 1975. Les organes lumineux a bacteries symbiotiques de quelques Teleosteens leiognathides. Arch. Zool. Exp. Genet. **116:** 359-373.

25. HASTINGS, J. W. 1971. Light to hide by: ventral luminescence to camouflage the silhouette. Science **173:** 1016-1017.

26. McFALL-NGAI, M. J. 1983. Adaptations for reflection of bioluminescent light in the gas bladder of *Leiognathus equulus* (Perciformes: Leiognathidae). J. Exp. Zool. **227:** 23-33.
27. RUBY, E. G. & K. H. NEALSON. 1976. Symbiotic association of *Photobacterium fischeri* with the marine luminous fish *Monocentris japonica:* a model of symbiosis based on bacterial studies. Biol. Bull. **151:** 574-586.
28. YAMADA, K., M. HAYGOOD & H. KABASAWA. 1979. On fertilization and early development in the pine-cone fish, *Monocentris japonicus.* Keikyu Aburatsubo Mar. Park Aquarium Annu. Rep. **10:** 31-38.
29. LEIS, J. M. & S. BULLOCK. 1986. The luminous cardinalfish *Siphamia* (Pisces, Apogonidae): development of larvae and the luminous organ. *In* Proceedings of the 2nd International Conference on Indo-Pacific Fishes. T. Uyemo, R. Arai, T. Taniuchi & K. Matsuura, Eds.: 703-714. Ichthyological Society of Japan. Tokyo, Japan.
30. REICHELT, J. L. & P. BAUMANN. 1973. Taxonomy of the marine, luminous bacteria. Arch. Mikrobiol. **94:** 283-330.
31. FITZGERALD, J. M. 1978. Studies on the taxonomy and bioluminescence of some luminous, marine bacteria. Ph.D. Dissertation, Monash University. Victoria, Australia.
32. DUNLAP, P. V. 1984. Physiological and morphological state of the symbiotic bacteria from light organs of ponyfish. Biol. Bull. **167:** 410-425.
33. NEALSON, K. H. 1979. Alternative strategies of symbiosis of some marine luminous fishes harboring light-emitting bacteria. Trends Biochem. Sci. **4:** 105-110.
34. TEBO, B. M., D. S. LINTHICUM & K. H. NEALSON. 1979. Luminous bacteria and light emitting fish: ultrastructure of the symbiosis. BioSystems **11:** 269-280.
35. KISHITANI, T. 1930. Studien uber die leuchtsymbiose in *Physiculus japonicus* Hilgendorf, mit der beilage der zwei neuen arten der leuchtbakterien. Sci. Rep. Tohoku Univ. Sect. 4 **5:** 801-823.
36. YASAKI, Y. & Y. HANEDA. 1935. On the luminescence of the deep-sea fishes, family Macrouridae. Oyo Dobutsu Zasshi Tokyo **7:** 165-177.
37. HANEDA, Y. 1938. Uber den leuchtfisch, *Malacocephalus laevis* (Lowe). Jpn. J. Med. Sci. **5:** 355-366.
38. NEWELL, S. Y. & R. R. CHRISTIAN. 1981. Frequency of dividing cells as an estimator of bacterial productivity. Appl. Environ. Microbiol. **42:** 23-31.
39. HAYGOOD, M. G., B. M. TEBO & K. H. NEALSON. 1984. Luminous bacteria of a monocentrid fish (*Monocentris japonicus*) and two anomalopid fishes (*Photoblepharon palpebratus* and *Kryptophaneron alfredi*): population sizes and growth within the light organs, and rates of release into the seawater. Mar. Biol. **75:** 249-254.
40. HASTINGS, J. W. & K. H. NEALSON. 1977. Bacterial bioluminescence. Annu. Rev. Microbiol. **31:** 549-595.
41. KARL, D. M. & K. H. NEALSON. 1980. Regulation of cellular metabolism during synthesis and expression of the luminous system in *Beneckea* and *Photobacterium*. J. Gen. Microbiol. **117:** 357-368.
42. NEALSON, K. H. & J. W. HASTINGS. 1977. Low oxygen is optimal for luciferase synthesis in some bacteria: ecological implications. Arch. Microbiol. **112:** 9-16.
43. McFALL-NGAI, M. J. 1981. New function for the teleost gas bladder: a source of oxygen for the bioluminescent bacteria in the leiognathid light organ. Abstr. West. Soc. Nat. 62nd Annu. Meet., Santa Barbara, Calif.: 31.
44. DUNLAP, P. V. 1985. Osmotic control of luminescence and growth in *Photobacterium leiognathi* from ponyfish light organs. Arch. Microbiol. **141:** 44-50.
45. HAYGOOD, M. G. & K. H. NEALSON. 1984. Effects of iron on bacterial growth and bioluminescence: ecological implications. *In* Current Perspectives in Microbial Ecology. M. J. Klug & C. A. Reddy, Eds.: 56-61. American Society of Microbiology. Washington, D.C.

# Molecular Interactions in Endosymbiosis between Legume Plants and Nitrogen-Fixing Microbes

DESH PAL S. VERMA AND JOHN STANLEY[a]

*Center for Plant Molecular Biology*
*Department of Biology*
*McGill University*
*1205 Docteur Penfield Avenue*
*Montreal, Quebec, Canada H3A 1B1*

Root-nodule symbiosis requires a highly regulated expression of genes in a eucaryotic host plant and its procaryotic microsymbiont. *Rhizobium* can act as a dimorphic infectious procaryote which changes from one form (free-living gram-negative rod) to another (pleomorphic bacteroid) as part of its life cycle, via a definable progression of gene expression. The legume host harvests procaryotic nitrogenase activity via a tightly regulated system of "contained infection," compartmentalizing rhizobia "outside" its own cytoplasm through a specialized membrane system. In this state, *Rhizobium* bacteroid resembles a nitrogen-fixing organelle, or "nitroplast."

The "endosymbiotic origin" model of eucaryotes postulates that mitochondrial genomes descend from those of once-free-living aerobic bacteria, while chloroplast genomes similarly originate from those of endosymbiont blue-green algae.[41] Although endosymbiosis is facultative for both rhizobia and legumes, it is possible that a similar trend towards cell fusion is a dynamic in the evolution of the root-nodule symbiosis. However, the basic constraints on the process of nitrogen fixation may always require the formation of an organ.

Both organogenesis of the root nodule and modified metabolism of the legume host result from its endosymbiotic interaction with *Rhizobium*. A set of plant genes involved in symbiosis directs synthesis of nodule-specific host proteins, nodulins[36] (see Reference 61). These have been characterized from soybean,[36] pea,[9] and alfalfa.[31] Coordinate expression of Sym genes in the microsymbiont directs the infection process and subsequently results in *de novo* synthesis of nodule-specific procaryotic proteins, bacteroidins, which are characteristic of the differentiated bacteroid state.[65] Ammonia, the product of the reduction of dinitrogen, is excreted from bacteroids to the host plant cell, and assimilated initially to glutamine: and thence to either ureides (allantoin, allantoate) or amides (asparagine). Tropical legumes are ureide producers, while temperate-zone legumes are amide producers.

Rhizobia fall into two markedly different groups. The fast-growing rhizobia are generally endosymbionts of temperate-zone legumes such as *Trifolium, Medicago,* or

[a] Present affiliation: University of Geneva, LBMPS, 1 Chemin de l'Imperatrice, 1292 Chambesy, Geneva, Switzerland.

*Pisum.* Slow-growing rhizobia, termed bradyrhizobia,[27] typically nodulate tropical legumes, and include the agronomically important *B. japonicum,* a group of (at least) two divergent slow-growing symbiotic bacteria of soybean.[57] Conserved genes essential for nodulation (*nod*) are encoded in a linkage group together with nitrogen fixation (*nif*) genes, located on large symbiotic plasmids in *R. trifolii, R. leguminosarum,* and *R. meliloti.* [4,38,19,55] However, they appear to be integrated in chromosome(s) of bradyrhizobia. In *B. japonicum,* structural *nif*HDK genes are organized in two[28] rather than the single transcriptional unit of fast-growing rhizobia.[54] The common nodulation genes (*nod*ABC) are found 10 kbp upstream of the *nif*A gene in *B. japonicum* genotype stI, (strain 110).[30]

# CONSTRAINTS ON SYMBIOSIS

In addition to the basic constraints on nitrogen fixation (i.e., the sensitivity to oxygen and the demand for high energy), there are other physiological constraints on this symbiosis. The development of this process, although a continuum of interactions, can be usefully divided into three phases: preinfection (I), infection (II), and nodulation function (III).

## Constraints on Preinfection and Infection

Constraints on preinfection and infection include the abiotic factors of combined nitrogen availability, the complex interactive phenomenon of competition for nodulation, and host specificity. Nitrate has a strong inhibitory effect on all stages of preinfection. Root-hair curling, infection-thread initiation and growth, and nodule initiation are all affected by nitrate either in rooting medium or by foliar application.

The plant host provides a rhizosphere that both supports and amplifies the saprophytic phase of the *Rhizobium* life cycle, and can evidently exert a selective effect on a mixed population of rhizobia.[38] The microsymbiont genome determines interstrain competition for nodulation via determinants for species cultivar specificity, bacteroidin production, or resistance to a variety of environmental factors such as combined nitrogen, pH, salinity, temperature, and dessication. Distinct *Rhizobium* genetic determinants govern the preinfection (phase I), infection/nodule formation (phase II), and nodule function (phase III). Interstrain competition determines the outcome of phase I; and as a consequence, a strain of *Rhizobium* more effective in phase III is usually displaced by relatively ineffective but more competitive indigenous rhizobial strains.[10]

## Constraints on Nodule Function

Symbiotic nitrogen fixation by the effectively nodulated legume is also subject to several major constraints—application of combined nitrogen results in oxidation of leghemoglobin (Lb) and the rapid senescence of nodule tissues. Ammonia is a nitrogenase inhibitor in free-living diazotrophic bacteria, though not in bacteroid rhizobia,[8] while nitrite inhibits nitrogenase activity in both bacteroid and asymbiotic *Bradyrhizobium.*[47] Nitrate is taken up rapidly and metabolized by root and nodule tissue. This provides a competitive sink for photosynthate, and diverts reductant in nodules to (host) nitrate assimilation rather than (bacteroid) nitrogen fixation. Hence, legumes exhibit a reciprocal systemic relationship between symbiotic nitrogen fixation and nitrate reduction, with photosynthate as the limiting resource for both.

The high energy requirement of the symbiosis is exemplified by the direct dependence of this process on photosynthesis; and thus, factors affecting photosynthesis, leaf area, or carbon dioxide concentration influence symbiotic nitrogen fixation (see Reference 49). The energy cost of symbiotic fixation as estimated (see References 40 and 63) includes the costs of nodule organogenesis, nodule maintenance, and the assimilation of fixed nitrogen.

An uptake hydrogenase system in some Hup$^+$ rhizobia and bradyrhizobia benefits nodule energy balance by scavenging oxygen and protecting the oxygen-sensitive nitrogenase, oxidizing hydrogen, and recovering the energy lost in hydrogen synthesis by nitrogenase.[17] The Hup$^+$ phenotype is affected by the plant, and various hosts produce Hup$^+$ or Hup$^-$ phenotypes with rhizobia of Hup$^+$ genotypes.[5]

## THE ROLE OF THE HOST IN FORMATION OF ENDOSYMBIOSIS

### Host Genes Defined by Classical Genetics

Classical genetics has defined a number of symbiotic loci in legumes (see Reference 63). For example, in soybean, *rj1 rj1,* a homozygous recessive locus, dictates nonnodulation via a root-specific factor, by most *B. japonicum* strains which normally infect this host.[68,12] However, the *rj1 rj1*-blocked nodulation phenotype can be overcome by *B. japonicum* strains with high capsular polysaccharide, high indoleacetic acid production, and inability to metabolize nitrate.[26] A number of dominant soybean genes condition Fix$^-$ (ineffective non-nitrogen-fixing) or abnormal nodules with the *B. japonicum* homologous for the wildtype plant. *Rj2* conditions small white nodules with *B. japonicum* strains C1 and 122,[16] *Rj3* gives similar nodules with *B. japonicum* strain 33, and *Rj4* conditions cortical proliferations with *B. japonicum* strain 61.[33] Similar symbiotic phenotypes can be generated by single transposon (Tn5) insertion in the rhizobial genome,[53,57] indicating that the development of the effective root nodule is an interdependent function of both procaryotic and eucaryotic genomes. A soybean nodulation-regulatory mutant *nts* 382 gives profuse nodulation (1000 nodules/plant) with or without nitrate in the rooting medium.[13] Expression of this Mendelian locus in contrast to *rj1* is demonstrated by grafting experiments to be shoot controlled.

A single Mendelian locus conditions ineffectiveness of nodulation of contemporary genetically improved soybean cultivars by fast-growing rhizobia from China which effectively nodulate cowpea and the unimproved Asian soybean var. Peking. Genetic analysis indicates that Peking carries the recessive allele governing effective symbiosis by these rhizobia while modern cultivars carry the dominant allele governing their ineffectiveness.[15]

## Host Genes and Their Function Defined by Molecular Biology

### Inter- and Intracellular Compartmentalization in Root Nodules

In determinate nodules of tropical legumes (reviewed in Reference 52), about half of the nodule cells are uninfected and appear to have a separate metabolic role in nitrogen assimilation. Nodulins are involved in root nodule organogenesis, maintenance, and function. In the soybean system, *in vitro* translation of polysomal messenger RNA of mature nodules followed by immunoprecipitation with nodule antibodies detects some 20 nodulins.[36] Up to 40 moderately abundant plant mRNAs are found to be nodule specific,[3] and cDNA cloning revealed that about 8% of the total transcript is comprised of four moderately abundant mRNA species, other than the superabundant Lb mRNAs.[22] TABLE 1 summarizes the current understanding of soybean nodulins. Analysis of the nodules from temperate legumes (e.g., pea)[24] indicated 21 nodulins comprising 8 major products included 4 Lbs and 13 minor nodulins.

### Centrality of the Subcellular Compartment Enclosing Bacteroids

A single infected cell of a legume root can contain up to 20,000 bacteroids,[14] each enclosed in peribacteroid membrane (pbm) sacs initially derived from plant plasma membrane during invasion. This subcellular compartment excludes these potentially pathogenic bacteria from the host cell cytoplasm. Thus, the peribacteroid membrane has a central role as the primary interface between micro- and macrosymbiont. Such "periendosymbiont" membranes are a general feature also in mycorrhizal and cyanobacterial (*Gunnera*) endosymbiosis. There is a very marked increase (up to 40-fold) in membrane proliferation in infected soybean cells over the 8-9 days of nodule initiation.[66] The synchronous increase in bacteroid number and pbm appears to be related to points of adhesion between bacteroid outer membrane and pbm.[50] The subsequent stability of pbm also depends on the endosymbiont. The *B. japonicum* Fix⁻ strain 61A24 produces stable bacteroids, but pbm is very unstable in this association and is lost early after nodule development. A similar loss of bacterial pbm-stabilizing function is observed in a number of other cases (Fix⁻ mutant or incompatible strains). It has been observed that in the nodules formed by strain 61A24, there is a 100-fold increase in the soybean phytoalexin, glyceollin I.[67] By compartmentalization with pbm, this host defense response to bacterial invasion may be avoided.

The phospholipid-to-protein ratio of pbm is higher than that of plasma membrane, and its phospholipid composition resembles that of endoplasmic reticulum.[51,43] Selective and directional transport functions of pbm must include the "importation" of ammonia to host cell cytoplasm and the "export" to bacteroids of substrates for their respiration. It can be assumed that hitherto uncharacterized molecules with signal- and symbiosis-specific physiological functions traverse pbm in both directions, facilitating "late" communication between host and bacteroids. Analysis of pbm, with reference to the plasma membrane of uninfected root cells by sodium dodecyl sulfate-polyacrilamide gel electrophoresis (SDS-PAGE) and Western blotting with a polyclonal nodule-specific antiserum, identifies pbm-bound nodulins.[21] One such nodulin could be the second form of choline kinase unique to soybean pbm.[43] The peribacteroid fluid (pbf) internal to pbm similarly contains nodulins. Both pbm and pbf nodulins can be precipitated from nodule polyA$^+$ mRNA *in vitro* translation products using respective antisera.[21]

One pbm nodulin that has been analyzed in detail is nodulin-24. DNA sequence analysis of the gene encoding this polypeptide (apparent molecular weight 24 kD) reveals a coding region interrupted by four introns.[29] The gene appears to have been generated by duplications of a unit resembling an insertion sequence. These exons encode three central hydrophobic domains in the protein. Nodulin-24 is shown to be a component specific to pbm, using immunogold electron microscopy with antiserum raised against a synthetic 18 amino acid peptide.[21] Immunocytochemistry using nodule-specific anti-pbm serum confirms that other membrane-bound nodulins are similarly specifically targeted to the pbm in root-nodule biogenesis.

TABLE 1. Nodule-Specific Host Genes and Their Products Identified from Soybean

| Nodulin | Gene Isolated | Subcellular Location | Function | References |
|---|---|---|---|---|
| Leghemoglobin | | | | |
| Lba | Yes | Infected cell cytoplasm[a] | Oxygen carrier | 35 |
| Lbc$_1$ | Yes | Infected cell cytoplasm[a] | Oxygen carrier | |
| Lbc$_2$ | Yes | Infected cell cytoplasm[a] | Oxygen carrier | |
| Lbc$_3$ | Yes | Infected cell cytoplasm[a] | Oxygen carrier | |
| Nodulin-35 | Yes | Uninfected cell[a] (peroxisomes) | Uricase | 45 |
| Nodulin-24 | Yes | Infected cell[a] (peribacteroid membrane) | Unknown | 29 21 |
| Nodulin-23 | Yes | Peribacteroid membrane[b] | Unknown | 42 and unpublished data |
| Nodulin-44 | Yes | Cytoplasm[b] | Unknown | unpublished data |
| Nodulin-27a | Yes | — | — | — |
| Nodulin-100 | No | — | — | — |

[a] Determined by immunocytochemistry.
[b] Determined by differential immunoprecipitation using antibodies against peribacteroid membranes.

*Role of Soluble Nodulins in Nodule Function*

Confinement of the aerobic bacterial endosymbiont at high local density in root nodules produces a constraint on oxygen supply, requiring an oxygen carrier. This requirement is further complicated by the sensitivity of the bacterial nitrogenase to free oxygen. The superabundant nodulin leghemoglobin (Lb) is found in all nitrogen-fixing nodules, as well as in fully differentiated Fix$^-$ nodules.[23] Efficient Lb-facilitated bacteroid respiration at extremely low free oxygen concentrations is a characteristic of all natural symbioses involving *Rhizobium*.[1] Consequently, Lb is a common (C-) nodulin.[63] Soybean contains four major Lb proteins, Lba, Lbc$_1$, Lbc$_2$, and Lbc$_3$, which differ slightly in amino acid sequence. They are induced at different times.[64] Leghemoglobins are found only in host-cell cytoplasm and not internal to pbm. They are, therefore, substantially physically separated from the bacteroid surface.[65] Although the plant origin of the Lb apoprotein is clear, the postulated bacteroid origin of the heme moiety (see Reference 62) remains unclear. The first committed step of heme synthesis in rhizobia is catalyzed by $\delta$-ALA synthetase, but transposon insertion mutants of *R. meliloti*[37] or *B. japonicum*[25] for the relevant gene produce Fix⁻ or Fix⁺ symbiosis on the alfalfa or soybean hosts respectively.

The presence of globin in plants capable of forming nitrogen-fixing symbiosis is an enigma.[35] Plant Lbs appear to have diverged from animal globins over a billion years ago and to have been present in a common ancestor of both kingdoms.[11] Intragenic organization of plant globin genes differs from animal genes in the presence of a third intron separating coding sequences for the protein structural units. In soybean, the Lba, $c_1$, and $c_3$ genes occur with one pseudogene in a major chromosomal locus while the $Lbc_2$ gene occurs in a minor locus with a second pseudogene. There are two other loci which carry only truncated genes.

The second most abundant soybean nodulin, nodulin-35, is a uricase[6] localized by immunocytochemistry in peroxisomes of the uninfected nodule cells.[45] This enzyme is central to the nitrogen assimilation pathway of ureide-producing legumes.[2] Uricase II (nodule-specific) activity replaces the "uricase I" activity of young uninfected soybean roots. There is no immunological cross-reaction with the uninfected tissue.[6] The cDNA clone for nodulin-35 mRNA does not hybridize with young roots or leaves confirming that the uricase I activity of these tissues is the product of a different gene. Messenger RNA of nodulin-35 appears a few days after those for Lb and nodulin-24 or nodulin-23.[45] Soybean, cowpea, and lima bean nodule cells contain xanthine dehydrogenase, a nodulin catalyzing oxidation of hypoxanthine to xanthine and thence to uric acid under microaerobic conditions. Uric acid is finally converted into allantoin. Initial expression of nodulin-35 is independent of nitrogen fixation but its level in nodules depends on the microsymbiont strain (e.g., References 36 and 58). It has been shown that in sterile soybean callus tissue, the specific uricase activity increases (by increased biosynthesis) when the oxygen concentration is lowered, and reaches a maximum at 4-5% oxygen.[34] Thus, the synthesis of this nodulin probably results from the lowered oxygen availability due to bacterial proliferation in the nodule. Another nodulin involved in primary nitrogen assimilation is the nodule-specific glutamine synthetase as shown in *Phaseolus*.[32] This nodulin is localized in infected cells of soybean nodules.[65]

*Nodulin Gene Regulation*

Induction of nodulin genes characterized to date occurs prior to and is independent of the commencement of nitrogen fixation in nodules. Certain developmentally regulated eucaryotic genes have been shown to possess 5' *cis*-regulatory upstream sequences capable of binding *trans*-activator molecules. The 5' flanking regions of genes for soybean $Lbc_3$ and nodulins 23, 24, and 35 have been sequenced.[42,45] Except for the nodulin-35 gene, these genes share three regions of homology on their 5' ends. Their transcripts are coordinately induced and appear both earlier than the uricase mRNA and in infected rather than the uninfected cells where uricase is produced. The orientation and spacing of three consensus sequences[42] indicate that these genes could be positively regulated by a common *trans*-activator molecule. A preliminary experiment, using isolated nuclei from uninfected tissue and factors from nodules, appears to produce transcripts corresponding to some nodulins (Mauro and Verma, unpublished data).

## INTERDEPENDENCE OF MACRO- AND MICROSYMBIONT

### *Plant Signals in Infection*

Several lines of investigation provide evidence of molecular communication between the legume host and microsymbiont throughout the endosymbiotic process. A mutant of *B. japonicum,* HS11,[56] which exhibits delayed initiation of nodulation of soybean, can be "phenotypically reversed" to wild-type nodulation kinetics by preincubation in soybean root exudate or seed lectin (G. Stacey, personal communication). Similarly, growth on the root exudate of *Vigna sinensis* enhances the nodulation kinetics of cowpea *Bradyrhizobium* 32H1, but root exudates from host plants grown in 5 mM ammonium chloride do not elicit this response.[7]

Random transcription fusions of Mu-d (*kan, lac*) into the genome of *R. japonicum* (*fredii*) USDA 201 identified three insertion mutants for rhizobial genes under host plant control. Promoters of genes upstream of the inserted *Lac* gene fusion are specifically induced by root exudate of *Glycine* or *Phaseolus* hosts of USDA 201, but no other legumes.[46] The inducer in root exudate is a low molecular weight organic compound.

A low molecular weight compound present in alfalfa root exudate is similarly responsible for activation of the essential nodulation genes, nod*DABC,* of *R. meliloti.*[44] This compound has been identified to be a flavonoid, luteolin, commonly present in plants (S. Long, personal communication). The *nod* genes have been shown by saturation transposon mutagenesis to be absolutely required for curling of root hairs and invasion of host cells via infection threads. The four genes concerned have been shown by interspecific complementation and sequence analysis to be conserved and represent allelic equivalents of *nod* loci in other rhizobia.

### *Coordinated Differentiation in Nodule Organogenesis*

Development of the nodule to its fully differentiated form of intracellular symbiosis is a tightly coordinated process requiring the expression of many genes in both partners. Mutations have been obtained in rhizobia that uncouple host plant and bacterial symbiotic differentiation "pathways." Some of these mutations, e.g., in exopolysaccharide synthesis, map at a locus within the large cryptic plasmid of *R. meliloti.* They abolish bacteriophage sensitivity and root-hair curling or infection-thread formation on alfalfa. The mutants, however, still induce cell proliferation typical of nodule initiation.[20] Though *Sym* mutations in rhizobia have classically been assigned to the categories Nod⁻ (Inf⁻, noninvasive on the host plant) or Fix⁻ (no nitrogen fixation), these two categories do not describe the level of nodule differentiation within Fix⁻ nodules. For instance, a transposon-insertion nodule development mutant of *R. japonicum* (*fredii*) USDA 191 induces pseudonodules on all its legume hosts which remain meristematic.[58] Neither nodulin-35 nor Lb, marker proteins for complete nodule differentiation, are induced by this mutant. These results confirm the existence of a class of nodule differentiation (Dif) rhizobial mutants, as opposed to Fix⁻ nitrogenase mutants like *B. japonicum* SM5. The latter gives nearly normal levels of nodulins by the time nitrogen fixation would normally reach a maximum.[23]

# SENESCENCE AND PATHOGENICITY

## *Nodule Senescence*

The duration of root nodules limits the availability of symbiotically fixed nitrogen to the host plant and release of the endosymbiont into the soil. It can be shown that *Bradyrhizobium* sp. RC3200 cannot simultaneously grow and fix nitrogen in suspension culture. On this basis, a model has been proposed whereby a stochastic switch in cell state occurs in some fraction of a cell population maintained under low oxygen and limiting nitrogen to a nongrowing state wherein nitrogen fixation then occurs exclusively.[39] These specialized nongrowing cells are formally similar to endosymbiotic bacteroids. The nodule contains rhizobia in various stages of differentiation into mature bacteroids, and the issue of rhizobial release into the environment remains to be clearly established. However, it seems unlikely that rhizobia enter the symbiotic state without a positive selection pressure for their own establishment in the soil.

Indeterminate nodules contain a zone of senescent tissue at the base of the nodules and are regenerable unless their corresponding meristematic zone degenerates. Determinate nodules constitute a nonreplaceable tissue whose senescence begins at the center. Senescence is generally related to the growth cycle of the specific host legume. Hence, nodules survive for a single growth season on herbaceous annual legumes and may survive for more than a year on perennial legumes. Incompatible strains of *Rhizobium*, e.g., 61A24 (see above), may induce premature nodule senescence.

Nodule senescence can be brought about by various environmental stresses of plant growth as well, notably temperature, water, or light stress (reviewed in Reference 60). Photosynthate deprivation affects nodule nitrogen fixation drastically, and combined-nitrogen addition (particularly as nitrate) similarly and rapidly induces senescence. The *in vivo* regulation of nodule senescence is a function of the integrated carbon-nitrogen metabolism of the nodulated legume. It is presently unknown whether it results directly from molecular signal(s) or whether it is a secondary result of the diversion of nutrients to other host plant tissues.

*Pathogenicity and Symbiosis*

There are a number of close analogies in the genetics and physiology of interaction between the *Rhizobium*-legume symbiosis and bacterial pathogenicity. This is not surprising since symbiosis and pathogenicity are evidently parallel evolutionary developments of the relationship between bacteria and plants, as ancient as plants themselves.

Plants possess preformed and induced resistance mechanisms. Preformed resistance, implicit in higher plant morphology (cuticle, epidermis, cell walls, etc.), is overcome generally by only a few bacterial pathogens (e.g., *A. tumefaciens* and *Erwinia carotovora*). The production of plant cell wall hydrolyzing enzymes by rhizobia remains to be fully elucidated. If present, it is likely to be only transiently induced, specifically during the symbiotic infection process. Rhizobial endosymbiosis occurs without either induction of hypersensitive necrosis of host tissue or induction of host plant phytoalexins. Only in certain cases such as *B. japonicum* 61A24 (which produces unstable pbm) are phytoalexins induced by endosymbiosis.

On the basis of symbiotic and pathogenic relationships, coevolution has occurred, leading to specialization of host or pathogen/endosymbiont. The development of races of the pathogen and resistant cultivars is based on gene-for-gene relationships between host resistance and pathogen avirulence (see Reference 48). Two independent complementary families of genes encode host resistance (R) and pathogen avirulence (P), interacting as "allele pairs." In the simplest case, a compatible reaction (susceptible host, virulent pathogen) occurs when one locus has a recessive allele. Race/cultivar-specific relationships have been demonstrated for the bacterial phytopathogens *Pseudomonas* on their respective hosts. Race-specificity genes have been cloned from race 6 of *P. syringase* pv. *glycinea,* and when introduced into pathovar races 1 or 4 these genes restrict the host range to that of race 6.[59] If these race-specificity bacterial genes are mutated or deleted, the pathogen can infect host plant cultivars that are normally resistant. Host-specific nodulation by rhizobia has evidently evolved similarly, in that transposon insertion mutations in nodulation gene region III of the clover-specific *R. trifolii* produce recessive phenotypes defined by nodulation of primitive and improved pea cultivars as well as clover.[18] Resistance phenotypes of the host may also be analogous since dominant Mendelian loci affecting specificity of symbiosis have been defined for soybean and pea (see above).[33]

# CONCLUDING REMARKS

The constraints of the nitrogen fixation process and the physical and metabolic pressures imposed by this endosymbiosis are dealt with in legume plants by forming a specialized organ, the root nodule. Inter- and intracellular compartmentalization in this organ allows various processes, seemingly incompatible, to occur within the same environment. A key feature of this, as well as of other facultative endosymbioses, is the formation and maintenance of a membrane enclosing the microsymbiont. In legumes, a number of plant genes have evolved whose products (nodulins) are responsible for structural and functional integrity of this novel compartment. Understanding the function of these genes and various signals that regulate the expression of these genes may shed light on the ways by which this association evolved and may provide means to manipulate it.

## REFERENCES

1. APPLEBY, C. A. 1984. Annu. Rev. Plant Physiol. **35:** 443-478.
2. ATKINS, C. A. 1984. Plant Soil **82:** 273-284.
3. AUGER, S. & D. P. S. VERMA. 1981. Biochemistry **20:** 1300-1306.
4. BANFALVI, Z., V. SAKANYAN, C. KONCZ, A. KISS, I. DUSHA & A. KONDOROSI. 1981. Mol. Gen. Genet. **184:** 318-325.
5. BEDMAR, E. J., S. A. EDIE & D. A. PHILIPS. 1983. Plant Physiol. **72:** 1011-1015.
6. BERGMANN, H., R. PREDDIE & D. P. S. VERMA. 1983. EMBO J. **2:** 2333-2339.
7. BHAGWAT, A. A. & J. THOMAS. 1984. Arch. Microbiol. **140:** 260-264.
8. BISSELING, T., R. C. VAN DEN BOS & A. VAN KAMMEN. 1978. Biochim. Biophys. Acta **539:** 1-11.
9. BISSELING, T., C. BEEN, J. KLUGKIST, A. VAN KAMMEN & K. NADLER. 1983. EMBO J. **2:** 961-966.
10. BROUGHTON, W. J., U. SAMREY, U. PREIFER & J. STANLEY. J. Bacteriol. (Submitted.)
11. BROWN, G. G., J. S. LEE, N. BRISSON & D. P. S. VERMA. 1984. J. Mol. Evol. **21:** 19-32.
12. CALDWELL, B. E. 1966. Crop Sci. **6:** 427-428.
13. CARROLL, B. J., D. L. MCNEIL & P. M. GRESSHOFF. 1985. Proc. Nat. Acad. Sci. USA **82:** 4162-4166.
14. DART, P. J. 1977. *In* A Treatise on Dinitrogen Fixation, R. W. F. Hardy & W. S. Silver, Eds.: 367-472. John Wiley and Sons, Inc. New York, N.Y.
15. DEVINE, T. E. 1984. Heredity **75:** 359-361.
16. DEVINE, T. E. & B. H. BREITHAUPT. 1980. Crop Sci. **20:** 394-396.
17. DIXON, R. O. D. 1972. Arch. Mikrobiol. **85:** 193-201.
18. DJORDJEVIC, M. A., P. R. SCHOFIELD & B. G. ROLFE. 1985. Mol. Gen. Genet. **200:** 463.
19. DOWNIE, J. A., Q. S. MA, C. D. KNIGHT, G. HOMBRECHER & A. W. B. JOHNSTON. 1983. EMBO J. **2:** 947-952.
20. FINAN, T. M., A. M. HIRSCH, J. A. LEIGH, E. JOHANSEN, G. A. KULDAU, S. DEEGAN, G. C. WALKER & E. R. SIGNER. 1985. Cell **40:** 869-877.
21. FORTIN, M., M. ZELECHOWSKA & D. P. S. VERMA. 1985. EMBO J. **4:** 3041-3046.
22. FULLER, F., P. W. KUNSTNER, T. NGUYEN & D. P. S. VERMA. 1983. Proc. Nat. Acad. Sci. USA **80:** 2594-2598.
23. FULLER, F. & D. P. S. VERMA. 1984. Plant Mol. Biol. **3:** 21-28.
24. GOVERS, S., T. GLOUDEMANS, M. MOERMAN, A. VAN KAMMEN & T. BISSELING. 1985. EMBO J. **4:** 861-867.
25. GUERINOT, M. L, & B. K. CHELM. 1985. *In* Nitrogen Fixation Research Progress. P. J. Bottomley & W. E. Newton, Eds.: 220. Martinus Nijhoff Publishers. Dordrecht, Holland.
26. HUBBELL, D. H. & G. H. ELKAN. 1966. Can. J. Microbiol. **13:** 235-241.
27. JORDAN, D. C. 1982. Int. J. Syst. Bacteriol. **32:** 136-139.
28. KALUZA, K., M. FUHRMANN, M. HAHN & H. HENNECKE. 1983. J. Bacteriol. **158:** 1168-1171.
29. KATINAKIS, P. & D. P. S. VERMA. 1985. Proc. Nat. Acad. Sci. USA **82:** 4157-4161.
30. LAMB, J. W. & H. HENNECKE. 1986. Mol. Gen. Genet. **202:** 512-517.
31. LANG-UNNASCH, N. & F. M. AUSUBEL. 1985. Plant Physiol. **77:** 833-839.
32. LARA, M., J. V. CULLIMORE, P. J. LEA, B. J. MIFLIN, A. W. B. JOHNSTON & J. W. LAMB. 1983. Planta **157:** 254-258.
33. LARUE, T. A., B. E. KNEEN & E. GARTSIDE. 1985. *In* Analysis of Plant Genes Involved in the Legume-Rhizobium Symbiosis. R. Marcellin, Ed.: 39-48. OECD, Paris. France.
34. LARSEN, K. & B. U. JOCHIMSEN. 1986. EMBO J. **5:** 15-19.
35. LEE, J. & D. P. S. VERMA. 1984. *In* Genetic Engineering: Principles and Methods. J. K. Setlow & A. Hollaender, Eds.: **6:** 49-69. Plenum Press. New York, NY.
36. LEGOCKI, R. P. & D. P. S. VERMA. 1980. Cell **20:** 153-163.
37. LEONG, S. A., G. S. DITTA & D. R. HELINSKI. 1982. J. Biol. Chem. **257:** 8724-8730.
38. LONG, S. R., W. J. BUIKEMA & F. M. AUSUBEL. 1982. Nature London **298:** 485-488.
39. LUDWIG, R. A. 1984. Proc. Nat. Acad. Sci. USA **81:** 1566-1569.
40. MAHON, J. D. 1983. *In* Nitrogen Fixation: Legumes. W. J. Broughton, Ed.: 299-325. Oxford University Press. Oxford, England.

41. MARGULIS, L. 1970. Origin of Eukaryotic Cells. Yale University Press. New Haven, Conn.
42. MAURO, V., T. NGUYEN, P. KATINAKIS & D. P. S. VERMA. 1985. Nucleic Acids Res. **13:** 239-249.
43. MELLOR, R. B., E. MORSCHEL & D. WERNER. 1984. Z. Naturforsch. **39c:** 123-125.
44. MULLIGAN, J. T. & S. R. LONG. 1985. Proc. Nat. Acad. Sci. USA **82:** 6609-6613.
45. NGUYEN, T., M. ZELECHOWSKA, V. FOSTER, H. BERGMANN & D. P. S. VERMA. 1985. Proc. Nat. Acad. Sci. USA **80:** 5040-5044.
46. OLSON, E., M. SADOWSKY & D. P. S. VERMA. 1985. Bio/Technology **3:** 143-149.
47. PAGAN, J. D., W. R. SCOWCROFT, W. F. DUDMAN & A. H. GIBSON. 1977. J. Bacteriol. **129:** 718-723.
48. PANOPOULOS, N. J. & R. C. PEET. 1985. Annu. Rev. Phytopathol. **23:** 381-419.
49. PATE, J. S. & C. A. ATKINS. 1983. *In* Nitrogen Fixation: Legumes. W. J. Broughton, Ed. **3:** 245-298. Clarendon Press. Oxford, England.
50. ROBERTSON, J. G. & P. LYTTLETON. 1984. J. Cell Sci. **69:** 147-157.
51. ROBERTSON, J. G., M. P. WARBURTON, P. LYTTLETON, P. M. FORDYCE & S. BULLIVANT. 1978. J. Cell Sci. **30:** 151-174.
52. ROLFE, B. G. & J. SHINE. 1984. *In* Genes Involved in Microbe-Plant Interactions. D. P. S. Verma & T. Hohn, Eds.: 95-128. Springer-Verlag. New York, N.Y.
53. ROSTAS, K., P. SISTA, J. STANLEY & D. P. S. VERMA. 1984. Mol. Gen. Genet. **197:** 230-235.
54. RUVKUN, G. B., V. SUNDARESAN & F. M. AUSUBEL. 1982. Cell **29:** 551-559.
55. SCHOFIELD, P. R., R. W. RIDGE, B. G. ROLFE, J. SHINE & J. M. WATSON. 1984. Plant Mol. Biol. **3:** 3-11.
56. STACEY, G., A. S. PAAU, K. D. NOEL, R. J. MAKER, L. E. SILVER & W. J. BRILL. 1982. Arch. Microbiol. **132:** 219-224.
57. STANLEY, J., G. G. BROWN & D. P. S. VERMA. 1985. J. Bacteriol. **163:** 148-154.
58. STANLEY, J., D. LONGTIN, C. MADRZAK & D. P. S. VERMA. 1986. J. Bacteriol. **166:** 623-634.
59. STASKAWICZ, B. J., D. DAHLBECK & N. T. KEEN. 1984. Proc. Nat. Acad. Sci. USA **81:** 6024-6028.
60. SUTTON, W. D. 1983. *In* Nitrogen Fixation: Legumes. W. J. Broughton, Ed. **3:** 144-212. Oxford University Press. Oxford, England.
61. VAN KAMMEN, A. 1984. Plant Mol. Biol. Rep. **2:** 43-45.
62. VERMA, D. P. S. & S. LONG. 1983. Int. Rev. Cytol. **14:** 211-245.
63. VERMA, D. P. S. & K. NADLER. 1984. *In* Genes Involved in Microbe-Plant Interactions. D. P. S. Verma & T. Hohn, Eds.: 57-93. Springer-Verlag. New York, N.Y.
64. VERMA, D. P. S., S. BALL, C. W. GUERIN & L. WANAMAKER. 1979. Biochemistry **18:** 476-483.
65. VERMA, D. P. S., M. G. FORTIN, J. STANLEY, V. P. MAURO, S. PUROHIT & N. MORRISON. 1986. Plant Mol. Biol. **7:** 51-61.
66. VERMA, D. P. S., V. KAZAZIAN, V. ZOGBI & A. K. BAL. 1978. J. Cell Biol. **78:** 919-936.
67. WERNER, D., R. B. MELLOR, M. G. HAHN & H. GRISEBACH. 1985. Z. Naturforsch. **40c:** 179-181.
68. WILLIAMS, L. F. & D. L. LYNCH. 1984. Crop Sci. **6:** 427-428.

# Transmission Modes and Evolution of the Parasitism-Mutualism Continuum[a]

PAUL W. EWALD

*Department of Biology*
*Amherst College*
*Amherst, Massachusetts 01002*

## BACKGROUND: THE COST/BENEFIT PERSPECTIVE

Although endosymbiotic associations have been a focus of interest since the discovery of microorganisms, we are just beginning to understand the influences on evolutionary change along the continuum from severe parasitism through mutualism.[b] The significance of this problem has been clouded by failure to consider the mechanism of natural selection. According to traditional thinking, commensalism is better for both host and parasite than parasitism because commensal organisms have a lower probability of destroying themselves through negative effects on their hosts; thus parasitism should eventually evolve towards commensalism.[1,2] Commensal relationships are then viewed as raw material for evolution of mutualisms.

The major problem with this view is that it does not differentiate effects of parasitism on the survival of species (or populations) of the endosymbionts from effects on the inclusive fitness[3] of the endosymbiont. Characteristics that harm hosts will spread through a population of symbionts so long as the beneficial effects on the spread of the symbiont's genetic instructions for such characteristics outweigh the detrimental effects. An understanding of where on the mutualism-parasitism continuum particular symbionts will evolve toward therefore must consider both the fitness costs and benefits to the symbiont associated with alternative levels of mutualism or parasitism.

This paper analyzes the entire span of this continuum, from highly beneficial obligate mutualistic relationships to parasitic relationships so severe that they are often considered a form of predation. My goal is to suggest ecological characteristics that should favor evolution towards particular portions of this continuum and then to draw upon data in the literature to evaluate these hypotheses.

[a] The ideas in this paper were developed with support from the Michigan Society of Fellows, a NATO/NSF fellowship awarded in 1983, a Biomedical Research Grant awarded by the National Institutes of Health to Amherst College, a symposium grant from the Wellcome Trust, and a Miner D. Crary fellowship from Amherst College.

[b] For the purposes of this paper endosymbiosis is defined broadly as the situation in which a parasite, mutualist, or commensal lives inside the body of a host organism. The net effect on the host's inclusive fitness of all harmful and beneficial characteristics is the theoretical criterion used to assign the labels parasitism, commensalism, or mutualism.

Mode of transmission is one aspect of host-symbiont relationships that should affect costs and benefits accrued by an endosymbiont at any particular point on the continuum. I begin by considering the parasitic part of the continuum.

## PATHOGENICITY AND TRANSMISSION BY VECTORS AND WATER

Parasite variants that reproduce more extensively within a host should accrue fitness benefits because there are more copies of their genetic instructions available for transmission; however, they should also accrue fitness costs because increased reproduction within a host generally should decrease the mobility of the infectious hosts and thereby decrease the frequency of contacts with susceptible hosts. The outcomes of host-parasite coevolution should depend upon the relative magnitudes of these costs and benefits. Parasites that can be transmitted to susceptible hosts from immobilized hosts should accrue lower costs at realistic levels of parasite reproduction. Such parasites should therefore be relatively severe.[4] This principle was supported by a cross-specific analysis of human pathogens other than those infecting the gastrointestinal tract. Specifically, pathogens transmitted by biting, terrestrial arthropods were significantly more severe than those not transmitted by such vectors.[4]

Like pathogens transmitted by biological vectors, pathogens transmitted from immobilized hosts to susceptible hosts by nonliving vehicles should accrue relatively low costs from immobilization of their hosts. The only existing test of this hypothesis involves the severity of bacterial pathogens of the human gastrointestinal tract. A review of all such pathogens regularly transmitted between humans revealed a statistically significant positive correlation between mortality and the degree of waterborne transmission.[5]

## VERTICAL TRANSMISSION AND MUTUALISM

These findings suggest that modes of transmission are related to the severity of parasitism. Might modes of transmission influence whether a host/symbiont association evolves toward mutualism as opposed to parasitism? Vertical transmission, defined as the direct transfer of infection from a parent organism to its progeny,[6] should favor evolution toward mutualism and benign parasitism. A vertically transmitted symbiont should accrue greater benefits than a horizontally transmitted symbiont in return for beneficial effects on the host because benefits to the host ultimately involve increased host reproduction. In contrast, horizontally transmitted symbionts do not benefit directly from increases in host reproduction.

This argument proposes that the presence of vertical transmission favors the evolution of mutualism. The general tendency in the literature, however, has been to emphasize the reverse mechanism: hosts benefit by evolving means of providing their offspring with mutualistic endosymbionts. This host bias apparently exists partly because of the presence of accessory structures facilitating vertical transmission, but such structures probably evolved after the presence of vertical transmission.

The explanatory power of this vertical-transmission-before-mutualism hypothesis depends on the strength of the relationship between vertical transmission and endosymbiotic mutualism and the prevalence of vertical transmission among parasites. The degree of association between vertical transmission and mutualism across species is a measure of the degree to which vertical transmission is sufficient to explain the existence of endosymbiotic mutualism. If a substantial proportion of endosymbiotic mutualisms are not vertically transmitted, then one must conclude that vertical transmission is not always necessary for the evolution and maintenance of mutualistic endosymbiotic associations.

Overviews of the frequencies of vertical transmission among endosymbiotic associations will not prove the degree to which vertical transmission is a cause or an effect of mutualism. Only experimental manipulation can accomplish that goal. Such overviews, however, should be a useful starting point for investigating the evolutionary role of vertical transmission and, specifically, for suggesting informative experiments. To provide such an overview of the prevalence of vertical transmission among mutualistic endosymbionts, I provide in TABLE 1 the modes of transmission for all of the mutualisms described by Buchner.[7]

TABLE 1 summarizes Buchner's work generally according to taxonomic families. In every case described by Buchner the symbionts were transmitted vertically, although the forms of vertical transmission varied considerably (see TABLE 1 and Buchner).[7] In only two cases was there a mode of horizontal transmission (TABLE 1). In each case the species also transmitted the symbiont vertically. Within each family Buchner generally provided illustrations from several species; the ubiquity of vertical transmission illustrated by TABLE 1 is therefore somewhat abbreviated. Although we need not invoke any causative factor other than vertical transmission (and of course potential mutual benefits) to explain the evolution of these mutualisms, the ubiquity of vertical transmission in Buchner's data could, of course, be explained as a consequence of mutualism rather than a cause. For this explanation to be viable, however, one must assume that all of these instances of mutualism have evolved past a nonvertically transmitted stage. Regardless of the evolutionary sequence, the data in TABLE 1 confirm that vertical transmission and mutualism are strongly linked.

# VERTICAL TRANSMISSION AND THE PARASITISM-TO-MUTUALISM TRANSITION

## Asexual Hosts

The probabilities of different phenotypic effects of genetic alterations and the associated costs and benefits to the parasite lend support for the idea that vertical transmission will generally precede mutualism, especially in asexually reproducing hosts. First, mutations yielding vertical transmission and no horizontal transmission in a symbiont population with both vertical and horizontal transmission should tend to occur relatively frequently. Any mutation by host or parasite that blocks any of the activities needed for horizontal transmission would be sufficient (e.g., production

**TABLE 1.** Transmission of Symbionts of Invertebrates Summarized from Buchner[a]

| Host | Symbiont | Mode of Transmission |
|---|---|---|
| **Insects** | | |
| Orthoptera | | |
|   Blattidae | b | transovarial |
| Isoptera | | |
|   Mastotermitidae | b | transovarial |
| Coleoptera | | |
|   Anobiidae | y | transovum |
|   Cerambycidae | y | transovum |
|   Lagriidae | b | transovarial |
|   Chrysomelidae | b | transovum |
|   Lyctidae | n | transovarial |
|   Curculionidae | b | transovum |
|   Cucujidae (Silvanidae) | n | transovarial |
|   Bostrichidae | n | transovarial |
|   Nosodendridae | b | transovarial |
|   Cimicidae | b | transovarial |
| Hymenoptera | | |
|   Siricidae | f | transovum |
|   Formicidae | b | larval |
| Diptera | | |
|   Tephritidae (Trypetidae) | b | transovarial, t |
|   Ceratopogonidae | b | transovarial |
|   Muscidae | b | larval |
|   Hippoboscidae | b | larval |
|   Nycterbiidae | n | larval |
| Hemiptera | | |
|   Triatomidae | b | transovum & fecal |
|   Cimicidae | b | transovarial |
| Homoptera | | |
|   Coccidae | b | transovarial |
|   Aphidae | b | transovarial |
|   Aleurodidae | n | transovarial |
|   Psyllidae | b | transovarial |
|   Cicadidae | y & n | transovarial |
|   Fulgoridae | n | transovarial |
| **Malophaga** | b | transovarial |
| **Anoplura** | b | transovarial |
| **Arachnoids** | | |
|   Gamasidae | b | transovarial |
|   Ixodidae | b | transovarial |
|   Argasidae | b | transovarial |
| **Annelids** | | |
|   Hirudinea | b | larval |
|   Lumbricidae | b | larval |
| **Tunicates** | | |
|   Pyrosomidea | b | larval |
|   Salpidea | b | larval |

[a] Families in parenthesis refer to the family name used by Buchner.[7] b, bacteria or bacterialike; f, filamentous fungi; y, yeast or yeastlike; n, symbionts not identified taxonomically. "Larval" refers to vertical transmission through parental fluids surrounding larva or embryo; t, tree to tree—the bacterium transmitted transovarially in the fly *Dacus oleae* is also transmitted from tree to tree by the fly, wind, and rain.

of dispersal stages, release from host cells, reproductive responses to the status of the host). With avenues of horizontal transmission nonexistent, all costs previously imposed on hosts for purposes of horizontal transmission are now costs to the parasite as well. For asexual hosts all descendants of such a mutant line of parasite are now doomed to extinction unless subsequent mutations or recombinations arise that reduce costs to the host (and/or increase the benefits to the host) to such a degree that the net effect of the parasite on its host is no longer negative.

Genetic alterations in the parasite resulting in benefits to the host should be less common than alterations resulting in reductions of costs because there are far more ways to terminate existing processes by random genetic change than to generate one of a few possible benefits to the host; however, genetic alterations beneficial to the host need not occur during the time period between loss of horizontal transmission and the evolution of a mutualistic association. Although a parasite by definition has a net negative effect on its host's fitness, it may have some positive effects; for example, bacteria infected with lysogenic viruses are often are less vulnerable to invasion by other viruses.[8] If such benefits exist, cost reduction would be sufficient as a mechanism for evolution of a mutualistic relationship from a solely vertically transmitted parasite.

Similarly, regeneration of horizontal transmission should be rare relative to alterations that terminate processes that were previously involved in horizontal transmission and were costly to the host. For such regeneration the genetic change would have to be reversed or a new mode of horizontal transmission would have to be developed. Genetic changes that result in loss of horizontal transmission in parasites of asexually reproducing hosts therefore should tend to result in either mutualism or extinction of the parasite lineage.

Each of the three possible sequences of events necessary for mutualism to precede vertical transmission seems less likely than the preceding argument. (1) For a free-living organism to become a mutualist, one must presume that the organism is pre-adapted to provide benefits to the host that are sufficiently strong to outweigh the inevitable costs associated with a new symbiotic existence. It is difficult to imagine associations other than perhaps photosynthetic endosymbionts[9] that could provide such benefits. (2) For a horizontally transmitted parasite to become a mutualist, one must propose genetic changes that are not just sufficiently beneficial to the host to permit the host to survive better than competing hosts. Rather, the symbiont must have a greater fitness from a mutualistic association than that of its competing parasites. Because the future return to a parasite from its contribution to the host should be less as the future association of that parasite lineage with that host decreases, such a threshold should be reached less frequently for a nonvertically transmitted parasite than for a vertically transmitted parasite. (3) A free-living organism that has just become a symbiont and has an initial net negative effect on the host must decrease its negative effect in a manner analogous to that of a vertically and horizontally transmitted parasite that has lost horizontal transmission, but it also must evolve a mechanism for vertical transmission. Without vertical transmission, such a symbiont's time period for generation of mutualism is restricted to one generation.

### Sexual Hosts

For sexually reproducing hosts, the situation is more complex because symbionts that are solely vertically transmitted can be maintained in a population even if they have a net negative affect on their host.[6] Specifically, according to the fundamental vertical transmission equation,[6] solely vertically transmitted parasites will be maintained if

$$\alpha\beta \ (d + v) > 1$$

Where $\alpha$ is the proportional effect on host fecundity (e.g., if infected hosts reproduce half as much as uninfecteds, $\alpha = 0.5$), $\beta$ is the proportional effect of infection on survival from egg through reproduction, $d$ is the probability that an infected mother will infect an average offspring (assuming the father is uninfected), and $v$ is the corresponding probability for the father.

To illustrate this point, if $\alpha = 0.9$, $\beta = 0.9$, $d = 0.8$, and $v = 0.5$ the above inequality holds and a parasitic symbiont can be maintained indefinitely. Here again, potential for regaining horizontal transmission should tend to be lost if characteristics needed for horizontal transmission impose fitness costs on the hosts. In contrast, costs imposed on hosts in order to increase $d + v$ or to otherwise increase the competitive ability of symbionts against other symbionts may increase. Evolution to mutualism will occur only if the advantage gained over symbiont competitors through parasitism is less than the increased number of infected offspring that result if the symbiont eliminates its net negative effects on the host. The potential for such competition is greater for sexual hosts than for asexual hosts because competitors may be introduced through reproduction. Balanced against this inhibition of evolution toward mutualism, however, is the fact that so long as the above inequality holds there is an unlimited amount of time for occurrence of mutations beneficial to the host.

If the above inequality is not met after a mutation eliminates horizontal transmission, additional genetic alterations must reduce the other costs to the host's fitness [or increase the efficiency of vertical transmission $(d + v)$] sufficiently to raise the quantity $\alpha\beta$ above $(d + v)^{-1}$. Substantial amounts of time for reduction of net cost to the host may occur; for example, according to the fundamental vertical transmission equation if $\alpha = 0.9$, $\beta = 0.9$, $d = 8$, $v = 0.4$, and 10% of a population of 10,000 hosts are infected, extinction of the symbiont will occur in 241 generations of the host with a total of 32,598 infected host·generations (i.e., the number of infected hosts surviving the $i$th generation summed over all generations) occurring before extinction. (An infection rate of 10% is reasonable if one assumes that infected hosts can increase in frequency during population growth, but not as fast as the uninfected population. The simulation assumes nonoscillatory stability at a carrying capacity of 10,000 hosts.)

If inheritance is only through the mother, the situation reverts back to that of asexual hosts. Only mutualists can be maintained indefinitely; for example, if in the preceding example $d = 1$, $v = 0$, extinction of the symbiont will occur in 35 generations of the host with a total of 4512 infected host·generations occurring before extinction.

## VERTICAL TRANSMISSION AND PATHOGENICITY IN VECTORS

The presence of well-developed vertical transmission of parasites is evidence that the raw material is available for such evolution of mutualistic associations. If the presence of vertical transmission favors reduction in pathogenicity one should see a correlation between the presence of vertical transmission in parasitic organisms and pathogenicity.

The general benignness of infections in vectors and the frequent presence of vertical transmission are not in themselves reliable support for the idea that vertical transmission reduces levels of severity; one expects benignness in vector-borne diseases through considerations of dispersal and resource use. Specifically, although greater use of host resources generally increases the amount of propagules generated per host it also generally reduces the dispersal ability of hosts. Because vertebrate hosts are larger than vectors, more host resources can be converted to pathogen reproduction before a given level of immobilization is produced; furthermore, immobilization of the vector generally should be associated with a greater loss in parasite fitness than immobilization of the vertebrate host (e.g., dispersal of malarial parasites is inhibited more by immobilized mosquitos than by immobilized humans). According to this dispersal/resource hypothesis, vector-borne parasites specialize on vertebrate hosts as a resource base and vectors as an agent of dispersal.

Rickettsial pathogens transmitted by arthropod vectors are one potential source of insight on the matter because they are often vertically transmitted and often benign. Information on rickettsial and rickettsia-like pathogens is presented in TABLE 2. The most severe disease is one of the few cases in which the parasite is not transmitted vertically: R. prowazekii in P. humanus; however, R. typhi is benign in its normal flea host but lethal in P. humanus,[11] suggesting that the lethal effects of R. prowazekii on P. humanus are more a function of host vulnerability than pathogen virulence. Several other rickettsial pathogens are not known to be transmitted vertically, but none of them are known to cause severe infections in their vectors;[12,15] thus none of the other known pathogens with insufficient data seem likely to generate a positive association between vertical transmission and benignness. If vertical transmission contributes to the general benignness of symbionts in vectors, the effect is presently too weak to detect among the rickettsial diseases.


# LOCATIONS ON THE CONTINUUM


### Divisions of the Continuum


Relationships between modes of transmission and the parasitism-mutualism continuum are diagrammed in FIGURE 1. Commensalism is represented as a mere point on the line dividing symbioses in which the net fitness benefit to each member is marginally positive from symbioses in which the net benefit to one member is marginally negative. As symbioses will almost always incur some costs or benefits on each partner true commensalism is probably less a reality than a heuristic concept used when such costs and benefits are too slight to measure. The parasitism-mutualism continuum can therefore be analyzed in terms of the magnitudes of the costs imposed by parasites on hosts and the degree to which mutualists approach the maximum possible mutual benefit.

For conceptual simplification, the parasitic end point in FIGURE 1 can be thought of as the point at which all infected hosts are killed; however, from an evolutionary viewpoint components other than percent mortality contribute to severity. For example complete castration of the host destroys its potential for further reproduction as completely as death. How the severity of castration is ranked relative to that of death depends on the interactions of a castrated host with other genetically related hosts;

for example, evolutionarily castration would be less severe if the castrated host has a net positive effect on propagation of its genetic material by giving aid to offspring and relatives.

## Vectors, Contact, and Waterborne Transmission

About half of the species of pathogens transmitted to and from humans by biting arthropod vectors are associated with death of more than 1% of infected but untreated

TABLE 2. Vertical Transmission and Severity of Rickettsial Infection in Vectors

| Rickettsial Pathogen | Arthropod Host | Vertical Transmission | Pathogenicity | Reference |
|---|---|---|---|---|
| Rickettsia prowazeki | Pediculus humanus | no | lethal | 12 |
| Rochalimaea quintana | Pediculus humanus | no | nonlethal | 13, 14 |
| Erlichia canis | Rhipicephalus sanguineus | no | mild | 15 |
| Rickettsia typhi | Xenopsylla cheopis | yes | mild | 11, 16 |
| Rickettsia rickettsi | Dermacentor species | yes | occasional le-thality & re-duced fecundity | 17 |
| Rickettsia tsutsugamushi | Leptotrombidum species | yes | mild | 18, 19 |
| Rickettsia akari | Lyponyssoides sanguineus | yes | mild | 12 |
| Rickettsia conori | Rhipicephalus sanguineus | yes | mild | 12 |
| Rickettsia sibirica | Dermacentor marginatus | yes | mild | 12, 15 |
| Wohlbachia pipientis | Culex pipiens | yes | mild | 20 |

human hosts.[4] All of the other half are sufficiently severe to incapacitate infected hosts. About one-tenth of the nondiarrheal human pathogens transmitted by host-to-host contact (including indirect contact via fomites) are associated with more than 1% mortality of untreated hosts.

All of the waterborne diarrheal bacteria (i.e., bacteria with about 50% or more outbreaks involving waterborne transmission) of humans are associated with mortality rates above 1%, but all of the nonwaterborne diarrheal bacteria (i.e., fewer than 25% of outbreaks involving waterborne transmission) have mortality below 0.1%.[5]

### Carnivorous Transmission

When a parasite is transmitted from a prey to a predator through predation (or to scavenger by scavenging), increased use of the prey's tissues should result in exceptionally low fitness costs to the parasite because parasites in weakened, less mobile prey can still be transmitted effectively to the predator. The point at which increased fitness benefits resulting from increased use of the prey's resources are outweighed by increased costs should therefore occur at a higher level of resource use, and consequently of pathogenicity, than in parasites not transmitted by predation. Although a thorough analysis of this idea has not yet been conducted, severe disease and mortality in prey hosts and mild or asymptomatic infestations in predator hosts are characteristic of helminths transmitted by predation.[21,22] The distribution in FIG-URE 1 is based on such helminths.

The arguments for vector-borne and predation-transmitted parasites are two special cases of a broader argument. In each case parasites should specialize on the feeding species for dispersal and on the fed-on species as a resource base. The fed-on species should therefore suffer high pathogenicity (e.g., human hosts of parasites transmitted by biting arthropods, prey hosts of helminths) while the feeding species should suffer relatively benign disease. The major difference between these two arguments is that in vector-borne transmission the consumer is much smaller than the consumee, while in predatory transmission the two are similar in size.

### Mobile Life History Stages

Parasites possessing life history stages that can move between hosts should have extremely low costs associated with extensive use of host resources. Even host death should impose little cost on such parasites so long as development to the mobile stage is possible. Although no systematic review of severity as a function of parasite mobility has been carried out, the most severe parasites tend to have mobile stages. Typical examples include flagellated coelomomycete fungi, myrmethid nematodes, and a vast variety of endoparasitic insects.[23] The severity of such parasites is so extreme that they are sometimes considered to be distinct from parasites, hence the term parasitoids; however, mobile parasites as a group are associated with mortality ranging relatively continuously from 100% to percentages frequently encountered among parasites possessing other modes of transmission. According to the theoretical framework of this paper, the selective pressures leading to their severity are similar to those leading to severity among nonmobile parasites (e.g., vector-borne parasites). They therefore seem most appropriately viewed as parasites possessing characteristics favoring evolution towards one end of the parasitism-mutualism continuum, rather than organisms fundamentally different from parasites.

### Vertical Transmission

The only mode of transmission implicated in favoring evolution of mutualism is vertical transmission. The location of a symbiosis on the mutualism side of the continuum should depend on the potential for genetic conflicts of interest between host and symbiont. Organelles appear to be the most extreme form of mutualist. Yet even many organelles may be slightly to the right of the mutualism end point if, for example, the genetic conflicts of interest between organelle and nucleus[24] result in one partner altering the cellular environment to better fulfill its own needs at expense to the other.

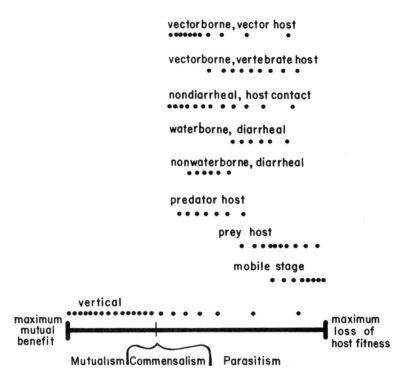

**FIGURE 1.** Relationships between modes of transmission and location of symbioses on the mutualism-parasitism continuum. The positioning of symbionts on the continuum represents an approximate ranking according to mortality, tissue damage, and degrees of immobilization in nonlethal infections, rather than a location according to an absolute scale of units. The dotted lines below each are given to indicate roughly how species in each class are distributed. These distributions are based on the information presented in the text.

## AVENUES OF FUTURE RESEARCH

The information presented in this paper draws attention to the need for comprehensive reviews of the literature to determine more precisely relationships between modes of transmission and locations on the parasitism-mutualism continuum; for example, the relative frequencies of vertical and horizontal transmission among mutualisms need to be analyzed on a broader scale. The resultant frequency of mutualisms that do not involve vertical transmission will provide a lower limit to the frequency with which vertical transmission is not necessary for the evolution of endosymbiotic mutualism.

This paper also draws attention to the need for experimental manipulations of transmission in the laboratory; for example, frequencies of vertical and horizontal transmission could be manipulated to determine whether increased frequencies of vertical transmission lead to mutualism and whether stable mutualisms can evolve in the absence of vertical transmission. The rapid evolution of mutualistic interdependence demonstrated in the laboratory (e.g., by Jeon)[25] indicates that such experiments are feasible within the time scales of laboratory experimentation.

# SUMMARY

An analysis of fitness costs and benefits associated with pathogenicity suggests that modes of transmission are key determinants of evolution toward severely pathogenic, benign, or mutualistic symbioses. Specifically, this approach suggests that (1) symbionts with mobile life history stages should evolve toward extremely severe parasitism, (2) vector-borne symbionts should evolve toward severe parasitism in vertebrate hosts and benign parasitism in the vectors, (3) waterborne symbionts should evolve toward severe parasitism, (4) symbionts transmitted by predation should evolve toward severe parasitism in prey hosts and benign parasitism in predator hosts, and (5) vertically transmitted symbionts should evolve toward benign parasitism and mutualism. Detailed reviews of the literature on human diseases support the hypothesized severity of vector-borne and waterborne transmission. Evaluation of the other associations is less detailed, but each association appears to be present. This framework draws attention to the need for (1) detailed reviews of relationships between transmission modes and the nature of symbiotic interactions, and (2) experimental manipulations of transmission.

## REFERENCES

1. HOEPRICH, P. D. 1977. *In* Infectious Diseases. P. D. Hoeprich, Ed. 2nd edit.: 34-45. Harper & Row. New York, N.Y.
2. DUBOS, R. 1965. Man Adapting. Yale University Press. New Haven, Conn.
3. HAMILTON, W. D. 1964. J. Theor. Biol. 7: 1-52.
4. EWALD, P. W. 1983. Annu. Rev. Ecol. Syst. 14: 465-485.
5. EWALD, P. W. 1986. Cultural vectors and the evolution of virulence among gastrointestinal bacteria. (Manuscript available from the author upon request.)
6. FINE, P. E. F. 1975. Ann. N.Y. Acad. Sci. 266: 173-194.
7. BUCHNER, P. 1965. Endosymbiosis of animals with plant microorganisms. Wiley Interscience. New York, N.Y.
8. BOTSTEIN, D. & M. M. SUSSKIND. 1974. *In* Mechanisms of Virus Disease. W. S. Robinson & C. F. Fox, Eds.: 57-108. W. A. Benjamin. Menlo Park, Calif.
9. TAYLOR, F. J. R. 1983. Int. Rev. Cytol. Suppl. No. 14: 1-28.

10.  JEON, K. W., Ed. 1983. Intracellular Symbiosis. International Review of Cytology, Supplement No. 14. Academic Press. New York, N.Y.
11.  ITO, S., J. W. VINSON & T. J. MCGUIRE, JR. 1975. N.Y. Acad. Sci. **266:** 35-60.
12.  BURGDORFER, W. 1975. *In* Diseases Transmitted from Animals to Man. W. T. Hubbert, W. F. McCulloch & P. R. Schnurrenberger, Eds. 6th edit.: 382-422. C. C. Thomas. Springfield, Ill.
13.  VINSON, J. W. 1975. Ann. N.Y. Acad. Sci. **266:** 3-5.
14.  WARREN, J. 1959. *In* Viral and Rickettsial Infections of Man. T. M. Rivers & F. L. Horsfall, Eds.: 918-920. J. B. Lippencott. Philadelphia, Pa.
15.  HARWOOD, R. F. & M. T. JAMES. 1979. Entomology in Human and Animal Health. 7th edit. Macmillan. New York, N.Y.
16.  FARHANG-AZAD, A., R. TRAUB & S. BAQAR. 1985. Science **227:** 543-545.
17.  BURGDORFER, W. & L. P. BRINTON. 1975. Ann. N.Y. Acad. Sci. **266:** 61-72.
18.  TRAUB, R. & C. L. WISSEMAN, JR. 1974. J. Med. Entomol. **11:** 237-303.
19.  TRAUB, R., C. L. WISSEMAN, JR., M. R. JONES & J. J. O'KEEFE. 1975. Ann. N.Y. Acad. Sci. **266:** 91-114.
20.  YEN, J. H. 1975. Ann. N.Y. Acad. Sci. **266:** 152-161.
21.  HOLMES, J. C. & W. M. BETHEL. 1972. *In* Behavioural Aspects of Parasite Transmission. E. U. Canning & C. A. Wright, Eds.: 123-149. Academic Press. London, England.
22.  OLSON, W. W. 1974. Animal Parasites, Their Life Cycles and Ecology. University Park Press. Baltimore, Md.
23.  ASKEW, R. R. 1971. Parasitic Insects. Heinemen. London, England.
24.  EBERHARD, W. G. 1980. Q. Rev. Biol. **55:** 231-249.
25.  JEON, K. W. 1983. Integration of bacterial endosymbionts in amoebae. Int. Rev. Cytol. Suppl. No. 14: 29-47.

# Coevolution in Lichens[a]

VERNON AHMADJIAN

*Department of Biology*
*Clark University*
*Worcester, Massachusetts 01610*

## INTRODUCTION

Lichens are good examples of coevolution because of their highly integrated nature and unique features. Lichens are associations of fungi and photosynthetic partners (photobionts) and they grow in many regions of the world, including habitats with extreme environments such as those of Antarctica and other deserts. Most lichen fungi are ascomycetes, but a few basidomycetes form lichens—this may reflect the older age of the Ascomycotina. Lichens are morphologically, chemically, and physiologically different from their component parts, that is, they are products of symbiotic interactions and not simply the sum of several parts. In many lichens, coevolution seems to be so far advanced that the symbionts can no longer live independently. For example, I doubt the independent existence of *Trebouxia,* by far the most common lichen photobiont. In my view, *Trebouxia* is an alga adapted for a symbiotic existence through a coevolutionary process. Its free-living ancestor may have been a soil alga, such as *Pleurastrum,* with which strong anatomical and taxonomic relationships have been made recently.[1] Lichen fungi also have changed from free-living fungi and there is a question as to whether the ascospores produced by lichen fungi in nature can form lichens, given the absence of potential, free-living photobionts. Most lichens propagate by means of asexual diaspores, such as soredia and isidia, or thallus fragments. The species-pair concept in lichenology states that an evolutionary process is occurring whereby the asexual propagation process in lichens is being selected over the sexual process of the symbionts.[2] Among lichens, many species pairs have been recognized, with one member of the pair being fertile and the other one sterile. It is assumed that reduction of sexuality among symbionts promotes genetic homogeneity and conservation of characters among those strains already well adapted to a symbiotic existence. From different pieces of evidence, therefore, it seems likely that the evolution of the fungal partner of a lichen has been closely linked with that of the photobiont resulting in reciprocal genetic changes in the symbionts.

Coevolution in lichens has not been considered by many scientists. Barrett discussed lichens in his chapter on plant-fungus symbioses in *Coevolution* by Futuyma and Slatkin (1983).[3] He concluded that there was little evidence that would unequivocally demonstrate that lichen symbionts have coevolved a mutualistic association. He may have been correct in assuming that lichens are not mutualistic associations but was incorrect, in my opinion, in presuming that coevolution between the symbionts has not occurred. Hawksworth used a coevolutionary approach to hypothesize that as-

[a] Research supported by the National Science Foundation.

comycetes, which include most of the lichen-forming fungi, arose from red algae that were parasitic on other algae.[4]

# ORIGIN OF LICHENS

When and how lichens arose and evolved are mostly a matter of speculation. The idea that they may be very ancient associations was proposed by Hallbauer[5] and Hallbauer et al.[6] in their observations of Witwatersrand conglomerates. Scanning electron microscope (SEM) observations of Witwatersrand fossils, which have been dated back to the Precambrian (2300 to 2700 million years), have revealed lichenlike organisms which the scientists have named *Thuchomyces lichenoides*. The lichens presumably extracted gold and other materials from their environment and deposited them within their thalli, thereby enabling them to become fossilized.

Among the ascomycetes only five orders consist exclusively of lichen species while the remaining taxonomic groups contain both lichenized and free-living species. There are about 14,000 species of lichen-forming fungi but there are many more associations of fungi and algae that are not lichens. There is a continuum of lichenization from associations that are loosely organized to ones that are highly structured and integrated. Why some fungal-algal relationships evolved into lichens and others did not is not clear. The lichen life-style seems to be evolving in different ascomycetes, appearing in some groups and disappearing in others.[7,8] Thus, lichen evolution may be more of a dynamic and ongoing process than earlier realized.

Many fungi that form lichens may have evolved from species that were once parasitic on plants. The evidence for this possibility is as follows: (1) laboratory studies have shown that lichen fungi are virulent parasites of algae;[9] (2) curious organelle-like structures called concentric bodies are present in the hyphae of both lichen fungi and pathogenic fungi. This common feature appears to be more than coincidental because concentric bodies have not been found in any other organism.

# FITNESS TRAITS DUE TO SYMBIOTIC INTERACTIONS
# THALLUS FORMATION

There are several different types of lichen thalli. A thallus forms largely because of the symbiotic transformation of the fungus. Some stimulus, undoubtedly received from the photobiont, triggers the fungus to form a thallus that will house the algal cells. According to Hawksworth and Hill, "these (thalli) can be interpreted as methods the mycobiont has evolved to display the photobiont in a manner designed to capture maximum irradiation and so ensure optimal photosynthesis."[8] Further, according to Kershaw, "the evolution of an effective lichen thallus thus appears to be a compromise between the provision of a moisture-holding structure for the maintenance of the normally aquatic algal component, and one which does not at the same time limit free gas exchange."[10]

Some loosely organized lichens such as *Collema* have their photobionts distributed randomly through the thallus. In most lichens, however, the photobionts are in a fairly well-defined layer of the thallus. In foliose and squamulose thalli, the photobiont layer is directly below the upper cortex and spread out horizontally like the photosynthetic cells in a leaf. In filamentous and fruticose thalli, the photobionts form a cylindrical zone below the outer cortex. The orientation of a thallus depends on the photobiont which stimulates the fungus to form cortical tissue. In lichen thalli that are turned upside down, the photobiont layer disintegrates and a new layer forms from errant photobiont cells located on the medulla or introduced onto the lower thallus. As these photobionts grow, they stimulate the fungus to form an overlying cortex.[11]

The evidence shows that a lichen thallus is a product of coevolution, i.e., the fungus forms a thallus in response to a specific stimulus from the photobiont which, in turn, is influenced by the fungus to release photosynthates. Why lichen fungi have evolved different kinds of thalli with similar types of photobionts is puzzling. There is a continuum of thallus forms from primitive crustose and leprose types to highly organized foliose and fruticose thalli. The different thalli, however, have basically the same layers, i.e., cortex, algal layer, and medulla.

## SECONDARY COMPOUNDS

Symbiotic interactions in lichens result not only in new morphological bodies, the thalli, but also in unique chemical substances, the "lichen acids." The inability of many mycobionts to synthesize in culture the compounds that they produce in symbiosis most likely is due to the absence of the photobiont which may play a role in the synthesis process. The major biosynthetic steps for the production of lichen compounds are those of the fungus and many of these compounds are the same or similar to those formed by nonlichenized fungi.[12] Lichens also synthesize unusual compounds such as depsides, depsidones, dibenzofurans, and usnic acids. These are aromatic compounds that are made up of two or three phenolic units joined together by oxidative coupling and ester linkages. About 75 of these compounds have been described.[12]

Culberson and Ahmadjian have proposed a hypothesis to explain how the alga helps to make the characteristic lichen products.[13] According to this hypothesis, lichen algae inhibit the decarboxylation of phenolic acid precursors produced by the mycobiont. Nonlichenized fungi use similar precursors to form different fungal products such as quinones that are toxic to algae. With that pathway blocked by the alga in a lichen, the phenolic acid precursors are shunted through an alternative pathway where they are esterified and coupled to form lichen compounds. The algal inhibitor of fungal decarboxylase must be a small molecule that is not specific for a particular alga. The evidence for this comes from synthesis studies that have shown that the mycobiont *Cladonia cristatella* produces the depside barbatic acid not only with different species of *Trebouxia* but also with the nonlichenized alga *Friedmannia israeliensis.*

There has been much speculation on the function of lichen substances.[14] According to the Culberson and Ahmadjian hypothesis, lichen substances are waste products that form because the alga protects itself by preventing the fungus from synthesizing toxic compounds.[13] In a similar fashion, lichen substances such as barbatic acid may slow the parasitism of the alga by the mycobiont. Such lichen compounds may be mycotoxic and may be produced as a result of the fungal parasitism on the alga. For example, many clones of *C. cristatella* mycobiont successfully lichenized the *Trebouxia erici* phycobiont.[9] Although algal cells from these successful syntheses contained fungal haustoria, squamules were formed and barbatic acid was synthesized, even at the earliest soredial stages of development. In contrast, one clone of the mycobiont formed soredial complexes with *T. erici* but did not produce squamules because it killed all the algal cells. Barbatic acid was not produced in these cultures. In the successful syntheses, barbatic acid may have protected the algal cells from being destroyed by the fungus.

## FUNGAL CONTROL OF PHOTOBIONT METABOLISM

It seems reasonable to assume that the mycobiont controls the metabolism of its photobiont. There must also be a feedback system that modifies the mycobiont's control when necessary, otherwise the continuous release of photosynthates would harm the photobiont and disrupt the balanced growth of the symbionts. The urease theory attempts to explain how the regulation of photobiont metabolism by the fungus occurs.

The urease theory is based on the ability of lichen fungi to produce urease in response to urea that accumulates in a thallus. Urea forms from the breakdown of arginine, which occurs as a free amino acid in lichens along with the enzymes that degrade it.[15,16] The breakdown of urea by urease produces ammonia and carbon dioxide. According to the urease theory, ammonia stimulates respiration of the photobiont, thereby increasing carbohydrate breakdown and the release of photosynthates while carbon dioxide stimulates photosynthesis thereby compensating for the loss of photosynthates. In the feedback system urease is inactivated by lichen substances such as usnic acid and furmarprotocetraric acid. Usnic acid binds with urease to form aggregates of high molecular weight,[17] while fumarprotocetraric acid inactivates urease by means of a depolymerization process.[18] Assuming a direct relationship between the amounts of photosynthates and lichen substances, increased quantities of the latter will cause greater inactivation of urease and thereby slow the release of photosynthates from the photobiont.

Urease activity has been demonstrated also in *Peltigera canina*.[19] This lichen has a blue-green photobiont and lacks the depsides and depsidones which inactivate urease in lichens with green phycobionts. The control mechanism that blue-green lichens use for nutrient regulation may be different from that of green lichens.

## REDUCED VIRULENCE

As in other host-parasite systems, one can detect among lichens the evolution of a reduced virulence of the fungal symbionts toward their algal hosts. Under natural conditions evolution has favored strains of fungi that did not kill their hosts quickly, since by doing so they would have jeopardized their own survival. The reduced virulence may also reflect a greater resistance of the algal host.

In laboratory syntheses of *Cladonia cristatella*, I found that the mycobiont was more virulent toward its algal symbiont than in the natural lichen.[20] In synthetic cultures, algal cells were deeply penetrated with large club-shaped haustoria whereas in natural thalli the haustoria were small and peg shaped and did not extend far beyond the algal cell wall. Clones of the mycobiont obtained from single ascospores varied widely with respect to growth rate and isozymes.[21] We found that some clones of the *C. cristatella* mycobiont were more virulent than others. For example, clone 50 did not progress beyond the soredial stages because the fungus killed the algal cells before a thallus (squamule) could be developed. Presumably, under natural conditions such virulent strains of the fungus would be selected against. *Lecanora dispersa* is a small, crustose lichen the fungus of which forms large haustoria inside cells of its *Trebouxia* photobiont. Despite such fungal penetrations, the algal cells remain viable and the symbionts form a thallus. In the laboratory, however, strains of the *L. dispersa* mycobiont that originated from ascospores rapidly parasitized and killed their own algal partner as well as other algal strains. The mycobionts that we obtain in the laboratory from ascospores have not undergone a period of natural selection and probably reflect what the early stages of lichen evolution were like.

There are three types of morphological relationships between lichen symbionts. The most frequent one is where the fungus penetrates the photobiont cells and forms haustoria. A second relationship is where differentiated fungal cells called appressoria press against the algal cells. In the third type of contact the photobiont cells are enveloped by fungal hyphae but are not penetrated and there are no specialized hyphal cells.

There have been conflicting opinions as to the frequency of haustoria among lichens. Collins and Farrar suggested that haustoria were rare in lichens.[22] I feel, however, that haustoria are common. Ahmadjian and Jacobs reported that in natural squamules of *C. cristatella* 57-65% of the algal cells contained haustoria.[9] The parasitic nature of lichen fungi, as determined from synthesis studies, suggests that the presence of haustoria in lichens is the rule rather than the exception. In many lichens haustoria are difficult to observe because the fungus does not penetrate deeply into the algal cell. For example, in *Endocarpon pusillum* only appressorial contacts between fungus and alga could be seen with the light microscope. The electron microscope, however, revealed the presence of tiny haustoria.[23] Cells of *Trebouxia* have chloroplasts that fill most of their cells and can conceal haustoria. Although a high percentage of photobiont cells in a lichen are penetrated by the fungus, some cells remain fungal free. These uninfected cells divide and maintain the photobiont population while those that are infected presumably succumb over time. How deeply fungal haustoria penetrate the cells of a photobiont varies among different lichens. In some species, haustoria penetrate to the center of the photobiont cell and often beyond. In other lichens the haustoria do not go far beyond the cell wall of the alga. Tschermak[24] and Plessl[25] tried to correlate the depth of haustoria penetration to the structural organization of a lichen. They assumed that lichens with highly structured thalli were phylogenetically more advanced than those with less structured thalli. Their results suggested that primitive lichens, i.e., those with poorly organized thalli, had photobionts that were more deeply penetrated by haustoria than were photobionts of advanced lichens, i.e., those with well-organized thalli. Although distinctions between advanced and primitive lichens based on the degree of thallus organization are tenuous, the general findings of these studies seem valid. Galun *et al.* also found that photobionts of structurally primitive lichens were more frequently infected with haustoria than were those of structurally advanced lichens.[26]

The degree of haustorial penetration in a lichen may vary depending on environmental conditions. When *Lecanora gangaleoides* grows under optimal conditions, i.e., on the exposed side of a boulder, its algal cells have only small haustoria. If the lichen grows under poor conditions, i.e., on the underside of an overhang where there is much shade, its algal cells have deeply penetrating haustoria.[27] The differences seen in physical contacts between symbionts of advanced and primitive lichens may be due to the degree of algal resistance to the fungus or to the fact that in more advanced lichens the algal cells divide rapidly and thus do not allow the fungus time for haustorial development. A photobiont cell that is penetrated by haustoria either dies or divides. Since fungal haustoria do not penetrate the plasma membrane of the photobiont, it can still divide and in the case of green phycobionts form either aplanospores or zoospores. These asexual propagules are then released from the sporangium which retains the fungal haustoria. Greenhalgh and Anglesea observed haustoria in aplanosporangia of the phycobiont of *Parmelia saxatilis* and felt that the fungus invaded the sporangia and separated the aplanospores by growing between them.[28] This facilitated the distribution of algal cells in the thallus. Some investigators have felt that haustoria stimulate the division of the photobiont[29] and that in some lichens growth of the fungus is synchronous with that of the photobiont; as a photobiont cell divides, its daughter cells become infected with haustoria.[30] Some photobiont cells, however, appear to be killed by the fungus since degenerating photobiont cells are seen that contain haustoria.

In some lichens, when a hyphal cell contacts the surface of an algal cell, it differentiates into a specialized cell called an appressorium. This cell loses its cylindrical shape and becomes flattened against the algal surface. What purpose it serves after it attaches to an algal cell is not known. Appressoria do not give rise to haustoria but rather are separate structures. The stimulus that causes a hyphal cell to form an appressorium and whether every hyphal cell can form these structures are not known. Appressoria can be formed by apical as well as intercalary cells of a hyphal filament. Appressoria may be used by the mycobiont to secure the photobiont cells necessary to start a lichen thallus or they may be a highly specialized means of obtaining nutrients without penetration of the photobiont cell. Appressoria are produced even in lichens that form haustoria.

Some lichens do not have haustoria or appressoria. There may be a thinning of the photobiont wall at the points of contact with the hyphae. Lichens that are in this category have *Coccomyxa* or *Myrmecia* as phycobionts. According to Brunner and Honegger, these phycobionts have sporopollenin in their cell walls which makes them resistant to fungal intrusions.[31] Sporopollenin is a lipidlike compound that is resistant to microbial decomposition. *Trebouxia* has cellulose walls without sporopollenin which may explain why their cells are easily penetrated by fungal haustoria.

## CONTROL OF PHOTOBIONT POPULATION BY THE MYCOBIONT

Control of the photobiont by the mycobiont involves not only carbohydrate release but also the population size of the photobiont inside a thallus. Hill considers lichen development to be a balanced codevelopment of the symbionts and a coordinated expression of the fungal and photobiont genomes.[32] According to Hill, there is a rapid multiplication of the photobiont along the growing margins or tips of a lichen thallus and a constant population, with some cell turnover, in the rest of the thallus.[32]

The necessity of a balanced growth between the symbionts of a lichen had been recognized earlier from synthesis studies that showed that unbalanced growth of either symbiont, because of nutritional and environmental conditions, would not result in a lichen.

During the coevolution of the symbionts, it would appear that mechanisms for the regulation of growth of both symbionts would have to be an early adaptation.

## LOSS OF SEXUALITY

One of the most perplexing problems in lichenology is our inadequate knowledge of how lichens develop in nature. Can lichens form *de novo* from ascospores and free-living algae? There are arguments for and against this possibility. The diversity found among lichen populations suggests that some type of genetic recombination occurs. Genetic diversity could be introduced into a population by means of ascospores, which are products of genetic recombination, forming new lichens. Arguments against this possibility include the belief by some lichenologists, including myself, that free-living populations of potential photobionts, at least those of the most common one *Trebouxia,* do not exist. It is possible, however, that the hyphae from germinated spores could associate with algae from the asexual propagules of other lichens and that fusions of hyphae between different lichens could occur and result in parasexual unions.

In 1970 Poelt proposed the concept of species pairs.[33] This concept states that many lichen species have two forms, one fertile, i.e., with ascocarps, and one sterile. The sterile strains are descendants of fertile forms and generally are more widely distributed. Poelt considers the fertile forms of the species to be relicts while the sterile forms have lost their evolutionary flexibility. According to Tehler, asexual propagules of sterile forms have an advantage over ascospores in dispersing the lichen and selection will favor the sterile form of the species which is more energy efficient in terms of reproduction.[2] Thus, there is a progressive sterilization among lichens.

## SLOW GROWTH OF THE SYMBIONTS

Another coevolved feature of lichens is the slow growth of the symbionts. Under natural conditions, most lichens grow only a few millimeters in radial growth each year and observations of the algal cells in a lichen thallus show only a small percentage of them in a stage of division. The symbionts also grow slowly, relative to free-living fungi and algae, in laboratory cultures. Individual lichens may live for hundreds or thousands of years.[34] The slow growth of both symbionts is an adaptation to the environmental habitats in which lichens grow. In these habitats lichens undergo frequent cycles of wetting and drying and as a result they have only brief periods in which they can metabolize. The amount of nutrient gain that the photobionts achieve during these brief periods of metabolism is small and will not support organisms with rapid growth rates.

## SUMMARY

Lichens are highly integrated and ancient symbiotic associations. Some lichenlike fossils have been dated back to the Precambrian. The lichen life-style is undergoing a dynamic period of evolution among different groups of ascomycetes. Unique features of lichens such as thallus formation and synthesis of unusual secondary compounds are collaborative results of the symbionts. Release of carbohydrates and other compounds from the photobiont and control of the photobiont cell population within a lichen thallus appear to be regulated by the mycobiont. Fitness traits due to the symbiotic interactions of fungi and algae also include reduced virulence of the fungus, loss of sexuality, and slow growth of the symbionts. These phenomena plus the absence of free-living populations of most lichen fungi and photobionts indicate that lichen symbionts have undergone long periods of coevolution.

## REFERENCES

1. MATTOX, K. R. & K. D. STEWART. 1984. Classification of the green algae: a concept based on comparative cytology. Syst. Assoc. Spec. Vol. **27:** 29-72.
2. TEHLER, A. 1982. The species-pair concept in lichenology. Taxon **31:** 708-717.
3. BARRETT, J. A. 1983. Plant-fungus symbiosis. *In* Coevolution. D. J. Futuyma & M. Slatkin, Eds.: 137-160. Sinauer Associates. Sunderland, Mass.
4. HAWKSWORTH, D. L. 1982. Co-evolution and the detection of ancestry in lichens. Hattori Bot. Lab. **52:** 323-329.
5. HALLBAUER, D. K. 1975. The plant origin of the Witwatersrand "carbon". Miner. Sci. Eng. **7:** 111-131.
6. HALLBAUER, D. K., H. M. JAHNS & H. A. BELTMANN. 1977. Morphological and anatomical observations on some precambrian plants from the Witwatersrand, South Africa. Geol. Rundschau **66:** 477-491.
7. HAWKSWORTH, D. L. 1978. The taxonomy of lichen-forming fungi: reflections on some fundamental problems. *In* Essays in Plant Taxonomy. H. E. Street, Ed.: 211-243. Academic Press. London, England
8. HAWKSWORTH, D.L. & D. J. HILL. 1984. The Lichen-Forming Fungi. Blackie. Glasgow & London.
9. AHMADJIAN, V. & J. B. JACOBS. 1981. Relationship between fungus and alga in the lichen *Cladonia cristatella*. Nature **289:** 169-172.
10. KERSHAW, K. A. 1985. Physiological Ecology of Lichens. Cambridge University Press. Cambridge, England.
11. GOEBEL, K. V. 1926. Induzierte Dorsiventralität bei Flechten. Flora **121:** 177-188.
12. CULBERSON, C. F. 1969. Chemical and Botanical Guide to Lichen Products. University of North Carolina Press. Chapel Hill, N.C.
13. CULBERSON, C. F. & V. AHMADJIAN. 1980. Artificial reestablishment of lichens. II. Secondary products of resynthesized *Cladonia cristatella* and *Lecanora chrysoleuca*. Mycologia **72:** 90-109.
14. RUNDEL, P. W. 1978. The ecological role of secondary lichen substances. Biochem. System. Ecol. **6:** 157-170.
15. LEGAZ, M. E., G. G. DE BUITRAGO & C. VICENTE. 1982. Exogenous supply of L-arginine modifies free amino acids content in *Evernia prunastri* thallus. Phyton **42:** 213-218.
16. VICENTE, C. & M. E. LEGAZ. 1982. Purification and properties of agmatine amidinohydrolase of *Evernia prunastri*. Physiol. Plant. **55:** 335-339.
17. VICENTE, C. & B. CIFUENTES. 1979. Reversal by L-cysteine of the inactivation of urease by L-usnic acid. Plant Sci. Lett. **15:** 165-168.
18. VICENTE, C. & L. X. FILHO. 1979. Urease regulation in *Cladonia verticillaris* (Raddi)Fr. Phyton **37:** 137-144.
19. BROWN, D. H., C. VICENTE & E. LEGAZ. 1982. Urease activity in *Peltigera canina*. Cryptogamie Bryol. Lichenol. **3:** 33-38.
20. AHMADJIAN, V. 1982. Algal/fungal symbioses. *In* Progress in Phycological Research. F. E. Round & D. J. Chapman, Eds. **1:** 179-233. Elsevier Biomedical Press. Amsterdam, Holland.
21. FAHSELT, D. 1985. Multiple enzyme forms in lichens. *In* Lichen Physiology and Cell Biology. D. H. Brown, Ed.: 129-143. Plenum Press. New York, N.Y.

22. COLLINS, C. R. & J. F. FARRAR. 1978. Structural resistance to mass transfer in the lichen *Xanthoria parietina.* New Phytol. **81:** 71-83.
23. AHMADJIAN, V. & J. B. JACOBS. 1970. The ultrastructure of lichens. III. *Endocarpon pusillum.* Lichenologist **4:** 268-270.
24. TSCHERMAK, E. 1941. Untersuchungen über die Beziehungen von Pilz und Alge in Flechtenthallus. Österr. Bot. Z. **90:** 233-307.
25. PLESSL, A. 1963. Über die Beziehungen von Haustorientypus und Organisationshohe bei Flechten. Österr. Bot. Z. **110:** 194-269.
26. GALUN, M., N. PARAN & Y. BEN-SHAUL. 1970. The fungus-alga association in the Lecanoraceae: an ultrastructural study. New Phytol. **69:** 599-603.
27. JAMES, P. W. 1970. The lichen flora of shaded rock crevices and overhangs in Britain. Lichenologist **4:** 309-322.
28. GREENHALGH, G. N. & D. ANGLESEA. 1979. The distribution of algal cells in lichen thalli. Lichenologist **11:** 283-292.
29. GEITLER, L. 1963. Über Haustorien bei Flechten und über *Myrmecia biatorellae* in *Psora globifera.* Österr. Bot. Z. **110:** 270-280.
30. TSCHERMAK-WOESS, E. 1951. Über wenig bekannte und neue Flechtengonidien II. Eine neue Protococcale, *Myrmecia reticulata,* als Algenkomponente von *Catillaria chalybeia.* Österr. Bot. Z. **98:** 412-419.
31. BRUNNER, U. & R. HONEGGER. 1985. Chemical and ultrastructural studies on the distribution of sporopolleninlike biopolymers in six genera of lichen phycobionts. Can. J. Bot. **63:** 2221-2230.
32. HILL, D. J. 1985. Changes in photobiont dimensions and numbers during co-development of lichen symbionts. *In* Lichen Physiology and Cell Biology. D. H. Brown, Ed.: 303-317. Plenum Press. New York, N.Y.
33. POELT, J. 1970. Das Konzept der Artenpaare bei den Flechten. Dtsch. Bot. Ges. **4:** 187-198.
34. LAWREY, J. D. 1984. The Biology of Lichenized Fungi. Praeger Scientific. New York, N.Y.

# Naturally Occurring and Artificially Established Associations of Ciliates and Algae[a]

## Models for Different Steps in the Evolution of Stable Endosymbioses

W. REISSER

*Department of Biology*
*Philipps University*
*Lahnberge*
*D-3550 Marburg, Federal Republic of Germany*

### INTRODUCTION

Among freshwater ciliates, specimens with a greenish color are frequently observed. Microscopic examination shows that the color originates from green algae which can be concentrated in digestive vacuoles or are distributed evenly throughout the cytoplasm of the protozoan. Whereas in the first case algae are digested and ciliates become colorless again under defined laboratory conditions,[1] in the second case ciliates remain green and seem to be permanently habituated by coccoid algae. There exists a tremendous variety of such associations between freshwater ciliates and coccoid green algae ranging from ephemeral systems which are only temporarily green to stable ones in which algae are individually enclosed in special vacuoles and are protected against the attack of hot lytic enzymes.[2] Since it is reasonable to assume that stable permanent associations require a higher degree of partner coordination than temporary ones they may be regarded as being evolutionarily more advanced. Thus the various naturally occurring associations of freshwater ciliates and green algae offer examples for different steps toward the evolution of stable endosymbiotic systems. Yet a systematic study has been seriously hampered by the fact that most of them are rather difficult to grow in sufficient amounts under defined laboratory conditions. Those difficulties have now been overcome by the development of techniques that allow the synthesis of so-called artificial ciliate-algae associations, e.g., associations between one ciliate host species and different algal partners that do not occur in the original system. Those artificial associations can be grown easily in the laboratory and show different degrees of stability. Thus they offer a unique possibility of studying the basic mech-

[a]This study has been supported by the Deutsche Forschungsgemeinschaft.

anisms and different steps in the evolution of integration and interaction between cells with genomes of different origin and the various parameters involved in symbiosis formation.

## MATERIALS AND METHODS

The green *Paramecium bursaria* Ehrbg. and its alga-free (colorless) counterpart were cultivated as has been described previously.[3] For synthesis of artificial systems with colorless *Paramecium bursaria*, algae taxonomically closely related to the original symbiotic *Chlorella* sp.[4] were obtained from the Sammlung von Algenkulturen Göttingen (SAG):[5] a free-living *Chlorella* species (SAG 211-30, *Chlorella lobophora* Andr.) and a *Chlorella* species isolated from *Trapelia coarctata* (Lichenes, SAG 3.80). Infection experiments were done according to Reference 6 and the new *Paramecium*/ algae associations were designated as Pb/30 and Pb/3.80 and will be called artificial. The original green *Paramecium bursaria* was isolated from a pond in the Old Botanical Garden of Göttingen University[7] and is designated as Pbg or as naturally occurring or original system. Its algae are called Pbi (*Paramecium bursaria* isolate).

Stock cultures of the original and artificial systems were maintained in Erlenmeyer flasks; growth experiments were run on depression slides. For photoaccumulation tests, oxygen measurements,[14] $CO_2$-fixation experiments, and electron microscopy specimen were taken from bacterized (*Enterobacter cloacae*) stock cultures.

Electron microscopy was done according to Reference 8. The technique of photoaccumulation experiments has been described previously.[9] For measurements of oxygen production see Reference 6. For $^{14}CO_2$-fixation experiments see References 4 and 8.

## RESULTS

The naturally occurring association of *Paramecium bursaria* Ehrbg. and its symbiotic *Chlorella* sp. is characterized by typical ultrastructural, physiological, and behavioral features that have been reported previously[2] and shall be used here as typical markers of a stable, i.e., evolutionarily advanced, ciliate-algae association. Accordingly those markers will be used for testing the degree of partner interaction in artificial systems in order to get information on their stability and hence their putative position in the course of evolution toward stable permanent systems.

### Ultrastructure

There are no principal differences in the ultrastructural organization between the naturally occurring (see also References 8 and 10) and the artificial systems (FIGURES 1-3) grown under identical conditions.

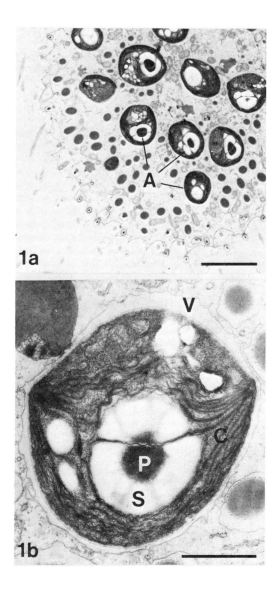

**FIGURE 1.** Original green *Paramecium bursaria.* (a) Algae (A) are enclosed in individual vacuoles which are mainly located in the periphery of the host cell. 3000×; bar, 5 μm. (b) Alga in individual vacuole (V). C, chloroplast; P, pyrenoid; S, starch; 20,000×; bar, 1 μm.

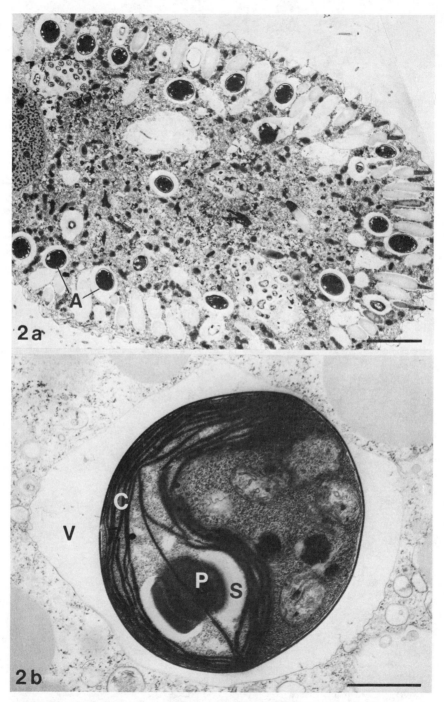

**FIGURE 2.** Artificial association of *Paramecium bursaria* with free-living *Chlorella* strain 211-30. (a) Algae (A) are enclosed in individual vacuoles which are mainly located in the periphery of the host cell. 3000×; bar, 5 μm. (b) Alga in individual vacuole (V); details as in FIGURE 1b; 20,000×; bar, 1 μm.

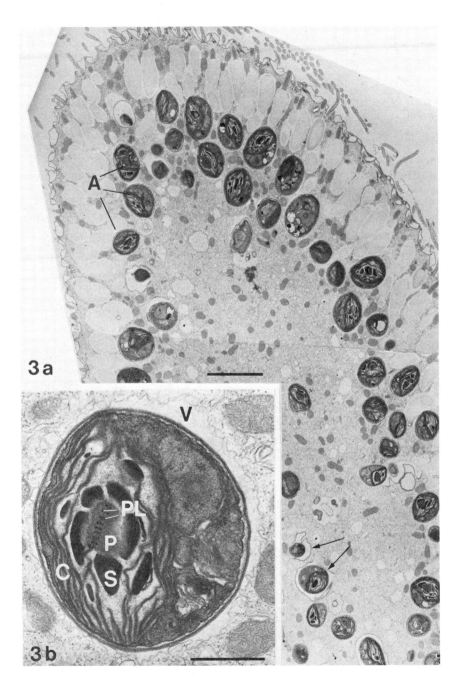

**FIGURE 3.** Artificial association of *Paramecium bursaria* with *Chlorella* strain 3.80 (isolated from *Trapelia coarctata,* Lichenes). (a) Algae (A) are enclosed in individual vacuoles which are mainly located in the periphery of the host cell. Note remnants of mother cell walls (arrows) together with algae in some vacuoles. 3000×; bar, 5 μm. (b) Alga in individual vacuole (V). PL, plastoglobuli; other details as in FIGURE 1b; 20,000×; bar, 1 μm.

Algae are enclosed in individual vacuoles which are mainly located in the periphery of the host cell. Chlorellae in those vacuoles are intact and show a prominent sheath of starch and a pyrenoid which is penetrated by a thylakoidal membrane (FIGURES 1b, 2b, and 3b). Algae divide in four autospores which are enclosed again in individual vacuoles. In Pb/3.80 frequently remnants of the mother cell wall can be observed together with enclosed algae (FIGURE 3b). In this system chlorellae often show a prominent accumulation of plastoglobuli which cannot be observed in the isolated algae grown in a nitrate-containing inorganic medium and thus this possibly indicates a deficiency of nitrogen supply *in situ*[11] (FIGURE 3b).

*Physiology*

In the naturally occurring green ciliates carbon metabolism is organized according to the animal-plant scheme, i.e., algal photosynthesis fixes respiratory $CO_2$ of the ciliate and produces oxygen which is used by host respiration and—at light intensities where the system works above its photosynthetic compensation point—is released into the medium. Therefore, in the artificial systems it had to be tested whether algae in individual vacuoles are not only morphologically but also physiologically intact and are able to meet the oxygen demand of their host. As is shown in FIGURE 4b there are no principal differences in gas metabolism between the original and the artificial systems. The photosynthetic compensation point of the artificial systems is higher (TABLE 1) but at about 3000 lx in all systems algal photosynthesis covers host respiratory oxygen demand and releases oxygen into the medium. In Pb/30 even more oxygen is produced than in the original system. Control experiments with chlorellae freshly isolated from their symbiotic unit show a prominent oxygen production (FIGURE 4a). Thus algae are physiologically intact. Their photosynthetic compensation points are much lower than the ones of the corresponding symbiotic associations (TABLE 1), e.g., oxygen released by algae *in situ* is fixed by host respiration.

Oxygen consumption by colorless *Paramecium bursaria* is about 30% higher than by the naturally green one in the dark on the basis of cell number.[2]

Obviously algae other than the original symbiotic ones are also able to exploit the host respiratory $CO_2$ supply and can build up a positive oxygen balance.

In order to get information on carbon fluxes in the artificial associations, green ciliates were labeled with $^{14}CO_2$ and activity was measured in the algal and in the *Paramecium* fraction. Data show (TABLE 2) that in the light, total $CO_2$ fixation in all systems is of the same order of magnitude. In the artificial associations it is even higher, e.g., plus 28% in Pb/30 and plus 30% in Pb/3.80. Hence algae other than the original ones are also able to maintain a high photosynthetic activity in the symbiotic habitat. Control experiments in darkness did not show any significant $^{14}CO_2$ fixation.

The predominant feature of carbohydrate metabolism in the original symbiotic system is the excretion of maltose by the algae[6,12] which can be measured as $^{14}C$ activity in the *Paramecium* fraction and amounts to about 36% of the total photosynthetically fixed material (TABLE 2). In the artificial systems the corresponding rates in the *Paramecium* fraction are about 5%. Since control experiments with isolated algae showed a comparably low activity in the supernatants of 211-30 and 3.80 assays but about 38% in the one of Pbi it is reasonable to conclude that the measured activity in the *Paramecium* fraction of artificial systems does not originate from the excretion

TABLE 1. Photosynthetic Compensation Points of Original and Artificial
*Paramecium bursaria-Chlorella* sp. Associations and of Algae Isolated from Them[a]

|          | Pbg  | Pbi | Pb/30 | 211-30 | Pb/3.80 | 3.80 |
|----------|------|-----|-------|--------|---------|------|
| PCP at lx | 1500 | 400 | 2600  | 600    | 2700    | 600  |

[a] For test conditions see Materials and Methods; PCP, photosynthetic compensation point; further abbreviations as in FIGURE 4.

of carbon by algae. Obviously the symbiotic milieu is not able to induce any excretion mechanisms in the free-living 211-30 or the 3.80. Remarkably, 3.80, in the appropriate milieu, e.g., as part of a lichen, excretes sugar alcohols. Low rates of activity in the *Paramecium* fraction of artificial systems also prove that algae in individual vacuoles cannot be digested in considerable amounts.

### Stability

A characteristic feature of stable ciliate-algae systems is the coordination of partner growth rates which guarantees that even under stress conditions acting selectively on only one partner the system is not broken up. Thus growth and stability of artificial associations had to be tested under various conditions stressing primarily either the ciliate (inorganic medium) or the algal partner (darkness) or imitating the natural milieu by feeding ciliates with bacteria.

As is demonstrated in TABLE 3 the naturally occurring system shows the highest division rates in the light when it is fed with bacteria. The division rate drops for about 50% without addition of bacteria but algal population size remains constant and there are no algae digested or expelled into the culture medium. When the association is cultured in continuous darkness with bacteria both division rate and algae number drop but again there is no digestion of algae. These observations confirm previously reported data showing that in the naturally occurring green *Paramecium bursaria* algal partners in their individual vacuoles are protected against the attack of host lytic enzymes.[13,14] The size of the algal population depends only on ecological parameters, e.g., on its ability to compete under autotrophic (light) or heterotrophic (darkness) conditions with the growth rate of the ciliate host (ecological regulation type).[2,13]

As is the case in the original system, in artificial associations the maximum population size is observed in the light and depends on the division rate of the host: a decreased host division rate results in a slight increase of the algal population size and vice versa. When specimens of Pb/3.80 with about 150 algae per ciliate are transferred from a bacterized to a bacteria-free medium the division rate of the host drops and algal number increases. Unlike chlorellae in the original system, algae in artificial associations show only a rather limited ability for heterotrophic growth. In darkness algal number rapidly decreases and in Pb/30 after about five days only alga-free ciliates are observed. Accordingly cultures of isolated 211-30 could not be grown under heterotrophic conditions.

A comparison of division rates of colorless *Paramecium bursaria* and naturally green ones shows that algae enhance the host division rate, e.g., sustain host growth.

This is obviously not the case in the artificial systems. Under comparable conditions their division rates are always lower than that of colorless *Paramecium bursaria*. Thus in these systems algae do not sustain host growth, either by excretion of sugars as is the case in the original system or by being digested as is excluded by the ultrastructural analysis, $^{14}CO_2$-fixation experiments, and cytochemical tests.[6,12,14]

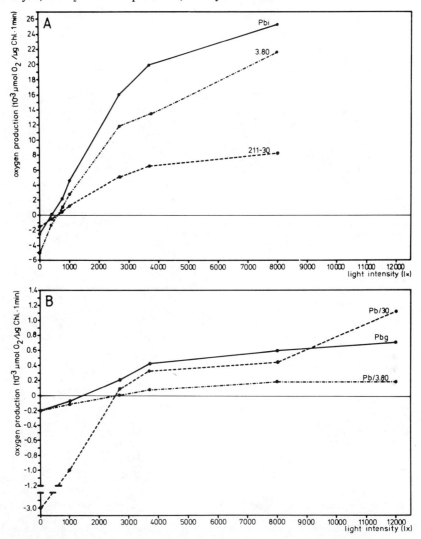

**FIGURE 4.** Photosynthetic oxygen production by original and artificial *Paramecium bursaria-Chlorella* sp. associations (B) and their freshly isolated algae (A). (For test conditions see Materials and Methods). Pbi, symbiotic *Chlorella* isolated from *Paramecium bursaria;* 3.80, symbiotic *Chlorella* isolated from *Trapelia coarctata* (Lichenes); 211-30, free-living *Chlorella lobophora* Andr.; Pbg, original green *Paramecium bursaria;* Pb/3.80, artificial association *Paramecium bursaria/Chlorella* strain 3.80; Pb/30, artificial association *Paramecium bursaria/Chlorella* strain 211-30.

TABLE 2. Photosynthetic Fixation of $^{14}CO_2$ by Original and Artificial *Paramecium bursaria–Chlorella* sp. Associations[a]

| | Pbg | Pb/30 | Pb/3.80 |
|---|---|---|---|
| Total fixation of whole association (cpm $\times$ 1 $\mu$g chlorophyll$^{-1}$ $\times$ 1 hr$^{-1}$ | 18733 | 23984 | 24367 |
| Activity (% of total fixation) | | | |
| In algal fraction | 64.3 | 95.2 | 94.5 |
| In ciliate fraction | 35.7 | 4.8 | 5.5 |

[a] Assay was run for 3 hours at 16°C and 13,000 lx (Philips white fluorescent bulbs); for further test conditions see Materials and Methods; abbreviations as in FIGURE 4.

In general, there is a remarkably higher inhomogeneity in the algal population size of artificial systems than in the original one. Besides green ciliates there are frequently observed colorless ones which are usually not found in cultures of the original system.

The upper limit of the algal population size in artificial systems is not given by available space. Even maximum algal numbers occupy a smaller proportion of the ciliate cell than algae in the natural system do (TABLE 3).

TABLE 3. Growth of Original and Artificial *Paramecium bursaria–Chlorella* sp. *Associations under Different Nutritional Conditions*[a]

| | Light-Dark Change | | Dark-ness |
|---|---|---|---|
| | +B | −B | +B |
| **Pbg** | | | |
| Divisions per day | 0.90 | 0.38 | 0.38 |
| Mean no. of algae in single vacuoles/P | 300–450 | 300–450 | 150–200 |
| Algae vol:vol of host minus algae | 1:11.5 | 1:11.5 | 1:26.6 |
| **Pb/3.80** | | | |
| Divisions per day | 0.28 | 0.15 | 0.25 |
| Mean no. of algae in single vacuoles/P | 120–160 | 120–180 | 10–25 |
| Algae vol:vol of host minus algae | 1:140.2 | 1:140.2 | 1:846.9 |
| **Pb/30** | | | |
| Divisions per day | 0.35 | 0.48 | 0 |
| Mean no. of algae in single vacuoles/P | 80–130 | 20–120 | 0 |
| Algae vol:vol of host minus algae | 1:29.8 | 1:29.8 | / |
| **Pbc** | | | |
| Divisions per day | 0.50 | 0 | / |

[a] Growth rate was monitored five days after start of experiment; light-dark change, 14:10 hours; 24°C; 3500 lx (Philips white fluorescent bulbs); for further test conditions see Materials and Methods; P, *Paramecium;* B, bacteria; Pbc, colorless *Paramecium bursaria;* further abbreviations as in FIGURE 4.

## Photobehavior

As has been demonstrated previously,[9] coordination of partners in ciliate-algae associations can result in a typical and qualitatively new type of behavior which is not shown by the isolated partners: green *Paramecium bursaria* show a step-down photophobic reaction and thus can gather in a light spot (photoaccumulation). This behavior depends on the photosynthetic activity of the algae and is an indicator of an intact algae population.

Measurements with artificial systems confirmed photoaccumulation for Pb/30 and Pb/3.80 (TABLE 4), although values of both thresholds and maxima are different. Threshold values for photoaccumulation in artificial systems are about 3500 lx for Pb/30 and about 400-500 lx in Pb/3.80 to about 600-1000 lx in the original system. In Pb/3.80 light intensities above 50000 lx lead to a decrease of the amount of accumulating specimen.

A step-up photophobic reaction is shown by all systems; probably it is a feature of the ciliate cell itself.

TABLE 4. Photoaccumulation of Original and Artificial *Paramecium bursaria–Chlorella* sp. Associations at Different Intensities of White Light[a]

| Light Intensity (lx) | Ciliates Accumulated in Light Spot after 2 Hours Illumination (%) | | |
|---|---|---|---|
| | Pbg | Pb/30 | Pb/3.80 |
| 100,000 | 95 | 75 | 10 |
| 10,000 | 70 | 35 | 75 |
| 5,000 | 25 | 0 | 40 |
| 500 | 0 | 0 | 5 |

[a] For test conditions see Materials and Methods; abbreviations as in FIGURE 4.

## DISCUSSION

There exist various reports on studies on the specificity of the *Paramecium bursaria-Chlorella* sp. association by infection of alga-free *Paramecium bursaria* with other than the original symbiotic algae.[6,15–20] Unfortunately most experiments were restricted to a phenomenological description of the eventually formed new associations. Ultrastructural characteristics as well as quantitative data on physiology and behavior are usually lacking so that it is rather difficult to draw any conclusions as to the efficiency of those systems in the natural habitat.

Nevertheless, experiments clearly demonstrate that the colorless *Paramecium bursaria* can take up not only its original symbiotic partners but also a limited number of other algal species and form with them "green" ciliates under special conditions. As could be shown here among these artificial systems the ones with *Chlorella* species 211-30 and 3.80 are fairly stable, i.e., can be cultured under appropriate conditions in large amounts in the laboratory. Both ultrastructural and physiological data show that the overwhelming portion of the algae is enclosed in individual vacuoles where they are protected against the attack of host lytic enzymes. Their integration in the host metabolism results in a photosynthetic activity even higher than the one of the original algae as well as in the unique phenomenon of photoaccumulation. Yet unless in the original system, algae cannot sustain host growth, either by excretion of carbohydrates or by being digested in significant amounts. As is the case in the original association the size of the algal population in artificial systems is also not limited by available space or by the membrane material needed for the formation of individual vacuoles.[13] The most important parameter limiting the number of algae is probably their ability for heterotrophic growth. Whereas Pbi remains within the system even after six weeks of continuous darkness[20] and accordingly can be grown heterotrophically on agar slants, 211-30 and 3.80 show only a limited capacity for heterotrophy and thus are lost from the system.

In general, it can be concluded that the number of both symbiotic and free-living algae taken up in individual vacuoles of *Paramecium bursaria* is not regulated primarily via digestion but by ecological parameters (ecological regulation type).[2,13] Besides these regulatory mechanisms, obviously additional factors exist that act on the algal number and are usually overlooked in the original system but are clearly expressed in the less stable artificial units: whereas in a culture of the original green *Paramecium bursaria* there are only rather few (less than 0.1%) individuals with fewer or even without algae, among the artificial systems the percentage of specimens with fewer than the average or without algae is much higher (up to 10-20%), at highest in Pb/30 where even in the light alga-free ciliates can be observed. Accordingly, in these cultures there are always algae growing on the bottom of the culture vessel that are physiologically intact and possibly have been expelled by paramecia or were set free following autolysis of some specimen. Probably this inhomogeneity of the ciliate population reflects some adaptation also of the host to its original symbiotic algal partner.

A comparison of original and artificial green *Paramecium bursaria* gives some interesting insights into the possible evolution of a stable ciliate-algae association. It is reasonable to assume that it is directed toward an increasing efficiency of the system. Accordingly, it must be possible to replace an existing algal partner by one that is better adapted to the symbiotic environment. A comparable mechanism has been observed when in a *Paramecium bursaria*-yeast association fungi were replaced by offered algae.[21] Thus the establishment of a stable *Paramecium*-algae association probably is a multistep process with different levels of partner selection. In the first step algae are taken up in special protective vacuoles. During this process algae are selected by size and mainly according to special cell wall surface structures,[22] which trigger the formation of a special vacuole membrane. As a result of this first step a relatively broad range of *Chlorella* species (SAG: 211-40c, 211-11b, 211-30, 3.80)[6,17] is taken up and can form "green" paramecia. In a second step algae are selected according to their ability to meet the ecophysiological requirements of the ciliate habitat, e.g., to maintain a growth rate that prevents them from being lost by mere "dilution." During this process possibly the ability to perform photosynthetic $CO_2$ fixation at a high level is only of minor importance. Probably more decisive are mechanisms that allow a heterotrophic growth and the exploitation of host N sources as is shown in Pb/3.80 where ultrastructural features of algae hint to a lack of nitrogen (accumulation of starch and plastoglobuli). In addition, those features of algae may be of evolutionary advantage that sustain host growth, e.g., carbohydrate excretion, and lead to a better growth of green ciliates as compared with alga-free ones.

Probably there is also a coevolution of the ciliate host which may contribute to the inhomogeneity of artificial cultures. Among most recent ciliate genera usually only one species forms a stable endosymbiotic unit with *Chlorella* sp. *Paramecium caudatum* digests every type of offered algae.[2]

As to the evolutionary advantages of symbiotic associations in comparison to colorless ciliates there exist only few data. Symbiotic units may have some selection advantages in habitats with less bacteria. An interesting observation was made as to the availability of green ciliates to predators: it could be shown that green *Paramecium bursaria* is significantly less attacked by *Didinium nasutum* than colorless *Paramecium bursaria* is, which never has been observed in the natural milieu. Perhaps, symbiotic algae release some "repelling metabolites."[23]

Thus the contemporary *Paramecium bursaria-Chlorella* sp. association results from a selection process that has led to a high specificity of both partners. A similar mechanism has been suggested for the *Hydra-Chlorella* sp. association.[24] In green *Paramecium bursaria* evolution is probably still at work on the level of ecophysiological adaptation. Preliminary investigations show that symbiotic chlorellae isolated from *Paramecium bursaria* of different collection sites show different ecophysiological features and lead in cross infections to artificial associations with different degrees of stability.[25]

# SUMMARY

Naturally occurring endosymbiotic associations of freshwater ciliates and coccoid green algae (*Chlorella* sp.) differ significantly in stability thus showing different levels of symbiotic integration. By the development of techniques to synthesize artificial endosymbiotic systems, i.e., associations of one ciliate species with other than the original symbiotic algae, it is possible to obtain associations that differ in stability and thus allow the study of those parameters of symbiosis formation that are responsible for the evolution of stable units.

Using the association of *Paramecium bursaria* Ehrbg. with its original symbiotic *Chlorella* sp., a free-living *Chlorella* (*Chlorella lobophora* Andr.), and a symbiotic *Chlorella* isolated from *Trapelia coarctata* (Lichenes) it can be shown by an analysis of ultrastructural, physiological (photosynthesis, respiration, growth) and behavioral (photobehavior) features, that the formation of a stable endosymbiotic unit is a multistep process including different levels of partner interaction. Suitable algae are recognized by the *Paramecium* by their size and certain cell wall surface structures which trigger the individual enclosure of algae in special vacuoles where they are protected against the attack of host lytic enzymes. Algae then can grow and populate their host according to their ability to meet the ecophysiological requirements of their special habitat, i.e., the amount of available organic and inorganic nutrients, light, and $CO_2$. Probably a less adapted algal population can be replaced by a better adapted one, and there are strong hints that also a coevolution of the ciliate host has taken place.

Probably in the *Paramecium bursaria-Chlorella* sp. system evolution is still going on since in associations collected from different localities different strains of *Chlorella* sp. have been found.

# ACKNOWLEDGMENTS

I wish to thank Dr. R. Meier, University of Heidelberg, for kindly providing FIGURES 2 and 3 and T. Klein for her help in preparing the manuscript.

## REFERENCES

1. REISSER, W. 1986. Endosymbiotic associations of freshwater protozoa and algae. Progr. Protistol. **1**: 195-214.
2. REISSER, W. & W. WIESSNER. 1984. Autotrophic eukaryotic freshwater symbionts. *In* Encyclopedia of Plant Physiology, New Series. Intercellular Interactions. H. F. Linskens & J. Heslop-Harrison, Eds. **17**: 91-112. Springer Verlag. Berlin, Heidelberg & New York.
3. NIESS, D., W. REISSER & W. WIESSNER. 1981. The role of endosymbiotic algae in photoaccumulation of *Paramecium bursaria*. Planta **152**: 268-271.
4. REISSER, W. 1984. The taxonomy of green algae endosymbiotic in ciliates and a sponge. Br. J. Phycol. **19**: 309-318.
5. SCHLÖSSER, U. G. 1982. Sammlung von Algenkulturen: List of strains. Ber. Dtsch. Bot. Ges. **95**: 181-276.
6. NIESS, D., W. REISSER & W. WIESSNER. 1982. Photobehavior of *Paramecium bursaria* infected with different symbiotic and aposymbiotic species of *Chlorella*. Planta **156**: 475-480.
7. REISSER, W. 1976. Die stoffwechselphysiologischen Beziehungen zwischen *Paramecium bursaria* Ehrbg. und *Chlorella* spec. in der *Paramecium bursaria*-Symbiose. I. Der Stickstoff- und der Kohlenstoffstoffwechsel. Arch. Microbiol. **107**: 357-360.
8. REISSER, W. 1976. Die stoffwechselphysiologischen Beziehungen zwischen *Paramecium bursaria* Ehrbg. und *Chlorella* spec. in der *Paramecium bursaria*-Symbiose. II. Spezifische Merkmale der Stoffwechselphysiologie und der Cytologie des Symbioseverbandes und ihre Regulation. Arch. Microbiol. **111**: 161-170.
9. REISSER, W. & D.-P. HÄDER. 1984. Role of endosymbiotic algae in photokinesis and photophobic responses of ciliates. Photochem. Photobiol. **39**: 673-678.
10. KARAKASHIAN, S. J., M. W. KARAKASHIAN & M. A. RUDZINSKA. 1968. Electron microscopic observations on the symbiosis of *Paramecium bursaria* and its intracellular algae. J. Protozool. **15**: 113-128.
11. MEIER, R. & W. REISSER. (In preparation).
12. MUSCATINE, L., S. J. KARAKASHIAN & M. W. KARAKASHIAN. 1967. Soluble extracellular products of algae symbiotic with a ciliate, a sponge and a mutant *Hydra*. Comp. Biochem. Physiol. **20**: 1-12.
13. REISSER, W., R. MEIER & B. KURMEIER. 1983. The regulation of endosymbiotic algal population size in ciliate-algae associations. An ecological model. *In* Endocytobiology II. W. Schwemmler & H. E. A. Schenk, Eds.: 533-543. W. de Gruyter & Co. Berlin, FRG.
14. KARAKASHIAN, S. J. & M. A. RUDZINSKA. 1981. Inhibition of lysosomal fusion with symbiont-containing vacuoles in *Paramecium bursaria*. Exp. Cell Res. **131**: 387-393.
15. OEHLER, R. 1922. Die Zellverbindung von *Paramecium bursaria* mit *Chlorella vulgaris* und anderen Algen Arb. Staatl. Inst. Exp. Ther. Georg-Speyer-Haus Frankfurt am Main **15**: 5-19.
16. PRINGSHEIM, E. G. 1928. Physiologische Untersuchungen an *Paramaecium bursaria*. Arch. Protistenk. **64**: 289-418.
17. BOMFORD, R. 1965. Infection of alga-free *Paramecium bursaria* with strains of *Chlorella, Scenedesmus* and a yeast. J. Protozool. **12**: 221-224.
18. KARAKASHIAN, S. J. & M. W. KARAKASHIAN. 1965. Evolution and symbiosis in the genus *Chlorella* and related algae. Evolution **19**: 368-377.
19. HIRSHON, J. B. 1969. The response of *Paramecium bursaria* to potential endocellular symbionts. Biol. Bull. **136**: 33-42.
20. MEIER, R., W. REISSER & W. WIESSNER. 1980. Cytological studies on the endosymbiotic unit of *Paramecium bursaria* Ehrbg. and *Chlorella* spec. II. The regulation of the endosymbiotic algal population as influenced by nutritional condition of the symbiotic partners. Arch. Protistenk. **123**: 333-341.
21. GÖRTZ, H.-D. 1982. Infections of *Paramecium bursaria* with bacteria and yeasts. J. Cell Sci. **58**: 445-453.

22. REISSER, W., A. RADUNZ & W. WIESSNER. 1982. The participation of algal surface structures in the cell recognition process during infection of aposymbiotic *Paramecium bursaria* with symbiotic chlorellae. Cytobios **33**: 39-50.

23. BERGER, J. 1980. Feeding behavior of *Didinium nasutum* on *Paramecium bursaria* with normal or apochlorotic zoochlorellae. J. Gen. Microbiol. **118**: 397-404.

24. RAHAT, M. 1985. Competition between chlorellae in chimeric infections of *Hydra viridis*: the evolution of a stable symbiosis. J. Cell Sci. **77**: 87-92.

25. REISSER, W. (In preparation.)

# Algal Symbiosis as the Driving Force in the Evolution of Larger Foraminifera[a]

JOHN J. LEE [b]

*Department of Biology*
*City College*
*City University of New York*
*Convent Avenue at 138th Street*
*New York, New York 10031*


PAMELA HALLOCK

*Department of Marine Science*
*University of South Florida*
*St. Petersburg, Florida 33701*

## INTRODUCTION

Foraminifera have been a very successful group of protozoans for over 500 million years. Loeblich and Tappan recognized 220 foraminiferal families;[1] species counts exceed 25,000. Though the majority of species are in the 0.1-1.0 mm size range, species whose members attain large sizes ( > 1 cm) are known both from modern seas and in the fossil record.

However, "larger" species do not occur throughout the history of foraminifera, but instead have arisen from time to time in evolutionary bursts from ordinary-sized ancestors. In fact, the term "larger foraminifera" was coined by paleontologists to refer to often unrelated species whose individuals reach large sizes with complex internal morphologies, have few external characteristics useful for identification, and typically occur in limestones, so study in thin section is necessary. Ross recognized larger representatives in 25 families,[2] but Loeblich and Tappan's taxonomic revisions nearly double that number.[1]

Larger foraminifera do not randomly appear in the fossil record. Their diversifications generally correspond with the "polytaxic" episodes in the geologic record defined by Fischer and Arthur,[3] and they diminish during "oligotaxic" episodes. For

[a] Portions of P. Hallock's work were supported by National Science Foundation Grants EAR-8026459 and EAR-8407781, and by a grant from the University of South Florida Faculty Creative Research Fund. Dr. J. J. Lee's work was supported by National Science Foundation Grant OCE 09266, U.S. Israel Binational Grant 3418/83, and PSC-CUNY Grant 666306.

[b] Also affiliated with the Department of Invertebrates, American Museum of Natural History.

example, diversifications of Orbitoididae and Pseudorbitoididae, as well as earliest diversifications of the important Cenozoic families Alveolinidae, Calcarinidae, and Nummulitidae, occurred during the polytaxic Cretaceous. Likewise, the Paleocene-Eocene saw diversifications of Lepidocyclinidae, Nummulitidae, Amphisteginidae, and others.

Perhaps the most interesting features of fossil larger foraminifera are recurring morphologic trends. For example, evolutionary trends towards similar shapes and geometries[2] can be seen in two unrelated fusiform taxa, Pennsylvanian fusulinids and Cretaceous-Cenozoic alveolinids. Other trends, including increasing adult and embryon size, increasingly complex internal morphologies, reduction in the number of juvenile chambers, and increasing surface-to-volume ratios, occur in larger foraminiferal lineages whether they are fusiform, planispiral, or discoid.

It was not until fairly recently that biological research on larger foraminifera showed its own "evolutionary burst" that solutions to some of the mysteries surrounding larger foraminifera began to appear. For the key to understanding the function of these morphologic peculiarities is algal symbiosis.

While micropaleontologists and specialists in foraminiferan symbiont biology may be familiar with the intricacies and complexities of larger foraminiferan cells and their cytoplasmic compartmentalization, most endosymbiologists have not yet gained an appreciation. Recently Hottinger, Hansen, and their collaborators have made significant advances in our ability to visualize easily in three dimensions the compartmentalization of cytoplasmic spaces within these highly evolved cells (e.g., References 4-8). The technique is simple; the foraminiferan shell is infiltrated by a plastic resin (Araldite), a vacuum is used to draw out any bubbles that might be trapped, and then the resin is polymerized. The shell is then dissolved away leaving a cast of the cytoplasmic spaces (see Reference 9 for technique). To pique intellectual curiosity we provide here a few illustrations [FIGURES 1(a-d), 2(2), and 3(4)]. An excellent, well-written, and clearly illustrated review on the subject is recommended as an introduction.[10]

# IDENTITY OF THE SYMBIONTS

The symbiotic relationships among the genera (e.g., *Peneroplis, Archaias, Sorites*) belonging to most of the superfamily Soritacea seem quite specific. *Archaias angulatus* is the host to a chlorophyte *Chlamydomonas hedleyii.*[11] A closely related foraminifera, *Cyclorbiculina compressa,* is the host to a closely related endosymbiotic alga, *Chlamydomonas provasolii.*[12] Neither alga has been found yet living free in the sea. Fine structural observations indicate that *Chlamydomonas hedleyii* may also be the endosymbiont of another larger foraminifer, *Peneroplis proteus* (=*Puteolina proteus*).[13] Another species of *Puteolina* from New Caledonia is also green, indicating that it may also harbor a chlorophyte.

*Peneroplis pertusus, P. planatus* [FIGURE 3(4)] and *Spirulina arietina* are hosts to a rhodophyte.[13,14] Fine structural study of one axenic isolate from *Peneroplis planatus* indicates that there are no morphological differences between it and an axenic reference strain (UTEX 637) of *Porphyridium purpureum* when they were both grown under identical conditions (Hawkins and Lee, in preparation).

All of the Soritinae (e.g., *Sorites* spp., *Amphisorus hemprichii, Marginopora vertebralis*) are hosts for endosymbiotic dinoflagellates.[13-16] These symbionts appear to

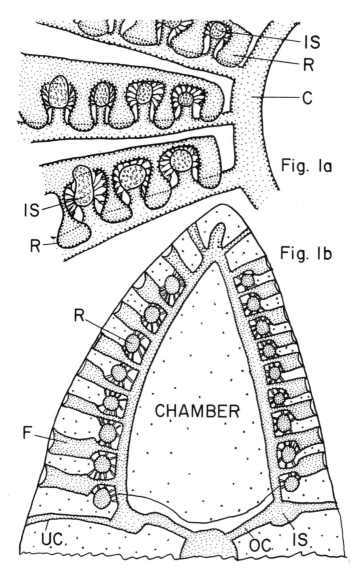

**FIGURE 1.** *Elphidium crispum* a foraminiferan with a well-developed canal system. This species temporarily husbands functioning chloroplasts derived from diatoms. (a) Simplified diagrammatic representation of a portion of the complex protoplasmic spaces found in this foraminifer. Representation is near umbilicus, perpendicular to axis, and shows umbilical spiral canal system (C), retral processes (R), and interlocular spaces (IS). Diagram based on SEM photographs made by Hottinger and Leutenegger (1980) and Hansen and Lykke-Anderson (1976) from casts. Magnification approximately 400×. (b) Simplified diagrammatic representation of the protoplasmic spaces when viewed on edge (side view). Diagram of the upper half of the animal in section and shows the location of the retral process (R), interlocular spaces (IS), fossette (F), vertical umbilical canal (UC), oblique vertical canal (OC), chamber or loculum. Magnification approximately 280×.

**FIGURE 1.** (*continued*). (c) SEM of a specimen collected by the authors near Mombasa on the Indian Ocean. 90×. (c) Enlargement of the above showing ponticulus (P) and fossette (F) intraseptal canal mouth. A pennate diatom frustule (d) illustrates the difference in size between the chloroplast donors and the host. 1700×.

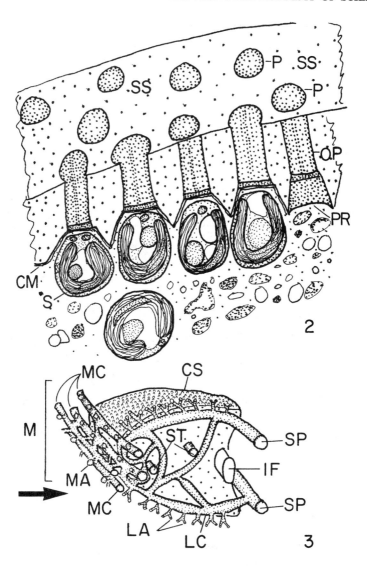

**FIGURE 2.** (2) Diagrammatic representation of the cortical region of *Amphistegina lobifera* showing outer shell surface (SS), pores (P), organic pore lining (OP), cell membrane (CM), inner shell surface-pore rim (PR), symbiotic diatom (S). An SEM of shell architecture is shown in FIGURE 3(6). Based on TEMS by Koestler *et al.*[58] and Leutenegger.[13] Magnification approximately 3200×. (3) Diagrammatic representation of canal system and chamber surface in a single operculinid chamber showing marginal cord (M), intercameral foramen (IF), lateral apertures (LA), lateral canal (LC), lateral chamber surface (CS), marginal aperatures (MA), marginal canal (MC), septal chamber surface (SS), stolo (ST), spiral umbilical canal (SP). Redrawn from Hottinger and Dreher (1974) who based their interpretation of SEM observations of casts made of the chamber spaces. A complete organism is shown in FIGURE 4. Imagine looking at FIGURE 4(9) and dissolving away the shell. The chambers would look like FIGURE 2(3) when viewed from the direction of the arrow. Approximately 200×.

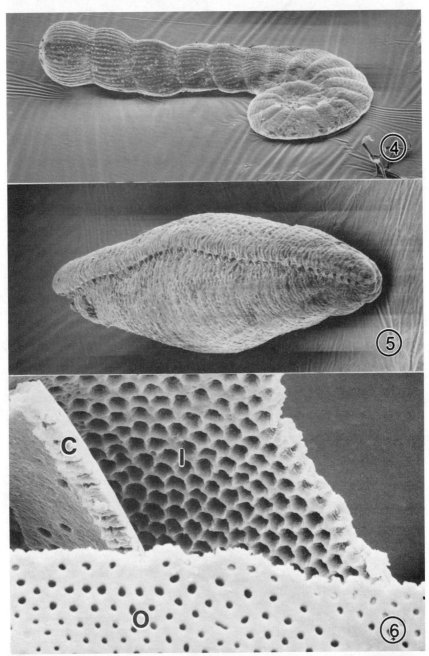

**FIGURE 3.** (4) *Peneroplis planatus* from 20 m at Wadi Taba on the Red Sea (SEM, 65×). (5) *Borelis schlumbergeri* from 20 m at Wadi Taba on the Red Sea (SEM, 85×). (6) SEM of a portion of a broken shell of *Amphistegina* Lobifera showing the outer surface (O), inner surface with pore rims (1), and a cross section (C) of the shell (1000×).

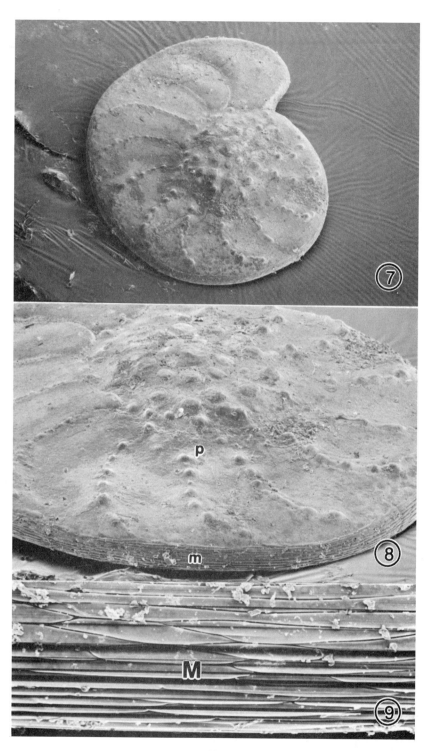

be similar, or even identical, to *"Symbiodinium microadriaticum,"* but definitive identification must wait until motile stages are examined and we know more about their
physiological ecology. Recent physiological and chromosomal studies indicate that
*"S. microadriaticum"* isolates from various invertebrate hosts are more diverse than
previous believed.[17-21] It is anticipated that further studies of the differentiation found
within *Symbiodinium* will lead to the erection of new specific epithets.[20,21] Perhaps
the endosymbiotic dinoflagellates of larger foraminifera are related to each other and
to symbionts of other invertebrates, but it is now clear that the comparisons will have
to be made only after very detailed studies.

Diatom endosymbionts have been found among the genera {e.g., *Borelis* [FIGURE
3(5)] and *Alveolinella*} of the imperforate family Alveolinidae and in many perforate
genera of foraminifera {e.g., *Calcarina, Amphisteqina* [FIGURE 3(6)], *Heterostegina,
Nummulites, Operculina* (FIGURE 4), *Heterocyclina, Cycloclypeus*} (e.g., References
14 and 22). Lee and co-workers have isolated and cultured diatom endosymbionts
from several of these genera and have used this material for study by light and electron
microscopy.[23-26]

For example, in July 1985, we collected several dozen specimens of *Calcarina
calcar,* a foraminifer whose symbionts were unknown, from a backreef habitat at
Nyali, Kenya (north of Mombasa). The samples were returned to New York where,
using aseptic technique, individual foraminifera were brushed, washed, and crushed
so that their endosymbiotic algae could be liberated and inoculated into media (e.g.,
Reference 23). The diatoms that grew were killed and cleaned with $H_2O_2$ so that
their frustules could be examined with the aid of a scanning electron microscope
(SEM). A single species, *Nitzschia frustulum* var. *symbiotica,* was found in all 24
specimens of *Calcarina* studied. Two of the 24 specimens had a second endosymbiotic
diatom species; one had *Nitschia panduriformis,* the other a *Navicula* sp. (similar to
an unknown new species collected from foraminifera from Elat on the Red Sea—Lee
and Reimer, in preparation).

Lee and co-workers have isolated and identified approximately 20 species of diatoms
from many of the above-mentioned genera of larger foraminifera using these methods.[23-26] Among the more commonly isolated endosymbionts are *Fragilaria shiloi,
Nitzschia frustulum* var. *symbiotica, N. frustulum* var. *subsalina, N. laevis,* and *N.
panduriformis* var. *continua* (References 23-26 and Lee *et al.,* in preparation).

Among diatom-bearing hosts, different populations (temporarily and spatially) of
the same host species can apparently harbor different species of endosymbiotic diatom.
This phenomenon has been widely documented by Lee and co-workers (References
23-26 and ms. in preparation). Hansen and Buchardt seem to have observed this
phenomenon in their transmission electron microscope (TEM) preparations.[27] Although Leutenegger, using TEM preparations of individual foraminifera, did not report
intraspecific variation in endosymbionts, she found four clearly distinct pyrenoid types
among endosymbiotic diatoms examined from various hosts.[22] The presence of other
differences such as cell size, chloroplast number, and minor differences in pyrenoid
structure indicated additional "subtypes."

Both direct TEM examination and isolation/culture have their limitations as
techniques for identification of diatom endosymbionts. Using TEM, the researcher
knows that the algal cells being examined were living within the host species. However,
preparative and examination time and cost limit the number of specimens that can

---

**FIGURE 4.** *Operculina ammonoides* from 38 m at Wadi Taba in the Red Sea. Note glassy
pustules (P) in (7) and (8) and marginal cord (M) in (8) and (9). Magnification: (7) 45×;
(8) 95×; (9) 525×.

be examined by TEM. For example, Leutenegger examined three specimens per species.[13] But diatom taxonomy is based on features of the frustule (outer envelope), an organelle not formed inside the host. Features observable in TEM, such as cell shape, size, chloroplast number and structure, and pyrenoids, vary among species, but have not been usefully applied to the taxonomy of the group. On the other hand, endosymbionts isolated from their hosts do form frustules in culture, and can thereby be identified taxonomically. However, there is always some question as to whether organisms isolated from within a foraminifer may be surviving ingested food rather than endosymbiotic. Therefore, Lee and associates (work in progress) are working on an immunological technique to identify endosymbionts *in situ.*

Axenic cultures of endosymbiotic *Fragillaria shiloi* and *Nitzschia panduriformis* were homogenized with the aid of a sterile Potter-Elvehjem tissue grinder. The cell envelope/frustule fraction of the cells was obtained by washing the homogenate in sterile seawater followed by centrifugation. After repeated washing and centrifugation, aliquots of the fraction were examined in the light microscope and filtered onto nucleopore membranes for examination in the SEM. We observed only frustule fragments and membrane debris in the preparations. Protein concentration of the fraction was determined by the Bio-Rad (Richmond, Calif.) microassay procedural modification of the Bradford method. Concentration was then adjusted until the protein level of the fraction was 1 mg/ml. A commercial vendor (Pocono Rabbit Farm and Laboratory, Canadensis, Pa.) prepared rabbit antisera for us. After absorbing the antisera with heterologous species, we incubated antisera with freshly collected, washed, and crushed *Amphistegina lobifera* and *Heterostegina depressa* from the Gulf of Elat. The specimens were washed in seawater and then incubated with anti-rabbit immunoglobulin G (IgG) (whole molecule) fluorescein isothiocyanate (FITC) conjugate. Axenic cultures of *F. shiloi* and *N. panduriformis* served as positive and negative controls respectively for each antiserum. The preparations were then examined with the aid of a Zeiss photomicroscope fitted with epifluoresecent optics. Part of the same collection of foraminifera was used to isolate, culture, and identify the endosymbionts using methods described previously (e.g., Reference 23). Although we did observe autofluorescence of the chloroplasts, we did not observe FITC fluorescence in any of the endosymbionts in the crushed animals. We did not find *F. shiloi* or *N. panduriformis* among the diatoms isolated from the other part of the collection after they grew and were studied. These negative and noncontradictory results encourage us to continue this line of research (Lee and co-workers, work in progress).

We were curious as to whether endosymbionts could be detected in representatives of some of the extinct larger foraminifera which were abundant in the Paleozoic. At the suggestion of two colleagues at the American Museum of Natural History, Drs. Norman Newell and Roger Batten, we examined samples of chert in the museum's collection. A thin section and small samples of a phosphoric chert from Three Forks Montana seemed quite promising.[28] We found within the very small sample available to us many hollow organic spherical remains within the extremely well preserved foraminifera (FIGURE 5). The spheres might have been endosymbiotic algae. These initial results encourage us to suggest that paleotoic cherts may some day yield more details of the putative symbionts of fusulinaceans.

## ADAPTATIONS OF THE HOST

To understand why algal symbiosis is such an important force driving morphologic evolution in larger foraminifera, two phenomena must be recognized. The first is the

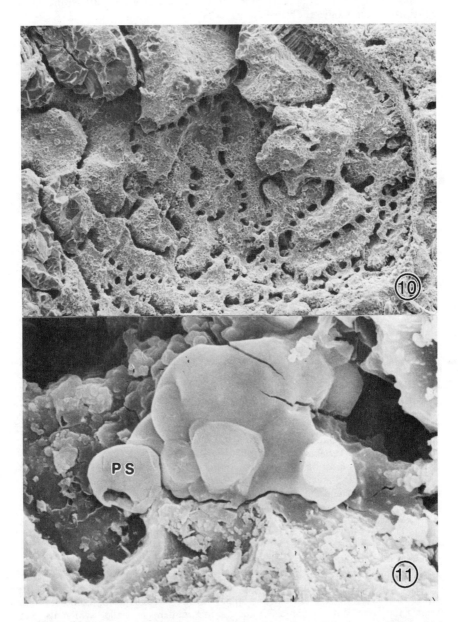

**FIGURE 5.** SEMs of a cut and etched *Pseudoschwagerina montanensis* from a sample of material used by Frenzel and Mundorff in their study of the Phosphoria formation of Montana.[28] (10) 85×. (11) Higher magnification (5000×) of the above showing putative symbiont (**PS**).

potential energetic benefit of algal symbiosis; the second is the climatic setting under which larger foraminiferal diversifications occurred.

Hallock showed mathematically that when nutrient and food supplies are scarce, algal symbiosis can theoretically provide the host-symbiont system with energetic benefit orders of magnitude over similar organisms lacking algal symbionts.[29] The relationship permits the host-symbiont system to function mixotrophically, photosynthesizing using scarce nutrients that the host has captured in their only concentrated form: particulate organic carbon. Thus, in an oligotrohpic environment, once the relationship is established, its energetic benefits can provide selective advantage to those members of a population adapted to best utilize their symbionts.

"Polytaxic" episodes in the geologic record coincide with high stands of sea level when climates were globally equable and rates of oceanic circulation diminished.[3] During such times, rates of nutrient recycling to surface waters were reduced, so organic productivity dropped as much as one to two orders of magnitude.[30] The resulting expanses of clear, oligotrophic, shallow seas probably provided ideal conditions for diversifications of larger foraminifera and other algal symbiont-bearing organisms.[31,32] Extinctions of larger foraminifera often occurred when those habitats were lost as seas regressed and oceanic circulation intensified.[33]

Among Cenozoic larger foraminifera, new characteristics or novelties in test morphology typically appeared suddenly (geologically speaking), while modification of characteristics such as embryon size generally occurred more gradually.[34] Hallock pointed out that in an algal symbiont-bearing lineage, a new characteristic that increased the advantage of the host-symbiont relationship or improved chances for juvenile survival probably spread rapidly under low nutrient conditions.[32] Likewise, morphologic characteristics prone to gradual modification, such as test shape or embryon size, would also be subject to strong selective pressures, for the energetic benefits of improving the symbiotic relationship would be substantial.

As for trends towards increasing adult and embryon size so commonly seen in larger foraminifera, Hallock showed that delayed maturation and growth to large sizes are only advantageous under relatively predictable environmental conditions where food resources are limited.[32] Both longevity and reproduction of relatively few, large embryos involve life history strategies that are simply not capable of responding rapidly to either changing conditions or abundant food resources. It is not surprising then that the largest, most highly specialized, modern larger foraminiferal species (e.g., *Marginopora vertebralis, Calcarina spengleri,* and *Cycloclypeus carpenteri*) are those living in the most oligotrophic environments.[33]

Haynes proposed that shape in larger foraminifera is a compromise between hydrodynamic factors and the metabolic requirements of algal symbiosis.[35] He suggested that in free-living forms, robust shells of maximum sphericity should indicate highenergy environments, while thin, fragile tests with maximum surface-to-volume ratios should be found in low-energy, low-light environments. The applicability of Haynes' hypothesis has consistently been illustrated by studies of both modern larger foraminiferal assemblages[36-40] and experimental studies.[38,41-43]

Increasing surface-to-volume ratios requires test elongation in fusiform lineages or development of very flat, thin tests in trochospiral, planispiral, or discoid lineages. In either case, the resulting foraminifera appear larger, even if protoplasmic volumes remain the same. With a chamber inside the foraminiferal test, algal symbionts tend to be concentrated around the outer walls of the chamber.[13] Flattening of the test and consequently flattening of the chambers provides more inner wall surface for the symbionts compared to the total volume of the chamber. Subdividing the chamber into chamberlets provides more wall surface, and it also strengthens the test, enabling outer walls to be thinner and more transparent. Thus, flatter, thinner-tested, more internally complex algal symbiont-bearing foraminifera can live at greater depths than thicker, more spheroid forms.[44]

The recurring trend towards increasing surface-to-volume ratios can be better understood in the context of oligotrophy. Hallock showed that euphotic habitats in clear oligotrophic waters provide the potential for many more niches for photosynthesizing organisms than do eutrophic waters where light penetration is limited by dense plankton stocks.[33] Many of those potential niches are below wave base at relatively low light levels, optimum conditions for large, flat algal symbiont-bearing foraminifera. In the geologic past, expansion of shallow seas during marine transgressions provided new niches for diversifying lineages. When seas subsequently regressed, many of those niches were lost.[33]

The relative transparency or opacity of the foraminiferan test, combined with the type of symbionts, may be important constraints influencing what euphotic niches certain lineages can fill. Rotaliine foraminifera with their lamellar, hyaline wall structure and diatom symbionts seem to have more potential to adapt to a wider variety of light levels than do milioline foraminifera with porcellaneous tests[13,38] or textularine foraminifera with agglutinated tests. Evolutionary developments in fusulinacean wall structure such as the transparent diaphanotheca and the honeycomblike keriotheca (e.g., Reference 45) may have provided more transparent, stronger walls.

The crystalline nature of the rotaliine test wall apparently provides the evolutionary potential to adapt to a wide spectrum of euphotic environments from very high light to very low light. Calcite crystals are secreted such that their C-axes are normal to the outer surface.[46] Therefore, light transmission is augmented and the shell is very transparent. The lamellar nature of the test mandates addition of layers throughout the life of the protist, but as experiments have shown,[43] the thickness of those layers, at least in some species, varies depending upon light and water motion. Thicker lamellae are deposited under higher light levels or moving water, while thinner lamellae are deposited under lower light or minimal water motion. Genera like *Calcarina* and *Baculogypsina*, which live in extremely clear, shallow high energy environments,[47] not only have nearly spheroid tests[38] with thick individual lamellae, but also have surface texture that increases the opacity of the shell, perhaps by scattering incoming light. Leutenegger suggested that light also might be scattered by their canal system or absorbed by thick organic linings.[13] *Amphistegina* spp. show the greatest intrageneric variation. Five species subdivide euphotic habitats, with subspheroid, thick-walled *A. lobifera* in the shallowest zone and flat, thin-walled *A. papillosa* living the deepest.[37,39] In the deepest dwelling algal symbiont-bearing taxa, e.g., *Operculina*, *Heterstegina*, and *Cycloclypeus*, which are also the thinnest and flattest forms, glassy pustules [e.g., FIGURE 4(7) and (8)] tend to grow on the outer walls at greater depth,[48] possibly forming lenses that concentrate incoming light.[49]

The miliolines, on the other hand, have a test wall constructed of a layer of randomly oriented calcite overlain by a bricklike layer in which the C-axes are parallel to the outer wall.[50] This structure scatters and reflects incoming light, giving dry tests a smooth, porcellaneous appearance. This wall structure probably provides very shallow-dwelling individuals with protection from ultraviolet radiation,[35] but its reflective properties may limit the amount of light that can enter the shell. Hallock found that both miliolines and rotaliines show similar shape trends with depth, but postulated that the reason the milioline tests are seldom as thick and that miliolines do not live as deep as rotaliines is because the milioline wall is more opaque.[38] Structural adaptations that permit light into the test include pits (*Archaias*, *Peneroplis*), grooves (*Alveolinella*, *Borelis*) [FIGURE 3(5)], or windows over chamberlets (*Amphisorus*, *Marginopora*, *Sorites*). Even with these adaptations, the deepest dwelling symbiont-bearing soritid in Hawaii, *Sorites marginalis*, was found living to 30-40 m, while rotaliine species were found as deep as 110 m.[47] The Alveolinelidae, with diatom symbionts, are the deepest dwelling larger porcellaneous species.[13]

Though restricted to shallower, brighter waters, the porcellaneous species tend to be more diverse in backreef and lagoonal habitats than the rotaliines.[33,51] This is best seen in the tropical western Atlantic, where there are three to four times as many algal symbiont-bearing porcellaneous species as analogous hyaline species (see, e.g., Reference 52). The taxonomic variety of symbionts may help account for this diversity. Chlorophyte-bearing archaiads and peneroplids tend to dominate quiet, more eutrophic lagoonal environments (e.g., Reference 53), while dinoflagellate-bearing soritids are prevalent in more open reef habitats (Hallock, unpublished). Chlorophyte-bearing species may also be restricted to shallower waters than other taxa, because chlorophytes lack accessory pigments that efficiently absorb light in the blue region of the spectrum.[54] Habitat preferences of rhodophyte-bearing peneroplids are poorly known.

An important predisposition for symbiosis in foraminifera seems to be the multiocular nature of most of the group. Subdivision of the cytoplasm into chambers restricts and channels internal flow patterns so that specialized subcellular regions are potentially created.[55] It has been suggested that the lateral position of endosymbiotic algae directly below outer chamber walls of many larger foraminifera [FIGURE 2(2)] avoids cytoplasmic currents that would sweep them away,[56,13] thus the separation of symbionts in chamber compartments while more rapidly moving cytoplasm flows through stolon or canal systems ("back doors" as Hottinger has colloquially called them).[5] The symbionts of Soritidae and Alveolinidae are kept in small lateral chamberlets and moving cytoplasm in younger parts of the stolon system.[13,15,57] Similarly, except in cases of experimental starvation or 3-(3,4-dichlorophenyl)-1,1-dimethyl urea (DCMU) treatment,[58] digestive and feeding activities are found in other regions of the cell. In *Amphestigina* spp. digestion begins outside and continues in the ventral medial regions of the youngest chambers.[15,58,59] Some foraminifera have canal systems within their test walls (e.g., Rotaliacean FIGURE 1). Röttger and others found that the canal system functions as a cytoplasm-transport system between the interior of the chamberlets and the environment in *Heterostegina depressa*.[60]

A radionuclide label ($^{14}$C) was used to study pore function in *Amphistegina lobifera*.[61] $^{14}CO_2$ was readily taken up through the pores of *A. lobifera* but ($^{14}$C) glucose was not. Similar tracer experiments with *Amphisorus hemprichii*, and imperforate porcellaneous species, showed that it could take up $^{14}CO_2$ through its thin lateral walls.[62] Exchange of small molecules through pores and pits could promote symbiont activities by preventing microhabit nutrient depletion (e.g., $CO_2$) or buildup (e.g., $O_2$).

## ADAPTATIONS AND REQUIREMENTS OF THE SYMBIONTS

Comparatively little is known of the biology of the endosymbiotic algae from foraminifera. No exceptional or unique characteristics are recognized as separating them from free-living algae from the same habitat. The nutrition of a variety of symbiont species isolated in axenic culture has been studied. *Chlamydomonas hedleyi* has no vitamin or organic requirements but its growth is increased threefold in the presence of thiamine and twofold in the presence of glutamic acid, histidine, and methionine. Urea was the best nitrogen source.[11] The growth of *Chlamydomonas provasoli* from *Cyclorbiculina compressa* and the dinoflagellate symbiont from *Sorites marginalis* were enhanced by more than two orders of magnitude by the presence of biotin, thiamin, and vitamin $B_{12}$.[12] The vitamin requirements of eight endosymbiotic diatom isolates have also been tested.[63] Except for isolates of *Nitzschia laevis* and *Navicula reissii,* algal clones required or were stimulated by biotin. The *Nitzschia frustulum* clone also required vitamin $B_{12}$. It is quite possible that as endosymbionts the algae could acquire these vitamins through the egis of their hosts feeding on microorganisms that would provide them in excess. There were three orders of magnitude difference between the optimum concentration of $NO_3^-$ required for growth of the diatoms and the $NO_3^-$ levels found in the Gulf of Elat where the symbionts were isolated.[63] This suggested that in their hosts the algae are constantly nitrogen limited or that the host/symbiont system sequesters and tightly recycles fixed nitrogen. A similar picture was obtained for phosphate.[63] Comparative nutritional studies of free-living diatoms from the same habitats as the hosts have not yet been done. Studies of the natural populations of diatoms in habitats where the hosts are abundant, however, indicate that diatom symbiont species are absent or very rare (work in progress). Perhaps this is an indication that the diatom symbiont species are not well adapted for survival as free-living forms.

The photosynthetic and growth responses of four species of endosymbiotic diatoms have also been studied.[63] Photosynthetic rates were photoinhibited at 625 $\mu W/cm^2$ but were reasonable at 175 $\mu W/cm^2$. Photocompensation points for all four species were 2% of the light available at 1 m depth at Elat (Red Sea) in the summer. The algae also grew better at relatively low light levels indicating that they are well adapted in this respect to photosynthesize at the depths where their hosts are found.

As a general rule endosymbiotic algae within invertebrate hosts do not have outer envelopes typical of their kind (reviewed by Cook and Taylor).[64,65] This is also the case in larger foraminifera. The mechanisms leading to the loss of or failure to form envelopes within various hosts are not understood. Recently we were able to show that something in a sterile homogenate prepared from the larger foraminifer *Amphestigina lobifera* affected the formation of new frustules of growing and dividing diatom endosymbiont species. It was inferred that host substances are probably responsible for the maintenance of the frustule-less state *in vivo.*[66]

A different picture has just emerged from a fine structural study of the red algal symbiont *Porphyridium purpureum* from two species of larger foraminifera, *Peneroplis planatus* and *Spirulina arietina* (Hawkins and Lee, in preparation). Axenic symbiotic *P. purpureum* have a thick fibrous sheath in axenic culture [FIGURE 6(12)]. The red algal symbionts are unusual in that they are not separated from their host's cytoplasm by symbiont vacuoles [FIGURE 6(13)]. At first glance it would appear as if the *P. purpurem* do not have, or form, a sheath in their hosts [FIGURE 6(13)]. Closer and more careful examination shows that there are fibers similar to those in the sheath scattered in the host's cytoplasm near the symbionts. In many cases cytoplasmic vesicles are in direct contact with the algal surface. Periodic acid–Schiff (PAS) histochemical preparations gave strong reaction in marginal vesicular packets, a sign that could be interpreted as production of carbohydrates. While it is difficult to infer dynamic properties from a static series of photographs it seems reasonable to suggest that these symbiotic red algae may be transferring some of their fixed carbon to their hosts via their sheath polysaccharides.

Research on modern larger foraminifera has the potential for helping us interpret the bursts of evolution that have taken place in the group in the past. It may also help us understand some of the fundamental properties in the biological design of these organisms that favor the development of protoctistan giants. Since we have barely scratched the surface of the potential biological studies of larger foraminifera, even more exciting discoveries undoubtedly are yet to come.

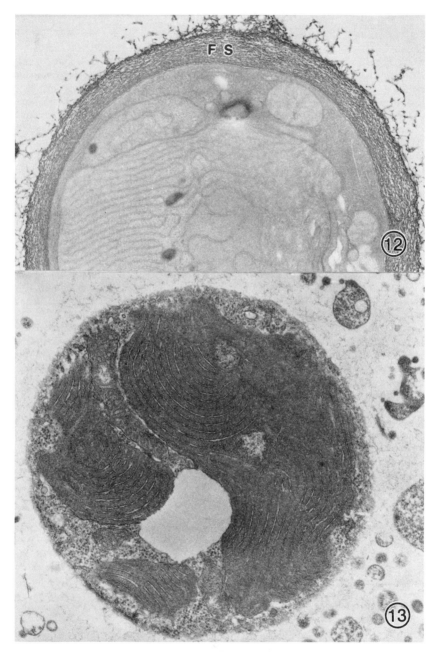

**FIGURE 6.** TEMs of *Porphyridium purpurem,* a red algal symbiont from *Peneroplis planatus.*
(12) From an axenic culture. Note thick fiberous sheath (fs). Magnification 25,300×. (13)
Symbiont within host. Note that the symbiont is not surrounded by a vacuole. Fibers that
resemble those of the sheath are scattered in the host cytoplasm. Magnification 25,200×. (Figure
reduced to 95%.)

## REFERENCES

1. LOEBLICH, A. R. & H. TAPPAN. 1982. Classification of the Foraminiferida. *In* Foraminifera—Notes for a Short Course. University of Tennessee Studies in Geology. T. W. Broodhead, Ed. **6:** 22-36. Knoxville, Tenn.

2. ROSS, C. A. 1979. The ecology of large, shallow-water, tropical foraminifera. Soc. Econ. Paleontol. Mineral. Short Course No. 6: 54-61. Tulsa, Okla.

3. FISCHER, A. G. & M. A. ARTHUR. 1977. Secular variations in the pelagic realm. Spec. Publ. Soc. Econ. Paleontol. Mineral. **25:** 19-50. Tulsa, Okla.

4. HANSEN, H. J. & A. L. LYKKE-ANDERSEN. 1976. Wall structure and classification of fossil and recent elphidiid and nonionid foraminifera. Fossils Strata **10:** 1-37.

5. HOTTINGER, L. 1982. Larger foraminifera, giant cells with a historical background. Naturwissenschaften **69:** 361-371.

6. HOTTINGER, L. & D. DREHER. 1974. Differentiation of Protoplasm in Nummulitidae (Foraminifera) from Elat, Red Sea. Mar. Biol. **25:** 41-61.

7. BILLMAN, H., L. HOTTINGER & H. OESTERLE. 1980. Neogene to recent rotaliid foraminifera from the Indopacific Ocean; their canal system, their classification and their stratigraphic use. Mem. Suisses Paleontol. **101:** 71-113.

8. HOTTINGER, L. & S. LEUTENEGGER. 1980. The structure of calcarinid foraminifera. Schweiz. Palaontol. Abhandl. **101:** 115-151.

9. HOTTINGER, L. 1979. Araldit als Helfer in der Mikropalaontologie. Aspekte (Ciba-Geigy) 3/1979 Publ. Nr. 24522.

10. HOTTINGER, L. 1978. Comparative anatomy of shell structures in selected larger foraminifera. *In* Foraminifera. R. H. Hedley & C. G. Adams, Eds. **3.** Academic Press. London, England.

11. LEE, J. J., L. J. CROCKETT, J. HAGEN & R. STONE. 1974. The taxonomic identity and physiological ecology of *Chlamydomonas hedleyi* sp. nov., algal flagellate symbiont from the foraminifer *Archaias angulatus.* Br. J. Phycol. **9:** 407-422.

12. LEE, J. J., M. MCENERY, E. KAHN & F. SCHUSTER. 1979. Symbiosis and the evolution of larger foraminifera. Micropaleontology **25:** 118-140.

13. LEUTENEGGER, S. 1984. Symbiosis in benthic foraminifera: specificity and host adaptations. J. Foraminiferal Res. **14:** 16-35.

14. LEUTENEGGER, S. 1977. Ultrastructure de foraminiferes perfores et imperfores ainsi que de Leurs symbiotes. Cahiers Micropaleontol. **3:** 1-52. (with English summary).

15. MULLER-MERZ, E. & J. J. LEE. 1976. Symbiosis in the larger foraminiferan *Sorites Marginalis* (with notes on *Archaias* spp.). J. Protozool. **23:** 390-396.

16. ROSS, C. A. 1972. Biology and ecology of *Marginopora vertebralis* (Foraminiferida), Great Barrier Reef. J. Protozool. **19:** 181-192.

17. SCHOENBERG, D. A. & R. K. TRENCH. 1980. Genetic variation in *Symbiodinium* ( = *Gymnodinium*) *microadriaticum* Freudenthal and specificity in its symbiosis with marine invertebrates. I. Isoenzyme and soluble protein patterns of axenic cultures of *S. microadriaticum.* Proc. R. Soc. London Ser. B 207: 405-427.

18. SCHOENBERG, D. A. & R. K. TRENCH. 1980. Genetic variation in *Symbiodinium* ( = *Gymnodinium*) *microadriaticum* Freudenthal and specificity in its symbionts, with marine invertebrates. II. Morphological variation in *S. microadriaticum.* Proc. R. Soc. London Ser. B 207: 429-444.

19. SCHOENBERG, D. A. & R. K. TRENCH. 1980. Genetic variation in *Symbiodinium* ( = *Gymnodinium*) *microadriaticum* Freudenthal and specificity in i-s symbiosis with marine invertebrates. III. Specificity and infectivity of *S. microadriaticum.* Proc. R. Soc. London Ser. B 207: 445-460.

20. BLANK, R. J. 1986. Unusual chloroplast structures in endosymbiotic dinoflagellates: a clue to evolutionary differentiation within the genus *Symbiodinium* (Dinophyceae). Plant Syst. Evol. **151:** 271-280.

21. BLANK, R. J. & R. K. TRENCH. 1986. Nomenclature of endosymbiotic dinoflagellates. Taxon **35:** 286-294.

22. LEUTENEGGER, S. 1983. Specific host-symbiont relationship in larger foraminifera. Micropaleontology **29**(2): 111-125.

23.  LEE, J. J. & C. W. REIMER. 1982. Isolation and identification of endosymbiotic diatoms
     from larger foraminifera of the Great Barrier Reef, Australia, Makapuu Tide Pool, Oahu,
     Hawaii, and the Gulf of Elat, Israel, with the description of new species, *Amphora
     rottgeri, Navicula hansenii* and *Nitzschia frustrulum* var. *symbiotica. In* Proceedings of
     the 7th International Symposium on Living and Fossil Diatoms. D. G. Mann, Ed. O.
     Koeltz. Koeingstein, FRG.
24.  LEE, J. J., M. MCENERY & C. W. REIMER. 1980. The isolation, culture and identification
     of endosymbiotic diatoms form *Heterostegina depressa* D'Orbigny and *Amphistegina
     lessonii* d'Orgigny (larger foraminifera) from Hawaii. Bot. Mar. **23:** 297-302.
25.  LEE, J. J., M. MCENERY, M. SHILO & Z. REISS. 1979. Isolation and cultivation of diatom
     endosymbionts form larger foraminifera (Protozoa). Nature **280:** 57-58.
26.  LEE, J. J., C. W. REIMER & M. E. MCENERY. 1980. The identification of diatoms isolated
     as endosymbionts from larger foraminifera from the Gulf of Elat (Red Sea) and the
     description of 2 new species, *Fragilaria shiloi* sp. nov. and *Navicula reisii* sp. nov. Bot.
     Mar. **23:** 41-48.
27.  HANSEN, H. J. & B. BUCHARDT. 1977. Depth distribution of *Amphistegina* in the Gulf of
     Elat. Utrecht Micropaleontol. Bull. **1:** 225-239.
28.  FRENZEL, H. & M. MUNDORFF. 1942. Fusulinidae from the phosphoria formation of
     Montana. J. Paleontol. **16:** 675-684.
29.  HALLOCK, P. 1981. Algal symbiosis: a mathematical analysis. Mar. Biol. **62:** 249-255.
30.  BRALOWER, T. J. & H. R. THEIRSTEIN. 1984. Low productivity and slow deepwater
     circulation in mid-Cretaceous oceans. Geology **12:** 614-618.
31.  HALLOCK, P. 1982. Evolution and extinction in larger foraminifera. *In* Proceedings of the
     3d North American Paleontological Convention **1:** 221-225. Department of Geology.
     University of Montreal. Montreal, Canada.
32.  HALLOCK, P. 1985. Why are larger foraminifera large? Paleobiology **11:** 195-208.
33.  HALLOCK, P. 1986. Paleonutrients and the evolution of larger reef foraminifera. *In* Second
     International Symposium on Paleoceanography. Woods Hole Oceanographic Institution.
     Woods Hole, Mass.
34.  ADAMS, C. G. 1983. Speciation, phylogenesis, tectonism, climate and eustacy: factors in
     the evolution of Cenozoic larger foraminiferal bioprovinces. System. Assoc. Spec. **23:**
     255-289.
35.  HAYNES, J. R. 1964. Symbiosis, wall structure and habitat in foraminifera. Cushman Found.
     Foraminiferal Res. Contrib. **16:** 40-43.
36.  LARSEN, A. R. 1976. Studies of Recent *Amphistegina:* taxonomy and some ecological
     aspects. Israel J. Earth-Sci. **25:** 1-26.
37.  LARSEN, A. R. & C. W. DROOGER. 1977. Relative thickness of the test in *Amphistegina*
     species of the Gulf of Elat. Utrecht Micropaleontol. Bull. **15:** 225-239.
38.  HALLOCK, P. 1979. Trends in test shape with depth in large, symbiont-bearing Foraminifera.
     J. Foraminiferal Res. **9:** 61-69.
39.  HALLOCK, P. & H. J. HANSEN. 1979. Depth adaptation in *Amphestigina:* change in lamellar
     thickness. Geol. Soc. Denmark Bull. **27:** 99-104.
40.  HOTTINGER, L. 1983. Processes determining the distribution of larger foraminifera in space
     and time. Utrecht Micropaleontol. Bull. **30:** 239-253.
41.  HALLOCK, P. 1981. Light dependence in *Amphistegina.* J. Foraminiferal Res. **11:** 40-46.
42.  KUILE, B. TER. & J. EREZ. 1984. *In-situ* growth rate experiments on the symbiont-bearing
     foraminifera *Amphistegina lobifera* and *Amphisorus hemprichii.* J. Foraminiferal Res. **14:**
     262-276.
43.  HALLOCK, P., L. B. FORWARD & H. J. HANSEN. 1986. Environmental influence of test
     shape in *Amphistegina.* J. Foraminiferal Res. **16:** 224-231.
44.  HALLOCK, P. & E. C. GLENN. 1986. Larger foraminifera: a tool for paleoenvironmental
     analysis of Cenozoic carbonate depositional facies. Palaios **1:** 55-64.
45.  LOEBLICH, A. R. & H. TAPPAN. 1964. Sarcodina, chiefly "Thecamoebians" and Fora-
     miniferida. *In* Treatise on Invertebrate Paleontology. Protista 2. R. C. Moore, Ed. **1-2.**
     Kansas University Press. Lawrence, Kansas.
46.  TOWE, K. M. & R. CIFELLI. 1967. Wall structure in the calcareous foraminifera: crystal-
     lographic aspects and a model for calcification. J. Paleontol. **41:** 742-762.

47. HALLOCK, P. 1984. Distribution of selected species of living algal symbiont-bearing foraminifera on two Pacific coral reefs. J. Foraminiferal Res. **14:** 250-261.

48. HOTTINGER, L. 1977. Distribution of larger Peneroplidae, *Borelis* and Nummulitidae in the Gulf of Elat, Red Sea. Utrecht Micropaleontol. Bull. **15:** 35-109.

49. DROOGER, C. W. 1983. Environmental gradients and evolutionary events in some larger foraminifera. Utrecht Micropaleontol. Bull. **30:** 255-271.

50. ROSS, C. A. & J. R. P. ROSS. 1978. Adaptive evolution in the soritids *Marginopora* and *Amphisorus* (Foraminiferida). Scanning Electron Microscopy / 1978 **2:** 53-60.

51. HALLOCK, P. 1984. Foraminifera of Indo-Pacific and Caribbean reefs. A comparison of selected algal symbiont-bearing taxa. *In* Advances in Reef Science: 50-51. University of Miami. Miami, Fla.

52. BOCK, W. D., G. W. LYNTS, S. SMITH, R. WRIGHT, W. W. HAY & J. I. JONES. 1961. A symposium of recent south Florida foraminifera. Memoir 1. Miami Geological Society. Miami, Fla.

53. HALLOCK, P., T. L. COTTEY, L. B. FORWARD & J. HALAS. 1986. Population biology and sediment production of *Archaias angulatus* (Foraminiferida) in Largo Sound, Florida. J. Foraminiferal Res. **16:** 1-8.

54. FUJITA, Y. 1970. Photosynthesis and plant pigments. Bull. Plankton Soc. Jpn. **17:** 20-31.

55. CAVALIER-SMITH, T. & J. J. LEE. 1985. Protozoa as hosts for endosymbioses, and the conversion of symbionts into organelles. J. Protozool. **32:** 376-379.

56. LEE, J. J. 1983. Perspective on algal endosymbionts in larger foraminifera. Int. Rev. Cytol. Suppl. **11:** 49-77.

57. LEE, J. J. & M. MCENERY. 1983. Symbiosis in foraminifera. *In* Algal Symbiosis. L. J. Goff, Ed.: 37-68. Cambridge University Press. Cambridge, England.

58. KOESTLER, R. J., J. J. LEE, J. REIDY, R. P. SHERYLL & X. XENOPHONTOS. 1985. Cytological investigation of digestion and reestablishment of symbiosis in the larger benthic foraminifer *Amphistegina lessonii*. Endocyt. Cell Res. **2:** 22-54.

59. MCENERY, M. E. & J. J. LEE. 1981. Cytological and fine structural studies of 3 species of symbiont-bearing larger foraminifera from the Red Sea. Micropaleontology **27:** 71-83.

60. RÖTTGER, R., M. SPINDLER, R. SCHMALJOHANN, M. RICHWIEN & M. FLADUNG. 1984. Functions of the canal system in the rotaliid foraminifer, *Heterostegina depressa.* Nature **309:** 789-791.

61. LEUTENEGGER, S. & H. HANSEN. 1979. Ultrastructural and radiotracer studies of pore-function in foraminifera. Mar. Biol. **54:** 11-16.

62. HANSEN, H. J. & P. DALBERG. 1979. Symbiotic algae in milioline foraminifera: $CO_2$ uptake and shell adaptions. Bull. Geol. Soc. Denmark **28:** 47-55.

63. LEE, J. J., M. MCENERY, M. J. LEE, J. REIDY, J. GARRISON & R. ROTTGER. 1980. Algal symbionts in larger foraminifera. *In* Endocytobiology. I. W. Schwemmler & H. E. A. Schenk, Eds.: 113-124. Walter de Gruyter and Co. Berlin, FRG.

64. COOK, C. B. 1983. Metabolic interchange in algae-invertebrate symbiosis. Int. Rev. Cytol. Suppl. **14.**

65. TAYLOR, D. L. 1971. Ultrastructure of the "Zooxanthellae" *Endodinium chattonii in situ.* J. Mar. Biol. Assoc. UK **51:** 227-234.

66. LEE, J. J., N. M. SAKS, F. KAPIOUTOU, S. H. WILEN & M. SHILO. 1984. Effects of host cell extracts on cultures of endosymbiotic diatoms from larger foraminifera. Mar. Biol. **82:** 113-120.

# Animals with Photosynthetic Symbionts

ROSALIND HINDE

*School of Biological Sciences*
*University of Sydney*
*Sydney, New South Wales 2006, Australia*

The various relationships between microscopic algae (both prokaryotic and eukaryotic) and aquatic invertebrates have been the subjects of a number of reviews during the last 20 years (for example, References 1 to 6), and another review does not seem justified. Since the theme of this volume is coevolution of endosymbionts and their hosts, I would like to discuss some aspects of these associations that will be important in understanding how they may have evolved. The discussion raises questions about the real nature of these symbioses, particularly about the status of the algae; it will, I hope, provide some stimulus to research in the neglected corners of this field.

It is generally held that associations between algae and invertebrates are mutualistic, provided that the algae retain their ability to photosynthesize. There is certainly abundant evidence that the animals benefit from their associations with algae. First, they obtain large amounts of organic carbon from the algae. In all the studies of carbon metabolism in associations between algae and invertebrates, substantial translocation of organic compounds from the algae to the animal has been observed. Estimates of the proportion of the products of photosynthesis that are used by the animal, rather than by the algae, vary among associations and also depend on the method used to make the estimate. Most published estimates are based on short-term experiments with the intact association or with freshly isolated symbiotic algae under various conditions, using $^{14}CO_2$ as a tracer.[1,6,7] In such experiments between 12% and 82% of the freshly fixed carbon is translocated from alga to animal in the first hour after the start of photosynthesis (with most values lying between 20% and 60%). Experiments using model systems involving isolated algae and host tissue homogenates give similar rates of translocation.[6] This is, however, only the beginning of the loss of photosynthetic products from the algae. These experiments measure the translocation of low-molecular-weight, water-soluble compounds. In addition, lipids (mostly triglycerides and phospholipids) that are synthesized by the algae move into the cells of the animal host.[8] Lipid is a major component of the $^{14}C$-labeled material in cnidarian hosts after photosynthesis in labeled $CO_2$, and there is evidence that much of it is synthesized in, and translocated from, the algae rather than being synthesized by the host from labeled low-molecular-weight precursors.[8] The proportion of total photosynthate eventually translocated as lipid does not seem to have been measured in long-term experiments. Little or no information is available about the possible translocation of other classes of compound (for example, oligosaccharides or polysaccharides secreted from the algal surface). The construction of carbon budgets for invertebrates that contain zooxanthellae suggests that, one way or another, more than 95% of the total carbon fixed during photosynthesis by the algae is used by the animal partner in the association.[9] The value of this to the animal will depend on the ratio between the biomass of the animal and the biomass of the plants, and on environmental factors that affect the rate of photosynthesis. In many cases, though, it is clear that photosynthesis can supply a very significant part of the animal's requirement for organic compounds. In the coral *Stylophora pistillata,* for example, translocation is equivalent to 143% of the daily respiratory needs in colonies living in well-lit areas;[9] the excess is presumably used for the animal's growth. In shaded habitats, translocation supplies 58% of the total respiratory carbon.[9] The flatworm *Convoluta roscoffensis* does not feed after infection with its symbiotic algae, and relies entirely on its symbionts and on the uptake of dissolved organic matter for nutrition.[11]

Many invertebrates also gain other advantages from their symbiotic algae. The presence of photosynthesizing algae in corals increases the rate at which they deposit skeletal calcium carbonate;[10] this presumably allows them to increase their biomass faster. Photosynthesis liberates oxygen into, and removes carbon dioxide from, the animals' tissues. There is, however, no evidence that the rate of either the supply of oxygen or the removal of $CO_2$ limits the growth of aposymbiotic animals in the relatively shallow waters in which these animal-algal associations live. In many associations, the algae can take up and use waste products of the animals' nitrogen metabolism. For example *Platymonas convolutae,* the symbiont of *Convoluta roscoffensis,* can grow with uric acid as the sole source of nitrogen, and the worms store uric acid only when they do not contain active symbiotic algae.[11] Organic nitrogen, in the form of amino acids, is translocated from the algae to the worms in this association.[11] Like most algae, zooxanthellae from corals can use ammonia as a source of nitrogen; ammonia is likely to be the major nitrogenous waste product of coral polyps.[12] Corals without zooxanthellae excrete more nitrogen than do corals with symbionts.[12,13] Another benefit to the animals has been demonstrated in some species of sponge, where the algae protect the delicate underlying cells of the sponge from excessive ultraviolet radiation.[14] Other animals with pigmented symbionts may benefit in the same way; certainly many of them live in shallow, tropical waters where light intensities may be high enough to damage unprotected cells.

Thus the animal hosts of a wide range of symbiotic algae derive considerable benefit from the presence of the algae. Little is known of the costs to the host, but presumably the animal expends some energy in maintaining its symbionts. For instance it may supply the algae with substrates for heterotrophic nutrition, or with vitamins or other essential nutrients; and it must synthesize control substances such as the active factors in host homogenates. These costs need to be investigated if complete models of the physiology of associations between algae and invertebrates are to be constructed. Nevertheless, the success of many of the associations in persisting through geological time and in dominating many habitats indicates their Darwinian fitness. The apparently dominant role of the host animals in the establishment and control of these associations (see below) suggests that if the presence of the algae were not beneficial to the animals, the associations would not persist.

The overall picture is less clear when benefits to the algae are considered. Some features of the associations are potentially beneficial to the plants. The host acts as a source of inorganic nutrients [for example, nitrogen (see above) and possibly phosphorus]. Since these nutrients are the ones likely to be limiting to marine and freshwater plants respectively, this improvement in their supply is almost certainly advantageous to the algae. Inorganic nutrients absorbed from the surrounding water by the host are also available to the algae. However, where there is a large population of unicellular algae within an animal, the total surface area of the algal cells must be much greater than the surface area of the parts of the host that are exposed to the water. The ability of most invertebrates to create currents of water over their respiratory surfaces would aid in the uptake of nutrients, but it seems unlikely that this would compensate for their relatively small surface area. Measurement of the rates at which nutrients are absorbed by suspensions of freshly isolated or cultured algae and by host animals would settle the question of whether symbiosis with an animal brings a net advantage (or disadvantage) to the algae with regard to the availability of exogenous inorganic nutrients.

Other advantages may flow from living in animals, but none of them has been fully investigated. There has been little work on the movement of organic compounds from animal hosts to the associated algae. Many free-living algae can grow heterotrophically when organic carbon is available, and some symbiotic algae are known to receive dissolved organic compounds via their hosts.[4,15] Little is known about heterotrophic growth in symbiotic algae, or about the extent of translocation from host to alga. It is also possible that the host may supply the algae with vitamins (for which many microalgae have absolute requirements)[16] or with other essential organic nutrients. Living within the tissues of the host may provide protection from herbivores which, obviously, do not normally attack animals. The fate of symbiotic algae that are eaten by predators feeding on their hosts does not seem to have been investigated, but some at least are likely to pass unharmed through the digestive system of most carnivores.[25] Unlike planktonic microalgae, symbiotic algae do not have to expend energy in order to stay in the photic zone, since they are found either adhering to their hosts (extracellular symbionts) or within the hosts' cells. The host's tissues and skeleton provide support and allow the algae to remain in one place indefinitely, without sinking. This means that they can adapt themselves closely to conditions in the host's habitat, which are likely to be more constant than those encountered by planktonic algae. In addition, the hosts tend to settle and grow in well-lit places.

In spite of these conditions, which appear to be ideal for algal growth, the growth rates of symbiotic algae are very low.[9,17] The huge losses of photosynthetically fixed carbon from the algae when they are in symbiosis could probably explain this, but there is also some evidence that supports the idea that the host actively controls both the size of the population of algae in its cells and the rate of growth or division of the algae, in addition to the rate of translocation.[17] In many symbiotic associations the number of algae per host cell is known to be regulated. For example, in zoanthids excess algae resulting from exposing the association to continuous light are expelled from the gastrodermis and passed out through the mouth;[18] in other symbioses involving zooxanthellae, excess algae are ejected or digested.[4] In the nudibranch *Pteraeolidia ianthina* the mitotic index of the zooxanthellae is much higher in animals with low densities of algae than in animals with large populations of algae.[19] Regulation of algal numbers by the host has been studied more fully in hydra than in other associations.[17] In hydra, neither ejection nor digestion of the algae is involved in regulation of the population under normal circumstances, although both can occur in the laboratory. Instead, the animal appears to inhibit division of the algae when the size of the population of algae is "suitable" for the conditions under which the association is growing.[17] Division of the algae and of the host cell seem to be linked (both in terms of numbers of divisions and of timing), but these links can be disrupted experimentally. The mechanism of regulation is not understood. McAuley has suggested that the animal may control algal growth and cell division by regulating the supply of one or more metabolites to the algae, or by regulating the extent of the stimulation of translocation of photosynthetic products from the algae.[17] Obviously such mechanisms are also likely to exist in associations involving zooxanthellae, particularly those that do not normally eject or digest their symbionts. There is no evidence that symbiotic algae can escape from the host's cells.

The host animal also seems to control the entry of symbionts into its tissues. A particular species of host collected from the wild will always contain algae of the same strain or, at most, algae of one of a small group of closely related strains.[20] While "foreign" algae that are taxonomically related to the normal symbionts can be introduced into some cnidarians,[20] *Paramecium,*[21] *Convoluta roscoffensis,*[22] and *Hydra,*[23] the resulting associations are unstable. The algae either fail to establish themselves and are ejected in a few days,[23] or they remain until the host is exposed to its usual symbionts and are then ejected and replaced by the "normal" symbionts.[20-22]

So the evidence as it stands at present is that in associations between algae and invertebrates, the invertebrate hosts, while they probably have to expend some energy to maintain their algal populations, receive a large part of their nutrition from their photosynthesizing partners. On the other hand, although the algae are living in a habitat in which they have good access to light and nutrients and are probably protected from herbivores (i.e., a habitat that should be more favorable to growth than the habitat of planktonic plants) they grow much more slowly than related free-living algae. Many symbiotic algae also grow faster in culture than in symbiosis. They are apparently prevented from retaining the greater part of their photosynthetic products, from growing and dividing at rates commensurate with their photosynthetic rates, and from controlling their entry to and exit from the host. It is the animal that seems to control all these aspects of the life of the algae. What does this control mean in terms of the nature of the associations, and their evolution?

The answers to these questions may be different for each association. They depend on whether the species of alga involved in the association also has free-living populations and, where applicable, on the relative fitness of the symbiotic and the free-living algae. It should be noted that these associations are almost always obligate for the host animal. Although in some cases aposymbiotic animals can be produced in the laboratory, they are rarely found in the field; when they do occur naturally they tend either to die or to become repopulated with algae.[19,24]

In each of the following three cases, symbiosis will have a different impact on the fitness of the algae: (1) The symbiotic algae may have no free-living stages in the life cycle and no free-living conspecifics. (2) There may be free-living algae that are conspecific with the symbionts and that, on average, have greater fitness than the symbionts. (3) The symbiotic algae may show greater fitness, on average, than the free-living forms.

1. If a given species of symbiotic alga does not have free-living populations or stages (that is, if the symbiosis is obligate for the alga), characteristics that decrease the likelihood of its remaining in the host will clearly decrease the fitness of the alga. In such an association there is likely to be a tendency for increasing interdependence of the partners and increasing accuracy of regulation of the interchange of metabolites and of growth rates. An association of this type must be regarded as mutualistic, no matter how much of the productivity of the alga is directed to maintaining the host.

2. Symbiosis may decrease the fitness of the algae compared to that of free-living members of the same species. In this case, selection will tend to produce populations dominated by individuals that can evade symbiosis. Such evasion might involve adaptations preventing the potential host from phagocytosing the alga, or adaptations increasing the chance of rejection after phagocytosis.

3. Symbiosis may have only a slight effect on net fitness, or the algae in symbiosis may tend to have greater fitness than free-living ones. In such cases both symbiotic and planktonic populations are likely to persist. If the host is normally infected by algae from the water column in each generation, there should be free genetic interchange between the symbiotic and planktonic populations. If

this interchange is restricted, for instance by the evolution of transovarial transmission of the symbionts (which is known to occur in a number of algal symbionts),[4] the populations might separate into two species, one more successful in symbiosis and the other in the plankton.

Unfortunately, we are in no position to assess the exact nature of any algal-invertebrate association in these terms. There are some data on the growth rates of algae in their hosts, but little is known of their mortality in any host. Even less is known about possible free-living conspecific algae. In most cases, we do not even know if such algae exist. In many symbioses between algae and invertebrates, the algae are transmitted during asexual reproduction by the host and are also passed on via the eggs or egg coverings during sexual reproduction.[4] Obviously some of these algae may never be planktonic, existing only as symbionts. In other cases the eggs of the host do not carry the symbionts, and the larvae or juveniles are infected by algae from the water column.[4] Even this is not watertight evidence of the existence of populations of actively growing, free-living algae of the infective species or strain. It is possible that new host individuals might be infected by algae that have been extruded from other hosts of the same species, or that had passed through the gut of a predator of the animal host.[25] Such algae may be able to survive in the plankton for some time without forming stable, reproducing populations there. For many of the well-known symbiotic algae there are simply no data about the occurrence of free-living forms.

In many cases it will probably be difficult to recognize the algae in the plankton—for instance, the zooxanthellae of most hosts are so similar in form that for a long time they have been placed in one morphological species, *Symbiodinium microadriaticum*, and their motile stages are similar to each other and to free-living gymnodinioid dinoflagellates. Many species of *Chlorella*, both free-living and symbiotic forms, are also very alike. Detailed immunological or genetic screening will probably be necessary to establish the identity of algal strains isolated from the plankton with any given symbiotic alga. *Platymonas convolutae*, the symbiont of *Convoluta roscoffensis*, does occur in the plankton. Its growth rate and mortality under natural conditions do not, however, appear to have been studied in the detail needed to make comparisons of its fitness with that of symbiotic *P. convolutae*. Many symbiotic algae have been cultured in isolation from their hosts. Measurements of fitness made in culture are, however, of little help in assessing the fitness of natural populations. Some types of symbiotic algae have not been cultured successfully; some of these species may turn out to be obligate symbionts. Difficulty in culturing them is not, of itself, sufficient evidence that the symbiosis is obligate for the algae.

At present, then, it is not possible to determine the effects of becoming a symbiont on the fitness of algae. Some associations between algae and invertebrates may in fact be parasitic relationships, with the "host" in this case being the smaller symbiont (the alga), whose photosynthetic ability is exploited by the animal partner. In other cases there may be no net change in average fitness of the algae (commensalism). And, of course, some of the associations are probably true mutualisms, where the fitness of both partners is enhanced. Until data are available about the occurrence, growth rates, and mortality of free-living populations, it is impossible to define the effects of symbiosis on the fitness of the algae. As long as this is so it will be difficult to support or refute models that seek to explain the evolution of the algae. Models involving coevolution of alga and host are even more complex, and therefore more difficult to test.

# SUMMARY

Associations between algae and invertebrates are commonly cited as prime examples of mutualistic symbioses. In this paper the known and hypothetical benefits of symbiosis to the animal and to the algae are outlined. In spite of some obvious advantages to the plants, it is impossible to decide whether the net outcome of symbiosis is beneficial to any given species of alga (i.e., whether its fitness is enhanced by association with the host). The precise benefits of symbiosis can only be clarified when sufficient data are available about the occurrence, growth rates, and survivorship of free-living populations of the species in question. Ignorance of the relative fitness of free-living and host-associated algae makes it difficult to erect useful models concerning the evolution (and putative coevolution) of these associations.

# ACKNOWLEDGMENTS

I am grateful to Dr. A. J. Underwood for his very helpful criticisms of an earlier draft of this paper.

## REFERENCES

1. SMITH, D. C., L. MUSCATINE & D. LEWIS. 1969. Carbohydrate movement from autotrophs to heterotrophs in parasitic and mutualistic symbiosis. Biol. Rev. **44**: 17-90.
2. TAYLOR, D. L. 1974. Symbiotic marine algae: taxonomy and biological fitness. *In* Symbiosis in the Sea. W. B. Vernberg, Ed.: 245-262. University of South Carolina Press. Columbia, S.C.
3. JENNINGS, D. H. & D. L. LEE, Eds. 1975. Symbiosis. Symp. Soc. Exp. Biol. **29**.
4. TRENCH, R. K. 1979. The cell biology of plant-animal symbiosis. Annu. Rev. Plant Physiol. **30**: 485-531.
5. GOFF, L., Ed. 1983. Algal Symbiosis—a Continuum of Interaction Strategies. Cambridge University Press. Cambridge & New York.
6. HINDE, R. 1986. Symbioses between aquatic invertebrates and algae. *In* Parasitology—Quo Vadit? M. J. Howell, Ed.: 383-390. Australian Academy of Science. Canberra, Australia.
7. SMITH, D. C. 1974. Transport from symbiotic algae and symbiotic chloroplasts to host cells. Symp. Soc. Exp. Biol. **28**: 485-520.
8. BATTEY, J. F. & J. S. PATTON. 1984. A reevaluation of the role of glycerol in carbon translocation in zooxanthellae-coelenterate symbiosis. Marine Biol. **79**: 27-38.
9. MUSCATINE, L., P. G. FALKOWSKI, J. W. PORTER & Z. DUBINSKY. 1984. Fate of photosynthetic fixed carbon in light- and shade-adapted colonies of the symbiotic coral *Stylophora pistillata.* Proc. Roy. Soc. London Ser. B **222**: 181-202.
10. TAYLOR, D. L. 1973. Algal symbionts of invertebrates. Annu. Rev. Microbiol. **27**: 171-187.
11. HOLLIGAN, P. M. & G. W. GOODAY. 1975. Symbiosis in *Convoluta roscoffensis.* Symp. Soc. Exp. Biol. **29**: 205-227.
12. MUSCATINE, L. 1980. Uptake, retention, and release of dissolved inorganic nutrients by marine algae-invertebrate associations. *In* Cellular Interactions in Symbiosis and Parasitism. C. B. Cook, P. W. Pappas & E. D. Rudolph, Eds.: 229-244. Ohio State University Press. Columbus, Ohio.

13. SZMANT-FROELICH, A. & M. E. Q. PILSON. 1984. Effects of feeding frequency and symbiosis with zooxanthellae on nitrogen metabolism of the coral *Astrangia danae*. Mar. Biol. **81:** 153-162.
14. WILKINSON, C. R. 1980. Cyanobacteria symbiotic in marine sponges. *In* Endocytobiology—Endosymbiosis and Cell Biology. W. Schwemmler & H. E. A. Schenk, Eds.: 553-563. Walter de Gruyter & Co. Berlin & New York.
15. SCHLICHTER, D. & B. P. KREMER. 1985. Metabolic competence of endocytobiotic dinoflagellates (zooxanthellae) in the soft coral, *Heteroxenia fuscescens*. Endocyt. Cell Res. **2:** 71-82.
16. PROVASOLI, L. & A. F. CARLUCCI. 1974. Vitamins and growth regulators. *In* Algal Physiology and Biochemistry. W. D. P. Stewart, Ed.: 741-787. Blackwell Scientific Publications. Oxford, England.
17. McAULEY, P. J. 1985. Regulation of numbers of symbiotic *Chlorella* in digestive cells of green hydra. Endocyt. Cell Res. **2:** 179-190.
18. REIMER, A. A. 1971. Observations on the relationships between several species of tropical zoanthids (Zoanthidea, Coelenterata) and their zooxanthellae. J. Exp. Mar. Biol. Ecol. **7:** 207-214.
19. HOEGH-GULDBERG, O., R. HINDE & L. MUSCATINE. 1986. Studies on a nudibranch that contains zooxanthellae. II. Contribution of zooxanthellae to animal respiration (CZAR) in *Pteraeolidia ianthina* with high and low densities of zooxanthellae. Proc. R. Soc. London Ser. B **228:** 511-521.
20. SCHOENBERG, D. A. 1980. An ecological view of specificity in algal-invertebrate associations of *Symbiodinium microadriaticum* and coelenterates. *In* Endocytobiology—Endosymbiosis and Cell Biology. W. Schwemmler & H. E. A. Schenk, Eds.: 145-154. Walter de Gruyter. Berlin & New York.
21. KARAKASHIAN, M. W. 1975. Symbiosis in *Paramecium bursaria*. Symp. Soc. Exp. Biol. **29:** 145-173.
22. DOUGLAS, A. E. 1983. Establishment of the symbiosis in *Convoluta roscoffensis*. J. Mar. Biol. Assoc. UK **63:** 419-434.
23. MUSCATINE, L., C. B. COOK, R. L. PARDY & R. R. POOL. 1975. Uptake, recognition and maintenance of symbiotic *Chlorella* by *Hydra viridis*. Symp. Soc. Exp. Biol. **29:** 175-203.
24. HARRIOTT, V. J. 1985. Mortality rates of scleractinian corals before and during a mass bleaching event. Mar. Ecol. Prog. Ser. **21:** 81-88.
25. SANTELICES, B., J. CORREA & M. AVILA. 1983. Benthic algal spores surviving digestion by sea urchins. J. Exp. Mar. Biol. Ecol. **70:** 263-270.

# Control of Translocation in Some Associations between Invertebrates and Algae[a]

ROSALIND HINDE

*School of Biological Sciences*
*University of Sydney*
*Sydney, New South Wales 2006, Australia*

## INTRODUCTION

Recent work in my laboratory has centered on four associations. The coral *Plesiastrea versipora,* the zoanthid *Zoanthus robustus,* and the nudibranch *Pteraeolidia ianthina* all contain zooxanthellae that are morphologically similar to *Symbiodinium microadriaticum.* All three are found in temperate waters; animals for experiments were collected from Port Jackson (Sydney). The sponge *Dysidea herbacea* contains the blue-green alga *Oscillatoria spongeliae;* most of the algae are extracellular but many of them are closely associated with host cells.[1] This sponge occurs in tropical waters, often on coral reefs. The work I wish to discuss is all concerned with the control of translocation of products of photosynthesis from the algae to the animals.

## CONTROL OF TRANSLOCATION FROM ALGA TO ANIMAL

### Associations with Zooxanthellae

Rates of photosynthesis and translocation were measured using $^{14}CO_2$ as a tracer. Algae were separated from animal tissue by homogenizing the association, and centrifuging the homogenate. For experiments with isolated zooxanthellae the algae were then washed three times with seawater and resuspended in seawater or host homogenate. Half-hour incubations were used (Hoegh-Guldberg and Sutton, Mar. Biol., submitted).

(i) In intact *P. versipora* there is short-term translocation of soluble photosynthetic products from the algae to the animal. About 10% of the total fixed carbon is translocated *in vivo,* mainly as glycerol, with small amounts of carbohydrates (Hoegh-

[a] This research has been supported by the Australian Research Grants Scheme, the Marine Sciences and Technologies Grants Scheme, and the University of Sydney Research Maintenance Fund.

Guldberg and Sutton, Mar. Biol., submitted). Zooxanthellae isolated from the coral translocate 5% to 10% of their total fixed carbon into seawater during incubation. Host homogenate increases the proportion translocated to between 25% and 48%.

Dark fixation by the zooxanthellae amounts to about 5% of the total fixation of $CO_2$ in the light, so that although the products of dark fixation "leak" into the medium much more rapidly than do products of photosynthesis, this cannot account for more than a small part of the material translocated into host homogenate. The compounds released into host homogenate are similar to the mobile compounds in the intact coral. In host homogenate most of the carbon released is in glycerol and a compound with $R_F$ similar to leucine. Succinic acid and glucose are the main compounds that "leak" into seawater (Hoegh-Guldberg and Sutton, submitted). These results are highly reproducible and the coral homogenates show activity at about the same rate throughout the year.

The activity of *P. versipora* homogenates cannot be mimicked by altering the pH or the concentrations of calcium, phosphate, or ammonia in the seawater in which the algae are suspended. The active factors in the homogenates are heat labile. Activity is lost slowly from frozen homogenates, and is destroyed by freeze-drying. Activity is abolished by dialyzing the homogenate; it can be restored by reconstituting the homogenate, that is, by mixing the retained material with the dialysate.

From the data outlined above I conclude that in *P. versipora* "host release factor" exists as two or more purifiable components of the homogenized host tissue, and that their activity is a normal component of the physiology of the association. At least one of these components has a molecular weight greater than about 8000 daltons, and at least one of less than 8000 D.

(ii) In intact *Zoanthus robustus* between 12% and 42% of total fixed carbon moves from the algae to the host during a 60-minute incubation in the light (Hoegh-Guldberg and Sutton, submitted). Nevertheless, homogenates of the zoanthid tissue have little or no effect on translocation by its algae or by algae isolated from *P. versipora*. Algae isolated from the zoanthid do respond to homogenates of the coral, usually giving rates of translocation almost as great as those from the corals' own zooxanthellae.

(iii) Translocation of 25% to 50% of total fixed carbon can be demonstrated in the intact association between the nudibranch *Pteraeolidia ianthina* and its zooxanthellae.[2] But algae isolated from *Pa. ianthina* do not respond strongly to homogenates of the nudibranch or of *P. versipora*; 2-3% of the total fixed carbon leaks into seawater, and 4 to 10% into host homogenates of either *Pa. ianthina* or *P. versipora* (Hoegh-Guldberg and Sutton, submitted.)

### Association between **Dysidea herbacea** and the Prokaryote **Oscillatoria spongeliae**

In this association the algae are mostly extracellular, although they are often closely associated with sponge cells.[1] In spite of the looser structural association of algae and host, there is substantial translocation of photosynthetic products in the intact association. In other symbiotic sponges 5% to 12% of total fixed carbon is translocated to sponge tissue as glycerol.[3] The mechanism by which this occurs is not yet known.

# DISCUSSION

## Associations Involving Zooxanthellae

The data outlined above suggest that each of these three associations has a different means of ensuring translocation of low-molecular-weight products of photosynthesis from the algae to the animal. *P. versipora* shows the "classical" response of the algae to host tissue; the other two associations clearly function differently. The direct translocation of lipids (see, for example, Reference 4) has not been studied in any of these associations. It may play a major role in translocation in the zoanthid and the nudibranch. Nevertheless, both of these associations show rapid translocation of low-molecular-weight compounds such as glycerol. There is no evidence of extensive necrotrophy in any of these associations.

Important future directions for research on translocation in associations involving zooxanthellae include the study of the unexpectedly diverse mechanisms of metabolite exchange in these symbioses, and the elucidation of the links between the loss of organic carbon from the algae and the control of their growth rates in symbiosis.

## Association of Animals and Prokaryotic Algae

There is, as yet, no strong evidence for or against stimulation of translocation in sponges, or in ascidians containing *Prochloron,* by host homogenates. This is largely because homogenizing the host tissue often releases toxic compounds which destroy the algae (as in *D. herbacea*).

Biotechnologists have found that immobilizing cells and growing them in a medium with relatively low concentrations of essential nutrients increases their accumulation of secondary metabolites.[5] In some cases these conditions are thought to promote the export of secondary metabolites.[6] Most symbiotic algae are immobilized; could this induce translocation? The effects of immobilization on algal symbionts should be investigated. They may be particularly important in the case of the extracellular, prokaryotic symbionts of sponges and ascidians, which seem to have less contact with the cells of their hosts than intracellular symbionts do.

# SUMMARY

Four associations are under investigation. In each the animal receives a substantial proportion of the photosynthetic products of the zooxanthellae when the association is intact. (a) Zooxanthellae from the coral *Plesiastrea versipora* show a classic response to homogenates of their hosts' tissues, translocating fixed carbon into the host homogenate at high rates. At least two components of the host homogenate are required for activity. Work on purification and characterization of the active compounds is continuing. (b) Zooxanthellae from the zoanthid *Zoanthus robustus* are not stimulated to translocate by zoanthid host homogenate but translocate almost as rapidly into *P. versipora* host homogenate as the symbionts of the coral do. (c) Zooxanthellae from the nudibranch *Pteraeolidia ianthina* do not respond to host homogenates of either *Pa. ianthina* or *P. versipora*. (d) The sponge *Dysidea herbacea* takes up photosynthetic products from its symbiotic blue-green algae, most of which are extracellular. The mechanisms involved are unknown. It is suggested that immobilization of the algae in the sponge may promote this translocation.

## ACKNOWLEDGMENTS

Ms. Ellen O'Brien's technical assistance has been invaluable throughout the project. I am grateful to Professor Arne Jensen for drawing to my attention the effects of immobilization of nonsymbiotic cells. Much of the as-yet unpublished work by O. Hoegh-Guldberg and D. Sutton was done at the Roche Research Institute of Marine Pharmacology, Sydney.

### REFERENCES

1.  BERTHOLD, R. J., M. A. BOROWITZKA & M. A. MACKAY. 1982. The ultrastructure of *Oscillatoria spongeliae,* the blue-green algal endosymbiont of the sponge *Dysidea herbacea.* Phycologia **21:** 327-335.
2.  HOEGH-GULDBERG, O. & R. HINDE. Studies on a nudibranch that contains zooxanthellae. I. Photosynthesis, respiration and the translocation of newly fixed carbon by zooxanthellae in *Pteraeolidia ianthina.* Proc. R. Soc. London Ser. B **228:** 493-509.
3.  WILKINSON, C. R. 1980. Cyanobacteria symbiotic in marine sponges. *In* Endocytobiology—Endosymbiosis and Cell Biology. W. Schwemmler & H. E. A. Schenk, Eds.: 553-563. Walter de Gruyter. Berlin & New York.
4.  BATTEY, J. F. & J. S. PATTON. 1984. A reevaluation of the role of glycerol in carbon translocation in zooxanthellae-coelenterate symbiosis. Mar. Biol. **79:** 27-38.
5.  LINDSEY, K. & M. M. YEOMAN. 1983. Novel experimental systems for studying the production of secondary metabolites by plant tissue cultures. *In* Plant Biotechnology. S. H. Mantell & H. Smith, Eds.: 39-66. Cambridge University Press. Cambridge, England.
6.  KOSARIC, N., A. WIECZOREK, G. COSENTINO, R. MAGEE & J. PRENOSIL. 1983. Ethanol fermentation. *In* Biotechnology. H. Dellweg, Ed. **3:** 257-386. Verlag Chemie. Weinheim, FRG.

# Change of Cellular "Pathogens" into Required Cell Components[a]

KWANG W. JEON

*Department of Zoology*
*University of Tennessee*
*Knoxville, Tennessee 37996-0810*

## INTRODUCTION

This is a progress report of a case study in which the transformation of harmful parasites into required cytoplasmic components has been followed under laboratory conditions. Cells, especially those living in natural environments, often act as suitable habitats for smaller microbes,[1-3] some of which are parasitic while others are innocuous without known functions.[4] A prolonged period of symbiosis may result in the partners becoming mutually dependent, interacting as a single functioning unit. Such a process of establishing stable symbiosis usually requires many cell generations and is not amenable to continuous observations or experimentation. The present model of ameba-bacteria symbiosis is an exception in that its history is known and the development of a host's dependence on symbionts can be readily reproduced under experimental conditions.

## INITIAL INFECTION

Cultures of a strain of *Amoeba proteus* that had been continuously kept under laboratory conditions for about 20 years became suddenly unhealthy in 1966. On examining individual amebae under the microscope, we found that they were heavily infected with rod-shaped gram-negative bacteria (called X-bacteria for their unknown origin), each ameba containing 60,000-150,000 bacteria.[5] Further comparisons between amebae of the normal (D) and infected (xD) strain revealed that the presence of infective bacteria had harmful effects on amebae, such as reduced cell size, fewer cytoplasmic crystals, slower cell growth with longer mean generation times, sensitivity to starvation, increased fragility, and a poor clonability. As a result of these harmful effects, most of the infected amebae died and extreme care was needed to maintain cultures of infected amebae. When bacteria from infected amebae were introduced into normal amebae either by exposure to a bacterial suspension or by intracellular injection,[5] the newly infected amebae died after dividing a few times, indicating the virulent nature of the newly introduced bacteria to host cells.

[a]The author's work was supported by grants from the National Science Foundation.

It had been known that some free-living amebae carried infective or symbiotic bacteria,[6-12] but the number of bacteria found in such amebae was usually small and there were no reports of harmful effects brought about by these infective bacteria. In only one case, harmful bacterial infection was described which led to the lysis of infected soil amebae within several days.[13] However, the host-parasite relationship was not studied further.

## MUTUAL ADAPTATION

### *Reduction of Parasites' Virulence*

Over a period of about a year, while the infected xD amebae were cultured with special care, it was noted that the adverse effects of infection diminished gradually.[5] The bacteria-bearing amebae were healthier and grew well. In contrast to earlier experiments, some newly infected amebae survived, indicating that the infective bacteria were less virulent. Further diminishment of harmful effects of infective bacteria was observed during the next few years. The reason for the reduction in virulence is not known, except that the number of bacteria per infected ameba decreased. At the present time, each xD ameba contains about 42,000 X-bacteria per cell and the bacteria appear to have the same morphology noted at the time of initial infection (FIGURE 1).

### *Development of Host's Dependence on Symbionts*

By 1971, the progeny of the originally infected amebae appeared to have become dependent on the bacteria.[14] Thus, nuclei of infected xD amebae were unable to form viable cells when combined with the cytoplasm of normal D amebae, unless cytoplasm from xD amebae containing bacteria was also transferred at the same time or following nuclear transplantation (see Reference 15 for later experiments). At that time, only a small proportion (7%) of xD nuclei transferred into D-ameba cytoplasm (i.e., $xD_nD_c$) was able to form viable clones without accompanying infective bacteria,[14] whereas over 90% of reciprocal nuclear transplants ($D_nxD_c$) were viable. In similar experiments carried out since 1980, we have found that xD nuclei rarely form viable cells with D cytoplasm.[16]

The dependence of infected amebae on their symbionts was further demonstrated by selectively removing symbiotic bacteria from adapted amebae either by treating them with chloramphenicol[17] or by culturing amebae at an elevated temperature causing digestion of symbionts.[16,18] The disappearance of symbionts was followed by death of aposymbiotic amebae within a week. However, these aposymbiotic amebae could be resuscitated by reintroducing symbiotic bacteria.[16]

Experimentally infected amebae also became dependent on their newly acquired symbionts. In one series of experiments, it was found that 90% of newly infected amebae became dependent on their symbionts in 18 months or 200 cell generations.[18]

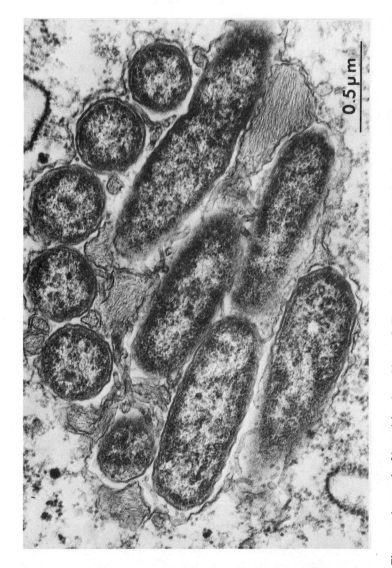

**FIGURE 1.** Electron micrograph of bacterial endosymbionts of xD amebae enclosed in a vesicle, showing the electron-dense cytoplasm and electron-lucent nuclear zones in the bacteria. Magnification, $\times$ 48,000.

## ROLES OF SYMBIONTS AS QUASI ORGANELLES

### Independent Genetic Machinery

Symbiotic bacteria contain their chromosomal DNA and two distinct plasmid DNAs[19] (FIGURE 2). The chromosomal DNA has not yet been characterized, but the sizes of plasmid DNAs have been found to be about 62 and 22 kilobases (kb), respectively. Although the role of these plasmid DNAs is not known, they are suspected to be involved in coding for cell-surface components conferring symbionts resistance to digestion or preventing lysosomal fusion with symbiont-containing vesicles.[20] In spite of the symbiotic relationship, X-bacteria have retained their ability to utilize exogenous thymine for DNA synthesis.[21]

### Initial Acceleration of Host Cell Growth

During the initial phase of experimental infection for up to 12 months, newly infected amebae grow faster than either established xD amebae or symbiont-free D amebae.[18] In one series of experiments, the mean generation time of newly infected amebae was about 1.8 days at 26°C, as compared to 2.1 days for symbiont-free amebae and 2.5 days for established xD amebae. Similar results have been obtained also at room temperature.

### Control of a Phenotypic Character, Temperature Sensitivity

Symbiont-bearing xD amebae cannot grow at temperatures above 26°C, one degree higher than their optimum growth temperature, and die within two weeks, while D amebae continue to grow at that temperature.[18] The death of amebae is preceded by the disappearance of symbionts, apparently by digestion,[16] which is in turn preceded by the disappearance of the plasmid DNAs normally present in symbionts.[22] Therefore, temperature sensitivity of symbionts determined by their plasmids has been conferred upon the host amebae.

### Control of Membrane Differentiation

Membranes of vesicles enclosing symbiotic bacteria are unique in that they do not fuse with lysosomes even though they share a common origin with phagosomes, being derived from the plasmalemma during phagocytosis.[15,24] These symbiont-enclosing vesicle membranes contain unique polypeptides ($M_r$ about 220,000)(FIGURE 3) not found in other ameba membranes.[23] The polypeptide group is located on the cyto-

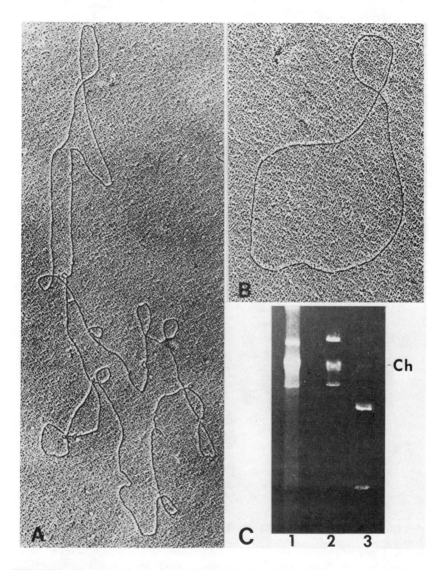

**FIGURE 2.** Chromosomal and plasmid DNAs of endosymbionts of xD amebae (after Han & Jeon.)[19] A and B, electron micrographs of two plasmid DNAs. C, agarose gel (0.7%) electrophoresis of chromosomal (Ch) and plasmid DNAs. Lane 1: DNA from X-bacterial lysate after partial removal of chromosomal DNA; two plasmid bands, one above Ch and another below, are visible. Lane 2: DNA of X-bacteria after further removal of chromosomal DNA by two cycles of CsCl-EtBr centrifugation. Lane 3: pBR322 and RSF2124 plasmids electrophoresed together for comparison.

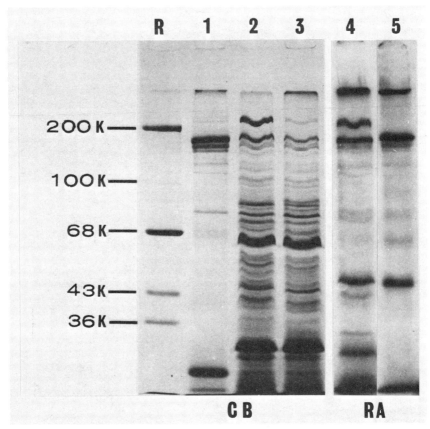

**FIGURE 3.** Coomassie blue-stained polyacrylamide gel of $^{125}$I-labeled and solubilized proteins of xD amebae (CB) and corresponding radioautograms (RA). Lane R, molecular-weight markers; lane 1, isolated plasmalemma; lane 2, symbiont-containing vesicles with X-bacteria; lane 3, isolated X-bacterial proteins; lanes 4 and 5, radioautograms of lanes 2 and 1, respectively. Note the presence of unique polypeptide band present in symbiont-containing vesicle membranes (lanes 2 and 4) which is not present in the plasmalemma (lane 1). (After Ahn and Jeon.)[23]

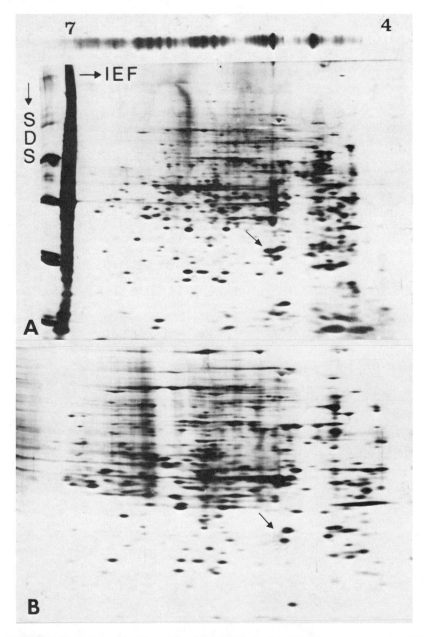

**FIGURE 4.** Two-dimensional electorphoretic gels of cytosol proteins from xD (A) and D (B) amebae. Note the presence of unique 29-kD polypeptides (arrow in A; pI, 5.5) in the cytosol of xD amebae (A). The gels were run by the usual method[36] and stained with silver.[37]

plasmic side of the vesicle membranes, as determined by labeling isolated vesicles with [125]I before solubilizing the membranes for polyacrylamide gel electrophoresis[23] (FIGURE 3). It is not yet known if these polypeptides are synthesized by the symbionts or by the host. These unique membrane proteins may play a role in the prevention of lysosomal fusion.

### Protein Synthesis

The symbiotic bacteria synthesize several polypeptides that are apparently transported across the vesicle membrane to the ameba cytosol. Among these, a prominent 29 kilo-Dalton component can be resolved by two-dimensional gel electrophoresis[25] (FIGURE 4). The 29-kD polypeptide appears to be a subunit of a native protein with a molecular weight of about 120,000[26] (FIGURE 5). This polypeptide is contained in the cytosol of established xD amebae and of newly infected D amebae. The bacterial origin of this polypeptide was first demonstrated indirectly by following its disappearance from the xD-ameba cytosol when symbiotic bacteria were removed by heat treatment[25] (FIGURE 6). When symbiont-free amebae were infected with X-bacteria, the 29-kD polypeptide appeared in the newly infected ameba cytosol[26] (FIGURE 7). Synthesis of this protein was suppressed when symbiont-bearing amebae were treated with chloramphenicol or rifampicin[27] (FIGURE 8). Cycloheximide failed to inhibit synthesis of the protein, suggesting symbiont origin of the polypeptides.

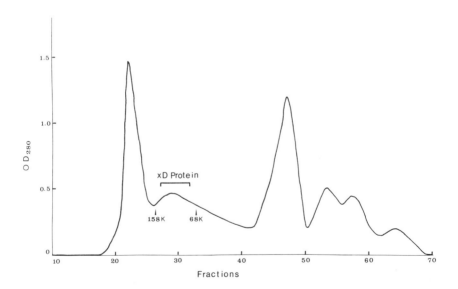

**FIGURE 5.** Elution pattern of xD-ameba cytosol proteins through a Sephacryl S-300 column, indicating the fractions for native proteins containing 29-kD polypeptides. (From Kim and Jeon.)[27]

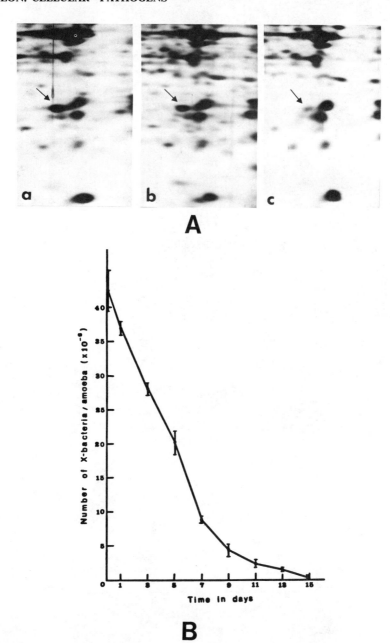

**FIGURE 6.** Decrease in the amount of 29-kD polypeptides from xD-ameba cytosol with the removal of symbionts. A; portions of two-dimensional electrophoretic gels of xD ameba cytosol, after 3 days (a), 9 days (b), and 14 days (c) of culture at 26°C, respectively. The 29-kD spot is significantly smaller after 14 days (arrow in c). B; a graph to show the decrease in the number of X-bacteria per xD ameba cultured at 26°C.

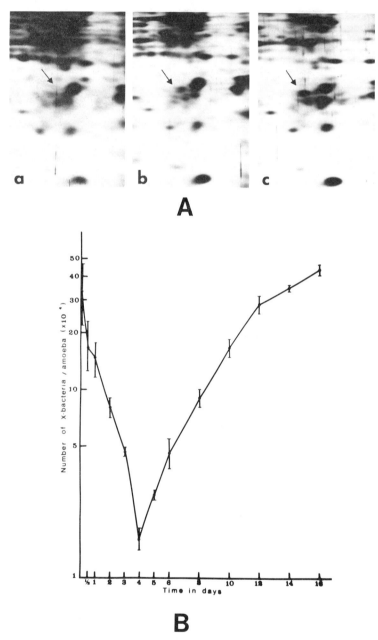

**FIGURE 7.** Appearance of 29-kD polypeptides in the cytosol of newly infected D amebae. A; portions of two-dimensional electrophoretic gels of cytosol proteins from newly infected amebae after 3 days (a), 6 days (b), and 9 days (c) of infection, respectively, to show the appearance of 29-kD polypeptides. The 29-kD spot (arrow) is detectable by the 6th day and on the 9th day it is only slightly smaller than that of the control xD cytosol (shown in other figures.) B; a graph to show changes in the number of symbionts per infected ameba with time.

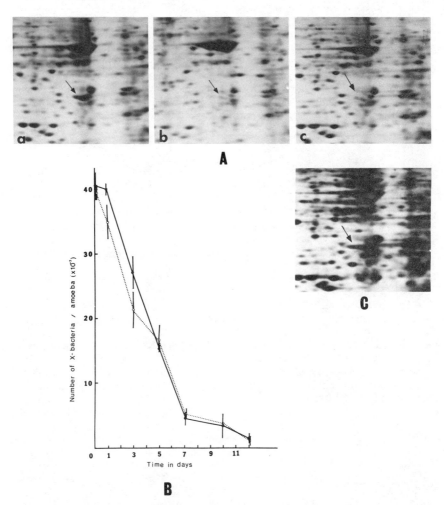

**FIGURE 8.** Effects of chloramphenicol and rifampicin on the synthesis of 29-kD polypeptides by symbionts of xD amebae. A; portions of radioautograms of two-dimensional electrophoretic gels of the cytosol proteins from xD amebae fed with [35]S-methionine-labeled *Tetrahymena*. Cytosol proteins were extracted 24 hours after feeding. a, control xD ameba cytosol, showing the presence of newly synthesized 29-kD polypeptides (arrow). b, cytosol of ameba treated with chloramphenicol (250 μg/ml) added at the time of feeding. c, cytosol of amebae treated with rifampicin (125 μg/ml). Note the absence of newly synthesized 29-kD polypeptides in b and c, while unlabeled 29-kD peptides are present in antibiotic-treated ameba cytosol (C). B; a graph to show changes in the average number of symbionts per xD ameba grown in chloramphenicol (solid line) and rifampicin (dotted line). Bars indicate standard deviations. C; a portion of silver-stained gel, from which the radioautogram in A,c was obtained. Note the presence of 29-kD peptides (arrow).

Chloramphenicol was originally used at concentrations of 0.5-1.5 mg/ml,[17] in an attempt to remove symbiotic bacteria from infected amebae. Recently, we have used chloramphenicol to selectively inhibit protein synthesis by the symbionts and to follow the effect of inhibition on the host cells.[27] When treated with chloramphenicol at 250 μg/ml, synthesis of the xD-specific polypeptides by the symbionts was immediately suppressed as determined by two-dimensional gel electrophoresis and radioautography and the average number of symbionts per ameba decreased rapidly (FIGURE 8). When symbiont-bearing and symbiont-free amebae were grown in a medium containing chloramphenicol, symbiont-bearing amebae died much sooner, suggesting that the symbionts' proteins play vital roles in the survival of host amebae.

## DISCUSSION

Our results show that initially harmful bacterial parasites have changed to required cell components of symbiont-bearing amebae. One unanswered question has been what the newly acquired symbionts do to induce host cell dependence. We now know that the symbionts produce at least one protein that is needed by host amebae, although it is not known where the protein is localized and what it does in amebae. We are in the process of obtaining monoclonal antibody to the 29-kD polypeptides to determine their site of action.

The symbionts do not grow *in vitro* and it is not possible to determine if they would independently synthesize 29-kD-containing protein by themselves or do so only under the influence of amebae. Recently Ishikawa found that protein synthesis by endosymbionts of aphids was closely controlled by their host, with one specific protein being synthesized by endosymbionts only when in the host cells.[28,29]

The serial endosymbiosis theory[30,31] for the origin of eukaryotic cell organelles appears to be gaining further support from recent molecular-biological studies. For example, ribosomal RNA sequencing data have been interpreted to support the origin of mitochondria from bacterial endosymbionts,[32,33,34] and the presence of DNA in the nucleomorph of cryptomonads has been viewed as evidence for the symbiotic origin of chloroplasts.[35] Our observation of transformation of pathogens into required cell components through mutual adaptations is concordant with this theory.

## ACKNOWLEDGMENTS

I thank Dr. M. A. Handel for her critical reading of the manuscript and Mr. H. B. Kim for his help in preparing the figures.

## REFERENCES

1. JENNINGS, D. H. & D. L. LEE, Eds. 1975. Symb. Symp. Soc. Exp. Biol. **29**.
2. RICHMOND, M. H. & D. C. SMITH, Eds. 1979. The Cell as a Habitat. Royal Society. London, England.

3. COOK, C. B., P. W. PAPPAS & E. D. RUDOLPH, Eds. 1980. Cellular Interactions in Symbiosis and Parasitism. Ohio State University Press. Columbus, Ohio.
4. PREER, J. R. 1971. Annu. Rev. Genet. **5:** 361-406.
5. JEON, K. W. & I. J. LORCH. 1967. Exp. Cell Res. **48:** 236-240.
6. BREED, R. S., E. G. D. & N. R. SMITH. 1957. In Bergey's Manual of Determinative Bacteriology. 7th edit.: 926-927. Williams & Wilkins. Baltimore, Md.
7. KUDO, R. R. 1957. J. Protozool. **4:** 154-164.
8. ROTH, L. E. & E. W. DANIELS. 1961. J. Biophys. Biochem. Cytol. **9:** 317-323.
9. DANIELS, E. W. 1964. J. Protozool. **11:** 281-290.
10. DANIELS, E. W., E. P. BREYER & R. R. KUDO. 1966. Z. Zellforsch. **73:** 367-383.
11. DANIELS, E. W. & E. P. BREYER. 1967. J. Protozool. **14:** 167-179.
12. CHAPMAN-ANDRESEN, C. 1971. Annu. Rev. Microbiol. **25:** 27-48.
13. DROZANSKI, W. 1963. Acta Microbiol. Pol. **12:** 9-24.
14. JEON, K. W. 1972. Science **176:** 1122-1123.
15. JEON, K. W. & M. S. JEON. 1976. J. Cell. Physiol. **89:** 337-347.
16. LORCH, I. J. & K. W. JEON. 1980. J. Protozool. **27:** 423-426.
17. JEON, K. W. & J. C. HAH. 1977. J. Protozool. **24:** 289-293.
18. JEON, K. W. & T. I. AHN. 1978. Science **202:** 635-637.
19. HAN, J. H. & K. W. JEON. 1980. J. Bacteriol. **141:** 1466-1469.
20. JEON, K. W. 1983. Int. Rev. Cytol. Suppl. **14:** 29-47.
21. FREESH, J. M. 1975. Master's Thesis. University of Tennessee. Knoxville, Tenn.
22. PARK, Y. M. 1983. Master's Thesis. University of Tennessee. Knoxville, Tenn.
23. AHN, T. I. & K. W. JEON. 1982. Exp. Cell Res. **137:** 253-268.
24. AHN, T. I. & K. W. JEON. 1979. J. Cell. Physiol. **98:** 49-58.
25. AHN, T. I. & K. W. JEON. 1983. J. Protozool. **30:** 713-715.
26. KIM, H. B. & K. W. JEON. 1985. J. Cell Biol. **101:** 493a.
27. KIM, H. B. & K. W. JEON. 1986. Endocyt. Cell Res. **3:** 299-309.
28. ISHIKAWA, H. 1984. Biosystems **17:** 127-134.
29. ISHIKAWA, H. 1985. Biosystems **17:** 327-335.
30. MARGULIS, L. 1970. Origin of Eukaryotic Cells. Yale University Press. New Haven, Conn.
31. TAYLOR, F. J. R. 1974. Taxon **23:** 229-258.
32. GRAY, M. R., D. SANKOFF & R. J. CEDERGREN. 1984. Nucleic Acids Res. **25:** 5837-5852.
33. STEWART, K. D. & K. R. MATTOX. 1985. J. Mol. Evol. **21:** 54-57.
34. YANG, D., Y. OYAIZU, G. J. OLSEN & C. R. WOESE. 1985. Proc. Nat. Acad. Sci. USA **82:** 4443-4447.
35. LUDWIG, M. & S. P. GIBBS. 1985. Protoplasma **127:** 9-20.
36. O'FARELL, P. H. 1975. J. Biol. Chem. **250:** 4007-4021.
37. MERRIL, C. R., D. GOLDMAN, S. A. SEDMAN & M. H. EBERT. 1980. Science **211:** 1437-1438.

# Compartmentation of Calcium and Energy Metabolic Pathways

## Implications for Eukaryote Evolution and Control of Cell Proliferation

J. BEREITER-HAHN

*Cinematic Cell Research Group*
*J. W. Goethe University*
*Senckenberganlage 27*
*D 6000 Frankfurt a.M., Federal Republic of Germany*

The endosymbiontic origin of plastids, mitochondria, and, with less probability, of centrioles organizing cilia and flagella is widely accepted (i.e., References 6, 12, 16, and 18). The closest recent relatives of the likely symbiotic candidates have been identified. Despite this impressive progress we still lack a continous deduction of the putative host cell and its properties. It is unclear whether this host was a eukaryotic or a prokaryotic cell[6] and how the main characteristics of a eukaryotic cell derived from prokaryotes.[5] As in all cases of reconstruction of historical events, theoretical considerations have to be combined with experimental evidence derived from modern organisms. Consequent research of this type can deepen our comprehension of the organization of living matter and reveal results that may delineate a convincing trace of evolutionary development.

## CHARACTERIZATION OF A PUTATIVE HOST CELL

Characteristics that can be ascribed to a putative host cell at the basis of eukaryote cell evolution are:

1. Wall-less protoplast with actin-based motility.
2. Motility regulation by $Ca^{++}$.
3. Energy supply system.
4. Aerophilia.

Assumption of a eukaryotic cell acting as host for prokaryotes becoming prospective organelles does not solve the problem of deducing such an organism from a prokaryote ancestor. Therefore the above-mentioned characteristics have either to be ascribed to a prokaryote or to be deduced from other prokaryote properties. There is no compelling reason why lysosomes or the development of a true nucleus should be essential at an early stage of a host-symbiont interaction.

### Wall-less Protoplast with Actin-Based Motility

Wall-lessness is a prerequisite for a protoplast, which is supposed to take up other organisms. How is wall-lessness achieved? Is it a primitive character of early prokaryotes or is it due to loss of wall structures?

With a few remarkable exceptions, prokaryotes possess cell walls. Their organization is very different from cell walls found in eukaryotes. The selective advantage of a wall lies in mechanical stabilization of a cell which is essential for living in turbulent water.[11] In addition, an osmotic imbalance between cytoplasm and the surrounding medium can be tolerated. Loss of bacterial cell walls by genetic defects is a well-known phenomenon leading to L-forms. These protoplasts form large protoplasmic layers containing multiple copies of chromosomes—chromosome replication and growth proceed; finally they are destroyed by mechanical forces resulting from agitation of their environment. This is the reason why penicillin is an effective antibiotic and why L-forms are a blind lane in evolution. Two ways can be assumed to lead out of this "blind lane"—the first is reduction of size combined with colonization of a niche of low turbulence, the second is the invention of internal reinforcements providing a minimum amount of mechanical stability. Nature tried both ways—the first resulted in evolution of mycoplasms,[19] the second in evolution of eukaryote cells, that is, of the nucleoplasmic part of eukaryote cells. Actin can be assumed to have formed the reinforcing structure. The evolutionary origin of actin still lies in the dark; however, all eukaryotic organisms seem to have actin, plants as well as animals or fungi. The molecule is highly conserved throughout evolution, which points to its extreme importance. The main traits of biochemical pathways have been invented by prokaryotes; therefore it is highly probable that actin also was first produced in a prokaryotic cell. In bacteria the selective benefit of this invention may be limited to L-forms, which became extinct due to competition with mobile eukaryotic cells living in the same ecological niche. Therefore no actin is found in recent bacteria.

Actin opened an evolutionary potential of extreme significance: together with microtubules, actin filaments are responsible for intracellular transport in eukaryotic cells, overcoming the diffusion limitation of size. An extreme example are characean cells or Acetabularia, huge cells with vigorous actin-based cytoplasmic streaming. Filamentous reinforcements of animal cells are fundamentals of multicellularity, which also means mechanical interaction of cells, and of animal mobility.

What could be the selective advantage of a protoplast stabilized by filaments in an environment with steadily increasing oxygen? In this situation abiotic formation of small organic molecules ceases, therefore food supply becomes a limiting factor for heterotrophs. Heterotrophs can solve this problem by several strategies:

1. By improvement of energy yield from food sources through perfection of the respiratory systems, acquiring the ability to use oxygen as final electron acceptor.
2. By becoming mobile to search chemotactically for new sources.
3. By extending food uptake from molecules dissolved in water to those absorbed on solid substrata.

The last strategy could have been followed by protoplasts which may cover relatively large areas of soil particles and dissolve organic matter by exoenzymes. The developmental potential of such protoplasts is limited, however, without the achievement of mechanical stabilization, which in addition allows slow creeping movements on the substratum. Actin alone is supposed to be sufficient to drive such locomotions forming a gel of high viscosity by polymerization and a fluid sol in the unpolymerized state. A model for a comparable type of locomotion has been developed by Taylor and Fechheimer to explain leukocyte motility.[22]

### Motility Regulation by $Ca^{++}$

Regulation of the events described above is mediated by $Ca^{++}$ and $Ca^{++}$-binding proteins. $Ca^{++}$ is involved in all types of actin-dependent mobilities so far investigated. However, the mode of control differs significantly, e.g., between skeletal muscle and Physarum plasmodia;[8] contraction as well as gelation/solation is $Ca^{++}$-controlled (comp. Reference 14). From these findings it follows that any cell, using actin, has to control intracellular free $Ca^{++}$ and to keep it at a low level. The specific type of interaction with actin-associated proteins, and later on also with microtubules, may have pullulated during evolution of eukaryote kingdoms.

In consequence of the proposal that $Ca^{++}$ control is a fundamental property of cells in the prokaryote/eukaryote transition zone, a certain degree of mitochondria-independent $Ca^{++}$ control should be expected in modern cells (comp. Compartmentation of $Ca^{++}$ and Relevance of the Experiments for Speculations on Cell Evolution and Growth Control).

### Energy Supply System

The general distribution of glycolysis leading from carbohydrates to pyruvate is a strong argument for the assumption that the cells giving rise to eukaryotes were equipped with this pathway. A high variety exists in the consecutive reactions. The products arise from the need to remove the active hydrogen from the $NADH_2$ formed in pyruvate production.[2] Lactic acid formation is the most widespread product under anaerobic conditions; therefore it might well be a very ancient process.

Interaction of a host prokaryote with prospective mitochondria was at its beginning not necessarily symbiotic, it could have been one of the many known syntrophies among bacteria. The consequence of this assumption is that not the exchange substrate/ATP but the flux of substrate from the prospective host to the prospective mitochondrion determine the physiological interaction; adenine-nucleotide exchange would emerge in an advanced stage of mutual interaction providing a high selective benefit.

According to this hypothesis, engulfment of bacteria with respiratory activities is preceded by syntrophic association. The bacteria may be covered by a sheet of protoplast cytoplasm and be engulfed by invaginating cortical cytoplasm of the protoplast (see Relevance of the Experiments). At this early stage true phagocytosis, involving protein synthesis on rough endoplasmic reticulum (rER) and further processing by dictyosomes might not have been elaborated. In this case no intracellular digestion would destroy the engulfed bacteria, which are resistant against exoenzymes of the protoplasts, because otherwise no syntrophy would have been established. Syntrophy rather than symbiosis initiated the biogenesis of mitochondria. Stabilization of intracellular syntrophy requires balance in substrate uptake and growth of the partners; regulation by phosphorylation potential or ATP would be of minor significance from the evolutionary point of view (comp. Glycolysis and Respiration).

### *Aerophilia*

Protoplasts and bacteria using oxygen as final electron acceptor can establish syntrophy only if the protoplasts are not aerophobic, otherwise they never would meet.

## EXPERIMENTS SUPPORTING THE PROTOPLAST-SYNTROPHY HYPOTHESIS

### *Compartmentation of $Ca^{++}$*

In myocytes cytoplasmic $Ca^{++}$ concentration is regulated by the sarcoplasmic reticulum and by mitochondria. These two systems have different affinities to $Ca^{++}$, as has been known for a long time.[1,3] Whether this type of control by two systems is valid for nonmuscle cells as well, and how these systems interact with each other has been investigated by Stolz and Bereiter-Hahn: endothelial cells of *Xenopus* tadpole hearts in culture (XTH-2 cells) were injected with $Ca^{++}$ at different velocities.[20] Speed and amount were controlled by regulating current during microiontophoretic injection. $Ca^{++}$ does not diffuse through the cytoplasm for more than a few micrometers (comp. Reference 15). Cell death occurs suddenly. The injection time required to kill the cells is used as a parameter indicating overtask of $Ca^{++}$ sequestration capacity. The results summarized in FIGURE 1 reveal the ER and mitochondria to be the compartments active in maintainance of low cytoplasmic $Ca^{++}$ levels. However, at slow injection speed ER stores become overtasked; cells die without any involvement of mitochondria in sequestering $Ca^{++}$. Mitochondria become active only if injection speed exceeds $20 \times 10^{-15}$ mol s$^{-1}$.

From these results we conclude that $Ca^{++}$ homeostasis is maintained primarily by the ER. In living cells, mitochondria are involved only in case of emergency.

### *Glycolysis and Respiration: Compartmentation and Interaction*

Location of respiratory enzymes on the inner mitochondrial membrane is a well-known fact. In addition the enzymes of glycolysis are not dissolved in the cytosol but rather are bound to F-actin.[7,13] This association is of high functional significance as it improves enzyme activity[7] and provides the energy for contraction and turnover of the actomyosin fibrils.[23]

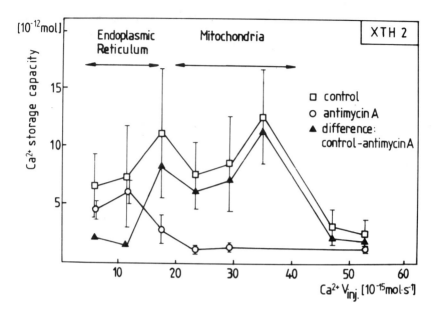

**FIGURE 1.** Relation of Ca$^{++}$-storage capacity of XTH-2 cells to injection speed; untreated cells (squares), and after inhibition of respiration by antimycin A (10 $\mu$M) (circles). The difference curve (triangles) shows the amount of Ca$^{++}$ sequestered by mitochondria. (Reproduced from Reference 20 with permission.)

Following intracellular ATP during cytochalasin D-initiated reactions of actin fibrils showed a high fluctuation of values without any change in oxygen consumption.[23] In a given cell type (XTH-2),[17] oxygen consumption and lactate formation are related to cell density (Münnich and Bereiter-Hahn, unpublished), and the ratio of lactate produced per oxygen consumed decreases considerably with increasing cell density (FIGURE 2). The only parameter that seems to be kept constant and that is independent from cell density is the fraction of reduced NADH$_2$ (Kajstura and Bereiter-Hahn, unpublished).

### Relevance of the Experiments for Speculations on Cell Evolution and Growth Control

The results show that even in modern cells:

1. Ca$^{++}$ homeostasis is provided by the nucleoplasmic part which is supposed to be derived from an actin-containing protoplast (see Motility Regulation by Ca$^{++}$).
2. A close relation between the mechanical system and glycolytic enzymes is maintained, as suspected by the endocytobiont hypothesis.

3. A certain degree of independence exists between glycolysis and respiration. The controlling link, which is kept constant, is the redox potential.

Retrieval of NAD can be assumed to represent the central goal of syntrophy for fermenting protoplasts. This avoids loss of pyruvate by formation of lactate, or other reduction products, which are not used for further syntheses. Kajstura and Korohoda draw attention to the importance of citric acid cycle intermediates for maintaining proliferative activity.[9] If supplied with some purines, mammalian cells are able to continue growth in the absence of respiration.[10] This shows that growth and proliferation of animal nucleocytoplasm are relatively independent from mitochondria. Concomitant with the establishment of mitochondria, their proliferation had to be balanced with that of the nucleocytoplasm. The balancing link could well be the redox potential. Disturbing this balance is supposed to be responsible for malignant growth of cancer.[21] This influence of redox potential deserves more detailed studies on malignant and nonmalignant cells.

## IMPLICATIONS OF HOST CELL CHARACTERISTICS ON THE ORIGIN OF THE NUCLEUS

According to Cavalier-Smith, the main advantage of nucleus formation has been the availability of the whole cell surface for phagocytosis.[4] This explanation does not regard that bacterial L-forms proceed with DNA replication without being able to divide. This problem could be overcome by internalization of plasmalemma with the

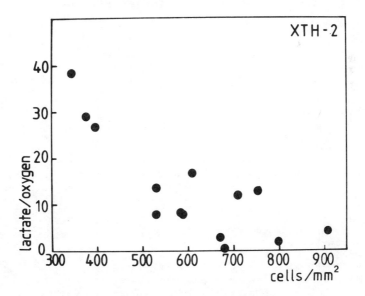

**FIGURE 2.** Dependence of the ratio of lactate production/oxygen consumption on density of XTH-2 cells (courtesy of A. Münnich).

attached chromosome (or chromosomes!) and formation of a vesicle, thus providing a stress-free space for separation of sister chromosomes. Membrane invaginations will easily appear as soon as fibrillar elements interconnect different parts of membranes and exert tension. This may well precede true phagocytosis and resemble the membrane foldings seen in the acellular slime mold Physarum.[24] Physarum plasmodia resemble most closely a primitive eukaryote stage as suggested by the hypothesis presented here or envisaged by Cavalier-Smith.[5]

## CONCLUDING REMARKS

Despite the fact that many of the deductions given here are very speculative, systematic investigation of the interaction of metabolic pathways gives some hints on their origin. Furthermore an evolutionary approach to regulation of cell metabolism is promising for deepening our comprehension of control of cell growth and proliferation.

### REFERENCES

1. BORLE, A. B. 1981. Control, modulation, and regulation of cell calcium. Rev. Physiol. Biochem. Pharmacol. **90:** 13-153.
2. BRODA, E. 1975. The Evolution of the Bioenergetic Processes. Pergamon Press. Oxford, England.
3. CARAFOLI, E. 1985. The homeostasis of $Ca^{2+}$ in heart cells. J. Mol. Cell Cardiol. **17:** 203-212.
4. CAVALIER-SMITH, T. 1975. The origin of nuclei and of eukaryotic cells. Nature **256:** 463-468.
5. CAVALIER-SMITH, T. 1981. The origin and early evolution of the eukaryotic cell. In Molecular and Cellular Aspects of Microbial Evolution. M. J. Carlile, J. F. Collins & B. E. B. Moseley, Eds.: 33-84. Cambridge University Press. Cambridge, England.
6. CAVALIER-SMITH, T. 1983. Endosymbiontic origin of the mitochondrial envelope. In Endocytobiology II. H. E. A. Schenk & W. Schwemmler, Eds.: 265-279. Walter de Gruyter. Berlin & New York.
7. CLARKE, F., P. STEPHAN, D. MORTON & J. WEIDEMANN. 1983. The role of actin and associated proteins in the organisation of glycolytic enzymes. In Actin. C. G. Dos Remedios & J. A. Barden, Eds.: 249-258. Academic Press. Sydney & New York.
8. HINSSEN, H. 1980. Regulation of actin transformation by a $Ca^{2+}$-dependent actin-modulating protein from the slime mould Physarum polycephalum. Eur. J. Cell Biol. **22:** 327.
9. KAJSTURA, J. & W. KOROHODA. 1983. Significance of energy metabolism pathways for stimulation of DNA synthesis by cell migration and serum. Eur. J. Cell Biol. **31:** 9-14.
10. LÖFFLER, M. 1985. Towards a further understanding of the growth-inhibiting action of oxygen deficiency. Exp. Cell Res. **157:** 195-206.
11. MÄRKL, H. & R. BRONNENMEIER. 1985. Mechanical stress and microbial production. Biotechnology **2:** 369-392.
12. MARGULIS, L. 1981. Symbiosis in Cell Evolution. Freeman. San Francisco, Calif.
13. MASTERS, C. 1984. Interactions between glycolytic enzymes and components of the cytomatrix. J. Cell Biol. **99:** 222s-225s.
14. MITTAL, A. K. & J. BEREITER-HAHN. 1985. Ionic control of locomotion and shape of epithelial cells. Cell Mot. **5:** 123-136.

15. ROSE, B. & W. R. LOEWENSTEIN. 1975. Calcium ion distribution in cytoplasm visualized by aequorin: diffusion in cytosol restricted by energized sequestering. Science **190:** 1204-1206.
16. SAGAN, L. 1967. On the origin of mitosing cells. J. Theor. Biol. **14:** 225-274.
17. SCHLAGE, W. & J. BEREITER-HAHN. 1981. Established Xenopus tadpole heart endothelium (XTH) cells exhibiting selected properties of primary cells. Eur. J. Cell Biol. **24:** 342.
18. SCHWEMMLER, W. 1985. Endocytobiosis formation: macromechanism of cell evolution. Endocytol. C. Res. **2:** 179-190.
19. STACKEBRANDT, E. & C. R. WOESE. 1981. The evolution of prokaryotes. Symp. Soc. Gen. Microbiol. **32:** 1-31.
20. STOLZ, B. & J. BEREITER-HAHN. 1987. Sequestration of iontophoretically injected calcium by living endothelial cells. Cell Calcium. (In press.)
21. SZENT-GYÖRGY, A. 1976. Electronic Biology and Cancer. A New Theory of Cancer. Marcel Dekker, Inc. New York & Basel.
22. TAYLOR, D. L. & M. FECHHEIMER. 1981. Cytoplasmic structure and contractility: the solation-contraction coupling hypothesis. Philos. Trans. R. Soc. London Ser. B **299:** 185-197.
23. TILLMANN, U. & J. BEREITER-HAHN. 1986. Relation of actin fibrils to energy metabolism of endothelial cells. Cell Tissue Res. **243:** 579-585.
24. WOHLFARTH-BOTTERMANN, K. E. 1974. Plasmalemma invaginations as characteristic constituents of plasmodia of *Physarum polycephalum.* J. Cell Sci. **16:** 23-27.

# Eubacterial, Eukaryotic, and Unique Features in the Mitochondrial Genome of *Oenothera*[a]

W. SCHUSTER, R. HIESEL, E. MANNA, W. SCHOBEL,
AND A. BRENNICKE[b]

*Department of Botany*
*University of Tübingen*
*Auf der Morgenstelle 1*
*D-74 Tübingen, Federal Republic of Germany*

## INTRODUCTION

According to the endosymbiotic hypothesis, mitochondria are thought to have evolved from prokaryotes accepted as endosymbionts by their host cells, which became eukaryotes by this and other steps of differentiation from the prokaryotic world.[1-4] The endosymbionts evolved after total dependence was established through mutations together with their hosts to the high degree of integration we observe for these cell organelles in the complex specialization through which tasks are distributed inside each cell.

From the time of merging cells, modern mitochondria could have retained some of the characteristics of a prokaryotic organism. Through their close cooperation with the evolving eukaryotic cytoplasm some of the newly established pathways and assemblies altered structures and functions also on the inside of the mitochondrial membranes and established similar activities and modes on both sides, in the eukaryotic cytoplasm and in the mitochondrion. Certain drifts however, probably occurring equally in all systems, could only manifest themselves in the increasingly smaller mitochondrial genetic systems and have thus remained developments unique to this compartment.

In our analysis of structure and function of the mitochondrial genome of the higher plant *Oenothera berteriana*, features are found supporting the view of influences from both pro- and eukaryotic specificities and are described here in context.

## MATERIALS AND METHODS

Mitochondrial DNAs were purified from mitochondria isolated by differential centrifugation as described.[5,6] Restriction, hybridization protection, and gel electro-

[a]This work was supported by grants from the Deutsche Forschungsgemeinschaft, a Graduiertenstipendium and a Heisenbergstipendium.

[b]To whom correspondence should be addressed.

phoresis experiments followed standard procedures.[7] Sequence analyses were performed by controlled chemical modification.[8] Construction of the cDNA clones and other clone libraries are described elsewhere.[9]

**TABLE 1.** Mitochondria Have Evolved Different Sets of Genetic Information in Different Species[a]

| Gene | | Plants | Chlamydomonas | Mammalia, Insects | Yeast | Neurospora | Aspergillus |
|------|------|--------|---------------|-------------------|-------|------------|-------------|
| COX | I | + | + | + | + | + | + |
| | II | + | + | + | + | + | + |
| | III | + | + | + | + | + | + |
| ATPase | 6 | + | ? | + | + | + | + |
| | 8 | ? | ? | + | + | + | + |
| | 9 | + | ? | − | − | − | + |
| | a | + | ? | − | − | − | − |
| ND | 1 | + | ? | + | − | + | + |
| | 2 | + | + | + | − | − | − |
| | 3 | ? | ? | + | − | − | − |
| | 4 | ? | ? | + | − | − | − |
| | 5 | + | + | + | − | − | − |
| tRNAs | | + | ? | + | + | + | + |
| 5S rRNA | | + | ? | − | − | − | − |
| s rRNA | | + | + | + | + | + | + |
| l rRNA | | + | + | + | + | + | + |

[a] Gene designations are COX for cytochrome oxidase subunits and ND for NADH dehydrogenase subunits. References are given in the text.

# RESULTS AND DISCUSSION

Recent work on structure and function of plant mitochondrial genomes has shown a surprisingly heterogenous constitution of these genetic entities.[10–12]

Mitochondria have retained different sets of genetic information in different species (TABLE 1). Once more is known about the mitochondrial genomes of more species, it might eventually be possible to reconstruct a timetable of development and answer the question of whether mitochondria have evolved several times in parallel or whether they originated from one event and spread through all eukaryotic cells. From the analyses of present-day organelles no clear line of development can yet be deduced. Comparison of mitochondrial genomes as they exist today shows remnants or secondary introduced specificities in a synthesis selected for functionality inseparable from each other in their contemporary status.

In the following we will discuss selected features of the plant mitochondrial genome and emphasize similarities with either the prokaryotic or the eukaryotic specificities and point out unique developments of higher plant mitochondria that we have deduced from analysis from the *Oenothera* mitochondrial genome. A summary of these comparisons is given in TABLE 2.

TABLE 2. Mitochondria Have Retained Many Prokaryotic Features, Evolved Some in Concert with Their Eukaryotic Hosts, and Combined These with Unique Characteristics[a]

| | Prokaryotic | Eukaryotic | Plant Mitochondria | Yeast Mitochondria | Mammalian Mitochondria |
|---|---|---|---|---|---|
| 5S rRNA | + | + | + | — | — |
| s rRNA | 1550 bp | 1800 bp | 1900 bp | 1700 bp | 1000 bp |
| l rRNA | 2900 bp | 3400–4700 bp | 3400 bp | 3300 bp | 1600 bp |
| Genetic code | standard | standard | CGG=Trp | UGA=Trp CTN=Thr | UGA=Trp AGA=Term ATA=Met AGG=Term |
| Introns | — | + | + | + | — |
| poly-A | — | + | — | — | + |
| mRNA cap | — | + | — | — | — |
| Transcription term | stem-loop (and others) | various sequences | stem-loop (others ?) | ? | D-loop region |
| tRNAfMet | + | — | + | — | — |

[a] Selected features of mitochondrial genomes from some species are compared with the respective characteristics in eubacteria and eukaryotes. References are given in the text.

*Translation*

*rRNAs*

Higher plant mitochondria contain a gene coding for a 5S rRNA unlike fungal and mammalian mitochondria.[10,13] The *Oenothera* 5S rRNA has in comparison with the respective wheat gene a deletion of four nucleotides in an A+T rich extra loop (helix 3) that has been analyzed in detail with possible phylogenetic implications.[14]

The sizes of both small (18S)[14] and large (26S)[15] rRNAs in *Oenothera* mitochondria with 1901 and 3268 nucleotides respectively appear to be very close to the sizes of the respective molecules in eukaryotic translation systems (TABLE 2). Sequence segments expanding the molecule lengths from the basic structure are localized in positions analogous to the insertion sites of additional nucleotides in the 18S and 28S rRNAs in eukaryotes. This comparable positioning of size enlargement could be due to a common expanding mechanism through mutually functional recombination activities or could be due to random recombinations having been selected through the functional constraints on these molecules that tolerate insertions only at certain positions. A detailed comparison of the positional analogies between the large rRNAs in *Oenothera* and eukaryotic molecules has been discussed,[15] and the existence of frequent rearrangements and sequence insertions in *Oenothera* mitochondria has been evidenced.[14,16,17] Antibiotic tolerance as deduced from nucleotide sequence analogies of the large rRNA indicates a sensitivity towards chloramphenicol as in wild-type bacteria, but a resistance against streptomycin different from the sensitivity of bacteria and chloroplasts.[14]

*tRNAfMet*

Plant mitochondria contain a specific tRNA sequence highly homologous to the initiator tRNA of *E. coli* shown to be encoded in the mitochondrial genomes of wheat,[18] maize,[19] and *Oenothera.*[20] This tRNAfMet is not found in the eukaryotic translation system and in fungal and mammalian mitochondria. This gene is thought to be of eubacterial origin for its conspicuous homology with the eubacterial translation initiator tRNA.[18]

*Ribosome Binding Sites*

Currently no experimental evidence is available on either cofactor or sequence requirements for ribosome binding in higher plant mitochondria. The sequence at the 3′-terminus of the small rRNA can usually be matched as well as sequence requirements in the *E. coli* system are within the range of nucleotides postulated. The putative ribosome binding sequence of the *Oenothera* ATPA gene,[17] however, is separated by only three nucleotides from the ATG codon, which has been observed as a functional alignment in *Corynebacterium*[21] but not yet in *Escherichia coli.* The proposed ribosome binding sequences are of course speculative until functional data show their importance in *in vitro* assays.

## Genetic Code

The genetic code in higher plants differs from the universal code by one alteration in codon translation, CGG is read as tryptophan instead of arginine.[22–24] This postulate is based solely on comparison of nucleotide sequence and therefrom deduced protein sequence data. The apparent equivalence of CGG and TGG in conservative tryptophan positions in maize and *Oenothera* protein coding genes has been deduced from the comparison of protein coding genes in these two species. The codon TGA is used in higher plants in common with the standard genetic code to terminate transcription as comparison of the cytochrome b and ATPA genes between maize[25,26] and *Oenothera*[24,17] has shown. Plant mitochondria differ here from the other mitochondrial systems investigated so far, in which TGA is assumed to specify the amino acid tryptophan. Only *Chlamydomonas* mitochondria use the universal code (see the paper by G. Gellissen and G. Michaelis, this volume).

## *Transcription*

### *Promotor Sequence Alignment*

A possible promotor sequence can be deduced in the region upstream of the cytochrome oxidase subunit II gene[23,27] from sequence alignment with promotor sequences shown to function in *E. coli*[28] (FIGURE 1). This comparison and deduction show an intriguing degree of homology between the two given examples. Similar homology of a region upstream of the maize cytochrome oxidase subunit I gene[29] with the yeast nonanucleotide unique as a promotor sequence in yeast mitochondria has been pointed out. As these data are all theoretical a more precise definition will need await further analyses. Interestingly the sequence upstream of the *Oenothera* COX II gene is perfectly conserved in a 40 nucleotide block in *Daucus carota* mtDNA (FIGURE 1) and hybridizes strongly to fragments containing the COX II gene in wheat and *Brassica* (R. Hiesel, A. Brennicke, B. Lejeune, M. Dron, and F. Quetier, unpublished results).

### *Terminators*

The precise mRNA 3′-termini have been determined for two *Oenothera* mitochondrial genes, COX II and ATPA.[30] The terminal region is identical for the last 50 nucleotides and hybridizes in restriction digests of total mitochondrial DNA to a

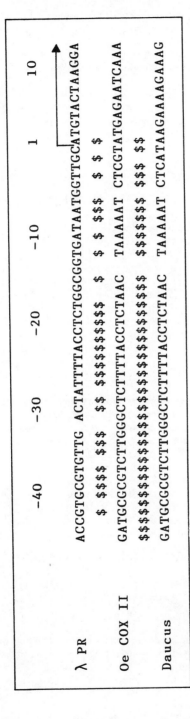

**FIGURE 1.** Alignment of the lambda promotor region [28] (PR) and the sequence preceding the mRNA start of the cytochrome oxidase subunit II (COX II) gene in *Oenothera* mitochondria and the region of homology in the *Daucus carota* mitochondrial genome. The 5'-termini of mRNAs in the phage promotor region and the *Oenothera* gene have been aligned; they both start at the ATG indicated by an arrow in the direction of transcription.

number of fragments in this plant mitochondrial genome (FIGURE 2). The significance of this homology is as yet unclear as other mRNA termini still have to be identified. Computer comparison with available sequence data from maize mitochondria shows a number of highly homologous sequences in regions where maize transcripts are thought to terminate. These terminal regions can be folded into two consecutive hairpin structures in both mitochondria, maize and *Oenothera,* similar to the simple eubacterial terminator structures in *E. coli.*[28] The last nucleotide of the mRNA resides at the distal foot of the second stem analogous to the bacterial terminators. The short run of T-residues observed at the terminus in *E. coli* is lacking in the mitochondrial terminator sequences however.

*Introns*

Sequence analysis of the locus encoding the NADH dehydrogenase subunit I (ND I) indicates the presence of an intron interrupting the regions of high homology with other known subunit sequences (D. B. Stern, personal communication). None of the other genes analyzed so far in the *Oenothera* mitochondrial genome contain an intron like the monocot mitochondrial genomes, where the COX II gene is split by intervening sequences of variable length.[22,31,32] Introns have not been found in eubacteria, where coding sequences are continuous reading frames.

*3'-Polyadenylation*

Plant mitochondrial mRNAs do not terminate in posttranscriptionally added poly-A tails,[10] although they can be retained on oligo-U columns. This binding might however be attributed to internal regions of elevated A proportions possibly in the transcribed, but noncoding, intergenic regions where blocks of nucleotides are found frequently in stretches larger than hexanucleotides alternating between pyrimidines and purines.

First strand synthesis priming with dT oligonucleotides for construction of the cDNA clone library of *Oenothera* mitochondrial transcripts was only successful after the *in vitro* addition of oligo A tails to the mRNA population.[9] Nucleotide sequence analysis of these cDNA clones then allows no conclusion of course on *in vivo* present short A tails. No poly-A addition signal sequence is observed in the *Oenothera* mRNA terminal sequences.

### Genome Organization

The mitochondrial genome of *Oenothera* is split into many different molecules each containing portions of the mitochondrial genetic information.[6] A master molecule sequence is generally postulated in higher plant mitochondria which generates several smaller molecules through various recombination processes. In the *Oenothera* mitochondrial genome, portions are split off as smaller circular molecules through site-

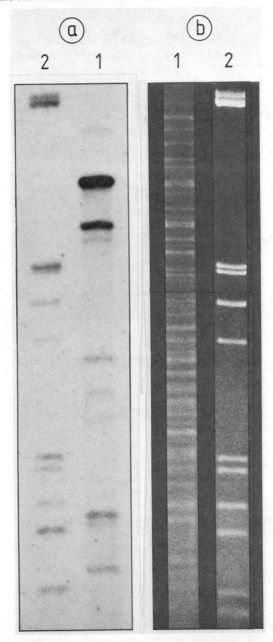

**FIGURE 2.** A Hind III restriction digest of mitochondrial DNA from *Oenothera* (lanes 1) has been probed with the sequence conserved at the 3′-termini of the mRNAs from the COX II and ATPase α-subunit loci in the *Oenothera* mitochondrial genome. A number of other regions of this mitochondrial genome show homology with this sequence. Part (a) shows the autoradiograph of the stained gel depicted in part (b) after hybridization with the terminator region. Lanes 2 show a phage-lambda Eco RI/Hind III size-marker digest with the sizes from top to bottom: 21.2 kilobases (lower band), 5.1 kb, 4.97 kb, 4.27 kb, 3.53 kb, 2.02 kb, 1.9 kb, 1.58 kb, 1.33 kb, 0.98 kb, 0.83 kb.

specific circularization processes[16] as well as by homologous recombination occurring via sequence repeats of 200 bp and more.[17] One site-specific event occurs within the gene encoding the large rRNA and leads to a small circular molecule with a truncated copy of the gene.[16] On this circular molecule only one allele of the tRNAfMet gene is present in full length. As we do not know whether this gene, located approximately 300 nucleotides behind the rRNA terminus, is transcribed independently from the rRNA, it is as yet unclear whether any product is synthesized from these small molecules at all. In Northern hybridization experiments no transcript can be detected from the smallest of these circular molecules.

Similarly are nontranscribed incomplete pseudogenes of the ATPase α-subunit (ATPA) gene found on different circular molecules.[17] The ATPA gene is encoded by mitochondria of higher plants, whereas mammalia and fungi synthesize this subunit in the nucleus (TABLE 2). The ATPase subunit 9, encoded in the nucleus of mammalia and most fungi, is specified in the mitochondrial genome of Neurospora[33] and higher plants as shown for maize and Oenothera.

The ATPase subunits encoded by Oenothera mitochondria are not clustered like the respective genes in bacteria or higher plant plastids,[34] where order is similar to E. coli, but coded in locations widely separated on the mitochondrial DNA.

Gene order is generally conserved among the smaller mitochondrial genomes of vertebrates, insects, and sea urchins, but very divergent in yeast and without any apparent conservation among the different plant mitochondrial genomes investigated so far.[10–12] Among higher plants not even rRNA gene order is in any way related between species, where each gene appears to exist in usually multiple copies within different contexts in the genomes investigated (wheat, Oenothera, and maize). The 5S rRNA gene in the Oenothera mitochondrial genome, for example, occurs in tandem behind the 18S rRNA gene in one allele and independent from the larger rRNA in several other genomic contexts. The 26S rRNA gene is far away from the other two rRNA genes in all alleles found.

*Variations in Noncoding Regions*

The intergenic noncoding regions in higher plant mitochondria are highly divergent between species. Genes appear to be accumulating mutations at an extremely slow rate compared to mitochondria from other species, particularly mammalia, with coding regions between the monocot Zea mays and the dicot Oenothera between 85% and 98% conserved. Homology extends occasionally to an additional sequence block of a few nucleotides outside of the coding regin and then ends completely. This high degree of divergence could be due to noninterfering nucleotide mutations or to the rapid continuous reorganization of the entire genome arrangement in higher plant mitochondria. The divergence of intergenic sequences could thus be due to relocation of the adjacent DNA into a different proximity. Analysis of recombination hotspots in the Oenothera mitochondrial genome locates many of them just outside of complete cistrons or, if within them, usually outside the coding regions.

The origin of the high sequence divergence in the noncoding regions of the plant mitochondrial genomes, where little homology between species is found, is as yet unclear. Possibly this apparent divergence is created by the movement of sequences in a way that leads to rapid recombination wherever tolerated. Then we would expect the noncoding region of one species outside of one gene to be found in another species only slightly diverged in its primary sequence in a different genomic location. Examples indicating such a derivation of noncoding sequences are found in the different degrees of conservation of sequence homologies between pseudogenes.[17]

# CONCLUSION

Research into structure and organization of the plant mitochondrial genomes has probably brought the strongest support of all the mitochondria investigated from different species so far for the endosymbiotic hypothesis, as many features observed in common with the prokaryotic kingdom are accounted for by the least complex model constructions when assumed to have been retained from an ancestral prokaryotic nature.

In an evaluation of the two different hypotheses proposed for the origin of mitochondria, the progressive filiation versus the endosymbiosis, arguments of progressiveness are meaningless. For example the loss of genes respectively transferred between organelles is an undisputed prerogative in both proposals. Evaluation can only concentrate on features that are retained from ancestral times for clues as near to proofs as possible for the most likely origin of these organelles.

Following this line of evidence it can never be positively proven that filiation was the way mitochondria arose, as the nucleus of the host cell has been in closest proximity and contact all the time of development and has influenced the mitochondrial genetic machinery in both hypotheses.

Thus the endosymbiotic hypothesis could only be positively proven if the original endosymbiont was a prokaryote that has retained some of its original specificities that are still recognizable in modern day prokaryotes as unique features for this group of organisms. And only if the two other possible sources of origin can be ruled out for such prokaryotic features, i.e., a prokaryotic influence on the genome from outside or events of convergent evolution.

From analysis of plant mitochondrial genomes we have no compelling evidence that plant mitochondria are positively derived from pro- or eukaryotes. The pro- and eukaryote-like features of the plant mitochondrial genome may well represent examples of convergent evolution on the surface of lengths of rRNAs for example.[15]

Notwithstanding where mitochondria came from, another still open question remains to be answered: Why do mitochondria maintain their own genetic machinery? To have survived with this rudimentary streamlined genetic apparatus, these organelles must have developed a selective advantage against the nuclear background, i.e., they must have been able to optimize and establish the encoded subunits just as quickly or quicker than the nucleus. This may be achieved by the high rate of fixed mutations in mammalian mitochondrial genomes and through the high rate of recombination between various pieces of DNA in the plant mitochondria, where the rate of fixed mutations is much slower as judged from the comparison of protein coding genes between the two best-studied plant mitochondrial genomes, maize and *Oenothera*.

Thus mitochondria might have found two different ways to keep pace with the rapidly optimizing nuclear genetic system: the first with a high mutation rate of single nucleotides is observed in mammalian mitochondria; the other in higher plant mitochondria consists mainly of an efficient recombination system that reshuffles and reorganizes sequence arrangements to try out new sequence domain combinations.

## ACKNOWLEDGMENTS

We thank D. B. Stern and C. J. Leaver for communicating results prior to publication, H. P. Mock and U. Seitz for their gift of carrot tissue culture and P. Blanz for discussions and his continuous interest. U. Christner did the *Oenothera* tissue culturing, C. Specht the photography, and V. Uhle-Schneider the artwork.

## REFERENCES

1. FREDRICK, J. F. 1981. Ann. N. Y. Acad. Sci. **361:** ix-x.
2. GILHAM, N. W. & J. E. BOYNTON. 1981. Ann. N. Y. Acad. Sci. **361:** 20-40.
3. GRAY, M. W. & W. F. DOOLITTLE. 1982. Microbiol. Rev. **46:** 1-42.
4. WALLACE, D. C. 1982. Microbiol. Rev. **46:** 208-240.
5. BRENNICKE, A. 1980. Plant. Physiol. **65:** 1207-1210.
6. BRENNICKE, A. & P. BLANZ. 1982. Mol. Gen. Genet. **187:** 461-466.
7. MANIATIS, T., E. F. FRITSCH & J. SAMBROOK. 1982. Molecular Cloning. Cold Spring Harbor Press. New York, N.Y.
8. MAXAM, A. & W. GILBERT. 1980. Methods Enzymol. **65:** 499-560.
9. HIESEL, R., W. SCHOBEL, W. SCHUSTER & A. BRENNICKE. 1987. EMBO J. **6:** 29-34.
10. GRAY, M. W. & C. J. LEAVER. 1982. Annu. Rev. Genet. **33:** 373-402.
11. LONSDALE, D. M. 1984. Plant Mol. Biol. **3:** 201-206.
12. QUETIER, F., B. LEJEUNE, S. DELORME & D. FALCONET. 1985. *In* Encyclopedia of Plant Physiology **18:** 25-36. Springer Verlag. Berlin, New York & Tokyo.
13. LEAVER, C. J. & M. A. HARMEY. 1976. Biochem. J. **157:** 257-277.
14. BRENNICKE, A., S. MÖLLER & P. BLANZ. 1985. Mol. Gen. Genet. **198:** 404-410.
15. MANNA, E. & A. BRENNICKE. 1985. Curr. Genet. **9:** 505-515.
16. MANNA, E. & A. BRENNICKE. 1986. Mol. Gen. Genet. **203:** 377-381.
17. SCHUSTER, W. & A. BRENNICKE. 1986. Mol. Gen. Genet. **204:** 29-35.
18. GRAY, M. W. & D. F. SPENCER. 1983. FEBS Lett. **161:** 323-327.
19. PARKS, T. D., W. G. DOUGHERTY, C. S. LEVINGS III & D. H. TIMOTHY. 1984. Plant Physiol. **76:** 1079-1082.
20. GOTTSCHALK, M. & A. BRENNICKE. 1985. Curr. Genet. **9:** 165-168.
21. ANDERSON, S., C. B. MARKS, R. LAZARUS, J. MILLER, K. STAFFORD, J. SEYMOUR, D. LIGHT, W. RASTETTER & D. ESTELL. 1985. Science **230:** 144-149.
22. FOX, T. D. & C. J. LEAVER. 1981. Cell **26:** 315-323.
23. HIESEL, R. & A. BRENNICKE. 1983. EMBO J. **2:** 2173-2178.
24. SCHUSTER, W. & A. BRENNICKE. 1985. Curr. Genet. **9:** 157-163.
25. DAWSON, A. J., V. P. JONES & C. J. LEAVER. 1984. EMBO J. **3:** 2107-2113.
26. ISAAC, P., A. BRENNICKE, S. M. DUNBAR & C. J. LEAVER. 1985. Curr. Genet. **10:** 321-328.
27. HIESEL, R. & A. BRENNICKE. 1985. FEBS Lett. **193:** 164-168.
28. ROSENBERG, M. & D. COURT. 1979. Annu. Rev. Genet. **13:** 319-353.
29. ISAAC, P. G., V. P. JONES & C. J. LEAVER. 1985. EMBO J. **4:** 1617-1623.
30. SCHUSTER, W., R. HIESEL, P. G. ISAAC, C. J. LEAVER & A. BRENNICKE. 1986. Nucleic Acids Res. **14:** 5943-5954.
31. BONEN, L., P. H. BOER & M. W. GRAY. 1984. EMBO J. **3:** 2531-2536.
32. KAO, T., E. MOON & R. WU. 1984. Nucleic Acids Res. **12:** 7305-7315.
33. MAHLER, H. 1981. Ann. N.Y. Acad. Sci. **361:** 53-75.
34. HENNIG, J. & R. G. HERRMANN. 1986. Mol. Gen. Genet. **203:** 117-128.

# Gene Transfer[a]

## Mitochondria to Nucleus

GERD GELLISSEN[b] AND GEORG MICHAELIS[c]

[b]Zoological Institute III
University of Düsseldorf
D-4000 Düsseldorf, Federal Republic of Germany
[c]Botanical Institute
University of Düsseldorf
D-4000 Düsseldorf, Federal Republic of Germany

The mitochondrial genomes of eukaryotic cells encode only a small proportion of the gene products necessary to constitute the organelle's functions, while the vast majority of proteins are encoded by the nucleus, translated in the cytosol, and imported into the organelle. The endosymbiotic theory of the origin of mitochondria implies that the genetic information of the presumptive endosymbiont, an aerobic prokaryote, has been reduced over evolutionary time by loss of genes and by a transfer of genes to the nucleus.[25,43] Resulting from this depletion, genome remainders persisted ranging in size from several hundred to 2400 kilobases in higher plants[39,48] and to circles of 14.5-20 kb in higher animals[3] where the mitochondrial DNA represents an extreme example of genetic economy. Following the endosymbiont theory, at least some nuclear genes for mitochondrial proteins could be considered as sequences transferred from mitochondria to the nucleus and modified by evolutionary adaptations to fulfill the requirements for expression and transport of the encoded peptides across the organelle membranes.[33]

This view is supported by the finding that at least two structural genes for mitochondrial proteins have different locations in various organisms. For example the gene for subunit 9 of mitochondrial ATPase is localized in the nucleus of mammals and *Drosophila*. The same gene maps on the mitochondrial genome of the higher plant *Zea mays*[17] and the yeast *Saccharomyces cerevisiae*.[47,56] In *Neurospora crassa*[6] and *Aspergillus nidulans*[8] an active nuclear gene has been found, but interestingly both organisms contain a second mitochondrial sequence which is silent during growth. Since these two sequences are formed by an open reading frame, there is a strengthened possibility that the mitochondrial gene has some biological significance unrelated to growth phenomena and that two functional genes are distributed between nucleus and mitochondrion. The second example concerns the α subunit of mitochondrial ATPase. The respective structural gene is absent from the mitochondrial genome of all orga-

[a]The authors' experimental work was supported by the Deutsche Forschungsgemeinschaft (to G. G.) and by "Forschungsmittel des Landes Nordrhein-Westfalen" (to G. M.). One of us (G. G.) gratefully acknowledges the receipt of a travel grant from "Verband der Chemischen Industrie."

nisms studied so far, and hence has to be a nuclear one, with the exception of higher plants for which a mitochondrial gene could be demonstrated recently.[7,28]

## THE UNIVERSAL GENETIC CODE IN THE MITOCHONDRIA OF *CHLAMYDOMONAS REINHARDTII*

The transfer of functional genes from mitochondria to the nucleus or vice versa is difficult to reconcile with the altered genetic code of animal mitochondria. Code alterations have been found in the mitochondria of all organisms studied so far (TABLE 1). The most striking difference concerns the UGA codon which is read as tryptophan in animal and fungal mitochondria but is recognized as stop codon in the cytosol of these organisms. After a presumptive gene transfer in such an organism, many mutations would be needed for the expression of a functional protein. However, many fewer alterations have been detected so far in the genetic code of higher plant mitochondria (TABLE 1). We have studied the mitochondrial genome of the unicellular green alga *Chlamydomonas reinhardtii*, a simple autotrophic organism with genetic information in the nucleus, the chloroplast, and the mitochondria.

One-third of this linear mitochondrial genome, about 16 kilobases (kb) in size, has been sequenced and not a single deviation from the universal genetic code could be detected so far.[57] Thus, in *Chlamydomonas reinhardtii* the same genetic code is used in the nucleus, the chloroplast, and the mitochondrion and, therefore, a hypothetic transfer of a functional gene from one compartment to another would not be discriminated against by different codon usage. This situation found in *Chlamydomonas*

TABLE 1. Anomalous Codon Usage in Mitochondria

| Organism | Codon[a] | Amino Acid | | References |
|---|---|---|---|---|
| | | Universal Code | Mitochondrial Code | |
| Mammals | UGA | Stop | Trp | 1 (human), 2 (bovine), |
| | AGR[+] | Arg | Stop | 5 (mouse) |
| | AUA | Ile | Met | |
| *Drosophila* | UGA | Stop | Trp | 12,13 (*D. yakuba*) |
| | AGA | Arg | Ser | |
| | AUA | Ile | Met | |
| *Neurospora* | UGA | Stop | Trp | 10 (*N. crassa*) |
| *Aspergillus* | UGA | Stop | Trp | 9 (*A. nidulans*) |
| *Saccharomyces* | UGA | Stop | Trp | 32 (*S. cerevisiae*) |
| | AUA | Ile | Met | |
| | CUN[++] | Leu | Thr | |
| Trypanosomes | UGA | Stop | Trp | 14 (*Leishmania tarentolae*) |
| | | | | 30 (*Trypanosoma brucei*) |
| Higher plants | CGG | Arg | Trp | 22 (*Zea mays*) |
| | | | | 31 (*Oenothera*) |

[a] R[+], A or G; N[++], A, G, C, or U.

**TABLE 2.** Fragment Diversity of Cloned Nuclear DNA Homologous to Mitochondrial rRNA Genes in *Locusta migratoria*

| Clone Number | EcoR I Fragments (size in kb)[a] | | |
|---|---|---|---|
| Lm  1 | *14.5* | 6.7 | |
| Lm  3 | *14.0* | | |
| Lm  6 | *15.5* | | |
| Lm 12 | *11.3* | | |
| Lm 14 | | 8.5 | *4.8* |
| Lm 21 | | 9.0 | *3.0* |

[a] Fragments bearing the homology are italicized.

*reinhardtii* seems to indicate a primitive characteristic originating from early eukaryotes.

## PROMISCUOUS MITOCHONDRIAL DNA IN
### *LOCUSTA MIGRATORIA*

Until a few years ago, the genetic systems of nucleus, chloroplasts, and mitochondria were considered to be discrete and separate in having completely distinct nucleotide sequences. A genomic lambda charon 4 library, constructed from partial EcoRI digests of *Locusta migratoria* fat body DNA, was screened with cDNA prepared against RNA from the fat body of reproductive females, adult males, and fifth instar larvae.

Six clones showing strong and 15 showing weak hybridization with all cDNA probes were isolated from 60,000 plaques and tested for the presence of sequences for constitutively expressed genes of serum proteins. They all showed some degree of homology to each other when tested with a plasmid subcloned probe, but were distinct when characterized by restriction enzyme analysis (TABLE 2). Four of these clones were hybridized on Northern blot with RNA of various locust tissues and showed homology with RNAs of 1.4 and 0.8 kb in all tissues tested. These RNA species were found to be of mitochondrial origin representing the small and the large ribosomal RNA. Because of the presence of poly (A) sequences these RNA molecules allowed the initiation of cDNA synthesis with an oligo (dT) primer. While being homologous to mitochondrial DNA (mtDNA) the cloned DNA is clearly of nuclear origin, since the clones contain sequences homologous to highly repeated nuclear DNA as well.[24] Thus, these findings demonstrate the existence of "promiscuous DNA,"[18] which means sequences common to both genetic systems, in *Locusta migratoria*. From the screening result, a number of several hundred highly diverged nuclear sequences homologous to mitochondrial rRNA genes can be estimated. Mitochondrial sequences other than the rDNA are not tested yet and might be represented in the nucleus as well. Nucleotide sequencing of one of the nuclear genomic homologues (lambda Lm12) revealed a structure corresponding to the 3' end of the large ribosomal RNA with a homology of more than 90%. The homology extends into an adjacent $tRNA_{CUN}^{Leu}$ gene neighboring the 3' end similar to the situation in *Drosophila yakuba* mtDNA.[12] Nucleotide exchanges introduce modifications into the secondary structures of the rRNA gene based on a general model of secondary structure[58] and shown to be valid for the respective mouse[19] and *Drosophila*[11] sequences. These changes, especially those of a loop that is 100% conserved between the human and *Drosophila* genes, provide some evidence that the sequence is possibly rendered into a pseudogene (FIGURE 1).

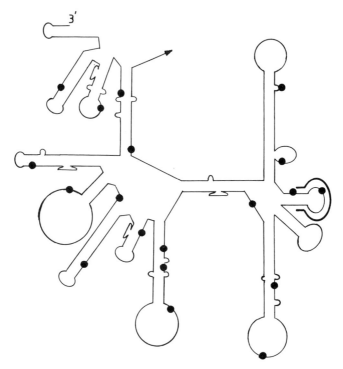

**FIGURE 1.** The nuclear pseudogene (Lm12) of *Locusta migratoria.* Base exchanges are indicated by dots in the model for the higher-order structure of the 3' end of mitochondrial 16S rRNA. The solid line surrounding the loop to the right indicates a region of 100% sequence homology between the respective human, *Locusta*, and *Drosophila* genes.

Sequencing of the other region of the DNA homologous to the small rRNA gene is not completed yet, but most likely it will reveal changes from a functionally intact sequence, as well.

## PROMISCUOUS MITOCHONDRIAL DNA IN OTHER ORGANISMS

As summarized in TABLE 3 promiscuity between nuclear and mitochondrial DNA has been demonstrated for species of a wide evolutionary range. In all animals analyzed so far, regions of the rRNA genes have homologous counterparts in the nucleus.

**TABLE 3.** Common Mitochondrial and Nuclear Sequences

| Organism | mtDNA Sequence Involved | Number of Homologous Sequences | Reference |
|---|---|---|---|
| *Saccharomyces cerevisiae* (yeast) | part of var 1 gene; 3' end of COB gene and part of ori/rep sequence | 1 | Farelly & Butow[21] (1983) |
| *Neurospora crassa* (Ascomycetous fungus) | gene for ATPase subunit 9 | 1[a] | van den Bogaart et al.[6] (1982) |
| *Aspergillus nidulans* (Ascomycetous fungus) | gene for ATPase subunit 9 | 1[a] | Brown et al.[8] (1984) |
| *Podospora anserina* (Ascomycetous fungus) | α event senDNA and β event senDNA plasmids | increasing during senescence; 1 in *P. anserina* mex 1[b] | Wright & Cummings[59] (1983) |
| *Strongylocentrotus purpuratus* (sea urchin embryo) | CO I gene; 3' end of 16S rRNA gene | 1 | Jacobs et al.[34] (1983) |
| *Locusta migratoria* (locust fat body) | rRNA genes | several hundred | Gellissen et al.[24] (1983) |
| *Rattus norvegicus* (rat liver) | rRNA genes and D-loop region | 1 | Hadler et al.[29] (1983) |
| *Homo sapiens* (human placenta) | rRNA genes; URF 2 and URF 4 genes; CO II gene | several hundred | Tsuzuki et al.[55] (1983); Fukuda et al.[23] (1985) |
| *Zea mays* (maize, corn) | S₁ and 1.94 kb plasmid 1.4 kb plasmid | few | Kemble et al.[36] (1983) |

[a] The nuclear gene is expressed during growth; the function of the mtDNA sequence is unknown, but it is possibly expressed.
[b] In *P. anserina* mex 1 the mitochondrial sequence is lost.

Additional sequence data are available for the sea urchin *Strongylocentrotus purpuratus,*[35] and *Homo sapiens.*[23,44] In FIGURE 2 these sequences are compared. In the sea urchin a sequence spanning 342 base pairs (bp) homologous to the 3' end of the large rRNA gene is flanked by a complete copy of the mitochondrial gene for subunit I of cytochrome oxidase (COI). Two-thirds of the large rRNA gene are absent from the 5' region. Instead, a stretch containing a sequence of middle repetitive nuclear DNA is present followed by a second shortened homologue of the COI gene. In *Homo sapiens* two sequences homologous to mitochondrial rRNA genes were characterized. One clone contains a 0.4 kb sequence corresponding to the 5' portion of the 16S rRNA gene, the second clone harbors 1.6 kb of homologous sequences comprising the 3' portion of the 12S rRNA gene, and an almost complete copy of the 16S rRNA. The two regions are interrupted by a tRNA$^{Val}$ gene as in the respective mitochondrial sequence. In all cases the homologous structures are shown to be pseudogenes, they never match exactly the ends for their expressed mitochondrial partners but extend into adjacent genes or represent shortened copies. Rearrangements and modifications are especially obvious in the case of the sea urchin. The overall homology of nucleotide

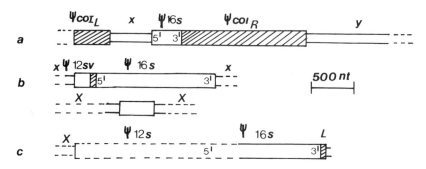

**FIGURE 2.** Comparison of sequenced nuclear pseudogenes for mitochondrial rRNA. (a) *Strongylocentrotus purpuratus* (adapted from Jacobs and Grimes, 1986).[35] (b) *Homo sapiens* (drawn after Nomiyama *et al.,* 1983).[44] (c) *Locusta migratoria* (Gellissen, unpublished). Abbreviations in (a) COI$_L$ = left (shortened) copy of the pseudogene for cytochrome oxidase subunit I; COI$_R$ = right (almost complete) pseudogene for cytochrome oxidase subunit I; 16S = pseudogene for the 16S rRNA; x = region of repetitive nuclear DNA; y = region of homology to a yet unknown part of mitochondrial DNA; in (b) 12S = pseudogene for 12S rRNA; 16S = pseudogene for 16S rRNA; v = tRNA$^{Val}$ gene; x = nonpromiscuous nuclear DNA; in (c) 12S = pseudogene for 12S rRNA; 16S = pseudogene for 16S rRNA; L = tRNA$^{Leu}$ sequence; x = nonpromiscuous nuclear DNA.

sequence is 80-84% in *Homo sapiens* and 92% in the sea urchin. For the nuclear sequence element of the sea urchin, a likely age of more than 30 million years is suggested. Comparison of human nuclear sequences homologous to URF 4 and URF 5 with those of mtDNA of several mammalian species revealed conservation of a part of the structures present in direct ancestral mtDNA. Thus, the mtDNA fragments seem to have been continuously integrated into mammalian nuclear DNA during evolution.[23] This might be the case in locusts, as well, where a comparable number of homologous sequences is found. In almost all examples listed in TABLE 3 and recently reviewed,[37] we are witnessing an evolutionary process.

In the ascomycetous fungus *Podospora anserina,*[59] however, DNA transfer from mitochondria and integration into the nucleus occur regularly during each life cycle.

Several fragments of mitochondrial DNA are excised and amplified during senescence. Examples are the $\alpha$ senDNA of 2.6 kb and the $\beta$ senDNA of 9.8 kb length containing parts of the genes for subunit I (COI) and subunit II (COII) of cytochrome oxidase, respectively. Both types of senDNA become integrated into the nuclear genome. The $\alpha$ senDNA represents the first intron of the COI gene,[45] an intron of group II with a reading frame homologous to the yeast RNA maturase. The mutant mex1, which was isolated as an outgrowth of senescent mycelia, has lost the mitochondrial $\alpha$ senDNA completely,[4] but it possesses respective sequences in the nucleus. Thus, *Podospora anserina* mex1 is the result of a very recent evolutionary process in which mitochondrial sequences are eliminated from the organelle genome and transferred into the nucleus.[59]

In many plant systems there are homologies between chloroplast and mitochondrial DNA[41,42,51] as well as homologies between chloroplast and nuclear DNA[52,54] extending the phenomenon of promiscuity to all three genetic compartments of a eukaryotic cell.[53] The results from the system describing homologies between chloroplast and nuclear DNA parallel the findings for gene transfer between mitochondria and nucleus. They provide evidence for movement and rearrangement of organelle sequences, which became stably integrated into the nuclear genome; all sequences found so far are pseudogenes. Since the vast majority of the gene products for both organelles are encoded by the nucleus and all above-described promiscuous sequences have lost their functions (possible exceptions are the mitochondrial genes for ATPase subunit 9 in *Neurospora* and *Aspergillus*), it seems that during evolution most of the genes were readily transferred and modified for expression. The duplicate genes were then subsequently deleted. *Podospora anserina* mex1 might be an example for this evolutionary process. On the other hand it is interesting to note that a regulatory sequence common to both genetic systems exists in yeast. The nuclear encoded genes for histone H4[49] and L(+) lactate cytochrome c oxidoreductase[27] are preceded by a nonanucleotide motive founded in putative origins of mitochondrial DNA replication and immediately upstream of several mitochondrial genes as a signal where initiation of transcription occurs.

# POSSIBLE MECHANISMS OF DNA TRANSFER

One can only speculate about the cause and mechanism of interorganellar gene transfer. According to a hypothesis by Hadler *et al.*, integration of mitochondrial DNA sequences into the nuclear genome occurs as a result of pressure on mitochondria by carcinogens and their metabolites similar to the integration of genetic material of an oncogenic virus.[29] Some significance is provided by the demonstration of a homology between a human bladder carcinoma gene product and a mitochondrial ATPase.[26] Intact mitochondrial DNA is claimed to be present in the nuclei of HeLa cells;[38] from another vertebrate cell line, the avian fibroblast cell line LSCC-H32, cells have been isolated after long-term exposure to ethidiumbromide, completely devoid of mtDNA molecules but still showing a faint hybridization reaction of its nuclear DNA when hybridized to mtDNA.[16] Thus, this system might offer the possibility of finding out whether sequences were transferred during exposure to the drug.

A mechanism similar to that of transposable elements has been proposed for integration into the nuclear DNA. In yeast, Ty elements were found to be present in the vicinity of the transferred sequences;[21] repetitive nuclear DNA neighboring the homologous structures is reported for locust,[24] sea urchin[34] and man.[55] Sequencing analysis, however, has revealed no characteristic structures such as direct or inverted repeats or duplication of nuclear sequences at the junctions commonly found to be associated with mobile elements, yet structures like that might have been lost during subsequent sequence modifications.

The transferred sequences may originate from mitochondrial RNA molecules converted into DNA with the aid of reverse transcriptase-like enzymes. Indeed, enzymes of that nature have been reported for *Podospora anserina* during senescence when α- and β-senDNA plasmids are created.[50] Similarly, a reverse transcriptase-like enzyme has been demonstrated in the developing macronucleus of the hypotrichous ciliate *Stylonychia*,[40] where in even more dramatic changes 90% of micronuclear sequences are eliminated.

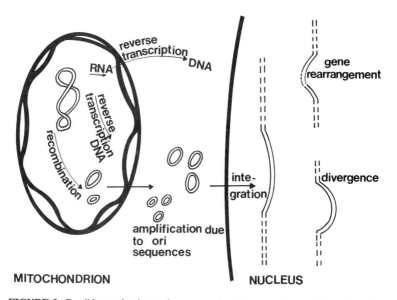

**FIGURE 3.** Possible mechanisms of gene transfer between mitochondria and nucleus.

On the other hand, recombination events leading to small circular molecules carrying portions of the total mitochondrial genome are quite common in yeast and plants,[46] and especially the rRNA genes are reported to be target sites for recombination.[20] The DNA sequence of the excision sites of α-senDNA in *Podospora anserina* is flanked by palindromic sequences which can be discussed within the context of site-specific recombination[15] as well.

A high copy number of the excised sequences might be an additional prerequisite for integration into the nuclear genome. This might be achieved by ori/rep sequences present on the transferred structures as reported for yeast.[21] In animal mitochondria ori sequences are in a close neighborhood of the rRNA genes having nuclear counterparts in many systems reported to date.

FIGURE 3 summarizes all possible mechanisms. We feel that the present state of research does not allow any preference. A detailed analysis of many of these exciting structures and an understanding of how they were created remain a fascinating subject of future research.

[**Note added in proof:** According to a recent report the integration of mitochondrial DNA into the nuclear genome of *Podospora anserina* during senescence has been questioned. (Koll, F. 1986. Does nuclear integration of mitochondrial sequences occur during senescence in *Podospora*? Nature **324:** 597-599.)]

## REFERENCES

1. ANDERSON, S., A. T. BANKIER, B. G. BARRELL, M. H. L. DE BRUIJN, A. R. COULSON, J. DROUIN, I. C. EPERON, D. P. NIERLICH, B. A. ROE, F. SANGER, P. H. SCHREIER, A. J. H. SMITH, R. STADEN & I. G. YOUNG. 1981. Sequence and organization of the human mitochondrial genome. Nature **290:** 457-465.
2. ANDERSON, S., M. H. L. DE BRUIJN, A. R. COULSON, I. C. EPERON, F. SANGER & I. G. YOUNG. 1982. Complete sequence of bovine mitochondrial DNA—Conserved features of the mammalian mitochondrial genome. J. Mol. Biol. **156:** 683-717.
3. ATTARDI, G. 1985. Animal mitochondrial DNA: an extreme example of genetic economy. Int. Rev. Cytol. **93:** 93-145.
4. BELCOUR, L. & C. VIERNY. 1986. Variable DNA splicing sites of a mitochondrial intron: relationship to the senescence process in *Podospora*. EMBO J. **5:** 609-614.
5. BIBB, M. J., R. A. VAN ETTEN, C. T. WRIGHT, M. W. WALBERG & D. CLAYTON. 1981. Sequence and gene organization of mouse mitochondrial DNA. Cell **26:** 167-180.
6. VAN DEN BOOGAART, P., J. SAMALLO & E. AGSTERIBBE. 1982. Similar genes for a mitochondrial ATPase subunit in the nuclear and mitochondrial genomes of *Neurospora crassa*. Nature **298:** 187-189.
7. BOUTRY, M., M. BRIQUET & A. GOFFEAU. 1983. The α subunit of a plant mitochondrial $F_1$-ATPase is translated in mitochondria. J. Biol. Chem. **258:** 8524-8526.
8. BROWN, T. A., J. A. RAY, R. B. WARING, C. SCAZZOCCHIO & R. W. DAVIES. 1984. A mitochondrial reading frame which codes for a second form of ATPase subunit 9 in *Aspergillus nidulans*. Curr. Genet. **8:** 489-492.
9. BROWN, T. A., R. B. WARING, C. SCAZZOCCHIO & R. W. DAVIES. 1985. The *Aspergillus nidulans* mitochondrial genome. Currr. Genet. **9:** 113-117.
10. BURGER, G., C. SCRIVEN, W. MACHLEIDT & S. WERNER. 1982. Subunit I of cytochrome oxidase from *Neurospora crassa:* nucleotide sequence of the coding gene and partial amino acid sequence of the protein. EMBO J. **1:** 1385-1391.
11. CLARY, D. O. & D. R. WOLSTENHOLME. 1985. The ribosomal genes of *Drosophila* mitochondrial DNA. Nucleic Acids Res. **113:** 4029-4045.
12. CLARY, D. O. & D. R. WOLSTENHOLME. 1985. The mitochondrial DNA molecule of *Drosophila yakuba*: nucleotide sequence, gene organization, and genetic code. J. Mol. Evol. **22:** 252-271.
13. CLARY, D. O., J. A. WAHLEITHNER & D. R. WOLSTENHOLME. 1984. Sequence and arrangement of the genes for cytochrome b, URF1, URF4L, URF4, URF5, URF6, and five tRNAs in *Drosophila* mitochondrial DNA. Nucleic Acids Res. **12:** 3747-3762.
14. DE LA CRUZ, V. F., N. NECKELMANN & L. SIMPSON. 1984. Sequences of six genes and several open reading frames in the kinetoplast maxicircle DNA of *Leishmania tarentolae*. J. Biol. Chem. **259:** 15136-15147.
15. CUMMINGS, D. J. & R. M. WRIGHT. 1983. DNA sequence of the excision sites of a mitochondrial plasmid from senescent *Podospora anserina*. Nucleic Acids Res. **11:** 2111-2119.
16. DESJARDINS, P., J.-M. DE MUYS & R. MORAIS. 1986. An established avian fibroblast cell line without mitochondrial DNA. Somatic Cell Mol. Genet. **12:** 133-139.

17. DEWEY, R. E., A. M. SCHUSTER, C. S. LEVINGS III & D. H. TIMOTHY. 1985. Nucleotide sequence of $F_o$-ATPase proteolipid (subunit 9) gene of maize mitochondria. Proc. Nat. Acad. Sci. USA **82:** 1015-1019.

18. ELLIS, J. 1982. Promiscuous DNA-chloroplast genes inside plant mitochondria. Nature **299:** 678-679.

19. VAN ETTEN, R. A., M. W. WALBERG & D. A. CLAYTON. 1980. Precise localization and nucleotide sequence of the two mouse mitochondrial rRNA genes and three immediately adjacent novel tRNA genes. Cell **22:** 157-170.

20. FALCONET, D., B. LEJEUNE, F. QUETIER & M. W. GRAY. 1984. Evidence for homologous recombination between repeated sequences containing 18S and 5S ribosomal RNA genes in wheat mitochondrial DNA. EMBO J. **3:** 297-302.

21. FARELLY, F. & R. A. BUTOW. 1983. Rearranged mitochondrial genes in the yeast nuclear genome. Nature **301:** 296-301.

22. FOX, T. D. & C. J. LEAVER. 1981. The *Zea mays* mitochondrial gene coding cytochrome oxidase subunit II has an intervening sequence and does not contain TGA codons. Cell **26:** 315-323.

23. FUKUDA, M., S. WAKASUGI, T. TSUZUKI, H. NOMIYAMA & K. SHIMADA. 1985. Mitochondrial DNA-like sequences in the human nuclear genome. Characterization and implications in the evolution of mitochondrial DNA. J. Mol. Biol. **186:** 257-266.

24. GELLISSEN, G., J. Y. BRADFIELD, B. N. WHITE & G. R. WYATT. 1983. Mitochondrial DNA sequences in the nuclear genome of a locust. Nature **301:** 631-634.

25. GRAY, M. W. & W. F. DOOLITTLE. 1982. Has the endosymbiont hypothesis been proven? Microbiol. Rev. **46:** 1-42.

26. GAY, N. J. & J. E. WALKER. 1983. Homology between human bladder carcinoma oncogene product and mitochondrial ATP-synthase. Nature **301:** 262-264.

27. GUIARD, B. 1985. Structure, expression and regulation of a nuclear gene encoding a mitochondrial protein: the yeast L(+)-lactate cytochrome c oxireductase (cytochrome $b_2$). EMBO J. **4:** 3265-3272.

28. HACK, E. & C. J. LEAVER. 1983. The α-subunit of the maize $F_1$-ATPase is synthesized in the mitochondrion. EMBO J. **2:** 1783-1789.

29. HADLER, H. I., B. DIMITRIJEVIC & R. MAHALINGAM. 1983. Mitochondrial DNA and nuclear DNA from normal rat liver have a common sequence. Proc. Nat. Acad. Sci. USA **80:** 6495-6499.

30. HENSGENS, L. A. M., J. BRAKENHOFF, B. F. DE VRIES. P. SLOOF, M. C. TROMP, J. H. VAN BOOM & R. BENNE. 1984. The sequence of the gene for cytochrome c oxidase subunit I, a frameshift containing gene for cytochrome c oxidase subunit II and seven unassigned reading frames in *Trypanosoma brucei* mitochondrial maxi-circle DNA. Nucleic Acids Res. **12:** 7327-7344.

31. HIESEL, R. & A. BRENNICKE. 1983. Cytochrome oxidase subunit II gene in mitochondria of *Oenothera* has no intron. EMBO J. **2:** 2173-2178.

32. HUDSPETH, M. E. S., W. M. AINLEY, D. S. SHUMARD, R. A. BUTOW & L. I. GROSSMAN. 1982. Location and structure of the var1 gene on yeast mitochondrial DNA: nucleotide sequence of the 40.0 allele. Cell **30:** 617-626.

33. HURT, E. C. & A. P. G. M. VAN LOON. 1986. How proteins find mitochondria and intramitochondrial compartments. Trends Biochem. Sci. **11:** 204-207.

34. JACOBS, H. T., J. W. POSAKONY, J. W. GRULA, J. W. ROBERTS, J. H. XIN, R. J. BRITTEN & E. H. DAVIDSON. 1983. Mitochondrial DNA sequences in the nuclear genome of *Strongylocentrotus purpuratus*. J. Mol. Biol. **165:** 609-632.

35. JACOBS, H. T. & B. GRIMES. 1986. Complete nucleotide sequences of the nuclear pseudogenes for cytochrome oxidase subunit I and the large mitochondrial ribosomal RNA in the sea urchin *Strongylocentrotus purpuratus*. J. Mol. Biol. **187:** 509-527.

36. KEMBLE, R. J., R. J. MANS, S. GABAY-LAUGHNAN & J. R. LAUGHNAN. 1983. Sequences homologous to episomal mitochondrial DNAs in the maize nuclear genome. Nature **304:** 744-747.

37. KEMBLE, R. J., S. GABAY-LAUGHNAN & J. R. LAUGHNAN. 1985. Movement of genetic information between plant organelles: mitochondria-nuclei. *In* Plant Gene Research. Genetic Flux in Plants. B. Hohn & E. S. Dennis, Eds.: 79-86 Springer. Wien & New York.

38. KRISTENSEN, T. & H. PRYDZ. 1986. The presence of intact mitochondrial DNA in HeLa cell nuclei. Nucleic Acids Res. **14:** 2597-2609.
39. LEAVER, C. J. & M. W. GRAY. 1982. Mitochondrial genome organization and expression in higher plants. Annu. Rev. Plant Physiol. **33:** 373-402.
40. LIPPS, H. J. 1985. A reverse transcriptase like enzyme in the developing macronucleus of the hypotrichous ciliate *Stylonychia.* Curr. Genet. **10:** 239-243.
41. LONSDALE, D. M., T. P. HODGE, C. J. HOWE & D. B. STERN. 1983. Maize mitochondrial DNA contains a sequence homologous to the ribulose-1.5-biphosphate carboxylase large subunit gene of chloroplast DNA. Cell **34:** 1007-1014.
42. LONSDALE, D. M. 1985. Movement of genetic material between the chloroplast and mitochondrion in higher plants. *In* Plant Gene Research. Genetic Flux in Plants. B. Hohn & E. S. Dennis, Eds.. Springer. Wien & New York.
43. MARGULIS, L. 1970. Origin of Eukaryotic Cells. Yale University Press. New Haven, Conn.
44. NOMIYAMA, H., M. FUKUDA, S. WAKASUGI, T. TSUZUKI & K. SHIMADA. 1985. Molecular structures of mitochondrial-DNA-like sequences in human nuclear DNA. Nucleic Acids Res. **13:** 1649-1658.
45. OSIEWACZ, H. D. & K. ESSER. 1984. The mitochondrial plasmid of *Podospora anserina:* a mobile intron of a mitochondrial gene. Curr. Genet. **8:** 299-305.
46. PALMER, J. D. & C. R. SHIELDS. 1984. Tripartite structure of the *Brassica campestris* mitochondrial genome. Nature **307:** 437-440.
47. PRATJE, E. 1984. Extrakaryotic inheritance: mitochondrial genetics. Prog. Bot. **46:** 226-240.
48. PRING, D. R. & D. M. LONSDALE. 1985. Molecular biology of higher plant mitochondrial DNA. Int. Rev. Cytol. **97:** 1-46.
49. SMITH, M. M.& O. S. ANDRESSON. 1983. DNA sequences of yeast H3 and H4 histone genes from two non-allelic gene sets encode identical H3 and H4 proteins. J. Mol. Biol. **169:** 663-690.
50. STEINHILBER, W. & D. J. CUMMINGS. 1986. A DNA polymerase activity with characteristics of a reverse transcriptase in *Podospora anserina.* Curr. Genet. **10:** 389-392.
51. STERN, D. B. & J. D. PALMER. 1984. Extensive and widespread homologies between mitochondrial DNA and chloroplast DNA in plants. Proc. Nat. Acad. Sci. USA **81:** 1946-1950.
52. TIMMIS, J. N. & N. S. SCOTT. 1983. Spinach nuclear and chloroplast DNAs have homologous sequences. Nature **305:** 65-67.
53. TIMMIS, J. N. & N. S. SCOTT. 1984. Promiscuous DNA: sequences homologous between DNA of separate organelles. Trends Biochem. Sci. **9:** 271-273.
54. TIMMIS, J. N. & N. S. SCOTT. 1985. Movement of genetic information between the chloroplast and nucleus. *In* Plant Gene Research. Genetic Flux in Plants. B. Hohn & E. S. Dennis, Eds.: 61-78. Springer. Wien & New York.
55. TSUZUKI, T., H. NOMIYAMA, C. SETOYAMA, S. MAEDA & K. SHIMADA. 1983. Presence of mitochondrial-DNA-like sequences in the human nuclear DNA. Gene **25:** 223-229.
56. TZAGOLOFF, A., G. MACINO & W. SEBALD. 1979. Mitochondrial genes and translation products. Annu. Rev. Biochem. **48:** 419-441.
57. VAHRENHOLZ, C., E. PRATJE, G. MICHAELIS & B. DUJON. 1985. Mitochondrial DNA of *Chlamydomonas reinhardtii:* sequence and arrangement of URF5 and the gene for cytochrome oxidase subunit I. Mol. Gen. Genet. **205:** 213-224.
58. WOESE, C. R., R. GUTNELL, R. GUPTA & H. F. NOLLER. 1983. Detailed analysis of higher order structure of 16S like ribosomal ribonucleic acids. Microbiol. Rev. **47:** 621-669.
59. WRIGHT, R. M. & D. J. CUMMINGS. 1983. Integration of mitochondrial gene sequences within the nuclear genome during senescence in a fungus. Nature **301:** 86-88.

# Nuclear Transfer from Parasite to Host[a,b]

## A New Regulatory Mechanism of Parasitism

LYNDA J. GOFF[c] AND ANNETTE W. COLEMAN[d]

[c]*Department of Biology and
Institute of Marine Sciences
University of California
Santa Cruz, California 95064*

[d]*Division of Biology and Medicine
Brown University
Providence, Rhode Island 02912*

Every theory of the course of events in nature is necessarily based on some process of simplification and is to some extent therefore a fairy tale.

Sir Napier Shaw
*Manual of Meteorology*

## INTRODUCTION

One of the most intriguing aspects of the phenomenon of host-parasite interactions is the mechanism by which a parasite elicits a particular set of responses in a host that results in the success of the parasitism. In necrotrophic interactions these mechanisms are obvious; the parasite simply destroys the tissue of the host via toxins and cell-degrading enzymes and then feeds on the remains. However, many parasites require substances produced only by living host cells. These biotrophic parasites must stimulate the production and release of required substances from the host and concomitantly ensure that the host is not destroyed in the process.[1,2] The mechanisms that mediate these most subtle biotrophic parasitic interactions are not at all clear.

Probably the greatest insight into how biotrophic parasites may regulate their hosts comes from the gene-for-gene concept of resistance and virulence which Flor proposed for the interaction of highly host-specific obligate fungi and their plant hosts.[3] This model suggests that parasite gene products involved in inducing the virulent reaction

[a]The authors dedicate this paper to Professor Janet R. Stein on the occasion of her retirement from the Department of Botany, University of British Columbia.

[b]Research support for these studies has been made available from the National Science Foundation (PCM-8313033, PRM-8211834, PCM-8003130, BSR-8415760 to Goff and PCM 8108122 and PCM 7923054 to Coleman). NSF's continued support of this research is gratefully acknowledged.

must be transferred from parasite to host across temporary intercellular bridges that form between the two cell types.[4] These inducers may bind to and inhibit the activity of host-produced resistance gene products, and concomitantly they may directly control the expression of other host genes. By these mechanisms parasite gene products could redirect (and frequently stimulate) the metabolism of the host, thereby providing necessary substances to feed the parasite, and compensate the host for the cost of the parasitism. Although this is an attractive hypothesis, it remains largely hypothetical as the transfer of gene products has not been documented,[5] nor has the conduit over which these gene products might be conveyed.[4]

One obvious evolutionary solution to the problem of moving gene products between host and parasite cells is to move the source of those gene products, the parasite nuclear genome, into the cytoplasm of the host. However, until very recently, no examples were known of such a direct transfer of nuclei from parasite to host cells.

Using quantitative fluorescence microscopy, we have recently determined that intercellular parasite-host nuclear transfer is characteristic of the many parasitic interactions that have evolved between red algal parasites and their red algal hosts.[6-8] The presence of parasite nuclei in the host's cytoplasm is directly correlated with changes in the morphology and metabolism of the host cell which result in the production of organic compounds required by the parasite. And, as we will discuss, the nuclei of some parasitic red algae replicate within their host's cytoplasm. The "transformed" host cell may then spread the intracellular parasite (reduced to genomic DNA) via cellular fusions and adventitious cellular connections to other host cells. In this paper we will examine the phenomenon and the consequences of intercellular nuclear transfer in the parasitic red algae, nuclear-organelle genome interactions that may occur in the heterokaryotic (parasite + host) cells, and the evolutionary implications of these intergeneric nuclear transfers.

## THE RED ALGAL PARASITES

Among the many peculiarities of red algae (Rhodophyta) is their propensity to exist exclusively as parasites of other organisms. Parasites are known from nearly every order of the major red algal class Florideophyceae and, in total, comprise nearly 20% of all known red algal genera.[9] Even more interesting is that these parasitic red algae employ only other red algae as their hosts, and most frequently the parasite and host are very closely related taxonomically. As the biology of these organisms has been reviewed recently,[9,10] it is only briefly summarized here.

Parasitic red algae are generally small (ca. < 0.5 cm) and morphologically simple, composed of branching filaments of cells which penetrate between the cells of the pseudoparenchymatous host, and a mass of tissue that protrudes from the host thallus and bears reproductive structures [FIGURE 1(1)]. The vegetative cells of most parasites have little or no photosynthetic pigment and consequently rely upon the photosynthetic host as a source of photosynthate.[11-14] Parasitic red algae are highly host specific, occurring only in association with a particular red algal host species or a particular physiological race of a specific host.[9]

The nature of the interaction of parasitic red algae and their hosts is most clearly seen by examining the development of the parasite and the responses it elicits within its host. Since these processes have been most thoroughly examined in two alloparasitic genera (i.e., parasites not closely related taxonomically to their host), Harveyella and

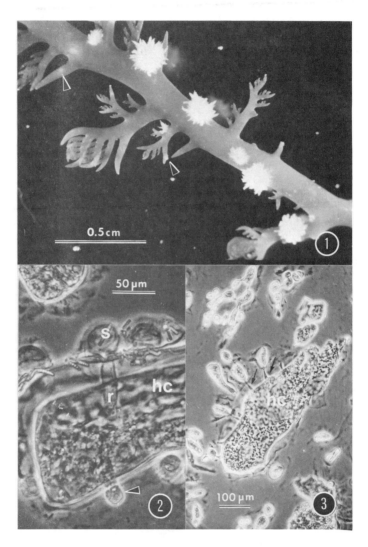

**FIGURE 1.** (1) Several reproductively mature individuals of the parasitic red alga *Plocamiocolax pulvinata* are seen growing from the tissues of the red algal host *Plocamium cartilagineum.* The arrowheads indicate very young infection regions where the parasite tissue has not emerged from the host tissue. (2) Spore germination by the parasitic red alga *Choreocolax* ( = *Leachiella*) upon attachment of the spore (s) to the outer wall of the host *Polysiphonia.* The spore produces an elongate rhizoid (r) which extends through the wall region separating host cells, and cuts off a cell (arrowhead). This cell has divided to form a small conjunctor cell which will fuse with and transfer a nucleus to the adjacent host cell. The cell will also divide to give rise to a filament of cells that will grow intrusively throughout the host. (Fixed and squashed tissue, phase microscopy). (3) A host cell (hc) of *Polysiphonia,* with many cells of the parasite *Choreocolax,* attached via secondary pit connections (arrows). (Fixed and squashed tissue, phase microscopy).

*Choreocolax* (syn = *Leachiella*),[7,15,16] their development is described first. Host infection by these taxa, as in all parasitic red algae, is initiated upon the attachment, germination, and penetration of parasite spores (haploid or diploid) on and into the tissues of the specific host [FIGURE 1(2)]. After extending into the host tissues, the rhizoid divides and branches to produce a network of cellular filaments that ramify between the cells of the host. These filaments may extend great distances from the original infection site, thereby spreading the infection throughout the host thallus.

Cells of these filaments are colorless [but possess plastid DNA (ptDNA) as do all parasitic red algae] and often contain conspicuous quantities of florideophycean starch. They are directly connected to host cells via connections termed "pit connections" (see discussion in Reference 7) [FIGURE 1(3)]. Electron microscopy reveals that these connections are not regions of host-parasite cytoplasmic confluence as once thought,[17] but rather they are plugged by a dense glycoprotein "pit plug."[7,16,18] The pit plug is extracellular, separated from the host and/or parasite cytoplasm by a membrane (FIGURE 2). However, host and parasite cells are interconnected by their plasma membranes which extend across this pit.

Host cells interconnected to parasite cells via pit connections undergo a range of responses that include (1) increase in production of carbohydrates and polymerization to form starch, (2) decrease in the size of the host cell vacuole and a concomitant increase in cytoplasmic volume, (3) increase in cytoplasmic organelles including ribosomes, mitochondria, and plastids, (4) increase in the ploidy of host nuclei, or in their number, (5) structural changes in the host cell wall, and (6) a dramatic increase in cell size (hypertrophy). All of these responses appear to reflect an increased activity in those host cells interconnected to parasite cells via pit connections.[7,16]

During later stages of the development of the parasite, parasite cells rapidly proliferate and coalesce into a pseudoparenchymatous mass of tissue which ruptures through the surface of the host [FIGURE 3(5) and (6)]. Numerous host cells, interconnected to parasite cells, appear to be caught up in these masses and carried up into the erumpent tissue where they continue to provide a nutrient source for the achlorotic parasite "sink" cells.[12] In other parasites examined (*Gardneriella, Gracilariophila*), localized proliferation of the parasite induces cell division in the overlying host's epidermis [FIGURE 3(7)]. In these taxa, the resultant tissue mass is therefore a mosaic of primarily photosynthetic host cells, host-parasite heterokaryons, and nonphotosynthetic parasite cells [FIGURE 3(8)].

Spores or gametes are produced from parasite cells in the peripheral regions of this tissue mass [FIGURE 3(5) and (6)]. Upon their release, the parasite and host tissues which comprise the erumpent thallus, plus some underlying cells, become necrotic. In the laboratory, most parasitic red algae complete this entire infection process (from spore germination-host penetration to parasite reproduction) within three to six weeks. In the absence of a suitable living host, parasite spores either fail to germinate[9] or develop only to a few-cell germling stage.[19]

The interaction of all parasitic red algae and their hosts is similar to the patterns seen in *Harveyella* and *Choreocolax*, although in some taxa the extent of intercellular parasite development may be much less extensive. In general these interactions are similar to those of many obligate biotrophic fungal parasites and their higher plant, or algal (in lichens), hosts. The parasite appears to stimulate the physiology of the host which results in an increased production of photosynthate, thereby compensating the host for the cost of the parasitism.

But how are these subtle controls effected? A clue to this mystery came from a realization that a feature that is inherent to all host-parasite interactions is the presence of pit connections between host and parasite cells.[7] In addition, studies of several parasitic taxa have revealed that the host cell responses listed earlier only occur in

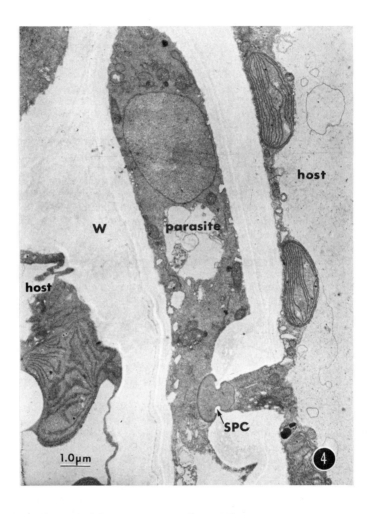

**FIGURE 2.** (4) A cell of the parasite *Harveyella mirabilis* is attached to an adjacent host cell via a secondary pit connection (SPC). The pit plug occludes the "connection" between the host and parasite cell (w = cell wall).

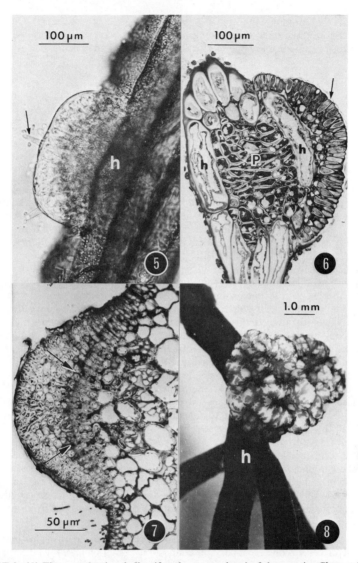

**FIGURE 3.** (5) The reproductive thallus (female gametophyte) of the parasite *Choreocolax* has emerged from the tissues of the host (h) *Polysiphonia.* The arrow indicates the receptive surface of the egg with several attached sperm. (6) A section through the reproductive thallus (tetrasporophyte) of *Choreocolax* reveals the presence of intercellular parasite (p) cells and several host cells that have become dislodged from one another. One host cell has been carried up into the proliferating parasite tissue mass. The arrow indicates a divided tetraspore (i.e., the site of meiosis) in the parasite. (Glycol methacrylate-embedded fixed tissue, stained with PAS and methylene blue). (7) During the development of the parasite *Gardneriella tuberifera* on its host *Sarcodiotheca,* the host epidermis (arrows) is induced by the parasite to divide, thereby forming a mass of tissue in which the parasite develops. [Fixed, embedded, and stained as in (6)] (8) A mature, reproductive plant of *Gardneriella.* The mosaic of red and white tissue in this tissue mass is due to the presence of pigmented host cells, and nonpigmented parasite cells. The host (h) is *Sarcodiotheca gaudichaudii.*

host cells that are interconnected to a parasite cell via a pit connection. Our attention was therefore directed to determining how these interconnections form between host and parasite cells, in the hope of elucidating how they may act.

## NUCLEAR TRANSFER FROM PARASITE TO HOST

A nearly century old paper by Rosenvinge provided the key to this puzzle.[20] In his elegant study of the development of the red alga *Polysiphonia,* Rosenvinge described the means by which two *Polysiphonia* cells that do not result from a common cell division become secondarily interconnected by a "secondary pit connection." According to his description, the nucleus in one cell divides and moves to the cell periphery where it is cut off into a small bud cell (conjunctor cell) (FIGURE 4a and b). The bud cell remains connected to the larger cell by a pit connection that results from incomplete septal formation during cytokinesis. The small, uninucleate bud cell fuses with an adjacent cell of the same individual. By this means, the two cells become interconnected by a secondary pit connection, and, more importantly for our present discussion, this process results in the transfer of a nucleus from one cell to another. Recently we reexamined this phenomenon using modern quantitative fluorescence microscopy and confirmed all of Rosenvinge's observations and conclusions.[21]

Thus, if secondary pit connection formation between host and parasite cells occurs in a manner similar to that described in *Polysiphonia,* then the red algae would have a mechanism for introducing a parasite genome into the cytoplasm of the host. Fine structural studies by Peyriére,[22] Wetherbee and colleagues,[23–26] and Goff and Coleman[7] of pit connection formation in several red algal host parasite associations showed a pattern of development similar to that in *Polysiphonia.* In these cases, the parasite produces the bud cell ("conjunctor cell") which subsequently fuses with an adjacent host cell (FIGURE 4c).

The ultrastructural studies by Peyriére and Wetherbee and colleagues indicated that the conjunctor cell contains a nuclear-like organelle; however these studies did not demonstrate that the organelle contained DNA nor could they determine its fate upon fusion of the conjunctor cell with the host. Both researchers speculated that if nuclear transfer did occur, the parasite nucleus must be subsequently eliminated[22] or degraded.[24,26]

To determine if a parasite nucleus is transferred from parasite to host cell via conjunctor cell fusion (secondary pit connection formation), and what its fate might be within host cytoplasm, we employed the DNA-specific fluorochromes DAPI (4',-6-diamidino-2-phenylindole) and mithramycin and quantitative fluorescence microscopy in a study of the alloparasitic interaction of *Choreocolax* (=*Leachiella*) and its host *Polysiphonia.*

The use of these fluorochromes clearly reveals that the double-membrane-bound body in the conjunctor cell contains DNA. And because these fluorochromes bind quantitatively to DNA, the fluorescence of a particular nucleus indicates its relative DNA content. Since host and parasite nuclei differed significantly in DNA content, this feature was used to "fingerprint" nuclei so that the identity of any could be established by its relative fluorescence which was determined using a microspectro-fluorometer.[15,21,27]

By this means, we determined that the parasite nucleus, which is cut off into the conjunctor cell, is "injected" into the cytoplasm of the host upon fusion of the conjunctor cell with the host. As the conjunctor cell also contains other cytoplasmic components (i.e., mitochondria, golgi, ribosomes, proplastids), these too enter the host cytoplasm upon the formation of host-parasite pit connections. And since

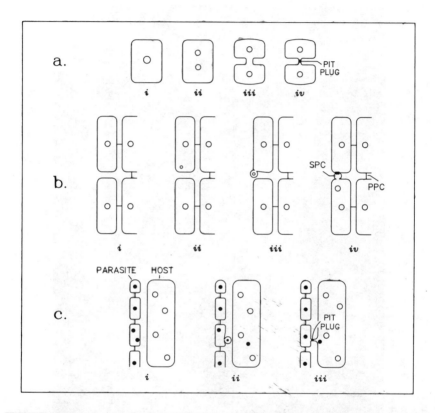

**FIGURE 4.** Pit connection formation in red algae. A primary pit connection (ppc) is formed between cells during cytokinesis (a). Septum formation is incomplete and the opening becomes plugged by a pit plug. A secondary pit connection (SPC) forms between cells that are not products of a single cytokinesis event (b). The process results in the formation of a pit connection, and transfer of a nucleus from one cell to another. Secondary pit connections may also form between parasite and host cells (c), and in the process, a parasite nucleus is injected into the host's cytoplasm.

hundreds of parasite cells may form connections to any single host cell, hundreds of parasite nuclei may come to reside in the host's cytoplasm [FIGURE 5(10)].

Upon transfer to the host cell, parasite nuclei are apparently unable to resume DNA synthesis; they remain in the $G_1$ stage (generally this is a very truncated portion of the cell cycle in red algae; most interphase is spent in $G_2$) and do not divide. Nor do the parasite nuclei degrade in the host's cytoplasm where they remain for the duration of its existence. Frequently a pairing of host and parasite nuclei is evident [FIGURE 6(11)] although it remains to be determined if, in the association of *Choreocolax* and *Polysiponia*, the nuclei ever fuse.

Host cells that receive parasite nuclei enlarge considerably, frequently increasing in volume by 20-fold. These enlarged, differentiated host cells undergo the typical

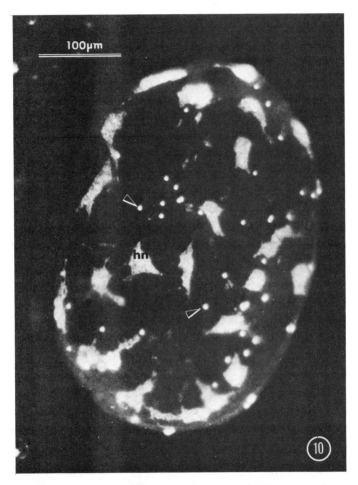

**FIGURE 5.** (10) An enlarged, pericentral cell of the host *Polysiphonia confusa*, infected by *Choreocolax*. The large host nuclei (hn) have become considerably polyploid in response to infection. Numerous, small parasite nuclei (arrowheads) are evident.

"stimulatory" reactions noted earlier. Starch accumulates and cytoplasmic organelles proliferate as the central vacuole decreases in size. Host nuclear DNA synthesis is stimulated in the infected host cell, resulting in an increase in the number of host nuclei, or an increase in DNA in each of the existing nuclei. These responses to parasite infection are highly localized, being restricted to host cells in which parasite nuclei are present. Even host cells that are adjacent, and pit connected (via primary

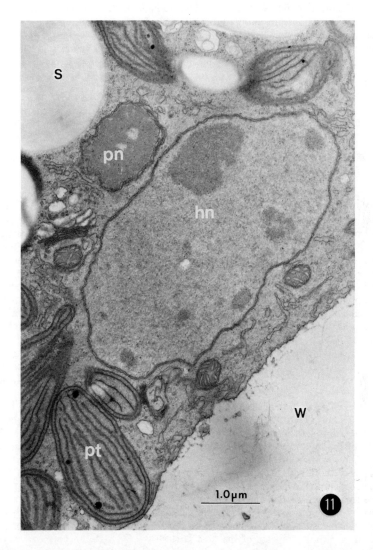

**FIGURE 6.** (11) A "pairing" of parasite nucleus (pn) (*Choreocolax*) and host nucleus (hn) is apparent in this infected *Polysiphonia* cell. A host plastid (pt) and starch (s) are indicated.

pit connections) to infected cells, show no responses if parasite nuclei are absent. Different regions of the same host cell may show different degrees of response, and this is directly correlated with the number of parasite nuclei within each region.

We still lack molecular evidence on whether the parasite nuclei are active in directing cellular activities within the host, though this is strongly suggested by the specificity of response in entire infected host cells, and in subregions of host cells containing parasite nuclei. At least some injected nuclei have nucleoli (evident in electron micrographs and acridine orange staining). Currently we are analyzing nuclear DNA to determine if the parasite and host have discernible differences, perhaps in their ribosomal DNA. If differences are found, we will use an appropriate probe for *in situ* localization of parasite ribosomal RNA. Adjacent uninfected host cells will provide the perfect control.

## HOST "CELLULAR TRANSFORMATION"

In our most recent studies of the phenomenon of parasite-host nuclear transfer, we turned our attention to the adelphoparasites (i.e., parasites that are very closely related taxonomically to their hosts), which comprise the majority (85-90%) of parasitic red algae. It immediately became obvious from our preliminary studies that the fate of the parasite nucleus in host cells in these adelphoparasites is clearly different than in the alloparasitic associations described earlier. This realization resulted in a reinterpretation of the development of these parasites that is vastly different from that accepted in the classic literature. It should be noted that our interpretations are based on research just now in progress in our laboratories, some of which is summarized below.

We began our investigation of the adelphoparasite *Janczewskia gardneri* on its host *Laurencia spectablis* (Ceramiales, Rhodomelaceae) by examining early (i.e., ca. 1 week after spore inoculation and host penetration) stages of the interaction of parasite and host cells. At the interface, filaments of colorless cells with large nuclei and copious quantities of ptDNA were observed ramifying between the photosynthetic subcortical cells of the host [FIGURE 7(12)]. Cells of the host could be readily distinguished from parasite cells by their pigment autofluorescence, and differences in nuclear DNA. Nuclei of the multinucleate host cells are substantially smaller and contain much less DNA than do parasite nuclei.

Cells of the intrusively growing parasite filaments divide to form conjunctor cells via a process similar to that described in the alloparasite *Choreocolax* [FIGURE 7(13)]. These small cells contain one or more nuclei (with DNA levels matching the "fingerprint" of *Janczewskia*), and evident organelle (plastid and/or mitochondrial) DNA. Upon fusing with an adjacent host cell, the contents of the conjunctor cell are injected into the cytoplasm of the host [FIGURE 7(14)]. Here, the parasite nucleus (which has a very conspicuous nucleolus) undergoes DNA synthesis and divides rapidly and frequently to form a population of parasite nuclei within the host cell [FIGURE 7(15)]. The parasite nuclei appear to divide synchronously within the host's cytoplasm and no nuclear division or DNA synthesis has been observed in the host nuclei of these cells.

Thus, from the formation of a single "secondary pit connection," a host cell may become populated with many parasite nuclei. Their rapid division, and the concomitant cessation of DNA synthesis and division of host nuclei in the enlarging host cells,

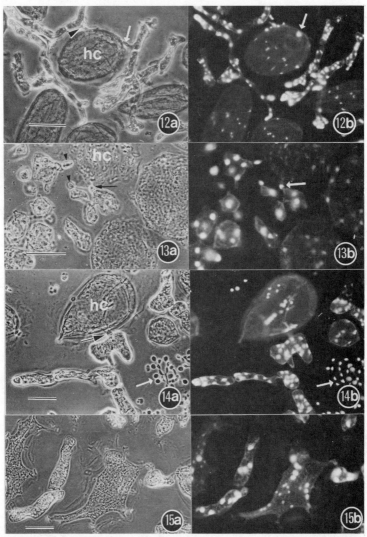

**FIGURE 7.** Early infection stages (ca. one week old) of the adelphoparasite *Janczewskia gardneri* (2N generation) on its host *Laurencia spectabilis* (1N, male gametophyte). All tissue was fixed, squashed, stained with DAPI and examined with epi-fluorescence microscopy. (12) Several filaments of *Janczewskia* surround and are pit connected (12a, arrows) to a host cell. The host and parasite nuclei are clearly distinguished by size and DNA content. Parasite nuclei are considerably larger than host nuclei. The arrow in (12b) indicates a parasite nucleus that has been injected, likely via the nearby pit connection, into the host cytoplasm. (13) Parasite cells form small uninucleate conjunctor cells (13a, arrowheads) that fuse with adjacent host cells (hc) (13a, arrow) and thereby transfer a parasite nucleus (13b, arrow) into the host cytoplasm. (14) A parasite cell has formed a pit connection with an adjacent host cell (hc) (14a). The host cytoplasm (14b) contains several large parasite nuclei, and smaller host nuclei. The arrows in (14) indicate sperm cells of the host which can be used as a DNA standard in microspectro-fluorometry measurements. (15) Several small host nuclei, and larger parasite nuclei occur in this heterokaryotic (host + parasite) cell.

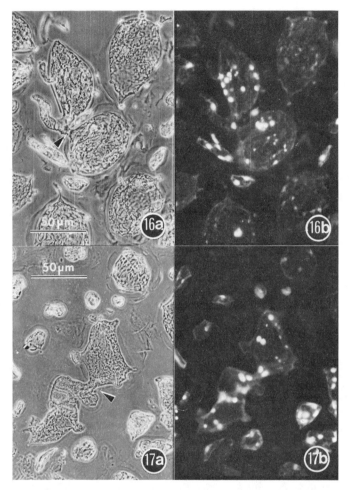

**FIGURE 7** (*continued*). (16) Two host cells, interconnected by a pit connection, have fused across that junction (16a, arrowhead) forming a syncytium. Parasite and host nuclei are clearly evident within the heterokaryotic cell (16b). (17) The pit connection between two host cells has increased considerably in diameter and is no longer plugged. Across this junction, parasite nuclei may be moved from one cell to another.

results in a progressive enrichment in the number of parasite vs. host nuclei; in some cells observed, parasite nuclei may outnumber host nuclei by 20 to 1.

The infected host cell subsequently undergoes developmental processes not normal in these differentiated cells. Pit plugs that occur within the pit connections between the infected host cell and contiguous uninfected cells appear to become dislodged into the cytoplasm of one of the cells. The pit connection greatly increases in diameter [FIGURE 7(16) and (17)] thereby forming a cellular syncytium (fusion cell) in which the parasite nuclei further replicate.

This mechanism results in the dispersal of the parasite genome locally, between contiguous, "pit connected" host cells [FIGURE 7(17)]. However, the genome may be dispersed to further regions of the host via cellular filaments that develop from the heterokaryotic (infected) cells. From localized regions of these heterokaryotic cells, parasite nuclei (and occasionally some host nuclei) are cut off into "bud" cells. These cells elongate and divide apically to form filaments of colorless cells that branch and grow intrusively through adjacent host tissues [FIGURE 8(18)]. Cells of these filaments form conjunctor cells which fuse with other host cells, thereby spreading the parasite genome throughout a region of host tissue.

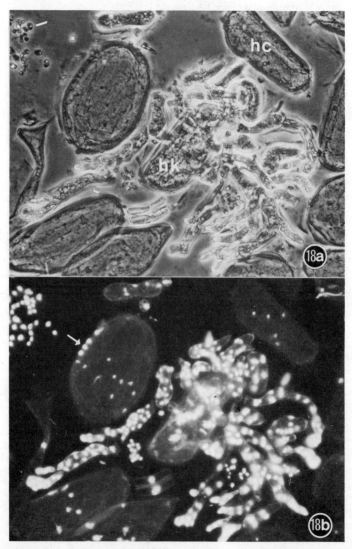

**FIGURE 8.** (18) Several branched filaments of cells are growing from this heterokaryotic cell (hk). Cells of these filaments contain primarily parasite nuclei. Fusion of these cells with other adjacent host cells (hc) spreads the parasite genome throughout localized regions of the host. The arrow in (18b) indicates several parasite nuclei within a host cell.

Within localized regions of infected host tissues, the filaments formed from the heterokaryotic cells coalesce to form a mass of tissue which pushes out through the overlying host cells. In the process, many of the enlarged, heterokaryotic cells, and previously uninfected host cells, may become dislodged and carried out into the erumpent tissue mass.

Initially this tissue is colorless and appears as an amorphous, tumorlike growth [FIGURE 9(19)]; however, as the tissue mass increases in size, two events occur, approximately simultaneously. The tissue mass becomes pigmented, changing in color from white to pink, and some cells of the tissue begin to express certain developmental patterns (i.e., polysiphonious development) which eventually result in a morphologically distinct tissue mass composed of radiating, pigmented branches.

**FIGURE 9.** (19) *Janczewskia gardneri* (2N) and its host *Laucrencia spectabilis*. Proliferation and coalescence of the filaments formed by heterokaryotic cells form a mass of unpigmented tissue that erupts through the surface of the host (upper arrow). As this tissue mass enlarges, pigmentation develops and the cell mass begins to express developmental patterns very similar to the host (lower arrow). The final result is a complex outgrowth composed of radiating branches which form spores, thereby dispersing the parasite genome into the environment.

Cells borne peripherally on the branches of *Janczewskia* will form reproductive cells characteristic of that generation, gametes (and from the fertilized egg, numerous copies of the zygote known as carpospores), and tetraspores (meiospores). Whether gametes or tetraspores are formed is determined by the ploidy (1N or 2N) of the initial parasite spore that established the infection.

Thus, for most of its life history, *Janczewskia gardneri* exists as an intracellular parasite of its host *Laurencia spectabilis*. In a very real sense, the pigmented mass of reproductive tissue is nothing more than a host gall whose cells are under the "control" of the parasite genome. In comparison with the alloparasite *Choreocolax*, the extent of exclusively parasite somatic tissue development in *Janczewskia* has been greatly reduced.

By examining spore germination and host penetration by *Janczewskia,* we are currently attempting to determine just how much somatic development of the parasite may occur before it assumes its "life" as an intracellular parasite. Evidence from other adelphoparasites (*Gardneriella* and *Gracilariophila*) indicates that the entire extent of parasite somatic development is the rhizoid produced by the germinating parasite spore [FIGURE 10(20)]. Upon penetrating the surface of the host, this rhizoid immediately fuses with, and injects a nucleus (and other cytoplasmic components) into the underlying epidermal or subepidermal cell. As in *Janczewskia,* the nucleus divides in this cell and is subsequently spread from the infection site via syncytium formation and the production of filaments that emanate from the heterokaryotic cell.

## NUCLEAR-ORGANELLE GENOME INTERACTIONS IN ADELPHOPARASITES

The final product of host cellular transformation by the parasite *Janczewskia,* a tumor of cells controlled by parasite nuclei, is capable of photosynthesis.[28] But are the functioning plastids in these tissues derived from the host or parasite?

If the plastids are host plastids, then this implies that the parasite nuclear genome may act in concert with the host plastid genome in regulating photosynthesis. It is also possible that pigment production is due to the development of parasite proplastids during the differentiation of the reproductive tissue. This latter possibility is suggested from observations of host cells in which parasite nuclei have only recently been injected. In these cells, discrete, discoid-shaped autofluorescent plastids (with DAPI-staining ptDNA), which are similar in morphology and autofluorescence to those in adjacent uninfected cells, are evident only in the region of cytoplasm occupied by host nuclei [FIGURE 11(21)]. In regions where parasite nuclei occur, the "host" plastids are absent. However, in this cytoplasmic region, extranuclear DAPI-staining DNA is evident. This DNA may represent the plastid DNA (and/or mitochondrial DNA) of the proliferating parasite plastids, and/or the remains of degenerating host plastids. Currently we are using chloroplast DNA endonuclease restriction patterns to determine whether host or parasite (or both) plastids end up populating the photosynthetic tissues of *Janczewskia.*

## EVOLUTIONARY CONSIDERATIONS

Based upon the morphological characteristics used to classify red algal taxa, the parasite *Janczewskia gardneri* is very closely related systematically to its host *Laurencia spectabilis.* Upon comparing several morphological features, Saito concluded that *J. gardneri* is more closely related to its host *L. spectabilis* than to other species of *Janczewskia.*[29] This may not be too surprising if *Janczewskia* is merely a gall of host tissue in which parasite nuclei reside. The expression of host developmental genes in tissues containing parasite nuclei may occur due to the (1) continued production of host gene products from host nuclei in the heterokaryotic cells. (2) expression of long-

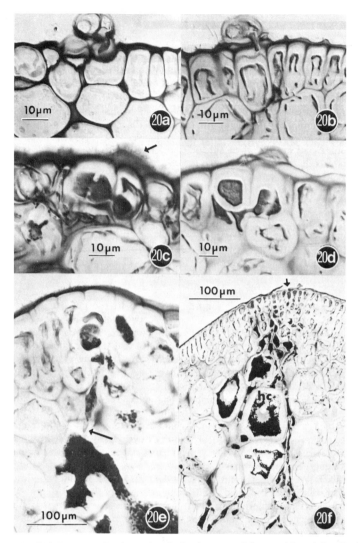

**FIGURE 10.** (20) Spore germination and early development of the parasitic red alga *Gardneriella tuberifera* on its host *Sarcodiotheca gaudichaudii.* (Fixed, embedded in glycol methacrylate, sectioned, and stained with PAS and methylene blue.) (20a) Eight hours after host inoculation, the parasite spore has germinated to produce one or more rhizoids each containing a nucleus. One rhizoid has penetrated the outer host wall surface. (20b) The parasite rhizoid has penetrated through the inner cell wall of the underlying host epidermal cell. (20c) Twenty-four hours after host inoculation, the underlying host cell, which was penetrated by the parasite rhizoid, has divided and is protruding (across the pit connection) into an adjacent host cell. The arrow indicates the region of parasite spore attachment. (20d) The host's epidermal cell, penetrated by the host, and several adjacent cells have been "transformed"; these cells, which are no longer vacuolate, contain large amounts of proteins. (20e) Four days after initial host infection, the infection has spread away from the epidermis via fusion of infected and uninfected host cells across pit connections, and by the development of filaments that intrusively grow into the host tissue and fuse with host cells (arrow). The granules in cells are PAS-stained starch grains. (20f) Seven days after initial host infection, "host"-derived rhizoidal cells, containing the parasite genome, have extended deep into the host tissue, and have transferred nuclei to other host cells. Note the large accumulation of starch in host cells (hc) and "parasite" cells. The arrow indicates a rupture through the outer host wall which was probably the site through which the parasite spore initially penetrated the host.

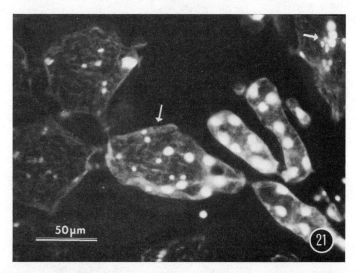

**FIGURE 11.** (21) This "host" cell of the host *Laurencia* has recently received nuclei from *Janczewskia*. Autofluroescent plastids, similar in morphology and autofluorescence to plastids of adjacent uninfected host cells, occur only in the cytoplasmic region still occupied by host nuclei (arrow). In cytoplasmic regions where the parasite nuclei are dividing, no plastids are evident, though higher magnification reveals the presence of organelle DNA (ptDNA?). Two host cells are pit connected to the left of the transformed host cell, and several "parasite" cells are associated at the right. The arrow indicates another heterokaryotic cell in which the parasite and host nuclei are closely associated.

lived host messenger RNA in host cells occupied by the parasite genome, (3) the expression of host-developmental genes that have, over evolutionary time, become incorporated into the parasite genome, or (4) the expression solely of parasite genes which are in fact very similar developmentally to those of the host.

How are the adelphoparasites and alloparasites related evolutionarily? Did adelphoparasites arise from alloparasites via a tremendous reduction in somatic tissue development, or are alloparasites really descendants of adelphoparasites that have radiated onto a different red algal host into which they are able to inject copies of their nuclear genome, but are unable to take over the host's cellular machinery necessary for host genome replication (FIGURE 12).

Lastly, how might these most sophisticated interactions have arisen evolutionarily? In many ways, the transfer of parasite nuclei to host cells, their replication within the host's cytoplasm, and the "packaging" of the parasite nuclei into host-cell derived dispersal units (i.e., spores and gametes) are extraordinarily similar to the processes involved in postfertilization development in most red algae. In most of these organisms, the diploid nucleus formed by karyogamy of the sperm and egg nucleus is transferred from the egg (carpogonium) to the cytoplasm of a predetermined vegetative (haploid) cell via cellular filament (ooblastema) or syncytium (i.e., fusion cell) formation. Within this cell (which is now heterokaryotic), the diploid nucleus replicates and copies are subsequently transferred to adjacent host cells upon pit plug dissolution or are cut off into cells that will form filaments of cells. These "diploidized" cells grow intrusively between cells of the haploid "host" plant (female gametophyte), frequently inducing hyperplasia in surrounding host tissue. The diploid cells may coalesce to form a mass

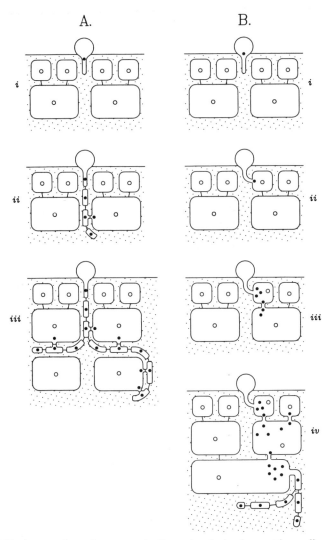

**FIGURE 12.** A comparison of spore germination and early development in an alloparasite (i.e., *Choreocolax, Harveyella*) and an adelphoparasite (i.e., *Janczewskia, Gardneriella*). The black nuclei are parasite nuclei.

of 2N tissue (gonimoblast), the peripheral cells of which differentiate into 2N spores (carospores). Alternately, cells of the intrusively growing diploid filaments may inject copies of the 2N genome into other haploid vegetative cells where the 2N genome is further replicated and eventually dispersed.

    Thus, in a way very analogous to red algal parasites, the diploid genome that results from fertilization is an intracellular parasite of haploid vegetative cells, utilizing these cells to replicate the 2N genome and disperse it into the environment. Do these remarkable similarities indicate that parasitic red algae may have originated from this complex sexual process, or are those similarities an example of convergent evolution, perhaps related to the ability of red algal cells to fuse with and transfer nuclei between genetically different cells.

## SUMMARY AND CONCLUSIONS

Parasitic red algae are highly sophisticated obligate biotrophic parasites that occur only within the tissues of other red algae. These parasites appear to elicit the particular set of required host responses by injecting a copy of the parasite nuclear genome (and other cytoplasmic components) into the host's cytoplasm. These specific responses are seen only in host cells containing parasite nuclei. In some interactions, the parasite nuclei do not undergo DNA synthesis or division in the host cytoplasm. Rather, the parasite genome is spread from parasite cells which grow intrusively throughout the host tissue and transfer nuclei to adjacent host cells.

In contrast, other parasitic red algae have little or no somatic tissue development of their own. Upon host penetration, these parasites inject their nuclear genome directly into the cytoplasm of underlying host cells. Here they rapidly divide and eventually outnumber the resident host nuclei. The infected, heterokaryotic host cell then produces cellular filaments which fuse with other host cells, thereby spreading the parasite genome intracellulary throughout localized regions of the host. Eventually the infected host tissue differentiates to form a tumorous mass of cells which disseminates the parasite genome, packaged in a "host-derived" cell.

This is a parasitism as total as a lysogenic viral infection but occurring only between red algae. The parasite, reduced to its genomic DNA (nuclear and perhaps organelle), utilizes the host's cellular machinery for its replication, packaging, and dispersal. The molecular regulatory mechanisms responsible for these most subtle and biologically sophisticated interactions remain to be discovered. Instead of being captured and enslaved by a cell, as has been postulated for the endosymbiotic theory of mitochondrial and plastid evolution, parasitic nuclei have become the invader, and conqueror, untamed.

## ACKNOWLEDGMENTS

The authors extend thanks to Dr. J. Michaelmore, University of California, Davis for his lively discussions of fungal biotropy and to the conference cochairs, Drs. John Lee and Jerome Fredrick, for extending the opportunity to present this paper.

## REFERENCES

1. LEWIS, D. H. 1973. Concepts in fungal nutrition and the origin of biotrophy. Biol. Rev. **48:** 261-278.

2. ANDREWS, J. H. & L. J. GOFF. 1985. Pathology. *In* Handbook of Phycological Methods—Ecological Field Methods: Macroalgae. M. M. Littler & D. S. Littler, Eds. Cambridge University Press. New York, N.Y.
3. FLOR, H. H. 1942. Inheritance of pathogenicity in *Melampsora lini.* Phytopathology **32:**
   653-669.
4. HESLOP-HARRISON, J. & H. F. LINSKENS. 1984. Cellular interactions: a brief conspectus.
   *In* Cellular Interactions. J. Heslop-Harrison & H. F. Linskens, Eds.: 2-17. Springer
   Verlag. New York, N.Y.
5. CLARK, A. E. & R. B. KNOX. 1979. Plants and immunity. *In* Developmental and Comparative Immunology. **3:** 571-589. Pergamon Press, Ltd. London, England.
6. GOFF, L. J. & A. W. COLEMAN. 1984. Transfer of nuclei from a parasite to its host. Proc.
   Nat. Acad. Sci. USA **81:** 5420-5424.
7. GOFF, L. J. & A. W. COLEMAN. 1985. The role of secondary pit connections in red algal
   parasitism. J. Phycol. **21:** 483-508.
8. LEWIN, R. 1984. New regulatory mechanism of parasitism. Science **226:** 427.
9. GOFF, L. J. 1982. The biology of parasitic red algae. Prog. Phycol. Res. **1:** 289-369.
10. EVANS, L. V., J. A. CALLOW & M. E. CALLOW. 1978. Parasitic red algae: an appraisal.
    *In* Modern Approaches to the Taxonomy of Red an Brown Algae. D. E. G. Irvine &
    J. H. Price, Eds.: 87-109. Academic Press. London, England.
11. EVANS, L. V., J. A. CALLOW & M. E. CALLOW. 1973. Structural and physiological studies
    of the parasitic red algae *Holmsella.* New Phytol. **72:** 393-402.
12. GOFF, L. J. 1979. The biology of *Harveyella mirabilis* (Cryptonemiales, Rhodophyceae).
    V1. Translocation of photoassimilated ¹⁴C. J. Phycol. **15:** 82-87.
13. CALLOW, J. A., M. E. CALLOW & L. V. EVANS. 1979. Nutritional studies on the parasitic
    red alga *Choreocolax polysiphoniae.* New Phytol. **83:** 451-462.
14. KREMER, B. P. 1983. Carbon economy and nutrition of the alloparasitic red alga *Harveyella
    mirabilis.* Mar. Biol. **76:** 231-239.
15. GOFF, L. J. & A. W. COLEMAN. 1984. Elucidation of fertilization and development in a
    red alga by quantitative DNA microspectrofluorometry. Dev. Biol. **102:** 173-194.
16. GOFF, L. J. 1976. The biology of *Harveyella mirabilis* (Cryptonemiales; Rhodophyceae).
    V. Host responses to parasite infection. J. Phycol. **12:** 313-328.
17. FELDMANN, J. & G. FELDMANN. 1958. Recherches sur quelques Floridees parasites. Rev.
    Gen. Bot. **65:** 49-128.
18. KUGRENS, P. & J. A. WEST. 1973. The ultrastructure of an alloparasitic red alga *Choreo-
    colax polysiphoniae.* Phycologia **12:** 175-186.
19. NONOMURA, A. M. & J. A. WEST. 1981. Host specificity of *Janczewskia* (Ceramiales,
    Rhodophyta). Phycologia **20:** 251-158.
20. ROSENVINGE, L. K. 1888. Sur la formation des pores secondaires chez *Polysiphonia.* Bot.
    Tidssk. **17:** 10-19.
21. GOFF, L. J. & A. W. COLEMAN. 1986. A novel pattern of apical cell polyploidy, sequential
    polyploidy reduction and intercellular nuclear transfer in the red alga *Polysiphonia.* Am.
    J. Bot. **73:** 1109-1130.
22. PEYRIÉRE, M. 1977. Ultra-structure d'*Harveyella mirabilis,* (Cryptonemiales, Rhodophy-
    cee) parasite de *Rhodomela confervoides* (Ceramiales, Rhodophycee): origine des synapses
    secondaires entre cellules de l'hote et entre cellules du parasite. C. R. Acad. Sci. Paris
    Ser. D **285:** 965-968.
23. QUIRK, H. M. & R. WETHERBEE. 1980. Structural studies on the host-parasite relationship
    between the red algae *Holmsella* and *Gracilaria.* Micron **11:** 511-512.
24. WETHERBEE, R. & H. M. QUIRK. 1982. The fine structure of secondary pit connection
    formation between the red algal alloparasite *Holmsella australis* and its red algal host
    *Gracilaria furcellata.* Protoplasma **110:** 166-176.
25. WETHERBEE, R. & H. M. QUIRK. 1982. The fine structure and cytology of the association
    between the parasitic red alga *Holmsella australis* and its red algal host *Gracilaria
    furcellata.* Protoplasma **110:** 153-165.
26. WETHERBEE, R., H. M. QUIRK, J. E. MALLETT & R. W. RICKER. 1984. The structure
    and formation of host-parasite pit connections between the red algal alloparasite *Harv-
    eyella mirabilis* and its red algal host *Odonthalia floccosa.* Protoplasma **119:** 62-73.

27.  COLEMAN, A. W., M. J. MAGUIRE & J. R. COLEMAN. 1981. Mithramycin- and 4',6-
     diamidino-2-phenylindole (DAPI)-staining for fluorescence microspectrophotometric
     measurement of DNA in nuclei, plastids, and virus particles. J. Histochem. Cytochem.
     **29:** 959-968.
28.  COURT, G. J. 1980. Photosynthesis and translocation studies of *Laurencia spectabilis* and
     its symbiont *Janczewskia gardneri* (Rhodophyceae). J. Phycol. **16:** 270-279.
29.  SAITO, Y. 1972. Two species of *Janczewskia* from Japan and their systematic relationships.
     *In* Proceeding of the Seventh International Seaweed Symposium. K. Nisizawa, Ed.:
     146-149. University of Tokyo Press. Tokyo, Japan.

# Evolution of Eukaryotic Cells via Bridge Algae

## The Cyanidia Connection

JOSEPH SECKBACH

*Department of Natural Sciences*
*School for Overseas Students*
*The Hebrew University of Jerusalem*
*Mount Scopus 91905, Israel*

Intelligent people ought to examine all different opinions.

> Maimonides
> "Guide to the Perplexed" (1190)

The transition of prokaryotic cyanobacteria (i.e., blue-green algae) into eukaryotic algae was an important event in the evolutionary history of the nucleated cell.[1] The origin of eukaryotic organisms and their organelles is still a biological mystery, because of the absence of any definite intermediate stage among the living organisms or in the fossil records. This proeukaryotic transformation is important also for the appearance of the nucleus and the organization of cell division as well as the concentration of all photosynthetic pigments of eukaryotic algae and plants into one organelle (i.e., the chloroplast) instead of being scattered around the cyanobacterial cell. Various suggestions have been proposed to cover the gap and to explain the big advance from anucleated cyanobacteria to nucleated Eukaryota. Two main theories for this evolutionary change have been proposed: the traditional scheme of direct filiation[2] and the endosymbiotic (intracellular symbiotic) hypothesis.[3] The former one considers the modern eukaryotic cell as derived during the Precambrian from a monophyletic (anucleated) source through autocompartmentalization by the folding of a prokaryotic cell membrane around cytoplasmic DNA fibers to form the nucleus, enveloping photosynthetic and respiratory membranes to evolve chloroplasts and mitochondria, respectively. On the other hand, the symbiotic proponents (supporting a theory that of late seems to be gains favor and wide popularity) consider the nucleated photosynthetic cell to be composed of a few once-primitive, free-living organisms dwelling intimately within the cell habitat and serving as organelles.[3] These endocytobiologists maintain, for example, that the chloroplasts represent the evolutionary residue of primitive single cyanobacteria, the mitochondria-aerobic bacteria that have somehow taken up permanent residence in a colorless cell and eventually become organelles symbiotic with their host. A few such partnerships are known even today within various organisms. Both evolutionary theories (the endosymbiotic polyphyletic origin and the autogenous monophyletic origin for the Eukaryota) are extremely hard to prove, since there are no records of the past that are unequivocally accepted today.

This article proposes to link the discontinuity between the Prokaryota and the

Eukaryota, in the blue-green and red algal line, with *Cyanidium caldarium*[4,5] cell types. Recent observations have shown that an apparently unialgal population of *Cyanidium caldarium* actually was a mixture of two or three related but different thermoacidophilic eukaryotic algae.[6-8] These algae have been isolated from different parts of the world[4,5,7,8] and recently also from Hamat Tiberias (Israel) hot springs.[9] Examinations of these *Cyanidia* species demonstrate (TABLE 1) possible orderly intraspecies progressional stages with changes in morphology and biological activities ranging from resemblance to primitive cyanobacteria to the more advanced unicells

**TABLE 1.** Features of Three Unicellular Acidothermophilic Algae (*Cyanidiaceae*)[a]

| Characteristic | *Cyanidioschyzon merolae* | *Cyanidium caldarium* | *Galdieria sulphuraria* |
|---|---|---|---|
| **Cell Morphology** | | | |
| Size (μm) | 1.5 × 3.5 | 2-6 | 3-11 |
| Shape | claviform to ovally | spherical | spherical |
| Division | binary fission | autospores | autospores |
| Number of spores | 2 | 4 | 4-32 |
| Mitotic & sexual reproduction | absent | absent | absent |
| **Microstructure** | | | |
| Chloroplast | polymorph | spheric or cup shape | multilobed (seen as several profiles) |
| Envelope of plastid | single | single | single |
| Mitochondria | one | one | several |
| Dictyosome | absent | absent | several primitive |
| Vacuole | absent | absent | several primitive |
| Pyrenoid | absent | absent | absent |
| **Activity & Bioproducts** | | | |
| Culturing | autotroph | autotroph | auto- & heterotroph |
| Halotolerance (% NaCl) | 3 | 3-4 | 10 |
| Linolenic acid | none | none | present |
| Storage glucan | phytoglycogen | phytoglycogen | amylopectin |
| Color with iodine | red | red | red-violet |
| Maximum absorption (nm) | 420 | 440 | 540 |

[a] After Seckbach *et al.*[6]

of Rhodophyceae. It seems that these organisms may serve as "bridges" linking the red with the blue-green algae thereby forming a continuous line of evidence in this evolutionary pathway. The idea that *Cyanidium caldarium,* an acid thermophilic photoautotroph, has a possible transitional status for bridging the hiatus between Pro- and Eukaryota has been discussed elsewhere.[1,2,10-15] However, by reconsidering the transitional model of *Cyanidium caldarium* together with the observations of the additional species in the Cyanidiaceae,[6-8] we feel that we have a good case for reestablishing the interkingdom bridge with these newly observed algal cells (FIGURE 1).

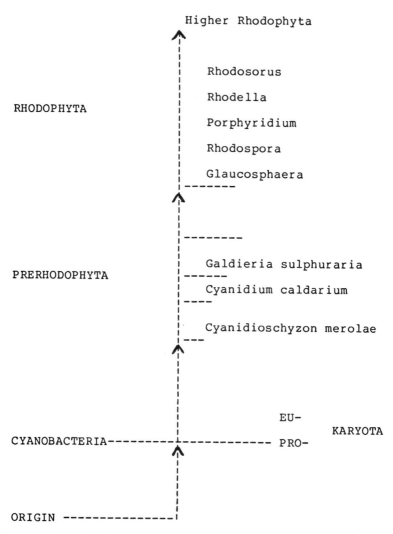

**FIGURE 1.** Schematic evolutional line of cyanobacteria through a group of "bridge algae" (classified in a new taxon, Prerhodophyta) towards the Rhodophyceae. Based on the morphology, fine structure, and biochemistry of these unicellular species. (See also Reference 6, Figure 2C.)

# THE TROUBLE WITH *CYANIDIUM* AND THE ELUCIDATION

*Cyanidium* is all things to all people
J. F. Fredrick (1984)

The organism known in the literature[1,4,5,10] as *Cyanidium caldarium* is a bluish-greenish eukaryotic microalga found in thermoacidic environments throughout the world.[4,5,7–9,16,17] It is the sole photoautotroph with an upper temperature over 50°C and a pH level less than 4. Almost every feature—such as cellular ultrastructure,[5,7,8,10,13,16,18–25] taxonomic position and phylogeny[1–5,10,12,15,20,25–29] biochemical content, and other aspects[1,2,4,5,11,15,28–30]—has been the subject of considerable controversy since its original collection by Tilden from the Yellowstone hot springs in 1898. For example, *Cyanidium* has been classified with practically every algal group[4,5,6,10,26–28] including the prokaryotic cyanobacteria. In addition, the nomenclature for *Cyanidium* has involved about a dozen names in the literature over time and the "merry-go-round" inflation in the naming list is not finished as yet.[4–8,16–18,21,30] It seems that each new analysis of this paradoxical alga brings only further trouble, serves to continue the systematic "musical chairs," and just clouds the picture.

Because of its cellular size and mode of division into autospores, it resembles *Chlorella,* but also differs from this green algal division with the photopigments.[4,5] *Cyanidium* contains chlorophyll *a* without chlorophyll *b,* C-phycocyanin, allophycocyanin, β-carotene, zeaxanthin, and lutein. This pigment spectrum is rather close to the cyanobacteria and at the same time to rhodophytes (e.g., *Porphyridium aerugineum*). The ultrastructure of *Cyanidium* shows a simple (FIGURES 2 and 4) and primitive cellular level; the plastids resemble the rhodoplasts (FIGURE 6) in their appearance and also the cyanobacterial cells (FIGURE 2). Phycobilisomes[19,20,25] are attached on the parallel or concentric running thylakoids, which are single and nonstacked. Several observations show a distinct girdle band[13,20,23,25,27] in this chloroplast which is enveloped by a single membrane[6–8,13,16,23] (FIGURES 4 and 6), although other reports claim to depict a double membrane.[23,25] Further investigations on the lipid content,[5,11,19,30,31] storage polysaccharides (the grains are on the outside of the chloroplast) and their enzymatic systems involved in the polyglucan biosynthesis,[1,6,12–15] ferredoxin sequence,[32] and primary photoassimilatory products[6,28,33–36,41] (low molecular sugars and floridosides) indicate the affinity of this controversial organism to the blue-green and red algae. With time all the "older" taxonomic proposals for this alga were shown to be invalid and were rejected and abandoned. We are left with three major hypotheses to account for the origin and relationships of this intriguing individualist:[6,27,33] (1) to place it within the Rhodophyceae; (2) to regard it as a symbiotic association (a cyanome), or (3) the transitional or "bridge" status, which considers it as an intermediate stage linking between pro- and eukaryotes.

The main reason for the contradictions and confusion in the results of studies concerning *Cyanidium caldarium* is that most of the workers have not been dealing with the same species but with one of the "strains" or with a mixture of two or three similar but different species of Cyanidiaceae mingled within their algal population.[6–8] The principle variations among these species are shown in TABLE 1. The electron micrographs (FIGURES 2-7) exhibit the ultrastructures of these three members of Cyanidiaceae. They were called provisionally[6,8,15] *Cyanidioschyzon merolae* (FIGURES 2 and 3), *Cyanidium caldarium* (FIGURE 4 and 5), and *Galdieria sulphuraria* (FIGURES 6 and 7). It is clear that there are some increasing developmental changes within the algal line. Further details on these algae are discussed below.

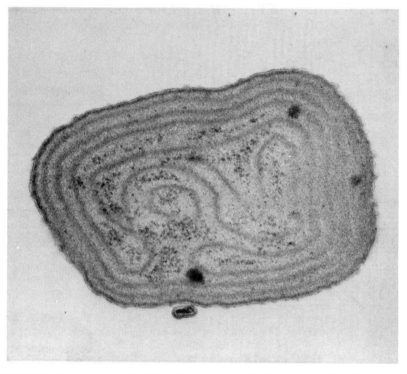

**FIGURE 2.** Electro micrograph of *Cyanidioschyzon merolae.* Note the primitive ultrastructure, i.e., only a few concentric thylakoids are visible. This cell resembles blue-green bacteria. Fixation with glutaraldehyde-osmium tetraoxide. 63500×.

## IS *CYANIDIUM* A RHODOPHYTAN?

A wolf changes his hair but not his nature.
Immanuel, Mahberot (c. 1300)

Seckbach *et al.* compared *Cyanidium* with other algal groups and indicated "phylogenetic scores" or various numerical considerations;[24] the highest similarity was obtained with rhodophytes (87%) and less with the cyanobacteria (57%). Additional studies, such as the ferredoxin sequence,[32] fine cellular structure,[7,13,15,16,20,21,23–25] accumulation of floridosides characteristic of red algae,[35,36] and analysis of sterol and lipid content[5,11,19,20,31] or of the chloroplast nucleoid of these algae[37] share wide agreement in the selection of the Rhodophyta as the host taxon for *Cyanidium* species. The multihierarchal systematic arrangement of *Cyanidium* species and further discussion have been published elsewhere.[6,26,27] There are, however, some arguments that may weaken the tendency of placing *Cyanidium* with the rhodophytes. For example, most of the lower unicell red algae possess some of the following components that are not shared with *Cyanidium:* phycoerythrin, true floridean starch, pyrenoid, and gelatinous sheath covering the cell. On the other hand, *Cyanidium* differs from the rhodophytes by its heavy proteinacious cell wall, its ability to thrive in thermoacidic environments, heterotrophic growth in darkness (with the loss of pigmentation), and the ability to grow in the absence of oxygen in an atmosphere if pure $CO_2$.[38] *Cyanidium* possesses some biochemical products, storage polysaccharides[15] or sterol content,[19] which are not totally characteristic to rhodophytans. These parameters and others suggest examining the other model for this paradoxical alga. In addition, due to the above arguments and others,[30,31] we are tempted to place *Cyanidium* species and related enigmatic organisms in a new taxon—Prerhodophyta (see FIGURE 1, and compare it to Reference 6, Figure 2C).

## EVALUATION OF *CYANIDIUM* AS ENDOCYANOME

We only see what we desire, only hear what we long for.

Luzzatto (1743)

Raven and co-authors used *Cyanidium caldarium* as a model for eukaryotic cells.[39] They considered it as an endosymbiotic association of a green alga harboring a

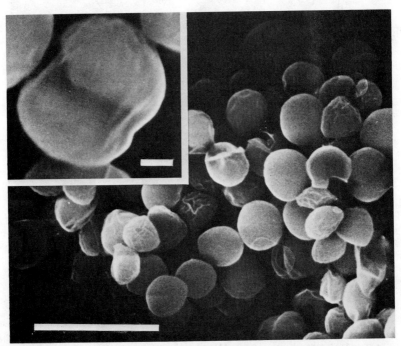

**FIGURE 3.** Cells of *Cyanidioschyzon merolae* (originally collected in Yellowstone National Park, Wyoming) as they appear in the scanning electron microscope (SEM). The algal cells are close to spherical in shape and a few are in the process of reproduction (bar = 10 $\mu$). The inset (bar = 1 $\mu$) demonstrates the two newly formed daughter cells in an oval cell before being discharged to the medium. This alga is considered to be the most primitive eukaryotic organism.

cyanobacterium (that had assumed the function of a chloroplast). Castenholz designed an ecophysiological model for this hot spring alga.[40] In his scheme, a thermocyanobacterium (tolerates high temperature up to 70°C in a neutral or alkaline environment) entered a colorless *Chlorella*-like (*Prototheca*) host which is adapted to acidic media in elevated temperature (up to ca. 42°C). The prokaryotic "resident" is protected

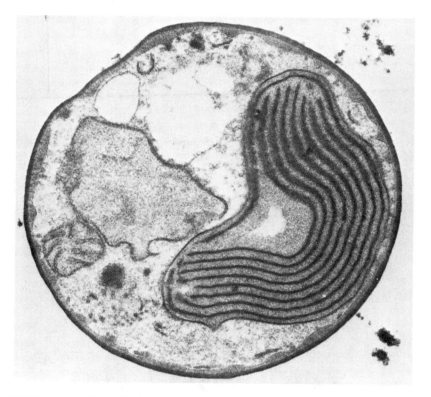

**FIGURE 4.** *Cyanidium caldarium* cross section. Three main organelles are shown: chloroplast with parallel running thylakoids which are limited by a single membrane envelope; a primitive mitochondrion; and between both of these organelle is the nucleus which has a slight projection. Fixed with KMnO₄. 28000×.

inside the host neutral cytoplasmic habitat against the extreme acidity, and this cyanobacterium bestows some "thermal resistant factors" upon its host alga. Thus, the upper temperature limit of this "combined organism" is elevated to 56°C—the highest limit of *Cyanidium*. Recently a relevant study by Seckbach and Fredrick noted that such a host candidate (colorless Chlorophyte *Prototheca zopfii*) has a starch and isozyme range involved in the polysaccharide synthesis that is typical of green algae and quite variable and dissimilar from *Cyanidium*.[1] This may add some dispute to the host nature and to the Castenholz endosymbiotic model.[40]

Most recently, Kremer and associates advocated that *Cyanidium* is a colorless eukaryotic alga of unknown affinities harboring a cyanobacterium as an endosymbiont.[28,33,34] They based their hypothesis on the comparative photoassimilatory products (e.g., free sugars, floridosides, or heterosides) of *Cyanidium* and related algae. They concluded,[28,33,34] because of absence of red algal floridosides together with the presence of free blue-green algal sugars (e.g., glucose and fructose) that *Cyanidium* is not a red alga but rather a cyanome, homologous to other known endosymbiotic organisms (e.g., *Cyanophora*). It seems that Kremer's reports do not overlap the articles by Nagashima and Fukuda[36,37] or those by De Luca and Moretti,[41] who reported the actual presence of rhodophytic floridosides in *Cyanidium,* while the low molecular sugars were absent.[36,37] Furthermore, Reed also reported, in contradiction to Kremer,[28,33,34] that *Cyanidium* accumulates such products in the presence of increasing salt concentrations.[29] In addition, the case of Kremer versus Nagashima and Fukuda and De Luca and Moretti is also weakened by the earlier report by Chapman, who rejected the symbiotic concept for *Cyanidium caldarium* because of ultrastructural and mode of division differences from cyanellae in *Cyanophora* and *Glaucocystis.*[26]

The symbiotic scheme for *Cyanidium* may suffer from additional arguments, for example, (1) the heavy proteinaceous cell wall is not an ideal path for symbiont penetration; (2) a lack of cyanobacterial cell wall traces as detected in cyanelle-bearing protists; (3) the chloroplast is freely embedded in cytoplasm and not in the host phagocytotic (food) vacuoles as in cyanellae; (4) if the chloroplast of *Cyanidium* is enveloped by a single membrane as observed by several workers[6-8,13,16,22] it causes phylogenetic questions concerning the symbiotic sources.[8,13] It seems from most of the data presented above that this concept is not supported enough by the available evidence. In fact, the symbiotic speculation for *Cyanidium* is not really new; earlier pioneers of this enigmatic alga already categorically rejected it on the basis of chemical content (Allen),[4] ultrastructure (Mercer, Bogorad, and Mullens),[20,24] or phylogenetic approach (Klein and Conquist).[10] One can only wonder if such a symbiotic scheme is possible for *Cyanidium* or whether it is just wishful thinking.


## *CYANIDIUM* SPECIES AT HAMAT TIBERIAS


More recently, Fredrick isolated for the first time *Cyanidium* cells from the hot spring of Hamat Tiberias (actually one mile south of modern Tiberias at the western bank of the Sea of Galilee, northern Israel). If further examinations will confirm the abundant presence of the cyanidiaceaen at Hamat Tiberias, then one can add the Middle East to the *Cyanidium* biogeographic distribution. At Hamat Tiberias the cells grow at their upper limit of acidity (pH 6) and at elevated temperature (ca. 35-40°C). The spring waters contain $H_2S$ and $CO_2$ and are rich in mineral (35 g salts per liter). The altitude of the site at the Sea of Galilee is 200 m *below* sea level. On the other hand, at Mount Shasta (California) *Cyanidium* has been found at elevations of 4316 m above sea level.[5] The initial analysis at Fredrick's laboratory indicated that three different cyanidiacean types occur in the premises of Hamat.

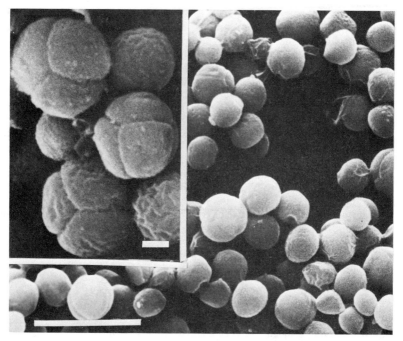

**FIGURE 5.** An SEM view of several *Cyanidium caldarium* cells (isolated originally at Naples, Italy) exhibiting stages of internal divisions into four spores (bar = 10 μ). At higher magnification (inset, bar = 1 μ), these tetrads are noticeable just before separating into four autospores.

## FOLLOW THE *CYANIDIA* BRIDGE

Without contrast, there can be no progress.

Marx (1847)

It is generally accepted that the prokaryotic cyanobacteria were the progenators of rhodophytes. To the present day, this evolutionary transformation from anucleated to nucleated cells is not fully understood and this step indicates an immense hiatus in the fossil records. Since the common characteristics of *Cyanidium* members are shared with the prokaryotic cyanobacteria and with the eukaryotic *Rhodophyta,* this makes them good candidates for covering the evolutionary gap between the two kingdoms. Indeed several authors have been intrigued by the idea that *Cyanidium caldarium* might serve as a bridge between the cyanobacteria and the red algae. Klein[2] and Seckbach and Fredrick[13] listed the requirements of such transitional organisms: they should (1) be eukaryotes; (2) show a primitive level of cellular organization (anlage of some organelles); (3) divide by autospores; (4) lack prokaryotic cell wall traces; and (5) their chloroplast, pigments, and storage glucans should be related and show similarity to both blue-green and red algae. If one adds the observations of lack of linolenic acid,[30,31] chloroplast DNA,[37] and the single membraned envelope[6,8,13,16,23] of *Cyanidium* chloroplast, then there is a good case of presenting these organisms upon the interface level or as "bridge algae." *Cyanidium* members fulfill all these criteria; therefore they seem to be the true transitional organisms between the anucleated and nucleated algae.

We have shown that the extension to support the symbiotic scheme for *Cyanidium* is untenable and has little foundation. So *Cyanidium* is not an endocyamone as proposed recently.[33] This enigmatic cell has an affinity to the lower (primitive) rhodophytes, it probably diverged from the remaining rhodophyceae at the Pro/Eukaryota interface. We propose to fit all *Cyanidium* types and other transitional algae into an intermediate category of "bridge" cells. The new taxonomic rank *Prerhodophyta* can group together the *Cyanidiaceae* and other related algae.

The transitional proponents consider direct filiation as the suitable pathway in the proeukaryotic cellular transformation. This "bridge" concept is now better and further supported with the separation of the three different species of *Cyanidium*. This "Italian style" *Cyanidium*[8] distinguishes between the simplest cell *Cyanidioschyzon merolae*, which resembles the cyanobacteria in its mode of division by binary fission, by its

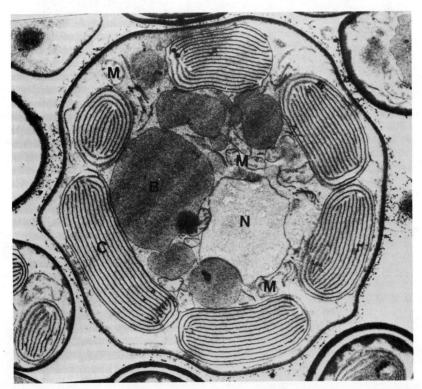

**FIGURE 6.** *Galdieria sulphuraria* cross section. Note the several profiles of the multilobed chloroplast (C); each one is encircled with a single membrane; the central nucleus (N); darker microbodies (B); several mitochondria (M) are in the cytoplasm. This cell is more closely related to the *Rhodophyta*. Fixed with $KMnO_4$. 12000×.

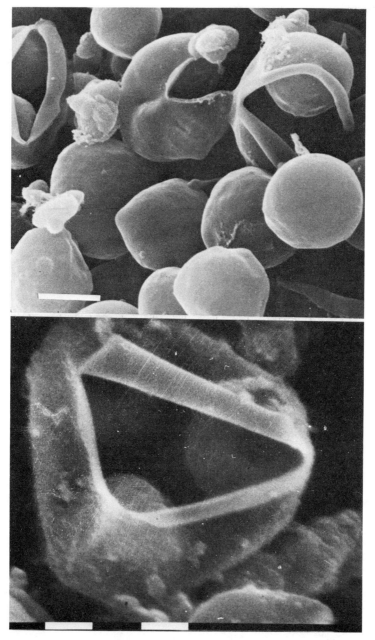

**FIGURE 7.** Top: observation of *Galdieria sulphuraria* cells with the SEM (source of algal culture as used in Reference 24). The cellular walls of several organisms curl during the release of the endospores (bar = 3 $\mu$). Bottom: a single cell of *G. sulphuraria* (isolated at Naples, Italy) during its reproduction and the release of autospores (bar = 1 $\mu$).

fatty acids,[30,31] and storage polysaccharides.[15] TABLE 1 and FIGURES 2 and 3 illustrate the characteristics and ultrastructure of this most primitive eukaryote.[15] The second member is *Cyanidium caldarium* (FIGURES 4 and 5, TABLE 1) which was used by several studies;[18-20] it indicated an intermediate status in the cyanidiaceaen line. The third cell, *Galdieria sulphuraria,* shows advanced features;[6,13,24] it has more developed organelles, a multilobed chloroplast[37] as in *Porphyridium,* and storage glucans closer to the lower *Rhodophyta,*[14,15] (FIGURES 6 and 7).

Prior to the discovery of the cyanidiaceaen members some authors had indicated the intermediate nature of *Cyanidium caldarium.* This proposal has been supported by the following studies: ferredoxin sequence analysis[32] (since this protein is not coded by the chloroplast DNA, it can be used as an evolutional marker for the whole cellular status); lipid composition;[11] storage polycarbohydrates and the enzymatic system involved in their biochemistry;[12] ultrastructure of the *Cyanidiaceae* and special reference to chloroplast architecture.[13] It is quite reasonable to conclude that these three links of Cyanidia can fit to cover the hiatus between the two kingdoms. These species show a gradual development within their mutual line from primitive blue-green (*Cyanidioschyzon* FIGURES 2 and 3, to *Cyanidium* FIGURES 4 and 5) towards further advanced red algae (*Galdieria,* FIGURES 6 and 7) in the following areas of biochemistry, e.g., lipids, polysaccharides, tolerance to salts,[6,17,29] (the higher red algae are marine) as well as micromorphology (see TABLE 1).

Further investigations, e.g., protein sequence analyses and microscopic observations, are required in order to support the "bridge" concept discussed in this article and for a better understanding of the possible answer to the key question in modern biology concerning the evolution of eukaryotic algae. Finally, one should keep in mind, as long as this line of proposal and debate between the various hypotheses (concerning the status of Cyanidia phylogenetic position) is kept in balance it may only enlighten the picture. After all, since these theoretical fields are not handed down to us on stone tablets,[6] they are part of man-made schemes and should be handled as such.

# SUMMARY

The transition from prokaryotic to eukaryotic cells was an important event in biological evolution. One possible pathway of the prokaryotic transformation is by internal gradual differentiation within the prokaryotic cell itself and with formation of intermediate (bridge) cells. *Cyanidium caldarium,* a thermophilic alga, may bridge the hiatus between the prokaryotic blue-green and eukaryotic red algae. *Cyanidium*'s phylogeny, taxonomic position, and cellular status have been widely disputed. There have been indications that apparently unicellular populations of *Cyanidium* actually contain three related but different algae. They were provisionally named *Cyanidioschyzon merolae, Cyanidium caldarium,* and *Galdieria sulphuraria.* More recently, Fredrick demonstrated for the first time that these *Cyanidia* members are also present in the hot springs of Hamat Tiberias (at the Sea of Galilee, Israel).[9] These Cyanidiacean species show progressive morphological and biochemical steps from cyanobacterial features (in *Cyanidioschyzon* and *Cyanidium*) to the rhodophytic affiliation (*Galdieria*) and may serve as transitional cells bridging the prokaryote to eukaryote missing connection.

## ACKNOWLEDGMENTS

The author is indebted to Prof. Paulo de Luca (University of Naples, Italy) for the algal cells (*Cyanidiaceae*). He gratefully acknowledges his colleagues Prof. David J. Chapman (University of California, Los Angeles) and Dr. Jerome F. Fredrick (Dodge Chemical Company, Bronx, N.Y.) for critical reading of the manuscript. Appreciation is due to Dr. Ezra Rahamim (Electron Microscope Unit of the Medical School, Hebrew University of Jerusalem) for his skillful assistance in the scanning electron microscope photos, to Mr. Shmuel Mantenband (then in New Jersey) for his technical help in the preparation of this manuscript, and to the New York Academy of Sciences for financial support.

## REFERENCES

1. SECKBACH, J. & J. F. FREDRICK. 1981. *In* Origin of Life. Y. Wolman, Ed.: 567-574. Reidel Publishing Company. Dordrecht, Holland.
2. KLEIN, R. M. 1970. Ann. N.Y. Acad. Sci. **175:** 623-633.
3. MARGULIS, L. 1981. Symbiosis in Cell Evolution. W. H. Freeman & Company. San Francisco, Calif.
4. ALLEN, M. B. 1959. Arch. Microbiol. **32:** 270-277.
5. BROCK, T. D. 1978. *In* Thermophilic Microorganisms and Life at High Temperatures: 255-302. Springer Verlag. New York, Heidelberg & Berlin.
6. SECKBACH, J., J. F. FREDRICK & D. J. GARBARY. 1983. *In* Endocytobiology. H. E. A. Schenk & W. Schwemmler, Eds. **2:** 947-962. Walter de Gruyter & Co. Berlin, FRG.
7. NAGASHIMA, H. & I. FUKUDA. 1981. Jpn. J. Phycol. **29:** 237-242.
8. MEROLA, A., R. CASTALDO, P. DE LUCA, R. GAMBARDELLA, A. MUSACCHIO & R. TADDEI. 1981. Giorn. Bot. Ital. **115:** 189-195.
9. SECKBACH, J. & J. F. FREDRICK. 1987. Mada **30:** 134-138. (In Hebrew.)
10. KLEIN, R. M. & A. CRONQUIST. 1967. Q. Rev. Biol. **42:** 105-296.
11. IKAN, R. & J. SECKBACH. 1972. Phytochemistry **11:** 1077-1082.
12. FREDRICK, J. F. 1976. Plant Cell Physiol. **17:** 317-322.
13. SECKBACH, J. & J. F. FREDRICK. 1980. Microbios **29:** 135-147.
14. FREDRICK, J. F. & J. SECKBACH. 1983. Phytochemistry **22:** 1155-1157.
15. FREDRICK, J. F. & J. SECKBACH. 1986. Phytochemistry **25:** 363-365.
16. DE LUCA, P., R. GAMBARDELLA & A. MEROLA. 1979. Bot. Gaz. **140:** 418-427.
17. DE LUCA, P., A. MUSACCHIO & R. TADDEI. 1981. Giorn. Bot. Ital. **115:** 1-9.
18. SECKBACH, J. 1971. Israel J. Bot. **20:** 302-310.
19. SECKBACH, J. & R. IKAN. 1972. Plant Physiol. **49:** 457-459.
20. SECKBACH, J. 1972. Microbios **5:** 133-142.
21. DE LUCA, P. & R. TADDEI. 1976. Webbia **30:** 197-218.
22. TÔYAMA, S. 1980. Cytologia **45:** 779-790.
23. UEDA, K. & J. YOKOCHI. 1981. Bot. Mag. Tokyo **94:** 159-164.
24. SECKBACH, J., I. S. HAMMERMAN & J. HANANIA. 1981. Ann. N.Y. Acad. Sci. **361:** 409-425.
25. FORD, T. W. 1984. Ann. Bot. **53:** 285-294.
26. CHAPMAN, D. J. 1974. Nova Hedwigia **25:** 673-682.
27. GARBARY, D. J., G. I. HANSEN & R. F. SCAGEL. 1980. Nova Hedwigia **33:** 145-166.
28. KREMER, B. P. 1982. Br. Phycol. J. **17:** 51-61.
29. REED, R. H. 1983. Phycologia **22:** 351-354.
30. MORETTI, A. & R. NAZZARO. 1980. Delpinoa **21:** 1-10.
31. BOENZI, D., P. DE LUCA & R. TADDEI. 1977. Giorn. Bot. Ital. **111:** 129-134.
32. HASE, T., S. WAKABAYASHI, K. WADA, H. MATSUBARA, F. JÜTTNER, K. K. RAO, I. FRY & D. O. HALL. 1978. FEBS Lett. **96:** 41-44.

33.  KREMER, B. P. 1983. *In* Endocytobiology. H. E. A. Schenk & W. Schwemmler, Eds. **2:** 963-970. Walter de Gruyter & Co. Berlin, FRG.
34.  KREMER, B. P. & G. B. FEIGE. 1979. Z. Naturforsch. **34c:** 1209-1214.
35.  NAGASHIMA, H. & I. FUKUDA. 1981. Phytochemistry **20:** 439-442.
36.  NAGASHIMA, H. & I. FUKUDA. 1983. Phytochemistry **22:** 1949-1951.
37.  NAGASHIMA, H., T. KUROIWA & I. FUKUDA. 1984. Experientia **40:** 563-564.
38.  SECKBACH, J., F. A. BAKER & P. M. SHUGARMAN. 1970. Nature **227:** 744-745.
39.  RAVEN, P. H., R. F. EVERT & H. CURTIS. 1981. Biology of Plants. Worth Publishing. New York, N.Y.
40.  CASTENHOLTZ, R. W. 1979. *In* Strategies of Microbial Life in Extreme Environments. M. Shilo, Ed.: 373-392. Verlag Chemie. Weinheim, FRG.
41.  DE LUCA, P. & A. MORETTI. 1983. J. Phycol. **19**(3): 368-369.

# Cyanidiophyceae

## Transition Stages in the Evolution of the Glucosyltransferase Isozymes

JEROME F. FREDRICK

*Research Laboratories*
*Dodge Chemical Company*
*3425 Boston Post Road*
*Bronx, New York 10469-2586*

### INTRODUCTION

There seems little doubt that the abiotic synthesis of hexoses and related sugars in primal atmospheres made these carbohydrates relatively abundant and available for use as primary energy sources and respirable substrates by the newly emergent cell forms.[1] Competition for these sugars, freely diffusible across the primitive cell membranes, probably caused the rapid depletion of these molecules in the rich chemical soup. Some of the hexose supply undoubtedly began to react intramolecularly: the reactive nature of the primary hydroxyl groups in glucose made for establishment of many alpha-1,6 glycosidic linkages, and hence, "branching" among the glucose molecules.[2]

As the free nonbranched sugar was depleted, only the branched (or alpha-1,6-linked) glucan remained as a potential source of energy-rich glucose. The new cell forms (whether as microspheres or coacervates) were faced with the need to "keep" these hexose molecules within their own entities. In order to do this, the branched, primitive glucan had to be broken down to more readily diffusible individual molecules of glucose.[3] These, after passing through the cell membranes, had to be *stored* within the confines of the cell for future use as energy source. Hence, the first *polysaccharides* appeared on the evolutionary scene.

Those primitive cells with the ability to form these storage sugars undoubtedly had survival advantage over their contemporaries, especially since carbohydrates became scarce in their immediate environments.[4] Therefore, until these cells developed their own internal machinery for producing hexoses via the complexity of photosynthesis and thylakoids, they were dependent upon the "tapping" of this internal storage glucan.[5]

The appearance of the first *glucosyltransferase* enzyme must have predated any internal formation of hexose via the process of photosynthesis. If the storage glucans,

438

invariably *branched* alpha-1,4-linked chains (branching via alpha-1,6 linkages),[6] were to be used in the ubiquitous Embden-Meyerhoff metabolic pathway as energy sources, there had to be conversion of the storage glucan to glucose-1-phosphate.[6] This is accomplished by the enzyme *phosphorylase* without the need for further expenditures of energy in the form of ATP.

The amino acid sequencing of phosphorylase has recently been accomplished.[7] It appears to be an ancient protein, in fact even older than cytochrome *c*. Its mutation rate has been calculated to be 1.1 to 1.7 amino acid changes per 100 residues per 100 million years (cytochrome *c* has about 3).[8] Undoubtedly, those other glucosyltrans-ferases associated with it are probably *old* as well.[8] In fact, since the formation of alpha-1,4 storage glucans in algae today appears to be the result of the interaction of phosphorylases (E.C. 2.4.1.1) glucan synthases (E.C. 2.4.1.11) and branching enzymes (E.C. 2.4.1.18) and these other enzymes are *related* to phosphorylase by similar sequences,[9] immune reactions,[10] and other physicochemical properties,[11] it has been postulated that all three groups of enzymes were once *one* catalytic polyfunctional protein[12] which later "fractionated" into multiple catalytic peptides.

The types of storage sugars found in prokaryotes (cyanobacteria, bacteria, and prochlorophytes) are glucans that consist of alpha-1,4-linked glucose residues with many alpha-1,6 branched points[13,14] such as glycogen and phytoglycogen.

Among eukaryotic algae, such as the rhodophytes, the storage glucan is still highly branched (floridean "starch"), but less so than the glycogen of cyanobacteria. The chlorophytes or green algae, like the higher plants that probably descended from them, form *true starch* (a mixture of *amylopectin,* a moderately branched glucan, and *amylose,* a relatively unbranched, exclusively alpha-1,4-linked glucose straight-chained poly-mer).

The three groups of enzymes responsible for the synthesis and degradation of storage glucans in algae[15] and higher plants[16,17] have been shown to exist in multiple molecular forms as isoenzymes ("isozymes"). Since both phosphorylases and syn-thases are involved only with the breakdown and formation, respectively, of alpha-1,4 glycosidic linkages, and the branching isoenzymes synthesize the alpha-1,6 linkages, the *branching* of the storage glucan ultimately depends upon the activity of these latter isozymes.[13,14]

The group of isozymes that synthesize alpha-1,6 glycosidic bonds between alpha-1,4-linked glucose chains are of two types. The *Q* branching isozymes can insert branched alpha-1,6 bonds into straight-chained amyloses, converting these unbranched sugars into moderately branched amylopectins.[13] At this point, these *Q* isozymes cannot further branch the amylopectin.

The *b.e.* branching isozymes, on the other hand, are capable of inserting further alpha-1,6 linkages into amylopectin (as well as amylose), and produce storage glucans that are highly branched, such as floridean "starch," phytoglycogen, and glycogen.[13,14] These storage glucans are found in prokaryotes and in very *primitive* eukaryotes.[18] More highly evolved eukaryotes all form less-branched amylopectins and amylose (true "starch"). (See FIGURE 1.)

TABLE 1 summaries our observation on the types of storage glucans formed by cyanobacteria, bacteria, and red and green algae. The branching glucosyltransferases present in those algae that form more highly branched storage glucans are of the *b.e.* type. These are also found in the bacteria and in the newly discovered *Prochloron,* which is thought to be on the direct evolutionary pathway to the chlorophytes.[19,20]

The interesting controversial eukaryotic alga *Cyanidium caldarium* has recently been shown to encompass *three* distinct thermoacidophilic algae.[21,22] These have been isolated and their morphology, ultrastructure,[23] and biochemistry reported.[22,24,25] These three distinct algae now are gathered together under the *Cyanidiophyceae* title.[21] The

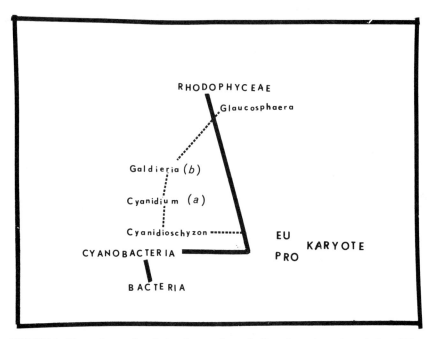

**FIGURE 1.** The pathway of evolution from prokaryotic Cyanobacteria to the red algae (Rhodophyceae) with "transition" forms (the Cyanidiophyceae) bridging the gap between prokaryotes and eukaryotes. The primitive *Glaucosphaera* (a rhodophyte) has many properties of *Galdieria*, particularly the branching isozymes (*Q* type) and storage glucan (Floridean starch).

original alga, first isolated in hot springs and called *Cyanidium calarium*, probably was the intermediate "type" and the report of its "conflicting" properties was due to the fact that the culture was not unialgal.[26] (See FIGURES 2-4.)

Some biological properties of the three algae forming this class, *Cyanidiophyceae*, are shown in TABLE 2. It should be noted that the thermoacidophilic alga having the most primitive properties (binary fission type of propagation, etc.) is *Cyanidioschyzon merolae*. This is, in all probability, the most closely related to the Cyanobacteria and hence has more prokaryotic properties than any other eukaryotic cell.

TABLE 3 shows studies performed on the three types of thermoacidophiles as far as the storage glucans they biosynthesize and the glucosyltransferases involved in their synthesis in each of the three types of Cyanidiophycean algae.

## THE STORAGE GLUCANS OF ALGAE

*Cyanidium caldarium* has variously been regarded as a primitive red alga,[27] as an endocyanome,[28] and as a transition or relic between the Cyanobacteria and the rhodophytes.[29] The storage glucan of *Cyanidioschyzon merolae*, as well as the glucosyltransferases (TABLE 3) of this alga, and its many biological properties, point to it as being the most primitive of this new group of hot-springs algae. As a result, *the* alga, originally studied as *Cyanidium caldarium,* appears to be a true transition form in the evolution of rhodophytes from cyanophytes. *Cyanidioschyzon,* with respect to its storage glucan, glucosyltransferase isozymes, etc., appears to be more cyanobacterial in its overall appearance than either *Cyanidium caldarium* or *Galdieria sulfuraria.*

The biosynthesis of storage glucans in both prokaryotic algae (cyanobacteria, *Prochloron,* etc.) and eukaryotic algae is dependent upon the three groups of isozymes described. However, the ultimate *form* of the glucan (its degree of branching or number of alpha-1,6 linkages) is determined by the activities of the alga's branching isozymes. If the *Q* isozymes are more active then the product will resemble amylopectin and, hence, contain fewer branch points than if the *b.e.* isozymes alone were present. Where the *b.e.* isozymes are the only branching enzymes present, the storage glucan is highly ramified and is usually *phytoglycogen.*

Since red algae form the storage glucan known as Floridean "starch" and since this glucan is much more highly branched than the amylopectin in the *true* starch stored by the green algae, it would seem that the *b.e.* isozymes exclusively present in the cyanobacteria and in *Cyanidioschyzon* and *Cyanidium* have undergone some change in progressing from *Cyanidium caldarium* to the unicellular red algae. This is apparent in the *most* eukaryotic of the Cyanidiaceae, *Galdieria sulfuraria.* TABLE 3 shows that this alga contains only *two b.e.* isozymes and *one Q* type. As in other red algae, however, even the presence of one *Q* isozyme radically influences the type of storage glucan synthesized.[30] Note that while *Cyanidium caldarium* has a branched storage glucan very much like that of *Cyanidioschyzon* (short average chain lengths), *Galdieria*'s storage glucan is much more like that found in red algae (much longer average chain lengths). This can be seen in TABLE 3.

It has been shown that the conversion of the *b.e.* type of branching isozyme to the *Q* type, may be due to replacement of two amino acid residues in the *b.e.* type.[31]

**TABLE 1.** Storage Glucans Formed by Bacteria and Algae

| Organism | Glucan Formed | Bonding | Structure |
|---|---|---|---|
| Bacteria | | | |
| *Escherichia coli* | glycogen | alpha-1,4 + alpha-1,6 | very highly branched |
| Cyanobacteria | | | |
| *Nostoc muscorum* | phytoglycogen | alpha-1,4 + alpha-1,6 | very highly branched |
| Rhodophytes | | | |
| *Porphyridium purpureum* | Floridean "starch" | alpha-1,4 + alpha-1,6 | Moderately branched |
| Chlorophytes | | | |
| *Chlorella pyrenoidosa* | starch | alpha-1,4[a] alpha-1,4 + alpha-1,6[b] | Unbranched Slightly branched |

[a] Amylose component.
[b] Amylopectin component.

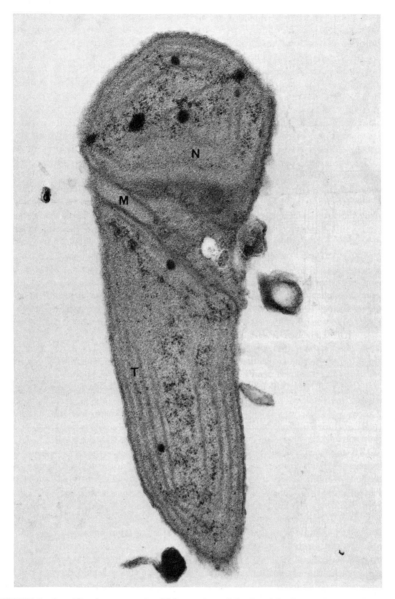

**FIGURE 2.** *Cyanidioschyzon merolae.* This member of the Cyanidiophyceae is most closely akin to the Cyanobacteria. It reproduces by binary scission as do bacteria and blue-green algae. Its biochemistry insofar as storage glucan formation is exactly similar to the blue-green algae. It has only *b.e.* branching isozymes, and it stores phytoglycogen. It has a single mitochondrion (M), a very diffuse nucleus (N), and peripherally located thylakoids (T) in a definitive chloroplast.

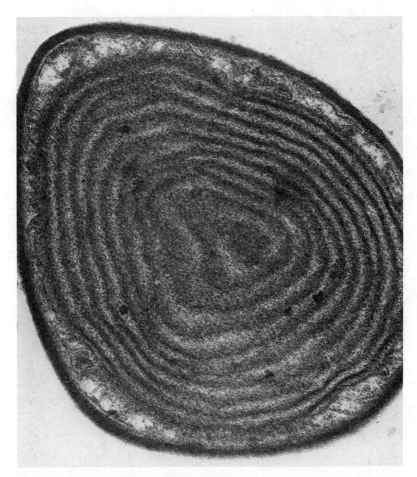

**FIGURE 3.** *Cyanidium caldarium*. This cell shows mainly the chloroplast. In addition, there is a single mitochondrion and nucleus. It forms the true bridge between the primitive *Cyanidioschyzon* and the more red-algal-like *Galdieria sulfuraria*. The average chain length of *Cyanidium*'s storage glucan appears to be longer than that of the cyanobacteria and that of *Cyanidioschyzon*, but shorter than that of *Galdieria*'s sugar (see TABLE 2).

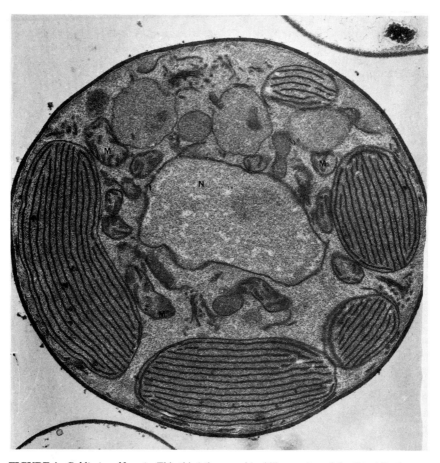

**FIGURE 4.** *Galdieria sulfuraria.* This third thermoacidophilic member of the Cyanidiophyceae is more like the red algae than any other member. It forms a Floridean-starch-like storage glucan and has the *Q* type of branching isozymes found in members of the red algae. It has many mitochondria and a sharply defined nucleus. While it resembles *Cyanidium,* the cell is much larger and has a more sharply defined chloroplast. Both *Galdieria* and *Cyanidium* reproduce by autospore formation (*Cyanidium* forms 4 autospores per cell, while *Galdieria* forms up to 32 per cell).

From the entire spectrum of branching in alpha-1,4-linked storage glucans in pro-karyotic and eukaryotic algae, it is possible to deduce two lines of possible evolution: one leading from the bacteria and cyanobacteria (where the glucan stored is glyco-genlike) to the Rhodophyceae and the Chlorophyceae (where amylopectins in the form of *Floridean* starch and *true* starch are the storage products). The line from the cyanobacteria to the red algae is *not* discontinuous, at least not in a biochemical sense with regard to the synthesis by the glucosyltransferase isozymes of storage glucans. This seems to form an orderly progression through the *Cyanidiceae* (from *Cyani-dioschyzon,* which is most like the bacteria and cyanobacteria even in its biological properties of mode of duplication by *binary fission,* through *Cyanidium,* which appears to be a real "transition" form, to *Galdieria,* the *most* eukaryotic of all three) to red algae.

## THE BIPHYLETIC PATH OF GLUCAN BIOSYNTHESIS

Since the basic glucan synthesizing and degrading enzymes must have predated the development of photosynthesis, it can be assumed that most, insofar as those that would survive, of the primordial cell entities had these enzymes. Subsequently, as some of these primitive structures developed porphyrins and, eventually, via chlorophyll(s) were able to form hexoses *in situ,* the newly-formed sugars could be "stored" immediately in such forms as *polymerized* glucose (glycogen, amylopectin, etc.).

There seems to be agreement that this *in situ* method of synthesizing a potentially energy-rich substance, eliminating the dependence on external sources for the substance, would confer a *selective advantage* on those primitive cells possessing this mechanism. Therefore, it seems altogether plausible that the Chlorophytes developed via the endosymbiotic route. Since all members of the green algae (Chlorophyceae) form their storage glucan *within* their chloroplast, and if the chloroplast originally constituted an endosymbiotic cyanobacteria with the ability to synthesize storage glucan *intra*thylakoidally, it seems logical to view the chloroplast structures in Chlorophytes as originating via this route. In fact, if one accepts the fact that *Prochloron,* the cyanobacterial precursor of the Chlorophycean chloroplast,[31] served as the ancestor for all green algae containing chlorophylls *a* and *b,* [32] then the next link in the evolution of this endosymbiotic relationship *had* to be an organism exactly like the newly discovered *Nanochlorum eucaryotum* which is definitely eukaryotic, forming its "true" starch within its chloroplast,[33] and yet has many cyanobacterial (or *prokaryotic*) traits as well (such as reproduction by binary fission as found in bacteria and blue-green algae).

Since the Cyanidiophycean algae (*Cyanidioschyzon, Cyandidium,* and *Galdieria*) and the Rhodophycean algae synthesize their highly branched storage glucans outside

TABLE 2. Biological Properties of the Acidothermophiles Comprising the Cyanidiaceae

| Feature | *Cyanidioschyzon merolae* | *Cyanidium caldarium*[a] | *Galdieria sulfuraria*[b] |
|---|---|---|---|
| Size of cell (nm) | $1.5 \times 4.0$ | 2–6 | 3–11 |
| Shape of cell | clublike | spherical | spherical |
| Vacuole | none | none | one or more |
| Mitochondrion | one | one | more than one |
| Chloroplast | one | one | many |
| Reproduction | binary fission | autospores (four usual) | autospores (more than four) |

[a] Old nomenclature: *Cyanidium caldarium forma A.*
[b] Old nomenclature: *Cyanidium caldarium forma B.*

TABLE 3. The Biosynthesis of Storage Glucans in the Cyanidiaceae

| Alga | Type of Glucan | Iodine Color (Maximum Absorption) | Average Chain Length[a] | Types of Branching Enzymes |
|---|---|---|---|---|
| Cyanidioschyzon merolae | glycogen | red (420 nm) | 6-7 | 3 (all b.e.) |
| Cyanidium caldarium | glycogen | red (440 nm) | 9-10 | 3 (all b.e.) |
| Galdieria sulfuraria | amylopectin (Floridean "starch"-like) | violet (540 nm) | 19-20 | 2 b.e. + 1 Q |

[a] Average chain length between branch (alpha-1,6) points.

of their chloroplasts, it seems likely that these algae represent a separate pathway in evolutionary progression from the Cyanobacteria.

For example, since the endosymbiont hypothesis assumes that while both *guest* and *host* algal cells possessed the ability for storage glucan formation but that one of the symbiotic partners gave up this ability,[34,35] it must be assumed that the *original* ability to synthesize branched glucans (as present in all "primitive" cells, see TABLE 1) was surrendered by the host algal cell in the Chlorophycean line. The storage glucan biosynthesis then remained exclusively the function of the guest alga (or *Prochloron*-like partner). Obviously, this ability from the standpoint of the *type* of storage glucan was altered during the evolution of the symbiotic course, or Chlorophytes would still be forming *phytoglycogens* rather than *true* starches.

In the progression or development of the Chlorophycean plastid from the Cyanobacterial "cyanelle" (or *Prochloron*), the next step was probably a cell of the *Nanochlorum* type. While *Prochloron* still forms highly branched phytoglycogens,[31] *Nanochlorum* seems to form *true starch* grains, at least insofar as revealed by freeze-fracture techniques.[33] True starch is a mixture of amylose and amylopectin. Amylose, the alpha-1,4-linked, relatively unbranched glucan, was also detected in Prochloron.[31]

The endosymbiotic explanation does not fit as well for the biphyletic pathway leading to the Rhodophycean line. In this pathway, the orderly "classical" slow evolutionary progression with many intermediates, leading from Cyanobacteria through the Cyanidiophyceae to the red algae, seems to be the rule. This is particularly evident from a study of the storage-glucan-synthesizing glucosyltransferase isozymes. There appears to be no discernible difference in the storage glucans formed by such blue-green algae as *Nostoc,* the Cyanidiceae (*Cyanidioschyzon* and *Cyanidium*). All are glycogenlike, highly ramified structures. These three algae also appear to have only the *b.e.* type of branching isozymes. The other member of this acidothermophilic *Cyanidiophyceae* class, *Galdieria sulfuraria,* has a branching isozyme that is of the *Q* type. This, because of its limited ability to branch only amylose to amylopectin, is probably responsible for the storage glucan formed by *Galdieria* (TABLE 3) which is similar to the Floridean starch formed by Rhodophytes. Indeed, our studies have revealed no differences between the storage glucan formed by unicellular red algae and that of this member of the Cyanidiophyceae. In all of these algae, the storage sugar is always formed *outside* of the chloroplast.

Hence, one can follow the evolution of the Rhodophyceae from the Cyanobacteria through the various "transition" forms (the Cyanidiophyceae) by a study of the storage glucans synthesized, and the glucosyltransferases responsible for their formation. The true relic, which appears to have very definitive prokaryotic, Cyanobacterial properties (TABLES 2 and 3) and very primitive eukaryotic properties, *Cyanidioschyzon merolae,* effectively can bridge the biological "discontinuity" claimed by many biologists to exist between the prokaryotic and eukaryotic forms of cellular structures.

# ACKNOWLEDGMENTS

We are indebted to Professor Paulo De Luca of the Botanical Institute, University of Naples, Italy for the cultures of *Cyanidium caldarium, Cyanidioschyzon merolae,* and *Galdieria sulfuraria.* We also thank our colleague Dr. Joseph Seckbach of the Hebrew University, Jerusalem, Israel for the electron microscopy.

## REFERENCES

1. FREDRICK, J. F. 1981. Ann. N.Y. Acad. Sci. *361:* 426-434.
2. PARTRIDGE, R. D., A. H. WEISS & D. TODD. 1972. Carbohydrate Res. **24:** 29-44.
3. DEWITT, W. 1977. Biology of the Cell. W. B. Saunders Company. Philadelphia, Pa.
4. CALVIN, M. 1969. Chemical Evolution. Oxford University Press. Oxford, England.
5. FREDRICK, J. F. 1962. Phytochemistry **1:** 152-157.
6. FREDRICK, J. F. 1968. Ann. N.Y. Acad. Sci. **151:** 413-423.
7. FLETTERICK, R. J. & N. B. MADSEN. 1980. Annu. Rev. Biochem. **49:** 31-61.
8. COHEN, P., J. C. SAARI & E. H. FISCHER. 1973. Biochemistry **12:** 5233-5241.
9. LARNER, J. & F. SANGER. 1965. J. Mol. Biol. **11:** 491-500.
10. FREDRICK, J. F. 1980. Phytochemistry **19:** 539-542.
11. FREDRICK, J. F. 1971. Phytochemistry **10:** 395-398.
12. FREDRICK, J. F. 1973. Ann. N.Y. Acad. Sci. **210:** 254-264.
13. FREDRICK, J. F. 1971. Physiol. Plant **24:** 55-58.
14. FREDRICK, J. F. & F. J. MULLIGAN. 1955. Physiol. Plant **8:** 74-83.
15. FREDRICK, J. F. 1975. *In* Isozymes IV. C. Markert, Ed.: 307-321. Academic Press. New York, N.Y.
16. DEFEKETE, M. A. R. 1968. Planta **79:** 208-213.
17. NELSON, O. E. 1973. Ann. N.Y. Acad. Sci. **210:** 113-121.
18. SECKBACH, J. & J. F. FREDRICK. 1980. Microbios **29:** 135-147.
19. FREDRICK, J. F. 1981. Phytochemistry **20:** 2353-2355.
20. LEWIN, R. A. 1984. Phycologia **23:** 203-208.
21. DE LUCA, P., R. TADDEI & L. VARANO. 1978. Webbia **33:** 37-44.
22. DE LUCA, P., R. GAMBARDELLA & A. MEROLA. 1979. Bot. Gaz. (Chicago) **140:** 418-427.
23. DE LUCA, P., A. MUSACCHIO & R. TADDEI. 1981. Giorn. Bot. Ital. **115:** 1-9.
24. FREDRICK, J. F. & J. SECKBACH. 1983. Phytochemistry **22:** 1155-1157.
25. MORETTI, A. & R. NAZZARO. 1980. Delpinoa **21:** 1-10.
26. FREDRICK, J. F. & J. SECKBACH. 1986. Phytochemistry **25:** 363-365.
27. SECKBACH, J. 1972. Microbios **5:** 133-142.
28. RAVEN, P. H. 1970. Science **169:** 641-646.
29. SECKBACH, J., J. F. FREDRICK & D. J. GARBARY. 1983. Endocytobiology II: 947-962. de Gruyter. Berlin, FRG.
30. FREDRICK, J. F. 1978. J. Therm. Biol. **3:** 1-4.
31. FREDRICK, J. F. 1980. Phytochemistry. **19:** 539-542.
32. FREDRICK, J. F. 1980. Phytochemistry. **19:** 2611-2613.
33. ZAHN, R. K. 1984. Origins of Life **13:** 289-303.
34. WHATLEY, J. M., P. JOHN & F. R. WHATLEY. 1979. Proc. R. Soc. London Ser. B. **204:** 165-187.
35. WHATLEY, J. M. 1981. Ann. N.Y. Acad. Sci. **361:** 154;165.

# Bacterial and Algal Effects on Metamorphosis in the Life Cycle of *Cassiopea andromeda*[a]

M. RAHAT

*Department of Zoology*
*The Hebrew University of Jerusalem*
*Jerusalem 91904, Israel*

D. K. HOFMANN

*Department of Zoology*
*Ruhr University*
*D-4630 Bochum, Federal Republic of Germany*

## INTRODUCTION

Considerable attention has been directed in recent years to the study of metamorphosis in marine invertebrate larvae (e.g., Reference 1). These studies were aimed at elucidating the factors inducing and releasing morphogenic processes. In the life cycle of some coelenterates two major events have been studied, i.e., the metamorphosis of sexually formed planula larvae and vegetatively formed buds into polyps, and the strobilation of medusae by such polyps.

The life cycle of the scyphozoan medusa *Cassiopea andromeda* comprises larval, polyp, bud, and medusal forms (FIGURES 1 and 2). The polyps feed on plankton and get infected by the dinoflagellate algae *Symbiodinium microadriaticum,* which are then hosted as symbiotic zooxanthellae in the polyps, buds, and medusae. *Cassiopea* can conveniently be cultured in the laboratory and is thus a suitable model for the study of morphogenic events in a life cycle.

Gohar and Eisawy noted that planulae of *C. andromeda* attached and metamorphosed into young scyphystomae only on solid surfaces but not on sandy or muddy bottoms.[2] Curtis and Cowden found that buds of *C. xamachana* settled and metamorphosed only on algae-covered surfaces but not on clean glass or plastic.[3] Hofmann *et al.* reported that settlement and metamorphosis of *C. andromeda* buds occurred in aged natural seawater but not in pasteurized or antibiotic-containing seawater.[4]

[a] Part of these studies was supported by the Deutsche Forschungsgemeinschaft.

On the basis of such observations, Hofmann *et al.* and Neuman suggested the involvement of exogenous factors of marine bacterial origin in the induction of metamorphosis in both buds and planulae larvae.[4-6]

It has also been claimed that the scyphozoan medusae *Mastigias papua*[7] and *Cassiopea andromeda*[8] will strobilate only if infected with zooxanthellae, apparently depending on a factor released by the latter.

Related phenomena have been reported for some other marine invertebrates. For the marine flatworms *Convoluta roscoffensis* it has been claimed that they are completely dependent on their symbiotic algae for "growth stimuli,"[9] and the closely related *Amphiscolops langerhansii* has been reported to achieve sexual maturity only if infected with algae.[10]

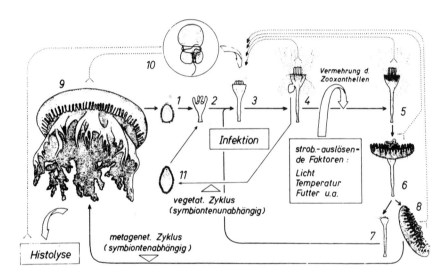

**FIGURE 1.** Life cycle of *Cassiopea andromeda*. (Reproduced from Reference 8 with permission from the publisher.)

It has long been known that external metabolites and dissolved organic material in the sea may affect marine life.[11,12] Nevertheless, the above experiments were done in natural seawater and with bacterized larvae. It is only recently that studies have been conducted considering these facts, and bacteria-free larvae in artificial seawater have been used.[13] Similarly, aposymbiotic polyps were used to reexamine the role of zooxanthellae in the life cycle of *C. andromeda*.[14]

In the following we describe studies that have critically examined effects of bacteria-derived metabolites and of endosymbiotic algae on the life cycle of *C. andromeda*. The question we asked was: Can *C. andromeda* complete its life cycle in the absence of exogenous organic matter or of endosymbiotic algae?

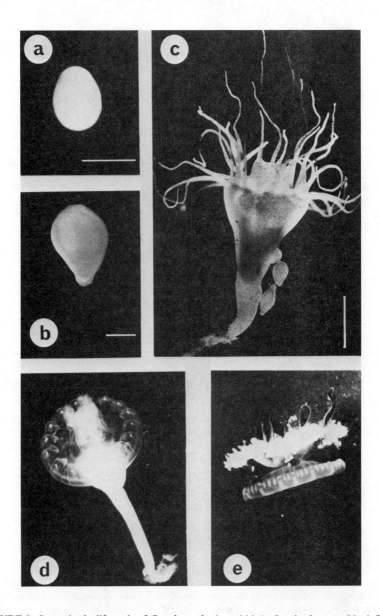

**FIGURE 2.** Stages in the life cycle of *C andromeda*. (a and b) A planulae larva and bud. Scale: 200 µm. (c) Budding polyp. Scale: 1 mm. (d) Strobilation of a medusae by a polyp. Scale: as in c. (e) Mature medusae in its normal "upside-down" position in the sea, exposing its algae-hosting oral lobes toward light. Diameter about 15 cm.

## METHODS

Sexually mature *Cassiopea* were collected in the gulf of Eilat, Israel, and egg masses were picked from between their oral lobes. Planula larvae usually hatched in the laboratory within 24-48 hours. Vegetatively formed buds were collected from polyps of *Cassiopea* cultured in the laboratory. The sexually formed larvae are free of infecting algae, and from them aposymbiotic polyps were obtained and cultured in sterilized seawater.

To critically investigate the effect of organics on the life cycle of *C. andromeda,* we used axenic planulae and buds of this medusa in synthetic sterile seawater.

Axenic planulae or buds were obtained by incubation in a mixture of antibiotics containing respectively 100 μg/ml or 500 μg/ml of neomycin penicillin and streptomycin prepared in natural seawater (NSW) or artificial seawater (ASW).

The media used were NSW, natural seawater from the gulf of Eilat; SNSW, NSW sterilized by autoclave; SASW, artificial seawater[15] adjusted with NaCl to 4.1% salinity, sterilized by autoclave. CTa to CTd were four different preparations of cholera toxin (CT) applied at 10-20 μg/ml. TSH was thyrotropine (Sigma), used at 300 μg/ml. CT, TSH, and a series of polypeptides were used as biochemical inducers. (Details of the procedures are given in References 13 and 14.)

## RESULTS

In natural seawater, 29-34% of the planulae and 7% of the buds metamorphosed into polyps within 30 days.

In SASW devoid of organic matter, axenic larvae or buds of *C. andromeda* did not metamorphose to form polyps; only partial or abnormal development was sometimes obtained. Addition of marine *Vibrio* bacteria extracts, cholera toxin, thyrotropine, and some low-molecular peptides to SASW induced 100% metamorphosis in larvae and buds respectively within 2-18 days (TABLES 1-3).

At normal seawater temperature, i.e., < 22°C, aposymbiotic polyps did not strobilate medusae. However, strobilation has been obtained at temperatures above 25°C (FIGURES 3-5).

## DISCUSSION

At first consideration the life cycles of all living organisms seem to be self-contained and self-supporting units. We intuitively assume that all information required by an organism to complete its life cycle is contained in its genome.

On closer examination, however, we find that all eukaryotic organisms are actually symbiotic communities, the cosymbionts showing various degrees of interdependence throughout their life cycle. Some participants of these symbioses we find to be completely and directly interdependent, one cosymbiont unable to live without the other, e.g., the herbivores and their intestinal flora. In other cases, however, the interde-

TABLE 1. Effect of CT and TSH on the Metamorphosis of *C. andromeda* Planulae

| Media Used | Planulae (n) | Days Watched[a] | Percent Complete Metamorphosis | Percent Partial Metamorphosis | Percent Abnormal Metamorphosis | Percent Swimming Planulae | Percent Dead Planulae |
|---|---|---|---|---|---|---|---|
| NSW | 500 | 30 | 29 | 0 | 0 | 49 | 22 |
| SNSW | 500 | 30 | 34 | 6 | 0 | 40 | 20 |
| SASW | 100 | 30 | 0 | 0 | 50 | 1 | 49 |
| SASW+CTa | 100 | 2 | 0 | 100 | 0 | 0 | 0 |
| SASW+CTb | 100 | 18 | 100 | 0 | 0 | 0 | 0 |
| SASW+CTc | 100 | 20 | 43 | 3 | 27 | 0 | 27 |
| SNSW+TSH | 70 | 3 | 11 | 89 | 0 | 0 | 0 |
| SASW+TSH | 100 | 2 | 0 | 79 | 0 | 24 | 0 |

[a] All experiments were followed for 30 days. When a lesser number is shown no change occurred after the given number of days.

TABLE 2. Effect of CT and TSH on Metamorphosis of Buds of *C. andromeda*

| Media Used | Buds (*n*) | Days Watched[a] | Percent Complete Metamorphosis | Percent Abnormal Metamorphosis | Percent Swimming Buds |
|---|---|---|---|---|---|
| NSW | 100 | 30 | 100 | 0 | 0 |
| SNSW | 100 | 30 | 7 | 93 | 0 |
| SASW | 100 | 30 | 0 | 100 | 0 |
| SASW + CTd | 100 | 5 | 24 | 0 | 76 |
| SNSW + TSH | 100 | 5 | 100 | 0 | 1 |

[a] All experiments were followed for 30 days. When a lesser number is shown no change occurred after the given number of days.

TABLE 3. Oligopeptides Tested for Their Capacity to Induce Metamorphosis in Planulae and Buds of *C. andromeda*[a]

| Peptides | Metamorphosis | |
|---|---|---|
| | Planulae | Buds |
| **Casomorphins** | | |
| Tyr-Pro-Phe | ND | − |
| Tyr-Pro-Phe-Pro | − | − |
| Tyr-Pro-Phe-Pro-Gly | + | + |
| Tyr-Ala-Phe-Pro-Gly | ND | + |
| Tyr-Ala-Phe-Pro-Met | ND | + |
| Tyr-Pro-Phe-Pro-Gly-Pro | − | − |
| Tyr-Pro-Phe-Pro-Gly-Pro-Ile | + | + |
| **Other Oligopeptides** | | |
| Gly-Pro-Ala | + | + |
| Gly-Gly-Ala | ND | − |
| Gly-Gly-Pro-Ala | + | + |
| Gly-Gly-Asp-Ala | ND | − |
| Z-Gly-Pro-Gly-Gly-Pro-Ala[b] | + | + |

[a] + = active; − = inactive; ND = no data; Z = carbobenzoxy. (Table compiled from References 17-19.)
[b] Active at $1.2 \times 10^{-5}$. Other peptides were used at various higher concentrations.

TABLE 4. Requirement of Exogenous Inducers for Metamorphoses in the Life Cycle of *C. andromeda*[a]

| Metamorphic Stage | Inducers | |
|---|---|---|
| | Bacterial/Organic | Zooxanthellae |
| Planula larvae to polyps (1) | Obligatory | Not required |
| Budding of polyps (3-11) | ND | Not required |
| Buds to polyps (11-2) | Obligatory | Not required |
| Polyp strobilation (6-8) | ND | Facilitated |
| Sexual maturation of medusae (9) | ND | ND |

[a] Numbers in parentheses indicate stages shown in FIGURE 1; ND = no data available.

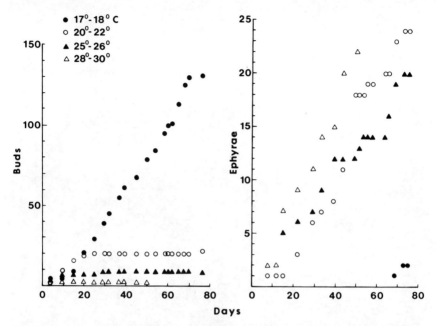

**FIGURE 3.** Effect of temperature on budding and strobilation of symbiotic *C. andromeda; n* = 7. (Reproduced from Reference 14 with permission from the publisher.)

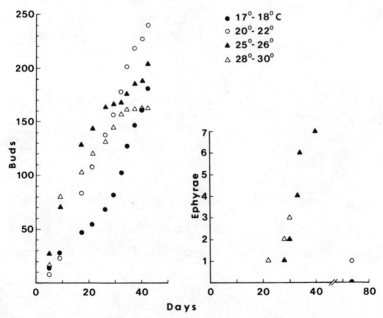

**FIGURE 4.** Effect of temperature on budding and strobilation of aposymbiotic *C. andromeda.* (Reproduced from Reference 14 with permission from the publisher.)

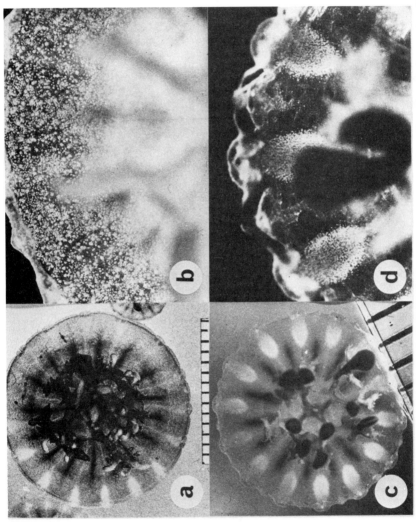

**FIGURE 5.** Symbiotic (a and b) and aposymbiotic (c and d) young medusae of *C. andromeda.* Milimetric scale in a and c. (b) Zooxanthellae in medusae are seen as white dots. (d) White dots are clusters of nematocytes. (Reproduced from Reference 14 with permission from the publisher.)

pendence is more subtle and to study it we have to completely separate the cosymbionts from each other and from the surrounding biosphere.

In spite of interdependences found, we must assume that originally all organisms had a full complement of genetic capacity that enabled them to complete their life cycle in a given environment. If a genetic capacity of a species however is not used or required in its environment, it might be lost by mutation and selection. Such a case of loss of genetic capacity has been reviewed by Fairbain.[16] We might envisage such a situation in the sea which is abundant with organisms that excrete their metabolites into the surrounding water, and these metabolites are then degraded by bacteria. Mutants of any organism that lose a factor required in their life cycle will not be selected against, if they can take up a replacing mimicking factor from the sea around them.

The variety of metabolites and oligopeptides inducing metamorphosis in *C. andromeda* show that it is not a specific inducer that is required. We do not know yet the limits of molecular variation that will induce metamorphosis.

As *C. andromeda* planulae or buds cannot metamorphose in the absence of exogenous metabolites, we may conclude that this medusa has lost a genetic capacity required for the completion of its life cycle, i.e., to induce or release metamorphosis in its planulae and buds.

Temperatures required for strobilation of aposymbiotic polyps of *C. andromeda* were higher than those required for strobilation of symbiotic polyps. We may thus assume that strobilation requires enhanced metabolism. Such enhancement is probably obtained through a factor supplied by the symbiotic zooxanthellae.

The facilitating effect of endosymbiotic zooxanthellae is by definition a "symbiotic" effect. Will we define as such also the exogenic effect of bacteria and their metabolites?

Our data leave open two questions: (1) How far can we generalize on our results showing that certain marine larvae have lost some genetic capacity and are completely dependent on nonnutritive exogenous factors abundant in the sea for the completion of their life cycles? 2. What are the receptors and by what pathway do the inducing factors affect metamorphosis?

Both questions warrant further research.

## ACKNOWLEDGMENTS

Drs. M. Wolk and W. K. Fitt and Ms. O. Adar participated in various stages of this study.

## REFERENCES

1. CHIA, F. S. & M. E. RICE, Eds. 1978. Settlement and metamorphosis of marine invertebrate larvae. Elsevier. Amsterdam, Holland.
2. GOHAR, H. A. F. & A. M. EISAWY. 1960. The development of *Cassiopea andromeda* (Scyphomedusae). Publ. Mar. Biol. Stn. Ghardaqua **11:** 148-190.
3. CURTIS, S. K. & R. R. COWDEN. 1971. Normal and experimentally modified development in buds of *Cassiopea*. Acta Embryol. Exp. **3:** 239-259.

4. HOFMANN, D. K., R. NEUMANN & K. HENNE. 1978. Strobilation and initiation of scyphystoma morphogenesis in the rhizostome *Cassiopea andromeda* (Cnidaria, Scyphozoa). Mar. Biol. **47:** 161-176.

5. NEUMANN, R. 1979. Bacterial induction of settlement and metamorphosis in the planula larva of *Cassiopea andromeda* (Cnidaria, Scyphozoa, Rhizostomae). Mar. Ecol. Prog. Ser. **1:** 21-28.

6. NEUMANN, R. 1980. Bakterielle Metamorphoseauslösung und kontrolle der Morphogenese bei Schwimmknospen und Planulalarven von *Cassiopea andromeda* (Cnidaria, Scyphozoa). Ph.D. Thesis. Cologne University. Cologne, FRG.

7. SUGIURA, Y. 1964. On the life history of rhizostome medusae. II. Indispensability of zooxanthellae for strobilation in *Mastigias papua*. Embryologia **8:** 223-233.

8. LUDWIG, F. D. 1969. Die Zooxanthellen bei *Cassiopea andromeda* Eschscholz 1829 (Polypstadium) und ihre Bedeutung fur die Strobilation. Zool. Jb. Abt. Anat. Ontog. Tiere **86:** 238-277.

9. PROVASOLI, L., T. YAMASU & I. MANTON. 1968. Experiments on resynthesis of symbiosis in *Convoluta roscoffensis* with different flagellate cultures. J. Mar. Biol. Assoc. U.K. **48:** 465-479.

10. TAYLOR, D. L. 1971. On the symbiosis between *Amphidinium Klebsii* (Dinophyceae) and *Amphiscolops langerhansii* (*Turbellaria: Acoela*). J. Mar. Biol. Assoc. U.K. **51:** 301-313.

11. LUCAS, C. E. 1949. External metabolites and ecological adaptation. Symp. Soc. Exp. Biol. **3:** 336-356.

12. LUCAS, C. E. 1961. On the significance of external metabolites in ecology. Symp. Soc. Exp. Biol. **15:** 190-206.

13. WOLK, M., M. RAHAT, W. K. FITT & D. K. HOFMANN. 1985. Cholera toxin and thyrotropine can replace natural inducers required for the metamorphosis of larvae and buds of the scyphozoan *Cassiopea andromeda*. Roux' Arch. Dev. Biol. **194:** 487-490.

14. RAHAT, M. & O. ADAR. 1980. Effect of symbiotic zooxanthellae and temperature on budding and strobilation in *Cassiopea andromeda* (Eschscholz). Biol. Bull. **159:** 394-401.

15. SPECHT, D. T. & W. E. MILLER. 1974. Modified Burkholder's artificial sea water. *In* Proceedings of a Seminar on the Methodology of Monitoring Marine Environments: 194-230. U.S. Environmental Protection Agency. Washington, D.C.

16. FAIRBAIRN, D. 1970. Biochemical adaptation and loss of genetic capacity in helminth parasites. Biol. Rev. **45:** 29-72.

17. BRAND, U. 1984. Bakterielle und enzymatische Abbauprodukte komplexer Proteine als Ausleser der Metamorphose von *Cassiopea andromeda*. Diplomarbeit. Bochum, FRG.

18. NAUST, G. 1985. Auslösung der Polypenmorphogenese bei *Cassiopea andromeda* (Scyphozoa): Zur Induktionswirkung von Caseinen und deren proteolytischen Spaltprodukte. Staatsexamensarbeit. Bochum, FRG.

19. FITT, W. K. & D. K. HOFMANN. 1985. Chemical induction of settlement and metamorphosis of planulae and buds of the reef-dwelling coelenterate *Cassiopea andromeda*. *In* Proceedings of the Fifth International Coral Reef Congress, Tahiti, **5:** 239-244.

# Cell Division Cycles and Circadian Clocks[a]

## Modeling a Metabolic Oscillator in the Algal Flagellate *Euglena*

LELAND N. EDMUNDS, JR.,[b] DANIELLE L. LAVAL-MARTIN,[b,c] AND KEN GOTO[b,d]

[b]*Department of Anatomical Sciences*
*School of Medicine*
*Health Sciences Center*
*State University of New York*
*Stony Brook, New York 11794*

[c]*Laboratory of Membrane Biology*
*University of Paris VII*
*75005 Paris, France*

[d]*Biological Laboratories*
*Obihiro University of Agriculture and Veterinary Medicine*
*Obihiro*
*Hokkaido 080, Japan*

Mechanisms proposed to control microbial cell division cycles (CDC) often include the notion of timers and clocks in the regulation of various pathways and sequences that culminate in mitosis and cell division.[1-6] Inasmuch as mitosis is a periodic event of short duration relative to the total length of the CDC, it is not surprising that various types of *autonomous* biochemical and macromolecular oscillators have been proposed to underlie this and other "landmarks" comprising the CDC.[4,7] These range from those of the relaxation type, in which a single continuous variable accumulates or declines in the cell, triggering some event in the CDC when it reaches a critical threshold and then resetting to a baseline so that the cycle starts again,[8] to those exhibiting limit cycle dynamics.[2,5,9,10] In the latter, a central clock, characterized by a self-sustained oscillation of at least two continuously varying biochemical species, would maintain stable periodic behavior and coordinate the timing of events comprising the CDC despite transient perturbations in phase; it would continue to function even if mitosis and division were blocked. Although the majority of these putative "clocks" have periods in the neighborhood of minutes to hours, they well may be relevant to circadian oscillators (having longer period lengths of about 20 to 28 hours) and to the regulation of circadian rhythms of cell division frequently observed in cell pop-

[a]Supported in part by National Science Foundation grants (PCM-8204368 and DCB-8501896) to L. Edmunds.

ulations.[7] Our long-range goal is to determine the biochemical nature of the oscillator(s) presumed to couple to and interact with the CDC and to generate these overt rhythmicities.

As a working hypothesis, based in part on results previously obtained by one of us[11,69] for the duckweed *Lemna gibba* G3 and on recent findings in the algal flagellate *Euglena gracilis* Klebs (strain Z), we have proposed that $NAD^+$, the mitochondrial $Ca^{2+}$-transport system, $Ca^{2+}$, calmodulin, and $NADP^+$ phosphatase and $NAD^+$ kinase together could constitute a light-entrainable, self-sustaining circadian oscillator.[12] These elements would constitute "gears" of the clock underlying the overt rhythm of cell division (and of other variables, or "hands") in eukaryotic microorganisms. This notion is supported also by much evidence that calmodulin and mitochondrial calcium transport play a pivotal role in cellular regulation.[13-18]

We have chosen *Euglena* (a facultative photoautotroph) to test this model so as not to have to confront the complexities of higher organisms and levels of organization. This eukaryotic, unicellular system is particularly useful in that it is well characterized physiologically and biochemically[19] and has been shown to display many other circadian periodicities,[20,21] including the extensively documented rhythm of cell division,[7] our primary overt reference cycle. We are thus in an advantageous position to study the nature and interaction of CDC oscillators and circadian clocks at both the formal and metabolic levels in a well-defined cellular system.

## CIRCADIAN CONTROL OF THE CELL DIVISION CYCLE IN *EUGLENA*

The CDCs of many algae, fungi, and protozoans exhibit persisting circadian rhythms of cell division, or "hatching."[4,22] Division of cells occurs only at a certain phase of the circadian cycle—often the times ("subjective" nights) in continuous darkness (DD) or illumination (LL) corresponding to the dark intervals in an environmentally synchronizing light-dark (LD) cycle. This "gating" phenomenon, reflecting an interaction between a circadian oscillator and the CDC, perhaps has been most intensively investigated in the eukaryote *Euglena*. The *formal* properties of circadian clocks—particularly entrainability, persistence, phase shiftability, and temperature compensation—have been found to characterize the circadian rhythm of cell division in this organism. Inasmuch as this periodicity will serve as the overt reference rhythm for our biochemical analysis of the underlying oscillator, we will now review the major lines of evidence that implicate an autonomous clock in the control of the CDC in this unicell.[7]

### *Entrainability*

Photoautotrophically grown cultures of wild-type *Euglena* (Z strain) can be routinely synchronized (entrained) by 24-hour LD cycles so that cell division synchrony is confined entirely to the dark intervals.[23,24] If one reduces the duration of the light interval (e.g., LD: 8,16) within a 24-hour framework, the amplitude of the division rhythm is reduced, and the average doubling time ($g$) of the culture is lengthened.[25]

Not every cell divides during each division burst; nevertheless, cell divisions, when the do occur, do so only during the dark interval each 24 hours, and the culture continues to be synchronized in the sense of event simultaneity,[22] in sharp contrast to the exponential growth curve obtained in LL having a g of 12-14 hours.[24]

Similarly, "skeleton" photoperiods comprising the framework of a full-photoperiod cycle (for example, LD: *3*, 6, *3*, 12) will also entrain the rhythm to a precise 24-hour period; divisions are confined to the main dark intervals.[25] Entrainment by non-24-hour LD cycles[25,26] (but only within certain limits) and by diurnal temperature cycles[27] may also occur. Finally, even more conclusive evidence for the role of a circadian oscillator in the control of the CDC has been obtained by synchronizing photosynthetic mutants (all obligate heterotrophs) of *Euglena*[22,28] with diurnal LD cycles, providing that the average g is longer than 24 hours. These studies, therefore, have effectively circumvented the problem of the dual use of imposed LD cycles: as an energy source, or "substrate," for growth, on the one hand, and as a timing cue (*Zeitgeber*) for the underlying clock, on the other.

### Persistence

A basic test for the existence of a circadian rhythm (*sensu stricto*) is to determine whether it will continue to free run for a number of cycles following transfer of the organism to conditions held constant with respect to the major environmental *Zeitgeber* (light, temperature); characteristically, the period ($\tau$) under such conditions only approximates 24 hours, as might be expected of an imperfect biological clock. Indeed, rhythmic cell division has been found to persist for many days in the Z strain of *Euglena* autotrophically grown in dim LL[29] and in photosynthetic mutants[30,31] organotrophically batch cultured or continuously cultured in DD or LL.

We have also observed "free-running" circadian rhythms of division in a variety of higher-frequency LD cycles[25,32] and even "random" illumination regimes,[7,33] although we do not exclude the possibility that such short-period cycles ($T \ll 24$ hours), providing no information to the cells with regard to 24-hour time, could modulate the period to some extent.[34] Those high-frequency regimens (see FIGURE 1) having symmetric photo-and scotophases (e.g., LD: 1,1 or LD: 3,3) and properly constructed random regimes have proved particularly useful in that they afford an amount of light during a 24-hour time span identical to that received in a full-photoperiod LD: 12,12 entraining reference cycle ($T = \tau = 24$ hours) yet elicit free-running circadian rhythms ($T \ll \tau \simeq 24$ hours) not only of cell division (see FIGURE 1), but also of motility[35] and of photosynthetic capacity and chlorophyll content.[32,36] With these exotic cycles, therefore, the total duration of light afforded for growth is held constant (reflected in the comparable step sizes of the growth curves), while the signaling information inherent in the light perturbations can be varied to manipulate the light-sensitive circadian oscillator.

### Phase Shiftability

Another basic property of circadian rhythms is that their phase can be reset (or shifted)—as the result of the lengthening or shortening of the period of one or more oscillations—by single light (or dark) or temperature signals. Characteristically, both the sign and magnitude of phase shifts ($\Delta\phi$'s) engendered by the signal are predictably dependent on the subjective circadian time (CT) at which the perturbations are applied. Similarly, the phenomenon of phase perturbation is inherent in most, if not all, models for CDC oscillators: the response of cells to external influences is often strongly dependent on the time of the CDC at which the agent is imposed.[37–43.]

Recently, we have derived a detailed, Type 0 ("strong") phase-response curve (PRC) for a *circadian* mitotic clock by utilizing photoautotrophic cultures of *Euglena* free running in a high-frequency LD: 3,3 cycle.[44] At different times throughout the 30.2-hour CDC, 3-hour light perturbations were imposed systematically during one of the intervals when dark would have fallen (FIGURE 1). Using the onset of division as the phase reference point ($\phi_r$), the net steady-state phase advance ($+\Delta\phi$) or delay ($-\Delta\phi$) of the rhythm was determined relative to an unperturbed control culture. Both $+\Delta\phi$ and $-\Delta\phi$ were found, with maximum values of about 11-12 hours being obtained at CT 22-23 (the "breakpoint"); little, if any $\Delta\phi$ occurred if the light signal were given between CT 6 and CT 12.

The CT at which delay or maximum $+\Delta\phi$'s were achieved corresponded, respectively, to the approximate position of the CDC during which division occurred (commencing at about CT 12) and to the first few hours of the $G_1$ phase when the free-running rhythm of photosynthetic capacity[32,36] displayed the lowest values. Although light is needed as a "substrate" for photosynthesis and the progression of the CDC in photoautotrophically cultured *Euglena*, it serves a quite distinct and separable function in phase shifting and entraining the circadian oscillator underlying the rhythm of cell division, whose cyclic sensitivity to light is reflected in the PRC. The $\Delta\phi$'s observed constitute true developmental advances (and delays) since a higher cell titer was attained before the same cell concentration was reached in the unperturbed control (while the converse held for phase delays).

### Singularity Point

The oscillatory motion of a biological clock can be represented on a plane as a stable trajectory (comprising two first-order differential equations with two state variables) that closes in on itself, termed a limit cycle.[45] Such a system, if disturbed, will always tend to return to an equilibrium configuration. Theoretical studies have further predicted that a circadian oscillator might be rendered arrhythmic—characterized by a phaseless, motionless state—by a critical pulse of a certain strength and duration given at a specific time (termed the "singularity point") in the circadian cycle[46] (see Winfree).[47] As stimulus strength is increased, the transition from Type 1 (weak pulse) to Type 0 (strong pulse) resetting is necessarily discontinuous at one special phase point ("breakpoint") corresponding to this unique singularity. This prediction has now been demonstrated for light perturbations in several multicellular circadian systems, and for critical pulses of anisomycin in the dinoflagellate *Gonyaulax*.[48]

Similarly, we have recently discovered that a 40- to 400-lux pulse given at CT 0.4 (the approximate location of the breakpoint, at about CT 23) during the free-running rhythm of cell division in *Euglena* induced arrhythmicity,[49] the population reverting to asynchronous, exponential growth. The intensity ($I^*$) of this annihilating pulse and the CT at which it was imposed were found to be quite specific: a 300-lux stimulus given at CT 21.5 merely generated a $-\Delta\phi$ of the same magnitude found for 7500-lux signals. Different degrees of asynchrony were observed as one approached the

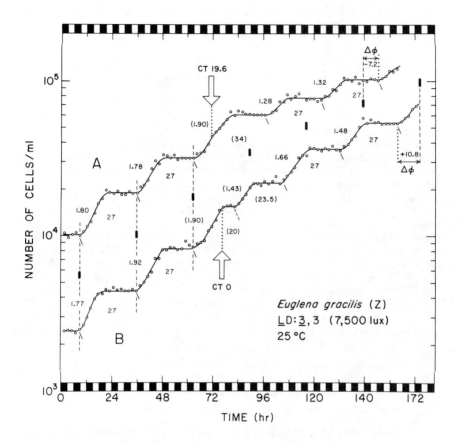

**FIGURE 1.** Examples and methodology used in the calculation of phase-shift delays ($-\Delta\phi$) and advances ($+\Delta\phi$) of the free-running circadian rhythm of cell division obtained in cultures of *E. gracilis,* grown photoautotrophically in LD: 3,3 (L = 7500 lux) at 25°C and perturbed by 2- to 3-hour light signals (7500 lux, given by deleting one of the normal dark intervals) or pulses of various chemicals given at different circadian times. Two in-phase, synchronously dividing cultures, at approximately the same titer (ordinate) and exhibiting a stable period ($\tau$) of 27 hours (vertical dashed lines) over a time span of at least 72 hours (abscissa), were subjected to a perturbation whose midpoint occurred at either hour 70.5 or hour 76.5, corresponding to circadian times (modulo 24) CT 19.6 or CT 24.0 (=CT 0), respectively. Onsets of cell division were taken as phase-reference points ($\phi_r$) and considered to fall at CT 12.0, corresponding to onset of darkness in LD: 12,12. After transients had subsided and free-running $\tau$ of 27 hours had been reestablished, the difference in phase ($\Delta\phi$, here given in real time) between the shifted rhythm (vertical dotted lines) and that projected (solid markers) for the rhythm before it was perturbed was measured and taken to be the steady-state phase shift engendered by the light signal or chemical pulse. (See References 7, 12, and 44.)

boundaries (CT,lux). Thus, the existence of this "critical pulse" and its corresponding singularity point not only further supports the hypothesis that a circadian oscillator regulates the CDC in *Euglena* but also suggests (though it does not demand) that the pacemaker may have limit cycle dynamics.

## Temperature Compensation

A final, and rather remarkable, property of circadian rhythms is that their period—but not their amplitude—is only slightly affected by the ambient steady-state temperature over the physiological range.[50] In fact, this is just what one would anticipate in a functional biological clock measuring astronomical time. In contrast, the duration of the CDC (that is, *g* ) is commonly thought to be highly temperature dependent, and, indeed, this is true for *Euglena* also.[27]

Several lines of evidence, however, suggest that the *period* of the endogenous oscillator hypothesized to underlie *rhythmic* cell division in a population of cells is conserved. For example, Klevecz and King have found that the $Q_{10}$ for cell division of the V79 line of Chinese hamster lung fibroblasts growing between 34°C and 40°C was between 1.15 and 1.26,[51] thus indicating that the mammalian CDC is temperature compensated over a span of 6°C to 7°C; and we have observed that $\tau$ of the free-running rhythm of cell division in the $P_4ZUL$ mutant of *Euglena* grown in DD is temperature compensated over a range of 7°C (14-21°C).[4] Finally, in a more extensive and rigorous comparative study of temperature compensation of $\tau$ in the Z and ZR strains of *Euglena* maintained in LD: 3,3, we have found a $Q_{10}$ of 1.05 in the former,[52] indicating that it is virtually unaffected by changes in temperature over a 10°C range (22-32°C).

## Other Considerations

This short review has demonstrated the key role of a circadian oscillator(s) in the control of the CDC in *Euglena,* taken to be representative of other eukaryotic microorganisms (and, perhaps, multicellular systems). Mitosis would not be an essential part of the oscillator but would lie downstream from it: blockage of cell division should not stop the system from oscillating.[4,7] Thus, cell division would be a "hand" of the underlying clock. We have recently tested this hypothesis in two ways: (1) if the division rhythm (free running in LD: 3,3) was stopped due to low initial levels of vitamin $B_{12}$, and if this inhibition subsequently was released by readdition of $B_{12}$, the cell division rhythm started up again in phase with an unperturbed control; and (2) if a pulse of lactate was given to a free-running culture, temporarily accelerating the CDC and overriding circadian oscillator controls,[30] the phase of the rhythm when it was finally restored after the substrate had been depleted was in phase with that of an unperturbed control. These results are consistent with those found previously for the in-phase restoration of rhythmicity in the $P_4ZUL$ mutant free running in LL by the addition of sulfur-containing compounds to the medium.[31]

Another question arises from the phase-shifting data reflected in the experimentally derived PRC for light signals in *Euglena:*[44] How are the observed shortenings and lengthenings of individual CDCs generated by a master (circadian) oscillator at the biochemical and molecular level? The evidence reviewed earlier *formally demands* that a clock of some sort predictably insert time segments into the CDC (time "dilation"), or delete them from it (time "contraction"). One way that the CDC might be "programmed" would be for a collection of timing loops of different lengths to couple together in various combinations to form a flexible timer, or "cytochron."[4] Temporal loci, or control points, would exist along the cytochron track at which decisions would be made with respect to the addition or deletion of time loops by interaction with a circadian oscillator. Thus, the cytochron and the circadian clock are posited to be functionally independent (although not necessarily entirely separate as to mechanism).

# BIOCHEMICAL ANALYSIS OF THE CIRCADIAN OSCILLATOR

In the previous subsection extensive evidence was documented implicating a circadian oscillator in the generation of the rhythm of cell division in *Euglena*—the overt rhythm that we have used as a reference point in our planned biochemical attack on the underlying mechanism. This section gives the background for this approach and summarizes the results obtained with this unicellular *Euglena* system.[12,53]

## General Background

The biochemical analysis of ultradian and circadian clocks[54] has proceeded along two lines of attack: (1) The assay of a variety of biochemical variables to ascertain whether they exhibit rhythmicity. The rationale here is to identify the immediate cause of an overt rhythm, such as luminescence or photosynthesis, and then to continue this analysis step by step in an attempt to thread one's way back through the biochemical pathways mediating the rhythm—the so-called transducing mechanisms[55]—until one arrives at their point of coupling to the oscillator. (2) To administer a variety of drugs or inhibitors (such as valinomycin, an ionophore, or cycloheximide, an inhibitor of 80S protein synthesis) to a rhythmic system to determine if they affect the oscillator, as reflected by a shift in the phase ($\Delta\phi$) of the overt rhythm or by an alteration in its period length. In this case, the rationale is that the known specific cellular target of the drug may be a part of the clock, or at least intimately associated with it.

With a few exceptions, however, most attempts to specifically perturb the oscillator and to elucidate its biochemistry have been confounded by the difficulties inherent in distinguishing between the so-called hands of the clock (clock-mechanism-irrelevant events) and the "gears" themselves (clock-relevant processes) due to possible side effects or chains of secondary effects induced by the perturbing agent.[56] Thus, a drug pulse might affect the clock only indirectly through some unknown pathway rather than directly via its known target; similarly, rather than being the direct result of drug inhibition, the observed effects of a perturbation on an overt rhythm might occur merely as a secondary consequence of inhibition. Some important criteria for ascertaining the role of a given biochemical component in circadian clock function have been discussed.[57,58] In a further attempt to circumvent the problem of secondary effects of drug pulses used to test our hypothesis, we have introduced the rationale that in order for the target of a perturbing drug (or other agent) to be designated an integral element of a circadian clock, it first must be demonstrated that transitorily and directly increasing or decreasing the level of the component (by activation or inhibition of some relevant process) should *each* cause a $\Delta\phi$ in overt rhythmicities driven by the clock.[12]

## The Rationale

An oscillator can be expressed mathematically as a set of differential equations comprising both state variables and parameters. The state variables characterize the state of the oscillation, with each set of values defining each phase of the oscillation. The parameters are constants constraining the manner in which the state variables change and determining the dynamics of the oscillations; a different set of parameter values gives a different solution of the rate equations. Any transitory alteration or perturbation of either the state variables or of the parameters can cause a permanent $\Delta\phi$ in an overt rhythm but has no permanent effect on its $\tau$; in contrast, permanent changes in parameter values can alter the $\tau$ of the oscillation.[59]

As was noted earlier, one of the most important drawbacks inherent in the phase-shift experiments widely used in determining a state variable or a parameter has been that we do not know whether the $\Delta\phi$ occurred as a result of an effect on the postulated primary target or rather on some other site secondarily affected by the drug. We have now operationally designated any element as a "gear" (G) of a circadian clock if it can be expressed as a state variable or a parameter; if not, it is a 'nongear" ($\sim$G), that is to say, a "hand" or other mechanism-irrelevant element. Together, the ensemble of gears would constitute a closed control loop, or oscillator.

Because an unperturbed $\sim$G in its normal or physiological oscillatory range would not be expected to regulate the operation of the G's themselves, its artificial perturbation within this range should not perturb circadian timekeeping (no steady-state $\Delta\phi$ in an overt rhythm should be observed). [A $\Delta\phi$ might occur if the level of the input of $\sim$G were so high or low (that is, outside of the normal range of its oscillation) that it limited the rate or otherwise affected the normal operation of the G's comprising the oscillator. In this latter case, even if both high- and low-level inputs of $\sim$G perturbed the overt circadian rhythm, they would probably be affecting different gear elements of the clock.] Consequently, if an experimental alteration in the level of a target within its normal range perturbs the clock and generates steady-state $\Delta\phi$'s, then that target is most likely a G (criterion A). It is conceivable, however, that the activation and resulting increase in the level of a $\sim$G might perturb timekeeping, whereas its inhibition would not (or vice versa). For example, although the inhibition of protein synthesis can perturb the clock in several organisms,[60] its activation in both *Neurospora*[61,62] and *Euglena*[63] does not; likewise, experimentally increasing (but not decreasing) the levels of cyclic AMP alters the period of the rhythm of conidiation in *Neurospora*.[64,65] To differentiate more stringently between G and $\sim$G, therefore, a further requirement for a target to be classified as a G is that both the direct activation as well as direct inhibition of the target must perturb the clock (criterion B).

Sometimes it will be impossible to attack a presumed target directly. Phase shifts generated by several drugs that only secondarily activate and inhibit a target, however, do not demonstrate necessarily that such a target is a G. Nevertheless, if a given target (B) is regulated by another target (A) and, in turn, regulates yet a third target

(C), and if it is probable that A and C are gears, then it is likely that B is one also. In contrast, if both targets A and C are known to be hands ($\sim$G) of the oscillator, then target B, of course, is also a $\sim$G. Only in the case where target A is shown to be a G and target C is proven to be a $\sim$G can target B not be classified.

Despite many experiments utilizing chemical perturbations, few, if any, targets that can be classified as a G on this rationale have been reported thus far. We now turn to our hypothesis that $NAD^+$, the mitochondrial $Ca^{2+}$-transport system, $Ca^{2+}$-calmodulin, $NAD^+$ kinase, and $NADP^+$ phosphatase together might constitute a self-sustained oscillating loop and will proceed to test this model in *Euglena* according to our criteria, utilizing the well-characterized rhythm of cell division as our primary reference cycle. In principle, this rationale is applicable to any set of processes thought to constitute an autonomously oscillatory clock.

### Oscillating Elements in Euglena

The *in vivo* levels of $NAD^+$, $NADP^+$, and NADPH, which were measured in synchronously dividing and in very slowly dividing cultures of *Euglena* photoautotrophically batch cultured at 25°C and free running in LD: 3,3 following transfer from LL (see FIGURE 1), oscillated with a circadian period (27 hours) but 180° out of phase with each other (FIGURE 2). The amplitude of the oscillations, showing as much as an 80% increase between minimum and maximum levels, was not directly related to the CDC inasmuch as similar values were found in cultures in which the entire population doubled during each circadian cycle and in cultures that had attained the stationary (infradian) growth phase, where only 14% (FIGURE 2) or less of the cells divided. Finally, a 3-hour light pulse applied at CT 18 to the free-running rhythm of cell division not only generated the expected $\Delta\phi$ in the division rhythm (see FIGURE 1) but also shifted the phase of the oscillation in total $NAD^+$ content by approximately the same extent.

There was also a circadian rhythm in the activity of $NAD^+$ kinase (peak at CT 0) in desalted extracts of *Euglena* with a phase relationship such that it could induce the rhythm in the *in vivo* level of $NAD^+$ (FIGURE 2). Taken together with the observation by Brinkmann (see Edmunds)[21] that neither the ratio of NADH to ($NAD^+$ + NADH) nor that of NADPH to ($NADP^+$ + NADPH) oscillates when the circadian clock is operating in *Euglena,* these findings suggest that the circadian oscillation in the *in vivo* level of $NAD^+$ could be ascribable, as for *Lemna,*[11] to that of the conversion between $NAD^+$ and $NADP^+$ or between NADH and NADPH (that is, to the activity of $NAD^+$ kinase or $NADP^+$ phosphatase, or both), but not to that of reduction-oxidation between $NAD^+$ and NADH.

### Are NAD⁺, NAD⁺ Kinase, and NADP⁺ Phosphatase "Gears"?

To determine whether $NAD^+$ constitutes a true clock G, small (25 ml) seed cultures displaying a free-running circadian rhythm of cell division were pulsed for 2 hours (see FIGURE 1) at various CTs with either $NAD^+$ (0.5 mM) or $NADP^+$ (0.2 mM)—treatments expected to elevate their own *in vivo* levels directly and then to increase or decrease the rate of the reactions catalyzed by $NAD^+$ kinase and

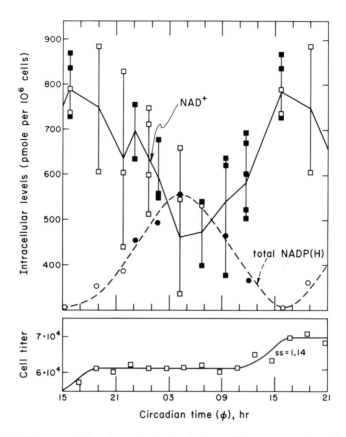

**FIGURE 2.** Upper panel: circadian variations in the intracellular content of NAD$^+$ and total NADP(H) (expressed as pmole per $10^6$ cells) in very slowly dividing autotrophic cultures of *E. gracilis* maintained in LD: 3,3 regimens. Two out-of-phase, free-running cultures (open and filled squares) were sampled at different circadian times, and their NAD$^+$ contents were spectrophotometrically measured using an enzymatic assay (2 to 5 determinations for each time point). The curve connects the mean values of all data points obtained at a given CT ($\phi$). Similarly, total NADP(H) was determined (open and filled circles). CT 0 indicates the phase point of a free-running rhythm that has been normalized to 24 hours and occurs at the onset of light in an LD: 12,12 reference cycle. Lower panel: representative growth curve for one of the two, very similar, analyzed cultures. Cell titer is plotted (on a log scale) as a function of CT ($\phi$). Note: (a) the very small factorial increase ($ss = 1.14$) indicating that only 14% of the free-running population divided during the cycle; (b) the amount of intracellular NAD$^+$ increased by 70% during this fission interval. (See Reference 12.)

NADP$^+$ phosphatase—or *p*-nitrophenylphosphate (pNPP, 2 mM), a competitive inhibitor of NADP$^+$phosphatase, and then resuspended in fresh medium (effectively terminating the pulse by dilution) for subsequent automated monitoring of the cell division rhythm. As shown in FIGURE 3 (top panel), pulses of each of these compounds were able to generate steady-state $\Delta\phi$'s of the division rhythm whose sign and magnitude were dependent upon the CT at which the pulse was applied.[12] It is not conceivable that these $\Delta\phi$'s were caused primarily by the perturbation of the CDC

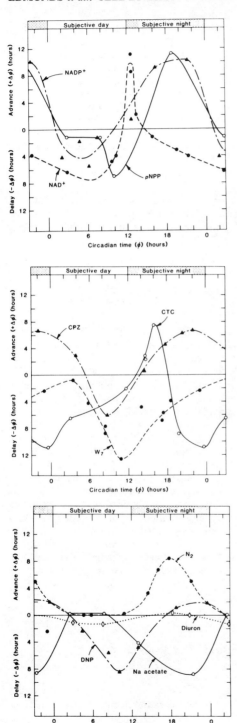

**FIGURE 3.** Phase response curves (PRCs) for the effect of pulses of different compounds on the free-running rhythm of cell division in photoautotrophic cultures of *E. gracilis* in LD: 3,3. Steady-state phase shift ($\pm \Delta\phi$) is plotted as a function of the midpoint of the perturbation (CT [$\phi$], normalized to 24 hours). The number of data points for a given PRC was varied from 5 to 10, depending on the complexity of its waveform; assays were clustered in instances of apparent sharp discontinuities (e.g., top panel: $NAD^+$, CT 12) or to insure that experiment-to-experiment variation was small (e.g., middle panel: W7, CTC). (Top: pulses of $NAD^+$ (0.5 mM, 2.3 hours) (filled circles), of $NADP^+$ (0.2 mM, 2 hours) (triangles), and of *p*-nitrophenylphosphate (pNPP; 10 mM, 2 hours) (open circles) applied during a light interval of the LD: 3,3 cycle. Middle: pulses of W7 (20 $\mu$M, 2.3 hours) (filled circles) and of chlorpromazine (CPZ; 50 $\mu$M, 2.3 hours) (triangles) applied during a light interval of the LD: 3,3 cycle, and of chlortetracycline (CTC; 200 $\mu$M, 3 hours) (open circles) applied during a dark interval. Bottom: pulses of nitrogen ($N_2$; 600 ml minute$^{-1}$, 3 hours) (filled circles) applied during a dark interval of the LD: 3,3 cycle; and of sodium acetate (10 mM, 2 hours) (open circles), of dinitrophenol (DNP; 100 $\mu$M, 2 hours) (triangles), and of diuron (DCMU; 10 $\mu$M, 2 hours) (diamonds) applied during a light interval. (Modified from Reference 12.)

itself, inasmuch as the latter has been shown to be driven, or phased, by the circadian clock.[7]

Furthermore, we have monitored the transitory change in the *in vivo* level of $NAD^+$, NADH, $NADP^+$, and NADPH at 30-minute intervals in a free-running culture of *Euglena* following a pulse of $NAD^+$ (0.5 mM, 2.5 hours) given at CT 21.7, which caused a 4-hour (delay) $\Delta\phi$ in the overt circadian division rhythm. Although the *in vivo* level of $NAD^+$ in an unperturbed, control culture decreased during this time span, the level in the pulsed cells increased over the next hour to a value slightly lower than that of the maximum level attained by the oscillation in unperturbed cultures, and finally decreased. Similar variations were found in $NADP^+$ and NADPH contents *in vivo,* all levels (as for $NAD^+$) falling within the normal range characterizing unperturbed control cultures, whereas that of NADH did not change significantly.

Taken together, these results satisfy criteria A and B which were established in our rationale and suggest that $NAD^+$ (or $NADP^+$, or NADPH, or any combination thereof), $NAD^+$ kinase, and $NADP^+$ phosphatase represent Gs of the underlying circadian oscillator.

### *Are $Ca^{2+}$ and $Ca^{2+}$-Calmodulin "Gears"?*

What, then, is the element that regulates $NAD^+$ kinase and $NADP^+$ phosphatase, already suggested to be Gs in themselves? If these elements are Gs, then the element regulating them should be a G also. $Ca^{2+}$-calmodulin activates $NAD^+$ kinase in many plants, including green algae,[66,67] and in the sea urchin[68] and the circadian rhythms in the activities of $NAD^+$ kinase and $NADP^+$ phosphatase appear to be generated by a rhythm in the *in vivo* level of this complex in *Lemna.*[11,69] Thus, $Ca^{2+}$-calmodulin would seem to be a likely candidate for this G in *Euglena.*

We attempted to cause a transitory decrease in $[Ca^{2+}]$ *directly* by means of 2- to 3-hour pulses of chlortetracycline (CTC, 200$\mu$M), a membrane-permeable chelator of $Ca^{2+}$, and to transitorily inhibit $Ca^{2+}$-calmodulin by similar short pulses of the calmodulin inhibitors W7 (20 $\mu$M) and chlorpromazine (CPZ, 50 $\mu$M).[12] These agents also yielded pronounced $\Delta\phi$'s of the cell division rhythm (FIGURE 3, middle panel). Although for technical reasons we have not been able to test the effects of a direct increase in $[Ca^{2+}]$, $\Delta\phi$'s in the overt rhythm were also obtained with the secondary increases in $[Ca^{2+}]$, and the ensuing tertiary activation of $Ca^{2+}$-calmodulin, resulting from nitrogen and dinitrophenol (DNP) pulses (see FIGURE 3, lower panel). These results suggest, therefore, that both cytosolic $Ca^{2+}$ and calmodulin probably constitute Gs. [Indeed, phase shifting by transitory perturbations (increases) of intracellular $Ca^{2+}$ have been reported in *Aplysia, Trifolium, Chlamydomonas,* and *Neurospora,* although pulses of the ionophore A23187 were ineffective in *Gonyaulax* (see Engelmann and Schrempf).][57]

### *Is the Mitochondrial $Ca^{2+}$-Transport System a "Gear"?*

There should be another G directly regulating $[Ca^{2+}]$. The main regulatory sites for many noncircadian systems are known to be the plasmalemma, the endoplasmic reticulum, and the mitrochondria. In *Lemna*,[70] $Ca^{2+}$ uptake across the plasmalemma does not oscillate (although that of $K^+$ does).

Although the pyridine nucleotides and $Ca^{2+}$ -calmodulin would be expected to affect general metabolism significantly, neither they nor $NAD^+$ kinase and $NADP^+$ phosphatase alone could constitute an autonomous oscillator because the requirement for negative cross-coupling[71] in the regulatory sequence $Ca^{2+}$ -calmodulin-$NAD^+$ kinase/$NADP^+$ phosphatase-$NAD^+$ is not satisfied. Although $Ca^{2+}$ -calmodulin can decrease the rate of net formation of $NAD^+$, the latter is not known to be able to increase directly the net formation rate of the former except through one or more additional steps. If it could be demonstrated in *Euglena* that $NAD^+$ could directly increase the rate of net $Ca^{2+}$ efflux, as observed in isolated rat liver mitochondria,[72] the resulting increase in the concentration of free $Ca^{2+}$ in the cytoplasm would cause an increase in the rate of net production of cytoplasmic $Ca^{2+}$ -calmodulin, and an oscillatory feedback loop could be constructed.

To test possibility that the mitochondrial $Ca^{2+}$ -transport system might be a G, the phase-shifting effects of short (2-or 3-hour) pulses of nitrogen, dinitrophenol (DNP) and sodium acetate on the cell division rhythm of *Euglena* were examined. Mitochondrial $Ca^{2+}$ -transport is closely related to electron transport or to ATP hydrolysis coupling to proton transport.[73] Nitrogen or DNP would be expected to enhance the rate of net mitochondrial $Ca^{2+}$ efflux by inhibiting energy-dependent $Ca^{2+}$ influx, whereas sodium acetate enhances $Ca^{2+}$ uptake via cotransport.[74] As is evident from the PRCs obtained (FIGURE 3, lower panel), all the perturbing agents were effective in generating steady-state $\Delta\phi$'s. The PRCs for nitrogen and sodium acetate, agents having opposite effects on transport, are almost mirror images of each other. Diuron [DCMU, 3- (3,4-dichlorophenyl)-1,1-dimethylurea], an inhibitor of photosynthesis, should not have any significant effect on mitochondrial $Ca^{2+}$ flux; it was shown to be ineffective for phase shifting (FIGURE 3, lower panel) at the concentration used. Therefore, the mitochondrial $Ca^{2+}$ -transport system would appear to be a G of the oscillator in *Euglena*.

## A Model for the Circadian Oscillator

Our findings suggest that $NAD^+$, the mitochondrial $Ca^{2+}$ -transport system, $Ca^{2+}$, calmodulin, $NAD^+$ kinase, and $NADP^+$ phosphatase represent clock "gears" that might constitute a self-sustained, circadian oscillating loop in *Euglena* and other eukaryotic microorganisms, as well as in higher plants and animals. This regulatory scheme, shown in FIGURE 4 together with the postulated primary actions of the drugs used in the present experiments, includes the following three steps: (1) $NAD^+$ would enhance the rate of net $Ca^{2+}$ efflux from the mitochondria (Mit), resulting in a maximal concentration of cytosolic (Cyt) $Ca^{2+}$ 6 hours (90°) later. Alternatively (1'), a photoreceptor (phytochrome, or perhaps a blue-light photoreceptor in *Euglena* and animal cells) stimulated by (red or blue) light would enhance either net $Ca^{2+}$ efflux from the Mit or net $Ca^{2+}$ influx across the plasmalemma into the Cyt. (2) $Ca^{2+}$ would immediately activate calmodulin by forming an activated $Ca^{2+}$ -calmodulin complex in the Cyt (maximal level at 90°). (3) This active form of $Ca^{2+}$ -calmodulin would decrease the *rate* of net production of $NAD^+$ by both activating $NAD^+$ kinase and inhibiting $NADP^+$ phosphatase in the Cyt so that the rate would become maximal 12 hours later (at 270°) when $Ca^{2+}$ -calmodulin reached its minimum level. After 6 more hours, when the *in vivo level* of NAD$^+$ then becomes lowest (at 0°), the regulatory sequence would be closed, and the cycle would repeat.

This proposed feedback loop can autonomously oscillate because the "cross-couplings" are always of opposite sign.[71] Thus, while the rise in NAD$^+$ causes increases in the rate of formation of the active Ca$^{2+}$-calmodulin complex by steps (1) and (2),

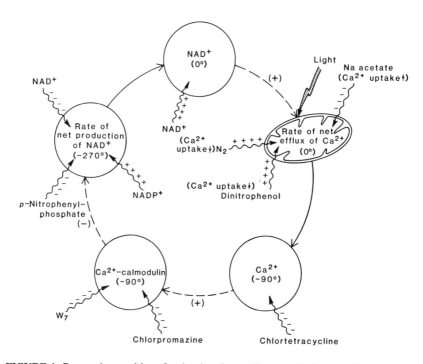

**FIGURE 4.** Proposed control loop for the circadian oscillator in *Euglena gracilis*. The pattern of regulation is indicated by both solid lines, which relate the reactions to their products, and dashed lines, which correspond to the sequence of steps. Each element oscillates with a circadian period (with peaks and troughs occurring 180° apart). The degrees in parentheses reflect the phases at which maximal values are attained. The maximal rate of a reaction precedes by 90° the maximal concentration of its product. The activation (+) or the inhibition (−) of each compound on the succeeding one is shown and can be considered to produce phase delays of 0° or 180°, respectively. The inhibitory effect of Ca$^{2+}$-calmodulin on the rate of net production of total NAD$^+$ derives from both its inhibitory effect on NADP$^+$ phosphatase activity and its activating effect on that of NAD$^+$ kinase. A number of different compounds and agents known to have either positive (activation: wavy arrows with plus signs) or negative (inhibition: wavy arrows with minus signs) effects on their targets are indicated. Their phase-shifting actions on the overt, circadian rhythm of cell division have been documented in FIGURE 3. (Reprinted, with permission, from Reference 12; copyright 1985 by the American Association for the Advancement of Science.)

the resulting rise in the level of active Ca$^{2+}$-calmodulin complex causes decreases in the rate of formation of NAD$^+$ by step (3). Step (1') would externally entrain (synchronize) the system. In order for this oscillator to display limit cycle dynamics, additional requirements would have to be satisfied.[71] The rise in NAD$^+$ must cause a decrease in the rate of net formation of $NAD^+$ (which is very likely). The rise in the level of active $Ca^{2+}$-calmodulin complex must cause an increase in its own rate of formation somewhere in the oscillatory region, perhaps by the cooperative binding of $Ca^{2+}$ to the four sites of each calmodulin molecule, which increases the binding affinity of calmodulin[17] for $Ca^{2+}$. Finally, the strength of this "self-coupling" must be weaker than that of the cross-coupling.

Our model does not purport to explain either the long period length or its steady-state temperature compensation characteristic of circadian rhythmicity, primarily because of a lack of hard data. A limit cycle could display a circa24-hour period given a particular set of parameter values or a network of interacting intracellular oscillators; alternatively, other time-consuming processes involving transcription, translation, or membrane-based ion transport may be important (see Engelmann and Schrempf;[57] Edmunds[75]). Similar explanations[12] could be invoked as compensatory mechanisms for the temperature-dependent steps (such as the $NAD^+$ kinase and $NADP^+$ phosphatase reactions) in our model. Finally, we have proposed a model—not for "the" clock—but for one oscillator in what is most probably a cellular "clock shop."

## REFERENCES

1. MITCHISON, J. M. 1974. *In* Cell Cycle Controls. G. M. Padilla, I. L. Cameron & A. Zimmerman, Eds.: 125-142. Academic Press. New York, N.Y.
2. KLEVECZ, R. R. 1976. Proc. Nat. Acad. Sci. USA **73:** 4012-4016.
3. WILLE, J. J., JR. 1979. *In* Biochemistry and Physiology of Protozoa. M. Levandowsky & S. H. Hutner, Eds. 2nd edit. **2:** 67-149. Academic Press, New York, N.Y.
4. EDMUNDS, L. N., JR. & K. J. ADAMS. 1981. Science **211:** 1002-1013.
5. GILBERT, D. A. 1981. S. Afr. J. Sci. **77:** 541-546.
6. EDMUNDS, L. N., JR. 1984. Cell Cycle Clocks. Marcel Dekker, New York, N.Y.
7. EDMUNDS, L. N., JR. & D. L. LAVAL-MARTIN. 1984. *In* Cell Cycle Clocks. L. N. Edmunds, Jr., Ed.: 295-324. Marcel Dekker. New York, N.Y.
8. SACHSENMAIER, W. 1976. *In* The Molecular Basis of Circadian Rhythms. J. W. Hastings and H.-G. Schweiger, Eds.: 409-420. Dahlem Konferenzen. Berlin, FRG.
9. KAUFFMAN, S. & J. J. WILLE. 1975. J. Theor. Biol. **55:** 47-93.
10. SHYMKO, R. M., R. R. KLEVECZ & S. A. KAUFFMAN. 1984. *In* Cell Cycle Clocks. L. N. Edmunds, Jr., Ed.: 273-293. Marcel Dekker. New York, N.Y.
11. GOTO, K. 1984. Z. Naturforsch. **39c:** 73-84.
12. GOTO, K., D. L. LAVAL-MARTIN & L. N. EDMUNDS, JR. 1985. Science **228:** 1284-1288.
13. CHEUNG, W. Y. 1980. Science **207:** 19-27.
14. KLEE, C. B., T. H. CROUCH & P. G. RICHMAN. 1980. Annu. Rev. Biochem. **49:** 489-515.
15. MEANS, A. R. 1981. Recent Prog. Hormone Res. **37:** 333-367.
16. MEANS, A. R., J. S. TASH & J. G. CHAFOULEAS. 1982. Physiol. Rev. **62:** 1-37.
17. RASMUSSEN, H. & D. M. WAISMAN. 1983. Rev. Physiol. Biochem. Pharmacol. **95:** 111-148.
18. WANG, J. H. & D. M. WAISMAN. 1979. Curr. Top. Cell. Regul. **15:** 47-107.
19. BUETOW, D. E., Ed. 1968 & 1982. The Biology of Euglena. **1, 2 & 3.** Academic Press. New York, N.Y.
20. EDMUNDS, L. N., JR. & F. HALBERG. 1981. *In* Neoplasms—Comparative Pathology of Growth in Animals, Plants and Man. H. E. Kaiser, Ed.: 105-134. Williams and Wilkins. Baltimore, Md.
21. EDMUNDS, L. N., JR. 1982. *In* The Biology of *Euglena.* D. E. Buetow, Ed. **3:** 53-142. Academic Press. New York, N.Y.
22. EDMUNDS, L. N., JR. 1978. *In* Aging and Biological Rhythms. H. V. Samis, Jr. & S. Capobianco, Eds.: 125-184. Plenum. New York, N.Y.
23. COOK, J. R. & T. W. JAMES. 1960. Exp. Cell Res. **21:** 583-589.
24. EDMUNDS, L. N., JR. 1965. J. Cell. Comp. Physiol. **66:** 147-158.

25. EDMUNDS, L. N., JR. & R. R. FUNCH. 1969. Planta **87:** 134-163.
26. LEDOIGT, G. & R. CALVAYRAC. 1979. J. Protozool. **26:** 632-643.
27. TERRY, O. W. & L. N. EDMUNDS, JR. 1970. Planta **93:** 106-127.
28. EDMUNDS, L. N., JR. 1975. Colloq. Int. CNRS (240): 53-67.
29. EDMUNDS, L. N., JR. 1966. J. Cell Physiol. **67:** 35-44.
30. JARRETT, R. M. & L. N. EDMUNDS, JR. 1970. Science **167:** 1730-1733.
31. EDMUNDS, L. N., JR., M. E. JAY, A. KOHLMANN, S. C. LIU, V. H. MERRIAM & H. STERNBERG. 1976. Arch. Microbiol. **108:** 1-8.
32. LAVAL-MARTIN, D. L., D. J. SHUCH & L. N. EDMUNDS, JR. 1979. Plant Physiol. **63:** 495-502.
33. EDMUNDS, L. N., JR. & R. R. FUNCH. 1969. Science **165:** 500-503.
34. ADAMS, K. J., C. S. WEILER & L. N. EDMUNDS, JR. 1984. *In* Cell Cycle Clocks. L. N. Edmunds, Jr., Ed.: 395-429. Marcel Dekker. New York, N.Y.
35. SCHNABEL, G. 1968. Planta **81:** 49-63.
36. EDMUNDS, L. N., JR. & D. L. LAVAL-MARTIN. 1981. *In* Photosynthesis. G. Akoyunoglou, Ed. **6:** 313-322. Balaban International Science Service. Philadelphia, Pa.
37. MIYAMOTO, H., L. RASMUSSEN & E. G. ZEUTHEN. 1973. J. Cell Sci. **13:** 889-900.
38. ZEUTHEN, E. 1974. *In* Cell Cycle Controls. G. M. Padilla, I. L. Cameron & A. M. Zimmerman, Eds.: 1-30. Academic Press. New York, N.Y.
39. SMITH, H. T. B. & J. M. MITCHISON. 1976. Exp. Cell Res. **99:** 432-435.
40. POLANSHEK, M. M. 1977. J. Cell Sci. **23:** 1-23.
41. KLEVECZ, R. R., J. KROS & S. D. GROSS. 1978. Exp. Cell Res. **116:** 285-290.
42. KLEVECZ R. R., G. A. KING & R. M. SHYMKO. 1980. J. Supramol. Struct. **14:** 329-342.
43. KLEVECZ, R. R., J. KROS & G. A. KING. 1980. Cytogenet. Cell Genet. **26:** 236-243.
44. EDMUNDS, L. N., JR., D. E. TAY & D. L. LAVAL-MARTIN. 1982. Plant Physiol. **70:** 297-302.
45. PAVLIDIS, T. 1968. *In* Lectures in Mathematics in Life Sciences. M. Gerstenhaber, Ed.: 88-112. American Mathematical Society. Providence, R.I.
46. WINFREE, A. T. 1970. J. Theor. Biol. **28:** 327-374.
47. WINFREE, A. T. 1980. The Geometry of Biological Time. Springer-Verlag. New York, N.Y.
48. TAYLOR, W., R. KRASNOW, J. C. DUNLAP, H. BRODA & J. W. HASTINGS. 1982. J. Comp. Physiol. **148:** 11-25.
49. MALINOWSKI, J. R., D. L. LAVAL-MARTIN & L. N. EDMUNDS, JR. 1984. J. Comp. Physiol. **B155:** 257-267.
50. SWEENEY, B. M. & J. W. HASTINGS. 1960. Cold Spring Harbor Symp. Quant. Biol. **25:** 87-104.
51. KLEVECZ, R. R. & G. A. KING. 1982. Exp. Cell Res. **140:** 307-313.
52. ANDERSON, R. W., D. L. LAVAL-MARTIN & L. N. EDMUNDS, JR. 1984. Exp. Cell Res. **157:** 144-158.
53. EDMUNDS, L. N., JR. 1987. *In* Algal Development. W. Wiessner & D. G. Robinson, Eds. Springer-Verlag. Heidelberg, FRG. (In press.)
54. HASTINGS, J. W. & H.-G. SCHWEIGER, Eds. 1976. The Molecular Basis of Circadian Rhythms. Dahlem Konferenzen. Berlin, FRG.
55. SWEENEY, B. M. 1969. Can. J. Bot. **47:** 299-308.
56. SARGENT, M. L., Rapporteur. 1976. *In* The Molecular Basis of Circadian Rhythms. J. W. Hastings & H.-G. Schweiger, Eds.: 295-310. Dahlem Konferenzen. Berlin, FRG.
57. ENGELMANN, W. & M. SCHREMPF. 1980. Photochem. Photobiol. Rev. **5:** 49-86.
58. ROEDER, P. E., M. L. SARGENT & S. BRODY. 1982. Biochemistry **21:** 4909-4916.
59. TYSON, J., Rapporteur. 1976. *In* The Molecular Basis of Circadian Rhythms. J. W. Hastings & H.-G. Schweiger, Eds.: 85-108. Dahlem Konferenzen. Berlin, FRG.
60. CORNELIUS, G. & L. RENSING. 1982. BioSystems **15:** 35-47.
61. NAKASHIMA, J., J. PERLMAN & J. F. FELDMAN. 1981. Science **212:** 361-362.
62. PERLMAN, J. & J. F. FELDMAN. 1982. Mol. Cell. Biol. **2:** 1167-1173.
63. BRINKMANN, K. 1973. *In* Biological and Biochemical Oscillators. B. Chance, E. K. Pye, A. K. Ghosh & B. Hess, Eds.: 523-529. Academic Press. New York, N.Y.
64. FELDMAN, J. F. 1975. Science **190:** 789-790.

65. FELDMAN, J. F. & J. C. DUNLAP. 1983. Photochem. Photobiol. Rev. **7:** 319-368.
66. MUTO, S. & S. MIYACHI. 1977. Plant Physiol. **59:** 55-60.
67. ANDERSON, J. M., H. CHARBONNEAU, H. P. JONES, R. O. MCCANN & M. J. CORMIER. 1980. Biochemistry **19:** 3113-3120.
68. EPEL, D., R. W. WALLACE & W. Y. CHEUNG. 1981. Cell **23:** 543-549.
69. GOTO, K. 1983. Plant Physiol. **72:** (suppl.): 86 (abstr.).
70. KONDO, T. 1982. Plant Cell Physiol. **23:** 901-908.
71. HIGGINS, J. 1967. Ind. Eng. Chem. **59:** 18-62.
72. PANFILI, E., G. L. SOTTOCASA, G. SANDRI & G. LIUT. 1980. Eur. J. Biochem. **105:** 205-210.
73. TEDESCHI, H. 1981. Biochim. Biophys. Acta Sci. **639:** 157-196.
74. LEHNINGER, A. L. 1974. Proc. Nat. Acad. Sci. USA **71:** 1520-1524.
75. EDMUNDS, L. N., JR. 1984. Adv. Microb. Physiol. **25:** 61-148.

# Endocytobiotic Coordination, Intracellular Calcium Signaling, and the Origin of Endogenous Rhythms

FRED KIPPERT

*Institute for Animal Physiology*
*University of Giessen*
*Wartweg 95*
*D-6300 Giessen, Federal Republic of Germany*

Elementary differences in the regulation of cellular physiology separate prokaryotic and eukaryotic organisms. As will be emphasized throughout this paper, fundamental disparities are already apparent at the unicellular level of eukaryotic organization. Regarding their universality in eukaryotes almost inevitably leads to the suggestion that some of these early acquisitions of the developing eukaryotic cell were of prominent evolutionary significance for the later emergence of eukaryotic complexity and diversity. The superior influence of interactions between cellular compartments on cellular evolution has been pointed out previously.[1] A more pervasive statement is the principal notion of this paper that those interactions culminating in mechanisms for the integration of today's semiautonomous organelles were even indispensable for the evolution of the eukaryotic cell out of an endocytobiotic organization. Hence, the endosymbiotic theory may not only account for the origin of mitochondria and plastids but equally for a substantial part of the developmental potency arising in conjunction with the first eukaryotic organisms. Two essential features of cell regulation ubiquitous in eukaryotes are suggested to have evolved as a consequence of mutual interactions between partners in a cellular symbiotic consortium developing towards the eukaryotic cell: endogenous rhythms and the calcium system of intracellular signaling. These two characters unique to eukaryotic cells may be intertwined, the calcium system being an integral part of the cellular clock mechanism.

## DIFFERENCES BETWEEN PROKARYOTIC AND EUKARYOTIC CELL REGULATION

The presence of a nucleus as the defining characteristic of eukaryotic cells is at the same time the most easy means to distinguish them from their prokaryotic counterparts. Equally eyecatching are their usually much larger size and the complex structure of the cytoplasm harboring various organelles. Differences in cellular morphology like these have been dealt with extensively; likewise those in essential metabolic pathways and the biochemistry of subcellular components. More recently, differences in molecular biology between the two basic subdivisions of life have constituted an additional focus of interest. From these fields of research emanate most of the features listed in earlier discussions on eukaryotic cell evolution as being exclusively eukaryotic.[1-4] Physiological features, if given at all, may be characterized as "steady-state physiology." In contrast, differences in dynamic cell physiology, especially aspects of cellular regulation, have so far been much neglected. TABLE 1 lists several features in addition to those given in previous reviews. Most significant, all of them are assumed to play essential roles in eukaryotic cell regulation already at the unicellular level. While ubiquitous in eukaryotic cells, they are either totally absent in prokaryotes or present only in some primordial form in a limited number of phylogenetic classes. Considering their universality among eukaryotes, from the most primitive microbes on, the evolutionary relevance of these features can hardly be overemphasized. It rather seems worth speculating that all these instruments of cellular regulation were of pivotal importance for the further evolution of eukaryotic complexity and diversity. Subsequent sections will be devoted to features suggested to have evolved as a consequence of interactions between those cellular compartments that are regarded as descendants of formerly autonomous organisms within an endocytobiotic consortium. Other features unique to eukaryotic cell physiology are described below.

**TABLE 1.** Essential Features of Cellular Regulation Suggested to Be Unique to and Universal in Eukaryotic Cells

Endogenous rhythms of the "cellular clock" type
"Calcium system of intracellular signaling"
Calmodulin
Calcium-calmodulin-dependent protein kinases
CAMP-dependent protein kinase
Protein kinase C
Phosphoinositide cascade
Regulatory GTP-binding proteins
Mono- and poly(ADP-ribosyl)ation
Dynamic cytoskeleton with regulatory properties

### Dynamic Cytoskeleton with Regulatory Properties

Besides actin and tubulin, generally recognized as universal components of eukaryotic cells, intermediate filament proteins represent a third category of cytoskeletal elements. Their recent discovery in very different cell types including unicellular organisms suggests that they, too, may be ubiquitous constituents of the eukaryotic cytoskeleton.[5] In addition to providing an intracellular skeleton, determining cell shape, and mediating motility, each of these cytoskeletal elements has been attributed some regulatory property. Actin does occur in the nucleus where it may be involved in the regulation of transcription.[6] Via the association of microfilaments with glycolytic enzymes and mitochondria, actin may also have a modulating influence on cellular energy metabolism.[7] A role for tubulin in the initiation of DNA synthesis has been discussed;[8] intermediate filaments were suggested to serve as nucleic acid binding proteins involved in signal transduction.[9] The cytoskeleton in toto has been implicated in the regulation of protein synthesis.[10] When considering these isolated reports in a joint context, the conclusion seems reasonable that cytoskeletal elements do in various ways participate in the regulation of eukaryotic cell physiology.

*Covalent Modifications*

Phosphorylation and dephosphorylation probably are the most elementary modifications of proteins. Although already employed in the regulation of prokaryotic physiology, this type of modification gained a totally different level of complexity in eukaryotes even at the unicellular level: (1) phosphorylating and dephosphorylating reactions are arranged in extended cascades, thereby forming a sophisticated network of interconverting enzymes; (2) an immense number of different protein kinases and phosphatases, frequently displaying stringent substrate specificity, are involved in virtually all domains of eukaryotic cell physiology. Among the major kinases apparently lacking in prokaryotes are: cAMP-dependent protein kinase (cAMP, a regulatory molecule in prokaryotes as well, has an entirely different mode of action in most eukaryotic cells by stimulating a protein kinase),[11] calcium-calmodulin-dependent protein kinase (to be discussed in more detail below), and protein kinase C[12] (implicated in the regulation of innumerable cellular functions, this kinase, too, is now being documented at all levels of eukaryotic organization).[13,14] ADP-ribosylation of both nuclear and cytoplasmic proteins[15,16] is just another example for a major covalent modification that in eukaryotic cells—again in contrast to the situation in prokaryotes—plays a variety of essential regulatory roles.

*Mediators in Cellular Regulation and Signal Transduction*

In conjunction with the advent of the eukaryotic cell evolved a remarkable diversity of pathways of signal transduction and fine tuning of cellular metabolism. For cells of multicellular organisms, which do anticipate signal input from a great variety of sources, a corresponding requirement is easily visualized. Most of the main signaling pathways, however, are found already in the simplest unicellular eukaryotes. So to say, evolution gave at hand to even the most primordial eukaryotic microbe a tremendous wealth of new instruments of cellular regulation. The developmental potency of eukaryotes, from the very beginning, has thus been of totally different quality from that of their prokaryotic counterparts. Besides calcium and calmodulin (to be the topic of a subsequent section), the hydrolysis of polyphosphoinositides seems to play a crucial role in eukaryotic signal transduction.[17] Originally discovered and investigated in vertebrate cells, this pathway apparently exists in plants[18] and unicellular organisms[19] as well, suggesting its ubiquitous distribution throughout the eukaryotic world. Another small molecule universally employed in both cellular regulation and signal transduction is GTP.[20] The number of documented regulatory GTP-binding proteins is increasing steadily;[21] the equally essential role they do play in multicellular and unicellular eukaryotes may be illustrated by the highly conserved nature of these G-proteins from yeast to man.[22]

Some features considered as being unique to eukaryotes, however, are found in at least some archaebacterial species. Members of this distinct class of prokaryotes are now widely assumed to resemble the ancestor of the eukaryotic cytoplasm. It is, therefore, not surprising that they display several characteristics typical for eukaryotes.[23] Findings relevant to this discussion of cellular regulation are yet restricted to reports on a calcium-regulated ameboid cytoskeleton[24] and a calmodulin-like protein[25] found in *Thermoplasma acidophilum*. Although it is at present not possible to assess how far the relationship between eukaryotic cells and these archaebacteria may go, the remaining gap seems at best only marginally diminished.

## TEMPORAL COORDINATION IN ENDOCYTOBIOSIS AND THE ORIGIN OF ENDOGENOUS RHYTHMS

One of the typical eukaryotic features is circadian rhythms. While absent in prokaryotes, they are widespread but not universal in eukaryotic cells. Circadian rhythms have long been estimated as *the* adaptation of cells to the geophysical cycle of the earth's rotation; a view challenged only infrequently.[26-28] In recent years, however, it has become apparent that they are just one member of a whole family of endogenous rhythms sharing a couple of striking features (summarized in TABLE 2) which make them unlike any other biological oscillations. Taking these peculiar properties as a criterion, a distinguished category "the cellular clock"[29] emerges which,

TABLE 2. Characteristic Properties of the Cellular Clock

| | |
|---|---|
| Ubiquity | Among eukaryotes, at every level of organization |
| Self-sustainment | No external stimuli required, persistence of circadian rhythms under constant conditions |
| Genetic determination | Innateness and endogeneity |
| General homeostasis | Period subject to active homeostatic regulation, insensitivity to minor perturbations |
| Entrainability | Of physiological significance for circadian rhythms only, synchronization to diurnal light and temperature cycles |
| Phase shiftability | By pulses of light, temperature, and chemicals, required for entrainment of circadian rhythms |

when considered in its entirety, is currently being realized as a unique *and* ubiquitous character of eukaryotic organisms. Besides circadian rhythms as the most prominent expression of cellular clocks, this class of oscillations includes rhythms whose periods are shorter (i.e., ultradian) or longer (i.e., infradian) than a day and do not correspond to any geophysical periodicity. The hypothesis of endogenous rhythms providing an adaptational advantage is severely challenged by this obvious lack of external correlates. Alternatively, with respect to their remarkable accuracy and homeostatic insensitivity to interference from the environment, it may be argued that cellular clocks primarily provide a mechanism to the eukaryotic cell whose functional role is *internal timekeeping.*[30] The question then arises why eukaryotic cells are in need of internal time sense while their prokaryotic counterparts obviously are not. A persuasive explanation for this further distinction between prokaryotes and eukaryotes is based on the endosymbiotic theory[31] of eukaryotic cell evolution: the most initial selective advantage of endogenous cellular clocks has been their providing an intrinsic device for temporal coordination to a host cell and its intracellular symbionts (to become the eukaryotic nucleocytoplasm and mitochondria, respectively).[28]

*On the Problem of Temporal Coordination in Endocytobiosis*
*and Eukaryotic Cells*

Following up on the question of why a device for temporal coordination should have provided a key advantage, one has to realize that only exceptionally did endocytobiotic associations give rise to a cellular entity; most likely because a symbiont does only in rare instances become tightly enough enmeshed in the internal economy of the cell to develop into a genuine cell organelle.[32] FIGURE 1 indicates three aspects of physiology for which coordination between the partners is definitely needed in order to achieve sufficiently tight integration. Metabolic coordination is the first step in any endocytobiosis with one partner supplying nutrients to (or removing waste products from) the other. A later step on the way toward a cellular entity might be epigenetic coordination; a multimeric protein for instance, whose subunits are encoded

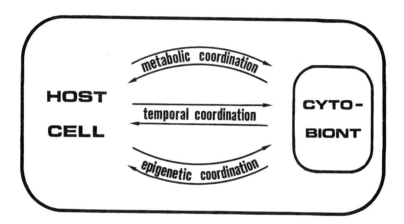

**FIGURE 1.** Scheme indicating domains of cellular physiology for which coordination between a host cell and its endocytobionts (and, likewise, between the eukaryotic nucleocytoplasm and its semiautonomous organelles) is required.

by different genomes, requires some regulatory coordination for its assembly. Still widely ignored is a third domain of cellular physiology with compelling demand for coordination: the temporal organization of the symbiotic association. A requirement for temporal coordination is easily conceivable for many aspects of internal cell economy; the discussion here will concentrate on a problem relevant to even the most primitive forms of endocytobiosis: attaining harmony between not only the division rates but also the division times of the partners.

Although the development of synchrony between endosymbiont division and host membrane division has been realized to represent a prerequisite for evolution from an endocytobiotic association toward the eukaryotic cell,[33] little information is available about strategies to meet this obvious demand for synchrony and, more basically, about mechanisms for regulation of symbiont populations within a host cell. Smith identifies ejection or digestion of surplus symbionts and restriction of symbiont division,[34] yet the elimination of symbionts is a crude mechanism unlikely to be found in any more advanced endocytobiosis. Rather does population size depend on a dynamic relationship between the division rates of the partners.[35] Suggestions for mechanisms involved in symbiont regulation are derived from studies on two systems only: the symbioses of Chlorella with Paramecium and Hydra. Reisser favors a concept of "ecological regulation" and views the symbiotic association as a microhabitat in which the equilibrium is primarily determined by factors both partners depend on equally.[36] Assuming some kind of feedback between upper and lower limits of symbiont number, this type of model is difficult to reconcile with the observation of synchronous symbiont development.[37] And it cannot be applied to associations with small symbiont number because the probability of a cell becoming devoid of its symbionts (or organelles) is biologically negligible only for symbiont number $> 10$.[38] The model of McAuley postulates that algal division is dependent on a "division factor" which is supplied by host cells only at the time of their own division.[39] Small and temporary changes in pH are assumed by Douglas and Smith to be the trigger for synchrony.[40] Regarding the pleiotropic responses frequently elicited by changes in pH or the concentrations of regulatory active cations, such a type of mechanism seems very appealing, yet experimental evidence is still limited for any of the models. Concerned with symbiotic systems comprising two eukaryotic partners and not being of an obligate nature for either of them, these studies are not very indicative for a discussion on those regulatory mechanisms that may have contributed to the evolution of the eukaryotic cell out of an endocytobiotic association. The number of reports on the population dynamics of prokaryotic cytobionts in relation to host cell cycle is vanishingly small, yet revealing the existence of (1) precise synchrony,[41,42] and (2) efficient mechanisms even competent to ensure a constant number of just a single pair of symbionts per host cell.[42]

A similar lack of data characterizes the state of knowledge about the regulation of mitochondrial populations within eukaryotic cells. Summing up a comprehensive review on mitochondrial dynamics, Lloyd and colleagues cannot but conclude that the degree remains unknown to which cycles of organelle growth and division are controlled.[43] A particularly conflicting picture emerges from reports concerned with the question as to whether the formation of mitochondria during the cell cycle is continuous or discontinuous.[43,44] This may at least in part be due to difficulties in obtaining synchronous cultures without imposing too much stress on the cells. It is therefore important to note that studies on naturally synchronous systems[45]—and those where synchrony is unharmfully achieved by entrainment to diurnal cycles[46]—provide good evidence for a distinct mitochondrial cycle tightly coupled to the nuclear cycle. Analysis of the available literature reveals that, while information is limited with respect to cell cycle related dynamics of mitochondria (and plastids),[47,48] virtually nothing is known about the mechanisms involved in temporal coordination between semiautonomous cellular compartments.

## On the Origin and Further Evolution of Endogenous Rhythms

The ubiquity of cellular clocks in eukaryotic organisms at all levels of organization substantiates the assumption that they have evolved concomitantly with the eukaryotic cell.[28] Considering the evolution of any character as fundamental as cellular clocks, only those selective forces prevalent at the early time of its origin are of relevance. This kind of reasoning brings about several arguments in contradiction to the long-held adaptational hypothesis for the origin of circadian rhythms.[28] Simple unicellular organisms can respond to changes in their natural environment (the light-dark cycle in this case) sufficiently fast and do not seem to be in need of any intrinsic timekeeping. Rather would stimulus-response coupling or at most some kind of hourglass timer be a fully adequate mechanism to modulate cellular functions in a daily fashion. On the other hand, although the postulated adaptational advantage should have been of equal significance for them, no circadian rhythms have ever been detected in prokaryotes. Interestingly, in the symbiotic prokaryote Prochloron, when directly isolated from its eukaryotic host, a diurnal rhythm of cell division can be found.[49] Another decisive argument against the adaptational hypothesis is, as mentioned above, the realization of circadian rhythms being just one member of a family of endogenous timing devices. For other members no external correlates will ever be found and, consequently, no adaptational advantage can be postulated.

Regarding these arguments, time was overdue for an alternative, unifying hypothesis for the origin of cellular clocks that had to account for their total lacking in prokaryotes. Bünning was the first to realize that the new spatial organization of eukaryotic cells made a new temporal organization indispensable.[50] Emphasizing the requirement of complex compartmentalized systems for an intricate balance of cellular functions, Paietta suggested that circadian rhythmicity may have been genetically internalized as a part of the organelle integration process.[51] Levandowsky presented the first "endosymbiotic hypothesis of circadian rhythms,"[52] implying that they may have developed as a consequence of interactions between host cells and their endocytobionts. The concept of endogenous rhythms providing temporal coordination for endocytobiotic associations[28] is the hypothesis for the origin of cellular clocks with the most far-reaching evolutionary implications. Besides indicating selective forces that could have urged the evolution of a whole array of endogenous rhythms without external correlates, this hypothesis offers a solution for the still open question of temporal coordination in endocytobiotic associations and eukaryotic cells.

One of the major fields where endogenous rhythms play a central role is identical to that domain of cellular physiology with the primary demand for temporal coordination. Interactions of cellular clocks with the eukaryotic cell cycle are a phenomenon now widely established (for a review see Edmunds).[53] Circadian rhythms have long been recognized to have a "gating" effect on cell cycle by restricting cell division and DNA synthesis to certain phases of the daily cycle. As a consequence, generation time (GT) is equal to, or an integer multiple of, 24 hours. An analogous relationship has more recently been documented for ultradian rhythms; these reports corroborate the concept of "quantized cell cycles"[30,54] postulating that the eukaryotic cell cycle is frequently, if not ubiquitously, a discrete multiple of subcycles of the cellular clock type: in slowly growing cells (GT $\geq 24$ hours) the subcycle is circadian, whereas in faster growing cells (GT $< 24$ hours) the underlying rhythm has an ultradian period.[30,43,54] Microorganisms seem capable of switching between the two modes, depending on growth conditions. In ciliates, for instance, several physiological processes were found to display oscillations with either a circadian period during slow growth and stationary phase, or an ultradian period during rapid proliferation. With most of the energy being generated in one compartment while utilized in a second compartment, energy metabolism is another domain of cellular physiology with compelling demand for temporal coordination. It has therefore been suggested that, besides temporal control of cell proliferation, temporal coordination of energy generation and utilization in different compartments contributed mainly to the selective advantage of endogenous rhythms.[28] Notably, in analogy to circadian organization, many of the ultradian rhythms involved in cell cycle regulation are at the same time rhythms of energy generating and utilizing processes.[30,43] An intimate interrelationship between cellular clocks, cell proliferation, and cellular energy metabolism apparently constitutes a fundamental characteristic of eukaryotic cell dynamics.

In particular with regard to cellular energy metabolism another aspect remains to be considered in a discussion on the selective advantage of cellular clocks. Complex physiological systems (such as the envisaged endocytobiotic association) comprising a network of feedback mechanisms will inevitably tend to break into oscillations.[27,55] There is a considerable likelihood of these oscillations becoming aperiodic, i.e., chaotic (the same outcome has to be expected when viewing the association as a set of mutually interacting oscillators).[27] Chaotic oscillations, however, are frequently associated with pathological conditions of physiological systems.[56] It seems highly probable that chaotic oscillations would also have destabilized the consortium, thereby preventing its development toward a cellular entity. A homeostatic, self-regulated oscillatory regime would thus have been of twofold advantage to the evolving eukaryotic cell by (1) providing a device for temporal coordination between still largely autonomous compartments, and (2) preserving the developing association from destabilizing chaotic fluctuations.

The aspect of energy is also of relevance considering the further evolution of endogenous rhythms. Although circadian rhythms should be regarded as just one member of the cellular clock family, they certainly played an outstanding role in eukaryotic evolution by contributing additional selective advantage of an adaptational nature to those organisms whose energy metabolism is in some way coupled to the daily cycle. That is why circadian rhythms have become the predominant expression of endogenous rhythmicity in animals (for which timing of their activities is important) and plants (which depend on the rhythmic energy supply from the sun). There are no similar constraints on nonphotosynthetic microorganisms and fungi. As a consequence, in these organisms noncircadian rhythms prevail, some of which may be more similar to a primordial clock. A scheme for the evolution of present-day circadian rhythms from primordial cellular clocks has been put forward recently[57] (FIGURE 2),

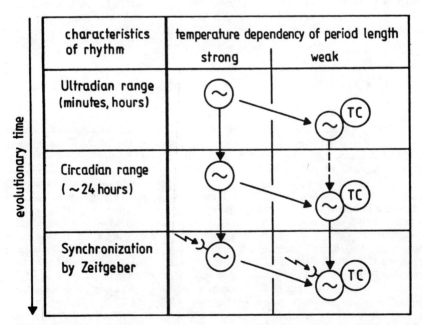

**FIGURE 2.** Proposed evolution of temperature-compensated and entrained circadian rhythms from a primordial clock (suggested to have been a less temperature-independent ultradian rhythm). (Reproduced with permission from Reference 57.)

TABLE 3. Examples for Different Expressions of the Cellular Clock (in Addition to Temperature-Compensated Circadian Rhythms Most Abundant in Eukaryotic Organisms)

| Organism | Period | Temperature Compensation | Reference |
|---|---|---|---|
| *Tetrahymena pyriformis* | ultradian (0.5 hours) | yes | 58 |
| *Acanthamoeba castellanii* | ultradian (1 hour) | yes | 59 |
| *Paramecium tetraurelia* | ultradian (1 hour) | yes | 60 |
| Mammalian cells | ultradian (4 hours) | yes | 61 |
| Tropical plant species | circadian | weak | 62 |
| *Thalassomyxa australis* | circadian | no | 63 |
| *Neuropora crassa* mutants | circadian | no | 64 |
| *Arthrobotrys oligospora* | infradian (40 hours) | yes | 65 |

pointing out that some of the properties generally ascribed to circadian rhythms have not necessarily existed from the very beginning of eukaryotic evolution. This has been stated above for entrainment to exogenous cycles and may be equally true for temperature compensation. TABLE 3 indicates that temperature compensation is found to be associated with endogenous rhythms from 0.5 to 40 hours, but on the other hand, it may be weak or even absent in otherwise fully competent circadian rhythms. At present it cannot be distinguished whether the nonubiquitous properties of endogenous rhythms represent secondary losses or acquisitions at later stages of clock evolution. Attempts to reveal (1) the minimum constituents of the cellular clock and (2) those properties that can be dissected from the basic clock mechanism will provide interesting perspectives for further chronobiological research.

# THE CALCIUM SYSTEM OF INTRACELLULAR SIGNALING

Calcium primarily is a cytotoxic substance. If cytoplasmic concentration reached the outside level (the natural habitat of unicellulars or the extracellular fluids of multicellular organisms), cell death would be the ultimate result. To prevent irreversible damage, all types of cells, prokaryotic and eukaryotic, have developed efficient transport systems maintaining cytoplasmic concentration about 1000-fold lower than that of the external milieu. In eukaryotic cells calcium ions are, in addition, a regulatory substance of outstanding importance. Slight changes in cytoplasmic $[Ca^{2+}]$, brought about by either increased influx across the plasma membrane or release from internal stores, exert a variety of regulatory functions.[66,67] These changes are in most cases translated into altered states of cellular physiology via regulatory calcium-binding proteins. Calmodulin (CaM) as the most abundant of these proteins has also been established as the primary physiological receptor for calcium in the eukaryotic cell.[68,69] From an evolutionary point of view most indicative are the ubiquity and remarkable conservation of CaM within the eukaryotic world.[68,70]

A raise in cytoplasmic $[Ca^{2+}]$ results in the formation of $Ca^{2+}/CaM$ complexes which will then activate or deactivate $Ca^{2+}/CaM$-dependent enzymes. Among those are the $Ca^{2+}/CaM$-dependent protein kinases and phosphatases which have a major mediating function in eliciting pleiotropic responses. Already in the unicellular eukaryote a broad spectrum of cellular processes (summarized in TABLE 4) are modulated or regulated by $Ca^{2+}/CaM$ (this abbreviation will in the further text denote the "calcium system of intracellular signaling," i.e., the total of $Ca^{2+}/CaM$-dependent pathways of cellular regulation). Most significantly, this record includes such elementary functions as cell proliferation or intermediary and energy metabolism, thereby highlighting the pivotal role of $Ca^{2+}/CaM$ in eukaryotic cell physiology.

## A Regulatory Role for Calcium in Prokaryotes?

Having seen this universality in eukaryotic cells, the question is tempting whether there are any comparable functions for $Ca^{2+}/CaM$ in prokaryotes. As far as CaM is concerned, there is at present no convincing evidence (based on immunological methods or cDNA hybridization) for its occurrence in bacterial cells. In contrast, a systematic search for CaM has proved unsuccessful.[68,70] With respect to a regulatory role of calcium outside the eukaryotic world, the presently available information suggests that it may be of importance in some specific instances of prokaryotic physiology only. Most thoroughly studied are the effects of calcium on bacterial motility and taxis.[71] In the cyanobacterium *Phormidium uncinatum* the opening of calcium-specific channels in the plasma membrane with subsequent influx of $Ca^{2+}$ is part of the phototactic response.[72] In both *Phormidium* and the archaebacterium *Halobacterium halobium* the internal $Ca^{2+}$ concentration seems to determine the rate of reversals,[71] what may be mediated by a direct effect of $Ca^{2+}$ on methylation and demethylation of membrane proteins.[73] Calcium accumulation appears to be associated with increased survival of thermophilic bacteria at higher temperatures, probably by virtue of a stabilizing effect on subcellular structures.[74] In the same way a requirement for calcium in sporulation of several species may be explained.[75] Cyanobacteria seem to require $Ca^{2+}$ for the functioning of photosystem II.[76] Since other cations can substitute for $Ca^{2+}$, a specific regulatory role has to be questioned; from an evolutionary point of view, however, it is important to note that $Ca^{2+}$ and CaM are involved in the regulation

TABLE 4. Cellular Processes Subject to Regulation by Calcium and Calmodulin Already at the Unicellular Level of Eukaryotic Organization

Motility, both ciliary and ameboid
Cytoplasmic transport
Endocytosis (including phagocytosis)
Exocytosis
Cyclic nucleotide metabolism
Intermediary metabolism
Respiration
Photosynthesis
Initiation of DNA synthesis
Mitosis
Extralysosomal protein degradation
Activity of $Ca^{2+}/CaM$-dependent protein kinases
Activity of $Ca^{2+}/CaM$-dependent phosphoprotein phosphatases
Activity of NAD-kinase and NADP-phosphatase
Activity of $Ca^{2+}$-ATPase

of chloroplast photosystem II.[77] Summarizing, the conclusion seems safe that, in contrast to its universal role in eukaryotic cells, $Ca^{2+}$ is regulatorily active only in rare and specific instances of prokaryotic physiology. It remains an interesting question whether $Ca^{2+}$/CaM will turn out to exert any more general functions in the cellular physiology of archaebacteria such as Thermoplasma.

### On the Origin of the Calcium System of Intracellular Signaling

Here again we realize a fundamental disparity between prokaryotes and even the simplest unicellular eukaryote. This observation is not restricted to $Ca^{2+}$/CaM itself but can be extended to the multitude of interactions between $Ca^{2+}$/CaM and other signal-transducing chains such as the cAMP pathway[78,79] or the phosphoinositide cascade with subsequent stimulation of protein kinase C.[17] And again, regarding the apparent universality among eukaryotes, it may be concluded that all these instruments of cellular signaling and regulation evolved concomitantly with the eukaryotic cell. It has to be argued, however, that this extent of a signaling machinery by far exceeds the need of a unicellular organism for pathways transducing external signals to the interior of the cell. And indeed, $Ca^{2+}$/CaM does regulate cellular functions independently of external stimuli in unicellular organisms. This holds, for example, for one of the most fundamental functions of any living cell: its own replication (in which both $Ca^{2+}$/CaM[80] and phosphoinositide metabolism[81] have been implicated).

Hence, the view of calcium being merely a second messenger involved in the transduction of external signals has to be called into question; and its most original function in early eukaryotes remains to be discussed. Levandowsky proposes an endocytobiotic origin of signal processing in eukaryotes.[82] Similarly, based on the assumptions that the outstanding role of $Ca^{2+}$/CaM in eukaryotic cells (1) evolved at the earliest unicellular level, (2) is essential for elementary functions of the single cell, and (3) in these cases does not necessarily depend on external stimuli, I would like to suggest that the $Ca^{2+}$/CaM system originally evolved as a device for signal transmission *between host cells and cytobionts*. These tools for intracellular signaling may have facilitated the mutual integration of their physiology within a symbiotic association developing toward the eukaryotic cell. And indeed, all of the functions for which temporal coordination between the partners has to be demanded are found again among the calcium-regulated processes. In analogy to the evolution of cellular clocks as discussed above, the symbiotic origin of the eukaryotic cell may likewise have necessitated the development of mechanisms transducing signals between the partners of the association. These pathways of intracellular signaling could subsequently be employed otherwise, thereby constituting some kind of preadaptation essential for the further evolution of eukaryotic complexity.

## THE MITOCHONDRIAL CALCIUM CYCLE MODEL
## FOR ENDOGENOUS RHYTHMS

Having suggested an analogy between the origin of the cellular clock and the origin of the calcium system of intracellular signaling, I will now, as a kind of synthesis of these considerations, advance a model for the cellular mechanism of endogenous rhythms. According to the mitochondrial calcium cycle model,[83] the "site of the clock" has to be expected at the interface between the nucleocytoplasmic and mitochondrial compartments. Since circadian rhythms generally attract most of the attention devoted to biological clocks, the model will be presented in its circadian version. Its basic features, however, will be the same for any other expression of the cellular clock mechanism. Membranes have been implicated in the generation of circadian rhythms for more than a decade;[84] the involvement of mitochondria,[85] calcium,[86,87] and even mitochondrial $Ca^{2+}$ transport[88,89] became apparent as a result of phase-shift experiments and genetic analysis. A recent model put forward by Edmunds and colleagues postulates changes in calcium concentration being an essential component of the clock mechanism.[90] The mitochondrial calcium cycle model is in general agreement with their suggestions but is more straightforward, attempting to explain the clock mechanism as just the consequence of counteracting calcium transport pathways between cytoplasm and mitochondrial matrix. By postulating a tight linkage between two fundamentally eukaryotic features of cellular regulation, the model hopefully contributes to the stimulation of future experimental approaches to the molecular basis of cellular clocks.

### The Mitochondrial Calcium Cycle Model

The model is based on the mitochondrial calcium cycle (MCC) most intensively studied in mammalian mitochondria:[91] calcium ions cycle continuously across the inner mitochondrial membrane, influx and efflux proceeding on two distinct pathways. Influx is via an electrophoretic uniporter driven by the membrane potential. The efflux mechanism is an electroneutral antiporter; depending on cell type either $2H^+$ or $2Na^+$ are exchanged for $1 Ca^{2+}$. These calcium-transport systems are, in their essential properties, similar to those found in most bacteria.[92] While the existence of the MCC is well established for vertebrate cells, information about other cell types is still equivocal; in particular the results obtained with plants[93] and fungi[94] are conflicting. However, recent repetition of earlier work employing improved experimental procedures[95] suggests that—despite some remaining differences to vertebrate mitochondria—an MCC does operate in other cell types as well. This assumption is substantiated by studies on *Tetrahymena pyriformis*[96] and *Physarum polycephalum*[97] revealing that already at this level of organization $Ca^{2+}$-transport pathways can be found analogous to those of vertebrate mitochondria.

The simultaneously active influx and efflux pathways constitute a futile cycle the physiological role of which is subject of an ongoing controversy. On the one hand, it is argued that the MCC does maintain and regulate cytoplasmic $[Ca^{2+}]$.[98] Another explanation suggests a role in the regulation of intramitochondrial $[Ca^{2+}]$ and, consequently, in the modulation of $Ca^{2+}$-sensitive matrix enzymes.[99] These two versions, however, are not mutually exclusive. The alternative interpretation of the cycle being involved in the regulation of both cytoplasmic and mitochondrial $[Ca^{2+}]$ is favored by the MCC model.

This model identifies the MCC as the key element of the cellular clock mechanism. In essence, it predicts that influx and efflux of $Ca^{2+}$ do not proceed at exactly the same rate; rather does one of the two pathways slightly predominate at any given time, resulting in a minute net flux into one direction. In its circadian version, the model postulates that efflux predominates during the day (FIGURE 3a); at night net flux will be into the reverse direction (FIGURE 3b). As a consequence—assuming that there is no net flux across the plasma membrane—the concentrations of $Ca^{2+}$ in cytoplasm and matrix will be subject to inverse changes (as outlined in FIGURE 4). These inverse changes are postulated to be the central element of the cellular timing mechanism and a corresponding course has to be assumed for any other expression of the cellular clock (i.e., ultradian and infradian rhythms like those listed in TABLE 3). Neither the experimental data on circadian rhythms accumulated over three decades of research nor the general information on the MCC is sufficient at the moment to permit the prediction of any further details. More important, however, seems the capacity of the model—even in this sketchy version— to provide an explanation for all the typical characteristics of the cellular clock.

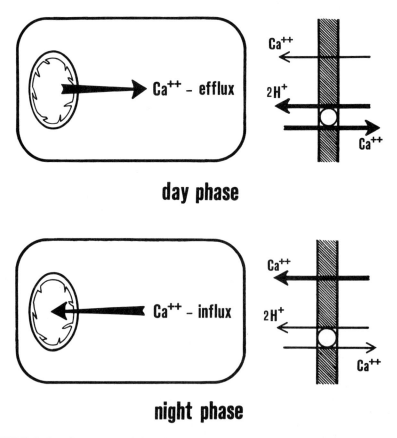

FIGURE 3. Postulated states of the mitochondrial calcium cycle at different phases of the circadian cycle. (a) During the day phase the efflux pathway (electrophoretic uniporter) predominates over the influx pathway (electroneutral antiporter), resulting in a net efflux of $Ca^{2+}$ from mitochondria (top panel). (b) During the night phase the influx pathway predominates over the efflux pathway, resulting in a net influx of $Ca^{2+}$ into mitochondria (bottom panel).

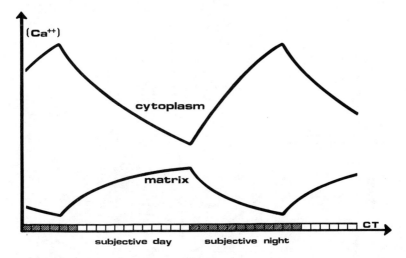

**FIGURE 4.** Postulated changes in the concentration of cytoplasmic and matrix calcium through a circadian cycle under constant conditions. Time scale normalized, 1 unit of CT (circadian time) corresponds to 1/24 of the free-running period.

## Explaining the Characteristic Features of the Cellular Clock

### Achieving a Long (24 Hour) Period

This is certainly one of the most difficult problems in modeling circadian rhythms. Chay presented a model for biological oscillations which seems adequate to explain long, even circadian, periods.[100] The molecular basis of her model are counteracting processes utilizing the proton gradient across the inner mitochondrial membrane. The presumptions made for the parameters of this feedback system, however, are equally applicable to the MCC. It may, therefore, be suggested that the MCC model can account for a circadian period without a requirement for any further assumptions. Yet, there are additional molecular mechanisms to attain a long period. Hysteresis is a phenomenon well known to be involved in the generation of oscillations with comparably low frequencies; notably, several calcium-transport systems do indeed display hysteresis.[79,101] In addition, there are recent reports of $Ca^{2+}$/CaM-dependent protein kinases undergoing autophosphorylation-dependent "molecular switches" which may last as long as the lifetime of the kinase molecule itself,[102] thereby a considerable time delay is introduced into the system. The incorporation of auxiliary mechanisms like these would definitely qualify the model of Chay to account for a 24 hour period without demanding an extreme value for any of the parameters.

### Entrainment and Phase Shifting

The inherent period of circadian rhythms, observed only under constant conditions, is different from 24 hours and has to be entrained, i.e., synchronized to the geophysical

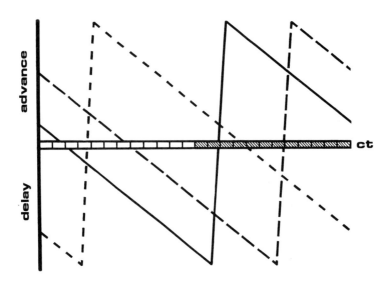

**FIGURE 5.** Simplified representation of the nonrandom distribution of phase response curves. Each of the lines stands for a cluster of phase response curves suggested to be associated with a distinct redistribution of cellular calcium: Solid line, general decrease in cellular calcium; long-dashed line, stimulated efflux of calcium from mitochondria; short-dashed line, general increase in cellular calcium.

cycle. For that reason the rhythm has to be accessible to phase shifting, a property widely used in the analysis of the circadian clock. Phase response curves (PRC) are experimentally established indicating the magnitude of a shift (advance or delay) after a perturbation placed at a certain time of the circadian cycle. Reanalysis of the numerous PRCs reported for *Neurospora crassa*[103] and *Aplysia californica,*[104] two objects of intensive circadian research, reveals a nonrandom distribution. Rather do the PRCs fall within clusters (indicated in FIGURE 5), and the interpretation of this pattern is in good agreement with the MCC model. Several equivalent PRCs result from treatments promoting $Ca^{2+}$ efflux from mitochondria; this cluster includes PRCs obtained with light pulses of a wavelength appropriate for entrainment in the respective organisms.[105] In plants, for example, $Ca^{2+}$ efflux was found to be stimulated by red light and phytochrome[106] (light at the same time triggered $Ca^{2+}$ uptake into chloroplasts).[107] Another cluster of PRCs is suggested to be associated with a general increase in cellular calcium. Virtually all of these PRCs are obtained with treatments inducing the heat shock response.[108] And indeed, cellular calcium does rise after heat shock[109] and other treatments equally effective in eliciting the heat shock response (Calderwood, personal communication) and phase shifting the circadian clock.

*General Homeostasis and Temperature Compensation*

In most phase-shift experiments high doses of the perturbing agent are required to obtain significant phase shifts. To only small perturbations the clock hardly reacts.

At first glance, there seems to emerge a difficulty in explanation since many external stimuli as well as normal cellular functions will cause transient changes of cytoplasmic $[Ca^{2+}]$. However, the MCC does not appear to be sensitive to such short-time transients. In heart muscle cells, for example, the MCC—because of its slow responses—does not interfere with beat-to-beat changes of $[Ca^{2+}]$.[110] There may be a similar insensitivity to transient changes of $[Ca^{2+}]$ accompanying processes such as ameboid movement, ciliary beating, endocytosis, etc. For temperature compensation several different explanations are available. Firstly, both transport pathways may have a very similar $Q_{10}$, the resulting net flux being only marginally altered at different temperatures. The alternative view of homeoviscous adaptation of membranes to altered temperatures[111] being responsible for steady-state temperature independence is favored by most of the membrane models for circadian rhythms.[84] And, of course, the involvement of both types of mechanisms in the temperature-compensation process is still another feasible possibility.

*Coupling of Hands to the Clock*

It is generally accepted that various overt rhythms are coupled to a central timing device. In that respect, the MCC model seems the ideal means to explain how very different physiological functions could be controlled by a single master clock. As an example, FIGURE 6 shows several functions that are subject to circadian modulation in the unicellular organism *Euglena gracilis*.[112] All of them are known to be sensitive to the actual calcium concentration and, therefore, can be suggested as being directly

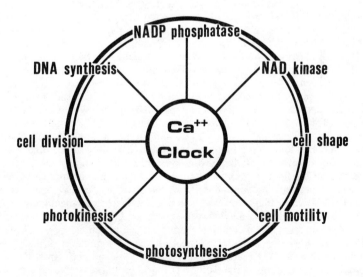

**FIGURE 6.** Physiological functions that in the unicellular *Euglena gracilis* display circadian rhythmicity and, being sensitive to the actual calcium concentration, are suggested to be directly coupled to a central "calcium clock."

coupled to a central "$Ca^{2+}$ clock." Moreover, the phase relationship of these processes to the daily cycle is also known and displays exactly the pattern predicted by the model: those processes requiring higher [$Ca^{2+}$] have their maximum toward the end of the day while those demanding lower [$Ca^{2+}$] peak during late night phase. Circadian—and other endogenous—rhythms do thus appear to provide temporal compartmentation of processes requiring different levels of [$Ca^{2+}$] in order to attain their optimal activity.

## CONCLUSIONS

The dominating theme of this paper has been the tremendous gap between prokaryotes and eukaryotes becoming evident when aspects of cellular regulation are considered. Several essential regulatory pathways ubiquitous in eukaryotes but absent in prokaryotes distinguish the two elementary subdivisions of life. Of preeminent significance for further eukaryotic evolution seem those that can be suggested to have evolved as a consequence of the symbiotic origin of the eukaryotic cell. Both endogenous clocks and calcium signaling appear to play pivotal roles in any process—such as development, growth, and differentiation—involved in the emergence of higher forms of eukaryotic organization. Not wanting to debase the relevance of features such as a separate nuclear compartment, sex, and meiosis, I would nevertheless like to end up with the provocative conclusion that those features resulting from a symbiotic origin of the eukaryotic cell contribute the more fundamental part to the quintessence of being eukaryotic.

[Note added in proof: Since submission of the manuscript several reports appeared concerned with two central aspects of this paper. A contribution to the discussion of the evolution of temporal structures are two recent papers demonstrating rhythms in cyanobacteria which persisted under continuous illumination (Grobbelaar, N., et al. 1986. FEMS Microbiol. Lett. 37: 173-177; Mitsui, A., et al. 1986: Nature 323: 720-722). Being close to 24 hours, these could be circadian rhythms; the data available to date do however permit alternative explanations. These studies were done on the activity of nitrogenase in $N_2$-fixing nonheterocystous species. In this type of cyanobacteria oscillations in nitrogenase activity have been reported earlier which, in continuous light, were far from 24 hours and partly irregular (Weare, N. M. & J. R. Benemann. 1974. J. Bact. 119: 258-265; Mullineaux, P. M., et al. 1981. FEMS Microbiol. Lett. 10: 245-247; Huang, T. C. & T. J. Chow. 1986. FEMS Microbiol. Lett. 36: 109-110). Since in these organisms two basically incompatible reactions, while not being separated spatially, have to be separated temporally, simple feedback could be responsible for the observed oscillations. In the species studied by Mitsui's group are these metabolic processes temporally integrated within the cell cycle (Mitsui, A., et al. 1987. Physiol. Plant. 69: 1-8) which under their experimental conditions is about 20 hours. Since no data for different growth conditions (light and temperature) are available (Mitsui, personal communication), no distinction can presently be made between an indication for the occurrence of circadian rhythms in prokaryotes and mere coincidence under a particular set of experimental conditions. In any case, new interesting systems for studying the evolution of timekeeping strategies seem to emerge. The second aspect is the role played by CaM in clock and coupling processes. "Molecular switch" mechanisms as described do now appear to be frequently employed in eukaryotic regulation (Lou, L. L., et al. 1986. Proc. Nat. Acad. Sci. USA 83:

9497-9501; Schworer, C. M., *et al.* 1986. J. Biol. Chem. **261:** 8581-8584). Given the requirement of various different overt rhythms being coupled to one central clock, it is important to note that this type of activity modulation is shown for the multifunctional $Ca^{2+}$/CaM-dependent protein kinase. Lonergan (1986. Plant Physiol. **82:** 226-229) demonstrated the involvement of CaM in the coupling pathway between the clock and the overt rhythm of photosynthesis. Taken together with the results presented by Edmunds (this volume), the picture of a "multifunctional" role of CaM in timekeeping emerges, providing additional support to the hypothesis of the $Ca^{2+}$/CaM system of intracellular signaling being an integral part of the cellular clock.]

## REFERENCES

1. CAVALIER-SMITH, T. 1981. Symp. Soc. Gen. Microbiol. **32:** 33-84.
2. TRIBE, M., A. MORGAN & P. WHITTAKER. 1982. The Evolution of Eukaryotic Cells. Arnold. London, England.
3. CARLILE, M. J. 1980. Symp. Soc. Gen. Microbiol. **30:** 1-40.
4. DODSON, E. O. 1979. Can. J. Microbiol. **25:** 651-674.
5. STEINERT, P. M. & D. A. D. PARRY. 1985. Annu. Rev. Cell Biol. **1:** 41-65.
6. SCHEER, U., H. HINSSEN, W. W. FRANKE & B. M. JOCKUSCH. 1984. Cell **39:** 111-122.
7. MASTERS, C. 1984. J. Cell Biol. **99:** 222s-225s.
8. OTTO, A. M. 1982. Cell Biol. Int. Rep. **6:** 1-18.
9. TRAUB, P. 1985. Ann. N.Y. Acad. Sci. **455:** 68-78.
10. NIELSEN, P., S. GOELZ & H. TRACHSEL. 1983. Cell Biol. Int. Rep. **7:** 245-254.
11. GANCEDO, J. M., M. J. MAZON & P. ERASO. 1985. TIBS **10:** 210-212.
12. NISHIZUKA, Y. 1984. Nature **308:** 693-697.
13. SCHÄFER, A., F. BYGRAVE, S. MATZENAUER & D. MARME. 1985. FEBS Lett. **187:** 25-28.
14. MACIVER, S. & C. KING. 1986. Eur. J. Cell Biol. **40**(suppl. 13): 12.
15. UEDA, K. & O. HAYAISHI. 1985. Annu. Rev. Biochem. **54:** 73-100.
16. GAAL, J. C. & C. K. PEARSON. 1986. TIBS **11:** 171-175.
17. BERRIDGE, M. J. 1986. J. Cell. Sci. Suppl. **4:** 137-153.
18. DROBAK, B. K. & I. B. FERGUSON. 1985. Biochem. Biophys. Res. Commun. **130:** 1241-1246.
19. EUROPE-FINNER, G. N. & P. C. NEWELL. 1986. Biochim. Biophys. Acta **887:** 335-340.
20. PALL, M. L. 1985. Curr. Top. Cell. Regul. **25:** 1-20.
21. BOURNE, H. R. 1986. Nature **321:** 814-816.
22. TATCHELL, K., L. C. ROBINSON & M. BREITENBACH. 1985. Proc. Nat. Acad. Sci. USA **82:** 3785-3789.
23. ZILLIG, W., R. SCHNABEL & K. STETTER. 1985. Curr. Top. Microbiol. Immunol. **114:** 1-18.
24. SEARCY, D. G. 1986. System. Appl. Microbiol. **7:** 198-201.
25. SEARCY, D. G. 1987. Ann. N.Y. Acad. Sci. (This volume.)
26. ENRIGHT, J. T. 1971. Annu. Rev. Ecol. System. **1:** 221-238.
27. WINFREE, A. T. 1980. The Geometry of Biological Time. Springer. New York, N.Y.
28. KIPPERT, F. 1986. J. Interdiscipl. Cycle Res. **16:** 77-84.
29. WILLE, J. J. 1979. *In* Biochemistry and Physiology of Protozoa. M. Levandowsky & S. H. Hutner, Eds.: 67-148. Academic Press. New York, N.Y.
30. LLOYD, D. & F. KIPPERT. 1987. Symp. Soc. Exp. Biol. **41.** (In press.)
31. MARGULIS, L. 1981. Symbiosis in Cell Evolution. Freeman. San Francisco, Calif.
32. CAVALIER-SMITH, T. & J. J. LEE. 1985. J. Protozool. **32:** 376-379.
33. WHATLEY, J. M. 1976. New Phytol. **76:** 111-120.
34. SMITH, D. C. 1980. *In* Endocytobiology I. W. Schwemmler & H. E. A. Schenk, Eds.: 317-332. de Gruyter. Berlin, FRG.

35. REISSER, W. 1980. *In* Endocytobiology I. W. Schwemmler & H. E. A. Schenk, Eds.: 97-104. de Gruyter. Berlin, FRG.
36. REISSER, W., R. MEIER & B. KURMEIER. 1983. *In* Endocytobiology II. H. E. A. Schenk & W. Schwemmler, Eds.: 533-543. de Gruyter. Berlin, FRG.
37. WEIS, D. S. 1977. Trans. Am. Microsp. Soc. **96:** 82-86.
38. BIRKY, C. W. & R. K. SKAVARIL. 1984. J. Theor. Biol. **106:** 441-447.
39. MCAULEY, P. J. 1985. J. Cell Sci. **77:** 225-239.
40. DOUGLAS, A. & D. C. SMITH. 1984. Proc. R. Soc. London Ser. B **221:** 291-319.
41. RUTHMANN, A. & G. NOLL-ALTMANN. 1980. *In* Endocytobiology I. W. Schwemmler & H. E. A. Schenk, Eds.: 361-370. de Gruyter. Berlin, FRG.
42. CHANG, K. P. & C. DAVE. 1980. *In* Endocytobiology I. W. Schwemmler & H. E. A. Schenk, Eds.: 349-359. de Gruyter. Berlin, FRG.
43. LLOYD, D., R. K. POOLE & S. W. EDWARDS. 1982. The Cell Division Cycle. Academic Press. London, England.
44. COTTRELL, S. F. 1982. Cell Biol. Int. Rep. **6:** 125-130.
45. KUROIWA, T., M. HIZUME & S. KAWANO. 1978. Cytologia **43:** 119-136.
46. DONNER, B. & L. RENSING. 1984. Eur. J. Cell Biol. **35:** 143-145.
47. WHATLEY, J. M. 1983. Int. Rev. Cytol. **Suppl. 14:** 329-373.
48. POSSINGHAM, J. V. & M. E. LAWRENCE. 1983. Int. Rev. Cytol. **84:** 1-56.
49. LEWIN, R. A., L. CHENG & J. MATTA. 1984. Phycologia **23:** 505-507.
50. BÜNNING, E. 1978. Arzneim. Forsch. Drug Res. **28:** 1811-1813.
51. PAIETTA, J. 1982. J. Theor. Biol. **97:** 77-82.
52. LEVANDOWSKY, M. 1981. Ann. N.Y. Acad. Sci. **361:** 369-374.
53. EDMUNDS, L. N., Ed. 1984. Clocked Cell Cycle Clocks. Marcel Dekker. New York, N.Y.
54. KLEVECZ, R. R., S. A. KAUFFMAN & R. M. SHYMKO. 1984. Int. Rev. Cytol. **86:** 97-128.
55. GOLDBETER, A., J. L. MARTIEL & O. DECROLY. 1984. *In* Dynamics of Biochemical Systems. J. Ricard & A. Cornish-Bowden, Eds.: 173-212. Plenum Press. New York, N.Y.
56. GLASS, L. & M. C. MACKEY. 1979. Ann. N.Y. Acad. Sci. **316:** 214-235.
57. SILYN-ROBERTS, H. & W. ENGELMANN. 1986. Endocyt. Cell Res. **3:** 239-242.
58. KIPPERT, F. 1985. J. Interdiscipl. Cycle Res. **16:** 272-273.
59. LLOYD, D., S. W. EDWARDS & J. C. FRY. 1982. Proc. Nat. Acad. Sci. USA **79:** 3785-3788.
60. KIPPERT, F. 1985. Eur. J. Cell Biol. **38**(Suppl. 9): 16.
61. KLEVECZ, R. R. & G. A. KING. 1982. Exp. Cell Res. **140:** 307-313.
62. MAYER, W. E. 1966. Planta **70:** 237-256.
63. SILYN-ROBERTS, H., W. ENGELMANN & K. G. GRELL. 1986. J. Interdiscipl. Cycle Res. **17:** 181-187.
64. MATTERN, D. L., L. R. FORMAN & S. BRODY. 1982. Proc. Nat. Acad. Sci. USA **79:** 825-829.
65. LYSEK, G. & B. NORDBRING-HERTZ. 1981. Planta **151:** 50-53.
66. CAMPBELL, A. K. 1983. Intracellular Calcium: Its Universal Role as Regulator. Wiley & Sons. Chichester, England.
67. HEPLER, P. K. & R. O. WAYNE. 1985. Annu. Rev. Plant Physiol. **36:** 397-439.
68. KLEE, C. B. & T. C. VANAMAN. 1982. Adv. Protein Chem. **35:** 213-316.
69. MEANS, A. R., J. S. TASH & J. G. CHAFOULEAS. 1982. Physiol. Rev. **62:** 1-39.
70. NAKAMURA, T., K. FUJITA, Y. EGUCHI & M. YAZAWA. 1984. J. Biochem. **95:** 1551-1557.
71. GLAGOLEV, A. N. 1984. Motility and Taxis in Prokaryotes. Harwood Academic Publications. Chur, Switzerland.
72. HÄDER, D. P. 1986. Biochim. Biophys. Acta **864:** 107-122.
73. SCHIMZ, A. 1982. FEBS Lett. **139:** 283-286.
74. STAHL, S. 1978. Arch. Microbiol. **119:** 17-24.
75. ROSEN, B. P. 1982. *In* Membrane Transport of Calcium. E. Carafoli, Ed: 187-216. Academic Press. London, England.
76. BLACK, C. C. & J. J. BRAND. 1986. *In* Calcium and Cell Function. W. Y. Cheung, Ed.: **6:** 327-355. Academic Press. New York, N.Y.
77. BARR, R., K. S. TROXEL & F. CRANE. 1982. Biochem. Biophys. Res. Commun. **104:** 1182-1188.

78. CHEUNG, W. Y. 1981. Cell Calcium **2**: 263-280.
79. RASMUSSEN, H. & P. Q. BARRET. 1984. Physiol. Rev. **64**: 938-984.
80. CHAFOULEAS, J. G., W. E. BOLTON, A. E. BOYD & A. R. MEANS. 1982. Cell **28**: 41-50.
81. SILLERS, P. J. & A. FORER. 1985. Cell Biol. Int. Rep. **9**: 275-282.
82. LEVANDOWSKY, M. 1987. Ann. N.Y. Acad. Sci. (This volume.)
83. KIPPERT, F. 1984. Poster presented at the Symposium "Temporal Order," Bremen.
84. ENGELMANN, W. & M. SCHREMPF. 1980. Photochem. Photobiol. Rev. **5**: 49-86.
85. BRODY, S., C. DIECKMANN & S. MIKOLAJCZYK. 1985. Mol. Gen. Genet. **200**: 155-161.
86. BOLLIG, I., K. MAYER, W. E. MAYER & W. ENGELMANN. 1978. Planta **114**: 225-230.
87. GOTO, K. 1984. Z. Naturforsch. **39c**: 73-84.
88. ESKIN, A. & G. CORRENT. 1977. J. Comp. Physiol. **117**: 1-21.
89. WOOLUM, J. C. & F. STRUMWASSER. 1983. J. Comp. Physiol. **151**: 253-259.
90. GOTO, K., D. L. LAVAL-MARTIN & L. N. EDMUNDS. 1985. Science **228**: 1284-1288.
91. NICHOLLS, D. & K. ÅKERMAN. 1982. Biochim. Biophys. Acta **683**: 57-88.
92. CARAFOLI, E., G. INESI & B. P. ROSEN. 1984. In Metal Ions in Biological Systems. H. Sigel, Ed. **17**: 129-186. Marcel Dekker. New York, N.Y.
93. MOORE, A. L. & K. E. O. ÅKERMAN. 1984. Plant Cell Environ. **7**: 423-429.
94. PITT, D. & U. O. UGALDE. 1984. Plant Cell Environ. **7**: 467-475.
95. MARTINS, I. E., E. G. S. CARNIERI & A. E. VERCESI. 1986. Biochim. Biophys. Acta **850**: 49-56.
96. KIM, J. V., L. J. KUDZINA, V. P. ZINCHENKO & J. V. EVTODIENKO. 1984. Cell Calcium **5**: 29-41.
97. HOLMES, R. P. & P. R. STEWART. 1979. Biochim. Biophys. Acta **545**: 94-105.
98. ÅKERMAN, K. E. O. & D. G. NICHOLLS. 1983. Rev. Physiol. Biochem. Pharmacol. **95**: 149-201.
99. McCORMACK, J. G. & R. M. DENTON. 1986. TIBS **11**: 258-262.
100. CHAY, T. R. 1981. Proc. Nat. Acad. Sci. USA **78**: 2204-2207.
101. SCHARFF, O. & B. FODER. 1986. Biochim. Biophys. Acta **861**: 471-479.
102. MILLER, S. G. & M. B. KENNEDY. 1986. Cell **44**: 861-870.
103. FELDMAN, J. F. & J. C. DUNLAP. 1983. Photochem. Photobiol. Rev. **7**: 319-368.
104. JACKLET, J. W. 1984. Int. Rev. Cytol. **89**: 251-294.
105. NINNEMANN, H. 1979. Photochem. Photobiol. Rev. **4**: 207-266.
106. SERLIN, B. S., S. K. SOPORY & S. J. ROUX. 1984. Plant Physiol. **74**: 827-833.
107. KREIMER, G., M. MELKONIAN & E. LATZKO. 1985. FEBS Lett. **180**: 253-258.
108. NOVER, L. 1984. Heat Shock Response of Eukaryotic Cells. Springer. Berlin, FRG.
109. STEVENSON, M. A., S. K. CALDERWOOD & G. M. HAHN. 1986. Biochem. Biophys. Res. Commun. **137**: 826-833.
110. CROMPTON, M. 1986. Curr. Top. Membranes Transp. **25**: 231-277.
111. COSSINS, A. R. 1983. In Cellular Acclimatisation to Environmental Change. R. Cossins & P. Sheterline, Eds.: 3-32. Cambridge University Press. Cambridge, England.
112. EDMUNDS, L. N. 1984. Adv. Microb. Physiol. **25**: 61-148.

# Endocytobionts and Mitochondria as Determinants of Leafhopper Egg Cell Polarity[a]

WERNER SCHWEMMLER

*Institute for Plant Physiology, Cell Biology and Microbiology*
*Free University of Berlin*
*1000 Berlin 33, Federal Republic of Germany*

## INTRODUCTION

The cytoplasm of probably all eggs is not uniformly structured, but exhibits regional differences in morphological as well as chemical properties. These differences occur in many eggs along a specific axis—the egg axis. This leads to the formation of different poles. This egg polarity subsequently influences development. The anterior pole region of oblong insect eggs normally differentiates, for example, to head-thorax segments; the posterior pole to abdominal segments.

The evidence from developmental physiological investigations indicates the existence of anterior and posterior organizational factors having counter-current concentration gradients, whose respective component ratios are supposed to determine and thereby control the segment formation of head, thorax, and abdomen.[1] The morphology and chemistry of the postulated determinants have not yet been elucidated, despite intensive research activities. In addition, the causal relationship is completely unknown, whereby such determinants play a role in the molecular mechanism of initiation and regulation of embryogenesis. This is counted among the major, still unsolved, puzzles of biology. With the solution of this specific problem of cellular differentiation, one also expects significant information on the molecular mechanism of cellular dedifferentiation like tumor formation. We have attempted to create the necessary conditions for commencing the chemical analysis of determinants, by elucidation of the basic morphological phenomena of oogenesis, blastoderm formation, and germ layer development using the small leafhopper *Euscelidius variegatus* as a model.

[a]The research was aided in part by funds from the Deutsche Forschungsgemeinschaft DFG (German Science Foundation) (Schw 175/10-2); above all, however, within the framework of a "research-project FPS—Structure and Evolution of the Cell" established by the Free University Berlin.

# MATERIAL AND METHODS

## Morphological Analysis

The experiments were conducted on growing oocytes, undeposited unfertilized eggs, as well as fertilized deposited eggs. The breeding of leafhoppers and eggs, preparation of chronologically dated eggs, vital observation, symbiont elimination attempts, as well as normal histology were conducted according to methods already published.[2,3] The techniques applied for the microstructure investigations have also already been published elsewhere.[4,5] The microorganization of the egg surface was analyzed with the usual techniques of scanning electron microscopy.[6,7]

## Physiochemical Analysis

The intracellular pH was measured in eggs that were treated in mature females by abdominal injection of specific fluorescing dyes (e.g., pyranine, DASPMI, and neutral red: 3 $\mu$M), or obtained by microinjection (Eppendorf 5242) into the egg and then evaluated by means of ultraspectral photometry, photography, and densitometry.[8] The total distribution and localization of ions ($K^+$, $Na^+$, $Ca^{2+}$, $Mg^{2+}$) were measured by the established methods of atomic absorption spectroscopy (AAS) and flame emission spectroscopy (FES),[9] as well as by x-ray microanalysis.[10] For x-ray microanalysis, the leafhopper eggs were cryogenically fixed (Freon, $-150°C$), then treated with 0.5% $OsO_4$/acetone, and embedded in epoxy resin.[11] Sections (0.5 $\mu$m thick) were mounted on 75 mesh Cu grids for evaluation. The physiological metabolic investigations were conducted by measurements in microrespiration measuring instruments according to Kalthoff,[12] by means of enzyme substrate tests,[13] as well as the application of fermentation inhibitors (monoiodoacetate) and respiration inhibitors (KCN, antimycin A, rotenone, benzohydroxamic acid).[14] Measurements of energy metabolism were conducted using bioluminescence methods.[15,16]

The experiments on gene expression were conducted with intact leafhoppers as well as with dissected symbiont organs and symbiont balls enriched for endocytobionts. For the isolation of host and endocytobiont DNAs, either radioactive labels were used followed by classical extraction procedures[17] or the CTAB method was applied to unlabeled material.[18] The method of Gause and Mikhailov was used for the isolation of mitochondrial DNA.[19] The characterization of the DNAs was determined by restriction digestion, hybridization, cloning, and sequencing. In order to obtain sufficient pure endocytobiont fractions from entire leafhoppers for DNA isolation, the improved methods of Ishikawa with a wider Ficoll gradient were employed as originally developed for the isolation of DNA-containing cell organelles.[20]

# RESULTS

## Morphology of the Determinants

### Oogenesis

As is well known, the polarity of the insect egg is already determined by its position in the ovary of the female. It was therefore interesting to examine certain aspects of oogenesis in *Euscelidius* to determine the extent to which it may eventually yield indications on the morphological nature of the determinants. The ovaries of leafhopper females are arranged in pairs and each consists of seven ovarioles which exit jointly through the vagina. The ovarioles belong to the meroistic telotrophic ovariole type.[21] In such ovarioles, the oocytes develop from the germ cells (oogonia) according to a predetermined divisional pattern; in addition the nutrient cells also develop from oogonia and remain in the nutrient chamber (germarium) of the terminal segment (FIGURE 1). The egg cells on the other hand migrate basally and penetrate into the egg chamber (vitellarium) where they are successively arranged in a follicle and gradually grow with an enormous increase of their cell bodies. During this growth process, the oocytes remain at the anterior pole united with the nutrient cells by nutrient strands. Nutrient-rich yolk (deutoplasm) arrives at the oocytes by means of the nutrient strands in the form of protein-, lipid-, and glycogen-containing granules, followed by ooplasmic organelles such as, among others, ribosomes, endoplasmic reticulum, and mitochondria. These structures proceeding from the anterior pole are then inhomogeneously incorporated in the oocyte either as yolk in the central deutoplasm or as organelles almost quantitatively in the peripheral periplasm, and a negligible remainder in the finely divided reticuloplasm of the deutoplasm or around the nucleus. Before the initial separation of the yolk membrane from the ooplasm (vitelline membrane) and then the egg shell from the follicle (chorion), about 200 bacterialike endocytobionts (protoplastoids) are finally embedded between the egg cell and egg membrane on the opposite posterior pole of the mature egg.

Considering the morphologically observable inhomogeneous incorporation of mitochondria and endocytobionts in the ooplasm, such structures appear to be suitable candidates for a direct or indirect relationship to putative anterior and posterior development factors.[5] In following the continuing egg development, special attention should therefore be directed to the further destiny of these two structures.

### Blastoderm Formation

The fertilization of the mature egg occurs during its passage through the vagina.[21] Thereby sperm from the receptaculum seminis penetrate into the egg interior through a structure of 13-15 $\mu$m diameter, a so-called micropyle (FIGURE 2), blocked by a lipid plug located on the egg anterior pole at about 75% of the egg length (EL). The sperm that is the first to reach the vicinity of the egg nucleus in the central or anterior portion of the egg discards its tail segment, develops a centriole from the central segment and a male pronucleus from the head segment; the other penetrating sperms

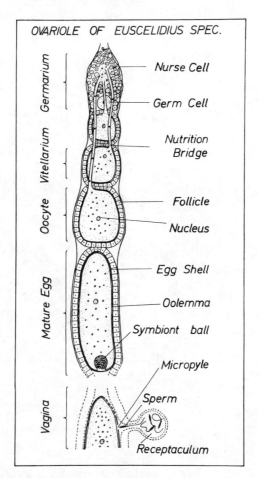

**FIGURE 1.** Oogenesis and egg fertilization in the small leafhopper *Euscelidius variegatus* based upon normal, semimicrohistology, and ultramicrohistology. The ovary type is meroistic telotrophic (partly with reference to 21).

are destroyed. During this event, the mall pronucleus fuses with the female pronucleus, which has previously divided according to a definite plan, to produce the zygote nucleus. The zygote nucleus begins the first stage of division in 4-5 hours at an exterior temperature of 24°C, following organization of the spindle apparatus from the male centriole. Both daughter nuclei, however, do not produce a cell membrane, but remain in a plasma area as so-called synergids which are interconnected with one another via plasma bridges. Subsequently, the daughter nuclei proceed through a series of synchronized nuclear division cycles, whereby the nuclei number is doubled every 1-2 hours without the formation of cell membranes (FIGURE 3). After the 7th division (128 nuclei), the synergids have finally been distributed over the entire egg area. Only after the 8th division (256 nuclei) do they begin to preferentially migrate into the egg cortex; the cortex of a leafhopper egg constitutes the outermost periplasm layer whose

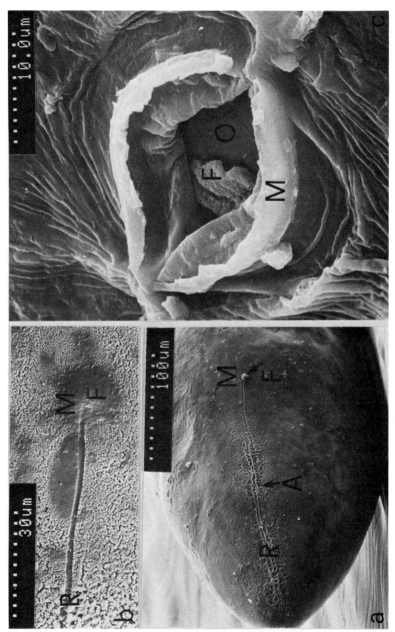

**FIGURE 2.** The micropyle of the leafhopper egg on the anterior pole as seen by scanning electron microscopy (micrographs, U. Gernert). (a) Performed rupture track for larval emergence in an unfertilized egg with the micropyle on the inner end sealed with a lipid plug. (b) Enlargement form a. (c) The micropyle of the fertilized egg opened by the penetrating sperm. A, aeropyle; F, lipid plug; M, micropyle; O, oolemma; R, scored track.

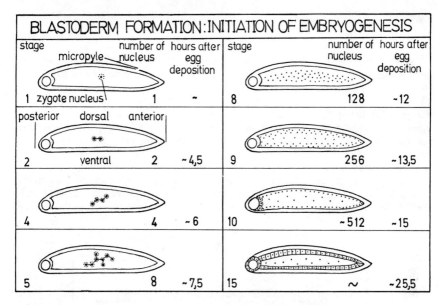

| BLASTODERM FORMATION: INITIATION OF EMBRYOGENESIS | | | | | | |
|---|---|---|---|---|---|---|
| stage | number of nucleus | hours after egg deposition | stage | | number of nucleus | hours after egg deposition |
| 1  zygote nucleus | 1 | ~ | 8 | | 128 | ~12 |
| 2 | 2 | ~4,5 | 9 | | 256 | ~13,5 |
| 4 | 4 | ~6 | 10 | | ~512 | ~15 |
| 5 | 8 | ~7,5 | 15 | | ~ | ~25,5 |

**FIGURE 3.** The first steps of division of the fertilized leafhopper egg until the blastoderm stage by means of vital preparation and normal histology. This initiation phase of embryogenesis is clearly different developmentally and physiologically from the succeeding phase of germinal layer formation. The time data are based upon a development temperature of 24°C.

organelles are difficult to displace even with centrifugation in comparison to the central deutoplasm. A few nuclei remain as vitellophages in the deutoplasm and multiply there further asynchronously. After the 9th division (circa 512 nuclei), the first onset of cell membrane formation occurs around certain synergids of the henceforth single-layer cortical nuclear stratum on the posterior pole. This process is completed at the 14th division, about 25-26 hours after egg deposition, by peripheral formation of the single-layer blastoderm.

The endocytobionts are taken up by the posterior-pole ooplasm during the blastoderm-formation phase and are there covered with a single layer of cells. The distribution of the mitochondria between the anterior and posterior cortex area shows negligible quantitative differences in the *Euscelidius* egg. However, small compact mitochondria dominate at the anterior pole region, whereas more oblong hyaline forms can be observed in the anterior pole region[5] (cp. also FIGURE 10).

At the conclusion of blastoderm formation, the mitochondria are found almost exclusively in the blastoderm or in the vitellophages. As we know, the determination and consequently the differentiation of the embryonic cells only begin after the blastoderm stage. Until then only a few genes are expressed (probably primarily for the formation of the cell membrane), since the egg cell is fully occupied with the doubling of its nuclear material[22] and is equipped for initial development performance with a "dowry" of diverse RNA types by the nutrient strands. Consequently egg development up to blastoderm formation is genetically as well as physiologically distinctly separate from the subsequent development of the embryonic germ layers.

*Germ Layer Development*

Development of the embryonic germ layers begins about 30 hours following the egg deposition with the segregation of the blastoderm into two cell areas, from which the area with smaller and numerous cells becomes the paired pre-germ anlage (FIGURE 4). About 40 hours following egg deposition, this pre-germ anlage contracts to form the germ anlage at the posterior pole; the germ anlage, in contact with the endocytobiont infection mass, begins to curl (anatrepsis) in the deutoplasm toward the anterior pole after the 2nd day. On the 3rd day the completely curled germ anlage becomes the germ band by segmentation, already allowing recognition of the basic body form. Following the caudal flexure on the 4th day, the germ band reaches its maximum length on the 5th day, but this is partially offset by embryonic shortening on the 6th day. On the 6th day the first signs of red eye pigment are indicated. After the 7th day, the embryo begins to uncurl (katatrepsis); from the 8th day it assumes

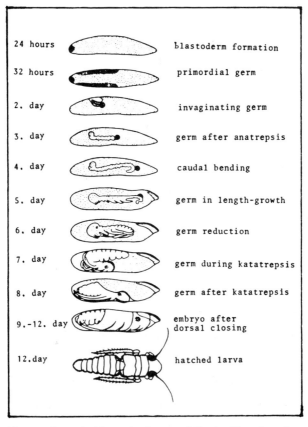

| | |
|---|---|
| 24 hours | blastoderm formation |
| 32 hours | primordial germ |
| 2. day | invaginating germ |
| 3. day | germ after anatrepsis |
| 4. day | caudal bending |
| 5. day | germ in length-growth |
| 6. day | germ reduction |
| 7. day | germ during katatrepsis |
| 8. day | germ after katatrepsis |
| 9.-12. day | embryo after dorsal closing |
| 12.day | hatched larva |

**FIGURE 4.** Diagram of germinal layer development following blastoderm formation in *Euscelidius variegatus* by means of vital observation and special histology.[3]

its final position in the egg following dorsal closure, whereby the differentiation of the organ anlage commences. From the 12th day, the embryo larva emerges.

Following the third day, the endocytobionts are incorporated according to a fixed predetermined program in special cells of the embryo (bacteriocytes), which later contract to form a common symbiont organ (bacteriome) finally yielding separate paired organs arranged left and right in the abdomen of the ready-to-emerge embryo. During embryonic development, the mitochondria show the spectrum of substructures typical for differentiated tissue. If one hinders the infection of the leafhopper egg with endocytobionts through chemical intervention (lysozyme), one obtains symbiont-free eggs from which head-thorax embryos lacking an abdomen develop following egg deposition (FIGURE 5).[4] Extensive functional elimination of the egg mitochondria

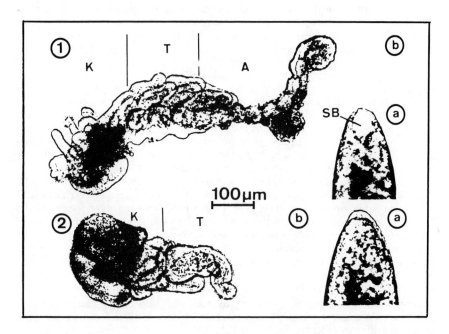

**FIGURE 5.** An approximately one-week-old leafhopper embryo. (1a) Egg posterior pole with symbiont ball. (1b) Normal embryo developed from 1a. (2a) Egg posterior pole without symbiont ball. (2b) Head-thorax embryo without abdomen developed from 2a.[4] A, abdomen anlage; K, head anlage; SB, symbiont ball; T, thorax anlage.

through antibiotics, on the other hand, yields malformations on "headless" abdomens.[23] It was thereby shown for the first time that endocytobionts directly or indirectly have an influence on the determinants controlling the abdomen, and possibly the mitochondria influence the determinants that control the head-development factors in the egg. Now that ooplasmic determinants could be narrowed down from the morphological aspects, it is interesting to attempt to characterize their chemical nature.

*Chemistry of the Determinants*

*Metabolism*

On the anterior pole of the leafhopper egg, a double track of irregularly arranged egg pores, the so-called aeropyles, extend between 80% and 100% of the EL and serve in the larval emergence process as a scored track and also function in the egg's gaseous exchange with the surrounding milieu (FIGURE 6). *Preliminary* measurements in microrespiration chambers yield a respiration quotient RQ = 0.92 ± 0.2[b] per medium dry weight of the unfertilized egg, and RQ = 0.63 ± 0.2[b] per medium dry weight of the fertilized egg after the third day. By artificially sealing the aeropyles and the micropyle opened upon penetration of the sperm, a developmental halt occurs, even if only after the third day. If one assumes that RQ values above 1 point to fermentation and values under 1 to respiration, these circumstances can be interpreted to mean that fermentation and respiration are both ongoing in an unfertilized egg—the measured RQ value of 0.92 could result from the averaging of a "glycolytic RQ value" of, for example, 1.2 and an "oxidative RQ value" of, for example, 0.6. In the fertilized egg, on the other hand, fermentation could be considerably reduced in favor of respiration.

For an exact examination of this presumption, various fermentation and respiration inhibitors were applied either externally or injected into the egg interior (cf. FIGURES 11 and 12). It was thus *preliminarily* shown that glycolytic and bypass inhibitors (e.g., monoiodoacetate to inhibit phosphofructokinase: $10^{-2}$ to $10^{-5}$ M; benzohydroxamic acid to specifically inhibit the electron transport chain: $1.5 \times 10^{-3}$ M) do not visibly disturb egg development. The same applies for short-term effects (3-7 hours) of respiration inhibitors (antimycin A, potassium cyanide, rotenone for specific inhibition of the electron transport chain, all at $10^{-3}$ M). If one however allows the respiration inhibitors to exert their effect over a long time, egg development totally breaks down, again about after the third day. This appears to confirm the initial presumption that the electron transport chain is not participating or only to a limited extent in the egg at the beginning of development: thus fermentation activities dominate, which are substantially reduced from the third day after egg deposition in favor of respiration activities. Now the question is consequently raised, Where exactly are such indirectly detectable fermentation and respiration activities localized in the egg?

*Preliminary* ion analyses, according to AAS and FES of egg plasm, egg shell (vitelline membrane + chorion), and symbiont masses of an unfertilized egg, yield an ion constitution (FIGURE 7):[24]

$$Na^+ \; 71\% > K^+ \; 18.7\% > Ca^{2+} \; 7.2\% > Mg^{2+} \; 2.9\%$$

There results from this, for example, an average concentration in the egg of 20 mM $Ca^{2+}$, which also is in good agreement with the x-ray microanalysis performed according to the EDAX method (FIGURES 8 and 9). Since it is known that 50% of the $Ca^{2+}$ of the egg cell is bound to protein, and the protein fraction increases from the posterior pole (10% EL: < 30%) to the anterior pole (90% EL: > 40%), based upon evaluated ultrathin sections,[5] it can be concluded that $Ca^{2+}$ gradient of the unfertilized egg has its highest concentration at the anterior pole. The endocytobionts of the egg posterior pole also show, in agreement with substrate tests, calcium-containing inclusion bodies which are partially transferred to the surrounding posterior-

---

[b] Mean value out of 1000 eggs, respectively.

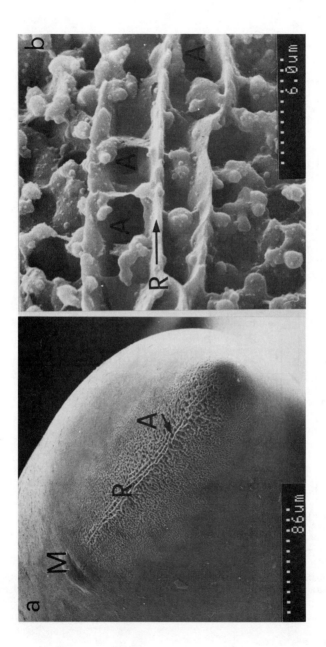

**FIGURE 6.** Irregular double track of aeropyles along the later scored track of the emerging leafhopper embryo on the egg anterior pole, using scanning electron microscopy (micrographs, U. Gernert). (a) Overview. (b) Detail. A, aeropyle; M, micropyle, R, rupture track.

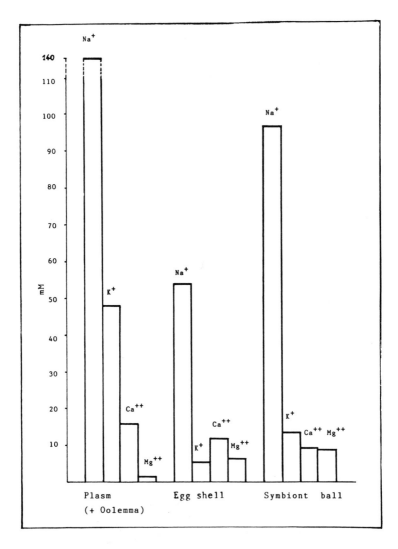

**FIGURE 7.** Distribution of the most important ions in various fractions of the unfertilized *Euscelidius* egg measured with AAS and FES;[24,25] 3 experimental series with 3 parallel tests, respectively.

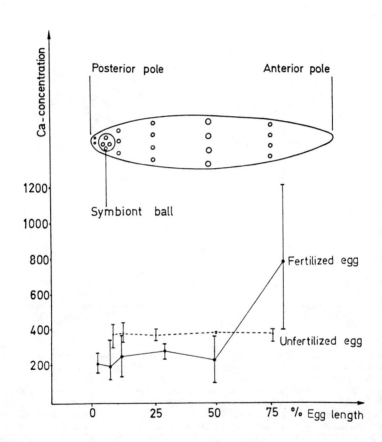

**FIGURE 8.** Data of total $Ca^{2+}$ concentration as counts per unit time in various substructures of the unfertilized and fertilized *Euscelidius* egg measured on semimicro sections by x-ray microanalysis (EDAX).[11] Each dot of the graph represents a mean value at least out of 4 measurements.

pole ooplasm during blastoderm formation, where they influence the developmental events.

By injection of fluorescing pH markers (pyranine, DASPMI, neutral red, 30 mM) into the abdomen of mature leafhopper females or directly into the deposited egg, with subsequent ultraspectral photometry or densitometry, the pH distribution in the egg interior can be determined (cf. FIGURES 10 and 11).[8] According to *preliminary* results, the anterior pole of unfertilized eggs exhibits pH values slightly under 7 while

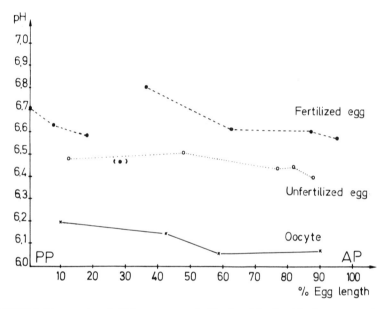

**FIGURE 9.** Topography of pH in oocytes, unfertilized and fertilized leafhopper eggs between the anterior pole (100% egg length) and posterior pole (0% egg length). Pyranine was the fluorescence probe having been abdominally injected into female leafhoppers. The fluorescent eggs were photographed at two excitation wavelengths ($\lambda_1 = 410$ nm, $\lambda_2 = 460$ nm, $\lambda$ emission $> 490$ nm) and the negatives were densitometrically evaluated (film: Agfapan Vario XL, 22-33 DIN). The determined pH values in the same cell as well as the differences between the three different cells are accurate within 0.05 pH units; however, there might be a general increase of the ordinate values by about 0.4 pH units, depending on the choice of the calibration curve used. The pH values are reproducible. PP = posterior pole, AP = anterior pole.

the posterior pole of such eggs has values above 7 (FIGURE 9). In addition, using bioluminescence methods, one finds higher ATP content at the anterior pole of leafhopper eggs than at the posterior pole (cf. FIGURE 10).[15,16] However, substrate tests reveal measurable lactate dehydroxygenase (LDH) and malate dehydroxygenase (MDH) activities only for the posterior pole and the endocytobionts, in contrast to an absence of succinate dehydrogenase (SDH) activities.

From all these physiochemical data, an assumption can be made that, initially, ooplasmic fermentation activities dominate in unfertilized eggs with an optimum at the egg's posterior pole, upon which the similarly glycolytically active endocytobionts have a direct or indirect influence.[5] However, in the fertilized eggs, the mitochondrial

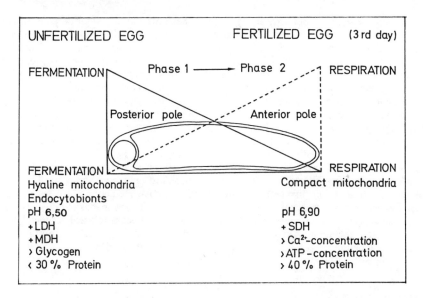

**FIGURE 10.** According to this hypothetical diagram, ooplasmic and endocytobiotic fermentation dominates with an optimum at the posterior pole of the unfertilized egg (phase 1); the fermentative activities are decisively reduced by oxidative metabolism of compact mitochondria forms at the anterior pole of the fertilized egg (phase 2) beginning at the third day after egg deposition (for details, see text).

**FIGURE 11.** By using the techniques of microinjection, it was possible to inject minute quantities of constant concentration at any location on the leafhopper egg. In this way, the fermentation and respiration inhibitors, the pH marker, as well as the substances to initiate the embryogenesis were applied.[24] Magnification ×60. (a) Injection of the pH marker fluorescein acetate (20 mM) in an unfertilized *Euscelidius* egg; automatic injection: 0.5 seconds, injection $P_i$ = 1400 hPa. (b) Injection of a double amount of the same substance.

respiration activities should gradually increase beginning at the anterior pole at the cost of glycolysis, and then dominate from the third day following egg deposition. Morphologically this change from fermentation to respiration is manifested at the anterior pole by the gradually beginning displacement of the large hyaline mitochondria by the small compact forms. To that extent, the morphologically confined determinants are in agreement with the characteristics physiochemically determined.

Should it now be possible for us to prove that the hyaline mitochondrial forms, in contrast to the compact forms, do not possess a complete respiratory chain,[25] the hypothesis formulated by my laboratory in 1977 and more precisely defined in 1980/83 would be experimentally verified in principle. This hypothesis states that the initiation, at least of leafhopper, embryogenesis is controlled by the local and chronologically accurate synchronized fluctuation between fermentation and respiration.

At present there are still no reliable conclusions to be drawn about the kinetics of the coupling of ion flux with glycolytic and oxidative metabolic activities of the egg. Experiments are therefore planned to determine ion flux by means of ion-sensitive microelectrodes or laser microanalysis, as well as experiments to localize glycolytic and oxidative metabolic activities in the leafhopper egg with the aid of immunofluorescence labeling. Based upon these experiments, an improved molecular model of the causes and mechanisms of egg polarity might be developed.

*Gene Expression*

Extensive density, sedimentation, and conformation analyses have shown that the leafhopper endocytobionts exhibit an apparent DNA molecular weight between 2.2 and $2.6 \times 10^7$ Daltons.[17,26] Should this be confirmed in current experiments, the leafhopper endocytobionts would have the lowest DNA molecular weight of all known prokaryotes. This would be a sensational finding to the extent that the endocytobiont DNA would consequently lie only slightly above the DNA molecular weight of leafhopper mitochondria of about $10^7$ Daltons. This could then only be explained if the endocytobiont genome either contains an overlapping gene expressivity unknown until now, or if a gene transfer from the symbionts into the host cell genome or gene loss from the symbiont genome has occurred.[27] The fact that the leafhopper endocytobionts cannot be reproduced *in vitro* extracellularly[28,29] rather indicates a gene loss or a gene transfer. A gene transfer could possibly be indicated by the demonstrated influence of endocytobionts on the germ-band development as well as the regulative coupling in energy metabolism, pH, and osmotic behavior between egg and symbiont system.[15,16,30,31] The leafhopper endocytobionts appear to be especially suitable for the investigation of the intermeshing of gene expression of egg and symbiont DNA, because they may have been conserved as a type of living fossil in the form of a "missing link" between mitochondria and archaebacteria, having had the protection of the leafhopper for millions of years of common coevolution.

The goal of the current experiments is therefore to examine whether gene transfer is indicated between egg and endocytobiont genomes similar to that between cell nucleus and mitochondrial DNA, and how these genetically regulate the initiation of embryonic cell growth.

## DISCUSSION

### Molecular Mechanism

According to preliminary model concepts,[32] external signals (S) are arranged as first messengers upon the surface of the egg cell on receptors (R), which activate phosphodiesterase ($PiP_2$-PDE) as a type of amplifier (FIGURE 12) via a G-protein (converter). $PiP_2$-PDE then cleaves phosphatidylinosit-4,5 diphosphate ($PiP_2$) into the secondary messengers inosittriphosphate ($IP_3$) and diacylglycerol (DG).[33] The water-soluble $IP_3$ diffuses into the egg plasm and there leads to the liberation of $Ca^{2+}$ out of the intracellular storage systems (symbionts, ooplasmic reticulum, oolemma, mitochondria?).[34] $Ca^{2+}$ itself stimulates the effector protein calmodulin, which in turn results in the directed phosphorylation of proteins.[35] The DG remaining in the membrane activities the protein C-kinase by means of phosphatidylserin (PS) which is also membrane bound. This effector then activates a membrane pump via the phosphorylation of various proteins, which exchanges $H^+$ for $Na^+$ in a still unexplained manner yielding an increase in the intracellular pH value as well as the $Na^+$ concentration.[36,37] Each of these changes in the $Ca^{2+}$ and $Na^+$ concentration and pH values could activate the machinery in the female egg nucleus (parthenogenesis) or zygote nucleus by various pathways, which finally lead to the cellular response of mitogenesis. The homeobox revealed by Gehring[22] has decisive significance for the genetic control of embryonic cell growth.

Possible interaction between these signal chains and ooplasm fermentation could exist via the liberation of $Ca^{2+}$, and with egg-mitochondria respiration by means of $O_2$ production through the mediation of C-kinase.[38] Such couplings could be decisively significant in the regulation of postulated fermentation and respiration gradients in the leafhopper.[39]

By microinjection of single components of the signal chain in the order of their probable occurrence, it can be examined to what extent the initiation of embryogenesis can be artificially induced, as has already been successfully achieved up to the artificial formation of a fertilization membrane in the sea urchin egg.[40] The initial evaluations following microinjection of $IP_3$ into an unfertilized leafhopper egg also allow the presumption of an incipient embryogenesis (FIGURE 11).

### Tumorgenesis

The exact analysis of the probable signal chain for the initiation of embryogenesis discussed above possibly also contributes decisively to the understanding of the signal chain in tumor formation. As we now know, many oncogenes are homologous with certain genes that initiate embryonic cell growth, so that they are designated in general as proliferation genes (cf. FIGURE 12).[41] In many ways, tumor formation can be even viewed as a reversal of embryogenesis. In the first case, differentiated cells become again undifferentiated; in the last case, an entire spectrum of histological differentiation arises from undifferentiated embryonic cells (blastoderm).[32]

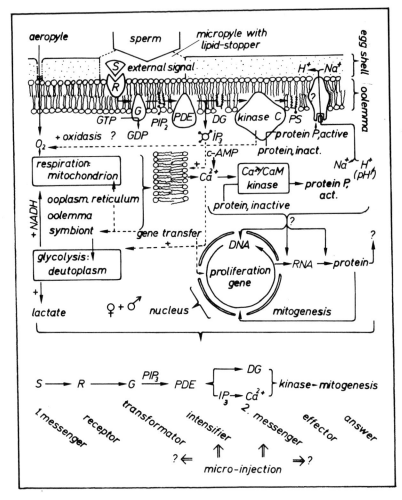

**FIGURE 12.** Hypothetical diagram of the signal chain for the initiation and regulation of embryogenesis in a leafhopper egg (with reference to 32).

## Endocytobiology

New experimental possibilities emerge from this presentation, which are all based upon the eukaryote cell being not an elementary organism, but rather representing an endocytobiosis. Whether the eukaryotic cell evolved out of three or four prokaryotic cells in the course of evolution is certainly an open question for the time being. This so-called endocytobiotic cell theory has finally achieved its experimental confirmation by the comparative sequence analyses of nucleic acids and proteins. For this new synthetic research area, Schwemmler in 1979 coined and defined the term "endocytobiology." In 1980/83 with H. Schenk, the term and definition were then officially introduced at the first and second International Colloquia for Endocytobiology in Tübingen, West Germany, in order to become established internationally in conjunction with the New York conference. Impulses for all experimental areas are expected from endocytobiology, wherein the interaction between nucleocytoplasm and DNA-containing cell organelles (mitochondria, plastids, endocytobionts: viruses, parasites, symbionts) is of principle importance; above all in the analysis of embryogenesis, tumorgenesis, and circadian rhythm.

## ACKNOWLEDGMENTS

I would like to thank all members of the project, especially Professors V. Erdmann (H. Toschka), G. Kraepelin (A. Brose), and A. Grässmann for their valuable assistance in the experiments of gene expression, respiration, and microinjection. I would also like to thank Dr. W. H. Schröder (Kernforschungszentrum, Jülich), Dr. N. Dencher (Biophysik, FU Berlin), and Dr. B. Behrens (Chemie, FU Berlin) as well as Dr. B. Schirpke and I. Peters of my laboratory for their effective cooperation.

## REFERENCES

1. LAWRENCE, P. A., Ed. 1976. Insect Development. (Symp. R. Entomol. Soc. London No. 8). Blackwell Scientific Publications. Oxford, London, Edinburgh & Melbourne.
2. SCHWEMMLER, W., J. L. DUTHOIT, G. KUHL & C. VAGO. 1973. Sprengung der Endosymbiose von *Euscelis plebejus* FALL. und Ernährung asymbiontischer Tiere mit synthetischer Diät (Hemiptera, Cicadina). Z. Morphol. Tiere **74:** 297-322.
3. ZABEL, U. & W. SCHWEMMLER. 1980. Das Experiment: Insekten-Embryogenese am Beispiel einer Kleinzikade. Biut **10**(4): 120-123.
4. SCHWEMMLER, W. 1974. Endosymbionts: factors of egg pattern formation. J. Insect Physiol. **20:** 1467-1474.
5. SCHWEMMLER, W. & G. KEMNER. 1983. Fine structural analysis of the egg cell of *Euscelidius* spec. (Homoptera, Cicadina). Cytobios **37:** 7-20.
6. HEINRICH, K. F. J. 1981. Electron Beam x-Ray Microanalysis. V.N.R. Company. New York, N.Y.
7. KLUG, W. S. & D. CAMPBELL. 1974. External morphology of the egg of *Drosophila melanogaster* (Diptera: Drosophilidae). Int. J. Insect Morphol. Embryol. 3(1): 33-40.
8. SCHWEMMLER, W. & N. DENCHER. (In preparation.)
9. WELZ, B. 1982. Atomspektrometrische Spurenanalytik. Verlag Chemie. Weinheim, Germany.
10. CHANDLER, J. A. 1977. X-Ray microanalysis in electron microscopy. *In* Methods of Practical Electron Microscopy. A. M. Glauert, Ed.: 319-541. North Holland Publishing Company. Amsterdam, New York & Oxford.
11. SCHWEMMLER, W., B. SCHIRPKE & W. H. SCHRÖDER. (In preparation.)
12. KALTHOFF, K., I. KANDLER-SINGER, O. SCHMIDT, D. ZISSLER & G. VERSEN. 1975. Mitochondria and polarity in the egg of *Smittia* spec. (Diptera, Chironomidae): UV irradiation, respiration measurements, ATP determinations and application of inhibitors. Wilhelm Roux's Arch. **178:** 99-121.
13. KNY, U. 1976. Histochemische Untersuchungen infektiöser und vegetativer Zikadenendocytobionten. Master's Thesis, Free University of Berlin. Berlin, FRG.

14. SCHWEMMLER, W. & I. PETERS. (In preparation).
15. SCHWEMMLER, W. & M. HERRMANN. 1979. Oszillationen im Energiestoffwechsel von Wirt und Symbiont eines Zikadeneies. I. Analyse möglicher stoffwechselphysiologischer Korrelationen beider Systeme. Cytobios **25**: 45-62.
16. SCHWEMMLER, W. & M. HERRMANN. 1980. Oszillationen im Energiestoffwechsel von Wirt und Symbiont eines Zikadeneies. II. Analyse möglicher endogener Rhythmen beider Systeme. Cytobios **27**: 193-208.
17. SCHWEMMLER, W., G. HOBOM & M. EGEL-MITANI. 1975. Isolation and characterization of leafhopper endosymbiont DNA. Cytobiologie **10**: 249-259.
18. TAYLOR, B. & A. POWELL. 1984. Isolation of plant DNA and RNA. Focus **4**: 4-6.
19. GAUSE, G. G. & V. S. MIKHAILOV. 1973. State of the DNA synthesizing system in isolated mitochondria from mature eggs of the loach (*Misgurus fossilis*). Biochim. Biophys. Acta **324**: 189-198.
20. ISHIKAWA, H. 1982. Isolation of the intracellular symbionts and partial characterization of their RNA species of the elder aphid, *Acyrthisiphon magnoliae*. Comp. Biochem. Physiol. **72**: 239-247.
21. WEBER, H. 1966. Grundriss der Insektenkunde. Gustav Fischer Verlag. Stuttgart, FRG.
22. GEHRING, W. J. 1985. The molecular basis of Development. Sci. Am. **253**: 137-149.
23. SCHWEMMLER, W. 1980. Endocytosymbiose: Modell zur molekularen Analyse von Circadianrhythmik, Eimusterbildung und Krebs. Naturwiss. Rundschau **33**: 52-59.
24. SCHWEMMLER, W., B. SCHIRPKE & B. BEHRENS. (In preparation.)
25. CHEN, P. S. 1971. Biochemical Aspects of Insect Development. Monogr. Dev. Biol. **3**.
26. SCHWEMMLER, W. 1971. Intracellular symbionts: a new type of primitive prokaryotes. Cytobiologie **3**: 427-429.
27. SCHWEMMLER, W. 1983. Analysis of possible gene transfer between an insect host and its bacteria-like endocytobionts. Int. Rev. Cyt. **14**: 247-266.
28. SCHWEMMLER, W. 1973. Beitrag zur Analyse des Endosymbiosezyklus von *Euscelis plebejus* F. (Hemip., Homop., Cicad.) mittels in-vitro-Beobachtung. Biol. Zbl. **92**: 749-772.
29. SCHWEMMLER, W. 1973. In vitro Vermehrung intrazellulärer Zikaden-Symbionten und Reinfektion asymbiontischer Mycetocyten-Kulturen. Cytobios **8**: 63-73.
30. SCHWEMMLER, W. & H. E. A. SCHENK, Eds. 1980. Endocytobiology I. Endosymbiosis and Cell biology. Synthesis of Recent Research. Walter de Gruyter. Berlin & New York.
31. SCHWEMMLER, W. 1984. Reconstruction of Cell Evolution: A Periodic System. CRC Press. Boca Raton, Fla.
32. BERRIDGE, M. & 1985. The molecular basis of communication within the cell. Sci. Am. **253**: 124-136.
33. ROSOFF, P. M. & L. C. CANTLEY. 1985. Lipopolysaccharide and phorbol esters induce differentiation but have opposite effects on phosphatidylinositol turnover and $Ca^{2+}$ mobilization in 70Z/3 pre-B lymphocytes. J. Biol. Chem. **260**: 9209-9215.
34. CLAPPER, D. L. & H. C. LEE. 1985. Inositol triphosphate induces calcium release from nonmitochondrial stores in sea urchin egg homogenates. J. Biol Chem. **260**: 13947-13954.
35. MCNEIL, P. L., M. P. MCKENNA & D. L. TAYLOR. 1985. A transient rise in cytosolic calcium follows stimulation of quiescent cells with growth factors and is inhibitable with phorbol myristate acetate. J. Cell Biol. **101**: 372-379.
36. GRINSTEIN, S., S. COHEN, J. GOETZ, A. ROTHSTEIN & E. W. GELFAND. 1985. Characterization of the activation of $Na^+/H^+$ exchange in lymphocytes by phorbol esters: change in cytoplasmic pH dependence of the antiport. Proc. Nat. Acad. Sci. USA **82**: 1429-1433.
37. CASSEL, D., B. WHITELEY, Y. X. ZHUANG & L. GLASER. 1985. Mitogen-independent activation of $Na^+/H^+$ exchange in human epidermoid carcinoma A431 cells regulation by medium osmolarity. J. Cell. Physiol. **122**: 178-186.
38. HOLIAN, A. & D. F. STICKLE. 1985. Calcium regulation of phosphatidylinositol turnover in macrophage activation by formyl peptides. J. Cell. Physiol. **123**: 39-45.
39. CHEUNG, W. Y. 1980-1983. Calcium and Cell Function **1-4**. Academic Press. New York, N.Y.
40. WHITAKER, M. 1984. Inositol-1,4,5-triphosphate microinjection activates sea urchin eggs. Nature **312**: 636-639.
41. KARLSON, P. 1982. Wie entstehen Krebszellen? Naturwiss. Rundschau **9**: 356-365.

# Tubulinlike Protein from
# *Spirochaeta bajacaliforniensis*[a]

D. BERMUDES,[b] S. P. FRACEK, JR.,[d] R. A. LAURSEN,[c]
L. MARGULIS,[b,e] R. OBAR,[c,f] AND G. TZERTZINIS [c]

[b]*Department of Biology*
*Boston University*
*2 Cummington Street*
*Boston, Massachusetts 02215*

[c]*Department of Chemistry*
*Boston University*
*590 Commonwealth Avenue*
*Boston, Massachusetts 02215*

## INTRODUCTION

Tubulin proteins are the fundamental subunits of all polymeric microtubule-based eukaryotic structures. Long, hollow structures each composed of 13 protofilaments as revealed by electron microscopy, microtubules (240 Å in diameter) are nearly ubiquitous in eukaryotes. These proteins have been the subject of intense biochemical and biophysical interest since the early 1970s[28] and are of evolutionary interest as well (e.g., References 27 and 29).

If tubulin-based structures (i.e., neurotubules, mitotic spindle tubules, centrioles, kinetosomes, axonemes, etc.) evolved from spirochetes by way of motility symbioses, tubulin homologies with spirochete proteins should be detectable (see Reference 4).

Tubulin proteins are widely thought to be limited to eukaryotes (e.g., Reference 36). Yet both azotobacters[1] and spirochetes[12,30] have shown immunological cross-reactivity with antitubulin antibodies. In neither of these studies was tubulin isolated nor any specific antigen identified as responsible for the immunoreactivity. Furthermore, although far less uniform in structure than eukaryotic microtubules, various cytoplasmic fibers and tubules (as seen by electron microscopy) have been reported in several types of prokaryotes (e.g., *Spirochaeta*;[12] large termite spirochetes;[18,30] tre-

[a]This work was supported by a National Science Foundation Graduate Fellowship (to D.B.), the Lounsbery Foundation (to L.M.), National Aeronautics and Space Administration Grant NGR-004-025 (to L.M.), NSF Grant DMB-8503940 (to R.A.L.), and the Boston University Graduate School.

[d]Present affiliation: Department of Biology, North Texas State University, Denton, Texas 76203.

[e]To whom correspondence should be addressed.

[f]Present affiliation: Worcester Foundation for Experimental Biology, 222 Maple Avenue, Shrewsbury, Massachusetts 01545.

515

ponemes;[19-22] cyanobacteria;[24] and *Azotobacter*.[35] This work forms a part of our long-range study of the possible prokaryotic origin of tubulin and microtubules.

Spirochetes are helically shaped gram-negative motile prokaryotes. They differ from all other bacteria in that the position of their flagella is periplasmic: their flagella lie between the inner and outer membranes of the gram-negative cell wall. Some of the largest spirochetes have longitudinally aligned 240 Å microtubules.[18] Unfortunately, in spite of many attempts, all of the larger spirochetes (family Pillotaceae) with well-defined cytoplasmic tubules and antitubulin immunoreactivity are not cultivable.[31]

However, a newly described spirochete species (*Spirochaeta bajacaliforniensis*)[12] possessing cytoplasmic fibers displays antitubulin immunoreactivity in whole-cell preparations. Since preliminary observations suggested that *Spirochaeta bajacaliforniensis* proteins may be related to eukaryotic tubulins, their characterization was undertaken.

Brain tubulin can be purified by utilizing its ability to polymerize at warm temperatures and to depolymerize in the cold.[43] After several cycles of sedimentation and redissolution the microtubule fraction is comprised of 75% tubulin and 20% high molecular mass microtubule-associated proteins (MAPs).[5,39] In this paper we report that components of cell lysates, prepared from a spirochete that contains cytoplasmic fibers (*Spirochaeta bajacaliforniensis*), also exhibit the property of temperature-dependent cyclical sedimentation. Additionally we report the identification and characterization of the polypeptide responsible for cross-reactivity with antitubulin antiserum.

## METHODS

### *Microscopy*

Light microscopic studies were performed using a Nikon Biophot microscope equipped with Nomarski differential interference optics. Electron microscopy was performed using a Jeol electron JEM 100B microscope on negative-stained material prepared by the method of Hovind-Hougen and Birch-Andersen.[21]

### *Bacterial Culture*

Spirochetes (*Spirochaeta bajacaliforniensis* ATCC 35968) were grown at 36°C in the medium reported in Fracek and Stolz.[12] Stock cultures were maintained in 10 × 150 mm screw-top culture tubes. Larger quantities were grown in 2-liter screw-top flasks.

### *Temperature-Dependent Cycling*

The method of Shelanski *et al.* for temperature-dependent sedimentation and redissolution of brain tubulin *in vitro* was employed.[40] Spirochetes (8-80 liters) were harvested by centrifugation at 21°C, using a Sorvall KSB flow-through system, in a Sorvall SS3 centrifuge with a SS-34 rotor at 12,000 rpm (approximately 17,000 $\times$ *g*). The pooled pellets were resuspended in the following buffer which was used for dilutions and resuspensions throughout the cycling: 0.1 M piperazine-*N,N'*-bis-(2-ethanesulfonic acid) (PIPES), 2 mM ethylene glycol bis (β-aminoethyl ether)-*N,N'*-tetraacetic acid (EGTA), 1 mM MgSO$_4$, 1mM dithiothreitol (DTT), 1 mM guanosine triphosphate (GTP) (see below), 1 mM phenylmethylsulfonylfluoride (PMSF), pH 6.9. Bacterial cells were lysed in a French pressure cell at a pressure greater than 10,000 psi in a one-inch cell precooled to 4°C. Cell debris was removed by centrifugation at 100,000 $\times$ *g* for 60 minutes at 4°C. The resulting (cold) supernatant (1CS) was diluted 1:1 with buffer warmed at 37°C and incubated at 37°C for 60 minutes. The aggregated material was then collected by centrifugation at 100,000 $\times$ *g* for 60 minutes at 37°C. The pellet (1WP) was resuspended to the original volume and incubated for 60 minutes at 4°C, then subjected to a second cold centrifugation. The supernatant (2CS) was used for characterization or subsequent cycles of sedimentation.

### Guanosine Triphosphate Dependence of Sedimentation

The procedure for temperature cycling of tubulins[40] was applied to the spirochete cell lysate in the presence or absence of GTP in the buffers. Glycerol was not present in any buffer. The French press lysate was divided into two equal fractions each of which was carried through two cycles of centrifugation; the two aliquots were treated identically except for the addition of GTP. The resultant second warm pellets, resuspended to form the third cold supernatants, were compared quantitatively for protein content by the use of the Bradford protein assay.[6]

### Isolation of Brain Tubulin

Tubulin was isolated from bovine brain by the method of Eipper.[8,9] This brain tubulin was used for comparative purposes.

### Sodium Dodecyl Sulfate-Polyacrylamide Gel Electrophoresis (NaDodSO₄-PAGE)

Polyacrylamide gels were run[42] on a Hoefer Model 220 slab gel apparatus. Proteins were stained with Coomassie Brilliant Blue R-250, silver stained by the method of Merril *et al.,*[32] or color stained.[38]

In preparative gels the locations of protein bands on the gels were determined by soaking the gels in an ice-cold solution of 3 M potassium acetate. The protein bands were excised from each gel in 3-mm slices and soaked in protein-denaturing solution (25% glycerol 1% NaDodSO₄, and 5% mercaptoethanol) for 15 minutes at 50°C.[42] The proteins were then electroeluted into dialysis tubing.[23]

*Isoelectric Point Determination*

The isoelectric points of the proteins were determined from one-dimensional iso-electric focusing (IEF) tube gels of purified proteins and from two-dimensional gels (first dimension IEF, second NaDodSO$_4$-PAGE) of crude material. The pI values obtained were corrected for primary medium cell effects and pH measuring cell effects.[13,14]

*Immunoblotting*

Polyacrylamide gels were blotted onto nitrocellulose.[41] The nitrocellulose blots were probed with guinea pig antitubulin antibody (a gift from Dr. George Bloom, Worcester Foundation). Visualization of bands was performed by the horseradish peroxidase method.[16]

## RESULTS AND DISCUSSION

Tubulin exists as a dimer of two distinct polypeptides ($\alpha$- and $\beta$-tubulin) as isolated from brain, the 9+2 axoneme of undulipodia (cilia, eukaryotic flagella), and the mitotic spindle. Each subunit has a molecular mass of 50 kilodaltons (kD) but can exhibit apparent molecular mass of 55 to 60 kD in NaDodSO$_4$ gels. The amino acid sequences of $\alpha$- and $\beta$-tubulin share approximately 40% sequence homology.[26]

When cell lysates of *Spirochaeta bajacaliforniensis* were subjected to cold-warm cycling under conditions used for brain tubulin purification, the lysate became enriched in two proteins, S1 and S2, which possess molecular masses of 65 kD and 45 kD, respectively. The results from two cold-warm cycles are shown in FIGURE 1. Additional cycles gave further enrichment, but material losses were severe because of diminishing yields of total protein. For this reason, most studies were done using the second cold supernatant (2CS) fraction. The S1 protein has been further purified by a combination of chromatographic methods[33] or preparative gel electrophoresis to give material that appeared to be homogeneous on NaDodSO$_4$-PAGE (FIGURE 2).

The warm-pelleted material contained fibers that tended to bundle together as seen by light microscopy; the smallest discernible fibers were approximately 0.5 $\mu$m in diameter (FIGURE 3). Warm-pelleted material is fibrous with strands varying in size from 0.15 to 1.5$\mu$m in diameter as revealed by electron microscopy of negative-stained preparations (FIGURE 4). The larger strands appeared to be aggregates of smaller ones. The 0.5 $\mu$m fibers apparently are composed of smaller fibrils approximately 5-10 nm in diameter. The size of the fibrils is about that of the protofilaments of eukaryotic microtubules (approximately 5 nm).[3] The diameter of the spirochete fibrils is also approximately that of muscle actin filaments (i.e., 5-7 nm).[25]

GTP promotes the assembly of brain tubulin subunits into microtubules by forming an active GTP-tubulin complex, but is not required for tubulin polymerization.[34] Colchicine, which inhibits assembly of brain tubulin,[44] had no effect on tubulinlike

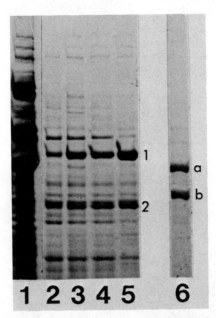

**FIGURE 1.** Spirochete cell lysate subjected to temperature-dependent cycling, 7.5% acrylamide-NaDodSO$_4$ gel. Lanes: (1) whole cell lysate, (2) first cold supernatant (1CS), (3) first warm pellet (1WP), (4) second cold supernatant (2CS), (5) second warm pellet (2WP), (6) brain tubulin; (a) $\alpha$-subunit and (b) $\beta$-subunit. Note that the 2WP material is largely enriched in the S1 and S2 proteins, indicated by "1" and "2" respectively.

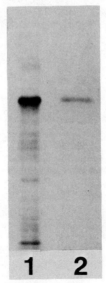

**FIGURE 2.** Silver-stained gel comparing purification methods. Lanes: (1) chromatographically purified S1; (2) electroeluted S1 from a sliced gel band.

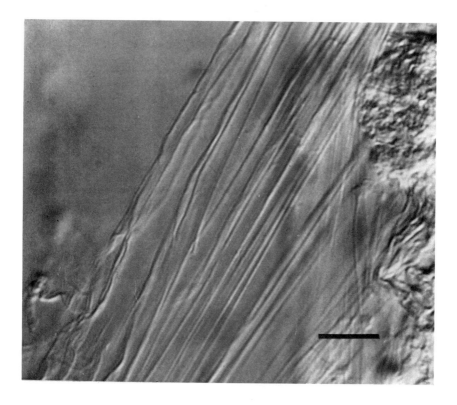

**FIGURE 3.** Warm-pelleted spirochete lysate. (Light micrograph, bar shown corresponds to 10 μm.)

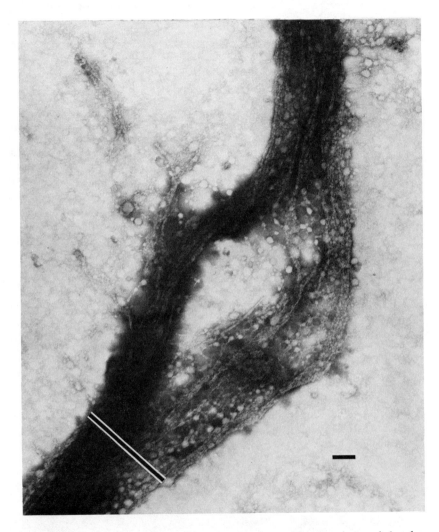

**FIGURE 4.** Negative-stained warm-pelleted spirochete lysate. (Electron micrograph, bar shown corresponds to 0.1 μm; black-on-white stripe corresponds to 0.45 μm.)

**TABLE 1.** Effect of GTP on Yield of Sedimented Protein[a]

| | Experiment | | |
|---|---|---|---|
| | 1 | 2 | 3 |
| + GTP | 0.014 | 0.213 | 0.092 |
| − GTP | 0.011 | 0.235 | 0.098 |

[a] Three independent temperature-cycling preparations with (+) and without (−) GTP. Protein concentration in the redissolved second warm pellets is given in mg/ml.

protein from *Spirochaeta bajacaliforniensis.*[11] In this study GTP had no apparent effect on the yield of spirochete tubulinlike protein during the cold-warm cycling process (TABLE 1). Colchicine and GTP sensitivities are not universally characteristic of tubulins, however. Brain tubulin can be polymerized in the absence of added nucleotides, although a microtubule-stabilizing agent such as glycerol is generally required.[40] Furthermore colchicine-insensitive tubulins are well known (e.g., *Chlamydomonas* mutants;[2,10] many fungi;[17] and protoctists such as *Physarum*[7]).

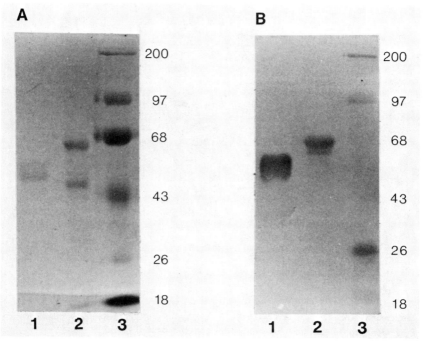

**FIGURE 5.** Immunological detection of the S1 protein with antitubulin antibody. Panel A: Coomassie Blue-stained 7.5% polyacrylamide gel; lanes: (1) brain tubulin, (2) second cold supernatant (2CS), (3) prestained molecular weight standard (Bethesda Research Laboratories). Panel B: Duplicate of gel in panel A, electrophoretically transferred to nitrocellulose and immunostained with antitubulin antibody as described in methods. Only tubulin and S1 protein bind the antibody. On the right of panels A and B, molecular weights are shown in kD.

An assessment of the similarity between brain tubulin and the spirochete proteins was made using immunochemical methods. Immunoblots of 2CS using guinea pig antitubulin ($\alpha$ and $\beta$) showed that at least one of the spirochete proteins, S1, is reactive with this antiserum (FIGURE 5). This cross-reactivity suggests that these two proteins share at least one epitope.

The isoelectric points of S1 and S2 were determined to be 5.9 and 5.5, respectively, which are similar to those of $\alpha$- and $\beta$- tubulin (approximately 5.4 and 5.3, respectively).[15] A two-dimensional map of a 2CS preparation, indicating the relative mobilities of S1 and S2 according to pI and size, is shown in FIGURE 6. Although

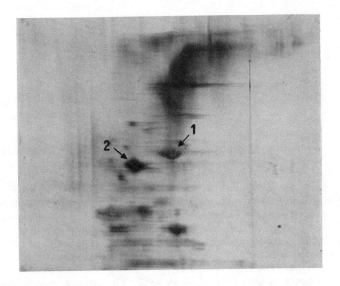

**FIGURE 6.** Two-dimensional gel of 2CS spirochete material stained with the Sammons color stain methods. First (horizontal) dimension, isoelectric focusing; second (vertical) dimension, NaDodSO$_4$-PAGE; pH increases from left to right: 1, S1; and 2, S2.

preparations of 2CS richer in S1 and S2 have been obtained, that shown in FIGURE 6 is typical. The absence of other major protein bands migrating with the same R$_f$ in the NaDodSO$_4$-PAGE dimension has facilitated purification of S1 and S2 by preparative gel electrophoresis.[33]

Proteins with temperature-dependent aggregation characteristics similar to tubulin have also been reported in *Mycoplasma* rho form,[37] but these have not been tested for immunological cross-reactivity with antitubulin antiserum. The relationship of these *Mycoplasma* proteins to those of spirochetes has not been determined.

## CONCLUSIONS

We have enriched two proteins by warm-cold cycling from the free-living cultivable anaerobic spirochete *Spirochaeta bajacaliforniensis,*[12] one 65 kD (S1) and the other 45 kD (S2). S1 cross-reacts with a polyclonal antibody made against calf brain tubulin. The purification of the S1 protein by chromatography and by excision from gels[33] should make possible protein sequencing for comparative studies.

The detection of tubulin in spirochetes would greatly strengthen the hypothesis of the symbiotic origin of the eukaryotic 9+2 motility organelle (undulipodium, cilium, flagellum) from prokaryotes.[4] Identification of S1 and S2 proteins is an important step in the verification or refutation of this hypothesis. Confirmation of the presence of tubulins in spirochetes, which by itself still will not prove the symbiotic spirochete hypothesis, will require at least protein or gene sequence analysis of S1 and S2. The existence of tubulinlike proteins in spirochetes has not, to our knowledge, been predicted by any other evolutionary hypotheses.[4]

## SUMMARY

Tubulin is the major protein component of the walls of microtubules, structures nearly ubiquitous in eukaryotic cells. The evolution of undulipodia (i.e., cilia, axonemes, and other tubulin-containing organelles that show a 9+2 array in transverse section) from motile prokaryotic symbionts has been proposed.[4,29] This hypothesis prompted our search for tubulin in spirochetes. We report here that *Spirochaeta bajacaliforniensis,* an anaerobic free-living bacterium, possesses at least two proteins, termed S1 and S2, which can be copurified by a temperature-dependent cycling method used for isolation of mammalian brain tubulin. S1 cross-reacts with brain antitubulin antiserum. Warm aggregates of spirochete-derived proteins contain fibrous material (as observed by light and electron microscopy). These structures form in either the presence or absence of added colchicine or GTP. The apparent relative molecular masses of S1 and S2 are 65 and 45 kD and their isoelectric points are 5.9 and 5.5, respectively. These results are consistent with the hypothesis of the origin of undulipodia from symbiotic spirochetes.

## ACKNOWLEDGMENTS

We thank George Bloom for the guinea pig antitubulin antiserum; Richard Vallee, George Bloom, and Frank Luca for assistance in the immunological detection; Giovanni Bosco and Barbara Dorritie for assistance in growing spirochetes; and Geraldine Kline for typing the manuscript.

## REFERENCES

1. ADAMS, G. M. W. 1983. Microtubule-like structures in the bacterium *Azotobacter vinelandii.* Abstract 794. J. Cell Biol. **97** (5, part 2): A209.
2. ADAMS, M. & J. R. WARR. 1972. Colchicine-resistant mutants of *Chlamydomonas reinhardtii.* Exp. Cell. Res. **71**: 473-475.
3. AMOS, L. A. 1979. Structure of microtubules. *In* Microtubules. K. Roberts & J. S. Hyams, Eds.: 1-64. Academic Press. London, England.
4. BERMUDES, D., L. MARGULIS & G. TZERTZINIS. 1987. Prokaryotic origin of undulipodia: application of the panda principle to the centriole enigma. Ann. N.Y. Acad. Sci. (This volume.)
5. BORISY, G. G., J. M. MARCUM, J. B. OLMSTED, D. B. MURPHY & K. A. JOHNSON. 1975. Purification of tubulin and associated high molecular weight proteins from porcine brain and characterization of microtubule assembly in vitro. Ann. N.Y. Acad. Sci. **253**: 107-132.
6. BRADFORD, M. M. 1976. A rapid and sensitive method for the quantitation of microgram quantities of protein utilizing the principle of protein-dye binding. Anal. Biochem. **72**: 248-254.
7. CLAYTON, L. & K. GULL. 1982. Tubulin and the microtubule organising centres of *Physarum polycephalum* myxamoebae. *In* Microtubules in Microorganisms. P. Cappuccinelli & N. R. Morris, Eds.: 179-201. Marcel Dekker, Inc. New York, N.Y.
8. EIPPER, B. A. 1972. Rat brain microtubule protein: purification and determination of covalently bound phosphate and carbohydrate. Proc. Nat. Acad. Sci. USA **69**: 2283-2287.
9. EIPPER, B. A. 1975. Purification of rat brain tubulin. Ann. N.Y. Acad. Sci. **253**: 239-246.
10. FLAVIN, M. & C. SLAUGHTER. 1974. Microtubule assembly and function in *Chlamydomonas:* inhibition of growth and flagellar regeneration by antitubulins and other drugs and isolation of resistant mutants. J. Bacteriol. **118**: 59-69.
11. FRACEK, S. P., JR. 1984. Tubulin-like proteins of *Spirochaeta bajacaliforniensis,* a new species from a microbial mat community at Laguna Figueroa, Baja California Norte, Mexico. Ph.D. Dissertation. Department of Biology. Boston University Graduate School. Boston, Mass.
12. FRACEK, S. P., JR. & J. F. STOLZ. 1985. *Spirochaeta bajacaliforniensis* sp. n. from a microbial mat community at Laguna Figueroa, Baja California Norte, Mexico. Arch. Microbiol. **142**: 317-325.
13. GELSEMA, W. J., C. L. DE LIGNY & N. G. VAN DER VEEN. 1978. Isoelectric focusing as a method for the characterization of ampholytes. III. Isoelectric points of carrier ampholytes and dissociation constants of some carboxylic acids and alkyl-substituted ammonium ions in sucrose-water, glycerol-water and ethylene glycol-water mixtures. J. Chromatogr. **154**: 161-174.
14. GELSEMA, W. J., C. L. DE LIGNY & N. G. VAN DER VEEN. 1979. Isoelectric points of proteins, determined by isoelectric focusing in the presence of urea and ethanol. J. Chromatogr. **171**: 171-181.
15. GEORGE, H. J., L. MISRA, D. J. FIELD & J. C. LEE. 1981. Polymorphism of brain tubulin. Biochemistry **20**: 2402-2409.
16. GEYSEN, J., A. DELOOF & F. VANDESANDE. 1984. How to perform subsequent or "double" immunostaining of two different antigens on a single nitrocellulose blot within one day with an immunoperoxidase technique. Electrophoresis **5**: 129-131.
17. HEATH, I. B. 1978. Experimental studies of mitosis in the fungi. A review. *In* Nuclear Division of the Fungi. I. B. Heath, Ed.: 89-176. Academic Press. New York, N.Y.
18. HOLLANDE, A. C. & I. D. GHARAGOZLOU. 1967. Infrastructural morphology of *Pillotina calotermitidis* nov. gen., nov. sp. spirochete of the intestine of *Calotermes praecox.* C. R. Acad. Sci. Paris **265**: 1309-1312.
19. HOVIND-HOUGEN, K. 1974. The ultrastructure of cultivable treponemes. Acta Pathol. Microbiol. Scand. B **82**: 329-344.
20. HOVIND-HOUGEN, K. 1976. Determination by means of electron microscopy of morphological criteria of value for classification of some spirochetes, in particular treponemes. Acta Pathol. Microbiol. Scand. B Suppl. No. 255: 1-41.

21. HOVIND-HOUGEN, K. & A. BIRCH-ANDERSEN. 1971. Electron microscopy of endoflagella and microtubules in *Treponema* Reiter. Acta Pathol. Microbiol. Scand. B **79**: 37-50.

22. HOVIND-HOUGEN, K., A. BIRCH-ANDERSEN & H. J. SKOVGAARD JENSEN. 1976. Ultrastructure of cells of *Treponema pertenue* obtained from experimentally infected hamsters. Acta Pathol. Microbiol. Scand. B **84**: 101-108.

23. HUNKAPILLER, M. W., E. LUJAN, F. OSTRANDER & L. E. HOOD. 1983. Isolation of microgram quantities of proteins from polyacrylamide gels for amino acid sequence analysis. Methods Enzymol. **91**: 227-236.

24. JENSEN, T. E. 1984. Cyanobacterial cell inclusions of irregular occurrence: systematic and evolutionary implications. Cytobios **39**: 35-62.

25. KORN, E. D. 1978. Biochemistry of actomyosin-dependent cell motility (a review). Proc. Nat. Acad. Sci. USA **75**: 588-599.

26. LITTLE, M. 1985. An evaluation of tubulin as a molecular clock. BioSystems **18**: 241-247.

27. LITTLE, M., G. KRÄMMER, M. SINGHOFER-WOWRA & R. F. LUDUEÑA. 1986. Evolutionary aspects of tubulin structure. Ann. N.Y. Acad. Sci. **466**: 8-12.

28. LUDUEÑA, R. F. 1979. Biochemistry of tubulin. *In* Microtubules. K. Roberts & J. S. Hyams, Eds.: 65-116. Academic Press. London, England.

29. MARGULIS, L. 1981. Symbiosis in Cell Evolution: life and its environment on the early earth. W. H. Freeman and Co. San Francisco, Calif.

30. MARGULIS, L., L. P. TO & D. CHASE. 1978. Microtubules in prokaryotes. Science **200**: 1118-1124.

31. MARGULIS, L., D. BERMUDES & D. CHASE. Morphology as a basis for taxonomy of large symbiotic spirochetes Pillotaceae Familia nova nom. rev.; *Pillotina calotermitidis* gen.n. sp.n. nom. rev.; *Diplocalyx calotermitidis* gen.n. sp.n. nom. rev.; *Hollandina pterotermitidis* gen.n. sp.n. nom. rev.; and *Clevelandina reticulitermitidis* gen.n. sp.n. nom. rev. Int. J. Syst. Bacteriol. (Submitted.)

32. MERRIL, C. R., D. GOLDMAN, S. A. SEDMAN & M. H. EBERT. 1981. Ultrasensitive stain for proteins in polyacrylamide gels shows regional variation in cerebrospinal fluid proteins. Science **211**: 1437-1438.

33. OBAR, R. 1985. Purification of tubulin-like proteins from a spirochete. Ph.D. Dissertation. Boston University Graduate School. Boston, Mass.

34. PANTALONI, D. & M.-F. CARLIER. 1986. Involvement of guanosine triphosphate (GTP) hydrolysis in the mechanism of tubulin polymerization: regulation of microtubule dynamics at steady state by a GTP cap. Ann. N.Y. Acad. Sci. **466**: 496-509.

35. POPE, L. M. & P. JURTSHUK. 1967. Microtubule in *Azotobacter vinelandii* Strain O. J. Bacteriol. **94**: 2062-2064.

36. PORTER, K. R. 1966. Cytoplasmic microtubules and their functions. *In* Principles of Biomolecular Organization. G. E. W. Wolstenholme & M. O'Connor, Eds.: 308-345. J. & A. Churchill Ltd. London, England.

37. RODWELL, A. W., J. E. PETERSON & E. S. RODWELL. 1975. Striated fibers of the *rho* form of *Mycoplasma:* in vitro reassembly, composition and structure. J. Bacteriol. **122**: 1216-1229.

38. SAMMONS, D. W., L. D. ADAMS, T. J. VIDMAR, C. A. HATFIELD, D. H. JONES, P. J. CHUBA & S. W. CROOKS. 1984. Applicability of color silver stain (GELCODE$^R$ system) to protein mapping with two-dimensional gel electrophoresis. *In* Two Dimensional Gel Electrophoresis of Proteins. J. E. Celis & R. Bravo, Eds.: 111-126. Academic Press. New York, N.Y.

39. SHEELE, R. B. & G. G. BORISY. 1976. Comparison of the sedimentation properties of microtubule protein oligomers prepared by two different procedures. Biochem. Biophys. Res. Comun. **70**: 1-7.

40. SHELANSKI, M. L., F. GASKIN & C. R. CANTOR. 1973. Microtubule assembly in the absence of added nucleotides. Proc. Nat. Acad. Sci. USA **70**: 765-768.

41. TOWBIN, H., T. STAEHELIN & J. GORDON. 1979. Electrophoretic transfer of proteins from polyacrylamide gels to nitrocellulose sheets: procedures and some applications. Proc. Nat. Acad. Sci. USA **76**: 4350-4354.

42. WEBER, K. & M. OSBORN. 1975. Proteins and sodium dodecyl sulfate: molecular weight determination on polyacrylamide gels and related procedures. *In* The Proteins. H. Neurath & R. L. Hill, Eds. 3rd edit.**1:** 179-223. Academic Press. New York, N.Y.
43. WEISENBERG, R. C. 1972. Microtubule formation in vitro in solutions containing low calcium concentrations. Science **177:** 1104-1105.
44. WILSON, L. & K. W. FARRELL. 1986. Kinetics and steady state dynamics of tubulin addition and loss at opposite microtubule ends: the mechanism of action of colchicine. Ann. N.Y. Acad. Sci. **466:** 690-708.

# Catalyst

JEROME F. FREDRICK

*Research Laboratories*
*Dodge Chemical Company*
*3425 Boston Post Road*
*Bronx, New York 10469-2586*

In December of 1969, the New York Academy of Sciences sponsored a conference: *Phylogenesis and Morphogenesis in the Algae.* Only one paper presented at that conference dealt with extranuclear DNA.[1] The rest of the papers, published as an Academy *Annal,*[2] completely ignored the seminal paper by Sager and Ishida on the DNA found in plastids.[3] Indeed, Sager's treatise, "Cytoplasmic Genes and Organelles," sums up the biological field's treatment of this event:

> Historically, the idea that chloroplasts and mitochondria might be autonomous organelles goes back to the early 1900's. However, genetic evidence of non-Mendelian genes accumulated slowly between 1910 and 1960, a time of great advances in genetics and molecular genetics. During this period, the study was fervently pursued by a few groups of investigators, but ignored or discounted with equal fervor by *most* geneticists. Indeed, the literature of cytoplasmic genetics was viewed as a *blot* on the escutcheon of Science rather than as a part of a more comprehensive genetic theory.[4]

But, the publication of a paper by Lynn Margulis (under her former name of Sagan)[5] on the origin of mitosing cells stirred up interest in this extranuclear DNA and caused the beginning of a "rift" between the so-called traditional biologists[6] and the new group of endosymbiontists (who believed that the extranuclear or cytoplasmic organellar DNA attested to the past when the organelles were symbiotic "guests" in a larger "host").

The amount of cellular research started to gain at this point. It was with the publication of Margulis' *Origin of Eukaryotic Cells*[7] that the schism in Biology prompted the first conference sponsored by the Academy and organized by Fredrick on examining the accumulating endosymbiont/traditionalist (or exogenous/endogenous) data in 1980.[8] The conference devoted sessions to the newly growing "endosymbiosis theory" and many papers now started to deal with extranuclear DNA. Two more major international symposia soon followed;[9,10] and today, as witnessed by the contents of this volume, there is a virtual exponential growth of papers devoted to the topic.

Certainly there is no doubt, Margulis' book acted as a catalyst, in the traditional sense!, to cause this stimulation of research. We are now learning facts about the cell that we never could have without this catalysis.

## REFERENCES

1. IWAMURA, T. 1970. Ann. N.Y. Acad. Sci. **175:** 488-510.
2. FREDRICK, J. F. & R. M. KLEIN, Eds. 1970. Ann. N.Y. Acad. Sci. **175:** 413-781.
3. SAGER, R. & M. R. ISHIDA 1963. Proc. Nat. Acad. Sci. USA **50:** 725-730.
4. SAGER, R. 1972. Cytoplasmic Genes and Organelles: 1-2. Academic Press. New York & London.
5. SAGAN, L. 1967. J. Theor. Biol. **14:** 225-275.
6. KLEIN, R. M. & A. CRONQUIST 1967. Q. Rev. Biol. **42:** 105-296.
7. MARGULIS, L. 1970. Origin of Eukaryotic Cells. Yale University Press. New Haven, Conn.
8. FREDRICK, J. F., Ed. 1980. Ann. N.Y. Acad. Sci. **361.**
9. SCHWEMMLER, W. & H. E. A. SCHENK, Eds., 1980. Endocytobiology. de Gruyter. Berlin, FRG.
10. SCHENK, H. E. A. & W. SCHWEMMLER, Eds. 1983. Endocytobiology II. de Gruyter. Berlin, FRG.

# Evolutionary Differentiation in Gymnodinioid Zooxanthellae[a]

R. J. BLANK

*Institute of Botany and Pharmaceutical Biology*
*University of Erlangen-Nürnberg*
*Staudtstrasse 5*
*8520 Erlangen, Federal Republic of Germany*

Described as the endosymbiotic dinoflagellate of the Caribbean jellyfish *Cassiopea* sp.,[1] *Symbiodinium microadriaticum* (*Dinophyceae*) became widely regarded as representing a single, pandemic species of gymnodinioid zooxanthellae harbored by a variety of marine protozoans and invertebrates.[2] Although differences were found as to morphological, behavioral, physiological, and biochemical characters of the symbionts present in different hosts,[3] the question of speciation remained unresolved for lack of genetic evidence. Quantitative studies based on three-dimensional reconstructions and morphometric analyses of serially sectioned zooxanthellae isolated from different hosts now uncovered fundamental differences established in the DNA-bearing organelles of these algae.

The dinoflagellates harbored by *Cassiopea frondosa* and *C. xamachana* are identical while those of the Hawaiian stony coral *Montipora verrucosa*, the Caribbean sea anemone *Heteractis lucida*, the Californian sea anemone *Anthopleura elegantissima*, the Caribbean zoanthid *Zoanthus sociatus*, and the Australian zoanthid *Zoanthus* sp. are not. The entirely discordant karyotypes are marked by differences in chromosome numbers, forms, and sizes (FIGURE 1). The chromosome numbers of the symbionts harbored by these hosts are $97 \pm 2$ ($n = 6$), $26 \pm 0$ ($n = 3$), $74 \pm 2$ ($n = 3$), $50 \pm 1$ ($n = 4$), $78 \pm 2$ ($n = 6$), and $65 \pm 2$ ($n = 2$), respectively. Different chromosomal and nuclear volumes are well documented in both $G_1$ and $G_2$ phases of the cell cycle.[4,5] The chromosome content calculated in percent of nuclear volume is $14.4 \pm 2.3$ ($n = 6$), $43.6 \pm 3.7$ ($n = 6$), $29.4$ pm $1.6$ ($n = 3$), $28.1 \pm 3.5$ ($n = 4$), $16.4 \pm 1.5$ ($n = 8$), and $18.6 \pm 0.3$ ($n = 2$), in order of the hosts listed above. Differences in the chromosome distribution within the nuclei of differing symbionts are even evident in single sections.[6]

Mitochondrial numbers seem to vary in the course of the dinoflagellate cell cycle,[6] while the volumes differ from host to host. The mitochondrial content in the zooxanthellae arranged by hosts as above is $1.5 \pm 0.2$ ($n = 3$), $5.1 \pm 1.0$ ($n = 3$), $2.8 \pm 0.5$ ($n = 3$), $4.7 \pm 0.9$ ($n = 2$), $1.3 \pm 0.1$ ($n = 3$), and $2.5$ ($n = 1$), calculated in percent of cell volume.

[a]Supported by grants B1 222/3-1, B1 222/3-2, and 477/23/86 of the German Research Association.

Plastidial numbers are stable. All the symbionts possess one chloroplast except for those of *H. lucida,* which contain an increased number of plastids.[6] Two stalks connect the pyrenoid with chloroplast lobes in the algae present in both *Cassiopea* species, *M. verrucosa,* and *Z. sociatus,* but higher numbers were found in the zooxanthellae of *H. lucida* and those of *A. elegantissima*[6] as well as in the dinoflagellates inhabiting *Zoanthus* sp. The plastidial content in the different symbionts is 22.2 ± 2.8 ($n$ = 3), 35.8 ± 2.4 ($n$ = 3), 18.5 ± 4.5 ($n$ = 3), 49.4 ± 0.6 ($n$ = 2), 36.1 ± 3.4 ($n$ = 3), and 23.4 ($n$ = 1), calculated in percent of cell volume in order of the symbionts harbored by both *Cassiopea* species, *M. verrucosa, H. lucida, A. elegantissima, Z. sociatus,* and *Zoanthus* sp., respectively. Readily observed in single

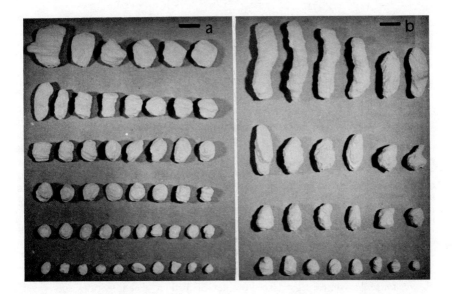

FIGURE 1. Differing karyotypes of the dinoflagellates harbored by two different host species of different geographical origins. (a) $G_1$ chromosomes of *Symbiodinium* sp. isolated from *A. elegantissima* (reproduced from Reference 5 with permission). (b) $G_1$ chromosomes of *Symbiodinium* sp. symbiotic with *M. verrucosa* (reproduced from Reference 9 with permission). Polystyrene-foam models built after three-dimensional reconstructions from 60 and 32 consecutive sections, respectively, through entire nuclei. Bars represent 0.5 μm.

sections and corroborated by freeze-fracture electron microscopy, the algae harbored by *A. elegantissima* and *Z. sociatus* are additionally distinguished by plastids with girdle thylakoids (FIGURE 2), a feature that they have in common with the symbionts of *Zoanthus* sp. which possess grana thylakoids in addition.[7] Polyhedral inclusion bodies resembling carboxysomes are frequently found in the chloroplasts of the algae symbiotic with *Cassiopea* sp., those of *M. verrucosa,* and the zooxanthellae of *Z. sociatus.*[8]

The three-dimensional cell architecture of the dinoflagellates inhabiting *M. verrucosa* was recently described.[9] Symbionts with hybrid organelles were never observed.

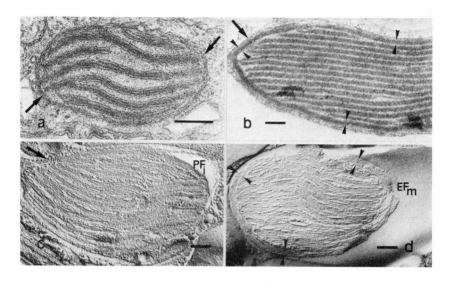

**FIGURE 2.** Differing thylakoid arrangements in chloroplasts of the dinoflagellates harbored by four different host species of different geographical origins. (a) Thin section through the chloroplast of *S. microadriaticum* harbored by *C. xamachana*. (b) Thin section through the chloroplast of *Symbiodinium* sp. isolated from *A. elegantissima*. (c) Freeze fracture of the chloroplast of *Symbiodinium* sp. harbored by *M. verrucosa*. (d) Freeze fracture of the chloroplast of *Symbiodinium* sp. symbiotic with *Z. sociatus* (reproduced from Reference 8 with permission). Triple-layered chloroplast envelope marked by arrows, peripheral thylakoid girdle completely surrounding the parallel thylakoid rows in the symbionts of *A. elegantissima* and *Z. sociatus* marked by arrowheads. $PF_i$ and $EF_m$ indicate fracture faces of inner and middle chloroplast envelope layers, respectively. Bars represent 0.2 $\mu$m.

It is concluded in the context of the evolutionary species concept[10] that the genus *Symbiodinium* encompasses a large complex of distinct genetic entities superficially resembling *S. microadriaticum* while this type species harbored by *Cassiopea* sp. is at the same time called by two incorrectly employed synonyms.[11]

# ACKNOWLEDGMENTS

Provision of the research subjects by Drs. R. K. Trench, D. J. Griffiths, and L. V. Thinh is gratefully appreciated.

# REFERENCES

1. FREUDENTHAL, H. D. 1962. J. Protozool. **9:** 45-52.
2. TAYLOR, D. L. 1984. *In* Encyclopedia of Plant Physiology. A. Pirson, M. H. Zimmermann,

H. F. Linskens & J. Heslop-Harrison, Eds. **17:** 75-90. Springer. Berlin, Heidelberg & New York.
3. TRENCH, R. K. 1987. *In* The Biology of Dinoflagellates. F. J. R. Taylor, Ed.: 530-570. Blackwell. Oxford, England.
4. BLANK, R. J. & R. K. TRENCH. 1985. Science **229:** 656-658.
5. BLANK, R. J. 1986. Arch. Protistenkd. **132:** 79-92.
6. BLANK, R. J. & R. K. TRENCH. 1986. *In* Proceedings of the Fifth International Coral Reef Congress, Tahiti. C. Gabrie, J. L. Toffart & B. Salvat, Eds. **6:** 113-117. Antenne du Museum National d'Histoire Naturelle et de l'Ecole Pratique des Hautes Etudes. Papeete, Tahiti.
7. THINH, L. V., D. J. GRIFFITHS & H. WINSOR. 1986. Phycologia **25:** 178-184.
8. BLANK, R. J. 1986. Pl. Syst. Evol. **151:** 271-280.
9. BLANK, R. J. 1987. Mar. Biol. **94:** 143-155.
10. WILEY, E. O. 1981. Phylogenetics. J. Wiley & Sons. New York, Chichester, Brisbane & Toronto.
11. BLANK, R. J. & R. K. TRENCH. 1986. Taxon **35:** 286-294.

# Cell Culture Models to Study Symbiosis[a]

PETER M. BRADLEY AND CLIFFORD S. DUKE

*Department of Biology*
*Northeastern University*
*360 Huntington Avenue*
*Boston, Massachusetts 02115*

Cell reconstruction techniques can be used as models for the putative formation of eukaryotic cells via endosymbiotic events. These events may have involved the uptake of bacteria into primitive cells. Mixed cell cultures have been used to show that free-living microalgae and cyanobacteria can form cellular associations with higher plant cells in culture,[1] and cellular interactions can be detected between the partners.[2] Work so far has introduced *Chlorella* into onion cells by microinjection, and demonstrated the uptake of *Chlorella* and cyanelles of *Glaucocystis nostochinearum* into isolated onion and carrot protoplasts using polyethyleneglycol[1] or oil drops.[3] Intercellular associations have also been produced. These used *Chlorella* and also several nitrogen-fixing cyanobacteria with carrot cells cocultured on medium with and without nitrate.[4] The cocultures were incubated at 20 and 25°C for seven weeks and were then fixed and sectioned. The percentage intact carrot cell results and full details of these experiments have been recorded previously.[2,4] The data have been examined further, and the new analysis is summarized in FIGURES 1 and 2. For example, some of the nitrate-free cocultures that contained the nitrogen-fixing cyanobacteria *Anabaena* and *Plectonema* contained more intact carrot cells than did control cultures. This suggests that some of the cultures obtained nitrogenous compounds from the microalgae for their continued survival. Other work has also shown that the nitrogen-fixing cyanobacterium *Anabaena* can successfully associate with cultured tobacco cells.[5] These results show that intracellular and intercellular associations can be established under controlled conditions and interactions can be detected. In addition, cocultures of *Chlorella* with *Amoeba proteus* were cultured successfully by M. C. Kaufman (personal communication) and viable *Chlorella* cells excreted from the *Amoeba* were encased in an additional membrane. The formation of vacuole membranes around *Chlorella* cells microinjected into *Amoeba* were observed in a transmission electron microscope (TEM) study.[6] In the future, we would like to construct a photosynthesizing cell by the uptake of *Chlorella*, cyanobacteria, or cyanelles into a host such as an albino plant cell or a protozoan. Data from these experiments may suggest how endosymbiosis could have occurred to form the early eukaryotic cells.

[a]This work was supported in part by National Aeronautics and Space Administration grant NSG-7154 to Dr. James Danielli, and the experimental work was performed at Worcester Polytechnic Institute, Massachusetts.

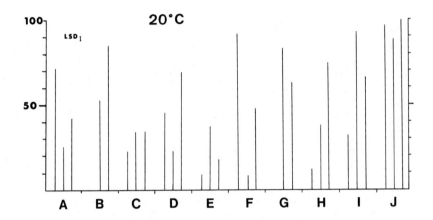

**FIGURE 1.** The percentage of intact carrot cells seen in sections of carrot callus that were cocultured with several microalgae at 20°C. Four microscope fields were counted to give means. Groups of three bars represent different culture concentrations of $KNO_3$: 0 (left), 0.015 mg/l (middle), 1.5 mg/l (right). Carrot callus is usually cultured on medium with 2.5 g/l $KNO_3$. Cultures labeled **A** were carrot callus controls that were not inoculated with a microalga. The other cultures were: **B,** *Chlorella* strain WPI-2; **C,** UV-killed WPI-2; **D,** autoclaved heat-killed WPI-2; **E,** a replicate *Chlorella* WPI-2; **F,** latex bead control; **G,** an alternate strain of *Chlorella;* and nitrogen-fixing cyanobacteria: **H,** *Gloeothece;* **I,** *Plectonema* and **J,** *Anabaena.* Results from an identical set of cocultures at 25°C are shown in FIGURE 2. Raw data of both sets were subjected to an analysis of variance. We are grateful to R. Alexander (University of Massachusetts) for his assistance with this. The least significant difference (LSD) is 2.6% which indicates the minimum difference between means necessary for statistical significance at the $p < 0.05$ level. The concentration of $KNO_3$, the temperature (20 or 25°C), and the microalga present in the cocultures all affected the percent intact carrot cells. A large part of the variation seen in this experiment is explained by interaction among these factors rather than by any individual factor. The alga present and its interaction with temperature and $KNO_3$ accounted for 74% of the variation, while temperature and its interaction with $KNO_3$ and the alga difference gave 43%, and $KNO_3$ with alga and temperature was 44%. The difference between the algal cultures had a greater influence on the carrot cells than did the $KNO_3$ concentration.

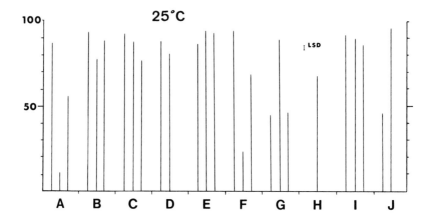

**FIGURE 2.** The percentage of intact carrot cells in 25°C cocultures. See legend to FIGURE 1 for details of these cultures. The analysis of variance showed that there was no effect of different concentrations of $KNO_3$ on the 25°C cultures as there was between the zero and highest concentration of $KNO_3$ at 20°C. The means (with 95% confidence limits) for all the cultures at a given temperature were 48.9 ± 6.0 for 20°C and 69.9 ± 6.0 for 25°C. The higher temperature gave greater intact carrot cells under all culture conditions. The means for each $KNO_3$ level were: 20°C lacking $KNO_3$ = 38.3 ± 12.3, 0.015 mg/l $KNO_3$ = 48.3 ± 0.2, and 1.5 mg/l $KNO_3$ = 60.2 ± 8.1. The corresponding means at 25°C were 72.6 ± 10.7, 71.8 ± 9.5, and 64.3 ± 11.9. At the higher temperature, the concentration of $KNO_3$ had no effect. FIGURES 1 and 2 suggest the following conclusions about these coculture experiments. The cultures having intact carrot cells greater than controls were *Anabaena* at 20°C and *Plectonema* at 25°C. WPI-2 at 25°C also gave high intact carrot cells but so did the killed WPI-2, suggesting that stored nitrogen in the algal cells could be responsible for these results. The $KNO_3$ concentration had an effect at 20° C but not at 25°C. Future coculture experiments should be performed at only one temperature but with fewer algal treatments and with more replicates. The optimum temperature to use should first be determined by a preliminary experiment for each alga/plant cell combination. It should now be possible to refine the conditions used for other experiments, and we also suggest that future work use $K^{15}NO_3$ to find out if any label is passed from the nitrogen-fixing cyanobacteria to the plant cells.

## REFERENCES

1. BRADLEY, P. M. 1980. *In* Endocytobiology I. W. Schwemmler & H. E. A. Schenk, Eds.: 515-522. de Gruyter. Berlin, FRG.
2. BRADLEY, P. M. 1984. PhD. Thesis. Worcester Polytechnic Institute. Worcester, Mass. (University Microfilm International 84-07038.)
3. BRADLEY, P. M. & A. LEITH. 1979. Naturwissenschaften **66:** 111-112.
4. BRADLEY, P. M. 1983. *In* Endocytobiology II. W. Schwemmler & H. E. A. Schenk, Eds.: 613-621. de Gruyter. Berlin, FRG.
5. GUSEV, M. V., T. G. KORZHENEVSKAYA, L. V. PYVOVAROVA, O. I. BAULINA & R. G. BUTENKO. 1986. Planta **167:** 1-8.
6. JEON, K. W. & I. J. LORCH. 1982. Exp. Cell Res. **141:** 351-356.

# Distribution and Transmission of Cytoplasmic DNA during Multiplication in *Netrium digitus* (Chlorophyceae)

MASAHIRO R. ISHIDA,[a] JIN HAMADA,[b] AND
MASAHIRO SAITO [a]

[a]*Department of Nuclear Biology*
*Research Reactor Institute*
*Kyoto University*
*Kumatoricho, Sennangun*
*Osaka 590-04, Japan*

[b]*Department of Community Medicine*
*Toyama Medical and Pharmaceutical University*
*Toyama 930-01, Japan*

The presence of DNA in the extranuclear organelles (chloroplasts and mitochondria) provides a new standpoint for studying the biogenesis of the organelles including DNA replication, transcription, and translation under the oligogenetic systems with a nuclear genome.[1,2] In the present study, the distribution and transmission of chloroplast DNA during multiplication using large algal cells were observed by epifluorescence microscopy and autoradiography.

## MATERIALS AND METHODS

*Netrium digitus* (Biebel 2)[3] was used throughout the present study. For epifluorescence microscopy, cells were fixed with carnoy's fluid or 1% glutaraldehyde, then stained with 4'6-diamidino-2-phenylindol (DAPI). For autoradiographic observation, cells were incubated with 15 $\mu$Ci/ml $^3$H-thymine in NPSY medium (0.1 g $KNO_3$, 0.01 g $K_2HPO_4$, 0.01 g $MgSO_4 \cdot 7H_2O$, 1 g yeast extract per liter) for 2.5 generations. For the chase study, the labeled cells were washed with the medium, they were spread on the medium with agar, and cultured for an additional 4 generations. Descendant cells from an original cell were mounted on a slide glass, dried, fixed, and washed with Carnoy's. Autoradiographic emulsion film was used.

## RESULTS

The cell contains two chloroplasts, one on each side of the nucleus. Epifluorescence microscopic observations after DAPI staining showed that the number of nucleoids per chloroplast was 150 to 200. These nucleoids were distributed throughout the chloroplast, and were mostly ovoid and rodlike in shape, and some were thread and network structures among thylakoid membranes. A ringlike association of chloroplast DNA was also found around the pyrenoids.

Autoradiograms of [3]H-thymine-incorporated cells for 2.5 generations showed no autoradiographic silver grain in the nuclei but many grains in the chloroplasts. A similar phenomenon was observed in *Spirogyra*.[4] This means that nuclei have no thymidine kinase. In the chase experiments, silver grains did not appear in the nuclei (FIGURE 1). The number of silver grains per chloroplast decreased during the 4

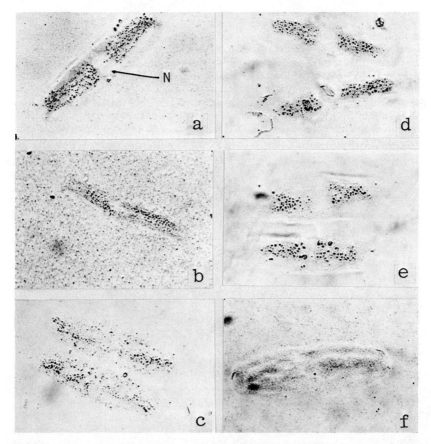

**FIGURE 1.** (a) Autoradiograms of the cells incubated with [3]H-thymine for 2.5 generations. After the treatment with radioactive precursor, it was chased for (b) 1 generation; (c) 2 generations; (d) 3 generations; and (e) 4 generations. (f) DNase-treated cells that have been incubated with [3]H-thymine for 2.5 generations.

TABLE 1. Change of Autoradiographic Grain Numbers in Descendant Cells of a Clone during the Chase Periods[a]

| Generations | Number of Cells per Clone | Number of Silver Grains | | Supply from Precursor Pool (%) |
| | | per Clone | per Cell | |
|---|---|---|---|---|
| 2.5 | 1 | 232.6 ± 62.9 | 232.6 | — |
| 2.5 + C 1[b] | 2 | 196.0 ± 67.3 | 98.0 | −15.8 |
| 2.5 + C 2 | 4 | 282.0 ± 97.2 | 70.5 | 21.2 |
| 2.5 + C 3 | 8 | 557.6 ± 143.7 | 69.7 | 139.7 |
| 2.5 + C 4 | 16 | 863.3 ± 159.2 | 54.0 | 271.2 |

[a] The original cells were incubated with $^3$H-thymine for 2.5 generations.
[b] The number of generations after the chase of the radioactive precursor.

generations; however, it was not in a geometrical progression. The numbers of silver grains were always higher than the theoretical ones in each generation (TABLE 1). This means that a large precursor pool is contained for the synthesis of chloroplast nucleoid DNA even in the fourth generation. The ratio (high value/lower value) of numbers of silver grains in two chloroplasts of a cell converged to 1. This means that the number of nucleoids is almost the same in the two chloroplasts for each cell and DNA synthetic activity is also the same level in the both chloroplasts. The nucleoid DNA in each chloroplast transmitted almost equally to the two chloroplasts of the descendant cells.

## REFERENCES

1. SAGER, R. & M. R. ISHIDA. 1963. Proc. Nat. Acad. Sci. USA 50: 725-730.
2. NASS, M. M. K. & S. NASS. 1963. J. Cell Biol. 19: 593-611.
3. STARR, R. C. 1978. J. Phycol. 14(suppl.): 47-100.
4. STOCKING, R. C. & E. M. GIFFORD, JR. 1959. Biochem. Biophys. Res. Commun. 1: 159-164.

# An Aphid Endosymbiont Has a Large Genome Rich in Adenine and Thymine

HAJIME ISHIKAWA

*Department of Biology*
*University of Tokyo*
*College of Arts and Sciences*
*Meguro-ku*
*Tokyo 153, Japan*

The intracellular symbionts in the mycetocyte of the aphid species (FIGURE 1) are typically prokaryotic in terms of their abilities for DNA, RNA, and protein synthesis.[1] Under *in vitro* conditions, the endosymbionts are able to synthesize at least several hundreds of protein species. In contrast, the endosymbiont *in situ* in the host insect seems to be under a stringent control and synthesizes only one protein species, symbionin.[2] For these reasons the aphid endosymbiont can be regarded as an evolutionary intermediate between the cell organelles, on the one hand, and inter and intracellular parasites such as mycoplasmas and rickettsiae, on the other. In this respect, it is interesting to determine basic properties of the aphid endosymbiont's genomic DNA.

Biologically active endosymbionts were isolated on a large scale from the pea aphid, *Acyrthosiphon pisum*, using a Percoll density gradient.[3] The DNA was extracted from the isolated symbiont and its nucleotide composition was estimated using three different methods: thermal denaturation, CsCl-density equilibrium centrifugation, and high performance liquid chromatography (HPLC) analysis of the P1 digest (TABLE 1).

All gave guanine + cytosine (G+C) contents of the symbiont DNA as low as 30%, which is reminiscent of the *Mycoplasma* species. However, the aphid endosymbiont is strikingly different from the parasitic bacteria with respect to the size of its genome.

Reassociation kinetics of the genomic DNA which was labeled by nick translation suggested that the endosymbiont genome is $1.4 \times 10^{10}$ daltons, about 5 and 18 times as large as those of *Escherichia coli* and *M. capricolum*, respectively. The results were confirmed by reassociation of the endosymbiont DNA which was labeled by incubation with ³H-thymidine in Grace's medium. The result contrasts sharply with the observation that the genome size of endosymbionts of the leaf hopper is even smaller than that of the chloroplast in higher plants.[4]

Though the aphid endosymbiont has, at least, several hundred protein genes, most of them are expressed only extracellularly. The fact that these gene products of the endosymbiont in *A. pisum* differ considerably from those of the one in *A. kondoi*, the closest relative of the former,[5] implies that these unexpressed genes are highly susceptible to mutation and evolve much faster than those in the host. In this respect, the endosymbiont's genes resemble the pseudogenes which undergo fast evolution with

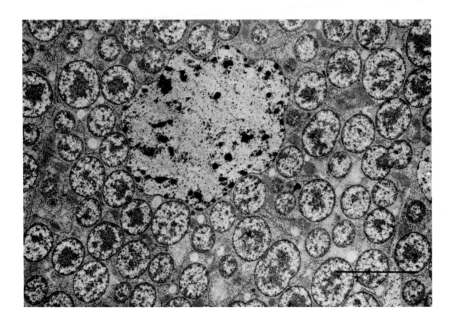

**FIGURE 1.** Intracellular symbionts of the pea aphid, *Acyrthosiphon pisum,* with the mycetocyte nucleus in the center. Bar represents 5 μm.

no functional constraint.[6] A high adenine + thymine (A+T) content of the endosymbiont genome is also reminiscent of the pseudogenes, where detected base substitutions are biased towards accumulation of A and T.[7]

If the aphid endosymbiont is destined, in the course of evolution, to specialize as a cell organelle through loss of much of its genetic coding capacity, the endosymbiont genome rich in A and T will be more amenable to such an evolutionary change than otherwise, and despite its presence the change has not yet happened to the large-sized genome of the aphid endosymbiont.

**TABLE 1.** Nucleotide Composition of Nuclease P1 Digest of Endosymbiont DNA (Determined by HPLC)

| DNA | Molar Percent | | | |
| --- | --- | --- | --- | --- |
| | C | T | G | A |
| Endosymbiont | 15.5 | 34.9 | 14.8 | 34.8 |
| *E. coli* | 25.5 | 24.8 | 25.0 | 24.7 |

## REFERENCES

1. ISHIKAWA, H. 1982. DNA, RNA and protein synthesis in the isolated symbionts from the pea aphid, *Acyrthosiphon pisum*. Insect Biochem. **12:** 605-612.
2. ISHIKAWA, H. 1984. Characterization of the protein species synthesized *in vivo* and *in vitro* by an aphid endosymbiont. Insect Biochem. **14:** 417-425.
3. ISHIKAWA, H. 1982. Isolation of the intracellular symbionts and partial characterizations of their RNA species of the elder aphid, *Acyrthosiphon magnoliae*. Comp. Biochem. Physiol. **72B:** 239-247.
4. SCHWEMMLER, W. 1980. Endocytobiosis: general principles. BioSystems **12:** 111-122.
5. ISHIKAWA, H. & S. M. YAMAJI. 1985. Protein synthesis by intracellular symbionts in two closely interrelated aphid species. BioSystems **17:** 327-335.
6. LI, W.-H., T. GOJOBORI & M. NEI. 1981. Pseudogenes as a paradigm of neutral evolution. Nature **292:** 237-239.
7. GOJOBORI, T., W.-H. LI & D. GRAUR. 1982. Patterns of nucleotide substitution in pseudogenes and functional genes. J. Mol. Evol. **18:** 360-369.

# Dynamics and Host Influence of Methane Production by Microbial Symbionts in a Lower Termite

MONICA J. LEE [a] AND ADAM C. MESSER [b]

[a]Department of Microbiology
[b]Department of Entomology
Cornell University
Ithaca, New York 14853

Symbiotic protozoa and bacteria in lower termites comprise a methane production system. Hydrogen and $CO_2$, resulting from anaerobic fermentation of cellulose by protozoa, are converted to $CH_4$ and $H_2O$ by methanogenic bacteria.

Methanogenic bacteria were found associated with three flagellated protozoa, *Trichomitopsis termopsidis*, *Hexamastix termopsidis*, and *Tricercomitus termopsidis*, symbiotic in the hindgut of *Zootermopsis angusticollis* termites.[1] Other protozoa did not carry symbiotic methanogens, and these bacteria were not observed free living in termite hindgut fluid, suggesting that these surface associations are specific and of an obligate nature.

Individual termites were placed in sealed butyl-rubber stoppered culture tubes and methane production was measured in both air and in 50% $H_2$ environments in self-paired experiments to test whether this anaerobic ecosystem is $H_2$ limited. Increased $CH_4$ production to 1950 nmoles hour$^{-1}$g$^{-1}$ termite fresh weight (U), 27% greater than paired control levels of 1550 U (p < 0.0001, t-test of means) suggested that although $H_2$ was rate limiting, it was close to saturation.

Hyperbaric $O_2$ treatments selectively eliminate protozoa from the termite hindgut. Exposure of *Z. angusticollis* termites to 300 kPa of 100% $O_2$ for one hour removed the cellulolytic *Trichonympha* and *Trichomitopsis* flagellates, but *Tricercomitus* and *Hexamastix* survived this treatment. With this treatment, methanogenesis was reduced 33-fold, to 50 U. Incubation of these partially defaunated termites in 50% $H_2$ resulted in a 7-fold increase in methanogenesis to 320 U (p<0.0001), suggesting that cellulolytic protozoa are important in methanogenic processes in the hindgut ecosystem.

A specific inhibitor of bacterial methanogenesis, 2-bromoethanesulfonic acid (BES), was fed to termites on filter papers. BES inhibited methane production to a point below the detection threshold. Incubation in 50% $H_2$ did not restore methanogenesis. Dissection and subsequent microscopic examination of gut contents of termites treated with BES for 5 days failed to reveal methanogens. All protozoan species were present, however.

The insect molting hormone, 20-hydroxyecdysone, was fed to termites to simulate the hormonal environment of molting, and to determine if methane production rates change during this process. Methane production was reduced to 250 U, a sevenfold reduction compared to untreated control levels. It was recoverable to 1830 U (p<0.0001), a level not statistically significantly different from the rate shown by normally faunated termites after incubation in 50% $H_2$ (p < 0.16). Dissections of these treated termites showed that the *Trichonympha* flagellates were no longer present, but all other protozoan species were observed. These observations suggest that host influence on methane production likely occurs when molting processes disrupt the flow of precursors from ecdysone-sensitive protozoa to methanogenic bacteria.

## REFERENCE

1. LEE, M. J., P. S. SCHREURS, A. C. MESSER & S. H. ZINDER. Association of methanogenic bacteria with flagellated protozoa from a termite hindgut. Current Microbiol. (In press.)

# Origins of Eukaryote Sensory Transduction

M. LEVANDOWSKY

*Haskins Laboratories*
*Pace University*
*41 Park Row*
*New York, New York 10038*

## THE EUKARYOTE PATTERN

In well-studied cases, eukaryotic sensory transduction involves a sequence of biochemical events that show a certain fundamental pattern:[1] (1) Stimulation of a receptor molecule. (2) Coupling of stimulated receptor with guanyl nucleotide binding proteins ("G proteins"). (3) Activation of phosphoinositidase C ("protein C"), causing hydrolysis of membrane inositol phospholipids, to yield second messengers: inositol trisphosphate, diacylglycerol, and other species. (4) Mobilization of external (noncalmodulin-associated) or internal (calmodulin-associated) $Ca^{++}$. (5) Activation of a protein kinase. This is a branch point, or node in sensory transduction, and can trigger a wide variety of responses. Numerous details vary, and complexities abound, but versions of this pattern are quite widespread, perhaps universal in eukaryote sensory transduction.

## THE PROKARYOTE PATTERN

Prokaryotes do not seem to follow the pattern just described. In eubacterial sensory transduction, by far the best-studied case being chemosensory transduction, there appears to be a great diversity of patterns:[2] (1) Gram-negative species (mainly *Escherichia coli* & *Salmonella typhimurium*). In many cases the signal is transported across the cytoplasmic membrane by methyl accepting proteins (MCPs). Some gram-negative responses, however, such as to oxygen ("aerotaxis"), do not involve MCPs, but rather a proton-motive force. (2) Gram-positive (i.e., *Bacillus subtilis*). Responses are associated with membrane depolarization and energy metabolism. (3) Spirochaetes. Chemoreception involves a $Ca^{++}$-dependent membrane depolarization. Work with Archaebacteria is in its infancy, and little is known yet about transduction in their chemosensory responses. Recently, however, an exciting report has appeared suggesting the presence of calmodulin in an archaebacterial species.[3]

# EVIDENCE OF PROKARYOTE HOMOLOGIES TO THE EUKARYOTE PATTERN

(1) Protein kinases. These were thought not to exist in bacteria, but have been found in recent years, although their physiological significance remains unknown. In most cases they phosphorylate at serine or threonine sites, rather than tyrosine as in many eukaryote kinases.[4] (2) G-protein homologies. It was found recently that G-protein has a great deal of primary structure homology with elongation factor in *E. coli.*[5] (3) Rhodopsin. (a) Relation of bacteriorhodopsin and eukaryote rhodopsin. Rhodopsin occurs in the archaebacterium, *Halobacterium.* There are many similarities. Both have 7 alpha helices spanning the membrane. Both have the amino terminus on the outside. In both, retinal is attached to the carboxyl terminal (7th) helix. There are similar regions in the 5th and 6th helices, and in a cytoplasmic loop. The central cores both contain high numbers of serines and threonines. (b) Functional homology of the rhodopsin-transducin system to the beta-adrenergic receptor-G protein system. The active beta-receptor agonist complex triggers GTP-GDP exchange in the G protein complex. Activated rhodopsin initiates GTP-GDP exchange in transducin. From this and many other similarities, it seems most likely that transducin and the G protein share structural domains. (c) Primary structure homology of rhodopsin and the beta-adrenergic receptor. This was recently discovered by Dixon *et al.*[6] (d) *E. coli* and other eubacteria (*Pseudomonas*) respond behaviorally to adrenergic signals. This was reported over a decade ago but has gone largely unnoticed.[7]

# TENTATIVE CONCLUSION

Vestiges of molecular homology to the eukaryote pattern can be seen in the two kinds of prokaryote lines. This suggests that the basic biochemical equipment used by eukaryotes may predate the divergence of the two prokaryote lines and the eukaryotes.

## REFERENCES

1. HELMREICH, E. J. M. & T. PFEUFFER. 1985. Regulation of signal transduction by beta-adrenergic hormone receptors. Trends Pharmacol. Sci. **6:** 438-443.
2. KOSHLAND, D. E., JR. 1980. Bacterial Chemotaxis as a Model Behavioral System. Raven. New York, N.Y.
3. SEARCY, D. G. 1987. Phylogenetic and phenotypic relationships between the eukaryotic nucleocytoplasm and thermophilic archaebacteria. Ann. N.Y. Acad. Sci. (This volume.)
4. COZZONE, A. J. 1984. Protein phosphorylation in bacteria. Trends Biochem. Sci. **9:** 400-403.
5. LOCHRIE, M. A., J. B. HURLEY & M. I. SIMON. 1985. Sequence of the alpha subunit of photoreceptor G protein: homologies between transducin, ras, and elongation factors. Science **228:** 96-99.
6. DIXON, R. A. F., *et al.* 1986. Cloning of the gene ans cDNA for mammalian beta-adrenergic receptor and homology with rhodopsin. Nature **321:** 75-79.
7. MARUYAMA, H. & H. AZUMA. 1974. Regulation of bacterial motility by cyclic nucleotides and effect of biogenic amines. J. Cyclic Nucleotide Res. **2:** 129-138.

# The Vesicle in the Bacteroid-Containing Cells of Root Nodules Is a Kind of Organelle

## A View on the Interrelation between Rhizobium and Legume Plants

LIANG OUYANG

*Jiangxi Agricultural University*
*Nachang, Jiangxi*
*China*

Usually microbiologists and agronomists recognize the root nodules of legume plants as symbiotic bodies participated in by the two organisms, the Rhizobium and legume plants. The bacteroids in these nodules exchange with their hosts the nitrogen nutrients for carbon nutrients.

However, there is a wealth of evidence to indicate that the relation between Rhizobium and its host is much closer than the generally recognized symbiotic one.

The root nodule cells of any variety of legume crops are composed of a strictly ordered ultrafine architecture. There are a great many vesicles in the bacterium-containing cells. The bacteria existing in the vesicles are differentiated into a special form known as bacteroids. The outermost layer of the vesicles is a plasma membrane of host plant origin. Leghemoglobin is located between the membrane and bacteroids. Such a fine structure is very similar to an organelle with an intracytoplasmic membrane system such as mitochondria and chloroplasts.

Morphology of the bacteroid is a special phase of Rhizobium. It is well known that free-living cells of Rhizobium, either in soil or in artificial medium, would not be differentiated into bacteroids; this occurs only in the vesicles of bacteria-containing cells of host root nodules. This seems to indicate that the conversion of Rhizobium vegetative cells into bacteroids is controlled by certain unknown mechanisms mediated by the host.

The vegetative cells differ from the bacteroids of Rhizobium not only in shape and size but also in cell composition. The DNA content of bacteroids is much less than that of vegetative cells. The former is only 40% of the latter. There are only a few ribosomes in bacteroids; the RNA content is only 13% that of vegetative cells. Also, bacteroids possess more mesosomelike structures than the vegetative cells. These important differences suggest that the nature of bacteroid is altered to a great extent.

Another feature of the bacteroid is its inability to divide into new daughter cells. The results of isolating single cells with microscopic manipulators have shown that only 0.02% of bacteroids are capable of forming colonies while 90% or more of vegetative cells form colonies. It should be accepted that loss of the ability to divide indicates the loss of the status of being an independent creature.

Recently, it has been shown that nodules can perform nitrogen fixation before the bacteria differentiate into bacteroids. Thus the latter process is helpful but not necessary for nitrogen fixation. One wonders why evolution selects bacteroids as the normal form in the mature nodule cells. The reason may not be for nitrogon fixation, at least not primarily for this. A more appealing reason may be for the integration of the nodules to make bacteria become inseparable from the whole finely organized functional cells. Such integration would achieve better coordination of the metabolism of bacteria and their host, and would also avoid damage to the plant. In this way, the nitrogen fixation would be enhanced greatly.

Mitochondria and chloroplasts were likely symbiotic procaryotes in eucaryotes one hundred million years ago; in the course of evolution they became organelles in plant cells. This hypothesis is widely accepted. In comparison with these, the vesicles in bacteroid-containing cells of root nodules seem to be similar in nature.

Leghemoglobin in legume root nodule cells serves a very important physiological function. It extensively absorbs oxygen to prevent nitrogenase from oxygen damage and releases it slowly to meet the requirement for oxidative phosphorylation in bacteroids. It is known that the genes coding for protein and heme moieties occur in legume plant cells and bacteroid respectively. Genes for such a physiologically important component are controlled by both Rhizobium and its host; this fact may only be explained if in the process of evolution the two organisms had been developing into an integrated organism.

The plants that fix nitrogen with root nodules differ in some aspects from those grown on combined nitrogen. Nitrogen-fixing plants increase the extent of their photosynthesis with increasing photon flux density, while those grown on combined nitrogen decrease their photosynthesis as the photon flux density exceeds a certain level. These results show that the nitrogen-fixing plants with root nodules should be regarded as a new type of organism with its metabolism comprehensively altered. It should not be regarded merely as a simple symbiont with Rhizobium.

The above observations lead the author to suggest that the root nodules of legume plants should be recognized as an organ of a new type of organism, and the vesicles in bacteroid-containing cells of a nodule as a kind of organelle. The author suggests that it may be named "nifixasome." Such an idea would be fitting for the true nature of noduled legumes and would be helpful for the elucidation of the internal mechanism of symbiotic nitrogen fixation. It would also provide some enlightenment into the creation of new eucaryotic nitrogen-fixing organisms.

# The Cyanelle Genome of
# *Cyanophora paradoxa*

## Chloroplast and Cyanobacterial Features

WOLFGANG LÖFFELHARDT,[a] HERMANN MUCKE,[a]
HEIMO BREITENEDER,[a] DAVE N. T. ARYEE,[a]
CHRISTIAN SEISER,[a] CHRISTINE MICHALOWSKI,[b]
MICHAEL KALING,[b] AND HANS J. BOHNERT[b]

[a]*Institute for General Biochemistry*
*University of Vienna and*
*Ludwig Boltzmann Laboratory for Biochemistry*
*A-1090 Vienna, Austria*

[b]*Department of Biochemistry*
*University of Arizona*
*Tucson, Arizona 85721*

*Cyanophora paradoxa*, a photoautotrophic protist, harbors so-called cyanelles which perform the functions of chloroplasts. The well-documented occurrence of peptidoglycan in the cyanelle envelope forbids the designation of cyanelles as a special type of chloroplasts, e.g., the rhodoplasts. However, when genome size only is considered these semiautonomous endosymbionts may be classified as chloroplastlike organelles. Investigation of cyanelle genome organization reveals dual properties, which is in accordance with the postulated position of cyanelles as intermediates between chloroplasts and free-living cyanobacteria in the sense of the endosymbiotic theory.

The inverted repeat (IR in FIGURE 1) containing two sets of rRNA genes is a common structural feature among chloroplast genomes from higher plants and algae. The distribution of tRNA genes on cyanelle DNA resembles that on several chloroplast DNAs.[1] The arrangement of some protein genes,[2,3] as summarized in TABLE 1, is shown in FIGURE 1. A number of gene clusters analogous to those found on chloroplast genomes were identified: two subunits of the cytochrome $b_6/f$ complex (petB, petD), the two photosystem I reaction center proteins (psaA, psaB), LSU and atpB (on opposite strands),[4] two subunits of the ATP synthase (atpA, atpH) and photosystem II reaction center and antenna proteins (psbC, psbD). Northern hybridization data (TABLE 1) indicate that most of them form transcription units as observed in chloroplasts.

On the other hand, cyanelle DNA contains genes not found on chloroplast DNAs from higher plants in the same arrangement as on cyanobacterial genomes. The large and the small subunit of ribulosebisphosphate carboxylase[4,5] as well as the $\alpha$ and $\beta$ subunits of phycocyanin and allophycocyanin[6] give rise to bicistronic mRNAs.

Two strains of *C. paradoxa* are known at present which differ in size and restriction

550

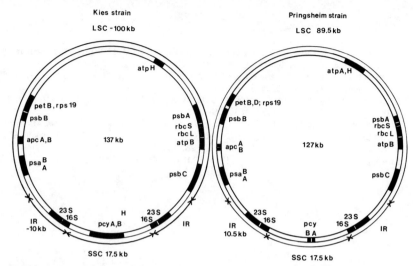

**FIGURE 1.** A comparison of protein gene loci on the cyanelle genomes from *C. paradoxa* LB 555 UTEX (Pringsheim strain) and from *C. paradoxa* 1555 (Kies strain).

pattern of cyanelle DNA and which possess similar but not identical nuclear rDNA units. However, as shown in FIGURE 1, the pattern of protein genes is remarkably conserved.

**TABLE 1.** Protein Genes Identified on Cyanelle DNA[a]

| Gene Symbol | Function/Description | mRNA size[b] (bases) |
|---|---|---|
| psbA | Herbicide binding protein | 1200 |
| atpA | Alpha subunit ATP synthase | |
| atpB | Beta subunit ATP synthase | 2800 |
| atpE | Epsilon subunit ATP synthase | |
| atpH | DCCD binding protein, $CF_o$-III | |
| psaA | PS I (P700) apoprotein | 6000 |
| psaB | PS I (P700) apoprotein | 6000 |
| psbB | PS II (55 Kd) apoprotein | |
| psbC | PS II (44 Kd) apoprotein | 3200 |
| rps19 | rbs protein S19 | 1500 |
| rpl2 | rbs protein L2 | 1500 |
| petA | Cytochrome f | 1600 |
| petB | Cytochrome b6/f complex, cyt. b6 | |
| petD | Cytochrome b6/f complex, SU4 | 1400 |
| rbcL | Rubisco, LSU | 2600 |
| rbcS | Rubisco, SSU | 2600 |
| pcyA/B | $\alpha$, $\beta$ subunits phycocyanin | |
| apcA/B | $\alpha$, $\beta$ subunits allophycocyanin | |

[a] Data from References 2 and 3.
[b] Determined from strain LB 555 UTEX.

## REFERENCES

1. KUNTZ, M., E. J. CROUSE, M. MUBUMBILA, G. BURKARD, J.-H WEIL, H. J. BOHNERT, H. MUCKE & W. LÖFFELHARDT. 1984. Mol. Gen. Genet. **194:** 508-512.
2. BOHNERT, H. J., C. MICHALOWSKI, S. BEVACQUA, H. MUCKE & W. LÖFFELHARDT. 1985. Mol. Gen. Genet. **201:** 565-575.
3. LAMBERT, D. H., D. A. BRYANT, V. L. STIREWALT, J. M. DUBBS, S. E. STEVENS & R. D. PORTER. 1985. J. Bacteriol. **164:** 659-664.
4. WASMANN, C. C. 1985. Ph.D. Thesis. Michigan State University. East Lansing, Mich.
5. STARNES, S. M., D. H. LAMBERT, E. S. MAXWELL, S. E. STEVENS, R. D. PORTER & J. M. SHIVELY. 1985. FEMS Microbiol. Lett. **28:** 165-169.
6. LEMAUX, P. G. & A. R. GROSSMAN. 1985. EMBO J. **4:** 1911-1919.

# Role of Light in the Symbiosis between Algae and Hydra

G. MULLER-PARKER[a] AND R. L. PARDY

*School of Biological Sciences*
*University of Nebraska*
*Lincoln, Nebraska 68588-0118*

Previous studies have shown that, when food is limiting, light promotes the growth of symbiotic green hydra.[1] When starved in the light, green hydra survive much longer than aposymbiotic hydra[1] and lose less weight than nonsymbiotic or aposymbiotic hydra.[2] These results and those of studies investigating the translocation of carbohydrates in the green hydra symbiosis[3] suggest that photosynthetic products of the symbiotic algae are of nutritional importance to the host hydra.

Since photosynthesis is dependent on irradiance, it was hypothesized that growth and survival of symbiotic hydra may be affected by the total photon flux density provided. To test this, green hydra from one clone were maintained under continuous light at four photon flux densities (5, 10, 15, and 30 $\mu E \times m^{-2} \times s^{-1}$) and either fed or starved. The following parameters were measured in green hydra acclimated to the four irradiances and fed $4\times$/week with brine shrimp nauplii: population growth rates, productivity of hydra at each culture irradiance, and host-algal biomass parameters. After acclimation to the four irradiances, hydra were subsequently starved and the following parameters analyzed: length of time green hydra survived starvation, and host-algal biomass parameters.

Culture irradiances ranging from 5 to 30 $\mu E \times m^{-2} \times s^{-1}$ had no significant effect on population growth rates of fed green hydra. As the productivity of hydra increased from 0.3 to 1.15 $\mu g$ $C \times mg$ protein$^{-1} \times h^{-1}$ with increase in culture irradiance, the productivity of symbiotic algae did not result in increased growth of the symbiotic association. These results indicate that the productivity of symbiotic algae does not affect the growth rate of green hydra when adequate food is provided. Biomass parameters of fed hydra indicated that algal density decreased in response to an increase in culture irradiance, although changes in the ratio of plant to animal biomass may have been less pronounced since algal cell volumes increased with increase in culture irradiance in both fed and starved hydra.

Survival of starved green hydra was not affected by culture irradiance. The decline in numbers of hydra was described by a second-order polynomial equation which predicts that 60 days without food results in a 50% reduction of the initial population. The protein biomass of green hydra starved for 28 days declined to about 10% of initial levels, whereas algal densities were twice those of initial levels. Since algal cell volumes from these hydra were much greater than those from fed hydra, the ratio of plant to animal biomass increased greatly with starvation of the host.

[a]Present affiliation: Chesapeake Biological Laboratory, University of Maryland, Center for Environmental and Estuarine Studies, Box 38, Solomons, Maryland 20688-0038.

## REFERENCES

1. MUSCATINE, L. & H. M. LENHOFF. 1965. Symbiosis of hydra and algae. II. Effects of limited food and starvation on growth of symbiotic and aposymbiotic hydra. Biol. Bull. **129:** 316-328.
2. COOK, C. B. & M. O. KELTY. 1982. Glycogen, protein, and lipid content of green, aposymbiotic, and nonsymbiotic hydra during starvation. J. Exp. Zool. **222:** 1-9.
3. MEWS, L. K. 1980. The green hydra symbiosis. III. The biotrophic transport of carbohydrate from alga to animal. Proc. R. Soc. London Ser. B **209:** 377-401.

# Major Acyl Lipids of Cyanelles
# from *Glaucocystis nostochinearum*

O. TACHEENI SCOTT

*Department of Biology*
*Northern Arizona University*
*Flagstaff, Arizona 86011*

## INTRODUCTION

*Glaucocystis nostochinearum* is an "alga" of indeterminate phylogenetic identity. Various structural and pigment characters have been used by investigators to suggest its taxonomic affinity to the prokaryotic cyanobacteria, and to members of eukaryotic phyla including the red algae, green algae, dinoflagellates, and the cryptomonads. In recognition of this combination of prokaryotic and eukaryotic characters in *Glaucocystis,* several investigators have considered it to be a unique eukaryotic organism which could be accommodated only by special taxonomic groupings, including Glaucocystideae[1] and Glaucophyta.[2]

The unsettled issue regarding the phylogeny of *Glaucocystis* persisted into the present decade following several unsuccessful attempts during the previous two decades to detect a remnant cyanobacterial cell wall in the cyanelles of *Glaucocystis.* Recently, however, Scott *et al.* reported that a peptidoglycan envelope existed in the cyanelle (there may be several to many cyanelles per host cell) of *Glaucocystis* and gave the first unequivocal evidence for the non-Archaebacterial ancestry of the cyanelles and, thus, an endosymbiotic explanation for *Glaucocystis nostochinearum.*[3] Furthermore, the argument can be made that *Glaucocystis nostochinearum* represents a unicellular plant system in which the organellar role of photosynthesis is being conducted by an intracellular cyanobacterium.

## MATERIALS AND METHODS

Axenic cultures of *Glaucocystis nostochinearum* (No. 229/1, Cambridge Collection of Algae and Protozoa, Cambridge University, Cambridge, United Kingdom) were grown in an unbuffered defined inorganic medium with added N-Z-Case and vitamin $B_{12}$, pH 7.1. The cultures were bubbled with filtered air in cotton-stoppered 1000-ml bubbler tubes under approximately 1700 lux of cool-white fluorescent illumination,

18°C. Intact *Glaucocystis* cells were harvested during midlate log-phase growth. Following centrifugation, the pellets were resuspended in culture medium before being subjected to mechanical disruption with a glass tissue homogenizer at 0°C. The brei was subjected to differential centrifugation resulting in pellets containing isolated cyanelles which were lyophilized. For use as positive controls, *Anacystis nidulans* and *Anabaena variabilis* were cultured in BG-11 in 5% $CO_2$-gassed carboys under 4000 lux of fluorescent illumination. Samples of the two cyanobacterial controls were harvested and lyophilized.

One hundred-milligram samples of lyophilized material were subjected to standard Bligh and Dyer extraction[4] and Folch wash[5] procedures. Following the "dry-down" step conducted in nitrogen, chloroform was used to adjust the final volume before 10 $\mu$l and 25 $\mu$l samples were spotted via Hamilton syringes onto hard, silica gel thin-layer chromotography (TLC) plates (Merck, No. 5769) before being dried immediately with a hand-held dryer-blower. The elution solvent consisted of acetone:benzene:water (91:30:8, by volume, total $= 129$ ml). Following an equilibration time of 2 hours, the TLC plates were introduced and permitted to run for 75 minutes before they were removed and air dried before analysis.

The following qualitative methods were used: (1) rhodamine, 1:100 dilution of stock (0.12%), a general stain for all lipids which was dispensed as a spray; (2) iodine (crystals in a covered chromatography jar, 90 minutes) used as a general stain for all lipids; (3) 5% sulfuric acid in methanol and ethanol, used over hot plate in a fume hood; (4) diphenylamine (method according to manufacturer, Applied Science Labs., Inc. State College, Pa.), dried over a very hot hotplate (5-10 minutes) elicited a blue color for glycolipids; and (5) molybdenum blue (Applied Science Labs., Inc.) used as a general indicator of phospholipids. The first three methods were used to indicate the presence of lipids with the rhodamine method, considered to be more sensitive than iodine, which in turn demonstrated greater sensitivity than the charring afforded by sulfuric acid. Authentic samples of digalactosyl diglyceride (4-6031) and phosphatidyl glycerol (4-6013), and nonpolar lipid mixes (4-7002 and 4-7007) and rhodamine were obtained from Supelco, Inc. (Bellefonte, Pa.).

## RESULTS

Initially, the total lipids for the two cyanobacterial controls were analyzed by the sequential applications of the first three assays described above. The results indicated similarities existed between the approximate $R_f$ values (mm) obtained for seven spots from each cyanobacterial sample, including 16, 20, 42, 90, 112, 122, and 135 for *Anacystis nidulans* and 14, 21, 37, 52, 94, 115, and 138 for *Anabaena virabilis*. Seven bands were obtained also for *Glaucocystis* cyanelles including 15, 40, 44, 50, 93, 122, and 136.

The spots that migrated the farthest generally occupied the largest areas and were slightly pigmented (green) and demonstrated fluorescence when illuminated with UV at 254 nm. The spots generally had $R_f$ values similar to those obtained for nonpolar lipids including cholesteryl palmitate, cholesteryl oleate, methyl oleate, oleic acid, tripalmitin, palmitic acid, triolein, and cholesterol.

The diphenylamine analysis for glycolipids revealed the spots at 37-40 mm to be digalactosyl diglyceride, as determined by the coelution of 50 $\mu$g ($10 = \mu$l sample of 5 mg/ml $CHCl_3$:MeOH, 2:1) of authentic digalactosyl diglyceride. The analysis revealed the 90-94 mm spots also to be a glycolipid as indicated by the blue reaction elicited by the heating of diphenylamine. Authentic monogalactosyl diglyceride was not coeluted, however, as a means of confirming the presence of this expected glycolipid in the cyanobacterial and cyanelle lipid samples.

The molybdenum blue analysis for phospholipids revealed the spot at 14-16 mm to be phosphatidyl glycerol, as determined by the coelution of 100 $\mu$g (10-$\mu$l sample of 10 mg/ml CHCl$_3$:CH$_3$OH, 2:1) of authentic phosphatidyl glycerol.

## DISCUSSION

According to Nichols, the cyanobacteria usually contain only one quantitatively important phospholipid (phosphatidyl glycerol) and three glycolipids (mono- and digalactosyl diglyceride and sulfoquinovosyl diglyceride).[6] Based on preliminary qualitative evidence for the presence of digalactosyl diglyceride, and perhaps monogalactosyl diglyceride, as the major glycolipid(s) and phosphatidyl glycerol as the major phospholipid in the cyanelles of *Glaucocystis nostochinearum,* the cyanelles appear to bear strong similarities to the cyanobacteria, including *Anacystis nidulans* and *Anabaena viriabilis.* Evidence for sulfoquinovosyl diglyceride, as suggested by Nichols, was not observed. It is quite possible, however, that this glycolipid might have been present in quantities below the resolution limit of the qualitative diphenylamine assay.

In conclusion, at least two, and perhaps three, of the four major acyl lipids of cyanobacteria were identified qualitatively as existing in the cyanelle of *Glaucocystis.* It is determined, then, that on the basis of lipids present in the cyanelle of *Glaucocystis,* it bears a strong resemblance to the cyanobacteria. The preliminary data indicate that more detailed studies must be conducted in order to strengthen further the notion that the cyanelle of *Glaucocystis* may indeed have retained its identity as a cyanobacterium, albeit as an obligately adapted intracellular photosynthetic endosymbiont in *Glaucocystis nostochinearum.*

## SUMMARY

*Glaucocystis nostochinearum* cells are characterized by a cellulosic cell wall, nucleus, mitochondria, internal flagella, "extrachloroplastidic" starch, autospore formation, and blue-green-pigmented plastidlike bodies endowed with a remnant peptidoglycan envelope. It was determined recently that *Glaucocystis* constitutes an endosymbiotic association involving what is apparently an achloroplastidic green algal host and the intracellular prokaryoticlike "cyanelle" which has been described either as (1) a "normal plastid," or (2) an intermediate form between the free-living unicellular cyanobacteria and the plastids (rhodoplasts) of eukaryotic red algae. In light of a recent hypothesis suggested for the multiple origins of plastids, including the so-called green and red lines of evolution, it was determined that the lipids from *Glaucocystis* cyanelles needed to be examined in the current efforts to trace its ancestry along the "red line of evolution." As a result of preliminary thin-layer chromatographic studies, it was determined that monogalactosyl diglyceride and digalactosyl diglyceride were the major glycolipids and phosphatidyl glycerol was the major phospholipid in the cyanelles of *Glaucocystis.* These lipids represent three of the four major acyl lipids normally found in cyanobacteria, suggesting that the *Glaucocystis* cyanelle may indeed have a prokaryotic ancestry traceable to the cyanobacteria.

## ACKNOWLEDGMENTS

I wish to thank Dr. Daniel Bershader, Director, Summer Faculty Program, Stanford/NASA-Ames Research Institute Aeronautics and Space Research Program, and Drs. Lawrence Hochstein and Linda L. Jahnke, Division of Extraterrestrial Research, for the opportunity to conduct research on *Glaucocystis* at the NASA-Ames Research Center, Moffett Field, California 94035.

## REFERENCES

1. WEST, G. S. 1904. The British Freshwater Algae. Cambridge University Press. Cambridge, United Kingdom.
2. LEE, R. E. 1980. Phycology. Cambridge University Press. Cambridge, United Kingdom.
3. SCOTT, O. T., R. W. CASTENHOLZ & H. T. BONNETT. 1984. Evidence for a peptidoglycan envelope in the cyanelles of *Glacucocystis nostochinearum* Itzigsohn. Arch. Microbiol. **139:** 130-138.
4. BLIGH, E. G. & W. J. DYER. 1951. Can. J. Biochem. Physiol. **37:** 911-917.
5. FOLCH, J. 1957. J. Biol. Chem. **266:** 497-509.
6. NICHOLS, B. W. 1973. Lipid Composition and Metabolism. *In* The Biology of the Blue-Green Algae. N. G. Carr & B. A. Whitton, Eds. University of California Press. Berkeley, Calif.

# Nucleotide Sequence of *Cyanophora paradoxa* Cellular and Cyanelle-Associated 5S Ribosomal RNAs

E. S. MAXWELL,[a] J. LIU,[a] AND J. M. SHIVELY [b,c]

[a]*Department of Biochemistry*
*North Carolina State University*
*Raleigh, North Carolina 27695*

[b]*Department of Biological Sciences*
*336 Long Hall*
*Clemson University*
*Clemson, South Carolina 29631*

*Cyanophora paradoxa* is a flagellated protozoan which has long been of particular interest because of its unique photosynthetic plastid or cyanelle. This cyanelle possesses morphological and genetic characteristics that indicate a close phylogenetic relationship with both prokaryotic cyanobacteria and eukaryotic higher plant chloroplasts.[1-3] It has therefore been postulated that these cyanelles might represent a potential intermediate in photosynthetic plastid evolution between cyanobacteria and plant chloroplasts.[4] In an effort to understand this evolutionary relationship better, we have isolated and sequenced both the cellular and cyanelle-associated 5S ribosomal RNAs.

*Cyanophora paradoxa* cellular and cyanellar 5S rRNAs were 119 and 118 nucleotides in length, respectively (FIGURE 1). Both exhibited the typical 5S rRNA secondary structures consisting of five helical regions. Analysis of conserved nucleotides, helix and loop sizes, and defined spacings demonstrated a eukaryotic 5S structure for the cellular species and a prokaryotic structure for the cyanellar species.[5] Sequence comparison of *Cyanophora paradoxa* cellular and cyanellar 5S rRNAs with other 5S rRNA sequences is shown in TABLE 1. Analysis of the cellular 5S rRNA revealed a relatively low degree of homology between *Cyanophora* and other eukaryotes, indicating a unique position for this organism in evolution. Highest homology was observed with *Euglena gracilis* 5S rRNA. The cyanelle-associated 5S rRNA exhibited high homology with both cyanobacterial and chloroplast 5S sequences. Highest homology was observed between the cyanellar sequence and the 5S rRNAs of the cyanobacterium *Synechococcus lividus*. Interestingly, *Synechococcus lividus* III has been postulated as a possible progenitor of plant chloroplasts because of its unique chloroplastlike 5S rRNA secondary structure.[6]

Sequence results obtained in these experiments are consistent with the idea that the cyanelle could represent an intermediate in chloroplast evolution from endocytosed cyanobacteria as postulated by the endosymbiont hypothesis.[7] It is impossible to say at this time whether the cyanelle is a direct precursor to chloroplasts (monophyletic event in endocytotic development of the chloroplast) or is the result of a similar but separate endocytotic event (polyphyletic) in photosynthetic plastid development.

[c]To whom correspondence should be addressed.

**A.** CELLULAR 5S RNA

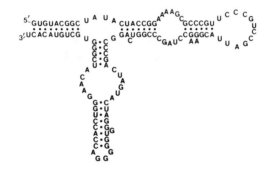

**B.** CYANELLAR 5S RNA

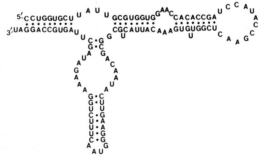

**FIGURE 1.** The primary sequence and predicted secondary structure of *Cyanophora paradoxa* cellular (A) and cyanellar (B) 5S rRNAs.

TABLE 1. *Cyanophora paradoxa* Cellular and Cyanellar 5S rRNA Homologies

| Organism | Cyanophora paradoxa | |
| --- | --- | --- |
| | Cellular 5S rRNA (% homology) | Cyanellar 5S rRNA (% homology) |
| *Euglena gracilis* | 79 | — |
| *Sargassum fulvellum* | 71 | — |
| *Chlorella pyrenoidosa* | 70 | — |
| *Acanthamoeba castellanii* | 70 | — |
| *Eisenia bicyclis* | 69 | — |
| *Tetrahymena thermophilia* | 67 | — |
| *Spinacia oleracea* | 65 | — |
| *Ulva pertusa* | 65 | — |
| *Porphyra tenera* | 62 | — |
| *Gracilaria compressa* | 60 | — |
| *Chlamydomonas reinhardtii* | 58 | — |
| *Synechococcus lividus* II | — | 78 |
| *Synechococcus lividus* III | — | 77 |
| *Marchantia polymorpha* chloroplast | — | 77 |
| *Prochloron* | — | 74 |
| *Anacystis nidulans* | — | 73 |
| *Nicotinia tobacum* chloroplast | — | 71 |
| *Spinacia oleracea* chloroplast | — | 71 |
| *Lemna minor* chloroplast | — | 70 |
| *Spirodela oligorhiza* chloroplast | — | 70 |
| *Bacillus acidocaldarius* | — | 67 |
| *Rhodospirillum rubrum* | — | 66 |
| *Chromatium* | — | 61 |
| *Euglena gracilis* chloroplast | — | 60 |
| *Chlamydomonas reinhardtii* chloroplast | — | 60 |
| *Rhodopseudomonas gelatinosa* | — | 60 |
| *Escherichia coli* | — | 59 |
| *Chlorobium* | — | 52 |

## REFERENCES

1. AITKEN, A. & R. Y. STANIER. 1979. J. Gen. Microbiol. **112:** 218-223.
2. HALL, W. T. & G. CLAUS. 1963. J. Cell Biol. **19:** 551-563.
3. LAMBERT, D., D. BRYANT, V. STIREWALT, J. DUBBS, E. STEVENS & R. D. PORTER. 1985. J. Bacteriol. **164:** 659-664.
4. HEINHORST, S. & J. M. SHIVELY. 1983. Nature **304:** 373-374.
5. DELIHAS, N. & J. ANDERSON. 1982. Nucleic Acids Res. **10:** 7323-7344.
6. DELIHAS, N., J. ANDRESINI, J. ANDERSON & D. BERNS. 1982. J. Mol. Biol. **162:** 721-727.
7. GRAY, M. & F. DOOLITTLE. 1982. Microbiol. Rev. **46:** 1-42.

# Chloroplast Retention by Marine Planktonic Ciliates[a]

DIANE K. STOECKER

*Biology Department*
*Woods Hole Oceanographic Institution*
*Woods Hole, Massachusetts 02543*


MARY W. SILVER

*Institute of Marine Science*
*University of California*
*Santa Cruz, California 95064*

## INTRODUCTION

Many marine planktonic ciliates in the family Strombidiidae, order Oligotrichida, are pigmented because they sequester algal chloroplasts.[1,2] When epifluorescence microscopy is used, the pigmented bodies fluoresce red or orange, indicating the presence of chlorophyll *a* and phycoerythrin, respectively.[3] In transmission electron micrographs, it is evident that these pigmented bodies are intact chloroplasts (FIGURE 1).

Chloroplast retention in ciliates had not been previously investigated experimentally. However, because the chloroplasts appear to be in good condition, they were thought to be photosynthetically functional.[1,2] We have isolated several pigmented ciliates, including *Laboea* sp. and several *Strombidium* spp., from coastal waters and have grown them in culture on a mixture of microalgae on a 14-hour L:10-hour D light cycle. Using cultured material, we are investigating (1) the specificity of chloroplast retention; (2) the life span of sequestered chloroplasts; (3) the growth of ciliates in the light and dark, in the presence and absence of food; and (4) the chlorophyll content and rate of photosynthesis in the ciliates.


## RESULTS AND CONCLUSIONS

The specificity of chloroplast retention varies among ciliates. For example, one *Strombidium* sp. (strain GR) only retains cryptophyte chloroplasts. In contrast, another *Strombidium* sp. (strain WHL-2) and a *Laboea* sp. (strain DYE) retain cryp-

[a] This research was supported by National Science Foundation grant OCE-8600765 to DKS.

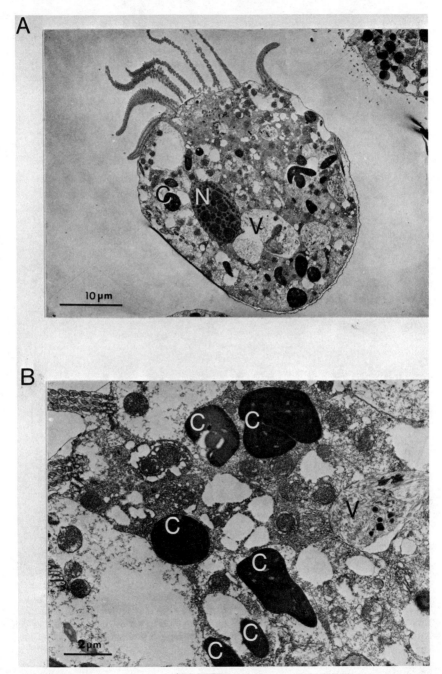

**FIGURE 1.** Transmission electron micrographs of *Strombidium* sp. (strain GR). Chloroplasts (C) derived from cryptophyte algae as well as cryptophytes in digestive vacuoles (V) of the ciliate are visible. N is the macronucleus of the ciliate.

tophyte chloroplasts and chloroplasts from other algal taxa. One *Strombidium* sp. (strain LB) retained chloroplasts from a prasinophyte, but not from cryptophytes, haptophytes, or dinophytes.

The chloroplasts are relatively long-lived in the ciliates. For example, *Strombidium* sp. (strain GR) retains cryptophyte chloroplasts for 14 days even when it is kept in the dark in the absence of food. *Laboea* sp. retains chloroplasts for at least 6 days. Darkness does not trigger digestion of chloroplasts by the ciliates.

In the presence of microalgal foods, the growth rate of the ciliates is slightly higher (although this difference is not always statistically significant) on a light:dark cycle than when the ciliates are kept continuously in the dark. In the absence of food, the difference between the growth of *Laboea* sp. in the light and dark is more pronounced;

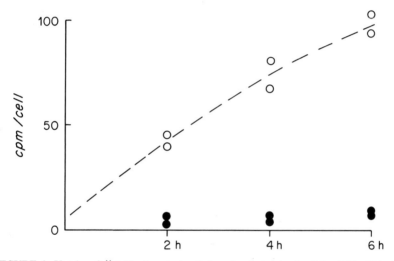

**FIGURE 2.** Uptake of $^{14}$C bicarbonate by *Laboea* incubated in the light (190 $\mu$E/m$^2$ per second) and in the dark at 15°C. Approximately 0.5 $\mu$Ci/ml of $^{14}$C bicarbonate (specific activity 53 mCi/mmol) was added to the seawater medium.

during the first two days without added food, *Laboea* continues to grow and divide when kept on a light:dark cycle, but then slowly declines. In continuous darkness, *Laboea* does not grow and declines rapidly after two days without added food.

*Laboea* sp. contains 126-219 pg chlorophyll *a*/cell (approximately 1.8 fg chlorophyll *a*/$\mu$m$^3$). The *Strombidium* spp. contain less chlorophyll than *Laboea* per cell and per unit biovolume. In *Laboea* sp. and in at least three pigmented *Strombidium* spp., the chloroplasts are photosynthetically functional. *Laboea* sp. fixes 292 ± 60 (standard deviation) pg carbon/cell per hour under our experimental conditions (FIGURE 2).

The relatively long life spans and functionality of algal chloroplasts in ciliates raise questions about the frequency of organelle transfer among protists.

# ACKNOWLEDGMENTS

Preliminary carbon fixation experiments were done with J. C. Goldman and M. R. Dennett. M. S. Roberts, while a Woods Hole Oceanographic Institution Summer Student Fellow, did the experiments on specificity of chloroplast retention and life span of chloroplasts in *Strombidium* sp. A. M. Michaels helped culture the ciliates. We thank them for their valuable advice and assistance.

## REFERENCES

1.  BLACKBOURN, D. J., F. J. R. TAYLOR & J. BLACKBOURN. 1973. Foreign organelle retention by ciliates. J. Protozool. **20:** 286-288.
2.  LAVAL-PEUTO, M. & M. FEBVRE. 1983. On chloroplast symbiosis on *Tontonia appendiculariformis* (Ciliophora Oligotrichina) (Abstract). Protistologica **19:** 464.
3.  MCMANUS, G. B. & J. A. FUHRMAN. 1986. Photosynthetic pigments in the ciliate *Laboea strobilia* from Long Island Sound, USA. J. Plank. Res. **8:** 317-327.

# Plant and Bacterial Cell Wall Deformations in the Functional Part of the Bacterial Leaf Nodule of Spearflower [*Ardisia crispa* (Thunb.) A.DC., Myrsinaceae]

URMAS SUTROP[a]

*Institute of Experimental Biology*
*Academy of Sciences of the Estonian SSR*
*203051 Tallinn-Harku*
*Estonian SSR, USSR*

During the coevolution of bacteria and host plants, the plant bacterial nodules (conventionally root nodules, leaf nodules, and tumors) are established. Plant bacterial leaf nodules (BLN) were first investigated in 1881 by F. R. von Höhnel,[1] but he misinterpreted BLN as a schizogenous albumin gland which contains "bakterienartigen" protein bodies.

Contemporary ultrastructural studies in spearflower have shown the fine structure and development of the BLN and endophyte,[2] which is identical with *Phyllobacterium myrsinacearum*.

BLN from both young and mature leaves of *Ardisia crispa* (Thunb.) A.DC. were routinely fixed (pH 7.3) and embedded in a mixture of Epon/Araldite. The samples were examined by the electron microscope Tesla BS 613.

Marginal BLN in spearflower consists of the functional part (FP) containing intercellular bacteria, as well as living and/or collapsed invasive plant cells, and of the covers that may regulate the endophyte population rate. The invasive cells form a three-dimensional network which compartments the FP. In the early stage of BLN ontogeny the intercellular spaces (compartments) take form lysigenously.[2] The compartments enlarge schizogenously in young BLN (FIGURE 1A), and rexigenously via the collapse and/or death of invasive cells in mature BLN (FIGURE 1B). Invasive cells in young BLN were found to be functionally active. In a mature BLN the invasive cells are deformed by thickening of the cell walls up to a hundred times in comparison with the mesophyll. Deformed walls consist of honeycomblike and denser layers with spherical cavities (FIGURE 1C and 1D).

Cell walls of the endophytic bacteria in the mature BLN are also deformed. The surface of the bacterial cells are covered with cell wall outgrowths (FIGURE 2A) which seem to be analogous with the microbodies formed via the bacterial cell wall protruding in the cytoplasm (FIGURE 2B).

[a] Present affiliation: Estonian Biocenter, Tähetorn Toomel, 202400 Tartu, Estonian SSR, USSR.

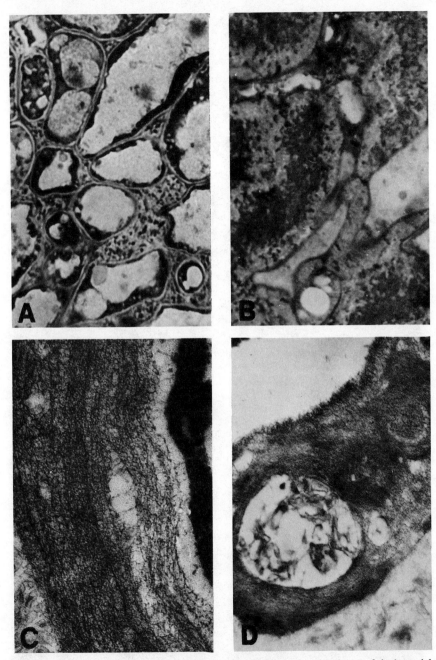

**FIGURE 1.** Deformation of the invasive plant cell walls in the functional part of the bacterial leaf nodule of *Ardisia crispa*. In a young nodule the intercellular spaces take form schizogenously (A), while the neighboring cell wall contours follow one another's tracks. In a mature nodule (B) the invasive cells form a three-dimensional network which compartments the bacterial population. Thickened invasive cell walls (C) consist of honeycomblike and denser layers with spherical cavities (D). Magnification: 1400× (A), 900× (B), 57,000× (C), 20,000× (D).

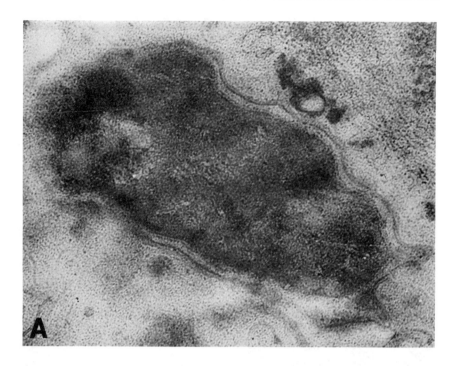

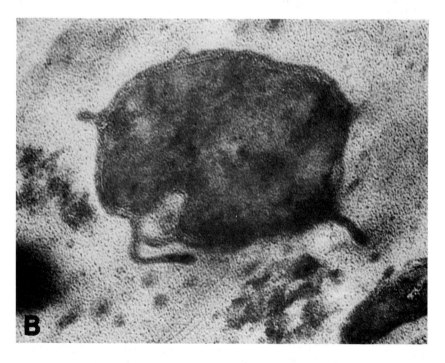

The interaction of bacteria and spearflowers is based on the bacterial cytokinin synthesis.[3] Cytokinins may be responsible for the invasive cell wall deformations. Since the bacterial cell cycle proceeds entirely in plants, the formation of bacterial cell wall outgrowths may correspond to the specific phase in the bacterial cell cycle or its degeneration. In the course of the intimate coevolution of endosymbionts and spearflowers, while the bacteria are carried from plant generation to generation in the seeds, the bacterial leaf nodules are established as bacteriomes with limited growth.

## REFERENCES

1. HÖHNEL, F. R. V. 1882. Sitzb. kaiser. Akad. Wiss. Mat.-naturw. Classe, Jg. **1881**: 565-603.
2. MILLER, I. A., I. C. GARDNER & A. SCOTT. 1983. Bot. J. Linnean Soc. **86**: 237-252.
3. RODRIGUES PEREIRA, A. S., P. J. W. HOUWEN, H. W. J. DEURENBERG-VOS & E. B. F. PEY. 1972. Z. Pflanzenphysiol. **68**: 170-177.

**FIGURE 2.** Deformation of the endophytic *Phyllobacterium myrsinacearum* cell walls in the functional part of the bacterial leaf nodule of *Ardisia crispa*. In a specific phase of bacterial cell cycle or its degeneration the cell wall outgrowths are formed (A). They are analogous with the microbodies formed via the bacterial cell wall protruding in the cytoplasm (B). Magnification: 36,500× (A), 105,000× (B).

# Condensation of Nucleoid Material in Leafhopper Endocytobionts

TOOMAS H. TIIVEL

*Institute of Chemical Physics and Biophysics*
*Estonian Academy of Sciences, Akadeemia 23*
*200026 Tallinn*
*Estonia, USSR*

The existence of genetic material in different intracellular microorganisms of insects has been a point of discussion.[1]

We have established the nucleoid region electron microscopically in all three types[2] of endocytobionts of the leafhopper *Philaenus spumarius* L. (FIGURES 1 and 2). The zone containing DNA seemed to be evenly distributed throughout the entire ribosome-free area of the microorganism. It had a grainy appearance with a finer fibrillar structure similar to that of various free-living bacteria[3-5] and other intracellular microorganisms.[6] Unlike the case with the endocytobionts in embryonic bacteriocytes,[7] filamentous structures of diameter of about 0.01-0.02 $\mu$ in the cytobionts of the second and third types were found. The nucleoid regions had often a central, more electron dense zone that could be up to 10 times bigger in diameter than that of the fibrils branching off that zone (FIGURES 1, 3, 4, 5, and 6). The nucleoid region was particularly clear-cut in the cells about to divide.

Among other intracytoplasmic structures,[2,8] there were a lot of those with homogeneous contents, surrounded by a membrane. They were round or irregular in shape, their size being 0.1-0.8 $\mu$. In several places a certain contact could be observed between them and the region of nucleoid with fibrils. This contact is revealed in the fibrils spreading from such membrane-bound structures—the nucleoid fibrils looked as if they were condensed around these configurations—due to which a certain periodicity was established in the surface layers (FIGURES 1 and 4).

No such condensation of nucleoid fibrils was observed in the endocytobionts of the first type. Their nucleoid zone was less clear-cut there, no electron-light area around the fibrils was noted either. In the endocytobionts of the third type the region of the cytoplasm containing nucleoid components was mostly situated in the center of the cells.

It might be suggested that in these cells of endocytobionts we have to do with certain "organizing centers" that form structures which are attached to the bacterial nucleoid fibrils. The variations in appearance of the nucleoid zones in endocytobionts (as well as in their different types) may be due to their different functional states and the different stages of their life cycles at the moment of fixation (with glutaraldehyde, $OsO_4$, or both), and also due to the differences in the condensation of DNA and proteins in various types of endocytobionts.

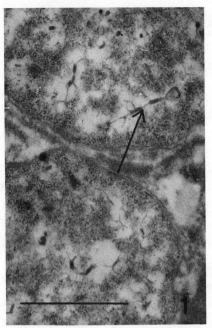

**FIGURE 1.** Fibrils of the nucleoid (arrow) in the cytoplasm of endocytobionts of the second type (glutaraldehyde and OsO₄ fixation). Bar = 0.5μ.

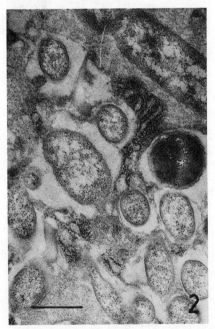

**FIGURE 2.** Cytobionts of the third type (glutaraldehyde fixation). Bar = 0.5μ.

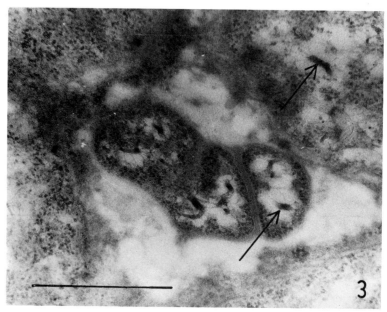

**FIGURE 3.** Fibrils of the nucleoid (arrow) in endocytobionts of the second type (glutaraldehyde and OsO$_4$ fixation). Bar = 0.5$\mu$.

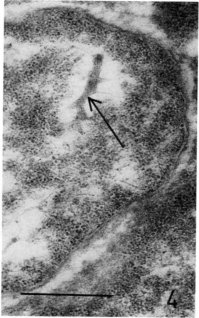

**FIGURE 4.** Structures connected with the nucleoid region (arrow) in the endocytobiont cytoplasm (glutaraldehyde fixation). Bar = 0.5$\mu$.

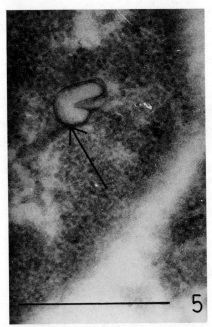

FIGURE 5. Structures connected with the nucleoid region (arrow) in the endocytobiont cytoplasm (glutaraldehyde fixation). Bar = 0.5μ.

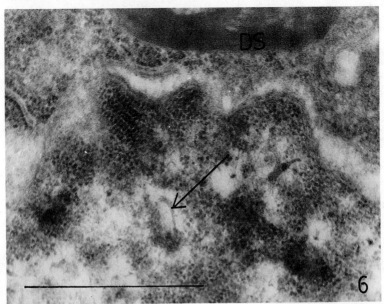

FIGURE 6. Different structures (arrow) in the endocytobiont cytoplasm. DA, structures connected with the degeneration of endocytobionts (glutaraldehyde and OsO₄ fixation). Bar = 0.5μ.

## REFERENCES

1. HOUK, E. J. & G. W. GRIFFITHS. 1980. Annu. Rev. Entomol. **25:** 161-187.
2. TIIVEL, T. 1984. Proc. Acad. Sci. Estonian SSR. Biol. **33**(4): 244-255.
3. MURRAY, R. G. E. 1978. *In* Essays in Microbiology. J. R. Norris & M. H. Richmond, Eds. **2:** 1-31. Wiley. Chichester & New York.
4. HOBOT, J. A., W. WILLIGER, J. ESCAIG, M. MAEDER, A. RYTER & E. KELLENBERGER. 1985. J. Bacteriol. **162**(3): 960-971.
5. DUBOCHET, J., A. W. McDOWALL, B. MENGE, E. N. SCHMID & K. G. LICKFELD. 1983. J. Bacteriol. **155**(1): 381-390.
6. CHANG, K. P. & A. J. MUSGRAVE. 1975. Can. J. Microbiol. **21:** 196-204.
7. TIIVEL, T. 1983. *In* Endocytobiology. Intracellular Space as Oligogenetic Ecosystem. H. E. A. Schenk & W. Schwemmler, Eds. **2:** 771-774. Walter de Gruyter. Berlin, FRG.
8. TIIVEL, T. & V. Mashansky. 1985. Endocytol. C. Res. **2**(2): 139-158.

# Cytochemical Determination of DNA Content in the Nucleus and Nucleoids of *Cyanidium* and Some Microbial Cells

IKUJIRO FUKUDA AND HIDEYUKI NAGASHIMA

*Department of Biology*
*Faculty of Science*
*Science University of Tokyo*
*Kagurazaka, Shinjuku-ku*
*Tokyo 162, Japan*

The evolution from prokaryote to eukaryote has been discussed for many years between the Endosymbiotic Theory and the Direct Filiation Theory. The present task on our recent trials could offer material for this discussion from a viewpoint of degree of eukaryotic cell structure. Among eukaryotes, *Cyanidium caldarium* RK-1, which possesses one each of a nucleus, chloroplast, and mitochondrion, is one of the most important organisms being situated in the lowest taxonomic position.[1-3]

In the present study, we used 4'6-diamidino-2-phenylindole (DAPI) staining and fluorescence microscopy including the video intensified microscopy (VIM) system, and determined the DNA content in the nucleus and chloroplast nucleoids. These data were compared with other prokaryotic and eukaryotic organisms such as virus, bacteria, fungi, Cyanophyta, Rhodophyta, Chlorophyta, and Euglenophyta. These results are shown in TABLE 1.

The ratios of nucleoids area to single cell size determined by fluorescence microscopy presented exceedingly high levels in blue-green algae. But the DNA contents per single cell determined by a photon counter presented values average among other microalgae. Photon numbers counted with the VIM system seem almost proportional to DNA content exactly. DNA content in the nucleus of *Cyanidium* cells counted by this method indicates considerably smaller amounts than those of other algal cells. But the DNA content per single cell presented about an average value according to their small cell sizes.

The comparison of DNA content of *Cyanidium* cells in nucleus and chloroplast nucleoids is shown in TABLE 2.

DNA ratio of organelle nucleoid to nucleus significantly suggests it to be an indicator of eukaryotic degree of evolution. *Cyanidium caldarium* RK-1 has an average value of 0.378 which includes from 0.198 (young cells) to 0.454 (aged cells). Otherwise the M-8 strain presented a value of 1.247 in light-grown young cells, and 0.308 in dark grown-young cells, respectively. We already presented the result that the RK-1 strain resembles a Cyanophyta type because of its rod-shaped chloroplast nucleoid (CN type), on the other hand, the M-8 strain resembles Phaeophyta or Bacillariophyta because of its ring-shaped chloroplast nucleoid (CL type).[4,5] The present data cal-

TABLE 1. Comparison of DNA Content of Nucleus (Nucleoids), Measured by a Photon Counter in Various Microorganisms

| Group | Species | Cell ($\mu M^2$) | Nucleus ($\mu M^2$) | Nucleus to Cell Ratio | Photon $\times 10^4$ in Nucleus | DNA/Cell |
|---|---|---|---|---|---|---|
| Virus | T$_4$-phage | — | — | — | 5.7 | — |
| Bacteria | Escherichia coli | 0.6 | 0.61 | — | — | — |
| Blue-green algae | Microcystis viridis | 26.6 | 12.10 | 0.456 | — | — |
| | Anabaena variabilis | 29.7 | 9.96 | 0.335 | 541 | 18.2 |
| Fungi | Saccharomyces cerevisiae | 18.7 | 1.21 | 0.065 | — | — |
| Red algae | Cyanidium caldarium RK-1 | 6.7 | 0.39 | 0.058 | 193 | 28.8 |
| | Cyanidium caldarium M-8[a] (light culture) | 16.9 | 0.40 | 0.023 | 185 | 10.9 |
| | Cyanidium caldarium M-8[a] (dark culture) | 12.9 | 0.27 | 0.021 | 141 | 10.9 |
| | Porphyridium cruentum | 69.5 | 1.49 | 0.021 | — | — |
| Green algae | Chlorella vulgaris | 21.9 | 0.47 | 0.022 | 348 | 15.8 |
| Euglenophyta | Euglena gracilis | 524.0 | 41.30 | 0.079 | 11,130 | 21.2 |

[a] Young cells.

TABLE 2. The Ratio of DNA Content of Chloroplast Nucleoid to Nucleus in *Cyanidium caldarium* Measured by a Photon Counter

| | DNA ($\times 10^4$ Photon) Average Value | | |
| --- | --- | --- | --- |
| | Cell Nucleus | Chloroplast Nucleoid | Ratio of Nucleoid to Nucleus |
| *Cyanidium caldarium* | | | |
| RK-1 type | 192.9 | 72.8 | 0.378[a] |
| M-8 type,[b] (young cells) | | | |
| Light culture | 185.0 | 230.8 | 1.247 |
| Dark culture | 140.8 | 43.3 | 0.308 |

[a] RK-1 contains young (0.198) to aged (0.454) cells.
[b] M-8 may be a different genus from *Cyanidium*.[1-3]

culated from DNA determinations suggest the evolutionary position of these algae from a viewpoint of eukaryotic degree which indicates the concentration of information contents to the nucleus.

## ACKNOWLEDGMENT

We are grateful to Prof. T. Kuroiwa for providing the opportunity to use the newest VIM system.

## REFERENCES

1. NAGASHIMA, H. & I. FUKUDA. 1981. Proc. Int. Bot. Congr. Sydney **13**: 288.
2. NAGASHIMA, H. & I. FUKUDA. 1981. Jpn. J. Phycol. **29**(4): 237-242.
3. SECKBACH, J., J. F. FREDRICK & D. J. GARBARY. 1983. Endocytobiology **2**: 947-962.
4. KUROIWA, T., T. SUZUKI, K. OGAWA & S. KAWANO. 1981. Plant Cell Physiol. **22**(3): 381-396.
5. NAGASHIMA, H., T. KUROIWA & I. FUKUDA. 1984. Experientia **40**(6): 563-564.

# Structural Homologies in the D-Loop-Containing Region of Vertebrate Mitochondrial DNA[a]

E. SBISÀ, M. ATTIMONELLI, AND C. SACCONE

*Center for Studies of Mitochondria and Energy Metabolism*
*of the CNR and*
*Department of Biochemistry and Molecular Biology*
*University of Bari*
*Bari, Italy*

The D-loop-containing region is the only major noncoding segment and, very likely, the only regulatory region of animal mitochondrial (mt) DNA. It spans between the tRNA-Phe and the tRNA-Pro genes and contains the site of initiation for heavy strand replication ($O_H$) and the promoters for heavy and light strand transcription (HSP and LSP). It is the most rapidly evolving part of the mitochondrial genome, varying among vertebrates (rat, mouse, cow, man, and Xenopus) in both base composition and length.

How the function in this region is preserved, in spite of such primary structural diversity, remains to be clarified. A detailed comparative analysis among the sequences so far available will be very useful for identifying the functional constraints present in this region.

Despite the high primary structural divergence that characterizes the D-loop-containing region of vertebrate mtDNA, we have discovered a number of features that are common to all mtDNAs. Among these, the most relevant is the presence of an open reading frame (ORF) in the four mammals. This is localized in a central region, rich in G, which is very conserved (sequence homologies: rat/mouse = 87%, rat/cow = 72%, rat/man = 72%, cow/man = 67%). The lengths of the ORFs are variable going from 32 amino acids in mouse to 62 in rat, 36 in man, and 110 in cow. They display significant amino acidic homology particularly when compared using functional and hydrophathic alphabets.

Another important common feature in the D-loop-containing region is the presence, at the 5' and 3' ends, of sequences that can be folded into tRNA-like structures.[1,2] These structures at the 5' end contain the CSB1[3] and the origin of replication of the heavy strand; the structures at the 3' end contain the TAS sequences.[4]

The potential of sequences at the initiation and termination sites for D-loop DNA synthesis to fold into tRNA-like secondary structures suggests regulation mechanisms similar to those found in other systems. In particular there are analogies with the Col E1 replication mechanism[5] and which has been recently described also for the replication of plasmid pRBH1.[6] It is obvious that only suitable *in vitro* systems, using

[a]This work has been partially supported by Progetto Finalizzato Ingegneria Genetica e Basi Molecolari delle Malattie Ereditarie and by Progetto Strategico Biotecnologie, CNR, Italy.

578

templates properly mutagenized according our theoretical analysis, will demonstrate the importance of the primary and secondary structural elements present in the regulatory region of mtDNA.

## REFERENCES

1. SACCONE, C., M. ATTIMONELLI & E. SBISÀ. 1985. *In*: Achievements and Perspectives of Mitochondrial Research. Biogenesis. E. Quagliariello, E. C. Slater, F. Palmieri, C. Saccone & A. M. Kroon, Eds. 2: 37-47. Elsevier Science Publishers. Amsterdam, Holland.
2. BROWN, G. G., G. GADALETA, G. PEPE, C. SACCONE & E. SBISÀ. (Submitted for publication.)
3. WALBERG, M. W. & D. A. CLAYTON. 1981. Nucleic Acids Res. 9: 5411-5421.
4. DODA, J. N., C. T. WRIGHT & D. A. CLAYTON. 1981. Proc. Nat. Acad. Sci. USA 78: 6116-6120.
5. TOMIZAWA, J. & T. ITOH. 1981. Proc. Nat. Acad. Sci. USA 78: 6096-6100.
6. ANO, T., T. IMANAKA & S. AIBA. 1986. Mol. Gen. Genet. 202: 416-420.

# Killer Particles in the Macronucleus of *Paramecium caudatum*

HELMUT J. SCHMIDT AND HANS-DIETER GÖRTZ

*Zoological Institute*
*University of Münster*
*D-4400 Münster, Federal Republic of Germany*

One of the earliest known examples of cytoplasmic inheritance is the killer trait in paramecium. For many years it was not known that such killer traits are determined by bacterial endosymbionts; and kappa, the first symbiont discovered in the *Paramecium aurelia* complex, was initially recognized only by its killing action on sensitive paramecia. Kappa and some other endosymbionts, all members of the genus *Caedibacter,* have two distinct morphological forms, one of which contains a large inclusion body known as a refractile body, or R body. Bacteria of this form are commonly called "brights" whereas symbionts without R bodies are referred to as "nonbrights." R bodies are proteinaceous ribbons, and several investigators have shown that the killing of sensitive paramecia is associated with such R bodies.[1]

All R body producing bacteria from *P. aurelia* live in the cytoplasm, whereas those observed in *P. caudatum* inhabit the macronucleus. Generally, many endosymbionts are endocytoplasmatic forms in *P. aurelia* host cells and endonuclear ones, living either in the micro- or macronucleus, if *P. caudatum* harbors them.

We discovered R body producing bacteria, looking similar to those described by Estève,[2] in the macronuclei of two *P. caudatum* lines which were isolated from a pond in Münster, and after their cultivation in the laboratory we observed that one of the two lines (C220) lost its ability to produce R bodies. Macronuclei of paramecia from this line now contain nonbrights only. They are uniform in size with a diameter of 0.4 $\mu$m. The average length is 1.0-1.5$\mu$m, but occasionally much longer rods can be found (FIGURES 1 and 2), which resemble a row of nonbrights sticking together end to end. We have been unable to infect symbiont-free *P. caudatum* cells. For such experiments either isolated nonbrights or homogenates from C 220 cells were used. We have also failed to induce R body production with the UV irradiation method.[3] The bacteria appear well adapted to their endonucleobiotic way of life. Rather than being evenly distributed within the macronucleus they are clustered. During cell division they get distributed with the dividing macronuclei of their hosts. Paramecium is obviously not harmed by the nonbright endonucleobionts.

Brights are always present in line C 221. Their numbers are low in rapidly growing cultures and increase considerably when such cultures starve. Brights are bigger than nonbrights, about 0.7 $\mu$m wide and 1.5-2.5$\mu$m long. C 221 host cells suffer and eventually die under certain conditions, e.g., in cultures that have starved for more than a week. Their death is probably caused by their endosymbionts.

We established two methods for the isolation and purification of nonbrights, brights, and R bodies. The first method involves ECTEOLA column chromatography of partly purified paramecium homogenates. High yields of pure nonbrights are obtained with this procedure (FIGURE 1). Brights, however, are lost irreversibly. They and their R bodies, which are unintentionally released to some extent during all isolation experiments, can be recovered with a PERCOLL density gradient centrifugation of pre-purified paramecium homogenates.

Although endosymbionts are very common among insects and protozoa and especially among ciliates, data concerning their biochemical properties are rather scarce.[4] This is mainly due to difficulties of growing sufficient amounts of host cells and harvesting enough endosymbionts for biochemical studies. The two methods mentioned above enable us to isolate sufficient numbers of symbionts. We have prepared DNA

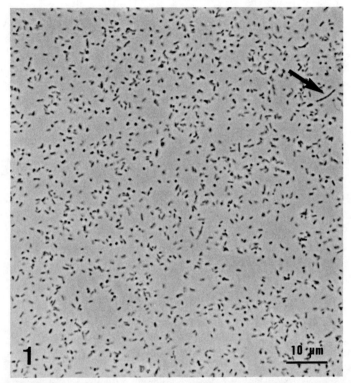

**FIGURE 1.** Nonbright symbionts, isolated and purified with ECTEOLA column chromatography, fixed with 2% glutaraldehyde, unstained, phase contrast. The arrow points out a long rod consisting of several symbionts sticking together end to end.

from frozen pooled symbionts. This DNA is currently used by Dr. R. L. Quackenbush to look for sequence homologies with other R body producing bacteria from the cytoplasm of several *P. aurelia* strains. He will also try to determine the degree of cross-reactions between isolated R bodies from *P. caudatum* and from *P. aurelia* endosymbionts. To facilitate this comparison we injected purified R bodies from *P. caudatum* symbionts into a rabbit and obtained an antiserum against them.

Preliminary results indicate substantial differences between R body producing endocytoplasmic bacteria and endonuclear forms. The GC content of the DNA from the latter seems to be significantly lower. Their R bodies do not unroll at low pH whereas most R bodies from *P. aurelia* kappas do. Plasmids which can carry the genetic determinant for R body synthesis[5] were never seen during DNA isolations from *P. caudatum* bacteria. Phagelike structures, however, which are also associated with R body production in other *P. aurelia* kappas, are routinely observed in brights from *P. caudatum* (FIGURE 2).

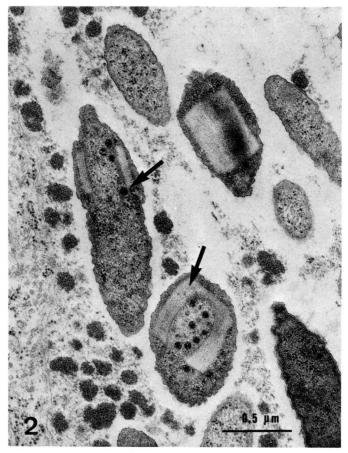

FIGURE 2. Electron micrographs of bright and nonbright endosymbionts. The arrows point out R bodies and phagelike structures.

Most of these differences were unexpected because of the morphological similarity of these macronuclear symbionts and their cytoplasmic counterparts. Obviously many of the endosymbionts living in paramecia are not as closely related as might be assumed on the basis of their biological and morphological resemblance.

# REFERENCES

1. PREER, J. R., JR., L. B. PREER & A. JURAND. 1974. Kappa and other endosymbionts in *Paramecium aurelia*. Bacteriol. Rev. **38**(2): 113-163.
2. ESTÈVE, J.-C. 1978. Une population de type killer chez *Paramecium caudatum* (Ehrenberg). Protistologica **14**: 201-207.
3. PREER, L. B., B. RUDMAN, J. R. PREER, JR. & A. JURAND. 1974. Induction of R bodies by ultraviolett light in killer paramecia. J. Gen. Microbiol. **80**: 209-215.
4. SCHMIDT, H. J. 1982. Isolation of omikron-endosymbionts from mass cultures of *Euplotes aediculatus* and characterization of their DNA. Exp. Cell Res. **140**: 417-425.
5. QUACKENBUSH, R. L. & J. A. BURBACH. 1983. Cloning and expression of DNA sequences associated with the killer trait of *Paramecium tetraurelia* stock 47. Proc. Nat. Acad. Sci. USA **80**: 250-254.

# Antigenic Properties of Erythrocyte Membrane from a Hypertensive Strain of Rats

M. FERRANDI, S. SALARDI, R. MODICA,
AND G. BIANCHI

*Carlo Erba Institute of Pharmalogical Research*
*Nerviano, Italy*
*and*
*Institute of Medical Science*
*Milan Institute of Studies*
*Milan, Italy*

## INTRODUCTION

Many "abnormalities" have been found in red blood cells (RBC) of Milan hypertensive rats (MHS) as compared with their normotensive controls (MNS): smaller cell volume, lower sodium content, faster Na-K cotransport.[1] The alteration of these parameters is genetically determined within the stem cell, and is correlated to blood pressure.[2] The same alterations of RBC were found in renal tubular cells, which are more directly involved in the development of hypertension.[1] The measurement of protein and lipid composition of RBC membranes showed only minor qualitative differences between the strains. Because a subtle alteration in protein structure, not detectable by conventional electrophoresis, cannot be excluded, we have cross-immunized the two strains with different antigens to see if one of them recognizes as "nonself" the other strain.

## METHODS

One-month-old rats were injected subcutaneously either with $2.10^9$ intact RBC or 1 mg of antigen diluted in 1 ml of saline+Freund's complete adjuvant (1:1). A second injection followed after two weeks, then the serum was tested for presence and specificity of antibodies two weeks later.

## RESULTS

FIGURE 1 shows the production and specificity of antibodies raised in a subgroup of animals against MNS and MHS RBC membrane proteins. Antibodies are mainly

584

against a 105 kD (kilodalton) protein which is present in RBC membrane of both strains. Antibodies are raised only in MHS and not in MNS and are absent in the preimmune sera of both strains.

TABLE 1 summarizes the results of immunization with different antigens in terms of production of antibody to the 105 kD protein. Antibodies are produced (1) mainly

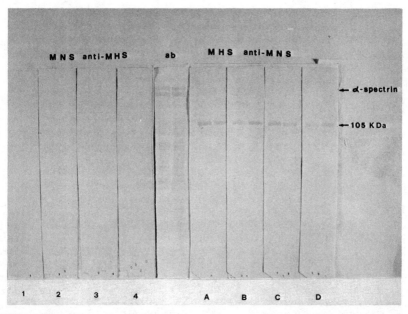

FIGURE 1. Specificity of antibodies raised in MHS rats upon immunization with MNS erythrocyte cytoskeletons; 1, 2, 3, 4 = immunoblot analysis of four different sera from MNS animals; A, B, C, D = the same for MHS animals; ab = amido-black staining of a parallel sample. Antibodies are mainly towards the 105 kD membrane protein. Sometimes these antibodies have a slight reactivity towards α-spectrin. Antigens were separated by sodium dodecyl sulfate-polyacrylamide gel electrophoresis with the Laemmli discontinuous system. After separation the gel was transferred to nitrocellulose membrane with the Trans-blot apparatus (3 hours, 90 volts). The nitrocellulose was incubated with block solution (8% bovine serum albumin), then with antiserum in BSA, then with sheep antirat antibody peroxidase conjugated and finally with the peroxidase substrates 6-chloro-1-naphtol and $H_2O_2$. Violet bands appear where the primary antibody binds to its antigen. Between incubations the sheets were washed with 500 mM NaCl+0.05% Tween 20. Comparison of peroxidase-stained sheets and amido-black stained ones allows determination of molecular weight of antigens. The intensity of color gives an idea of the specificity of the antiserum.

in MHS rats; (2) only upon immunization with antigens from young MNS rats (1-month-old); (3) the antigenic information is conserved in the low ionic strength extract but is lost in its pellet. Other characteristics of the 105 kD protein are (1) isoelectric point ~ 5.7 (by bidimensional electrophoresis); (2) abundance—about 0.2% of RBC membrane proteins which roughly corresponds to 6000 copies per cell; (3) it is found

**TABLE 1.** Immunogenic Properties of Various Antigen Preparations from MHS and MNS Rats[a]

| Antigen | Donor | | Recipient | | Production of Antibody towards 105 kD Protein | |
| --- | --- | --- | --- | --- | --- | --- |
| | Strain | Age (month) | Strain | Number of Rats | Positive | Negative |
| Intact red blood cell | MNS | 1 | MHS | 3 | — | 3 |
| | MHS | 1 | MHS | 3 | — | 3 |
| | MHS | 1 | MNS | 3 | — | 3 |
| | MNS | 1 | MNS | 3 | — | 3 |
| Red blood cell ghost | MNS | 1 | MHS | 12 | 12 | — |
| | MNS | 2 | MHS | 5 | — | 5 |
| | MHS | 1 | MNS | 7 | — | 7 |
| Low ionic strength extract | MNS | 1 | MHS | 15 | 14 | 1 |
| | MHS | 1 | MHS | 5 | — | 5 |
| | MHS | 1 | MNS | 10 | — | 10 |
| | MNS | 1 | MNS | 6 | 3 | 3 |
| Low ionic strength pellet | MNS | 1 | MHS | 3 | — | 3 |
| | MHS | 1 | MHS | 3 | — | 3 |
| | MHS | 1 | MNS | 3 | — | 3 |
| | MNS | 1 | MNS | 3 | — | 3 |

[a] The parameter evaluated is the presence or absence of antibody towards 105 kD protein (as determined by immunoblotting procedure) in sera of immunized animals. All antigens were prepared from a pool of MNS or MHS red blood cells. Recipient rats were one month old at the beginning of the immunization.

in the brain with expressions at 105 and 50 kD (both enriched in the cytosol), and in the kidney with an expression at 78 kD (enriched in the membrane). However cross-reactivity with other proteins cannot be excluded.

## CONCLUSIONS

The data show that a 105 kD protein from cytoskeletons of MHS and MNS RBC has different immunogenic properties, without evident qualitative difference according to the strain of origin. Moreover, if we considered that (1) the differences in RBC function between MHS and MNS are lost when RBC are deprived of the cytoskeleton (inside-out vesicles); (2) the RBC functional alterations of MHS are genetically determined and correlated to hypertension;[1] (3) similar alterations are present in RBC and renal tubular cells of MHS; and (4) the 105 kD protein is cytoskeletal in nature and probably expressed in the kidney, we can postulate that the 105 kD protein could be a candidate for the primary molecular alteration that underlies hypertension in this animal model.

### REFERENCES

1. BIANCHI, G., *et al.* 1985. Hypertension **7**: 319.
2. TRIZIO, D., *et al.* 1983. J. Hypertension **1**(suppl.2): 6.
3. BIANCHI, G., *et al.* 1984. *In* Handbook of Hypertension. W. de Jong, Ed. **4**: 328. Elsevier. Amsterdam, Holland.

# Index of Contributors

589